TOPICS IN CONTEMPORARY MATHEMATICS

TOPICS IN CONTEMPORARY MATHEMATICS

FOURTH EDITION

JACK R. BRITTON
University of South Florida

IGNACIO BELLO
Hillsborough Community College

DELLEN PUBLISHING COMPANY
San Francisco, California
Divisions of Macmillan, Inc.

COLLIER MACMILLAN PUBLISHERS
London

© Copyright 1989 by Dellen Publishing Company,
a division of Macmillan, Inc.

Printed in the United States of America

Permissions: Dellen Publishing Company
               400 Pacific Avenue
               San Francisco, California 94133

Orders: Dellen Publishing Company
               c/o Macmillan Publishing Company
               Front and Brown Streets
               Riverside, New Jersey 08075

Collier Macmillan Canada, Inc.

Library of Congress Cataloging-in-Publication Data

Britton, Jack Rolf
    Topics in contemporary mathematics/Jack R. Britton, Ignacio Bello. — 4th ed.
        p.   cm.
    Includes index.
    ISBN 0-02-308061-2
    1. Mathematics—1961–   I. Bello, Ignacio.   II. Title.
QA39.2.B77   1989
510–dc19   88-31777
                        CIP

Printing: 2 3 4 5 6 7 8 9   Year: 9 0 1 2 3

ISBN 0-02-308061-2

CONTENTS

* Sections marked by an asterisk can be omitted without loss of continuity.

TABLES

PREFACE

In this fourth edition, as in the previous ones, our purpose is to introduce the student to many of the interesting mathematical concepts that are used in our contemporary world. We have tried to bring out the basic ideas and techniques as simply and clearly as possible and have related these ideas to other areas—such as sociology, psychology, and business—that will be attractive to the reader. Whenever feasible, elementary applications are given; these will be found throughout the text as well as in the sets of problems.

The more abstract and theoretical aspects of the subject matter have been de-emphasized. Instead, emphasis has been placed on the understanding and use of the various concepts that are introduced. An important aid to this goal will be found in the exercises, which include many hundreds of problems ranging from the necessary routine drill to challenges for the better students. The reader will find much help in the nearly 450 worked-out examples. Material marked with an asterisk may be omitted without harm to an understanding of the remainder of the book.

Important Features

▶ The book has been written with the student in mind and has been class-tested with gratifying results at Hillsborough Community College and the University of South Florida.

▶ Most of the chapters are preceded by a short biography of a person who made important contributions to the area being discussed. This shows the reader that mathematicians are human beings and places the ideas developed in their proper historical perspective.

▶ Most of the sections are followed by a short problem section entitled "Using Your Knowledge." This section gives an immediate application of the topic under discussion. We have included these sections as answers to that often asked question, "Why do I have to learn this, and what good is it?"

▶ At the end of many of the sections, we have also included a "Discovery" section. These are brief excursions into related topics, extensions, generalizations, and sometimes just interesting problems. This material is more challenging and is geared toward the better students.

▶ A Summary is included at the end of each chapter.

▶ A Practice Test is included at the end of each chapter.

Suggested Courses Using this Book

The book is quite flexible, a large selection of topics being available to suit various courses. The entire book can be covered easily in a full year's course, while different suitable choices can be made for a two-quarter or a one-semester course. Here are some of the courses for which the book is suggested:

▶ General education or liberal arts mathematics (the book follows most of the CUPM (Committee on the Undergraduate Program in Mathematics) recommendations for Mathematics E)

▶ Topics in contemporary mathematics or introduction to mathematics

▶ College mathematics or survey of mathematics courses

▶ Introduction to mathematics or applications of mathematics courses

Supporting Materials and Supplements

This text has an extensive supporting package that includes:

1. An **instructor's manual** containing commentary and suggestions for each chapter, suggested course syllabi, answers to the even-numbered problems in the exercises, three tests similar to the Practice Tests at the end of each chapter and three multiple-choice versions of these tests, as well as the answers to all test questions in the manual
2. A **solutions manual** containing complete solutions to all odd-numbered problems in the exercises
3. A competency-based **study guide** containing the CLAST (College Level Academic Skills Competencies) competencies required by the State of Florida
4. A **computer-generated random test system** utilizing the IBM PC™ or the Apple™ computer
5. A disk containing the **computer programs** given in Chapter 16 and suitable for interactively working the appropriate problems in the text

What Is New in the Fourth Edition?

We have followed the valuable suggestions of users of previous editions and the many reviewers contributing to this fourth edition to clarify the exposition, expand the coverage, and, in general, improve the book. To this end, we have:

▶ Completely redesigned the format of the book

▶ Divided each section into subsections labeled A, B, C, and so on, and keyed the topics in these subsections to the exercises

▶ Where appropriate, added sections called "Calculator Corner" and/or "Computer Corner" following the exercise sections in the book

▶ Provided about 100 new worked-out examples

▶ Revised the exercises, adding nearly 600 new problems, to bring the total to over 2700

▶ Added 50 new "Using Your Knowledge" sections, to make a total of over 300, and 20 "Discovery" problems, to bring the total to 236

▶ Expanded the chapter on numeration systems (Chapter 3) to include the binary, octal, and hexadecimal systems used by computers

▶ Expanded the chapter dealing with the rational numbers (Chapter 4) to include more material on terminating and repeating decimals, as well as scientific notation

▶ Added a new chapter dealing with the real and the complex numbers (Chapter 5); this chapter includes a section on operations with radicals and a section on number sequences

▶ Completely revised the geometry chapter (Chapter 9), which now includes more material on angle measurements, the classification of plane figures, and similar triangles

▶ Revised the chapter on mathematical systems (Chapter 11), placing it after number systems, algebra, geometry, and matrices as a unifying chapter

▶ Shortened the chapter dealing with calculators and computers (Chapter 16), since "Calculator Corner" and "Computer Corner" sections now appear throughout the book

▶ Included a set of programs (Appendix to Chapter 16) that can be used to work the problems in the appropriate sections of the book

▶ Added a Summary at the end of each chapter

▶ Expanded the Practice Tests provided at the end of each chapter

Acknowledgments

We wish to thank the following reviewers for their many valuable suggestions:

First Edition: Calvin Lathan, Monroe Community College
                      Edward T. Ordman, Memphis State University

Second Edition: Homer Austin, Salisbury State College
                      Benjamin Divers
                      Gary Tikriti, University of South Florida

Third Edition: Gloria Arnold, University of Arkansas
John Christy, University of South Dakota
Lucy Dechéne, Fitchburg State College
Calvin Lathan, Monroe Community College
James O. Morgan, Southeastern Louisiana University
Curtis Olson, University of South Dakota
Carol B. Ottinger, Mississippi University for Women
Jean T. Sells, Sacred Heart University
Fay Sewell, Montgomery County Community College
John Vangor, Housatonic Community College
Harlie White, Jr., Campbellsville College

Fourth Edition: James Bagby, Virginia Commonwealth University
Barbara Burrows, Santa Fe Community College
Barbara Cohen, West Los Angeles College
David Cusick, Marshall University
Duane Deal, Ball State University
Philip Downum, State University of New York
Margaret Hackworth, School Board, Pinellas County, Florida
Jim Hodge, College of Lake County
Charles Klein, Midland College
Charles C. Miles, Hillsborough Community College
Sunny Norfleet, St. Petersburg Junior College
Josephine Rinaldo, School Board, Hillsborough County, Florida

We wish to express our particular appreciation to Dr. Heriberto Hernandez, who gave us invaluable help with the problems relating to medicine; Bill Albrecht, who created the computer programs appearing in Chapter 16; Barbara Burrows, our tireless reviewer, who offered invaluable comments and suggestions and kept us straight; Joe Clemente, who checked the accuracy of the problems while only complaining about his health; Gary Etgen, our final answer and page reader, who was both accurate and timely; Josephine Rinaldo, who proofread galleys and pages overnight and only doubted the accuracy of one of our probability problems; Prakash Sach, who did a wonderful job with most of the photos in the book (and refrained from criticizing our own photos); and to the following colleagues for all their helpful criticisms and suggestions: Diana Fernandez, James Gard, George Kosan, Chester Miles, Donald Clayton Rose, and Donald Clayton Rose, II. We would also like to thank our tireless leader, Don Dellen, for his encouragement and for providing us with superb reviewers and an outstanding production supervisor, Phyllis Niklas. Last, but not least, we wish to express our sincere thanks to the many users — students and instructors alike — of the previous editions. We hope that this edition will continue to please them.

CHAPTER 1

SETS

Georg F. L. P. Cantor made the first successful attempts to answer certain difficult questions concerning infinite sets. Born in Russia in 1845, Cantor moved to Germany at the age of 11. At 15 he entered the Wiesbaden Gymnasium (a preparatory school), where he developed an abiding interest in mathematics. His father tried to persuade him to study engineering, a much more profitable profession than teaching, but to no avail. Cantor continued in mathematics, received his Ph.D. from the University of Berlin in 1867, and ended by making important contributions to some of the most abstract areas of mathematics.

Cantor's first works, excellent though they were, gave little hint of the great inventiveness of his mind. It was in the papers published between 1874 and 1884 that his most important contributions appeared. In these papers, he attacked the basic questions of infinite sets — questions that concern the very heart of mathematical analysis. The entirely new methods he employed and the outstanding results he obtained marked him as a creative mathematician of most unusual originality.

Unfortunately, Cantor received almost no recognition during this highly productive period of his life.

He never attained his desire for a professorship at Berlin. Instead, he spent his entire professional career at the much lower rated University of Halle, and even there he did not become a full professor until 1879.

As often happens with innovative and original ideas, Cantor's work was rewarded by ridicule and abuse from many of his most famous contemporaries. Among these critics was the important mathematician, Leopold Kronecker, who had been one of Cantor's instructors at the University of Berlin. The bitterness and severity of Kronecker's criticism were undoubtedly one of the main reasons for Cantor's failure to obtain an appointment to Berlin. As a result of the lack of recognition and acceptance of his work, Cantor suffered a series of breakdowns and eventually died in a mental institution in 1918.

It was only in the later years of his life that Cantor's ideas gained a measure of recognition from his colleagues. The importance of his contribution lay in his perception of the significance of the one-to-one correspondence principle and its logical consequences. Today, we know that much of the foundation of mathematics rests directly on Cantor's work.

GEORG F. L. P. CANTOR (1845–1918)

Courtesy of David Eugene Smith Collection, Rare Book and Manuscript Library, Columbia University

Intuition (male, female, or mathematical) has been greatly overrated. Intuition is the root of all superstition.

E. T. BELL

SETS

A set of dishes
Courtesy of PSH

The concept of a **set** is simply a generalization of an idea that is familiar in everyday life. For example, a collection of dishes in the kitchen is often called a **set** of dishes, and a collection of mechanics' tools is a tool **set**. The volumes of an encyclopedia form a **set** of books. A molecule of a substance is a certain **set** of atoms. In each case, the collection of dishes, tools, books, or atoms is considered to be a unit.

Sets need not consist of physical objects; they may well consist of abstract ideas. For instance, the *Ten Commandments* is a **set** of moral laws. The *Constitution* is the basic **set** of laws of the United States.

A. Well-Defined Sets and Notation

We study sets in this book not only because much of elementary mathematics can be based on this concept, but also because many mathematical ideas can be stated most simply in the language of sets.

▶ **Definition 1.1a**

> A **set** is a well-defined **collection** of objects called **elements,** or **members,** of the set.

The main characteristic of a set in mathematics is that it is **well-defined.** This means that given any object, it must be clear whether that object is a member of the set. Thus, if we consider the set of even whole numbers, we know that every even whole number, such as 0, 2, 4, 6, etc., is an element of this set, but nothing else is. Thus, the set of even whole numbers is well-defined. On the other hand, the set of funny comic sections in the daily newspaper is *not* well-defined, because what one person thinks is funny may not agree with what someone else thinks is funny.

EXAMPLE 1   Which of the following descriptions define sets?

a. Interesting numbers   **b.** Multiples of 2
c. Good writers   **d.** Current directors of General Motors
e. Numbers that can be substituted for x so that $x + 4 = 5$

Solution   Descriptions b, d, and e are well-defined. Descriptions a and c are not well-defined, because people do not agree on what is "interesting" and what is "good." ◀

We use capital letters, such as A, B, C, X, Y, and Z, to denote sets, and lowercase letters, such as a, b, c, x, y, and z, to denote the elements of the set. It is customary, when practical, to list the elements of a set in braces and to separate these elements by commas. Thus, $A = \{1, 2, 3, 4\}$ means that "A is the set consisting of the elements 1, 2, 3, and 4." To indicate the fact that "4 is an element of the set A" or "4 is in A," we write $4 \in A$ (read "4 is an element of the set A"). To indicate that 6 is not an element of A, we write $\mathbf{6 \notin A.}$

EXAMPLE 2   Let $X = \{$Eva, Mida, Jack, Janice$\}$. Which of the following are correct statements?

    a. Mida $\in X$ **b.** Jack $\notin X$ **c.** Janice $\in \{$Eva, Mida, Jack, Janice$\}$
    d. $E \in X$ **e.** $X \in X$

Solution   Parts a and c are the only correct statements. ◀

B.   Describing Sets

Sets may be defined in three ways:

> 1. By giving a **verbal description** of the set
> 2. By **listing** the elements of the set
> 3. By using **set-builder notation**

Description	List
The set of counting numbers less than 5	$\{1, 2, 3, 4\}$
The set of natural Earth satellites	$\{$Moon$\}$
The set of even counting numbers	$\{2, 4, 6, \ldots\}$ The three dots mean the list goes on without end.
The set of odd counting numbers less than 15	$\{1, 3, 5, \ldots, 13\}$ The three dots mean the odd numbers after 5 and before 13 are in the set but not listed.

In set-builder notation, we use a defining property to describe the set. A **vertical bar** is used to mean "such that." Thus, the preceding sets can be written as follows:

{x|x is a counting number less than 5}	Read, "The set of all elements x, such that x is a counting number less than 5."
{x|x is a natural Earth satellite}	Read, "The set of all elements x, such that x is a natural Earth satellite."
{x|x is an even counting number}	Read, "The set of all elements x, such that x is an even counting number."
{x|x is an odd counting number less than 15}	Read, "The set of all elements x, such that x is an odd counting number less than 15."

EXAMPLE 3 Write a verbal description for the following sets:

a. {a, b, c, . . . , z} **b.** {1, 3, 5, . . .} **c.** {3, 6, 9, . . . , 27}

Solution **a.** The set of letters in the English alphabet
b. The set of odd counting numbers
c. The set of counting numbers that are multiples of 3 and less than 28 (or 29, or 30) ◄

EXAMPLE 4 Write the following sets using the listing method and using set-builder notation:

a. The set of digits in the number 1896
b. The set of odd counting numbers greater than 6
c. The set of counting numbers greater than 0 and less than 1
d. The set of counting numbers that are multiples of 4

Solution

List	**Set-Builder Notation**
a. {1, 8, 9, 6}	{x|x is a digit in the number 1896}
b. {7, 9, 11, . . .}	{x|x is an odd counting number and $x > 6$} The symbol ">" means "greater than."
c. { }	{x|x is a counting number and $0 < x < 1$} The symbol "<" means "less than."
d. {4, 8, 12, . . .}	{x|x is a counting number that is a multiple of 4} ◄

A set with no elements, as in part c of Example 4, will be denoted by the symbol { } or ∅.

▶ **Definition 1.1b** | The symbol { } or ∅ represents the **empty,** or **null,** set. |

Note that the set {∅} is not empty, because it contains the element ∅.

C. Equality of Sets

Clearly, the order in which the elements of a set are listed does not affect membership in the set. Thus, if we are asked to write the set of digits in the year in which Columbus discovered America, we may write {1, 4, 9, 2}. Someone else may write {1, 2, 4, 9}. Both are correct! Hence, {1, 4, 9, 2} = {1, 2, 4, 9}. Similarly, {a, b, c, d, e} = {e, d, c, b, a}.

▶ **Definition 1.1c**

In general, two sets are **equal,** denoted by $A = B$, if they have the same members (not necessarily listed in the same order).

For example, {1, 3, 2} = {1, 2, 3} and {20, $\frac{1}{2}$} = {$\frac{1}{2}$, 20}.

Notice also that repeated listings do not affect membership. For example, the set of digits in the year in which the Declaration of Independence was signed is {1, 7, 7, 6}. This set also may be written as {1, 6, 7}. Therefore, {1, 7, 7, 6} = {1, 6, 7}. In the same way, {a, a, b, b, c, c} = {a, b, c}, because the two sets have the same elements. By convention, we do not list an element more than once.

EXERCISE 1.1

A.
1. State whether each of the following sets is well-defined:
   a. The set of grouchy people
   b. The set of retired baseball players with a lifetime batting average of .400 or over
   c. {x|x is an odd counting number}
   d. {x|x is a good college course}
2. State whether each of the following sets is well-defined:
   a. The set of good tennis players in the United States
   b. The set of students taking mathematics courses at Yale University at the present moment
   c. {x|x is an even counting number}
   d. {x|x is a bad instructor}
3. Let A = {Desi, Gidget, Jane, Dora}. Which of the following are correct statements?

   a. $D \in A$ **b.** Desi $\in A$ **c.** $A \in$ Jane
   d. $D \notin A$ **e.** Jane $\notin A$
4. Let X = {a, b, x, y}. Fill in the blanks with $\in$ or $\notin$ so that the result is a correct statement.

   a. a ___$\in$___ X **b.** x ___$\in$___ X **c.** X ___$\notin$___ X
   d. A ___$\notin$___ X **e.** {bay} ___$\notin$___ X

B.
5. Write a verbal description for each of the following sets:
   a. {a, z}   **b.** {m, a, n}

c. {Adam, Eve}

d. {Christopher Columbus}

e. {7, 2, 6, 3, 5, 4, 1}

f. {2, 6, 12, 20, 30}

6. Write a verbal description for each of the following sets:

a. {1, 3, 5, . . . , 51}

b. {3, 6, 9, 12, . . . , 36}

c. {1, 4, 7, 10, . . . , 25}

d. {1, 6, 11, . . . , 31}

In Problems 7–16, list the elements in the given set.

7. {x|x is a counting number less than 8}

8. {x|x is a counting number less than 2}

9. {n|n is a counting number less than $7\frac{1}{2}$}

10. {n|n is a counting number less than $8\frac{1}{4}$}

11. {x|x is a counting number between 3 and 8}

12. {x|x is a counting number between 2 and 7}

13. {n|n is a counting number between 6 and 7}

14. {n|n is a counting number between 8 and 10}

15. {x|x is a counting number greater than 3}

16. {x|x is a counting number greater than 0}

In Problems 17–24, a set is specified by certain conditions. List the elements in the set. In each of these problems, n is a counting number—that is, $n \in \{1, 2, 3, \ldots\}$.

17. $\{x|x = 5n\}$

18. $\{n|3 < n < 7\}$

19. $\{n|n^2 < 0\}$

20. $\{n^2|0 \leq n \leq 4\}$ (The symbol "≤" means "less than or equal to.")

21. $\{n|n^3 < 15\}$

22. $\{n|4 < n^2 < 40\}$

23. $\{n|1 < n < 10$, n is an even number$\}$

24. $\{x|x = 2n - 1\}$

Problems 25–28 refer to the stock listing in the margin. Use the information in this list of stocks for each of these problems.

Name	Volume	High	Low	Last	Chg.
WangB	1,593,900	11⅝	10½	11¼	. . . .
Gull	1,493,800	25½	12	15	−10
WhrEnf	1,254,200	13¾	13	13⅝	+2½
HomeSh	1,152,200	6⅛	5¼	5½	−⅜
EchBg s	928,300	24	22½	23¾	+1¼
FAusPr	904,000	8½	7½	7⅝	−⅛
LorTel	896,500	9¾	8⅜	9¼	+⅜
ENSCO	778,800	3⅛	2½	3⅛	+½
WDigitl	776,100	17¼	15⅞	16¾	+1
TexAir	767,100	12⅜	11¼	11¾	−½

This listing gives the sales volume; high, low, and closing price; and net change from last week of this week's ten most active stocks traded on the American Stock Exchange (AMEX) for more than $1.

25. List the elements in the following sets:

a. Stocks whose sales volume was greater than 1,300,000

b. Stocks whose sales volume was greater than 1,000,000

c. Stocks whose sales volume was less than 800,000

d. Stocks whose price went up more than $\frac{1}{2}$ point

26. List the elements in the following sets:

a. Stocks that increased in price by more than 1 point

b. Stocks that increased in price by less than $\frac{1}{2}$ point

c. Stocks that increased in price by exactly $\frac{1}{2}$ point

In Problems 27 and 28, study the "Last" column and use set-builder notation to describe the following sets:

27. a. {HomeSh, FAusPr, LorTel, ENSCO}

b. {EchBg, WDigitl}

28. a. {EchBg}

b. {ENSCO}

29. List the elements in each of the following sets:
  a. The different letters in the word MISSISSIPPI
  b. The planets of our solar system
  c. All positive fractions having 1 for a numerator and a counting number as denominator
  d. The U.S. astronauts who have landed on the planet Pluto

30. Which of the sets in Problem 29 are empty?

C.

31. In each of the following, state whether the sets A and B are equal:
  a. $A = \{2n + 1 | n$ a counting number$\}$,
  $B = \{2n - 1 | n$ a counting number$\}$
  b. $A = \{4n | n$ a counting number$\}$,
  $B = \{2n | n$ a counting number$\}$
  c. $A = \{1, 1, 2, 2, 3\}$, $B = \{1, 2, 3\}$
  d. $A = \{x | x$ is a cow that has jumped over the moon$\}$,
  $B = \{x | x$ is an astronaut who has landed on Pluto$\}$

32. Let $A = \{5\}$, $B = \{f, i, v, e\}$, $C = \{e, f, v, i\}$, and D be the set of letters in the word *repeat*. Find:
  a. The set containing 5 elements $\{\ \}$
  b. The set equal to B $\{f, i, v, e\}$
  c. The set of letters in the word *five* $\{f, i, v, e\}$

33. Let $A = \{1, 2, 3, 4\}$, $B = \{4, 3, 2, 1\}$, and $C = \{4, 3, 2, 1, 0\}$. Fill in the blanks with "=" or "≠" to make a true statement:
  a. A _____ B **b.** A _____ C **c.** B _____ C

34. Let $A = \{x | x$ is a counting number between 4 and 5$\}$, $B = \varnothing$, and $C = \{\varnothing\}$. Fill in the blanks with "=" or "≠" to make a true statement:
  a. A __=__ B **b.** A __≠__ C **c.** B __≠__ C

35. Let A be the set of astronauts who have landed on Pluto, $B = \{0\}$, C be the empty set, and $D = \{\varnothing\}$. Write *true* or *false* for each of the following:
  a. $A = B$ **b.** $A = C$ **c.** $B = C$
  d. $A = D$ **e.** $B = D$ **f.** $C = D$

Here are two barbers who do not shave themselves.

USING YOUR KNOWLEDGE 1.1

Gepetto Scissore, a barber in the small town of Sevilla, who was naturally called the Barber of Sevilla, decided that as a public service he would shave all those men and only those men of the village who did not shave themselves. Let $S = \{p | p$ is a man of the village who shaves himself$\}$ and $D = \{p | p$ is a man of the village who does not shave himself$\}$.

1. If g represents Gepetto:
  a. Is $g \in S$?   **b.** Is $g \in D$?

The preceding problem is a popularization of the **Russell paradox,** named after its discoverer, Bertrand Russell. In studying sets, it seems that one can classify sets as those that are members of themselves and those that are not members of themselves. Suppose that we consider the two sets of sets: $M = \{X | X \in X, X \text{ is a set}\}$ and $N = \{X | X \notin X, X \text{ is a set}\}$.

2. Answer the following questions:
   a. Is $N \in M$?   **b.** Is $N \in N$?
   Think about the consequences of your answers!

DISCOVERY 1.1

You should find the following paradox amusing, puzzling, and perhaps even thought-provoking: Let us define a **self-descriptive word** to be a word that makes good sense when put into both blanks of the sentence, "_____ is a(an) _____ word." Two simple examples of self-descriptive words are "English" and "short." Just try them out!

   Now define a **non-self-descriptive word** to be a word that is not self-descriptive. Most words will fit into this category. Try it out again. Now consider the following questions:

1. Let S be the set of self-descriptive words, and let S' be the set of non-self-descriptive words. How would you classify the word non-self-descriptive? Is it an element of S? Or is it an element of S'? You should get into difficulty no matter how you answer these questions. Think about it!

2. Let $A = \{1, 2, 3, \ldots\}$ and $B = \{\ \}$. Then proceed as follows:
   a. Take the numbers 1 and 2 from A and place them in B.
   b. At $\frac{1}{2}$ hr before noon, remove the largest number, 2, from B so that the number 1 remains.
   c. Take the numbers 3 and 4 from A and place them in B.
   d. At $\frac{1}{4}$ hr before noon, remove the largest number, 4, from B so that 1 and 3 remain.
   e. Take the numbers 5 and 6 from A and place them in B.
   f. At $\frac{1}{8}$ hr before noon, remove the largest number, 6, from B so that 1, 3, and 5 remain.
   If this procedure is continued, what will B consist of at noon?

3. In Problem 2, amend steps b, d, and f by replacing "largest number" with "smallest number." Note that at $\frac{1}{2}$ hr before noon, the smallest number left in B is 2; at $\frac{1}{4} = (\frac{1}{2})^2$ hr before noon, the smallest number left in B is 3; at $\frac{1}{8} = (\frac{1}{2})^3$ hr before noon, the smallest number left in B is 4; and so on. Can you discover what the smallest number left in B is at $(\frac{1}{2})^n$ hr before noon?

SUBSETS

It sometimes happens that all the elements of a set A are also elements of another set B. For example, if A is the set of all students in your class, and B is the set of all students in your school, every element of A is also in B (because every student in your class is a student in your school). In such cases, we say that the set A is a **subset** of the set B. We denote this by writing $A \subseteq B$ (read, "A is a subset of B").

▶ **Definition 1.2a** | The set A is a **subset** of B (denoted by $A \subseteq B$) if every element of A is also an element of B.

Thus, if $A = \{a, b\}$, $B = \{a, b, c\}$, and $C = \{b\}$, then $A \subseteq B$, $C \subseteq A$, and $A \subseteq A$.

It is a consequence of the definition of a subset that $A = B$ when both $A \subseteq B$ and $B \subseteq A$. Furthermore, for any set A, since every element of A is an element of A, we have $A \subseteq A$.

The definition of a subset may be restated in the following form:

▶ **Definition 1.2b** | The set A is a **subset** of B if there is no element of A that is not an element of B.

From this it follows that $\varnothing \subseteq A$, because there is no element of $\varnothing$ that is not in A. This means that **the empty set is a subset of every set.**

A set A is said to be a **proper subset** of B, denoted by $A \subset B$, if A is a subset of B and there is at least one element of B that is not in A. In other words, $A \subset B$ means that all elements of A are also in B, but B contains at least one element that is not in A. For example, if $B = \{1, 2\}$, the proper subsets of B are $\varnothing$, $\{1\}$, and $\{2\}$, but the set $\{1, 2\}$ itself is not a proper subset of B.

A. Finding Subsets

In everyday discussions, we are usually aware of the "universe of discourse," that is, the set of all things we are talking about. In dealing with sets, the universe of discourse is called the **universal set.**

▶ **Definition 1.2c** | The **universal set** $\mathcal{U}$ is the set of all elements under discussion.

Thus, if we agree to discuss the letters in the English alphabet, then $\mathcal{U} = \{a, b, c, \ldots, z\}$ is our universal set. On the other hand, if we are to discuss counting numbers, our universal set is $\mathcal{U} = \{1, 2, 3, \ldots\}$.

EXAMPLE 1   Find all the subsets of the set $\mathcal{U} = \{a, b, c\}$.

Solution   We have to form subsets of the set $\mathcal{U}$ by assigning some, none, or all of the elements of $\mathcal{U}$ to these subsets. We organize the work as follows:

Form all the subsets with no elements: $\emptyset$.
Form all the subsets with 1 element: $\{a\}, \{b\}, \{c\}$.
Form all the subsets with 2 elements: $\{a, b\}, \{a, c\}, \{b, c\}$.
Form all the subsets with 3 elements: $\{a, b, c\}$.   ◀

Notice that the set in this example has 3 elements and $2^3 = 2 \times 2 \times 2 = 8$ subsets. Similarly, a set such as $\{a, b\}$, containing 2 elements, has $2^2 = 2 \times 2 = 4$ subsets, namely, $\emptyset, \{a\}, \{b\}$, and $\{a, b\}$. We shall show later that:

A set of n elements has 2^n subsets.

EXAMPLE 2   If $A = \{a_1, a_2, \ldots, a_8\}$, how many subsets does A have?

Solution   Because the set A has 8 elements, A has $2^8 = 256$ subsets.   ◀

EXAMPLE 3   List all the proper subsets of the set $\mathcal{U} = \{1, 2, 3\}$.

Solution   The proper subsets are $\emptyset, \{1\}, \{2\}, \{3\}, \{1, 2\}, \{1, 3\}$, and $\{2, 3\}$.   ◀

Note that a set with n elements has $2^n - 1$ proper subsets.

B.   Applications

The ideas presented in the preceding examples can be applied to many practical problems. For example, if a police dispatcher has 3 patrol cars, a, b, and c, available, the possibilities for handling a particular call correspond to those given in Example 1. Thus, on a given call the dispatcher has the following options:

Send no cars: $\emptyset$

Send 1 car: $\{a\}, \{b\}, \{c\}$.

Send 2 cars: $\{a, b\}, \{b, c\}, \{a, c\}$.

Send 3 cars: $\{a, b, c\}$.

The total number of choices is $2^3 = 8$.

EXAMPLE 4   The Taste-T Noodle Company has 6 employees. The payroll department classifies these employees as follows:

P: Part-time employees who work 20 hr or less per week

F: Full-time employees who work more than 20 hr per week

H_5:   Employees paid at the rate of $5 per hour

H_6:   Employees paid at the rate of $6 per hour

O: Employees working over 40 hr per week and receiving a time-and-a-half rate for all hours worked over 40 per week

List the elements in the sets P, F, H_5, H_6, O, and $\mathcal{U}$ using the employee numbers and the data in the chart.

WEEK ENDING ⟶

NAME	EMPL. NO.	Status	Exemptions	HOURS							TOTAL HOURS	RATE	EARNINGS		
				SUN.	MON.	TUES.	WED.	THURS.	FRI.	SAT.			REGULAR	OVERTIME	OTHER
1 Amy Able	01	S	1		7	7	7	7	7	0	35	$6.00	$210 00		
2 Bernie Baker	02	M	2		8	8	8	8	8	4	44	$5.00	$200 00	$30 00	
3 Cindy Chan	03	M	2		0	0	0	0	0	8	8	$5.00	$40 00		
4 Donald Dellen	04	S	1		8	0	8	0	8	0	24	$5.00	$120 00		
5 Emilio Estéves	05	M	3		8	8	8	8	8	2	42	$6.00	$240 00	$18 00	
6 Felicia Fellini	06	S	1		0	4	4	4	4	0	16	$6.00	$96 00		
7															
8															
9															
10															
TOTALS															

Solution $P = \{03, 06\}$

$F = \{01, 02, 04, 05\}$

$H_5 = \{02, 03, 04\}$

$H_6 = \{01, 05, 06\}$

$O = \{02, 05\}$

$\mathcal{U} = \{01, 02, 03, 04, 05, 06\}$ ◀

A. 1. List all the subsets of the following sets:
   a. $\mathscr{U} = \{a, b\}$ **b.** $\mathscr{U} = \{1, 2, 3\}$
   c. $\mathscr{U} = \{1, 2, 3, 4\}$ **d.** $\mathscr{U} = \{\varnothing\}$

2. List all the proper subsets of the sets given in Problem 1.

3. List all the subsets of the following sets:
   a. $\{1, 2\}$ **b.** $\{x, y, z\}$

4. List all the proper subsets of the sets given in Problem 3.

5. How many subsets does the set $A = \{a, b, c, d\}$ have? *16*

6. How many proper subsets does the set $\{1, 2, 3, 4\}$ have? *15*

7. If $A = \{\frac{1}{1}, \frac{1}{2}, \frac{1}{3}, \ldots, \frac{1}{10}\}$, how many subsets does A have? *1024*

8. How many proper subsets does the set A of Problem 7 have? *1023*

9. A set has 32 subsets. How many elements are there in this set? *5*

10. A set has 31 proper subsets. How many elements are there in the set? *5*

11. A set has 64 subsets. How many elements are there in the set? *6*

12. A set has 63 proper subsets. How many elements are there in the set? *6*

13. Is $\varnothing$ a subset of $\varnothing$? Explain. *Yes*

14. Is $\varnothing$ a proper subset of $\varnothing$? Explain. *No*

15. If A is the set of numbers that are divisible by 2 and B is the set of numbers that are divisible by 4, is $A \subseteq B$? Is $B \subseteq A$?

16. Give an example of a set P and a set Q such that $P \in Q$ and $P \subseteq Q$.

In Problems 17 and 18, fill in the blanks with $\in$, $\notin$, $\subset$, or $\not\subset$ so that the result is a correct statement.

17. **a.** $\{2, 3\}$ _____$\subset$_____ $\{3, 5, 2\}$ **b.** 5 _____$\notin$_____ $\{2, 4\}$
    c. $\{5\}$ _____$\not\subset$_____ $\{2, 4\}$

18. **a.** 31 _____$\notin$_____ $\{1, 3, 5, \ldots\}$
    b. $\{2, 4, 6, \ldots\}$ _____$\subset$_____ $\{1, 2, 3, \ldots\}$
    c. $\{1, 3, 5, \ldots\}$ _____$\not\subset$_____ $\{1, 3, 5\}$

B. 19. The sign shown in the margin says you can use any combination of nickels, dimes, and quarters. If you have three coins — a nickel, a dime, and a quarter — how many different sums of money can you select? Use set notation to list all the choices you have. How many choices are there if you must use at least one coin with each choice?

20. If you have five coins — a penny, a nickel, a dime, a quarter, and a half-dollar — how many different sums of money can you select? If you must use at least one coin, how many different sums can you select?

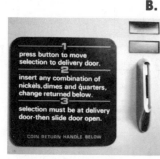

press button to move selection to delivery door.
—2—
insert any combination of nickels, dimes and quarters, change returned below.
—3—
selection must be at delivery door-then slide door open.

COIN RETURN HANDLE BELOW

USING YOUR KNOWLEDGE 1.2

The idea of a universal set and its subsets can be used in many applications. For example, the figure on the next page shows the human species as the

Do Odd Problems

15. $^A\{2,4,6,8,10,12,14,16,\ldots\}$
    $^B\{4,8,12,16\}$

16. $^P\{a\}$
    $^Q\{a,b,c\}$

universal set. As you can see from the diagram, there are 3 major racial stocks: Caucasian (C), Mongoloid (M), and Negroid (N).

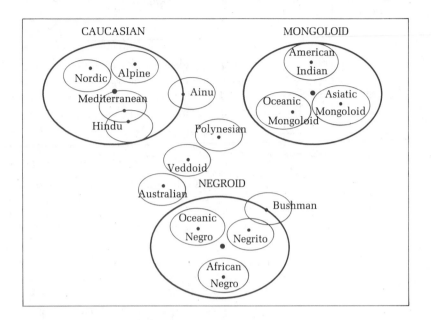

1. Name all the subraces (subsets) of the Caucasian race.
2. Name all the subraces (subsets) of the Mongoloid race.
3. Name all the subraces (subsets) of the Negroid race.
4. Name the 3 subraces (subsets) that do not belong to any of the 3 major races.
5. Name the subsets of the Caucasian race that have some members in common.
6. Name the subsets of the Negroid race that have some members in common.

Another application of subsets is as follows: Suppose that television station WICU has 3 commercials available, c_1, c_2, and c_3.

7. How many choices are available if the station director may choose to have none, 1, 2, or 3 different commercials run, and is not concerned with the order in which they are run? [Hint: How many subsets does the set $\{c_1, c_2, c_3\}$ have?]
8. How many choices are available if the director decides to run either none or else all of the commercials?
9. How many choices are available if the director decides to have exactly 1 commercial?
10. How many choices are available if the director decides to run exactly 2 commercials?

n \ k	0	1	2	3	4	5	t
0	1						$1 = 2^0$
1	1	1					$2 = 2^1$
2	1	2	1				$4 = 2^2$
3	1		3	1			$8 = 2^3$
4	1						
5							

The empty set has only one subset (itself). A 1-element set has two subsets, the empty set and the 1-element set itself. As we saw in this section, a 2-element set has four subsets, and a 3-element set has eight subsets. In the table, the column under n shows the number of elements in the given set, the row labeled k gives the number of elements in the subset, and the last column, t, shows the total number of subsets for a given n-element set.

The other entries in the body of the table show the number of k-element subsets for a given n-element set. For example, to find the number of 2-element subsets for a given 3-element set, we go down the column under n to the number 3 and then across until we are under the entry k = 2, where we read 3. This tells us that a 3-element set has three 2-element subsets.

1. Fill in the missing items in the table.
2. It is possible to construct each row (after the first) of the table from the preceding row. For example, to go from the row n = 2 to the row n = 3, imagine that you have a set of 2 elements {a, b}. You know that this gives one empty subset, two 1-element subsets, and one 2-element subset: ∅, {a}, {b}, {a, b}. Now let us add another element, c, to the given set to form a 3-element set, {a, b, c}. How can we form the 2-element subsets of the new set from the subsets of the old set? The answer is that we can use the old 2-element subset {a, b} as one of the required subsets, and we can adjoin the new element, c, to each of the old 1-element subsets, {a} and {b}, to form the required subsets {a, c} and {b, c}. The three subsets obtained in this way are all the 2-element subsets of the new 3-element given set. Thus, we see that the number of 2-element subsets of a 3-element given set is the sum of the number of 1-element subsets and the number of 2-element subsets of a 2-element given set. Can you discover how this idea works in general so that you can build up the row n = 4 from the row n = 3?
3. Suppose that the symbol $\binom{n}{k}$ stands for the number of k-element subsets of an n-element given set. For instance, $\binom{3}{2}$ stands for the number of 2-element subsets of a 3-element given set. Can you discover a formula for $\binom{n+1}{k}$ in terms of the symbols for an n-element given set? [*Hint:* In Problem 2, we found that $\binom{3}{2} = \binom{2}{1} + \binom{2}{2}$.]
4. Can you now discover how many 3-element subsets a 6-element given set has?

The triangular array of numbers starting with the column under 0 in the table given above is known as **Pascal's triangle** in honor of Blaise Pascal (1623–1662). This triangle is usually arranged as shown in the figure in the margin. Compare the entries in the table with those in the figure.

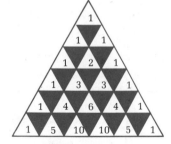

Pascal's triangle

▶ **Computer Corner 1.2** *We have shown you how to find all the subsets of a set of 2 or 3 elements. But can you find all the subsets of a set with 5 elements? A computer can! The* computer program on page 742 will find the 32 subsets in no time at all. The set consisting of all the subsets of a given set is called the **power set** of the original set.

1. Use this program to check Problems 1, 3, and 5 in Exercise 1.2.

SET OPERATIONS

It is often important to ascertain which elements two given sets have in common. For example, the sets of symptoms exhibited by patients with too little sugar in the blood (hypoglycemia) or too much sugar in the blood (hyperglycemia) are as given in Table 1.3a.

Table 1.3a

TOO LITTLE SUGAR IN THE BLOOD	TOO MUCH SUGAR IN THE BLOOD
Nausea (n)	Headache (h)
Visual disturbances (v)	Stomach cramps (s)
Trembling (t)	Nausea (n)
Headache (h)	Rapid breathing (r)

A. Intersections and Unions

We can clearly see that the set of symptoms *common* to both sets listed in Table 1.3a is $\{n, h\}$. The set $\{n, h\}$ is called the **intersection** of the two given sets.

▶ **Definition 1.3a**

If A and B are sets, the **intersection** of A and B, denoted by $A \cap B$ (read, "A intersection B"), is the set of all elements that are common to both A and B. That is,

$$A \cap B = \{x | x \in A \text{ and } x \in B\}$$

Thus, if $A = \{a, b, c, d\}$ and $B = \{b, d, e\}$, then $A \cap B = \{b, d\}$.

If we list all the symptoms mentioned in Table 1.3a, we obtain $\{n, v, t, h, s, r\}$. This set is called the **union** of the two given sets.

▶ **Definition 1.3b**

If A and B are sets, the **union** of A and B, denoted by $A \cup B$ (read, "A union B"), is the set of all elements that are either in A or in B or in both A and B. That is,

$$A \cup B = \{x | x \in A \text{ or } x \in B\}$$

Note that we use the **inclusive or,** that is, $x \in A$, $x \in B$, or x may be in *both* A and B.

Hence, if $A = \{1, 3, 4, 6\}$ and $B = \{3, 6, 7\}$, then $A \cup B = \{1, 3, 4, 6, 7\}$.

EXAMPLE 1   Let $A = \{a, b, c, d, e\}$ and $B = \{a, c, e, f\}$. Find:

a. $A \cap B$ **b.** $A \cup B$

Solution   **a.** $A \cap B$ is the set of all elements common to both A and B. That is, $A \cap B = \{a, c, e\}$.

b. $A \cup B$ is the set of all elements in A or B (or both). That is, $A \cup B = \{a, b, c, d, e, f\}$. Note that a, c, and e occur in both sets, yet we list each of these elements only once.   ◄

B. Complement of a Set

We are often interested in the set of elements in the universal set under discussion but *not* in some specified set.

Definition 1.3c

> Let $\mathcal{U}$ be the universal set, and let A be a subset of $\mathcal{U}$. The complement of A, denoted by A' (read, "A prime" or "A complement"), is the set of elements in $\mathcal{U}$ but not in A. That is, $A' = \{x | x \in \mathcal{U} \text{ and } x \notin A\}$. This set is also symbolized by $\mathcal{U} - A$.

For example, if $\mathcal{U}$ is the set of students in your school, and A is the set of those students who have taken algebra, then A' is the set of those students in your school who have *not* taken algebra. Similarly, if $\mathcal{U}$ is the set of all letters in the English alphabet, and $A = \{s, l, y\}$, then A' is the set of all letters in the alphabet *except* s, l, and y.

EXAMPLE 2   Let $\mathcal{U} = \{1, 2, 3, 4, 5, 6\}$, $A = \{1, 3, 5\}$, $B = \{2, 4\}$. Find:

a. A' **b.** B' **c.** $A' \cap B'$ **d.** $A \cap B$

Solution   **a.** $A' = \{2, 4, 6\}$ **b.** $B' = \{1, 3, 5, 6\}$ **c.** $A' \cap B' = \{6\}$

d. $A \cap B = \varnothing$, because there are no elements common to these two sets.   ◄

Note that, because $\varnothing \subseteq \mathcal{U}$, and no element of $\mathcal{U}$ is an element of $\varnothing$, it follows that $\varnothing' = \mathcal{U}$ and $\mathcal{U}' = \varnothing$.

EXAMPLE 3   If A, B, and $\mathcal{U}$ are the same as in Example 2, find:

a. $(A \cup B)'$ **b.** $(A \cap B)'$ **c.** $A' \cup B'$ **d.** $A \cup (A \cup B)'$

Solution   **a.** $A \cup B = \{1, 3, 5\} \cup \{2, 4\} = \{1, 2, 3, 4, 5\}$

Hence, $(A \cup B)' = \{6\}$.

[Note that in order to find $(A \cup B)'$, we first find $A \cup B$ and then take its complement.]

b. $A \cap B = \{1, 3, 5\} \cap \{2, 4\} = \emptyset$

Hence, $(A \cap B)' = \mathcal{U} = \{1, 2, 3, 4, 5, 6\}$.

c. $A' \cup B' = \{1, 3, 5\}' \cup \{2, 4\}'$

$\qquad = \{2, 4, 6\} \cup \{1, 3, 5, 6\}$

$\qquad = \{1, 2, 3, 4, 5, 6\}$

d. $A \cup (A \cup B)' = \{1, 3, 5\} \cup \{6\} = \{1, 3, 5, 6\}$ ◀

Notice that the answers to parts b and c are identical. Also, notice that the answers to Example 2c and Example 3a are identical. We can show for any sets A and B that $(A \cap B)' = A' \cup B'$ and $(A \cup B)' = A' \cap B'$. (See Exercise 1.4, Problems 14a and 14b.)

It is also possible to form intersections, unions, and complements using more than two sets, as in the next example.

EXAMPLE 4   Let $\mathcal{U} = \{a, b, c, d, e, f\}$, $A = \{a, c, e\}$, $B = \{b, e\}$, and $C = \{a, b, d\}$. Find $(A \cup B) \cap C'$.

Solution   Since $A \cup B$ is in parentheses, we find $A \cup B$ first:

$$A \cup B = \{a, c, e\} \cup \{b, e\} = \{a, b, c, e\}$$

Then,

$$C' = \{c, e, f\}$$

Hence,

$$(A \cup B) \cap C' = \{a, b, c, e\} \cap \{c, e, f\}$$
$$= \{c, e\}$$ ◀

C. Difference of Two Sets

In some cases, we might be interested in only part of a given set. For example, we might want to consider the set of all nonpoisonous snakes. If we let S be the set of all snakes and P be the set of all poisonous snakes, we are interested in the set of all snakes *except* (excluding) the poisonous ones. This set will be denoted by $S - P$.

▶   **Definition 1.3d** | If A and B are two sets, the **difference** of A and B, denoted by $A - B$ (read, "A minus B"), is the set of all elements that are in A and not in B. That is, $A - B = \{x | x \in A \text{ and } x \notin B\}$.

Notice that

1. The definition of A' is a special case of Definition 1.3d, because $\mathcal{U} - A = A'$. (See Definition 1.3c.)
2. $A \cap B' = A - B$, because $A \cap B'$ is the set of all elements in A and not in B, and this is precisely the definition of $A - B$.

EXAMPLE 5   Let $\mathcal{U} = \{1, 2, 3, 4, 5, 6\}$, $A = \{1, 2, 3, 4\}$, and $B = \{1, 2, 5\}$. Find:

   a. $\mathcal{U} - A$ **b.** A' **c.** $A - B$ **d.** $B - A$

Solution   **a.** $\mathcal{U} - A$ is the set of all elements in $\mathcal{U}$ and not in A; that is, $\{5, 6\}$.
   b. A' is the set of all elements in $\mathcal{U}$ and not in A; that is, $\{5, 6\}$.
   c. $A - B$ is the set of all elements in A and not in B; that is, $\{3, 4\}$.
   d. $B - A$ is the set of all elements in B and not in A; that is, $\{5\}$.   ◀

D. Applications

EXAMPLE 6   A small rural electric company has 10 employees who are listed by number as 01, 02, 03, . . . , 10. The company classifies these employees according to the work they do:

   P: The set of part-time employees
   F: The set of full-time employees
   S: The set of employees who do shop work
   O: The set of employees who do outdoor field work
   I: The set of employees who do indoor office work

The payroll department lists these employees as follows:

   $\mathcal{U} = \{01, 02, 03, 04, 05, 06, 07, 08, 09, 10\}$

   $P = \{01, 02, 05, 07\}$

   $F = \{03, 04, 06, 08, 09, 10\}$

   $S = \{01, 04, 05, 08\}$

   $O = \{03, 04, 06, 09\}$

   $I = \{02, 05, 07, 10\}$

Find and describe each of the following sets:

   a. $P \cap S$ **b.** $O \cup S$ **c.** $F \cap I$ **d.** P'
   e. $F \cap (S \cup O)$ **f.** $S' \cup (O' \cap I')$

Solution   **a.** $P \cap S = \{01, 05\}$; the set of part-time employees who do shop work
   b. $O \cup S = \{01, 03, 04, 05, 06, 08, 09\}$; the set of employees who do out-
      side field work or shop work
   c. $F \cap I = \{10\}$; the set of full-time employees who do indoor office work
   d. $P' = \{03, 04, 06, 08, 09, 10\}$; the set of employees who are not part-time
      — that is, the set of full-time employees
   e. $F \cap (S \cup O) = \{03, 04, 06, 08, 09\}$; the set of full-time employees who do
      shop work or outside field work
   f. $S' \cup (O' \cap I') = \{01, 02, 03, 06, 07, 08, 09, 10\}$; the set of employees who
      do not do shop work combined with the set of employees who do
      neither outdoor field work nor indoor office work   ◀

A.　**1.** Let $A = \{1, 2, 3, 4, 5\}$, $B = \{1, 3, 4, 6\}$, and $C = \{1, 6, 7\}$. Find:

　　a. $A \cap B$　　　　　**b.** $A \cap C$　　　　　**c.** $B \cap C$

　　d. $A \cup B$　　　　　**e.** $A \cup C$　　　　　**f.** $B \cup C$

　　g. $A \cap (B \cup C)$　　　　　**h.** $A \cup (B \cap C)$

　　i. $(A \cap B) \cup C$　　　　　**j.** $(A \cap B) \cup (A \cap C)$

2. If A, B, and C are the same as in Problem 1, find:

　　a. $A \cup (B \cup C)$　　　　　**b.** $(A \cup B) \cap (A \cup C)$

　　c. $A \cap (B \cap C)$　　　　　**d.** $(A \cup B) \cap C$

3. Let $A = \{\{a, b\}, c\}$, $B = \{a, b, c\}$, and $C = \{a, b\}$. Find:

　　a. $A \cap B$　　**b.** $A \cap C$　　**c.** $A \cup B$　　**d.** $A \cup C$

4. Let $A = \{\{a, b\}, \{a, b, c\}, a, b\}$ and $B = \{\{a, b\}, a, b, c, \{b, c\}\}$. Which of the following statements are correct?

　　a. $\{b\} \subseteq (A \cap B)$　　　　　**b.** $\{b\} \in (A \cap B)$

　　c. $\{a, b\} \subseteq (A \cap B)$　　　　　**d.** $\{a, b\} \in (A \cap B)$

　　e. $\{a, b, c\} \subseteq (A \cup B)$　　　　**f.** $\{a, b, c\} \in (A \cup B)$

　　g. $3 \subseteq (A \cap B)$　　　　　**h.** $3 \in (A \cap B)$

B.　The sets $\mathscr{U} = \{a, b, c, d, e, f\}$, $A = \{a, c, e\}$, $B = \{b, d, e, f\}$, and $C = \{a, b, d, f\}$ will be used in Problems 5–7.

5. Find:

　　a. A'　　　　　**b.** B'　　　　　**c.** $A' \cap B'$

　　d. $(A \cup B)'$　　　　　**e.** $A' \cup B'$　　　　　**f.** $(A \cap B)'$

　　g. $(A \cup B) \cap C'$　　　**h.** $(A \cap B) \cup C'$　　　**i.** $C \cup (A \cap B)'$

6. Find:

　　a. $A' \cup B$　　　　　**b.** $A \cup B'$　　　　　**c.** $A' \cap B$

　　d. $A \cap B'$　　　　　**e.** $A' \cap (A \cup B')$　　　**f.** $A \cup (A \cap B')$

　　g. $C' \cup (A \cap B)'$　　　**h.** $C' \cup (A \cup B)'$　　　**i.** $(C \cup B)' \cap A$

C.　**7.** Find:

　　a. $\mathscr{U} - A$　　　**b.** $\mathscr{U} - B$　　　**c.** $A - B$　　　**d.** $B - A$

8. Let $\mathscr{U} = \{1, 2, 3, 4, 5\}$, $A = \{2, 3, 4\}$, and $B = \{1, 4, 5\}$. Find:

　　a. B'　　　**b.** $\mathscr{U} - B$　　　**c.** $A - B$　　　**d.** $B - A$

In Problems 9–18, $\mathscr{U}$ is some universal set of which A is a subset. In each case find the indicated set in terms of A, $\mathscr{U}$, or $\varnothing$ alone.

　　9. $\varnothing'$　　　　**10.** $\mathscr{U}'$　　　　**11.** $A \cap \varnothing$　　　　**12.** $A \cap A$

13. $A \cap \mathscr{U}$　　　**14.** $A \cup \varnothing$　　　**15.** $A \cap A'$　　　**16.** $A \cup A'$

17. $(A')'$　　　**18.** $A \cup A$

19. If $A = \{1, 2, 3\}$, $B = \{2, 3, 4\}$, and $C = \{1, 3, 5\}$, find the smallest set that will serve as a universal set for A, B, and C.

D.　Problems 20–25 refer to the following data: What traits do men and women like in each other? In an attempt to analyze factors in popularity, not only between members of the same sex but between men and women, a psychologist asked 676 college men and women to indicate a few persons whom they liked and to tell why they liked those persons.

TRAITS MEN LIKED IN WOMEN (M_w)	TRAITS WOMEN LIKED IN MEN (W_m)	TRAITS MEN LIKED IN MEN (M_m)	TRAITS WOMEN LIKED IN WOMEN (W_w)
Beauty	Intelligence	Intelligence	Intelligence
Intelligence	Consideration	Cheerfulness	Cheerfulness
Cheerfulness	Kindliness	Friendliness	Helpfulness
Congeniality	Cheerfulness	Congeniality	Loyalty

20. Find the smallest set that will serve as a universal set for M_w, W_m, M_m, and W_w.
21. Find the set of traits that are mentioned only once.
22. Find $M_w \cap W_w$.
23. Find $M_w \cap M_m$.
24. What set of traits is common to M_m and M_w?
25. Name the traits that are common to all four of the sets; that is, find $M_w \cap M_m \cap W_w \cap W_m$.

26. *Boss behavior.* In an article in the *Harvard Review*, 606 participants reported on 17 specific changes in their boss's behavior from one year to the next. Here are some of the traits that were most frequently mentioned in each of these years:

FIRST YEAR	SECOND YEAR
Encourages suggestions	Is self-aware
Sets goals with me	Listens carefully
Gets me to have high goals	Follows up on action
Listens carefully	Gets me to have high goals
Is aware of others	Encourages suggestions
Is self-aware	Sets goals with me

Let S_1 be the set of traits mentioned in the first year, and let S_2 be the set of traits mentioned in the second year.
a. Find $S_1 \cap S_2$, the set of traits mentioned in both years.
b. What traits were mentioned only once?
c. Find the smallest set that will serve as a universal set for S_1 and S_2.
d. Find S_1' relative to the universal set found in part c.
e. Find S_2' relative to the universal set found in part c.

In Problems 27–30, let

$\mathcal{U}$ be the set of employees of a company,

M be the set of males who are employees,

F be the set of females who are employees,

D be the set of employees who work in the data-processing department,

T be the set of employees who are under 21,

S be the set of employees who are over 65, and

∅ be the empty set.

27. Find a single letter to represent each of the following sets:
   a. $M \cup F$ **b.** $M \cap F$ **c.** M'
   d. F' **e.** $T \cap S$

28. Describe verbally each of the following sets:
   a. $M \cap D$ **b.** $F \cap T$ **c.** $M \cap T'$
   d. $(T \cup S)'$ **e.** $(D \cap T)'$

29. Find a set representation for the set of:
   a. Employees in data processing who are over 65
   b. Female employees who are under 21
   c. Male employees who work in data processing

30. Write out in words the complement of each of the sets in Problem 29.

For Problems 31 and 32, refer to the data in Example 6. Find and describe the sets in each problem.

31. a. $F \cap S$ **b.** $P \cap (O \cup I)$
32. a. $P \cap I$ **b.** $P \cap O' \cap S'$

USING YOUR KNOWLEDGE 1.3

The ideas of sets, subsets, unions, intersections, and complements are used in zoology and in other branches of science. Here are some typical applications.

1. A zoology book lists the following characteristics of giraffes and okapis:

GIRAFFES	OKAPIS
Tall	Short
Long neck	Short neck
Long tongue	Long tongue
Skin-covered horns	Skin-covered horns
Native to Africa	Native to Africa

Let G be the set of characteristics of giraffes, and let O be the set of characteristics of okapis.
   a. Find $G \cap O$.
   b. What set of characteristics is common to okapis and giraffes?
   c. Find the smallest set $\mathcal{U}$ that will serve as a universal set for G and O.
   d. Find G'.   **e.** Find O'.

The Snake Families

Problems 2 and 3 refer to the table below.

2. Let B be the set of characteristics of blind snakes, T be the set of characteristics of thread snakes, P be the set of characteristics of pipe snakes, and BP be the set of characteristics of boas and pythons. Find:
   a. $B \cap T \cap P$ **b.** $B \cap T \cap P \cap BP$

3. If TL is the set of snakes with two lungs and SS is the set of snakes with spiny shields on the tail, find $TL \cap SS'$.

				FAMILY	NUMBER OF SPECIES	COMMON NAME
With limb remnants	Burrowers	All teeth in upper jaw		Typhlopidae	200	Blind snakes
		All teeth in lower jaw		Leptotyphlopidae	40	Thread snakes
		Eyes with spectacle		Anilidae	10	Pipe snakes
	Surface dwellers			Boidae	100	Boas and pythons
No limb remnants	Two lungs	Spiny shield on tail		Uropeltidae	40	Shieldtails
				Xenopeltidae	1	Sunbeam snakes
	One lung	No poison fangs or back fanged		Colubridae	2500	Typical snakes
		Rigid fangs	Land	Elapidae	150	Cobras
			Sea	Hydrophidae	50	Seasnakes
		Folding fangs		Viperidae	80	Vipers

DISCOVERY 1.3

Let the sets B_1 and B_2 be defined as follows:

a. Place the numbers 1 and 2 in B_1 and also in B_2.

b. At $\frac{1}{2}$ hr before noon, take the smallest number—that is, 1—from B_1 and the largest number—that is, 2—from B_2. At this time, $B_1 \cap B_2 = \varnothing$.

c. Place the numbers 3 and 4 in B_1 and also in B_2.

d. At $\frac{1}{4}$ hr before noon, take the smallest number—that is, 2—from B_1 and the largest number—that is, 4—from B_2. At this time, $B_1 \cap B_2 = \{3\}$.

e. Place the numbers 5 and 6 in B_1 and also in B_2.

f. At $\frac{1}{8}$ hr before noon, take the smallest number—that is, 3—from B_1 and the largest number—that is, 6—from B_2. At this time, $B_1 \cap B_2 = \{5\}$.

Continue this process.

1. Verify that at $\frac{1}{16}$ hr before noon, $B_1 \cap B_2 = \{5, 7\}$.

2. Verify that at $\frac{1}{32}$ hr before noon, $B_1 \cap B_2 = \{7, 9\}$.

3. Verify that at $\frac{1}{64}$ hr before noon, $B_1 \cap B_2 = \{7, 9, 11\}$.

Note that at $\frac{1}{16} = (\frac{1}{2})^4$ hr before noon, the largest number in $B_1 \cap B_2$ was 7, and that at $\frac{1}{32} = (\frac{1}{2})^5$ hr before noon, the largest number in $B_1 \cap B_2$ was 9.

4. Can you discover what will be the largest number in $B_1 \cap B_2$ at $(\frac{1}{2})^n$ hr before noon?

5. Can you discover what the smallest number in $B_1 \cap B_2$ will be at $(\frac{1}{2})^n$ hr before noon?

1.4

VENN DIAGRAMS

The ideas of sets, subsets, and the operations used to combine sets can be illustrated graphically by the use of diagrams called **Venn diagrams,** after John Venn (1834–1923), an English mathematician and logician. In these diagrams, we represent the universal set $\mathcal{U}$ by a rectangle, and we use regions enclosed by simple curves (usually circles) drawn inside the rectangle to represent the sets being considered. For example, if A is a subset of a universal set $\mathcal{U}$, we can represent this universal set by the set of points in the interior of the rectangle shown in Fig. 1.4a. The interior of the circle represents the set of points in A, while the set of points inside the rectangle and outside the circle represents the set A'. Obviously, closed figures other than circles may be used to represent the points of the set A. Figure 1.4b shows a Venn diagram in which A is represented by the points inside a triangle.

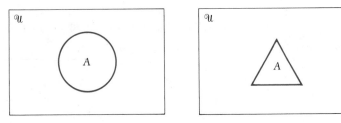

Figure 1.4a   **Figure 1.4b**

A. Drawing Venn Diagrams

We illustrate the idea of Venn diagrams in the following examples.

EXAMPLE 1   Let $\mathcal{U} = \{a, b, c, d, e\}, A = \{a, b, c\}$, and $B = \{a, e\}$. Draw a Venn diagram to illustrate this situation.

Solution
We draw a rectangle whose interior points represent the set $\mathscr{U}$ and two circles whose interior points represent the points in A and B. The completed diagram appears in Fig. 1.4c. We note that a is in both A and B, because $A \cap B = \{a\}$. Also, d is the only element that is not in A or in B, so $(A \cup B)' = \{d\}$. ◄

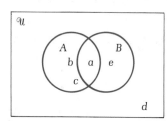

Figure 1.4c

Intersections and unions of sets can be represented by Venn diagrams. For example, given two sets A and B, we can draw a Venn diagram to represent the region corresponding to $A \cap B$. We proceed as follows:

1. As usual, the points inside the rectangle represent $\mathscr{U}$, and the points inside the two circles represent A and B (Fig. 1.4d). Note that A and B overlap to allow for the possibility that A and B have points in common.
2. We shade the set A using vertical lines (Fig. 1.4e).
3. We shade the set B using horizontal lines (Fig. 1.4f). The region in which the lines intersect (cross-hatched in the diagram) is the region corresponding to $A \cap B$.

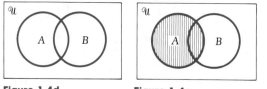

Figure 1.4d   **Figure 1.4e**   **Figure 1.4f** $A \cap B$

EXAMPLE 2   Draw a Venn diagram to represent the set $A' \cap B$.

Solution   We proceed as in Example 1:

1. We draw a rectangle and two circles as in Fig. 1.4g.
2. We shade the points of A' (the points outside A) with vertical lines.
3. We shade the points of B with horizontal lines. Then $A' \cap B$ is represented by the cross-hatched region. ◄

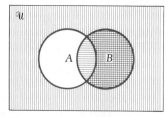

Figure 1.4g $A' \cap B$

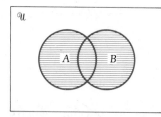

Figure 1.4h $A \cup B$

We may associate with the formation of $A \cap B$ a command to shade the region representing the set to the left of the symbol $\cap$ one way and the region representing the set to the right of the symbol another way. For example, in finding $A \cap B$, because A is to the left of $\cap$, we shade region A vertically, and because B is to the right of the symbol $\cap$, we shade region B horizontally. As before, $A \cap B$ is represented by the region in which the lines intersect. (If $A \cap B = \varnothing$, A and B are said to be **disjoint**.) On the other hand, the operation $\cup$ may be thought of as a command to shade the regions representing the sets to the left and right of the symbol $\cup$ with the same type of lines (horizontal or vertical). Thus, in finding $A \cup B$, we shade A with, say, horizontal lines and shade B in the same way. The union of A and B will be the entire shaded region (see Fig. 1.4h).

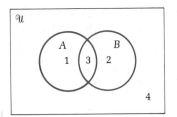

Figure 1.4i

A somewhat simpler procedure can be adopted in the construction of Venn diagrams. As before, we start with a rectangle and two circles representing the sets $\mathcal{U}$, A, and B, respectively. We then number the four regions into which the universal set is divided (see Fig. 1.4i). The numbering of the regions is completely arbitrary. By referring to the figure, we can identify the various sets as follows:

$A \cap B$ is the set of elements common to A and B—represented by region 3.

$A' \cap B$ is the set of elements that are not in A and that are in B—represented by region 2.

$A \cap B'$ is the set of elements in A and not in B—represented by region 1.

$A \cup B$ is the set of elements in A or in B, or in both A and B—represented by regions 1, 2, 3.

A' is the set of elements that are not in A—represented by regions 2, 4.

B' is the set of elements that are not in B—represented by regions 1, 4.

B. Verifying Equality

Venn diagrams are convenient for analyzing problems involving sets as long as there are not many subsets of $\mathcal{U}$ to be considered. For example, referring to Fig. 1.4i, we note that $A \cap B$ is the set of points in A and in B (region 3); but $B \cap A$ is the set of points in B and in A (region 3). Hence, these two sets refer to the same region, and we can see that $A \cap B = B \cap A$.

EXAMPLE 3   If A, B, and C are subsets of $\mathcal{U}$, use the preceding method to show that $A \cap (B \cup C) = (A \cap B) \cup (A \cap C)$.

Solution   We draw the rectangle and the circles representing the sets $\mathcal{U}$, A, B, and C, and number the regions 1, 2, 3, 4, 5, 6, 7, 8, as shown in the figure in the margin. Note that when we had 2 sets, we used $2^2 = 4$ regions. In this example, we have 3 sets; hence, we need $2^3 = 8$ regions.

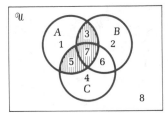

We first consider $A \cap (B \cup C)$:

a. A is represented by the set of regions $\{1, 3, 5, 7\}$.
b. $B \cup C$ is represented by the set of regions $\{2, 3, 4, 5, 6, 7\}$.
c. $A \cap (B \cup C)$ is therefore represented by the set of regions common to the two sets in parts a and b—that is, by the set of regions $\{3, 5, 7\}$.

We next consider $(A \cap B) \cup (A \cap C)$:

d. $(A \cap B)$ is represented by the set of regions common to the circles representing A and B—that is, by the set of regions $\{3, 7\}$.
e. $(A \cap C)$ is represented by the regions common to the circles representing A and C—that is, by the set of regions $\{5, 7\}$.

f. $(A \cap B) \cup (A \cap C)$ is therefore represented by the union of the two sets of regions found in parts d and e — that is, by the set of regions {3, 5, 7}.

Because $A \cap (B \cup C)$ and $(A \cap B) \cup (A \cap C)$ are both represented by the set of regions {3, 5, 7}, we see that

$$A \cap (B \cup C) = (A \cap B) \cup (A \cap C) \qquad \blacktriangleleft$$

C. Applications

EXAMPLE 4   There are three antigens in human blood, A, B, and Rh. A person can be A, or B, or AB, depending on which of the antigens the person has, with type O having neither A nor B. A person is Rh positive if the person has the Rh antigen and Rh negative otherwise; plus and minus signs are used to indicate this. For example, AB⁺ means that the person has all three antigens, and AB⁻ means that the person has the A and B antigens, but not the Rh antigen. Draw a Venn diagram and identify each of the areas.

Solution   We draw the rectangle and circles representing the sets $\mathcal{U}$, A, B, and Rh, as shown in the margin. The eight types of blood are A⁻, A⁺, B⁻, B⁺, AB⁻, AB⁺, O⁻, O⁺, where A⁺ means that the person has both A and Rh antigens, A⁻ means that the person has antigen A but not Rh, and similarly for the remaining symbols. Note that the circle labeled A represents the set of persons having A⁺, A⁻, AB⁺, or AB⁻ blood type, and likewise for the other two circles. $\qquad \blacktriangleleft$

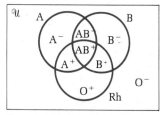

Human blood types

EXAMPLE 5   The ideas presented in this section have been used in recent years by forecasters in the National Weather Service. For example, in the accom-

panying map, rain, showers, snow, and flurries are indicated with different types of shadings. Find the states in which:

a. It is warm and there are showers.   **b.** It is warm and raining.
c. It is warm only.   **d.** It is snowing.
e. There are snow flurries.

Solution   **a.** Florida   **b.** Texas   **c.** California, Arizona
d. North Dakota, Minnesota   **e.** Minnesota   ◄

EXERCISE 1.4

A.   1. Let $\mathcal{U} = \{1, 2, 3, 4, 5\}$, $A = \{1, 2\}$, and $B = \{1, 3, 5\}$. Draw a Venn diagram to illustrate the relationship among these sets.
2. Do the same as in Problem 1 for the sets $\mathcal{U} = \{a, b, c, d, e, f\}$, $A = \{a, b, c\}$, and $B = \{d, e, f\}$.
3. Draw a Venn diagram to represent each of the following:
   a. $A \cap B'$ **b.** $A' \cup B'$ **c.** $(A \cup B) - (A \cap B)$
4. Draw a Venn diagram to represent each of the following:
   a. $A \cup B'$ **b.** $A' \cap B'$ **c.** $(A \cup B) - A$

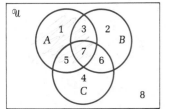

5. By using the numbered regions of the figure shown in the margin, identify each of the following sets:
   a. $A - (B \cup C)$ **b.** $C \cap (A \cup B)$
   c. $(A \cap B \cap C) - (A \cap B)$ **d.** $(A \cap B') \cup (A \cap C')$
   e. $(A \cup B') \cap C$
6. By using the numbered regions of the figure in the margin, identify each of the following sets:
   a. $(A \cup B) - C$ **b.** $(A \cap B') \cup C$
   c. $B \cap C' \cap A$ **d.** $(A \cup B \cup C)'$
7. Draw a Venn diagram to illustrate each of the following sets:
   a. $\{x | x \in A \text{ or } x \in B\}$ **b.** $\{x | x \notin A \text{ or } x \notin B\}$
   c. $\{x | x \in A \text{ and } x \notin B\}$
   d. $\{x | x \in A \text{ and } (x \in B \text{ and } x \notin C)\}$
8. Draw a Venn diagram to illustrate each of the following sets:
   a. $\{x | x \in A \text{ and } x \in B\}$ **b.** $\{x | x \notin A \text{ and } x \in B\}$
   c. $\{x | x \in A \text{ or } (x \in B \text{ or } x \in C)\}$
   d. $\{x | x \notin A \text{ and } (x \in B \text{ and } x \in C)\}$

B.   9. Draw a Venn diagram such that:
   a. $A \cap B = \emptyset$ **b.** $A \cap B = B$
   c. $(A \cup B) \cap C = \emptyset$ **d.** $A \cap (A \cap B) = A$

In Problems 10–14, use the numbered regions in the figure for Problems 5 and 6 to verify the given equalities.

10. **a.** $A \cup B = B \cup A$

b. $A \cap B = B \cap A$

(These two equations are called the **commutative laws** for set operations.)

11. **a.** $A \cup (B \cup C) = (A \cup B) \cup C$

    b. $A \cap (B \cap C) = (A \cap B) \cap C$

    (These two equations are called the **associative laws** for set operations.)

12. $A \cup (B \cap C) = (A \cup B) \cap (A \cup C)$

    [This equation and the equation $A \cap (B \cup C) = (A \cap B) \cup (A \cap C)$, which was verified in Example 3, are known as the **distributive laws** for set operations.]

13. **a.** $A \cup A' = \mathcal{U}$ **b.** $A \cap A' = \varnothing$

    c. $A - B = A \cap B'$

14. **a.** $(A \cup B)' = A' \cap B'$ **b.** $(A \cap B)' = A' \cup B'$

    (These two equations are known as **De Morgan's laws.**)

15. Referring to the figure for Problems 5 and 6, the set of regions $\{3, 7\}$ represents which of the following?

    a. $A \cap B$ **b.** $A \cap B \cap C$ **c.** $(A \cup B) \cap C$

    d. $(A \cap B) \cup C$ **e.** None of these

16. Referring to the figure for Problems 5 and 6, the set of regions $\{1, 2, 3\}$ represents which of the following?

    a. $(A \cup B) \cap C$ **b.** $(A \cup B) \cap C'$

    c. $(A \cap B) \cup C$ **d.** $(A \cap B) \cup C'$

    e. None of these

17. Given $A \cap B = \{a, b\}$, $A \cap B' = \{c, e\}$, $A' \cap B = \{g, h\}$, and $(A \cup B)' = \{d, f\}$, use a Venn diagram to find:

    a. A, B, and $\mathcal{U}$ **b.** $A \cup B$ **c.** $(A \cap B)'$

18. Given $A \cap B = \{b, d\}$, $A \cup B = \{b, c, d, e\}$, $A \cap C = \{b, c\}$, and $A \cup C = \{a, b, c, d\}$, use a Venn diagram to find:

    a. A, B, and C **b.** $A \cap B \cap C$ **c.** $A \cup B \cup C$

19. Draw a Venn diagram representing the most general situation for four sets A, B, C, and D. [Hint: There should be $2^4 = 16$ regions.]

20. Referring to the map in Example 5, find the states in which it is warm.

21. Referring to the map in Example 5, find the states in which it is warm and/or it is raining.

22. Referring to the map in Example 5, find the states in which it is warm and it is raining or there are showers.

USING YOUR KNOWLEDGE 1.4

In Example 4 of this section, blood was classified into eight different types. In blood transfusions, the recipient (the person receiving the blood) must have all or more of the antigens present in the donor's blood. For instance,

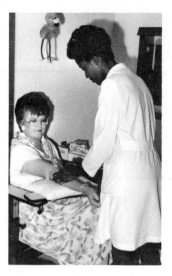

an A^+ person cannot donate blood to an A^- person, because the recipient does not have the Rh antigen; but an A^- person can donate to an A^+ person. Refer to the Venn diagram in Example 4 and:

1. Identify the blood type of universal recipients. *AB^+*
2. Identify the blood type of universal donors.
3. May an AB^- person give blood to a B^- person? *no*
4. May a B^- person give blood to an AB^- person?
5. May an O^+ person give blood to an O^- person? *no*
6. May an O^- person give blood to an O^+ person?

DISCOVERY 1.4

John Venn, the English logician who discovered the diagrams introduced in this section, used them to illustrate his work, Symbolic Logic. The Swiss mathematician Leonhard Euler (1707–1783) also used similar diagrams to illustrate his work. For this reason, Venn diagrams are sometimes called **Euler circles.**

We have seen in the preceding examples that if we have one set, the corresponding Venn diagram divides the universe into two regions. Two sets divide the universe into four regions, and three sets divide it into eight regions.

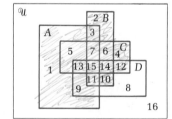

1. Can you discover the maximum number of regions into which four sets will divide the universe? *16*
2. The diagram for a division of the universal set into the 16 regions corresponding to four given sets may look like that at left. Can you guess the maximum number of regions into which n sets will divide the universe?
3. Referring to the diagram for Problem 2, find the regions corresponding to:

   a. $A \cap B \cap C' \cap D$
   {11}

   b. $(A \cup B \cup C)'$
   {8, 16}

1.5

THE NUMBER OF ELEMENTS IN A SET

One of the simplest counting techniques involves the counting of elements in a given set. If A is any set, the number of elements in A is denoted by **$n(A)$**. For example, if $A = \{g, i, r, l\}$, then $n(A) = 4$. Likewise, if $B = \{@, \#, \$\}$, then $n(B) = 3$. We shall be interested here in counting the number of elements in sets involving the operations of union, intersection, and taking complements.

A. Counting the Elements of a Set

EXAMPLE 1

When registering for the Graduate Record Examination (GRE), students must fill in the appropriate box in the form shown in the margin. Suppose that 30 students filled in the first box, 50 students filled in the second box, and 20 students filled in the third box. If G is the set of students registering for the GRE and M is the set of students registering for the Minority Locater Service, find:

a. $n(G)$ **b.** $n(M)$ **c.** $n(G \cup M)$

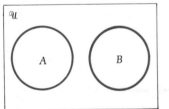

Solution

We first draw a Venn diagram showing the given information:

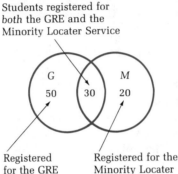

Students registered for *both* the GRE and the Minority Locater Service

Registered for the GRE *only*

Registered for the Minority Locater Service *only*

From the diagram above, we can see that:

a. $n(G) = 30 + 50 = 80$
b. $n(M) = 30 + 20 = 50$
c. $n(G \cup M) = 30 + 50 + 20 = 100$

Note that

$$n(G \cup M) = n(G) + n(M) - n(G \cap M)$$

that is,

$$100 = 80 + 50 - 30 \qquad \blacktriangleleft$$

We now examine the problem of finding the number of elements in the union of two sets in a more general way. Let us assume that A and B are any two given sets. We must consider two possibilities:

1. $A \cap B = \emptyset$ (see Fig. 1.5a): In this case, $n(A \cup B) = n(A) + n(B)$.
2. $A \cap B \neq \emptyset$ (see Fig. 1.5b): In this case, we note that $A \cup B$ includes all the elements in A and all the elements in B, but each counted only once. It is thus clear that $n(A) + n(B)$ counts the elements in $A \cap B$ twice and so exceeds $n(A \cup B)$ by $n(A \cap B)$. Therefore, $n(A \cup B) = [n(A) + n(B)] - n(A \cap B)$, or

Figure 1.5a $A \cap B = \emptyset$

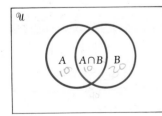

Figure 1.5b $A \cap B \neq \emptyset$

$$n(A \cup B) = n(A) + n(B) - n(A \cap B) \qquad (1)$$

Notice that equation (1) is correct even if $A \cap B = \varnothing$, because in that case $n(A \cap B) = 0$.

EXAMPLE 2   If $n(A) = 20$, $n(B) = 30$, and $n(A \cap B) = 10$, find $n(A \cup B)$.

Solution   Using equation (1), we have

$$n(A \cup B) = n(A) + n(B) - n(A \cap B) = 20 + 30 - 10 = 40 \qquad \blacktriangleleft$$

It is possible to develop a formula similar to equation (1) for the case where three or more sets are considered. However, we will rely on the use of Venn diagrams to solve such problems.

B. Applications

Venn diagrams can be used to study surveys, as in the following examples.

EXAMPLE 3   A survey of students at Prince Tom University shows that:

   29 liked jazz.
   23 liked rock.
   40 liked classical music.
   10 liked classical music and jazz.
   13 liked classical music and rock.
    5 liked rock and jazz.
    3 liked rock, jazz, and classical music.

If there was a total of 70 students in the survey, find:

a. The number of students who liked classical music only
b. The number of students who liked jazz and rock, but not classical music
c. The number of students who did not like jazz, rock, or classical music

Courtesy Ellis Richman

Solution   Let J be the set of students who liked jazz, R be the set of students who liked rock, and C be the set of students who liked classical music. Since the data indicate some overlapping, we draw a Venn diagram as shown in the figure in the margin. We fill in the numbers in this figure as follows: Since 3 students liked all three types of music, the region common to J, R, and C ($J \cap R \cap C$) must contain the number 3. Next, we see that 5 students liked jazz and rock, so that the region common to J and R ($J \cap R$) must contain a total of 5. But we already have 3 in $J \cap R \cap C$, a portion of $J \cap R$, so we must put 2 in the remainder of $J \cap R$. (See the figure, step 1.)

Similarly, since 13 students liked rock and classical music, the region common to R and C ($R \cap C$) must contain a total of 13. We already have 3 in

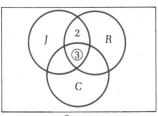

Step 1

$J \cap R \cap C$, a portion of $R \cap C$, so we must put 10 in the remainder of $R \cap C$ (step 2).

Since 10 students liked classical music and jazz, the region common to C and J ($J \cap C$) must contain a total of 10. We already have 3 in a portion of this region ($J \cap R \cap C$), so the remainder of the region must contain $10 - 3 = 7$ (step 3).

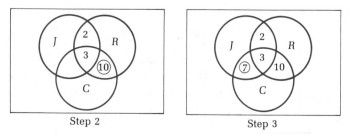

Step 2   Step 3

Next, we see that 40 students liked classical music, and we have already accounted for $7 + 3 + 10 = 20$ of these. Thus, the remainder of region C, outside of regions J and R, must contain $40 - 20 = 20$ (step 4).

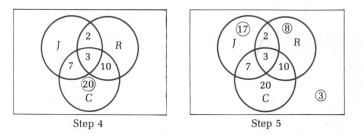

Step 4   Step 5

By proceeding in the same way, we can fill in the numbers for the remaining regions in the figure. From the completed diagram (step 5), we see that:

a. 20 students liked classical music only.
b. 2 students liked jazz and rock, but not classical music.
c. Since the numbers inside the circles add up to 67, 3 students did not like jazz, rock, or classical music. ◀

EXAMPLE 4

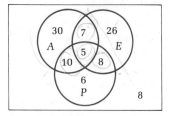

In a survey of 100 students, the numbers taking algebra (A), English (E), and philosophy (P) are as shown in the figure in the margin.

a. How many students were taking algebra or English, but not both?
b. How many students were taking algebra or English, but not philosophy?
c. How many students were taking one or two of these courses, but not all three?
d. How many students were taking at least two of these courses?
e. How many students were taking at least one of these courses?

Solution   **a.** Here, we want all the elements in A or E, but not in both. We read the
numbers from the Venn diagram to get

$$30 + 10 + 26 + 8 = 74$$

Note that we have taken all the numbers in $A \cup E$ except those in
$A \cap E$. $\{7,5\}$

b. Here, we need all the numbers in A or E that are not in P. From the
diagram, we find

$$30 + 7 + 26 = 63$$

c. The required number here is the number in the entire universal set,
100, minus the number taking all three courses or none of these
courses. Thus, the result is

$$100 - 5 - 8 = 87$$

d. The number wanted here is the number in $(A \cap E) \cup (A \cap P) \cup (E \cap P)$,
which is

$$7 + 5 + 10 + 8 = 30$$

e. We can get the required number by taking the number in the universal
set minus the number taking none of these courses. Thus, we find the
result to be

$$100 - 8 = 92 \qquad \blacktriangleleft$$

EXAMPLE 5   The Safe-T Auto Insurance Company has classified a group of drivers as
indicated in the table. Find the number of persons who:

a. Are low risks under 21   **b.** Are not high risks and are over 35
c. Are under 21   **d.** Are low risks

	UNDER 21	21–35	OVER 35
Low risk	15	20	35
Average risk	25	15	10
High risk	50	10	30

Solution   **a.** We see that there are 15 persons under the column labeled "Under 21"
and in the row of "Low risk."

b. There are 30 persons who are over 35 and high risks, so the rest of the
persons in the column labeled "Over 35" are not high risks. This number is $10 + 35 = 45$.

c. We count all the persons in the column labeled "Under 21." The sum is
$15 + 25 + 50 = 90$.

d. We count the persons in the row labeled "Low risk." The sum is $15 +$
$20 + 35 = 70$. $\blacktriangleleft$

A.

1. Suppose that $n(A) = 15$, $n(B) = 20$, and $n(A \cap B) = 5$. Find $n(A \cup B)$.
2. Suppose that $n(A) = 12$, $n(B) = 6$, and $n(A \cup B) = 14$. Find $n(A \cap B)$.
3. Suppose that $n(A) = 15$, $n(A \cap B) = 5$, and $n(A \cup B) = 30$. Find $n(B)$.
4. There are 50 students in an algebra (A) class and 30 students in a chemistry (C) class. Find:
    a. $n(A)$ **b.** $n(C)$
    c. The total number of students taking either algebra or chemistry, if it is known that none of the students is taking both courses
    d. The number of students taking algebra and/or chemistry if it is known that 10 students are taking both courses

5. Upon checking with 100 families, it was found that 75 families subscribe to *Time*, 55 to *Newsweek*, and 10 to neither magazine. How many subscribe to both?
6. If, on checking with 100 families, it was found that 83 subscribe to *Time*, 40 to *Newsweek*, and 30 to both magazines, how many subscribe to neither?

B.

7. In a survey of 100 students, the numbers taking various courses were found to be as follows: English, 60; mathematics, 40; chemistry, 50; English and mathematics, 30; English and chemistry, 35; mathematics and chemistry, 35; courses in all three areas, 25.
    a. How many students were taking mathematics but neither English nor chemistry?
    b. How many were taking mathematics and chemistry but not English?
    c. How many were taking English and chemistry but not mathematics?
8. Mr. N. Roll, the Registrar at Lazy U, has observed that, of the students:
    45% have a 9 A.M. class.
    45% have an 10 A.M. class.
    40% have an 11 A.M. class.
    20% have a 9 and a 10 A.M. class.
    10% have a 9 and an 11 A.M. class.
    15% have a 10 and an 11 A.M. class.
    5% have a 9, a 10, and an 11 A.M. class.
    a. What percent of the students have only a 9 A.M. class at these times?
    b. What percent of the students have no classes at these times?
9. The table at the top of the next page shows the distribution of employees at the Taste-T Noodle Company.
    a. How many employees are there in the Purchasing Department?
    b. How many skilled employees are there in the factory?
    c. How many of the skilled employees are in the Janitorial Department?

Personnel Distribution

DEPARTMENT	ADMIN-ISTRATOR (A)	CLERICAL (C)	OTHER (O)	SKILLED (SK)	SEMI-SKILLED (SS)	UNSKILLED (U)
Purchasing (P)	1	14	7	0	0	0
Quality Control (Q)	11	7	6	21	53	11
Sales (S)	8	8	40	0	0	0
Manufacturing (M)	5	7	0	9	23	37
Janitorial (J)	3	0	0	6	8	11

10. Use the table in Problem 9 to find the number of persons in the following sets:

   a. $A \cap S$ b. $S \cup P$

   c. $M \cap A' \cap SK'$ d. $S \cap A' \cap C'$

11. The table gives the estimated costs for a proposed computer system in 2-year intervals, projected over 10 years (figures in thousands of dollars).

Cost Projections

ITEM	YEAR				
	1–2	3–4	5–6	7–8	9–10
Data-processing equipment	215	240	260	295	295
Personnel	85	85	95	95	105
Materials	120	65	35	35	40
All others	90	85	90	95	120

   a. How much money would be spent on materials in the first 2 years?
   b. What would be the total cost at the end of the second year?
   c. What would be the cost of the data-processing equipment over the 10-year period?

12. In a survey of 100 investors, it was found that:

   5 owned utilities stock only.
   15 owned transportation stock only.
   70 owned bonds.
   13 owned utilities and transportation stock.
   23 owned transportation and bonds.
   10 owned utilities and bonds.
   3 owned all three kinds.

   a. How many investors owned bonds only?
   b. How many investors owned utilities and/or transportation stock?
   c. How many investors owned neither bonds nor utilities?

13. In a recent survey of readers of the *Times* and/or the *Tribune*, it was found that 50 persons read both the *Times* and the *Tribune*. If it is

known that 130 persons read the *Times* and 120 read the *Tribune*, how many people were surveyed?

14. In a survey conducted in a certain U.S. city, the data in the table were collected.

INCOME	WHITE (*W*)	BLACK (*B*)	OTHER (*O*)
Over $10,000 (*H*)	50	15	10
$7,000–$10,000 (*M*)	40	25	15
Under $7,000 (*L*)	30	35	20

Find the number of people in:
a. M **b.** M' **c.** $(O \cup B) \cap W'$
d. $L \cup O'$ **e.** $H \cap B'$

15. In a survey of 100 customers at the Royal Hassle Restaurant, it was found that:

40 had onions on their hamburgers.
35 had mustard on their hamburgers.
50 had catsup on their hamburgers.
15 had onion and mustard on their hamburgers.
20 had mustard and catsup on their hamburgers.
25 had onion and catsup on their hamburgers.
5 had onions, mustard, and catsup on their hamburgers.

a. How many customers had hamburgers with onions only?
b. How many customers had plain hamburgers (no condiments)?
c. How many customers had only one condiment on their hamburgers?

16. A survey of 900 workers in a plant indicated that 500 owned their homes, 600 owned cars, 345 owned boats, 300 owned cars and houses, 250 owned houses and boats, 270 owned cars and boats, and 200 owned all three.
a. How many of the workers did not own any of the three items?
b. How many of the workers owned only two of the items?

17. A coffee company was willing to pay $1 to each person interviewed about his or her likes and dislikes on types of coffee. Of the persons interviewed, 200 liked ground coffee, 270 liked instant coffee, 70 liked both, and 50 did not like coffee at all. What was the total amount of money the company had to pay?

18. In a recent survey, a statistician reported the following data:

15 persons liked brand *A*.
18 persons liked brand *B*.
12 persons liked brand *C*.
8 persons liked brands *A* and *B*.
6 persons liked brands *A* and *C*.
7 persons liked brands *B* and *C*.

2 persons liked all three brands.

2 persons liked none of the three brands.

When the statistician claimed to have interviewed 30 persons, he was fired. Can you explain why?

19. In Problem 18, a truthful statistician was asked to find out how many people were interviewed. Can you tell what this statistician's answer was?

20. In an experiment, it was found that a certain substance could be of type x or type y (not both). In addition, it could have one, both, or neither of the characteristics m and n. The table gives the results of testing several samples of the substance. Let M and N be the sets with characteristic m and n, respectively, and let X and Y be the sets of type x and y, respectively. How many samples are in each of the following sets?

a. $M \cap X$

b. $(X \cup Y) \cap (M \cup N)$

c. $(Y \cap M) - (Y \cap N')$

d. $(X \cup Y) \cap (M \cup N')$

	m only	n only	m and n	Neither m nor n
x	6	9	10	20
y	7	11	15	9

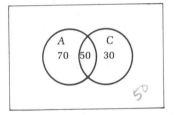

21. The number of students taking algebra (A) or calculus (C) is shown in the figure in the margin. Find:

a. $n(A)$

b. $n(C)$

c. $n(A \cap C)$

22. Referring to the figure in the margin, find $n(A \cup C)$.

23. If the total number of students surveyed to obtain the data of Problems 21 and 22 is 200, find:

a. $n(A')$

b. $n(C')$

c. $n(A' \cap C')$

24. With the total number of students as in Problem 23, find:

a. $n(A' \cup C)$

b. $n(A \cup C')$

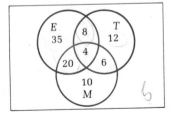

25. Upon checking with 100 investors to see who owned electric company stock (E), transportation stock (T), or municipal bonds (M), the numbers shown in the figure in the margin were found.

a. How many investors owned electric company or transportation stock but not both?

b. How many owned electric company or transportation stock but not municipal bonds?

c. How many had one or two of these types of investment but not all three?

d. How many had at least two of these types of investment?

e. How many had none of these?

26. A number of people were interviewed to find out who buys products $A, B,$ and C regularly. The results are shown in the figure in the margin on the next page.

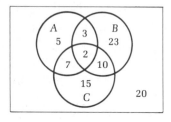

a. How many buy product A?
b. How many buy product A but not B?
c. How many buy product B or C but not A?
d. How many do not buy product C?
e. How many people were interviewed?

USING YOUR KNOWLEDGE 1.5

The cartoon shown here seems to indicate that it is impossible to have the morale statistics as follows:

58% want out (WO).
14% hate his guts (HG).
56% plan to desert (PD).
  8% are undecided (UD) (do not plan to do any of the above).

CROCK by Rechin, Parker, & Wilder. © Field Enterprises, Inc., 1976. Courtesy of Field Newspaper Syndicate.

However, a new statistician is hired and finds that in addition to the original information, the following statements are also true:

12% want to do only one thing—hate his guts.

36% want to do exactly two things. Of these, 34% want out and plan to desert, and 2% hate his guts and want out.

Of course, nobody in his right mind would do all three things.

1. Based on all the information, both old and new, draw a Venn diagram and show that it is possible to have the statistics quoted in the cartoon.

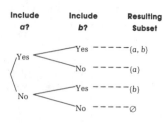

DISCOVERY 1.5

In Section 1.2 we discussed the subsets of a given set. It is interesting to diagram the formation of such subsets. We imagine that the elements of the given set are listed, and we look at each element in turn and decide whether or not to include it in the subset. For example, suppose that the given set is

Include a?	Include b?	Include c?	Resulting Subset

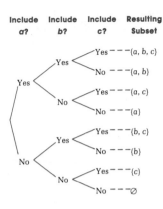

{a, b}. Then our diagram has two steps, as shown at the bottom of page 38. This diagram makes it clear that there are 2 × 2, or 4, subsets in all.

The diagram at the left is for the three-element set {a, b, c}.

Diagrams like these are called **tree diagrams.** The second tree diagram shows that a 3-element set has 2 × 2 × 2, or 8, subsets. Can you discover an easy way to explain this?

1. Can you discover the tree diagram for a 4-element set and its subsets?
2. Can you discover how to count the number of subsets of a 4-element set?
3. Can you now discover how to explain why an n-element set has
$$\underbrace{2 \times 2 \times 2 \times \cdots \times 2}_{n \text{ twos}}$$
or 2^n subsets?

SUMMARY

Section	Item	Meaning	Example
1.1A	{ }	Set braces	{1, 2, 3} is a set
1.1A	∈	Is an element of	$2 \in \{1, 2, 3\}$
1.1A	∉	Is not an element of	$4 \notin \{1, 2, 3\}$
1.1B	(1, 2)	List notation	
1.1B	{x|x has property P}	Set-builder notation	{x|x is a counting number}
1.1B	{ } or ∅	Empty, or null, set	The set of words that rhyme with "orange"
1.1C	$A = B$	A equals B	$\{1, 2\} = \{2, 1\}$
1.2	$A \subseteq B$	A is a subset of B	$\{1, 2\} \subseteq \{1, 2, 3\}$
1.2	$A \subset B$	A is a proper subset of B	$\{a\} \subset \{a, b\}$
1.2A	$\mathcal{U}$	Universal set	
1.3A	∩	Intersection	$\{1, 2, 3\} \cap \{2, 3, 4\} = \{2, 3\}$
1.3A	∪	Union	$\{1, 2, 3\} \cup \{2, 3, 4\} = \{1, 2, 3, 4\}$
1.3B	A'	Complement	If $\mathcal{U} = \{1, 2, 3, 4, 5\}$ and $A = \{1, 2\}$, then $A' = \{3, 4, 5\}$.
1.3C	$A - B$	Set difference	If $A = \{1, 2, 3, 4, 5\}$ and $B = \{1, 2\}$, then $A - B = \{3, 4, 5\}$.
1.5	$n(A)$	Number of elements in set A	If $A = \{a, b, c\}$ and $B = \{d, e\}$, then $n(A) = 3$ and $n(B) = 2$.

1. List the elements of the set:

   {x|x is a counting number between 2 and 10}

2. Describe the following sets verbally and using set-builder notation:
   a. {a, e, i, o, u}   **b.** {2, 4, 6, 8}

3. List all the proper subsets of the set {$, ¢, %}.

4. Complete the following definitions by filling in the blanks with the symbol $\in$ or $\notin$:
   a. $A \cup B = \{x|x$ _____ A or x _____ $B\}$
   b. $A \cap B' = \{x|x$ _____ A and x _____ $B\}$

5. Complete the following definitions by filling in the blanks with the symbol $\in$ or $\notin$:
   a. $A' = \{x|x$ _____ $\mathcal{U}$ and x _____ $A\}$
   b. $A - B = \{x|x$ _____ A and x _____ $B'\}$

6. Let $\mathcal{U} = \{$Ace, King, Queen, Jack$\}$, $A = \{$Ace, Queen, Jack$\}$, and $B = \{$King, Queen$\}$. Find:
   a. A' **b.** $(A \cup E)'$ **c.** $A \cap B$
   d. $\mathcal{U} - (A \cap B)'$

7. If, in addition to the sets in Problem 6, we have $C = \{$Ace, Jack$\}$, find:
   a. $(A \cap B) \cup C$ **b.** $(A' \cup C) \cap B$

8. Draw a pair of Venn diagrams to show that $A - B = A \cap B'$.

9. Draw a Venn diagram to illustrate the set $A \cap B \cap C'$.

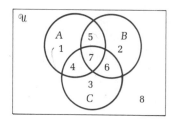

10. Find the sets of numbered regions in the figure in the margin that represent the following sets:
    a. $(A \cup B) \cap C'$ **b.** $A' \cup (B' \cap C)$

11. Use the numbered regions in the figure in the margin to verify that $(A \cap B) \cup C = (A \cup C) \cap (B \cup C)$.

12. Use the numbered regions in the figure in the margin to verify that $(A \cap B)' = A' \cup B'$.

For Problems 13–15, let $\mathcal{U} = \{1, 2, 3, 4, 5, 6, 7, 8\}$, $A = \{1, 3, 5, 7\}$, $B = \{2, 4, 6, 8\}$, and $C = \{1, 4, 5, 8\}$.

13. Fill in the blanks with $\in$ or $\notin$ to make correct statements:
    a. 2 _____ $A \cap B$ **b.** 4 _____ $A \cup (B \cap C)$
    c. 4 _____ $A \cap (B \cup C)$

In Problems 14 and 15, fill in the blanks with $=$ or $\neq$ to make correct statements.

14. **a.** $n(A \cup C)$ _____ 6   **b.** $n(B \cap C)$ _____ 3

15. **a.** $(A \cup C) \cap B$ _____ $\{4, 8\}$
    b. $(A \cap C) \cup B$ _____ $\{1, 2, 3, 5, 7, 8\}$

16. Let $n(A) = 25$ and $n(B) = 35$. Find $n(A \cup B)$ if:
    a. $A \cap B = \varnothing$ **b.** $n(A \cap B) = 5$

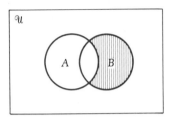

17. Let $n(A) = 15$, $n(B) = 25$, and $n(A \cup B) = 35$. Find:
    a. $n(A \cap B)$ **b.** $n(\mathcal{U})$ if $n(A' \cap B') = 8$

18. In the figure in the margin, the rectangular region represents the universal set $\mathcal{U}$, and the circular regions represent the subsets A and B of $\mathcal{U}$. Find an expression for the shaded region in the diagram.

19. Upon checking 200 students, it is found that 70 are taking French, 40 are taking German, 75 are taking Spanish, 10 are taking French and German, 30 are taking French and Spanish, 15 are taking German and Spanish, and 70 are taking no language. If it is known that no students are taking all three languages, draw a Venn diagram to determine the answers to the following questions:
    a. How many are taking two languages?
    b. How many are taking Spanish and no other language?
    c. How many are taking Spanish and not French?

20. A survey of people to determine who buys products A, B, and C regularly gave the numbers shown in the figure in the margin.
    a. How many people were surveyed?
    b. How many buy product A but not B?
    c. How many buy product B or C but not A?
    d. How many buy both products B and C but not A?
    e. How many do not buy either product B or product C?

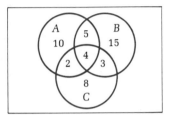

CHAPTER 2

LOGIC

George Boole, the son of an impoverished shopkeeper, was born in Lincoln, England, on November 2, 1815. Even as a small boy, Boole's overriding ambition was to lift himself out of the social mire into which he had been born. With the mistaken idea that a knowledge of Latin and Greek was the key to upper-class society, he taught himself both languages.

By the age of 16, Boole was educated enough to become an assistant teacher in an elementary school. The wage was pitifully small, and the poverty of his parents forced him to continue a miserable existence for about 4 years. At 20, he opened an elementary school of his own.

Boole had received early instruction in the rudiments of mathematics from his father, who had by his own efforts been able to educate himself beyond the level of the common schools. This was essentially Boole's training in mathematics when he discovered in his own school how miserable were the mathematics textbooks of the time. He became deeply interested in learning more mathematics, which he again did on his own, even making his first original contribution to the subject with no aid from others.

Now his self-training in Latin and Greek proved its value. He sent a paper to the influential Scottish mathematician D. F. Gregory, who was so impressed with the style and content that he had the paper published. With Gregory's friendship, Boole was assured of the publication of his future work.

During the next few years, Boole also became a friend and admirer of Augustus De Morgan (1806–1871), a vigorous mathematician and a great logician. Greatly influenced by De Morgan, Boole published a pamphlet, *The Mathematical Analysis of Logic,* in 1848. At the age of 34, Boole was recognized by being appointed Professor of Mathematics at Queen's College in Cork, Ireland.

Then, with an assured income and adequate time, he began to work earnestly on completing his masterpiece. In 1854, he published it: *An Investigation of the Laws of Thought, on Which Are Founded the Mathematical Theories of Logic and Probabilities.* Boole was 39.

In 1855, Boole married Mary Everest, niece of the professor of Greek in Queen's College. With her, he lived a happy few years, honored and growing in fame. He died of pneumonia in December 1864.

Since Boole's original work, his great invention has been improved and extended in many directions. Today, symbolic (or mathematical) logic is indispensable for anyone who wishes to understand the nature of logic or of mathematics.

GEORGE BOOLE (1815–1864)
The Bettmann Archive

Others before Boole, notably Leibniz and De Morgan, had dreamed of adding logic to the domain of algebra; Boole did it.

E. T. BELL

STATEMENTS

The word "logic" is derived from the Greek word *logos*, which may be interpreted as "reason" or "discourse." The ancient Greeks are usually credited with initiating the study of our reasoning process. The principles discovered by the Greeks were first systematized by Aristotle (ca. 384–322 B.C.), and Aristotle's type of reasoning constitutes the traditional logic that has been studied and taught from his time to the present day. A simple illustration of Aristotelean logic goes as follows:

All men are mortal.
Socrates was a man.
Therefore, Socrates was mortal.

This is a typical argument that is known as a "syllogism."

Although the modern study of symbolic logic (the subject of this chapter) finds its roots in the works of such men as the German mathematician Leibniz (1647–1716), modern symbolism and algebraic-type operations were first systematically applied to logic in the works of George Boole (see the introduction to this chapter).

The study of logic has developed at an accelerated pace since Boole's analysis, and considerable progress toward an understanding of logical truth has been made. Whether we are trying to solve a problem, taking part in a debate, or working a crossword puzzle, we are engaged in a mental activity called "logical reasoning." This reasoning is usually expressed in terms of declarative sentences, and it is to the study of these sentences that we now turn our attention.

A. Recognizing Statements

In this and the following sections, we shall be concerned with certain types of declarative sentences called **statements** and the manner in which we can combine such sentences and arrive at valid conclusions.

> In general, a **statement** is a declarative sentence that can be classified as true or false, but not both simultaneously.

This capability of being classified as true or false makes statements different from questions, commands, or exclamations. Questions can be asked, commands given, and exclamations shouted, but only statements can be classified as true or false. The sentences in Example 1 are illustrations of statements.

EXAMPLE 1   **a.** Boston is the capital of Massachusetts.
            b. 2 is even and less than 20.

c. There are 5 trillion grains of sand in Florida.

d. Either you study daily or you get an F in this course.

e. If 2 is even, then 2 + 2 is even.   ◀

Note that the truth or falsity of the first statement in Example 1 can be determined by a direct check, while the third one is true or false, even though there are no immediate or practical methods to determine its truth or falsity.

In contrast with the statements in Example 1, the following are illustrations of **nonstatements.**

EXAMPLE 2   **a.** What time is it?

b. Dagwood for president!

c. Good grief, Charlie Brown!

d. Close the door.

e. This statement is false.   ◀

The sentences in Example 2 are not statements. Notice that if we assume that sentence e is true, then it is false, and if we assume that it is false, then it is true. Hence, the sentence cannot be classified as either true or false, so it is not a statement.

B.   Conjunction, Disjunction, and Negation

Having explained what is meant by a statement, we now turn our attention to various combinations of statements. In Example 1, for instance, statements a and c have only one component each (that is, each says only one thing); while statement b is a combination of two components, namely, "2 is even," and "2 is less than 20." Statements a and c are **simple,** while statement b is **compound.**

As a further example, "John is 6 ft tall" is a simple statement. On the other hand, the statement, "John is 6 ft tall, *and* he plays basketball," is a compound statement, because it is a combination of the two simple statements, "John is 6 ft tall" and "he plays basketball."

As the reader may realize, there are many ways in which simple statements can be combined to form compound statements. Such combinations are formed by using words called **connectives** to join the statements. Two of the most important connectives are the words "and" and "or." Suppose we use the letters p and q to represent statements as follows:

p: It is hot today.

q: The air-conditioner in this room is broken.

Then we can form the following compound sentences:

p *and* q: It is hot today, *and* the air-conditioner in this room is broken.

p *or* q: It is hot today, *or* the air-conditioner in this room is broken.

In the study of logic, the word "and" is symbolized by $\land$ and the word "or" by $\lor$. Thus,

p **and** q is written $p \land q$

p **or** q is written $p \lor q$

▶ **Definition 2.1a**

If two statements are combined by the word **and** (or an equivalent word), the resulting statement is called a **conjunction.** If the two statements are symbolized by p and q, respectively, then the conjunction is symbolized by $p \land q$.

EXAMPLE 3   Symbolize the following conjunctions:

a. Tom is taking a math course, and Mary is taking a physics course.
b. Ann is passing math, but she is failing English.

Solution   **a.** Let m stand for "Tom is taking a math course" and p stand for "Mary is taking a physics course." Then the given conjunction may be symbolized by $m \land p$.

b. Let p stand for "Ann is passing math" and f stand for "she is failing English." The given conjunction is then symbolized by $p \land f$. Here, the word "but" is used in place of "and." ◀

▶ **Definition 2.1b**

If two statements are combined by the word **or** (or an equivalent word), the result is called a **disjunction.** If the two statements are symbolized by p and q, respectively, then the disjunction is symbolized by $p \lor q$.

EXAMPLE 4   Symbolize the disjunction, "We stop inflation, or we increase wages."

Solution   Letting p stand for "We stop inflation" and q stand for "we increase wages," we can symbolize the disjunction by $p \lor q$. ◀

Another construction important in logic is that of negating a given statement.

▶ **Definition 2.1c**

The **negation** of a given statement is a statement that is false whenever the given statement is true, and true whenever the given statement is false. If the given statement is denoted by p, its negation is denoted by $\sim p$. (The symbol $\sim$ is called a "tilde.")

The negation of a statement can always be written by prefixing it with a phrase such as, "it is not the case that." Sometimes, the negation can be obtained simply by inserting the word "not" in the given statement. For example, the negation of the statement, "Today is Friday," can be written as,

"It is not the case that today is Friday."

Or as,

"Today is not Friday."

Similarly, if p stands for "It is hot today," then ~p (read, "not p") may be written either as, "It is not hot today," or as, "It is not the case that it is hot today."

In the preceding illustrations, we have negated simple statements. We often have to consider the negation of compound statements, as in the next example.

EXAMPLE 5   Let p be "the sky is blue," and let q be "it is raining." Translate the following statements into English:

a. ~(p ∧ q)   **b.** ~p ∧ ~q   **c.** ~q ∧ p

Solution   **a.** It is not the case that the sky is blue and it is raining. Another form of the negation is: The sky is not blue or it is not raining.
b. The sky is not blue and it is not raining.
c. It is not raining and the sky is blue.   ◄

The two forms of the solution to part a of Example 5 illustrate the fact that the negation of p ∧ q can be written either as ~(p ∧ q) or as ~p ∨ ~q. Thus,

~(p ∧ q)   means   ~p ∨ ~q

Similarly,

~(p ∨ q)   means   ~p ∧ ~q

because the statement p or q is false when and only when p and q are *both* false. Thus, we have **De Morgan's laws:**

~(p ∧ q)   means   ~p ∨ ~q
and
~(p ∨ q)   means   ~p ∧ ~q

EXAMPLE 6   Consider the two statements:

p: Sherlock Holmes is alive.
q: Sherlock Holmes lives in London.

Write the following statements in symbolic form:

a. Sherlock Holmes is alive, and he lives in London. p ∧ q
b. Either Sherlock Holmes is alive, or he lives in London. p ∨ q
c. Sherlock Holmes is neither alive, nor does he live in London. ~p ∧ ~q
d. It is not the case that Sherlock Holmes is alive and he lives in London.
~(p ∧ q)

Drawing by Angrave

Solution   **a.** $p \wedge q$ **b.** $p \vee q$ **c.** $\sim p \wedge \sim q$ **d.** $\sim(p \wedge q)$ ◀

Be sure to notice the use of parentheses to indicate which items are to be taken as a unit. Thus, in part a of Example 5 and in part d of Example 6, $\sim(p \wedge q)$ means the negation of the entire statement $(p \wedge q)$. It is important to distinguish $\sim(p \wedge q)$ from $\sim p \wedge q$. The latter means that only the statement p is negated. For example, if p is "John likes Mary," and q is "Mary likes John," then $\sim(p \wedge q)$ is, "It is not true that John and Mary like each other." But $\sim p \wedge q$ is, "John does not like Mary, but Mary likes John."

EXAMPLE 7   Let p be the statement, "Tarzan likes Jane," and let q be the statement, "Jane likes Tarzan." Write in words:

a. $\sim(p \wedge q)$ **b.** $\sim p \vee \sim q$ **c.** $\sim p \wedge \sim q$

Solution   **a.** It is not the case that Tarzan likes Jane and Jane likes Tarzan. That is, it is not the case that Tarzan and Jane like each other.
b. Either Tarzan does not like Jane or Jane does not like Tarzan.
c. Tarzan does not like Jane, and Jane does not like Tarzan. That is, Tarzan and Jane dislike each other.   ◀

Some words require special care when dealing with negation. Examples of such words are **all, some,** and **none.** In general, the two statements, "all . . . are . . ." and "some . . . are not . . . ," are negations of each other. For instance, the following are negations of each other:

    p: All dogs are fierce.
    $\sim p$:   Some dogs are not fierce.

Note that another form for $\sim p$ is, "Not all dogs are fierce." However, the sentence, "All dogs are not fierce," is not the negation of p, because it says that there are no fierce dogs rather than simply negating the part, "*All* dogs are . . . ."

Similarly, two statements of the form, "some . . . are . . ." and "no . . . are . . . ," are negations of each other. For example, these two sentences are negations of each other:

    p: Some of us like football.
    $\sim p$:   None of us like football.

EXAMPLE 8   Write the negation of:

a. All of us like pistachio nuts.
b. Nobody likes freezing weather.
c. Some students work part-time.

Solution   **a.** Some of us do not like pistachio nuts. An alternate form is: Not all of us like pistachio nuts.

b. Somebody likes freezing weather.

c. No student works part-time.

◀

EXERCISE 2.1

A. *In Problems 1–8, determine whether the sentence is a statement. Classify each sentence that is a statement as simple or compound. If it is compound, give its components.*

1. Circles are dreamy.
2. Lemons and oranges are citrus fruits.
3. Jane is taking an English course, and she has four themes to write.
4. Apples are citrus fruits.
5. Do you like mathematics?
6. Walk a mile.
7. Students at Ohio State University are required to take either a course in history or a course in economics.
8. Today is Sunday, and tomorrow is Monday.

B. *In Problems 9–16, write the given statement in symbolic form using the letter in parentheses to represent the corresponding component.*

9. This is April (a), and income tax returns must be filed (f).
10. Logic is a required subject for lawyers (r) but not for most engineers ($\sim e$).
11. Dick Tracy is a detective (d) or a fictitious character in the newspaper (f). $d \vee f$
12. Snoopy is not an aviator ($\sim a$), or the Sopwith Camel is an airplane (p).
13. Violets are blue (b), but roses are pink (p).
14. The stock market goes up (u); nevertheless, my stocks stay down (d).
15. I will take art (a) or music (m) next term.
16. I will not drive to New York ($\sim d$); however, I shall go by train (t) or by plane (p).
17. Let p be, "Robin can type," and let q be, "Robin takes shorthand." Write the following statements in symbolic form:
    a. Robin can type and take shorthand. $p \wedge q$
    b. Robin can type, but does not take shorthand. $p \wedge \sim q$
    c. Robin can neither type nor take shorthand. $\sim(p \vee q)$ OR $\sim p \wedge \sim q$
    d. It is not the case that Robin can type and take shorthand. $\sim(p \wedge q)$ OR $\sim p \vee \sim q$
18. Let p be, "Dagwood loves Blondie," and let q be, "Blondie loves Dagwood." Give a verbal translation of each of the following statements:
    a. $p \vee \sim q$ **b.** $\sim(p \vee q)$ **c.** $p \wedge \sim q$
    d. $\sim p \wedge \sim q$ **e.** $\sim(p \wedge q)$

19. Write the negation of each of the following sentences:
  a. It is a long time before the end of the term.
  b. Bill's store is making a good profit.
  c. 10 is a round number.

20. Let the given statement be denoted by p. Write out the statement ~p.
  a. My dog is a spaniel.
  b. Your cat is not a Siamese.
  c. I do not like to work overtime.

21. Determine whether the statements p and q are negations of each other.
  a. p: Sally is a very tall girl. *no*
    q: Sally is a very short girl.
  b. p: All squares are rectangles. *yes*
    q: Some squares are not rectangles.
  c. p: All whole numbers are even. *yes*
    q: At least one whole number is not even.

22. Give the negation of each of the following statements:
  a. All men are mortal.
  b. Some women are teachers.
  c. Some basketball players are not 6 ft tall.
  d. Some things are not what they appear to be.

23. State the negation of each of the following statements:
  a. Either he is bald or he has a 10-inch forehead.
  b. Nobody does not like Sara Lee.
  c. Some circles are round
  d. Some men earn less than $5 an hour, and some men earn more than $50 an hour.

24. State the negation of each of the following statements:
  a. Somebody up there loves me.
  b. Nothing is certain but death and taxes.
  c. Everybody likes to go on a trip.

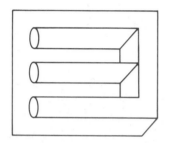

Some things are not what they appear to be.

2.2

TRUTH VALUES OF STATEMENTS

We regard *T* (for true) and *F* (for false) as the possible **truth values** of a statement. Thus, the statement, "George Washington was the first president of the United States" has the truth value *T*; and the statement, "The moon is made of green cheese" has the truth value *F*.

One of the principal problems in logic is that of determining the truth value of a compound statement when the truth values of its components are known. In order to attack this problem, we must first assign appropriate truth values to such statements as $p \wedge q$, $p \vee q$, and $\sim p$. Although the symbols $\wedge$, $\vee$, and $\sim$ were introduced in Section 2.1, they were not completely defined there. We shall complete their definitions by assigning appropriate truth values to statements involving these symbols.

A. The Conjunction

Suppose you are offered a position in a firm that requires that:

p: The applicant be at least 18 years of age

and

q: The applicant be a college graduate

We can easily see that to be eligible for this position you must meet *both* requirements; that is, both p and q must be true. Thus, it seems desirable to say that for a conjunction to be true, both components must be true. Otherwise, the conjunction is false.

If we have a conjunction with two components, p and q, we have four possible pairs of truth values for these statements, namely:

1. p true, q true   2. p true, q false
3. p false, q true   4. p false, q false

As in the preceding example, it seems reasonable to assign the value T to the statement $p \wedge q$ only when both components are true. The assignment of these truth values can be summarized by means of a **truth table,** as shown in Table 2.2a. This truth table is to be regarded as the definition of the symbol $\wedge$, and it expresses the fact that:

Table 2.2a Conjunction ($\wedge$)

1	2	3
p	q	$p \wedge q$
T	T	T
T	F	F
F	T	F
F	F	F

> The conjunction $p \wedge q$ is true when p and q are both true; otherwise, the conjunction is false.

Recall from Section 2.1 that other English words such as "but," "nevertheless," "still," "however," and so on, are sometimes used in place of the connective "and." Thus, the statement, "Mary is young but not under 15" is a conjunction. On the other hand, the statement, "Mary and Sue are sisters" is not a conjunction (unless they are sisters in a sorority).

B. The Disjunction

As we shall see, there are two types of disjunctions, and we shall illustrate these by examples.

EXAMPLE 1   Let p be, "I will pass this course."
Let q be, "I will flunk this course."

Form the disjunction of p and q, and discuss its truth values.

Solution The statement, "I will pass this course, or I will flunk it" is the desired disjunction. This statement will be true only when *exactly one* of the components is true; it will be false otherwise. ◄

EXAMPLE 2 Consider the two statements:

m: I will study Monday.
s: I will study Saturday.

Form the disjunction of m and s, and discuss its truth values.

Solution The statement, "I will study Monday, or I will study Saturday" is the required disjunction. It will be false only when both components are false; it will be true otherwise. ◄

Table 2.2b Disjunction (∨)

1	2	3
p	q	p ∨ q
T	T	T
T	F	T
F	T	T
F	F	F

If we compare the disjunctions in Examples 1 and 2, it is clear that in the statement contained in Example 1, only one of the two possibilities can occur: I will either pass the course or not. However, in Example 2, I have the possibility of studying Monday or Saturday, or even both. The meaning of the second usage is clarified by replacing the word "or" by "and/or." Instead of arguing which usage should be called the disjunction of the two statements, we shall refer to the "or" used in Example 1 as the **exclusive or**, or the **exclusive disjunction**. On the other hand, the "or" used in Example 2 will be called the **inclusive or**, or the **inclusive disjunction**, and will be denoted by ∨. The truth table defining the symbol ∨ appears in Table 2.2b and expresses the important fact that:

> The inclusive disjunction p ∨ q (read, "p or q") is false only when p and q are both false; otherwise, it is true.

In the remainder of this chapter it will not be necessary to use the exclusive disjunction.

C. Negations

Finally, we shall consider the **negation** of a statement. In the English language, a negation is usually formed by inserting a "not" into the original statement. Of course, this is not the only way to negate a statement. A column by Bill Gold, which appeared in the *Washington Post*, tells the story of a woman who offered a bus driver a $5 bill and asked for tokens. His reply? "I am sorry, lady, but there ain't no bus driver got no change or no tokens, no time, nowhere, no more."

EXAMPLE 3 Let p be the statement, "I will be drafted." Express the statement ~p verbally, and discuss its truth values.

Solution | Either of the statements, "I will not be drafted" or "It is not the case that I will be drafted," is the negation of the statement p and may be symbolized by ~p. Because every statement is either true or false (and not both), we see that ~p must be false whenever p is true, and ~p must be true whenever p is false. ◄

Table 2.2c Negation (~)

1	2
p	~p
T	F
F	T

Table 2.2c defines the negation symbol ~. This table expresses the following definition:

> ~p is false whenever p is true, and ~p is true whenever p is false.

EXAMPLE 4 | Your driver's license will be renewed if you are a safe driver, have no physical disability, and are not addicted to drugs or intoxicants. Let s be the statement, "You are a safe driver," p be the statement, "You have a physical disability," and q be the statement, "You are addicted to drugs or intoxicants." Write a statement in symbolic form whose truth will guarantee that your driver's license will be renewed.

Solution | $s \land \sim p \land \sim q$ ◄

EXERCISE 2.2

1. Let p be, "Today is Friday," and let q be, "Today is Monday."
   a. Write in words the disjunction of the two statements.
   b. Write in words the conjunction of the two statements.
   c. Write in words the negation of the statement p.
   d. Which of the statements in parts a, b, and c always has the truth value F?
2. Let g be, "He is a gentleman," and let s be, "He is a scholar." Write in words:
   a. The disjunction of the two statements
   b. The negation of the statement g
   c. The conjunction of the two statements
3. Use the two statements g and s of Problem 2, and write in symbolic form:
   a. He is not either a gentleman or a scholar.
   b. He is a gentleman and a scholar.
   c. He is neither a gentleman nor a scholar.
4. Consider the statements p and q:

   p: It is raining.
   q: I will go to the beach.

Write the statements in parts a and b in symbolic form.

a. It is raining, but I will go to the beach.

b. It is raining, or I will go to the beach.

c. Assume that p is true and q is false. Find the truth values of the statements given in parts a and b.

5. Let p be, "Mida is cooperative," and let q be, "Desi is uncooperative." Write the statements in parts a, b, c, and d in symbolic form.

a. Mida and Desi are both cooperative. $p \wedge \sim q$

b. Neither Desi nor Mida is uncooperative. $\sim q \wedge p$

c. It is not the case that Mida and Desi are both uncooperative. $\sim(\sim p \wedge q)$

d. Either Mida is cooperative or Desi is uncooperative.

e. Assume that Mida is cooperative and Desi is uncooperative. Which of the statements in parts a, b, c, and d are true?

6. Suppose that p is true and q is false. Write each of the following statements in symbolic form, and find its truth value:

a. Either p or q **b.** Either p or not q

c. Neither p nor q **d.** p or q but not both

e. Not q and not p

7. Consider the statements:

   g: I go to college.

   j: I join the army.

Suppose that g is false and j is true. Write each of the following statements in symbolic form, and find its truth value:

a. Either I go to college or I join the army.

b. I go to college, or I do not join the army.

c. I neither go to college, nor do I join the army.

d. I go to college or I join the army, but not both.

8. *An application to law.* A lawyer who specializes in damage suits arising out of automobile accidents knows that:

   1. The court will decide in favor of his client if his client was not negligent and the other driver was negligent.

   2. The court will decide against his client if both drivers were negligent or if neither was negligent.

Let c be the statement that the client was negligent, and let d be the statement that the other driver was negligent.

a. Use statement 1 to write a compound statement in symbolic form whose truth guarantees that the court will decide in favor of the client.

b. Use statement 2 to write a compound statement in symbolic form whose truth guarantees that the court will decide against the client.

9. A person is considered to have established his or her age for Social Security benefits if the person furnishes one of the following:

1. Birth certificate (b)

2. Church baptismal record (c), giving the date of birth

3. Early school record (s) and an employment record (e), both giving the date of birth

Write in symbols (using the suggested abbreviations) the conditions under which a person has established his (or her) age.

10. *An application to medicine.* In diagnosing diseases, it is extremely important to recognize the symptoms that distinguish one disease from another (usually called the "differential diagnosis"). The *Diagnosis Treatment* handbook* states: "Many specific infectious diseases present initial manifestations indistinguishable from those of common respiratory disease. Vigilance is required to avoid diagnostic errors of omission." Next are listed some symptoms that may be present in a patient with a respiratory disease:

s_1: Patient has a high white blood cell count.

s_2: Patient has fever.

s_3: Patient has nasal discomfort.

s_4: Patient has a sore throat.

s_5: Patient has a cough.

s_6: Patient has a headache.

s_7: Patient has a low white blood cell count.

s_8: Patient has the influenza virus.

In a certain hospital, it was found that:

1. If s_8 was false, but s_1 and either s_4 or s_5 was true, the diagnosis was a cold.

2. If s_8 was true, the diagnosis was influenza.

3. If s_8 was false, but s_7, s_6, s_4 and either s_2 or s_3 was true, the diagnosis was influenza.

a. State in symbolic form the statement whose truth implies that the diagnosis was a cold.

b. Do the same for influenza.

11. Here is a sign that appeared on a football stadium: "Students must present a valid student ID card (i) and agree to sit in the student section (a), or else purchase a general admission ticket (p)." Write in symbols, using the suggested abbreviations, the conditions under which a student would *not* be admitted to the stadium.

12. A person can check books out of a certain county library only if that person has a valid ID card (c), and, in addition, either is a resident of

---

* H. Brainerd, M. Chatton, and S. Margen, *Diagnosis Treatment.* Los Altos, Ca.: Lange Medical Publications, 1962, p. 102.

the county (r) or else pays a $12 annual fee (p). Use the suggested symbols to write a statement in symbolic form whose truth would *not* permit a person to check books out of this library.

13. The Florida "Intangible Tax Return" states, "A Florida beneficiary having one or more property rights in a trust must file a return and pay a tax unless a Florida Intangible Tax Return was filed by the trustee." The state did not receive a return from Sam Slick, a Florida beneficiary with property rights in a trust. May the state assume that Sam was breaking the law?

Problems 14–18 give some income tax applications.

14. The 1040 Federal Income tax instructions state: "If you do not file a joint return, you may claim an exemption for your spouse only if your spouse had no income from U.S. sources and is not the dependent of another taxpayer." Ms. Mulberry filed a separate return from her spouse, who had no income at all, claiming him as an exemption. What additional true statement can you make about Ms. Mulberry's spouse?

15. In a recent set of income tax forms, in Schedule B, Interest and Dividend Income, there appeared the following statement: "If you are required to list interest in Part I (i) or dividends in Part II (d), OR if you had a foreign account (f) or were a grantor of (g) or a transferor to a foreign trust (t), you must answer both questions in Part III." Use the suggested abbreviations and write in symbolic form the condition that would require the taxpayer to answer both questions in Part III. [*Hint:* You need two sets of parentheses.]

16. Refer to the statement in Problem 15.
    a. What is the minimum number of statements that have to be true in order for the taxpayer to have to answer both questions in Part III?
    b. Name the five different statements such that if any one of them is true, the taxpayer must answer both questions in Part III.

17. The interest income section of Schedule D, Form 1040, states: "If you received interest that actually belongs to another person (i), or you received (r) or paid (p) interest on securities transferred between interest payment dates, see page 24." Write in symbolic form the condition that requires the taxpayer to see page 24.

18. Refer to the statement in Problem 17, and write a statement in symbolic form—without using parentheses—that excuses the taxpayer from referring to page 24.

19. The Higher Education Act states that for a student to be eligible to apply for a loan, that student must be:
    1. Enrolled and in good standing, or accepted for enrollment, in an eligible school
    2. Registered for at least one-half of the normal full-time work load as determined by the school

3. A citizen or national of the United States, or in the United States for other than a temporary reason

Which of the following students are eligible to apply for a loan?

a. Sally has applied for admission to an eligible school and intends to register for a full-time load. She is a citizen of the United States.

b. Pedro has been accepted for enrollment in an eligible school and is registered for a full-time load leading to a Master's degree. Pedro is a citizen of Brazil and intends to return there after he earns the Master's degree.

c. Boris has been accepted for enrollment in an eligible school and is registered for a full-time load. Boris is a Russian refugee and has a permit to reside in the United States permanently.

d. Susan, who is a citizen of the United States, is enrolled and in good standing in an eligible school. Susan has a full-time job, so is registered for only 7 credit hours. The full-time load at her school is 15 credit hours.

20. Problem 14 gave the IRS instruction: "If you do not file a joint return, you may claim an exemption for your spouse only if your spouse had no income from U.S. sources and is not the dependent of another taxpayer." Assume that all of the following filed separate returns. Which of them are allowed to claim the spouse as an exemption?

a. Mr. Ambrose, whose wife had no income of her own but is the dependent of her father, who is a taxpayer

b. Mrs. Brown, whose husband has a large income from U.S. Treasury bonds

c. Mr. Cary, whose wife has some income from an investment in Switzerland, but is otherwise dependent on her husband

d. Mrs. Dolan, whose husband has no separate income, but is the dependent of his mother, who lives in Ireland and does not pay U.S. taxes

USING YOUR KNOWLEDGE 2.2

Here is an application of the material we have studied. The Higher Education Act states that any student is eligible to apply for a loan, provided the student is:

a. Enrolled (e) and in good standing (g), or accepted for enrollment (a), at an eligible school

b. Registered for at least one-half of the normal full-time work load as determined by the school (h)

c. A citizen (c) or national (n) of the United States, or in the United States for other than a temporary purpose ($\sim t$).

1. Translate requirements a, b, and c into symbolic form.
2. Can you discover the general compound statement whose truth implies that the student may apply for a loan?
3. A three-component statement whose truth implies that the student may apply for a loan is $a \wedge h \wedge c$. Can you discover two others?

Consider the four statements:

a. $g \wedge s$ *b.* $g \vee s$ *c.* $\sim g \vee \sim s$ *d.* $\sim g \wedge \sim s$

1. Make a table with the headings g, s, $g \wedge s$, $g \vee s$, $\sim g \vee \sim s$, and $\sim g \wedge \sim s$, and fill in all the possible combinations of truth values for the four statements.
2. Can you discover which of the four statements can be simultaneously true?
3. If you assume that two of the statements are true, can you discover the status (true or false) of the other two?
4. If you assume that two of the statements are false, can you discover the status (true or false) of the other two?
5. Can you discover a rule that gives the status (true or false) of the four statements in every possible case?

2.3

TRUTH TABLES

In many cases, it is convenient to construct truth tables to determine the truth values of certain compound statements involving the symbols $\wedge$ (and), $\vee$ (or), and $\sim$ (not). These symbols were defined by truth tables in the preceding section. It is important to keep in mind that a conjunction $p \wedge q$ is true when p and q are both true and is false otherwise; a disjunction $p \vee q$ is false when p and q are both false and is true otherwise; and if p and $\sim p$ are negations of each other, then $\sim p$ is false whenever p is true, and $\sim p$ is true whenever p is false.

A. Making Truth Tables

EXAMPLE 1 Construct the truth table for the statement $\sim p \vee q$.

Solution First, we break the statement down into its components to see what headings we need for the truth table. The statement $\sim p \vee q$ has the components $\sim p$ and q. We regard p and q as the primitive components, and we

1	2	3	4	
p	**q**	**~p**	**∨**	**q**
T	T	F	T	
T	F	F	F	
F	T	T	T	
F	F	T	T	

get the truth values of ~p from those of p. This breakdown suggests that the proper headings are p, q, ~p, and ~p ∨ q, where the last two items are obtained from the preceding ones. We now construct the table in the margin, where the numbers at the top give the order in which the required statement is put together from p and q. We proceed as follows:

1. We write in columns 1 and 2 the four possible pairs of truth values for p and q.
2. Using column 1 as a reference, we negate the statement p to get the entries in column 3. Note that we simply write F in the rows where we wrote T for p, and we write T in the rows where we wrote F for p.
3. We combine columns 3 and 2 using the disjunction "or," denoted by ∨. Recall (see Table 2.2b) that a disjunction is false only when both its components are false; it is true otherwise. Thus, we write F in the second row of column 4, where both components, ~p and q, are false, and we write T in the other rows. This completes the table. ◄

EXAMPLE 2   Let p be, "I lie," and let q be, "I would tell you." When will the statement, "I do not lie or I would tell you," be false?

Solution   The statement under consideration can be symbolized as ~p ∨ q. From the table in Example 1, we see that ~p ∨ q is false when p is true and q is false. Hence, the given statement is false when "I lie" and "I would not tell you." ◄

EXAMPLE 3   Construct the truth table for the statement ~(p ∧ ~q).

Solution   As in Example 1, we first break the given statement down into its primitive components. The statement ~(p ∧ ~q) is the negation of p ∧ ~q, which is the conjunction of the components p and ~q. We can write the truth values of ~q from those of q, and p is itself a primitive component. Thus, the headings of the truth table will be p, q, ~q, p ∧ ~q, and ~(p ∧ ~q). The table is filled out in the following steps, where the numbers at the top of the table are the column numbers:

1	2	5	4	3
p	**q**	**~(p**	**∧**	**~q)**
T	T	T	F	F
T	F	F	T	T
F	T	T	F	F
F	F	T	F	T

1. We write in columns 1 and 2 the four possible pairs of truth values for p and q.
2. Using column 2 as reference, we negate q to get column 3.
3. To get column 4, we combine columns 1 and 3, using the conjunction "and," denoted by ∧. Recall (see Table 2.2a) that a conjunction is true only when both its components are true and is false otherwise. Thus, we write T in the second row of column 4, where p and ~q are both true, and we write F in the other rows.
4. We negate the truth values in column 4 to get those in column 5. The statement ~(p ∧ ~q) has the truth values shown in column 5 of the table. ◄

B. Equivalent Statements

Notice that the statements given in Examples 1 and 3 have exactly the same truth values, TFTT; hence, the two statements must have the same meaning (say the same thing).

▶ **Definition 2.3a**

Two statements p and q that have identical truth values are said to be **equivalent** (denoted by p ⇔ q).

Accordingly, from Examples 1 and 3, we may write

$$\sim p \vee q \Leftrightarrow \sim(p \wedge \sim q)$$

EXAMPLE 4

Show that $p \wedge (q \vee r) \Leftrightarrow (p \wedge q) \vee (p \wedge r)$.

Solution

To show this equivalence, we need the truth values of the two given statements, so we make a truth table for each of them. Because we have three primitive statements (p, q, and r), each of which has two possible truth values, there are $2 \times 2 \times 2 = 8$ possible cases. Thus, the truth tables must have eight lines.

First, we examine the statement $p \wedge (q \vee r)$, which we break down into the components p and $q \vee r$. The second of these breaks down into the components q and r. Hence, the truth table must have columns for p, q, r, $q \vee r$, and $p \wedge (q \vee r)$, as shown.

The table is filled out in the same manner as the tables for our previous examples:

1. In the first three columns, we write the possible truth values for p, q, and r. In column 1, we enter four T's and four F's; in column 2, we enter two T's, two F's, two T's, and two F's; in column 3, we enter alternately one T and one F. This gives all the possible combinations of T's and F's for the three statements.

2. We combine columns 2 and 3 with a disjunction ∨ to obtain the truth values of $q \vee r$ in column 4. Because $q \vee r$ is false only when both q and r are false (rows 4 and 8), we enter F in rows 4 and 8 and T's in the remaining rows.

3. We combine columns 1 and 4 with a conjunction ∧ to obtain the truth values of $p \wedge (q \vee r)$. Since a conjunction is true only when both components are true, we complete column 5 by writing T's in the first three rows and F's in the other rows. The given statement has the truth values shown in column 5 of the table.

The second statement, $(p \wedge q) \vee (p \wedge r)$, has the components $p \wedge q$ and $p \wedge r$, and these two have the components p, q, and r. Thus, the truth table should have headings p, q, r, $p \wedge q$, $p \wedge r$, and $(p \wedge q) \vee (p \wedge r)$. The table

1	2	3	5	4
p	q	r		$p \wedge (q \vee r)$
T	T	T	T	T
T	T	F	T	T
T	F	T	T	T
T	F	F	F	F
F	T	T	F	T
F	T	F	F	T
F	F	T	F	T
F	F	F	F	F

1	2	3	4	6	5
p	q	r			$(p \wedge q) \vee (p \wedge r)$
T	T	T	T	T	T
T	T	F	T	T	F
T	F	T	F	T	T
T	F	F	F	F	F
F	T	T	F	F	F
F	T	F	F	F	F
F	F	T	F	F	F
F	F	F	F	F	F

is filled out in the same manner as above. Details are left to the student. The statement $(p \land q) \lor (p \land r)$ has the truth values shown in column 6 of the table.

Since the final columns of the two tables are identical, the given statements are equivalent. ◄

EXERCISE 2.3

A. In Problems 1–14, construct a truth table for the given statement.

1. $p \lor \sim q$ 2. $\sim(p \lor q)$ 3. $\sim p \land q$

4. $\sim p \lor \sim q$ 5. $\sim(p \lor \sim q)$ 6. $\sim(\sim p \lor \sim q)$

7. $\sim(\sim p \land \sim q)$ 8. $(p \lor q) \land \sim(p \land q)$

9. $(p \land q) \lor (\sim p \land q)$ 10. $(p \land \sim q) \land (\sim p \land q)$

11. $p \land (q \lor r)$ 12. $p \lor (q \land r)$

13. $(p \lor q) \lor (r \land \sim q)$ 14. $[(p \land q) \lor (q \land \sim r)] \lor (r \land \sim s)$

15. Let p be, "Mida is blonde," and let q be, "Mida is 6 ft tall."
   a. Under what conditions is the statement, "Mida is blonde and 6 ft tall" true?
   b. Under what conditions is the statement, "Mida is blonde and 6 ft tall" false?
   c. Under what conditions is the statement, "Mida is blonde or 6 ft tall" true?
   d. Under what conditions is the statement, "Mida is blonde or 6 ft tall" false?

16. Let p be, "Eva is a high school graduate," and let q be, "Eva is over 16 years old."
   a. Under what conditions is the statement, "Eva is neither a high school graduate nor over 16 years old" true?
   b. Under what conditions is the statement in part a false?
   c. Under what conditions is the statement, "Either Eva is a high school graduate or she is over 16 years old" true?
   d. Under what conditions is the statement in part c false?

B. 17. In each of the following, use truth tables to show that the two statements are equivalent.
   a. $p \lor (q \land r)$ and $(p \lor q) \land (p \lor r)$
   b. $\sim(p \lor q)$ and $\sim p \land \sim q$
   c. $\sim(p \land q)$ and $\sim p \lor \sim q$

18. Use truth tables to show the following equivalences:
   a. $(p \land q) \lor \sim p \Leftrightarrow q \lor \sim p$
   b. $(p \lor q) \land (\sim p \lor \sim q) \Leftrightarrow (p \land \sim q) \lor (\sim p \land q)$

19. a. Verify the entries in the following table:

p	q	p ∧ q	p ∧ ~q	~p ∧ q	~p ∧ ~q
T	T	T	F	F	F
T	F	F	T	F	F
F	T	F	F	T	F
F	F	F	F	F	T

b. Look at the last four columns of this table. Each of these columns has one T and three F's, and there is exactly one T on each line. This T occurs on the line where both components of the corresponding column heading are true. The headings of these last four columns are called **basic conjunctions.** By using these conjunctions, we can write statements having any given four-entry truth table. For instance, a statement with the truth table TFFT is (p ∧ q) ∨ (~p ∧ ~q). Explain this.

c. By forming disjunctions of the basic conjunctions, we can write statements with given truth tables as noted in part b. Write a statement having truth table FTTF. Do the same for FTTT. Can you write a simpler statement for the truth table FTTT?

20. The ideas in Problem 19 can be generalized to statements with any given number of components. Thus, the statement p ∧ ~q ∧ ~r would have a truth table with a T on the line corresponding to p true, q false, and r false, and F's on the other seven lines. What would be the truth table for (p ∧ ~q ∧ r) ∨ (~p ∧ ~q ∧ ~r)?

21. Describe the circumstances under which the following statements have truth value T. [*Hint:* First write the statement in symbolic form.]

a. Billy goes to the zoo and feeds peanuts to the elephants or the monkeys.

b. I file my income tax return and pay the tax, or go to jail.

22. Describe the circumstances under which the following statements have truth value T (see the hint in Problem 21):

a. John gets up before 7:00 A.M. and has cereal or pancakes for breakfast.

b. I do not drive over the speed limit and I obey the traffic signals, or I get a traffic ticket.

23. Let the connective * be defined by the table in the margin. Construct the truth tables for:

a. (p ∧ q) * p   **b.** (p ∧ ~q) * q   **c.** (p ∨ q) * ~p

24. Rework Problem 23 by replacing the connective * by @, where @ is defined by the lower table in the margin.

25. Using the method given in Problem 19 and the truth tables in the margin, find a statement that has the same truth table as:

a. p * q   **b.** ~(p @ q)   **c.** (p * q) @ ~p

p	q	p * q
T	T	F
T	F	F
F	T	F
F	F	T

p	q	p @ q
T	T	F
T	F	T
F	T	T
F	F	T

26. The truth table for a statement compounded from two statements has $2^2 = 4$ rows, and the truth table for a statement compounded from three statements has $2^3 = 8$ rows. How many rows would the truth table for a statement compounded from four statements have? How many for five? For six? For n?

USING YOUR KNOWLEDGE 2.3

In medical practice, a method commonly used to diagnose certain food allergies is the food diary. This method uses a record form listing the foods eaten each day, as shown in the table.

	DATE													
	1	2	3	4	5	6	7	8	9	10	11	12	13	14
Indigestion occurred	×						×							×
Food														
Coffee	×	×	×	×	×	×	×	×	×	×	×	×	×	×
Eggs	×		×	×	×	×			×	×		×	×	×
Chicken	×						×	×			×		×	
Fish	×						×							×
Pork			×				×			×				

Since coffee was drunk every day by the person with this diary, and indigestion did not occur every day, coffee can be eliminated as a suspect.

Let e be, "The person ate eggs."
Let c be, "The person ate chicken."
Let f be, "The person ate fish."
Let p be, "The person ate pork."

Write a symbolic statement telling what the person ate:

1. On day 1
2. On day 7
3. On day 14
4. On all three days when indigestion occurred
5. Based on your answer to Problem 4, which food do you think gives this person indigestion?

DISCOVERY 2.3

In Logictown, there live four men, Mr. Baker, Mr. Carpenter, Mr. Draper, and Mr. Smith. One is a baker, one a carpenter, one a draper, and one a smith, but none follows the vocation corresponding to his name. A logician tries to find out who is who, and he obtains the following partially correct information:

a. Mr. Baker is the smith.
b. Mr. Carpenter is the baker.
c. Mr. Draper is not the smith.
d. Mr. Smith is not the draper.

1. If it is known that three of the four statements are false, who is the carpenter? [*Hint:* Consider the four possible sets of truth values given for the statements in the table in the margin.]

STATEMENT	I	II	III	IV
a	T	F	F	F
b	F	T	F	F
c	F	F	T	F
d	F	F	F	T

▶ **Computer Corner 2.3** *In this section, you learned how to write a statement with a given truth table (Problems 19 and 20). The computer can construct a statement corresponding to a given truth table, provided you enter the table. The program appears on page 742.*

1. Use it to check the answer to Problem 19c in Exercise 2.3.

2.4

THE CONDITIONAL AND THE BICONDITIONAL

IF YOU STOP
HERE - YOUR
PAIN WILL TOO
6510

It is sometimes necessary to specify the conditions under which a given event will be true. For example, one might say, "If the weather is nice, then I will go to the beach." If we let *p* stand for "the weather is nice" and *q* stand for "I will go to the beach," then the preceding compound statement is of the form "If *p*, then *q*." Statements of this kind are called **conditional statements** and are symbolized by $p \rightarrow q$. (Read, "if *p*, then *q*" or "*p* arrow *q*" or "*p* conditional *q*.") In $p \rightarrow q$, the statement *p* is sometimes called the **antecedent** and *q* the **consequent.**

A. The Conditional

Table 2.4a

p	*q*	$p \rightarrow q$
T	T	T
T	F	F
F	T	T
F	F	T

Table 2.4a shows that if *p* and *q* are both true, then $p \rightarrow q$ is true, and if *p* is true and *q* is false, then $p \rightarrow q$ is false. In the last two lines of the table, *p* is false so that it would be incorrect to say that $p \rightarrow q$ is false. Since we want a complete truth table, we have assigned the value *T* to $p \rightarrow q$ in these two lines. See Problem 23, Exercise 2.4.

To understand the truth table for the conditional, consider the sign in the margin. It promises, "If you stop here, your pain will too." Under what circumstances is this promise broken? Obviously, only if you *do* stop here and your pain *does not* stop. Thus, we should write *F* for $p \rightarrow q$ if *p* is true and *q* is false; otherwise, we should write *T*. Table 2.4a expresses the fact that:

The **conditional statement** $p \to q$ ("if p, then q") is false only when p is true and q is false; otherwise, it is true.

B. The Biconditional

In certain statements, the conditional is used twice, with the antecedent and the consequent of the first conditional reversed in the second conditional. For example, the statement, "If money is plentiful, then interest rates are low, and if interest rates are low, then money is plentiful," uses the conditional twice in this manner. It is for this reason that such statements, which can be written in the form $(p \to q) \land (q \to p)$, are called **biconditionals.** The biconditional is usually symbolized by the shorter form $p \leftrightarrow q$, so that, by definition,

$$p \leftrightarrow q \leftrightarrow (p \to q) \land (q \to p)$$

Table 2.4b

1	2	3	5	4
p	**q**	**(p**	**→ q) ∧ (q**	**→ p)**
T	T	T	T	T
T	F	F	F	T
F	T	T	F	F
F	F	T	T	T

Table 2.4b is the truth table for the statement $(p \to q) \land (q \to p)$. This table is filled out in the usual way:

1. In columns 1 and 2, we write the four possible pairs of truth values for p and q.
2. We combine columns 1 and 2 with the conditional ($\to$) to form column 3. Since $p \to q$ is false only when p is true and q is false, we write F in the second line and T in the other lines.
3. We combine columns 2 and 1 — *in that order* — with the conditional ($\to$) to form column 4. Because $q \to p$ is false only when q is true and p is false, we write F in the third line and T in the other lines.
4. We combine columns 3 and 4 with the conjunction ($\land$) to form column 5. Since the conjunction is true only when both components are true, we write T in the first and fourth lines and F in the other two lines. This completes the table.

Table 2.4b shows that:

The **biconditional** $p \leftrightarrow q$ is true when and only when p and q have the same truth values; it is false otherwise.

EXAMPLE 1   Give the truth value of each of the following:

a. If Tuesday is the last day of the week, then the next day is Sunday.
b. If Tuesday is the third day of the week, then the next day is Sunday.
c. If Tuesday is the third day of the week, then Wednesday is the fourth day of the week.
d. If Tuesday is the last day of the week, then the next day is Wednesday.

Solution   All these statements are of the form $p \rightarrow q$, where p is the antecedent and q is the consequent. (Recall that the antecedent is the "if" part, and the consequent is the "then" part.)

a. Because p is false, the statement $p \rightarrow q$ is true.
b. Because p is true and q is false, the statement $p \rightarrow q$ is false.
c. Because p and q are both true, the statement $p \rightarrow q$ is true.
d. Because p is false, the statement $p \rightarrow q$ is true.

(The moral is that if you start off with a false assumption, then you can prove anything!)   ◀

EXAMPLE 2   Is the statement $(3 + 5 = 35) \leftrightarrow (2 + 7 = 10)$ true or false?

Solution   This statement is of the form $p \leftrightarrow q$, where p is "$3 + 5 = 35$" and q is "$2 + 7 = 10$." Since the biconditional $p \leftrightarrow q$ is true when p and q have the same truth value, and the p and q in this example are both false, the given statement is true.   ◀

EXAMPLE 3   Let p be, "x is a fruit," and let q be, "x is ripe." Under what conditions is the statement $p \rightarrow q$ false?

Solution   The statement $p \rightarrow q$ is a conditional statement, and thus is false only when p is true and q is false; hence, the given statement is false if x is a fruit that is not ripe.   ◀

EXAMPLE 4   Show that the statements $p \rightarrow q$ and $\sim p \vee q$ are equivalent; that is, $(p \rightarrow q) \leftrightarrow (\sim p \vee q)$.

Solution   The truth table for $(p \rightarrow q)$ was given in Table 2.4a and that for $(\sim p \vee q)$ was given in Example 1 in Section 2.3 (page 59). The two statements have identical truth tables, so the statements are equivalent.   ◀

The equivalence in Example 4,

$$(p \rightarrow q) \leftrightarrow (\sim p \vee q)$$

is of great importance, because it allows us to handle a conditional statement in terms of the logical symbols for negation and disjunction. This equivalence will be used in several of the problems in Exercise 2.4 and later in this chapter. Note that since

$$(p \rightarrow q) \leftrightarrow (\sim p \vee q)$$

the negation of $p \rightarrow q$ is equivalent to the negation of $\sim p \vee q$; that is, the negation of a conditional statement may be written:

$$\sim(p \rightarrow q) \Leftrightarrow \sim(\sim p \lor q)$$

See Problem 17 in Exercise 2.4 for the simplification of the right-hand side of this equivalence.

EXERCISE 2.4

A.
1. Show that the statement $\sim q \rightarrow \sim p$ is equivalent to $p \rightarrow q$.
2. Use truth tables to show the following equivalences:

   a. $p \rightarrow \sim q \Leftrightarrow \sim(p \land q)$ **b.** $\sim p \rightarrow q \Leftrightarrow p \lor q$
3. Give the truth value of each of the following statements:

   a. If $2 + 2 = 22$, then $22 = 4$.

   b. If $2 + 2 = 4$, then $8 = 5$.

   c. If $2 + 2 = 22$, then $8 = 4 + 4$.

   d. If $2 + 2 = 22$, then $4 = 26$.

In Problems 4–7, find all the number replacements for x that make the given sentence true.

4. If $2 + 2 = 4$, then $x - 2 = 5$.
5. If $2 + 2 = 22$, then $x - 2 = 5$.
6. If $x + 2 = 6$, then $3 + 2 = 5$.
7. If $x + 2 = 6$, then $2 + 2 = 32$.

8. Let p be, "I kiss you once," and let q be, "I kiss you again." Under what conditions is the statement $p \rightarrow q$ false?
9. Under what condition is the statement, "If you got the time, we got the beer" false?
10. Construct truth tables for the following statements. Note the importance of the parentheses and the brackets to indicate the order in which items are grouped.

    a. $[(p \rightarrow q) \rightarrow p] \rightarrow q$ **b.** $(p \rightarrow q) \leftrightarrow (p \lor r)$

    c. $(p \rightarrow q) \leftrightarrow (p \rightarrow \sim q)$

B.
11. Construct truth tables for:

    a. $p \rightarrow (q \land r)$ **b.** $(p \rightarrow q) \land (p \rightarrow r)$

    c. Are the statements in parts a and b equivalent?
12. Let p be, "I will buy it," and let q be, "It is a poodle." Translate into symbolic form:

    a. If it is a poodle, then I will buy it.

    b. If I will buy it, then it is a poodle.

    c. It is a poodle if and only if I will buy it.

    d. If it is not a poodle, then I will not buy it.

e. If I will not buy it, then it is not a poodle.

f. If it is a poodle, then I will not buy it.

13. Let $\sim s$ be, "You are out of Schlitz," and let $\sim b$ be, "You are out of beer." Translate into symbolic form:

a. If you are out of Schlitz, you are out of beer.

b. If you are out of beer, you are out of Schlitz.

c. Having beer is equivalent to having Schlitz.

14. Write each of the following statements in symbolic form using $\sim$ and $\vee$. Also write the corresponding verbal statement. [*Hint:* $p \rightarrow q$ is equivalent to $\sim p \vee q$.]

a. If the temperature is above 80° (a), then I will go to the beach (b).

b. If Mida is home by 5 (h), then dinner will be ready by 6 (r).

c. If Eva has a day off (o), then she will go to the beach (g).

15. In Example 4 it was shown that $p \rightarrow q$ is equivalent to $\sim p \vee q$. Use this equivalence to write the following as disjunctions:

a. If you work, you have to pay taxes.

b. If you got the time, we got the beer.

c. If you find a better one, then you buy it.

16. The statement, "All even numbers are divisible by 2" can be translated as, "If it is an even number, then it is divisible by 2." In general, the statement, "All . . . are . . ." can be translated as, "If it is a . . . , then it is a . . . ." Using this idea, write each of the following in the if–then form:

a. All dogs are mammals.

b. All cats are felines.

c. All men are created equal.

d. All prime numbers greater than 2 are odd numbers.

e. All rectangles whose diagonals are perpendicular to each other are squares.

17. Because $p \rightarrow q$ is equivalent to $\sim p \vee q$ (see Example 4), the negation of $p \rightarrow q$ should be equivalent to the negation of $\sim p \vee q$. Show that the negation of $\sim p \vee q$ is $p \wedge \sim q$; that is, show that $\sim(\sim p \vee q)$ is equivalent to $p \wedge \sim q$.

18. From Problem 17 it is clear that the negation of $p \rightarrow q$ is equivalent to $p \wedge \sim q$. Verify this by means of a truth table.

19. Problem 18 verified that the negation of $p \rightarrow q$ is $p \wedge \sim q$. This means that to negate an "if . . . , then . . ." statement, we simply assert the *if clause* and deny the *then clause*. For instance, the negation of the statement, "If you are out of Schlitz, you are out of beer" is the statement, "You are out of Schlitz, but you are not out of beer." Write in words the negation of each of the following:

a. If you earn much money, then you pay heavy taxes.

b. If Johnny does not play quarterback, then his team loses.

c. If Alice passes the test, then she gets the job.

20. Refer to Problem 19 and then write out the negation of each of the following:

a. If I kiss you once, I kiss you again.

b. If Saturday is a hot day, I will go to the beach.

c. Evel Knievel will lose his life if he is careless.

21. From Problem 18 we can see that $p \wedge \sim q$ has truth values *FTFF*. If you know that $p \wedge \sim q$ is the negation of $p \rightarrow q$, how can you define the truth table for $p \rightarrow q$?

22. Write each of the following statements in the if–then form:

a. Johnny does not play quarterback or his team wins.

b. Alice fails the test or she gets the job.

c. Joe had an accident or he could get car insurance.

23. In defining $p \rightarrow q$, it is easy to agree that if p is true and q is true, then $p \rightarrow q$ is true; also, if p is true and q is false, then $p \rightarrow q$ is false. Assuming that the entries in the first two rows in the table in the margin are *TF*, respectively, we have four possible definitions for $p \rightarrow q$, as listed in the table.

a. Show that if we use Definition 1, then $p \rightarrow q$ and $p \wedge q$ have the same truth table.

b. Show that if we use Definition 2, then $p \rightarrow q$ and $p \leftrightarrow q$ have the same truth table.

c. Show that if we use Definition 3, then $p \rightarrow q$ and q have the same truth table.

Thus, the table shows that if we wish $p \rightarrow q$ to be different from $p \wedge q$, $p \leftrightarrow q$, and q, then we must use Definition 4.

		DEFINITION OF $p \rightarrow q$			
p	**q**	**1**	**2**	**3**	**4**
T	T	T	T	T	T
T	F	F	F	F	F
F	T	F	F	T	T
F	F	F	T	F	T

24. a. A mother promises her child, "If you eat the spinach and the liver, then you may go out to play." The child eats only the spinach, but the mother lets him out to play. Has she broken her original promise?

b. In place of the statement in part a, suppose the mother says, "If you do not eat the spinach and liver, then you may not go out to play." If the child eats only the spinach and the mother lets him out to play, has she broken her promise?

25. The Score Report Request Form for the Graduate Record Exam includes the following statements:

1. If you entered a Future Test Date (f), your score will be reported after scores from that date become available (a).

2. If you entered a Previous Test Date (p), your scores will be reported from 2 to 4 weeks after this request is received (t).

Use the suggested abbreviations to write in symbolic form:

a. Statement 1   **b.** Statement 2

USING YOUR KNOWLEDGE 2.4

A certain credit union issues the following memorandum with its monthly statement of account: Please examine carefully the enclosed statement.

Report all differences to the Auditing Division. If no differences are reported in 10 days, we shall understand that the balance is correct as shown.

Let d be, "A difference is found."
Let r be, "A report is made in 10 days."
Let a be, "The credit union makes the adjustment."

1. Write in symbols: If a difference is found, then a report is made in 10 days.
2. Write in symbols: If a report is made in 10 days, the credit union makes the adjustment.
3. Write in symbols: If a difference is found, then a report is made in 10 days and the credit union makes the adjustment.
4. Does the statement in Problem 3 indicate that the credit union will make no adjustment if a late report of differences is made?

DISCOVERY 2.4

A logician is captured by a tribe of savages, whose chief makes the following offer: "One of these two roads leads to certain death and the other to freedom. You may select either road after asking any one question of one of these two warriors. I must warn you, however, that one of them is always truthful and the other always lies."

Let p be, "The first road leads to freedom."
Let q be, "You are telling the truth."

1. What should the question be? [Hint: We are to construct a question so that if p is true the answer is "Yes," and if p is false the answer is "No." Complete the table, and then refer to Problem 19, Exercise 2.3, to find the desired question.]

p	q	ANSWER	TRUTH TABLE OF QUESTION TO BE ASKED
T	T	Yes	
T	F	Yes	
F	T	No	
F	F	No	

2.5

VARIATIONS OF THE CONDITIONAL

In the preceding section, we observed that equivalent statements have identical truth tables and may be considered different forms of the same statement. In this section, we shall be concerned with some of the different forms in which the conditional statement $p \rightarrow q$ can be expressed.

A. Converse, Inverse, and Contrapositive

The conditional differs from conjunctions, disjunctions, and biconditionals in that the two components may *not* be interchanged to give an equivalent statement. Thus, $p \vee q \leftrightarrow q \vee p$,* $p \wedge q \leftrightarrow q \wedge p$, but $p \rightarrow q$ is *not* equivalent to $q \rightarrow p$. If we attempt to discover a statement that is equivalent to $p \rightarrow q$ (that is, that has an identical truth table) and involves p and q and the conditional, or $\sim p$, $\sim q$, and the conditional, we find the following possibilities:

$q \rightarrow p$	**Converse** of $p \rightarrow q$
$\sim p \rightarrow \sim q$	**Inverse** of $p \rightarrow q$
$\sim q \rightarrow \sim p$	**Contrapositive** of $p \rightarrow q$

Table 2.5a shows the truth tables for these statements.

Table 2.5a

p	q	Conditional $p \rightarrow q$	Converse $q \rightarrow p$	Inverse $\sim p \rightarrow \sim q$	Contrapositive $\sim q \rightarrow \sim p$
T	T	T	T	T	T
T	F	F	T	T	F
F	T	T	F	F	T
F	F	T	T	T	T

Notice that $p \rightarrow q$ is equivalent to its contrapositive, $\sim q \rightarrow \sim p$ (because they have identical truth tables).

The contrapositive of a statement is used in proving theorems in which a direct proof is difficult, but the proof of the contrapositive is easy. The next example illustrates this idea.

EXAMPLE 1   Prove that if n^2 is odd, then n is odd.

Solution   Let p be, "n^2 is odd," and let q be, "n is odd." We have to prove that if p is true, q is always true. Because the statement $p \rightarrow q$ is equivalent to $\sim q \rightarrow \sim p$, it is sufficient to prove the equivalent statement, "If $\sim q$ is true, then $\sim p$ is true." To this end, we assume that $\sim q$ is true; that is, we assume that n is even. If n is even, then $n = 2k$, where k is an integer. Then $n^2 = (2k)^2 = 4k^2 = 2(2k^2)$, which is also even. Hence, when $\sim q$ is true, $\sim p$ is always true. We have thus proved that, "If n^2 is odd, then n is odd." ◄

* Recall that p and q are equivalent (denoted by $p \leftrightarrow q$) if p and q have identical truth tables.

EXAMPLE 2 Let s be, "You study regularly," and let p be, "You pass the course." Write the converse, contrapositive, and inverse of the statement $s \rightarrow p$, "If you study regularly, then you pass this course."

Solution *Converse* If you pass this course, then you study regularly.

Contrapositive If you do not pass this course, then you do not study regularly.

Inverse If you do not study regularly, then you do not pass this course.

◀

B. Conditional Equivalents

Frequently in mathematics, the words **necessary** and **sufficient** are used in conditional statements. To say that p is sufficient for q means that when p happens (is true), q will also happen (is also true). Hence, "p is sufficient for q" is equivalent to "If p, then q."

Similarly, the sentence, "q is necessary for p" means that if q does not happen, neither will p. That is, $\sim q \rightarrow \sim p$. The statement $\sim q \rightarrow \sim p$ is equivalent to $p \rightarrow q$, so the sentence, "q is necessary for p" is equivalent to, "If p, then q."

Finally, "p only if q" also means that if q does not happen, neither will p, that is, $\sim q \rightarrow \sim p$. The statement $\sim q \rightarrow \sim p$ is equivalent to $p \rightarrow q$, so the sentence, "p only if q" is equivalent to $p \rightarrow q$. The equivalences discussed, together with the variation, "q, if p" are summarized in Table 2.5b. To aid you in understanding this table, notice that in the statement $p \rightarrow q$, p is the sufficient condition (the antecedent) and q is the necessary condition (the consequent).

Table 2.5b

STATEMENT	EQUIVALENT FORMS
If p, then q.	p is sufficient for q
	q is necessary for p
	p only if q
	q, if p

From Table 2.5b, we see that the statements

p is necessary and sufficient for q
q is necessary and sufficient for p
q if and only if p

are all equivalent to the statement "p if and only if q," and may be symbolized by $p \leftrightarrow q$.

EXAMPLE 3 Let s be, "You study regularly," and let p be, "You pass this course." Translate the following statements into symbolic form:

a. You pass this course only if you study regularly.
b. Studying regularly is a sufficient condition for passing this course.
c. To pass this course it is necessary that you study regularly.
d. Studying regularly is a necessary and sufficient condition for passing this course.

e. You do not pass this course unless you study regularly. [*Hint: p* unless *q* means $\sim q \rightarrow p$.]

Solution **a.** $p \rightarrow s$

b. Because *s*, studying regularly, is the sufficient condition, we write $s \rightarrow p$.

c. Since *s* is the necessary condition, we write $p \rightarrow s$.

d. $p \leftrightarrow s$ or $s \leftrightarrow p$

e. Since not studying regularly is a sufficient condition for not passing, we write $\sim s \rightarrow \sim p$. ◄

EXERCISE 2.5

A.   **1.** Using a technique similar to the one employed in Example 1, prove that if n^2 is even, then *n* is even.

2. Let *p* be, "You brush your teeth with Clean," and let *q* be, "You have no cavities." Write the converse, contrapositive, and inverse of the statement $p \rightarrow q$, "If you brush your teeth with Clean, then you have no cavities."

B.   **3.** Let *p* and *q* be defined as in Problem 2. Translate the following statements into symbolic form:

a. You have no cavities only if you brush your teeth with Clean.

b. Having no cavities is a sufficient condition for brushing your teeth with Clean.

c. To have no cavities, it is necessary that you brush your teeth with Clean.

4. Write in the if–then form:

a. If I kissed you once, I will kiss you again.

b. To be a mathematics major it is necessary to take calculus.

c. A good argument is necessary to convince Eva.

d. A two-thirds vote is sufficient for a measure to carry.

e. To have rain, it is necessary that we have clouds.

f. A necessary condition for a stable economy is that we have low unemployment.

g. A sufficient condition for joining a women's club is being a woman.

h. Birds of a feather run together.

i. All dogs are canines.

5. Use a truth table to show that, in general, the converse and the inverse of the statement $p \rightarrow q$ are equivalent (have identical truth tables).

6. Let *p* be, "I will pass this course," and let *s* be, "I will study daily." Write the following statements in symbolic form:

a. Studying daily is necessary for my passing this course.

b. A necessary and sufficient condition for my passing this course is studying daily.

c. I will pass this course if and only if I study daily.

7. Write the converse, inverse, and contrapositive of the following statements:

a. If you do not eat your spinach, you will not be strong.

b. If you eat your spinach, you will be strong.

c. You will be strong only if you eat your spinach.

8. Which statements in Problem 7 are equivalent?

9. Write the converse of each of the following statements. Which of these converses is/are always true?

a. If an integer is even, then its square is divisible by 4.

b. If it is raining, then there are clouds in the sky.

c. In order to get a date, I must be neat and well-dressed.

d. If M is elected to office, then all our problems are over.

e. In order to pass this course, it is sufficient to get passing grades on all the tests.

10. Write the contrapositive of each of the following:

a. In an equilateral triangle, the three angles are equal.

b. If the research is adequately funded, we can find a cure for cancer.

c. Black is beautiful.

d. All radicals want to improve the world.

11. The Score Report Request Form for the Graduate Record Exam includes the following statement: Use this box (u) only if your most recent scores were earned after October 1 (a).

a. Use the suggested abbreviations to write the given statement in symbols.

b. Write the given statement in words using the "if . . . , then . . ." form.

12. The following statement appeared in an IRS Form 1040A: "If you want IRS to figure your tax (f), please stop here (h) and sign below (s)."

Step 7	If You Want IRS To Figure Your Tax, See Page 24 of the Instructions.		
Figure your tax, credits, and payments (including advance EIC payments)	**Caution:** If you are under age 14 and have more than $1,000 of investment income, see page 24 of the instructions and check here . . . . ▶ ☐		
	18 Find the tax on the amount on line 17. Check if from: ☐ Tax Table (pages 32–37); or ☐ Form 8615, Computation of Tax for Children Under Age 14 Who Have Investment Income of More Than $1,000.	18	
	19 Credit for child and dependent care expenses. Complete and attach Schedule 1, Part I.	19	
	20 Subtract line 19 from line 18. Enter the result. (If line 19 is more than line 18, enter -0- on line 20.) This is your **total tax**. ▶	20	

a. Use the given abbreviations to write the statement in symbolic form.

b. Write, in symbols, the contrapositive of the statement in part a.

13. Here is a tip that may save you money on your tax return: If you rent

your vacation home for less than 15 days a year (f), you do not need to report the income ($\sim r$).

   a. Write the given statement in symbolic form using the suggested abbreviations.

   b. Write, in symbols, the contrapositive of the statement in part a.

   c. Write, in words, the contrapositive of the statement in part a.

14. The Driver's License Bureau of a certain state includes the following statement in its form: "If you do not enclose the correct fee ($\sim e$), your request will be returned (r)."

   a. Use the given abbreviations to write the statement in symbolic form.

   b. Write, in symbolic form, the contrapositive of the given statement.

15. Here is a statement found outside a certain establishment: "Under 18 not admitted without parent or guardian."

   a. Write the statement in the "if . . . , then . . ." form.

   b. Write the contrapositive of the statement in part a.

16. Let d be, "The postal service delivers your letter," and let p be, "You use the proper postage stamps." Write in symbols:

   a. For the postal service to deliver your letter, it is necessary that you use the proper postage stamps.

   b. Using the proper postage stamps is a sufficient condition for the postal service to deliver your letter.

   c. A necessary and sufficient condition for the postal service to deliver your letter is that you use the proper postage stamps.

$u \rightarrow \sim a$

$\sim(\sim a) \rightarrow \sim u$

In Problems 17–20, use the suggested abbreviations to write the given statements in symbolic form.

17. a. To keep wood from rotting (k), it is necessary to use a preservative (p). $K \rightarrow P$

   b. Using a preservative is a necessary and sufficient condition to keep wood from rotting. $P \leftrightarrow K$

   c. A sufficient condition to keep wood from rotting is that you use a preservative. P

18. a. To avoid a late penalty ($\sim p$), you must file your income tax return by April 15 (f).

   b. Filing your income tax return by April 15 is enough to avoid a late penalty.

   c. A necessary and sufficient condition for avoiding a late penalty is that you file your income tax return by April 15.

19. a. A necessary and sufficient condition for snow to be beautiful (b) is that it be clean (c).

   b. Snow must be clean in order for it to be beautiful.

   c. Being clean is a sufficient condition for snow to be beautiful.

20. a. A triangle is equilateral (has three equal sides) (s) if and only if it is equiangular (has three equal angles) (a).

Mt. Everest has one of the cleanest snows on earth.

Courtesy of PSH

b. A rectangle is a square (s) only if its adjacent sides are equal (e).

c. A necessary and sufficient condition for a quadrilateral (a four-sided plane figure) to be a rectangle (r) is that its angles are all right angles (a).

DISCOVERY 2.5

The following properties are used by mathematicians and logicians. In Problems 1–5, express the statements in symbols, and explain why they are true.

1. The contrapositive of the statement $\sim q \rightarrow \sim p$ is equivalent to $p \rightarrow q$.
2. The inverse of the inverse of $p \rightarrow q$ is equivalent to $p \rightarrow q$.
3. The contrapositive of the inverse of $p \rightarrow q$ is equivalent to $q \rightarrow p$.
4. The statement $r \vee s \vee \sim p \vee \sim q$ is equivalent to the contrapositive of $(p \wedge q) \rightarrow (r \vee s)$.
5. The statement $(\sim r \wedge \sim s) \vee (p \vee q)$ is equivalent to the converse of $(p \vee q) \rightarrow (r \vee s)$.

	d	*c*	*p*	*i*
d	d	c	p	i
c	c	d	i	p
p	p	i	d	c
i	i	p	c	d

Some of the properties in Problems 1–5 are summarized in the table in the margin, where d stands for direct statement, c for converse, p for contrapositive, and i for inverse. Using the table, find:

6. The contrapositive of the contrapositive
7. The inverse of the inverse
8. The converse of the contrapositive
9. The inverse of the converse
10. The inverse of the contrapositive

*2.6

IMPLICATION

In Definition 2.3a, two statements p and q were defined to be equivalent $(p \leftrightarrow q)$ if they have identical truth tables. An alternate definition states that p is equivalent to q $(p \leftrightarrow q)$ if the biconditional $p \leftrightarrow q$ is always true. (Can you see why these definitions are really the same?)

A. Tautologies and Contradictions

▶ **Definition 2.6a**

A statement that is always true is called a **tautology.** A statement that is always false is called a **contradiction.**

* Material marked with an asterisk is optional and may be omitted.

EXAMPLE 1   Show by means of a truth table that the statement $p \lor \sim p$ is a tautology.

Solution   The truth table for $p \lor \sim p$ is given below.

p	$\sim p$	$p \lor \sim p$
T	F	T
F	T	T

We note that in every possible case, $p \lor \sim p$ is true; therefore, the statement $p \lor \sim p$ is a tautology.   ◀

EXAMPLE 2   Show by means of a truth table that the statement $p \land \sim p$ is a contradiction.

Solution   The truth table for $p \land \sim p$ is given in the margin. We note that in every possible case, $p \land \sim p$ is false; therefore, $p \land \sim p$ is a contradiction (always false).   ◀

p	$\sim p$	$p \land \sim p$
T	F	F
F	T	F

It is easy to restate the definition of equivalence in terms of a tautology.

▶ **Definition 2.6b**

The statement p is equivalent to the statement q ($p \leftrightarrow q$) if and only if the biconditional $p \leftrightarrow q$ is a tautology.

EXAMPLE 3   Show that the biconditional $\sim(p \land q) \leftrightarrow \sim p \lor \sim q$ is a tautology.

Solution   We can do this by an easy check.

1. If the left side, $\sim(p \land q)$, is true, then $p \land q$ is false. Thus, at least one of p and q is false, so that at least one of $\sim p$ and $\sim q$ is true. Therefore, the right side, $\sim p \lor \sim q$, is also true. ·
2. If the left side, $\sim(p \land q)$, is false, then $p \land q$ is true. Thus, both p and q are true, so that both $\sim p$ and $\sim q$ are false. Therefore, the right side, $\sim p \lor \sim q$, is also false.

Because both sides always have the same truth value, the biconditional is always true, that is, it is a tautology. (This also shows that the two sides of the biconditional are equivalent.)   ◀

B.   Implications

Another relationship between statements that is used a great deal by logicians and mathematicians is that of **implication.**

▶ **Definition 2.6c**

Implication

The statement p is said **to imply** the statement q (symbolized by $p \Rightarrow q$) if and only if the conditional $p \rightarrow q$ is a tautology.

EXAMPLE 4   Show that $[(p \rightarrow q) \wedge p] \Rightarrow q$.

Solution   *First method* By Definition 2.6c, we must show that $[(p \rightarrow q) \wedge p] \rightarrow q$ is a tautology. A conditional is true whenever the antecedent is false, so we need to check only the cases where the antecedent is true. Thus, if $(p \rightarrow q) \wedge p$ is true, then $p \rightarrow q$ is true and p is true. But if p is true, then q is also true (why?), so both sides of the conditional are true. This shows that the conditional is a tautology and thus, $(p \rightarrow q) \wedge p$ implies q.

Second method A different procedure, which some people prefer, uses truth tables to show an implication. In order to show that $a \Rightarrow b$, we need to show that $a \rightarrow b$ is a tautology. But this only means that the truth tables for a and b (in this order) must not have a line with the values TF (in the same order), because this is the only case in which $a \rightarrow b$ is false. Thus, we may simply examine the given truth table for $(p \rightarrow q) \wedge p$ and q to show the implication. Notice that in columns 2 and 3 of the table there is no row with TF (in this order). Thus, $[(p \rightarrow q) \wedge p] \Rightarrow q$.

		1	2	3
p	q	$p \rightarrow q$	$(p \rightarrow q) \wedge p$	q
T	T	T	T	T
T	F	F	F	F
F	T	T	F	T
F	F	T	F	F

◀

EXAMPLE 5   Identify the combinations of truth values that may arise to determine whether $p \Rightarrow q$ or $q \Rightarrow p$, or neither, where p and q are as follows:

p: The fruit is an apple.
q: The fruit is not a pear.

Solution   We make a truth table and give the verbal interpretation of each line:

	p	q
The fruit is an apple and not a pear (possible).	T	T
The fruit is an apple and also a pear (impossible).	T	F
The fruit is not an apple and is not a pear (possible).	F	T
The fruit is not an apple and is a pear (possible).	F	F

Thus, we see that the only line of the truth table that cannot occur is the TF line. Hence, $p \Rightarrow q$.   ◀

A. 1. Show by means of a truth table that the statement $(p \wedge q) \to p$ is a tautology. This demonstrates that $(p \wedge q) \Rightarrow p$.

2. Show by means of a truth table that the statement

$$[(p \to q) \wedge (q \to r)] \to (p \to r)$$

is a tautology. This demonstrates that $[(p \to q) \wedge (q \to r)] \Rightarrow (p \to r)$.

3. Show by means of a truth table that the statement $p \leftrightarrow \sim p$ is a contradiction.

4. Classify each of the following statements as a tautology, a contradiction, or neither:

   a. $p \leftrightarrow p$

   b. $(p \to q) \leftrightarrow (p \wedge \sim q)$

   c. $(p \to q) \leftrightarrow (\sim p \vee q)$

   d. $(p \wedge \sim q) \wedge q$

B. 5. Find all implications that exist between the statements u, v, and w with truth values shown in the table in the margin.

u	v	w
T	T	F
F	T	F
F	T	T
F	F	F

In Problems 6–14, identify the combinations of truth values that may arise to determine whether $p \Rightarrow q$ or $q \Rightarrow p$, or neither.

6. p: The number is positive.
   q: The number is negative.

7. p: The number is less than 0.
   q: The number is greater than or equal to 0.

8. p: The number is negative.
   q: The number is less than or equal to 0.

9. p: The book is a dictionary.
   q: The book is useful.

10. p: The animal is a dog.
    q: The animal is not a cat.

11. p: Jennie is tall or blonde.
    q: Jennie is either tall or blonde, but not both.

12. p: It rains every weekend.
    q: It never rains except on weekends.

13. p: To pass this course, I must average at least 70% on the tests.
    q: I will fail this course if and only if I do not average at least 70% on the tests.

14. p: To make money on the stock market, one must be willing to gamble.
    q: If one is willing to gamble, one will make money on the stock market.

15. Select the maximum number of consistent statements from the following set. [*Note:* A set of statements is **consistent** if it is possible for all of them to be true at the same time.

a. D is stupid.   **b.** D is careless.
c. D is careless but not stupid.
d. If D is stupid, then D is careless.
e. D is careless if and only if D is stupid.
f. D is either stupid or careless, but not both.

In Problems 16–22, two statements are given. Determine whether they are equivalent or if one implies the other, or neither.

16. $\sim(p \vee q)$; $\sim p \wedge \sim q$ **17.** $\sim p \wedge q$; $p \rightarrow q$
18. $\sim p \rightarrow \sim q$; $\sim p \rightarrow q$ **19.** $p \vee (p \wedge q)$; p
20. $p \wedge (p \vee q)$; p **21.** $\sim p \vee \sim q$; $p \wedge \sim q$
22. $(p \wedge q) \rightarrow r$; $\sim p \vee \sim q \vee r$

USING YOUR KNOWLEDGE 2.6

In describing sets using set-builder notation, the reader probably observed the close connection between a set and the statement used to define that set. If we are given a universal set $\mathcal{U}$, there is often a simple way in which to select a subset of $\mathcal{U}$ corresponding to a statement about the elements of $\mathcal{U}$. For example, if $\mathcal{U} = \{1, 2, 3, 4, 5, 6\}$ and p is the statement, "The number is even," the set corresponding to this statement will be $P = \{2, 4, 6\}$; that is, P is the subset of $\mathcal{U}$ for which the statement p is true. The set P is called the **truth set** of p. Similarly, P′ is the truth set of $\sim$p.

Let $\mathcal{U} = \{a, b, c, d, e\}$. Then, let p be the statement, "The letter is a vowel"; let q be the statement, "The letter is a consonant"; and let r be the statement, "The letter is the first letter in the English alphabet."

P, the truth set of p, is {a, e}.
Q, the truth set of q, is {b, c, d}.

Can you find:

STATEMENT LANGUAGE	SET LANGUAGE
p	P
q	Q
$\sim$p	P′
$\sim$q	
p $\vee$ q	
p $\wedge$ q	
p $\Rightarrow$ q	
p $\Leftrightarrow$ q	
t, a tautology	
c, a contradiction	

R, the truth set of r?
R′, the truth set of $\sim$r?

Because p and q are statements, $p \vee q$ and $p \wedge q$ are also statements; hence, they must have truth sets. To find the truth set of $p \vee q$, we select all the elements of $\mathcal{U}$ for which $p \vee q$ is true (that is, the elements that are vowels or consonants). Thus, the truth set of $p \vee q$ is $P \cup Q = \{a, b, c, d, e\} = \mathcal{U}$. Similarly, the truth set of $p \wedge q$ is the set of all elements of $\mathcal{U}$ that are vowels and consonants; that is, the truth set of $p \wedge q$ is $P \cap Q = \varnothing$.

1. With this information, complete the table in the margin.

2. If P, Q, and R are the truth sets of p, q, and r, respectively, find the truth sets of the following statements:
   a. $q \wedge \sim r$ **b.** $(p \wedge q) \wedge \sim r$
3. As in Problem 2, find the truth sets of the following statements:
   a. $p \wedge \sim (q \vee r)$ **b.** $(p \vee q) \wedge \sim (q \vee r)$

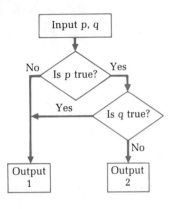

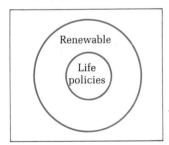

DISCOVERY 2.6

1. Here is a diagram for a machine that is set up to decide whether the statement $p \rightarrow q$ is true or false. What are the output decisions labeled Output 1 and Output 2?
2. Can you discover a similar diagram for a machine that will decide whether the statement $p \wedge q$ is true or false?
3. Can you discover a diagram for a machine that will decide whether the statement $p \vee q$ is true or false?

2.7

EULER DIAGRAMS

Figure 2.7a

Sometimes statements in logic involve relationships between sets. Thus, a renewal provision of a life insurance policy states, "All life policies are renewable for additional term periods." This statement is equivalent to the following two statements:

1. If it is a life policy, then it is renewable for additional term periods.
2. The set of life policies is a subset of the set of all policies renewable for additional term periods.

Statements 1 and 2 can be visually represented by an Euler diagram (another name for a type of Venn diagram), as shown in Fig. 2.7a.

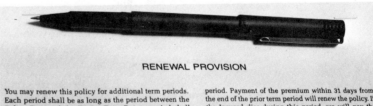

RENEWAL PROVISION

You may renew this policy for additional term periods. Each period shall be as long as the period between the Policy Date and the Expiry Date. But no period shall run past the Final Expiry Date. The Table of Insurance and Premium Amounts shows the premiums for each period. Payment of the premium within 31 days from the end of the prior term period will renew the policy. If the Insured dies during this period, we will pay the insurance proceeds less any premium that has not been paid.

A. Drawing Euler Diagrams

In this section, we shall study the analysis of arguments by using Euler diagrams, a method that is most useful for arguments containing the words "all," "some," or "none." In order to proceed, we must define what is meant by an **argument**.

▶ **Definition 2.7a**

> An **argument** is a set of statements, the **premises**, and a claim that another statement, the **conclusion**, follows from the premises.

We can represent four basic types of statements in Euler diagrams; these are illustrated in Figs. 2.7b, c, d, and e.

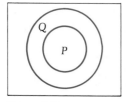

Figure 2.7b
All *P*'s are *Q*'s

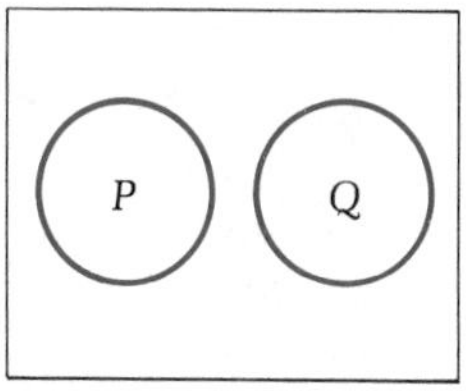

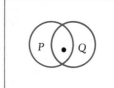

Figure 2.7c
No *P*'s are *Q*'s

Figure 2.7d
Some *P*'s are *Q*'s

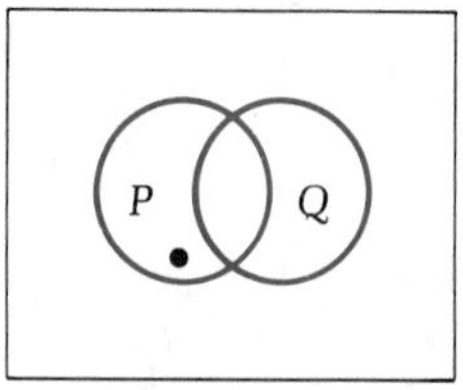

Figure 2.7e
Some *P*'s are not *Q*'s

The following examples discuss some simple arguments. Note that the premises are written on individual lines above a horizontal line; the conclusion is written below this line. This conclusion is preceded by the symbol ∴, which is read "therefore."

EXAMPLE 1   Consider the following argument:

> All men are mortal.
> Socrates is a man.
> ∴Socrates is mortal.

a. Identify the premises and the conclusion.
b. Make an Euler diagram for the premises.

Solution   a. The premises, which appear above the horizontal line, are "All men are mortal" and "Socrates is a man." The conclusion is "Socrates is mortal."

b. To diagram the first premise, we begin by drawing a region to represent "mortals." Since all men are mortal, the region for "men" appears inside the region for mortals, as shown in Fig. 2.7f. The second premise, "Socrates is a man," indicates that "Socrates" goes inside the region representing "men." If s represents "Socrates," the diagram showing both premises appears in Fig. 2.7g.

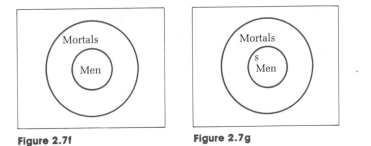

Figure 2.7f   **Figure 2.7g**   ◀

B. Valid Arguments

If we use the premises in an argument to reach a conclusion, we would like the resulting argument to be **valid.**

▶ **Definition 2.7b**   An argument is **valid** if whenever all the premises are true, the conclusion is also true. If an argument is not valid, it is said to be **invalid.**

Thus, the argument shown in Fig. 2.7g is valid because if Socrates is in the set of all men and the set of all men is inside the set of mortals, it must follow that Socrates is mortal.

EXAMPLE 2   Use an Euler diagram to test the validity of the following argument:

  All foreign cars are expensive.
  My car is expensive.
  ∴My car is a foreign car.

Solution   The first premise means that the set of all foreign cars is a subset of the set of expensive cars, while the second premise places "my car" (represented by m) within the set of expensive cars without specifying exactly where, as shown in the figure.

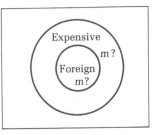

Since the information given is not enough to determine which alternative is correct—that is, m can be foreign or not—we say the argument is invalid.   ◀

EXAMPLE 3   Test the validity of the following argument:

> Some students are dangerous.
> All dangerous people are crazy.
> ∴ Some students are not crazy.

Solution   The diagram in Fig. 2.7h shows the premise, "Some students are dangerous" by two intersecting circles, with at least one student (represented by
s) included in both circles. The second premise, "All dangerous people are
crazy," can be shown by enclosing the set of dangerous people inside the
set of crazy people, as in Fig. 2.7i or as in Fig. 2.7j. Since we do not know
which of these two drawings is correct, we cannot conclude that "some
students are not crazy." Thus, the argument is invalid.

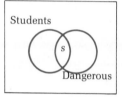

Figure 2.7h

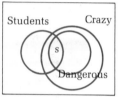

Figure 2.7i

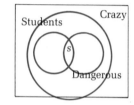

Figure 2.7j   ◀

EXAMPLE 4   Test the validity of the following argument:

> Some students are intelligent.
> All intelligent people are snobs.
> ∴ Some students are snobs.

Solution   The first premise indicates that the set of students and the set of intelligent
people have at least one element in common, represented by s (see the
figure). Since all intelligent people are snobs, the set of intelligent people
appears inside the circle of snobs. In this case, we can conclude that some
students are snobs (that is, there is at least one student, s, who is also a
snob). The argument is valid.

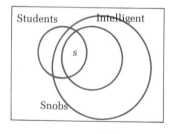

◀

EXAMPLE 5   Determine the validity of the following argument:

All persons taking the test outside the United States should not register.

All persons not registered should write to the company.

∴Persons taking the test outside the United States should write to the company.

Solution   Let O be the set of all persons taking the test outside the United States, let NR be the set of all persons not registered for the test, and let W be the set of all persons that should write to the company. The first premise is diagramed by placing O inside NR, while the second premise indicates that NR is inside the set W (see the figure). It follows from the diagram that the set O is inside the set W. Thus, the argument is valid.

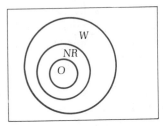

◀

EXERCISE 2.7

In Problems 1–6, state the premises and the conclusion for the given argument.

1.   No misers are generous.
     Some old persons are not generous.
     ∴Some old persons are misers.

2.   No thieves are honest.
     Some dishonest people are convicted.
     ∴Some thieves are convicted.

3.   All diligent students make A's.
     All lazy students are not successful.
     ∴All diligent students are lazy.

4.   All students like logic.
     Robin likes logic.
     ∴Robin is a student.

5.   No kitten that loves fish is unteachable.
     No kitten without a tail will play with a gorilla.
     ∴No unteachable kitten will play with a gorilla.

6.   No birds are proud of their tails.
     Some birds cannot sing.
     ∴Peacocks cannot sing.

In *Problems 7–20, use Euler diagrams to determine the validity of the given argument.*

7. All professors are wise.
   Ms. Brown is a professor.
   ∴Ms. Brown is wise.

8. All students are studious.
   Mr. Smith is studious.
   ∴Mr. Smith is a student.

9. No drinkers are healthy.
   No joggers drink.
   ∴No joggers are healthy.

10. All students are dedicated.
    All wealthy people are students.
    ∴All wealthy people are dedicated.

11. All men are funny.
    Joey is a man.
    ∴Joey is funny.

12. All football players are muscular.
    Jack is muscular.
    ∴Jack is a football player.

13. All felines are mammals.
    No dog is a feline.
    ∴No dogs are mammals.

14. Some students drink beer.
    All beer drinkers are dangerous.
    ∴All students are dangerous.

15. No mathematics teacher is wealthy.
    No panthers teach mathematics.
    ∴No panthers are wealthy.

16. All hippies have long hair.
    Some athletes are hippies.
    ∴Some athletes have long hair.

17. All mathematics teachers have publications.
    Some Ph.D.'s have publications.
    ∴Some Ph.D.'s are mathematics teachers.

18. All beer lovers like Schlitz.
    All people who like Schlitz get drunk.
    ∴All beer lovers get drunk.

19. All heavy cars are comfortable to ride in.
    No car that is comfortable to ride in is shoddily built.
    ∴No heavy car is shoddily built.

20. Some Datsun owners save money.
    Some fast drivers save money.
    ∴Some fast drivers are Datsun owners.

In Problems 21 and 22, use Euler diagrams to determine which (if any) of the given arguments are valid.

21. a. All bulldogs are ugly. This dog is ugly. So it must be a bulldog.
   b. All peacocks are proud birds. This bird is not proud. Therefore, it is not a peacock.
   c. All students who make A's in mathematics are drudges. This student made a B in mathematics. Hence, this student is not a drudge.
22. a. No Southerners like freezing weather. Joe likes freezing weather. Therefore, Joe is not a Southerner.
   b. Some fishermen are lucky. Fred is unlucky. Therefore, Fred is not a fisherman.

USING YOUR KNOWLEDGE 2.7

At the beginning of this chapter, we mentioned a type of argument called a "syllogism." The validity of this type of argument can be tested by using the knowledge you obtained about Venn diagrams in Chapter 1. We shall first diagram the four types of statements involved in these syllogisms.

As you recall, when we make a Venn diagram for two sets that are subsets of some universal set, we divide the region representing the universal set into four different regions, as shown in Figs. 2.7k, l, m, and n.

Figure 2.7k represents the statement, "All P's are Q's." Note that region 1 must be empty, because all P's are Q's.

Figure 2.7l represents the statement, "No P's are Q's." Note that region 3 must be empty, because no P's are Q's.

Figure 2.7m represents the statement, "Some P's are Q's." The dot in region 3 indicates that there is at least one P that is a Q.

Figure 2.7n represents the statement, "Some P's are not Q's." The dot in region 1 indicates that there is at least one P that is not a Q.

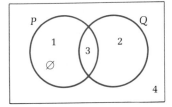

Figure 2.7k
All P's are Q's.

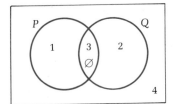

Figure 2.7l
No P's are Q's.

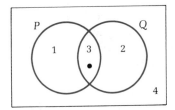

Figure 2.7m
Some P's are Q's.

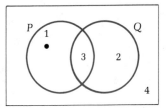

Figure 2.7n
Some P's are not Q's.

Similar considerations govern diagrams involving three sets, as you will see in the following illustration, where we examine the syllogism:

All kangaroos are marsupials.
All marsupials are mammals.
Therefore, all kangaroos are mammals.

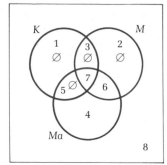

Figure 2.7o

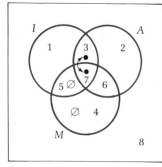

Figure 2.7p

To diagram this argument, we draw a rectangle and three circles, and label the circles K (kangaroos), M (marsupials), and Ma (mammals). As before, these circles divide the rectangle (the universal set) into eight regions. The statement, "All kangaroos are marsupials" makes regions 1 and 5 empty (see Fig. 2.7o); whereas the statement, "All marsupials are mammals" makes regions 2 and 3 empty. In order for the argument to be valid, regions 1 and 3 must be empty, because we wish to conclude that all kangaroos are mammals. Since the diagram shows that this is the case, the conclusion follows and the argument is valid.

Using a similar technique and the diagram in Fig. 2.7p, we can show that the following argument is invalid:

Some intelligent people are attractive.
All models are attractive.
Therefore, some intelligent people are models.

To diagram the statement, "Some intelligent people are attractive," we put dots in both regions 3 and 7, with a curved arrow symbol joining them. This indicates that the dot may be in region 3 or in region 7, or there may be dots in both, but we do not know which is the case; there is no statement in the argument that prevents region 7 from being empty. Thus, the conclusion could be false so the argument is invalid.

1.–10. Use these ideas to examine the validity of the arguments in Problems 11–20 of Exercise 2.7.

Use the same ideas to solve the following two problems:

11. Given the premises:

No student who does not study will get an A in this course.
Some students in this course do not study.

Which of the following conclusions (if any) are valid?
a. Some students in this course will not get A's.
b. No students in this course will get A's.
c. All students in this course who do study will get A's.

12. Given the premises:

Unless it rains, the grass will not grow.
If the grass grows, I will cut it.
It rains.

Which of the following conclusions (if any) are valid?
a. I will cut the grass.
b. The grass grows.
c. If the grass does not grow, I will not cut it.

TRUTH TABLES AND VALIDITY OF ARGUMENTS

In the preceding section, a valid argument was defined as follows: An argument is **valid** if whenever all the premises are true, the conclusion is also true. If an argument is not valid, it is said to be **invalid.** This definition suggests that a truth table can be used to check the validity of an argument. In order to construct such a truth table efficiently, the argument must be in symbolic form. The following examples illustrate the idea.

EXAMPLE 1   Write the following argument in symbolic form:

> If today is Sunday, then I will go to church.
> Today is Sunday.
> ∴ I will go to church.

Solution   Let s be, "Today is Sunday," and let c be, "I will go to church." Then the argument is symbolized as follows:

$$s \rightarrow c$$
$$\underline{s\qquad}$$
$$\therefore c$$

◀

EXAMPLE 2   Symbolize the following argument:

> A whole number is even or odd.
> This whole number is not even.
> ∴ This whole number is odd.

Solution   Let e be, "A whole number is even," and let o be, "A whole number is odd." Then the argument is symbolized as

$$e \vee o$$
$$\underline{\sim e\qquad}$$
$$\therefore o$$

◀

EXAMPLE 3   Determine the validity of the argument in Example 2.

Solution   We make a truth table for this argument with a column for each premise and a column for the conclusion as shown:

e	o	PREMISE $e \vee o$	PREMISE $\sim e$	CONCLUSION o	
T	T	T	F	T	
T	F	T	F	F	
F	T	T	T	T	← True premises True conclusion
F	F	F	T	F	

According to the definition of a valid argument, we need to examine only those rows of the table where all the premises are true. Consequently, we need to check only the third row of the truth table. (In rows 1, 2, and 4, at least one of the premises is false, and so these rows are crossed out.) For the argument to be valid, the remaining items in the conclusion column must all be T's. Thus, the table shows that the argument is valid.

◄

EXAMPLE 4   Determine the validity of the following argument:

> Either the puppy is cute or I will not buy it.
> The puppy is cute.
> ∴ I will buy it.

Solution   By writing p for "The puppy is cute" and b for "I will buy it," we can symbolize the argument in the form

$$p \lor \sim b$$
$$\underline{p}$$
$$\therefore b$$

Now, we construct the table for this argument:

			PREMISE	PREMISE	CONCLUSION
p	b	$\sim b$	$p \lor \sim b$	p	b
T	T	F	T	T	T
T	F	T	T	T	F
F	T	F	(F)	(F)	T
F	F	T	T	(F)	F

← True premises
False conclusion

Next, we cross out all the rows where an F occurs in a premise column (rows 3 and 4 in the table). In the remaining rows, the premises are all true, so the remaining items in the conclusion column must all be T's if the argument is to be valid. Since there is an F in the second row of this column, the argument is invalid.

◄

EXAMPLE 5   Determine the validity of the following argument: You will not get overtime pay unless you work more than 40 hr per week. You work more than 40 hr per week. So you will get overtime pay.

Solution   To symbolize the argument, let p be, "You get overtime pay," and let m be, "You work more than 40 hr per week." Then, the argument is as follows:

$$p \to m$$
$$\underline{m}$$
$$\therefore p$$

Notice that "You will not get overtime pay unless you work more than 40 hr per week" means that "If you get overtime pay, then you work more than 40 hr per week." The truth table for this argument is:

p	m	PREMISE $p \to m$	PREMISE m	CONCLUSION p	
T	T	T	T	T	
T	F	Ⓕ	Ⓕ	T	
F	T	T	T	F	← True premises False conclusion
F	F	T	Ⓕ	F	

In the first and third rows of the table, both premises are true; but in the third row, the conclusion is false. Hence, the argument is invalid. ◀

EXAMPLE 6   Determine the validity of the following argument:

Dictionaries are useful books.
Useful books are valuable.
∴Dictionaries are valuable.

$d \to u$
$u \to u$
$\therefore d \to v$

Solution   We first write the argument in symbolic form. The statement, "Dictionaries are useful books" is translated as "If the book is a dictionary (d), then it is a useful book (u)." "Useful books are valuable" means, "If the book is a useful book (u), then it is valuable (v)." "Dictionaries are valuable" is translated as, "If the book is a dictionary (d), then it is valuable (v)." Thus, the argument is symbolized as

$d \to u$
$u \to v$
$\therefore d \to v$

From the table, we see that whenever all the premises are true (rows 1, 5, 7, and 8), the conclusion is also true. Hence, the argument is valid.

d	u	v	PREMISE $d \to u$	PREMISE $u \to v$	CONCLUSION $d \to v$	
T	T	T	T	T	T	←
T	T	F	T	Ⓕ	F	
T	F	T	Ⓕ	T	T	
T	F	F	Ⓕ	T	F	
F	T	T	T	T	T	←
F	T	F	T	Ⓕ	T	
F	F	T	T	T	T	←
F	F	F	T	T	T	←

True premises True conclusion

◀

EXAMPLE 7   Translate the argument in the cartoon into symbolic form and check its validity.

© 1977 United Feature Syndicate, Inc. Reprinted by permission of UFS, Inc.

Solution   Let r be, "You do not know how to read."
Let w be, "You cannot read *War and Peace*."
Let h be, "Leo Tolstoy will hate you."

Then the argument can be translated as follows:

$$r \to w$$
$$\underline{w \to h}$$
$$\therefore r \to h$$

Since the form of this argument is identical to that in Example 6, this argument is also valid.   ◄

Examples 6 and 7 are illustrations of the fact that an argument of the form

$$p \to q$$
$$\underline{q \to r}$$
$$\therefore p \to r$$

is always valid.

In using a truth table to check the validity of an argument, it is to be emphasized that we need to examine only those rows where the premises are all true. This points out a basic logical principle: **If there is no case where the premises are all true, then the argument is valid regardless of the conclusion.**

This idea is used in the next example.

EXAMPLE 8   Determine the validity of the following argument:

It is raining now.
It is not raining now.
$\therefore$ It is raining now.

Solution   We let p be, "It is raining now," and symbolize the argument to get

p	~p	p ∧ ~p
T	F	F
F	T	F

p
$\underline{\sim p}$
$\therefore p$

From the table in the margin, we see that there is no case in which the conjunction of the premises, $p \wedge \sim p$, is true. Thus, the argument is valid.

◀

In many cases, we are confronted with arguments of the form

1. $p \rightarrow q$
2. $q \rightarrow r$
3. $\underline{r \rightarrow s}$
4. $\therefore p \rightarrow s$

Four statements are involved in this argument, so the truth table to determine its validity will have $2^4 = 16$ rows. To construct such a table would be a tedious task, indeed. For this reason, we develop an alternative method for establishing the validity of such arguments. We first consider the premises $p \rightarrow q$ and $q \rightarrow r$. From these we can conclude (see Example 6) that $p \rightarrow r$. Hence, substituting our conclusion in the given argument, we have

2. $p \rightarrow r$
3. $\underline{r \rightarrow s}$
   $\therefore p \rightarrow s$

But this new argument again has the form of the one in Example 6, so we know it is valid.

In many cases, instead of determining the validity of an argument, we have to supply a valid conclusion for a given set of premises. Using the idea in Example 6, we can see that if the premises are $(p \rightarrow q)$, $(q \rightarrow r)$, . . . , $(x \rightarrow y)$, $(y \rightarrow z)$, a valid conclusion is $p \rightarrow z$. We illustrate this idea in the following examples.

EXAMPLE 9   Supply a valid conclusion using all the following premises:

1. $p \rightarrow q$
2. $q \rightarrow r$
3. $\sim s \rightarrow \sim r$

Solution   The third premise may be rewritten as $r \rightarrow s$, because a statement and its contrapositive are equivalent. Hence, the entire argument may be written as

1. $p \rightarrow q$
2. $q \rightarrow r$
3. $r \rightarrow s$

Thus, a valid conclusion using all the premises is $p \rightarrow s$.   ◀

EXAMPLE 10   Suppose you know it to be true that:

1. If Alice watches TV, then Ben watches TV. $A \to B$

2. Carol watches TV if and only if Ben watches. $B \to C$

3. Don never watches TV if Carol is watching. $\sim C \to \sim D$

4. Don always watches TV if Ed is watching. $D \to E$

Show that Alice never watches TV when Ed is watching.

Solution   Let:

    a be "Alice watches TV" $q \to b$

    b be "Ben watches TV" $b \leftrightarrow c$

    c be "Carol watches TV"

    d be "Don watches TV"

    e be "Ed watches TV"

The argument above can be symbolized as follows:

1. $a \to b$

2. $c \leftrightarrow b$ or, equivalently,   **2′.** $b \leftrightarrow c$

3. $c \to \sim d$

4. $e \to d$

If these premises are arranged in the order 1, 2′, 3, and the contrapositive of 4, we obtain

1. $a \to b$

2′. $b \leftrightarrow c$

3. $c \to \sim d$

4. $\sim d \to \sim e$

Consequently, we may conclude that $a \to \sim e$; that is, "If Alice watches TV, then Ed does not watch TV," or, equivalently, "Alice never watches TV when Ed is watching."   ◀

In our day-to-day reasoning, we do not make truth tables or check our arguments in any formal fashion. Instead, we (perhaps unconsciously) learn a few argument forms, which we use as we need them. The most commonly used of these argument forms are as follows:

Modus Ponens	Modus Tollens	Hypothetical Syllogism	Disjunctive Syllogism
$p \to q$	$p \to q$	$p \to q$	$p \lor q$
p	$\sim q$	$q \to r$	$\sim p$
$\therefore q$	$\therefore \sim p$	$\therefore p \to r$	$\therefore q$

The names of the first two of these are derived from the Latin and mean, respectively, "a manner of affirming" and "a manner of denying" (the parts of a conditional). The modus ponens and the two types of syllogism have already been discussed in this section; see Examples 1 and 2 and Example 6. It is left for the reader to show the validity of the modus tollens in Problem 20 of Exercise 2.8.

EXERCISE 2.8

In Problems 1–16, symbolize the argument using the suggested abbreviations. In each case, determine the validity of the given argument.

1. If you eat your spinach (e), you can go out and play (p).
   You did not eat your spinach.
   Therefore, you cannot go out and play.

2. If you eat your spinach (e), you can go out and play (p).
   You cannot go out and play.
   Therefore, you did not eat your spinach.

3. If you study logic (s), mathematics is easy (e).
   Mathematics is not easy.
   Therefore, you did not study logic.

4. I will learn this mathematics (m), or I will eat my hat (e).
   I will not eat my hat.
   Therefore, I will learn this mathematics.

5. The Good Taste Restaurant has good food (g).
   Hence, the Good Taste Restaurant has good food, and I will recommend it to everyone (r).

6. If prices go up (u), management will scream (s).
   If management screams, then supervisors will get tough (t).
   Hence, if prices go up, supervisors will get tough.

7. If I work (w), then I have money (m).
   If I don't work, I have a good time (g).
   Therefore, I have money or a good time.

8. Babies (b) are illogical (i).
   Nobody is despised (d) who can manage a crocodile (m).
   Illogical persons are despised.
   Hence, babies cannot manage crocodiles.

9. If you have the time (t), we got the beer (b).
   You have the time.
   So we got the beer.

10. Bill did not go to class this morning (~g), because he wore a red shirt (r), and he never wears a red shirt to class.

11. Where there is smoke (s) there is fire (f).
    There is smoke.
    Hence, there is fire.

12. If you are enrolled (e) or have been accepted half-time at a college (h), you may apply for a loan (a).
You have not been accepted half-time at a college.
Hence, you may not apply for a loan.

13. You will be eligible for a grant (e) if you meet all the criteria (m).
You do not meet all the criteria.
So you are not eligible for a grant.

14. We will pay for collision loss (p) only if collision coverage is afforded (a).
Collision coverage is not afforded.
Hence, we will not pay for collision loss.

15. If spouse is also filing (f), give spouse's Social Security number (s).
Spouse is not filing.
Hence, do not give spouse's Social Security number.

16. Additional sheets of paper will not be attached (~a) unless more space is needed (m).
More space is needed.
Hence, additional sheets of paper are attached.

An argument is given in each of Problems 17–21. Determine whether each argument is valid or invalid.

17. $p \lor q$
$\underline{\sim p}$
$\therefore q$

18. $p \to q$
$\underline{p}$
$\therefore q$

19. $p \to q$
$\underline{\sim p}$
$\therefore q$

20. $p \to q$
$\underline{\sim q}$
$\therefore \sim p$

21. $p \to q$
$\underline{q \to r}$
$\therefore \sim r \to \sim p$

In Problems 22–31, find a valid conclusion using all the premises.

22. $p \to q$
$q \to r$
$\underline{r \to \sim s}$

23. $p \to q$
$\sim q \lor r$
[Hint: $\sim q \lor r \Leftrightarrow q \to r$]

24. $p \to q$
$s \to \sim r$
$t \to r$
$q \to u$
$\sim u \lor t$

25. $p \to q$
$q \to r$
$\sim s \to \sim r$
$\underline{p}$

26. $p \to \sim q$
$r \to q$
$\underline{r}$

27. $\sim p \to q$
$\sim p \lor r$
$\underline{\sim r}$

28. If it rains, then the grass will grow.
A sufficient condition for cutting the grass is that it will grow.
The grass is cut only if it is higher than 8 inches.

29. The only books in this library that I do not recommend are unhealthy.
    All bound books are well-written.
    All romances are healthy in tone.
    I do not recommend any of the unbound books.

30. All ducks can fly.
    No land bird eats shrimp.
    Only flightless birds do not eat shrimp.

31. If you are not patriotic, then you do not vote.
    Aardvarks have no emotions.
    You cannot be patriotic if you have no emotions.

32. Classify the arguments in Problems 2, 4, 6, and 11 as modus ponens, modus tollens, hypothetical syllogism, or disjunctive syllogism.

DISCOVERY 2.8

Consider the following premises taken from Symbolic Logic, a book written by Lewis Carroll (logician, mathematician, and author of Alice's Adventures in Wonderland):

No kitten that loves fish is unteachable.
No kitten without a tail will play with a gorilla.
Kittens with whiskers always love fish.
No teachable kitten has green eyes.
Kittens that have no whiskers have no tails.

1. Find a valid conclusion using all the premises. [*Hint:* "No _____ is . . ." is translated as "If it is a _____, then it is not a. . . ."

*2.9

SWITCHING NETWORKS

The theory of logic discussed in this chapter can be used to develop a theory of simple switching networks. A **switching network** is an arrangement of wires and switches that connect two terminals. A **closed** switch permits the flow of current, while an **open** switch prevents the flow. One may also think of a switch as a valve that controls the flow of water through a pipe, or as a drawbridge over a river controlling the flow of traffic along a road (see Fig. 2.9a, page 98).

Two switches may be connected in **series** (in a line from left to right), as in Fig. 2.9b. In this network, the current flows between the **terminals** A and B only if both switches P and Q are closed.

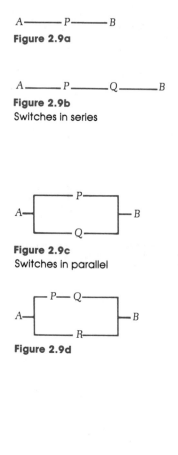

A———P———B

Figure 2.9a

A———P———Q———B

Figure 2.9b
Switches in series

Figure 2.9c
Switches in parallel

Figure 2.9d

To see how the preceding sections can be used in connection with switching networks, we let p be the statement, "Switch P is closed" and q be the statement, "Switch Q is closed." Then in Fig. 2.9a, for example, current will flow if and only if p is true (switch P is closed). On the other hand, in Fig. 2.9b, current will flow if and only if both p and q are true (switch P is closed and switch Q is closed); that is, if and only if $p \wedge q$ is true. For this reason, the network in Fig. 2.9b is associated with the statement $p \wedge q$. If we interpret the network in Fig. 2.9a as a drawbridge over a river controlling traffic between points A and B, it is clear that we can go from A to B only when the bridge is down (closed). Similarly, to go from A to B in Fig. 2.9b, we need both bridges down (closed).

Two switches also can be connected in **parallel** (as shown in Fig. 2.9c). It is clear that in Fig. 2.9c the current will flow if either P or Q is closed; that is, if the statement $p \vee q$ is true. For this reason, the network in Fig. 2.9c is associated with the statement $p \vee q$, where p is the statement, "Switch P is closed," and q is the statement, "Switch Q is closed."

These two simple networks (series and parallel) can be combined to form more complicated networks as in Fig. 2.9d. In this figure, the current will flow if P and Q are both closed or if R is closed. For this reason, the network in Fig. 2.9d is associated with the statement $(p \wedge q) \vee r$. If we interpret Fig. 2.9d as drawbridges controlling traffic between A and B, we can see that we are able to go from A to B when both P and Q are down (closed) or when R is down (closed).

Any switch P can be coupled with a switch (denoted by P') having the opposite effect of P; that is, if P is open, P' is closed, and vice versa. Thus, if p is the statement that corresponds to the switch P, then $\sim p$ is the statement that corresponds to the switch P'. The switches P and P' are called **complementary.** In the event that we have two switches that open and close simultaneously, these switches will be represented by the same letter and will be called **equivalent** switches.

In the same manner in which we have associated the statements $p \vee q$ and $(p \wedge q) \vee r$ with the networks shown in Figs. 2.9c and 2.9d, respectively, every compound statement can be represented by a switching network. The next example illustrates this idea.

EXAMPLE 1   Construct a network corresponding to the statement $(p \vee r) \wedge (q \vee r)$.

Solution   The network associated with the given statement appears in Fig. 2.9e. The parallel circuit containing the switches P and R corresponds to the statement $p \vee r$. Similarly, the parallel circuit with the switches Q and R corresponds to the statement $q \vee r$. These two parallel circuits are connected in series to correspond to the connective $\wedge$, which joins the two statements $p \vee r$ and $q \vee r$.   ◄

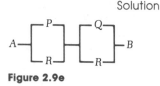

Figure 2.9e

Notice that, in the networks of Figs. 2.9d and 2.9e, current will flow if P and Q are closed or if R is closed. For this reason, these two networks are said to be equivalent. In general, two networks are **equivalent** if their corresponding statements are equivalent. For example, the statement corresponding to the network in Fig. 2.9d is $(p \wedge q) \vee r$ and that of Fig. 2.9e is $(p \vee r) \wedge (q \vee r)$. The truth tables of these two statements are identical, so the networks of Figs. 2.9d and 2.9e are equivalent.

Finally, we shall consider the design of certain networks having specified properties. An equivalent problem is that of constructing a compound statement having a specified truth table. The procedure used will involve the basic conjunctions given in Table 2.9a. For example, a network associated with a statement having truth table TTFF will be the one corresponding to the statement $(p \wedge q) \vee (p \wedge \sim q)$. (See Exercise 2.3, Problems 19 and 20.)

Table 2.9a

p	q	BASIC CONJUNCTION
T	T	$p \wedge q$
T	F	$p \wedge \sim q$
F	T	$\sim p \wedge q$
F	F	$\sim p \wedge \sim q$

EXAMPLE 2

A toy designer plans to build a battery-operated kitten, with arms that can be lowered or raised and a purring mechanism. He wants his kitten to purr only when the *right* arm or *both* arms are raised; with any other arrangement, the purring mechanism is to be off. Construct a switching circuit that will do this.

Solution

Let p be the statement, "The right arm is raised," and let q be the statement, "The left arm is raised." The desired truth table is Table 2.9b. We note that the cat will purr when p is true (rows 1 and 2). Hence, our network will correspond to a statement having truth table TTFF. We have just seen that one such statement is $(p \wedge q) \vee (p \wedge \sim q)$, so the network associated with this statement (see Fig. 2.9f) is a possible network for the toy kitten. ◀

Table 2.9b

p	q	DESIRED TRUTH TABLE
T	T	T
T	F	T
F	T	F
F	F	F

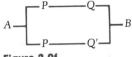

Figure 2.9f

Figure 2.9g

Notice that, in the network of Fig. 2.9f, current will flow when P is closed (because if Q is closed, current will flow through the top branch, and if Q is open, current will flow through the bottom branch). Hence, an equivalent network is the one given in Fig. 2.9g.

As indicated in Problems 19 and 20 of Exercise 2.3, we can always write a statement corresponding to a given truth table as follows:

1. For each row with a T in the final column, write a conjunction using each variable with a T in its column and the negation of each variable with an F in its column.
2. Write the disjunction of these conjunctions.

For example, suppose that two rows of the given truth table are as given below and that all other rows end with an F.

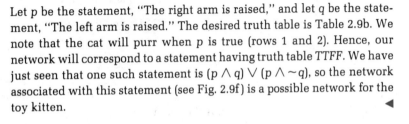

p	q	r	FINAL COLUMN	DESIRED CONJUNCTION
T	T	F	T	$p \wedge q \wedge \sim r$
T	F	F	T	$p \wedge \sim q \wedge \sim r$

Then the desired conjunctions are as shown in the table, and the final desired statement is

$$(p \wedge q \wedge \sim r) \vee (p \wedge \sim q \wedge \sim r)$$

Of course, this procedure does not always give the simplest result for the given truth table. Thus, the statement obtained can be simplified to $p \wedge \sim r$, as you can verify.

EXERCISE 2.9

In Problems 1–6, construct a network corresponding to the given statement.

1. $(p \wedge q) \vee p$ **2.** $p \vee (q \wedge r)$
3. $(\sim p \wedge q) \vee (p \wedge \sim r)$ **4.** $(p \vee q) \wedge \sim r$
5. $[(p \vee \sim q) \vee q \vee (\sim p \vee q)] \vee q$
6. $[(p \vee \sim q) \vee q \vee (\sim p \vee q)] \wedge q$

In Problems 7–11, find the compound statement corresponding to the network given.

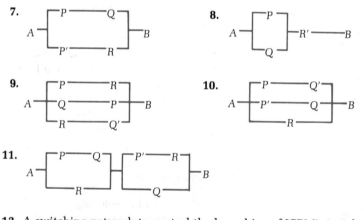

7.

8.

9.

10.

11.

12. A switching network to control the launching of ICBM's is to be designed so that it can be operated by three generals. For democracy's sake, the Defense Department assumes that in order to fire the missile, two of the three generals will have to close their switches. Design a network that will do this.

In Problems 13–15, draw a pair of switching networks to indicate that the given statements are equivalent.

13. $p \vee p$ and p
14. $p \vee (q \wedge r)$ and $(p \vee q) \wedge (p \vee r)$
15. $(p \wedge q) \vee p$ and p

In Problems 16–20, simplify the given network.

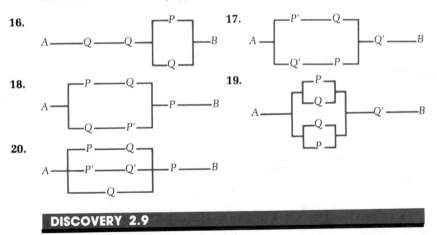

16.

17.

18.

19.

20.

DISCOVERY 2.9

Computer Gates

Digital computers have circuits in which the flow of current is regulated by **gates,** as shown below. These circuits respond to high (1) or low (0) voltages, and can be described by **logic statements.**

AND gate (p ∧ q)   P ⎯⎯⎯ Q ⎯⎯⎯   Output 1 if all input voltages (P and Q) are 1. Output 0 otherwise.

OR gate (p ∨ q)   P ⎯⎯⎯ Q ⎯⎯⎯   Output 1 if at least one input (P or Q) is 1. Output 0 otherwise.

NOT gate (~p)   P ⎯⎯⎯   0 if the input P is 1. 1 if the input P is 0.

Keep in mind that 1 and 0 correspond to high and low voltages, respectively.

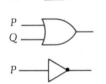

Logic diagrams show how the gates are connected. The outputs in these diagrams can be obtained in the same way in which truth tables are constructed, with 1's replacing the T's and 0's replacing the F's. For example, we can list the outputs for all possible inputs for the logic diagram shown in the margin as follows: The output of A is symbolized by p ∧ q and the output of B by ~q. Thus, the inputs of C correspond to p ∧ q and ~q. Since C is an OR gate, the final output can be symbolized by the statement (p ∧ q) ∨ ~p. If we construct the truth table for this statement using 1 for T and 0 for F, we obtain the table given in the margin. The table shows that the final output (column 5) is always a high (1) voltage except in the third row, where the input voltage corresponding to P is low (0) and that corresponding to Q is high (1). Since we have shown earlier that the statement (p ∧ q) ∨ ~q is equivalent to p ∨ ~q, the given circuit can be simplified to one corresponding to the latter statement. The diagram for this circuit is

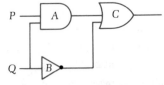

1	2	3	5	4
p	**q**	**(p ∧ q)**	**∨**	**~q**
1	1	1	1	0
1	0	0	1	1
0	1	0	0	0
0	0	0	1	1

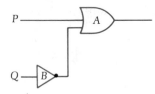

P ———[A >————

Q —[B>·——|

In Problems 1–3, see if you can discover the final outputs for all possible inputs for the given diagrams.

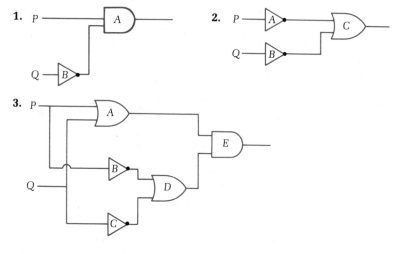

1. P ————[A >————

Q —[B>·——|

2. P —[A>·—————[C >———

Q —[B>·——|

3. P ———[A >————————[E >———

Q ————[B >——[D >——

———[C >——

+Test

Section	Item	Meaning	Example
2.1B	p, q, r, etc.	Statements	p: Today is Monday. q: The sky is blue.
2.1B	$p \wedge q$	Conjunction (p **and** q)	Today is Monday, **and** the sky is blue.
2.1B	$p \vee q$	Disjunction (p **or** q)	Today is Monday, **or** the sky is blue.
2.1B	$\sim p$	Negation (**not** p)	Today is **not** Monday.

2.2A, B, C						
	p	**q**	**p ∧ q**	**p ∨ q**	**~p**	Truth tables for the statements $p \wedge q$, $p \vee q$, and $\sim p$
	T	T	T	T	F	
	T	F	F	T	F	
	F	T	F	T	T	
	F	F	F	F	T	

Section	Item	Meaning	Example
2.3B	$p \Leftrightarrow q$	The statements p and q are equivalent; that is, they have identical truth tables.	$\sim(p \vee q) \Leftrightarrow \sim p \wedge \sim q$
2.4A	$p \rightarrow q$	p conditional q	**If** today is Monday, **then** I will go to school.
2.4B	$p \leftrightarrow q$	p biconditional q	Today is Monday **if and only if** I go to school.

Section					Meaning	Example
2.4A, B	p	q	$p \rightarrow q$	$p \leftrightarrow q$	Truth tables for the conditional and the biconditional	
	T	T	T	T		
	T	F	F	F		
	F	T	T	F		
	F	F	T	T		

Section	Item	Meaning	Example
2.5A	$q \rightarrow p$ $\sim p \rightarrow \sim q$ $\sim q \rightarrow \sim p$	Converse of $p \rightarrow q$ Inverse of $p \rightarrow q$ Contrapositive of $p \rightarrow q$	
2.5B	p is **sufficient** for q q is **necessary** for p p **only if** q q **if** p	Statements equivalent to "If p, then q."	
*2.6A	A tautology	A statement that is always true	2 is even. $p \vee \sim p$
*2.6A	A contradiction	A statement that is always false	2 is odd. $p \wedge \sim p$
*2.6B	$p \Rightarrow q$	p implies q; the conditional $p \rightarrow q$ is a tautology.	The animal is a dog implies that the animal is a mammal.
2.7A	$p \rightarrow q$ $\underline{p}$ $\therefore q$	An argument (a set of statements, the premises, and a claim that another statement, the conclusion, follows from the premises)	
2.7B	Valid argument	An argument is valid if whenever all the premises are true, the conclusion is also true.	All men are mortal. <u>Socrates is a man.</u> $\therefore$ Socrates is mortal.
2.8	Modus ponens	$p \rightarrow q$ $\underline{p}$ $\therefore q$	

Section	Item	Meaning
2.8	Modus tollens	$$\begin{array}{r} p \to q \\ \sim q \\ \hline \therefore \sim p \end{array}$$
	Hypothetical syllogism	$$\begin{array}{r} p \to q \\ q \to r \\ \hline \therefore p \to r \end{array}$$
	Disjunctive syllogism	$$\begin{array}{r} p \vee q \\ \sim p \\ \hline \therefore q \end{array}$$

PRACTICE TEST 2

1. Which of the following are statements?
   a. Green apples taste good.
   b. 1989 is a leap year.
   c. No fish can live without water.
   d. Some birds cannot fly.
   e. If it rains today, my lawn will get wet.
   f. Can anyone answer this question?
2. Identify the components and the logical connective in each of the following statements. Write each statement in symbolic form using the suggested abbreviations.
   a. If the number of a year is divisible by 4 (d), then the year is a presidential election year (p).
   b. I love Bill (b), but Bill does not love me ($\sim m$).
   c. A candidate is elected president of the United States (e) if and only if he receives a majority of the electoral college votes (m).
   d. Janet can make sense out of symbolic logic (s), or she fails this course (f).
   e. Janet cannot make sense out of symbolic logic ($\sim s$).
3. Let g be, "He is a gentleman," and let s be, "He is a scholar." Write in words:
   a. $\sim(g \wedge s)$ b. $\sim g \wedge s$
4. Write the negation of each of the following statements:
   a. I will go to the beach or to the movies.
   b. I will stay in my room and do my homework.
   c. Pluto is not a planet.
5. Write the negation of each of the following statements:
   a. All cats are felines.
   b. Some dogs are well-trained.
   c. No dog is afraid of a mouse.

6. Write the negation of each of the following statements:
   a. If Joey does not study, he will fail this course.
   b. If Sally studies hard, she will make an A in this course.
7. In the table below, identify each entry under a, b, c, d, and e by matching it with the appropriate one of the following statements: $\sim p$, $p \rightarrow q$, $p \wedge q$, $p \vee q$, $p \leftrightarrow q$.

p	q	a.	b.	c.	d.	e.
T	T	T	T	F	T	T
T	F	F	F	F	F	T
F	T	F	F	T	T	T
F	F	T	F	T	T	F

8. Construct a truth table for the statement

   $(p \vee q) \wedge (\sim p \vee \sim q)$

9. Construct a truth table for the statement

   $(p \vee q) \rightarrow \sim p$

10. Which of the following statements is equivalent to $\sim p \vee q$?
    a. $\sim p \wedge \sim q$ b. $\sim(p \wedge \sim q)$
11. Under what conditions is the following statement true? "Sally is naturally beautiful, or she knows how to use makeup."
12. Is the following statement true or false? If $2 + 2 = 5$, then $2 \cdot 3 = 6$.
13. Construct the truth table for the statement

    $(p \rightarrow q) \leftrightarrow (q \vee \sim p)$

14. Write:
    a. The converse   b. The inverse   c. The contrapositive
    of the statement, "If you make a golf score of 62 once, you will make it again."
15. Let p be, "You get overtime pay," and let m be, "You work more than 40 hr per week." Symbolize the following statements:
    a. If you work more than 40 hr per week, then you get overtime pay.
    b. You get overtime pay only if you work more than 40 hr per week.
    c. You get overtime pay if and only if you work more than 40 hr per week.
16. Let b be, "You get a bank loan," and let c be, "You have a good credit record." Symbolize the following statements:
    a. For you to get a bank loan, it is necessary that you have a good credit record.
    b. Your having a good credit record is sufficient for you to get a bank loan.
    c. A necessary and sufficient condition for you to get a bank loan is that you have a good credit record.

a	b	c
T	T	T
T	F	T
F	F	F
T	F	F

17. The table in the margin gives the truth values for three statements, a, b, and c. Find all the implications among these statements.

18. Which of the following statements are tautologies?

    a. $p \wedge \sim p$ **b.** $p \vee \sim p$

    c. $(p \rightarrow q) \leftrightarrow (\sim q \vee p)$

19. Use an Euler diagram to check the validity of the following argument:

> All students study hard.
> John is not a student.
> ∴John does not study hard.

20. Use an Euler diagram to check the validity of the following argument:

> No loafers work hard.
> Sally does not work hard.
> ∴Sally is a loafer.

21. Use a truth table to check the validity of the argument in Problem 19.

22. Use a truth table to check the validity of the argument in Problem 20.

23. Use a truth table to check the validity of the following argument:

> If you win the race, you are a good runner.
> You win the race.
> ∴You are a good runner.

24. Show that the following argument is invalid:

> $p \rightarrow q$
> $\sim q \rightarrow \sim r$
> ∴ $p \rightarrow \sim r$

25. Construct a switching network that represents the statement $(p \rightarrow q) \wedge (\sim p \rightarrow \sim q)$.

CHAPTER 3

NUMERATION SYSTEMS

It seems fairly certain that some ancient peoples used a base of 2 for counting: 1, 2, 2-1, 2-2, 2-2-1, and so on. Others used a base of 3 in a similar way. As these people became builders and farmers, they had to deal with larger numbers, and many used their own fingers and toes as counters. This led to new numbers, all the way to 20. Number names based on 20 still occur, as in the French words for 80 and 90, *quatre-vingt* and *quatre-vingt-dix*, which literally mean "four-twenty" and "four-twenty-ten," and in the English word "score," which Lincoln used in his famous Gettysburg address, "Four score and seven years ago. . . ."

Our present-day written symbols for the digits 1, 2, 3, 4, 5, 6, 7, 8, and 9 originated with the Hindus. These numbers were designed for a base 10, or decimal, system of counting, so named after the Latin word *decima*, which means "tenth."

Why do we use a base of 10? There really is no good reason except perhaps the number of fingers on a pair of human hands. We could just as well use a base of 5 or 12 or 20. Until about the fifteenth century, fractions were written by a complicated positional system based on the number 60. This sexagesimal (base 60) system had been developed by the ancient Mesopotamians more than 3000 years ago.

The great disadvantage of the sexagesimal system was the need to represent 59 numerals. No one wanted to memorize 59 separate symbols, and to get around this difficulty, the ancients used various combinations of two symbols, one representing our number one and the other representing our number ten.

It was not until about the year 500 that the Hindus devised a positional notation for the decimal system. They discarded the separate symbols for numbers greater than nine, and standardized the symbols for the digits from 1 through 9. It was not until much later that the very important 0 symbol came into use.

The first person who attempted to popularize the decimal system was an Arab mathematician, al-Khowârizmî of Baghdad. In the early part of the ninth century, he wrote a book about the Hindu numerals and recommended their use to merchants and mathematicians everywhere. However, it took about 200 more years for these new numerals to reach Spain, and it was not until the late thirteenth century that their use became at all widespread. As you might imagine, it was mostly the merchants who were won over by the ease of calculating in the new system and who were mainly responsible for its wide adoption.

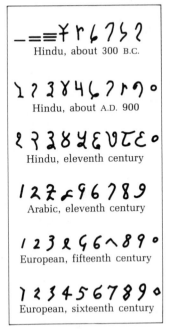

Hindu, about 300 B.C.

Hindu, about A.D. 900

Hindu, eleventh century

Arabic, eleventh century

European, fifteenth century

European, sixteenth century

EVOLUTION OF OUR NUMERALS

Note the absence of a 0 in the early Hindu and Arabic numerals.

In this chapter we study various numeration systems, starting with the additive system used by the ancient Egyptians. This was a system in which the position of a numeral had no significance; number values simply were added. We also look briefly at the positional system of the Babylonians. This appears to be the oldest system known in which the value of a numeral depends on its placement. We then look at the Roman numeral system, which uses base 10 as the Egyptians did and constructs numerals by means of simple addition and subtraction. The most efficient system for everyday use, our present decimal system, also called the Hindu–Arabic system, receives the main emphasis and is discussed in Section 3.2. In the remainder of the chapter, we study systems with bases other than 10, because these help us to understand the decimal system better.

3.1

EGYPTIAN, BABYLONIAN, AND ROMAN NUMERATION SYSTEMS

During the period of recorded history (beginning about 4000 B.C.), people began to think about numbers as abstract concepts. That is, they recognized that two fruits and two arrows have something in common—a quantity called *two*—which is independent of the objects. The perception of this quantity was probably aided by the process of tallying. In different civilizations, different tallying methods have been found. For example, the Incas of Peru used knots in a string or rope to take the census, the Chinese used pebbles or sticks for computations, and the English used tally sticks as tax receipts. As a result of human efforts to keep records of numbers, the first **numerals,** reflecting the process of tallying, were developed. (A **numeral** is a symbol that represents a number. For example, the numeral 2 represents the number two.) Table 3.1a shows three ancient sets of numerals. The property shared by these three numeration systems is that they are **additive;** that is, the values of the written symbols are added to obtain the number represented.

Table 3.1a   Ancient Numerals

	EGYPTIAN	SUMERIAN	MAYAN
1	I	I	•
2	I I	I I	••
3	I I I	I I I	•••
4	I I I I	I I I / I	••••
5	I I I / I I	I I I / I I	—
6	I I I / I I I	I I I / I I I	•̶
7	I I I I / I I I	I I I / I I I / I	••̶
8	I I I I / I I I I	I I I / I I I / I I	•••̶
9	I I I / I I I / I I I	I I I / I I I / I I I	••••̶

A.   The Egyptian System

Let us look at the Egyptian system in greater detail. The Egyptians used hieroglyphics (sacred picture writing) for their numerals. The first line of Table 3.1b (page 110) shows these symbols and their probable numerical values. Convenient names for the hieroglyphics are given in the margin.

As you can see from Table 3.1b, the Egyptians used a **base of 10.** That is, when the tallies were added and they reached 10, the 10 tallies were replaced by the symbol ∩. The Egyptians generally wrote their numbers from right to left, although they sometimes wrote from left to right, or even from top to bottom! Thus, the number 12 could have been represented by I I ∩ or by ∩ I I . For this reason, we say that the Egyptian system is not a

Table 3.1b Numerals

EGYPTIAN, ABOUT 3000 B.C.

1	10	100	1000	10,000	100,000	1,000,000

BABYLONIAN, ABOUT 2000 B.C.

0	1	10	12	20	60	600

EARLY GREEK, ABOUT 400 B.C.

1	5	10	50	100	500	5000

MAYAN, ABOUT 300 B.C.

1	5	7	10	12	15	20

TAMIL, EARLY CHRISTIAN ERA

1	2	3	5	6	10	1000

HINDU–ARABIC, CONTEMPORARY

0	1	2	3	4	5	6

|	Stroke
∩	Heelbone
♀	Scroll
⚘	Lotus flower
⌒	Pointing finger
☜	Fish
⚡	Astonished man

positional system. In contrast, our own **decimal system** (called **decimal** because we use a base of 10) is a **positional system.** In our system, the numerals 12 and 21 represent different numbers.

Computation in the Egyptian system was based on the **additive principle.** For example, to add 24 to 48, the Egyptians proceeded as follows:

```
           Exchange                    ∩
    24       |||∩∩              ∩∩
  +48     ||||||∩∩∩∩       ||∩∩∩∩
    72                     ||∩∩∩∩∩∩∩
```

As you can see, before the computation was done, 10 of the strokes were replaced by a heelbone (see the names given in the margin).

The Egyptians performed subtraction in a similar manner. Thus, to subtract 13 from 22, they proceeded as follows:

$$
\begin{array}{r}
22 \\
-13 \\
\hline 9
\end{array}
$$

The procedure for Egyptian multiplication is explained in the Rhind papyrus (an ancient document bought in Egypt by A. Henry Rhind and named in his honor). The operation was performed by **successive duplications.** Multiplying 19 by 7, for example, the Egyptians would take 19, double it, and then double the result. Then they would add the three numbers, thus:

\1		19
\2		38
\4		76
Total	7	133

The symbol \ is used to designate the submultipliers that add up to the total multiplier, in this case 7. In the Rhind papyrus, the problem 22 times 27 looks like this:

1		27
\ 2		54
\ 4		108
8		216
\16		432
Total	22	594

Again, the numbers to be added are only those in the lines with the symbol \ .

EXAMPLE 1   Problem 79 of the Rhind papyrus states: "Sum the geometrical progression of five terms, of which the first term is 7 and the multiplier 7." It can be shown that the solution of the problem is obtained by multiplying 2801 by 7. Use the method of successive duplications to find the answer.

Solution

\1		2,801
\2		5,602
\4		11,204
Total	7	19,607

◀

It is said that in later years the Egyptians adopted another multiplication technique generally known as **mediation and duplation.** This system consists of halving the first factor and doubling the second. For example, to find the product 19×7, we may successively halve 19, discarding remainders at each step, and successively double 7:

19 is odd →	19	(7)
9 is odd →	9	(14)
	4	28
	2	56
1 is odd →	1	(112)

Notice that half of 19 is regarded as 9, because all the remainders are discarded. The process is completed when a 1 appears in the left-hand column. Opposite each number in the left-hand column there is a corresponding number in the column of numbers being doubled. The product 19×7 is found by adding the circled numbers—those opposite the odd numbers in the column of halves. (Can you see why this works?) Thus, $19 \times 7 = 7 + 14 + 112 = 133$.

EXAMPLE 2
Solution

Use the method of mediation and duplation to find the product 18×43.

	18	43
9 is odd →	9	(86)
	4	172
	2	344
1 is odd →	1	(688)

$18 \times 43 = 86 + 688 = 774$

◄

Additive numeral systems were devised to keep records of large numbers. As these numbers became larger and larger, it became evident that tallying them was difficult and awkward. Thus, numbers began to be arranged in groups and exchanged for larger units, as in the Egyptian system in which 10 tallies were exchanged for a heelbone (∩). The scale used to determine the size of the group to be exchanged (10 in the case of the Egyptians) was the **base** for the system. These additive systems were advantageous for record-keeping operations, but computation in these systems was extremely complicated. As problems became even more complex, a new concept evolved to help with computations, that of a **positional numeral system.** In such a system, a numeral is selected as the base, then symbols ranging from 1 to the numeral that is one less than the base are also selected. Numbers are then represented by placing the symbols in a specified order. For example, the Babylonians used a sexagesimal (base 60) system. In their system, a vertical wedge ▼ was used to represent 1 and the symbol ◄ represented 10. (These symbols first appeared on the clay tablets of the Sumerians and Chaldeans, but were later adopted by the Babylonians; see Fig. 3.1a.)

Figure 3.1a
Sumerian clay tablet

Wide World Photos

B. The Babylonian System

The Babylonian numerals, which may look odd to you, were simply wedge marks in clay. Figure 3.1b shows a few numerals. Notice that the same symbols are used for the numerals 1 and 60. To distinguish between them, a wider space was left between the characters. Thus, ▼▼▼ represents the numeral 3, while ▼ ▼▼ is 62.

Figure 3.1b
Babylonian numerals

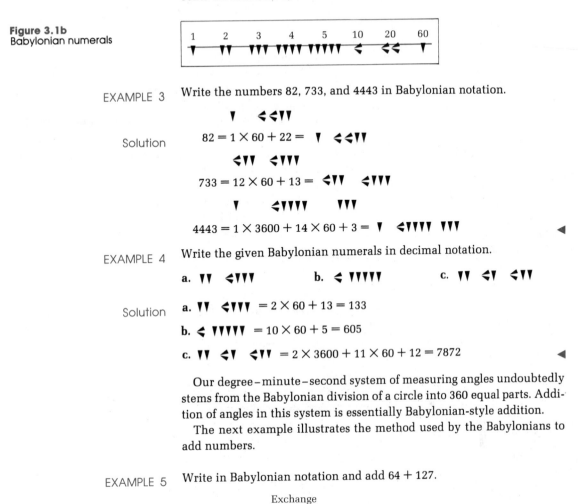

EXAMPLE 3 Write the numbers 82, 733, and 4443 in Babylonian notation.

Solution

$$82 = 1 \times 60 + 22 = \text{▼ ◀◀▼▼}$$

$$733 = 12 \times 60 + 13 = \text{◀▼▼ ◀▼▼▼}$$

$$4443 = 1 \times 3600 + 14 \times 60 + 3 = \text{▼ ◀▼▼▼▼ ▼▼▼}$$

EXAMPLE 4 Write the given Babylonian numerals in decimal notation.

a. ▼▼ ◀▼▼▼   **b.** ◀ ▼▼▼▼▼   **c.** ▼▼ ◀▼ ◀▼▼

Solution

a. ▼▼ ◀▼▼▼ $= 2 \times 60 + 13 = 133$

b. ◀ ▼▼▼▼▼ $= 10 \times 60 + 5 = 605$

c. ▼▼ ◀▼ ◀▼▼ $= 2 \times 3600 + 11 \times 60 + 12 = 7872$

Our degree–minute–second system of measuring angles undoubtedly stems from the Babylonian division of a circle into 360 equal parts. Addition of angles in this system is essentially Babylonian-style addition.

The next example illustrates the method used by the Babylonians to add numbers.

EXAMPLE 5 Write in Babylonian notation and add $64 + 127$.

Solution

Exchange

$$191 = 3(60) + 11$$

C. The Roman System

The Roman numeral system is still used today — for example, on the faces of some clocks, for chapter numbers in books, and for construction dates on buildings. How did the Roman system work? The Roman symbol for the number one was I, which was repeated for two and three. Thus, II was

Table 3.1c

NUMBER	ROMAN NUMERAL
1	I
5	V
10	X
50	L
100	C
500	D
1000	M

Courtesy of PSH

2 and III was 3. This is similar to the Egyptian system. However, unlike the Egyptians, the Romans introduced a special symbol for the number five. They then used another special symbol for ten, and repeated this symbol for 20 and 30. Other special symbols were used for 50, 100, 500, and 1000, as shown in Table 3.1c.

Although both the Roman and Egyptian systems used the addition principle, the Romans went one step further and used the **subtraction principle** as well. For instance, instead of writing IIII for the number four, the Romans wrote IV with the understanding that the I (one) is to be *subtracted* from the V (five). In the Roman system, the value of a numeral is found by starting at the left and adding the values of the succeeding symbols to the right, unless the value of a symbol is less than that of the symbol to its right. In the latter case, the smaller value is *subtracted* from the larger one. Thus, $XI = 10 + 1 = 11$, but $IX = 10 - 1 = 9$. Only the numbers 1, 10, and 100 were allowed to be subtracted, and these only from numbers not more than two steps larger. For example, I could be subtracted from V to give $IV = 4$, or from X to give $IX = 9$, but could not be subtracted from C or L. Here are some other examples:

Addition Principle

$LX = 50 + 10 = 60$
$CX = 100 + 10 = 110$
$MC = 1000 + 100 = 1100$

Subtraction Principle

$XL = 50 - 10 = 40$
$XC = 100 - 10 = 90$
$CM = 1000 - 100 = 900$

EXAMPLE 6   Write the following Roman numerals in decimal notation:

a. DCXII   **b.** MCMXLIX

Solution   **a.** Since the value of each symbol is larger than that of the one to its right, we simply add these values:

$DCXII = 500 + 100 + 10 + 1 + 1 = 612$

b. Here we must use the subtraction principle and write

$$M(CM)(XL)(IX) = 1000 + (1000 - 100) + (50 - 10) + (10 - 1)$$
$$= 1000 + 900 + 40 + 9$$
$$= 1949 \qquad \blacktriangleleft$$

The largest number that can be written using Roman numerals without using either the bar or the subtraction principle is

MMMDCCCLXXXVIII

What is this number in decimal notation?

Another way in which the Roman system went further than the Egyptian system is the use of a **multiplication principle** for writing larger numbers. A multiplication by 1000 was indicated by placing a bar over the entire numeral. Thus,

$\overline{X} \quad = 10 \quad \times 1000 = 10,000$
$\overline{LI} \quad = 51 \quad \times 1000 = 51,000$
$\overline{DC} \quad = 600 \quad \times 1000 = 600,000$
$\overline{M} \quad = 1000 \times 1000 = 1,000,000$

EXAMPLE 7   Write in Roman numerals:

   a. 33,008   **b.** 42,120

Solution   **a.** 33 is written as XXXIII in Roman numerals, so $33{,}000 = \overline{\text{XXXIII}}$. To write 33,008, we need to add 8 more. Thus,

$$33{,}008 = 33{,}000 + 8 = \overline{\text{XXXIII}}\text{VIII}$$

b. $42 = \text{XLII}$, so $42{,}000 = \overline{\text{XLII}}$. Since $120 = \text{CXX}$, we see that

$$42{,}120 = \overline{\text{XLII}}\text{CXX}$$ ◄

EXERCISE 3.1

A.   In Problems 1–6, use the symbols given in Tables 3.1a and 3.1b to write each number in Egyptian notation.

   1. 24   **2.** 54   **3.** 142
   4. 1247   **5.** 835   **6.** 11,209

In Problems 7–12, translate the Egyptian numerals into decimal notation.

   7. Ͻ∩||| **8.** ⌡ϽϽϽϽϽϽϽϽϽ||||

   9. ϽϽϽ∩∩|| **10.** ∩∩∩∩|||||

   11. 𝒞⌡ϽϽ∩∩∩|| **12.** ϽϽϽ∩∩∩|||
                                              ϽϽϽ∩∩ ||

In Problems 13–16, write the numbers in Egyptian notation and perform the indicated operations.

13. 34	**14.** 148	**15.** 432	**16.** 1203
+23	+ 45	−143	− 502

Use the Egyptian method of successive duplications to find the product in each of Problems 17–20.

   17. 15×40 **18.** 25×15 **19.** 22×51 **20.** 21×63

Use the Egyptian method of mediation and duplation to find the product in each of Problems 21–24.

   21. 18×32 **22.** 15×32 **23.** 12×51 **24.** 40×61

B.   In Problems 25–34, write the given numbers in Babylonian notation.

   25. 6   **26.** 24   **27.** 32   **28.** 64
   29. 123   **30.** 144   **31.** 258   **32.** 192
   33. 3733   **34.** 3883

In Problems 35–40, write the given Babylonian numbers in decimal notation.

35. ▼ ◄◄◄▼▼   36. ◄▼▼▼ ◄▼▼

37. ▼▼▼ ◄▼▼   38. ◄▼▼▼▼ ▼▼▼

39. ▼ ◄▼▼ ▼▼   40. ▼▼ ◄▼ ▼▼▼▼

In Problems 41–44, write the numbers in Babylonian notation and do the addition Babylonian style.

41.	32	42.	63	43.	133	44.	242
	+43		+81		+ 68		+181

C. In Problems 45–50, write the numbers in decimal notation.

45. CXXVI   46. DCXVII   47. $\overline{\text{XLII}}$
48. $\overline{\text{XXXDCI}}$ 49. $\overline{\text{XCCDV}}$ 50. $\overline{\text{LDDC}}$

In Problems 51–56, write the numbers in Roman numerals.

51. 72   52. 631   53. 145
54. 1709   55. 32,503   56. 49,231

USING YOUR KNOWLEDGE 3.1

The Rhind papyrus is a document that was found in the ruins of a small ancient building at Thebes. The papyrus was bought in 1858 by a Scottish antiquary, A. Henry Rhind, and most of it is preserved in the British Museum where it was named in Rhind's honor. The scroll was a handbook of Egyptian mathematics containing mathematical exercises and practical examples. Many of the problems were solved by the **method of false position.**

For example, one of the simple problems states: "A number and its one-fourth added together become 15. Find the number." The solution by false position goes like this: Assume that the number is 4. A number (4) and its one-fourth ($\frac{1}{4}$ of 4) added become 15; that is,

$$4 + \frac{1}{4}(4) \text{ must equal } 15$$

But

$$4 + \frac{1}{4}(4) = 5$$

and we need 15, which is three times the 5 we got. Therefore, the correct answer must be three times the assumed answer; that is, 3×4, or 12.

See if you can use the method of false position to solve the following problems.

A portion of the Rhind papyrus

1. A number and its one-sixth added together become 21. What is the number?
2. A number, its one-half, and its one-quarter add up to 28. Find the number.
3. If a number and its two-thirds are added, and from the sum one-third of the sum is subtracted, then 10 remains. What is the number?

A Babylonian tablet giving the values of $n^3 + n^2$ for $n = 1$ to 30 was discovered a few years ago. The decimal equivalents of the first few entries in the table can be found as follows:

For $n = 1$, we have $1^3 + 1^2 = 2$.
For $n = 2$, we have $2^3 + 2^2 = 12$.

(Recall that $2^3 = 2 \times 2 \times 2 = 8$ and $2^2 = 2 \times 2 = 4$.)

For $n = 3$, we have $3^3 + 3^2 = 36$.

Complete the following:

$4^3 + 4^2 = \Box$, $5^3 + 5^2 = \Box$, $6^3 + 6^2 = \Box$, $7^3 + 7^2 = \Box$,
$8^3 + 8^2 = \Box$, $9^3 + 9^2 = \Box$, $10^3 + 10^2 = \Box$

1. Using the preceding information, find the solution of
   a. $n^3 + n^2 - 810 = 0$ b. $n^3 + n^2 - 576 = 0$
2. There are many equations in which this method does not seem to work. For example, $n^3 + 2n^2 - 3136 = 0$. However, a simple transformation will reduce the sum of the first two terms to the familiar form $(\ )^3 + (\ )^2$. For example, let $n = 2x$. Now try to solve the following equation:

$$n^3 + 2n^2 - 3136 = 0$$

3.2

THE HINDU–ARABIC (DECIMAL) SYSTEM

In this section, we study our familiar **decimal system,** which is also called the **Hindu–Arabic system.** This numeration system is a positional system with 10 as its base, and it uses the symbols (called **digits**) 0, 1, 2, 3, 4, 5, 6, 7, 8, 9. Furthermore, each symbol in this system has a **place value;** that is, the value represented by a digit depends on the position of that digit in the numeral. For instance, the digit 2 in the numeral 312 represents two ones, but in the numeral 321 the digit 2 represents two tens.

A. Expanded Form

To illustrate the idea further, we can write both numbers in **expanded form.**

$$312 = 3 \text{ hundreds} + 1 \text{ ten} + 2 \text{ ones}$$
$$= (3 \times 100) + (1 \times 10) + (2 \times 1)$$
$$321 = 3 \text{ hundreds} + 2 \text{ tens} + 1 \text{ one}$$
$$= (3 \times 100) + (2 \times 10) + (1 \times 1)$$

These numbers can also be written using exponential form, a notation introduced by the French mathematician René Descartes. As the name indicates, **exponential form** uses the idea of exponents. An **exponent** is a number that indicates how many times another number, called the **base,** is a factor in a product. Thus, in 5^3 (read, "5 cubed" or "5 to the third power") the exponent is 3, the base is 5, and $5^3 = 5 \times 5 \times 5 = 125$. Similarly, in 2^4, 4 is the exponent, 2 is the base, and $2^4 = 2 \times 2 \times 2 \times 2 = 16$. Based on this discussion, we state the following definition.

Definition 3.2a

> If a is any number and n is any counting number, then a^n (read, "a to the nth power") is the product obtained by using a as a factor n times; that is,
>
> $$a^n = \underbrace{a \times a \times \cdots \times a}_{n \text{ } a\text{'s}}$$
>
> For any nonzero number a, we define $a^0 = 1$.

Thus, in expanded form,

$$312 = (3 \times 100) + (1 \times 10) + (2 \times 1)$$
$$= (3 \times 10^2) + (1 \times 10^1) + (2 \times 10^0)$$

The exponent 1 usually is not explicitly written; we understand that $10^1 = 10$ and, in general, $a^1 = a$. Also, we used $10^0 = 1$ in the last term. With these conventions, we can write any number in expanded form.

EXAMPLE 1   Write 3406 in expanded form.

Solution $3406 = (3 \times 10^3) + (4 \times 10^2) + (0 \times 10) + (6 \times 10^0)$

Notice that we could have omitted the term (0×10), because $0 \times 10 = 0$. Using the same ideas, we can convert any number from expanded form into our familiar decimal form.   ◀

EXAMPLE 2   Write $(5 \times 10^3) + (2 \times 10) + (3 \times 10^0)$ in ordinary decimal form.

Solution $(5 \times 10^3) + (2 \times 10) + (3 \times 10^0) = (5 \times 1000) + (2 \times 10) + 3$
$$= 5023$$ ◀

B. Operations in Expanded Form

The idea of expanded form and place value can greatly simplify computations involving addition, subtraction, multiplication, and division. In the following examples, we present the usual way in which these operations are performed together with the steps depending on place value.

EXAMPLE 3   Add 38 and 61 in the usual way and in expanded form.

Solution

$$
\begin{array}{r}
38 \\
+61 \\
\hline
99
\end{array}
\qquad
\begin{array}{l}
(3 \times 10) + (8 \times 10^0) \\
(6 \times 10) + (1 \times 10^0) \\
\hline
(9 \times 10) + (9 \times 10^0)
\end{array}
$$

◀

EXAMPLE 4   Subtract 32 from 48 in the usual way and in expanded form.

Solution

$$
\begin{array}{r}
48 \\
-32 \\
\hline
16
\end{array}
\qquad
\begin{array}{l}
(4 \times 10) + (8 \times 10^0) \\
-(3 \times 10) + (2 \times 10^0) \\
\hline
(1 \times 10) + (6 \times 10^0)
\end{array}
$$

◀

Before illustrating multiplication and division, we need to determine how to multiply and divide numbers involving exponents. For example, $2^2 \times 2^3 = (2 \times 2) \times (2 \times 2 \times 2) = 2^5$, and

$$a^m \times a^n = \underbrace{(a \times a \times a \times \cdots \times a)}_{m\ a\text{'s}} \times \underbrace{(a \times a \times a \times \cdots \times a)}_{n\ a\text{'s}}$$

$$= a^{m+n}$$

In order to divide 2^5 by 2^2 we can proceed as follows:

$$\frac{2^5}{2^2} = \frac{2 \times 2 \times 2 \times 2 \times 2}{2 \times 2} = 2 \times 2 \times 2 = 2^3$$

In general, if $m > n$,

$$a^m \div a^n = \frac{a^m}{a^n} = a^{m-n}$$

Laws of Exponents

$$a^m \times a^n = a^{m+n} \qquad \text{and} \qquad a^m \div a^n = \frac{a^m}{a^n} = a^{m-n}, \quad m > n$$

EXAMPLE 5   Perform the indicated operation and leave the answer in exponential form.

a. $4^5 \times 4^7$ **b.** $3^{10} \div 3^4$

Solution   **a.** $4^5 \times 4^7 = 4^{5+7} = 4^{12}$ **b.** $3^{10} \div 3^4 = 3^{10-4} = 3^6$

◀

EXAMPLE 6   Multiply 32 and 21 in the usual way and in expanded form.

Solution

$$
\begin{array}{r}
32 \\
\times\ 21 \\
\hline
32 \\
64 \\
\hline
672
\end{array}
$$

$$
\begin{array}{r}
(3 \times 10) + 2 \times 10^0 \\
\times (2 \times 10) + 1 \times 10^0 \\
\hline
(3 \times 10) + 2 \times 10^0 \\
(6 \times 10^2) + (4 \times 10) \\
\hline
(6 \times 10^2) + (7 \times 10) + 2 \times 10^0
\end{array}
$$

◄

EXAMPLE 7   Divide 63 by 3 in the usual way and in expanded form.

Solution

$$
3\overline{)63} \quad \begin{array}{c} 21 \end{array}
$$

$$
3\overline{)(6 \times 10) + 3 \times 10^0} \quad \begin{array}{c} (2 \times 10) + 1 \times 10^0 \end{array}
$$

◄

EXERCISE 3.2

A.   In Problems 1–6, write the given number in expanded form.

1. 432   **2.** 549   **3.** 2307
4. 3047   **5.** 12,349   **6.** 10,950

In Problems 7–15, write the given number in decimal form.

7. 5^0
8. $(3 \times 10) + (4 \times 10^0)$
9. $(4 \times 10) + (5 \times 10^0)$
10. $(4 \times 10^2) + (3 \times 10) + (2 \times 10^0)$
11. $(9 \times 10^3) + (7 \times 10) + (1 \times 10^0)$
12. $(7 \times 10^4) + (2 \times 10^0)$
13. $(7 \times 10^5) + (4 \times 10^4) + (8 \times 10^3) + (3 \times 10^2) + (8 \times 10^0)$
14. $(8 \times 10^9) + (3 \times 10^5) + (2 \times 10^2) + (4 \times 10^0)$
15. $(4 \times 10^6) + (3 \times 10) + (1 \times 10^0)$

B.   In Problems 16–19, add in the usual way and in expanded form.

16. $32 + 15$ **17.** $23 + 13$ **18.** $21 + 34$ **19.** $71 + 23$

In Problems 20–23, subtract in the usual way and in expanded form.

20. $34 - 21$ **21.** $76 - 54$ **22.** $45 - 22$ **23.** $84 - 31$

In Problems 24–31, perform the indicated operation and leave the answer in exponential form.

24. $3^5 \times 3^9$ **25.** $7^8 \times 7^3$ **26.** $4^5 \times 4^2$ **27.** $6^{19} \times 6^{21}$
28. $5^8 \div 5^3$ **29.** $6^{10} \div 6^3$ **30.** $7^{15} \div 7^3$ **31.** $6^{12} \div 6^0$

In Problems 32–35, multiply in the usual way and in expanded form.

32. 41×23 **33.** 25×51 **34.** 91×24 **35.** 62×25

In Problems 36–39, divide in the usual way and in expanded form.

36. $48 \div 4$ **37.** $64 \div 8$ **38.** $93 \div 3$ **39.** $72 \div 6$

DISCOVERY 3.2

The Rhind papyrus contains a problem that deals with exponents. Problem 79 is very difficult to translate, but historian Moritz Cantor formulates the problem as follows: "An estate consisted of seven houses; each house had seven cats; each cat ate seven mice; each mouse ate seven heads of wheat; and each head of wheat was capable of yielding seven hekat measures of grain: Houses, cats, mice, heads of wheat, and hekat measures of grain, how many of these in all were in the estate?" Here is the solution:

Houses	$7 = 7^1$
Cats	$49 = 7^2$
Mice	$343 = 7^3$
Heads of wheat	$2,401 = 7^4$
Hekat measures	$\underline{16,807} = 7^5$
Total	$19,607$

Because the items in the problem correspond to the first five powers of 7, it was at first thought that the writer was introducing the terminology houses, cats, mice, and so on, for first power, second power, third power, and so on!

1. A similar problem can be found in *Liber abaci*, written by Leonardo Fibonacci (1170–1250). The problem reads as follows:

> There are seven old women on the road to Rome. Each woman has seven mules; each mule carries seven sacks; each sack contains seven loaves; with each loaf are seven knives; and each knife is in seven sheats. Women, mules, sacks, loaves, knives, and sheats, how many are there in all on the road to Rome?

Can you find the answer?

2. A later version of the same problem reads as follows:

> As I was going to St. Ives
> I met a man with seven wives;
> Every wife had seven sacks;
> Every sack had seven cats;
> Every cat had seven kits.
> Kits, cats, sacks, and wives,
> How many were going to St. Ives?

[Hint: The answer is not 2800. If you think it is, then you did not read the first line carefully.]

NUMBER SYSTEMS WITH BASES OTHER THAN 10

As you learned in Section 3.1, in a positional number system a number is selected as the base, and objects are grouped and counted using this base. In the decimal system, the base chosen was 10, probably because the fingers are a convenient aid in counting.

A. Other Number Bases

As we saw earlier, it is possible to use other numbers as bases for numeration systems. For example, if we decide to use 5 as our base (that is, count in groups of 5), then we can count the 17 asterisks below in this way:

and we write 32_{five}.

If we select 8 as our base, the asterisks are grouped this way:

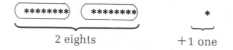

and we write 21_{eight}. Thus, if we use subscripts to indicate the manner in which we are grouping the objects, we may write the number 17 as follows:

$$17_{\text{ten}} = 32_{\text{five}} = 21_{\text{eight}}$$

Using groups of seven, we can indicate the same number of asterisks by

$$23_{\text{seven}} = 2 \text{ sevens} + 3 \text{ ones}$$

Note: **In the following material, when no subscript is used it will be understood that the number is expressed in base 10.**

EXAMPLE 1

Group 13 asterisks in groups of eight, five, and seven, and write the number 13 in:

a. Base 8   **b.** Base 5   **c.** Base 7

Solution

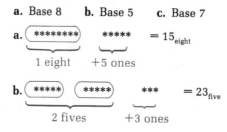

c. $\overbrace{(\texttt{******})}^{\text{1 seven}}\; \underbrace{\texttt{******}}_{+6\text{ ones}} = 16_{\text{seven}}$ ◀

B. Changing to Base 10

Recall that $a^0 = 1$ for $a \neq 0$. Thus, $10^0 = 1$ and $2 \times 10^0 = 2$.

Now consider the problem of "translating" numbers written in bases other than 10 into our decimal system. For example, what number in our decimal system corresponds to 43_{five}? First, recall that in the decimal system, numbers can be written in expanded form. For example, $342 = (3 \times 10^2) + (4 \times 10) + (2 \times 10^0)$. As you can see, when written in expanded form, each digit in 342 is multiplied by the proper power of 10 (the base being used). Similarly, when written in expanded form, each digit in the numeral 43_{five} must be multiplied by the proper power of 5 (the base being used). Thus,

$$43_{\text{five}} = (4 \times 5^1) + (3 \times 5^0) = 23$$

EXAMPLE 2   Write the following numbers in decimal notation:

a. 432_{five} **b.** 312_{eight}

Solution   **a.** $432_{\text{five}} = (4 \times 5^2) + (3 \times 5) + (2 \times 5^0) = 117$

b. $312_{\text{eight}} = (3 \times 8^2) + (1 \times 8) + (2 \times 8^0) = 202$ ◀

In the seventeenth century, the German mathematician Gottfried Wilhelm Leibniz advocated the use of the **binary system** (base 2). This system uses only the digits 0 and 1, and the grouping is by twos. The advantage of the binary system is that each position in a numeral contains one of just two values (0 or 1). Thus, electric switches, which have only two possible states, *off* or *on*, can be used to designate the value of each position. Electronic computers utilizing the binary system have revolutionized technology and the sciences by speedily performing calculations that would take humans years to complete. Hand-held calculators operate internally on the binary system.

A binary counter

Creative Publications, Inc.

Now consider how to convert numbers from base 2 to decimal notation. It will help you to keep in mind that in the binary system, numbers are built up by using blocks that are powers of the base 2:

$$2^0 = 1, \quad 2^1 = 2, \quad 2^2 = 4, \quad 2^3 = 8, \quad 2^4 = 16, \; \ldots$$

When we write 1101 in the binary system, we are saying in the yes/no language of the electronic computer, "a block of 8, yes; a block of 4, yes; a block of 2, no; a block of 1, yes." Thus,

$$\begin{aligned}
1101_{\text{two}} &= (1 \times 2^3) + (1 \times 2^2) + (0 \times 2^1) + (1 \times 2^0) \\
&= 8 + 4 + 0 + 1 \\
&= 13
\end{aligned}$$

EXAMPLE 3   Write the number 10101_{two} in decimal notation.

Solution   We write

$$10101_{two} = (1 \times 2^4) + (0 \times 2^3) + (1 \times 2^2) + (0 \times 2^1) + (1 \times 2^0)$$
$$= 16 + 4 + 1$$
$$= 21$$ ◄

So far, we have used only bases less than 10, the most important ones being base 2 **(binary)** and base 8 **(octal)**, which are used by electronic computers. Bases greater than 10 also are possible, but then new symbols are needed for the digits greater than 9. For example, base 16 **(hexadecimal)** also is used by electronic computers, with the "digits" A, B, C, D, E, and F used to correspond to the decimal numbers 10, 11, 12, 13, 14, and 15, respectively. We can change numbers from hexadecimal to decimal notation in the same way as we did for bases less than 10.

EXAMPLE 4   Write the number $5AC_{sixteen}$ in decimal notation.

Solution
$$5AC_{sixteen} = (5 \times 16^2) + (10 \times 16) + (12 \times 16^0)$$
$$= (5 \times 256) + (160) \quad + 12$$
$$= 1280 \quad + 160 \quad + 12$$
$$= 1452$$ ◄

C. Changing from Base 10

Up to this point, we have changed numbers from bases other than 10 to base 10. Now we shall change numbers from base 10 to another base. A good method for doing this depends on successive divisions. For example, to change 625 to base 8, we start by dividing 625 by 8, obtaining 78 with a remainder of 1. We then divide 78 by 8, getting 9 with 6 remaining. Next, we divide 9 by 8, obtaining a quotient of 1 and a remainder of 1. We diagram these divisions as follows:

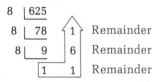

The answer is read upward as 1161_{eight} (see the arrow).

Why does this method work? Suppose we wish to find how many eights there are in 625. To find out, we divide 625 by 8. The quotient 78 tells us that there are 78 eights in 625, and the remainder tells us that there is 1 left over. Dividing the 78 by 8 (which is the same as dividing the 625 by $8 \times 8 = 64$) tells us that there are 9 sixty-fours in 625, and the remainder tells us that there are 6 eights left over. Finally, dividing the quotient 9 by 8 gives a new quotient of 1 and a remainder of 1. This tells us that there is 1 five hundred twelve ($8 \times 8 \times 8$) contained in 625 with 1 sixty-four left over. Thus, we see that

$$625 = (1 \times 8^3) + (1 \times 8^2) + (6 \times 8) + (1 \times 8^0)$$
$$= 1161_{\text{eight}}$$

EXAMPLE 5   Change the number 33 to:

a. Base 2   **b.** Base 5

Solution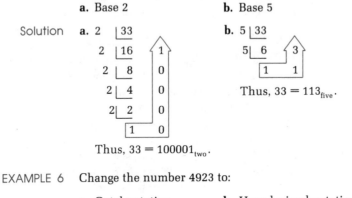

Thus, $33 = 100001_{\text{two}}$.   ◄

EXAMPLE 6   Change the number 4923 to:

a. Octal notation   **b.** Hexadecimal notation

Solution   **a.** 8 | 4923

8	615	3
8	76	7
8	9	4
1	1

Thus, $4923 = 11473_{\text{eight}}$.   ◄

b. 16 | 4923

16	307	11
16	19	3
1	3

Thus, $4923 = 133B_{\text{sixteen}}$.   ◄

EXERCISE 3.3

A. In Problems 1–4, write numerals in the bases indicated by the manner of grouping.

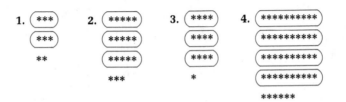

In Problems 5–8, draw a diagram as shown in Problems 1–4, and then write the given decimal number in the given base.

5. 15 in base 8   **6.** 15 in base 5

7. 15 in base 7   **8.** 15 in base 12

B. In Problems 9–16, write the given number in decimal notation.

9. 42_{five} **10.** 31_{five} **11.** 213_{eight}

12. 563_{eight} **13.** 11011_{two} **14.** 101001_{two}

15. 123_{sixteen} **16.** $\text{ACE}_{\text{sixteen}}$

C. In Problems 17–32, use the method of successive divisions.

17. Write the number 15 in base 5 notation.
18. Write the number 27 in base 5 notation.
19. Write the number 28 in binary notation.
20. Write the number 43 in binary notation.
21. Write the number 25 in hexadecimal notation.
22. Write the number 121 in hexadecimal notation.
23. Write the number 25 in base 6 notation.
24. Write the number 38 in base 6 notation.
25. Write the number 64 in base 7 notation.
26. Write the number 123 in base 7 notation.
27. Write the number 38 in octal notation.
28. Write the number 135 in octal notation.
29. Write the number 1467 in hexadecimal notation.
30. Write the number 145,263 in hexadecimal notation.
31. Write the number 73 in binary and in octal notation.
32. Write the number 87 in octal and in hexadecimal notation.

USING YOUR KNOWLEDGE 3.3

Here is a trick that you can use to amaze your friends. Write the numbers from 1 to 7 in binary notation. They look like this:

Decimal	Binary
1	1
2	10
3	11
4	100
5	101
6	110
7	111

Now label three columns A, B, and C. In column A, write the numbers that have a 1 in the units place when written in binary notation (see below).

In column B, write the numbers that have a 1 in the second position from the right when written in binary notation; and in column C, write the numbers with 1's in the third position from the right.

A	B	C
1	2	4
3	3	5
5	6	6
7	7	7

Ask someone to think of a number and tell you in which columns the number appears. Say the number is 6 (which appears in columns B and C). You find the sum of the numbers at the top of columns B and C. This sum is $2 + 4 = 6$, and you have the desired number!

1. Can you explain why this works?
2. If you extend this trick to cover the first 15 numbers, how many columns do you need?
3. Can you discover how to do the trick with 31 numbers?

▶ **Calculator Corner 3.3** **Converting from Base *b* to Base 10**

For convenience and to save space, we shall write the base *b* as an ordinary decimal rather than spelling it out. Thus, for example, 47_8 means exactly the same thing as 47_{eight}. Suppose that 4735_8 is to be converted to base 10. We know that

$$4735_8 = (4 \times 8^3) + (7 \times 8^2) + (3 \times 8) + 5$$

Since the three quantities in parentheses are all divisible by 8, we may rewrite the expression to get

$$4735_8 = 8 \times [(4 \times 8^2) + (7 \times 8) + 3] + 5$$

Next, we see that the two quantities in parentheses inside the brackets are divisible by 8, so we may rewrite again to get

$$4735_8 = 8 \times [8 \times \{(4 \times 8) + 7\} + 3] + 5$$

Now we can evaluate the last expression by a simple step-by-step procedure. Start with the innermost parentheses, multiply 4 by the base 8 and add 7 to the result. Then, multiply the last result by 8 and add 3 to the product. Finally, multiply the preceding result by 8 and add 5 to the product. The final sum is the required answer.

The arithmetic can be done on the calculator by keying in the following:

$$\boxed{4}\;\boxed{\times}\;\boxed{8}\;\boxed{+}\;\boxed{7}\;\boxed{=}\;\boxed{\times}\;\boxed{8}\;\boxed{+}\;\boxed{3}\;\boxed{=}\;\boxed{\times}\;\boxed{8}\;\boxed{+}\;\boxed{5}\;\boxed{=}$$

The calculator will show the result 2525.

Notice that we multiplied the first octal digit, 4, by the base, 8, and added the next octal digit, 7. Then we multiplied the result by the base and added the next octal digit, 3. Finally, we multiplied by the base again and added the last octal digit, 5.

This procedure holds for any base. Thus, to convert from any base b to base 10:

1. Multiply the first digit of the number by b, and add the second digit to the result.
2. Multiply the preceding result by b, and add the third digit of the given numeral.
3. Continue the same procedure until you have added the last digit of the given numeral.

The calculator will show the final answer. Be sure that you work from left to right in using the digits of the given numeral.

As another example, let us convert the hexadecimal numeral $BF3_{16}$ to base 10. Recall that the hexadecimal digits B and F are the numbers 11 and 15, respectively, in decimal notation, so you key in the following:

Your calculator should show the result 3059.

Convert each of the following to decimal notation:

1. 1101_2 2. 231_4 3. 423_5 4. 752_8
5. 3572_8 6. 873_9 7. $A3C_{16}$ 8. $93DC_{16}$

▶ **Computer Corner 3.3**   We have already learned how to convert from one base to another using a calculator. But there is a faster way! Use a computer. We have three different programs that will convert numbers from one base to another. The first program (page 742) converts numbers from base 10 to any other base. The second program (page 743) converts numbers from any base to base 10, and the third program (page 743) converts from **any** base to **any** other base, as long as both bases are less than 11.

3.4

BINARY, OCTAL, AND HEXADECIMAL CONVERSION

The octal and hexadecimal systems are used by computers to store data and instructions. Sometimes, operators must convert between these two systems and the binary system.

A. Binary and Octal Conversion

For the octal system, the procedure is simplified by using Table 3.4a. This table tells us, for example, that $101_2 = 5_8$ and that $6_8 = 110_2$. Notice that

Table 3.4a

BINARY	OCTAL
000	0
001	1
010	2
011	3
100	4
101	5
110	6
111	7

three-digit binary numbers correspond to one-digit octal numbers throughout. Accordingly, to convert from binary to octal, we simply break the given binary number into groups of three, starting from the right, and then write the octal equivalent for each group of three digits. (Consult Table 3.4a when doing this.) Thus, to convert 11010_2 to octal, 11010_2 is broken into groups of three digits (a 0 is attached at the left), and the octal equivalent for each group is written as follows:

$$\text{We attached this } 0 \rightarrow \underset{\underset{3}{\downarrow}}{011} \quad \underset{\underset{2}{\downarrow}}{010}$$

We then have $11010_2 = 32_8$.

EXAMPLE 1   Convert 11101110_2 to octal notation.

Solution   We break the given number into groups of three and write the octal equivalent of each group (a 0 was attached to the leftmost group):

$$\underset{\underset{3}{\downarrow}}{011} \quad \underset{\underset{5}{\downarrow}}{101} \quad \underset{\underset{6}{\downarrow}}{110}$$

Thus, $11101110_2 = 356_8$.   ◄

To change from octal to binary, the procedure can be reversed as in the next example.

EXAMPLE 2   Convert 5732_8 to binary notation.

Solution   Using Table 3.4a, we write the binary equivalents of 5, 7, 3, and 2. Thus,

$$\underset{\underset{101}{\downarrow}}{5} \quad \underset{\underset{111}{\downarrow}}{7} \quad \underset{\underset{011}{\downarrow}}{3} \quad \underset{\underset{010}{\downarrow}}{2}$$

and $5732_8 = 101111011010_2$.   ◄

B.   Binary and Hexadecimal Conversion

Conversions between hexadecimal and binary are done in the same way, except that we need a table giving the relationship between binary and hexadecimal numbers (Table 3.4b, page 130). This table tells us, for example, that $1001_2 = 9_{16}$ and that $D_{16} = 1101_2$. Note that each four-digit binary number in Table 3.4b corresponds to a one-digit hexadecimal number. Thus, the conversions will use groups of four binary digits for each hexadecimal digit, as shown in the following examples.

EXAMPLE 3   Convert 1101110_2 to hexadecimal notation.

Table 3.4b

BINARY	HEXADECIMAL
0000	0
0001	1
0010	2
0011	3
0100	4
0101	5
0110	6
0111	7
1000	8
1001	9
1010	A
1011	B
1100	C
1101	D
1110	E
1111	F

Solution   We divide the digits in the binary numeral into groups of four, starting from the right, and write under these groups the corresponding hexadecimal digits. (Consult Table 3.4b to do this.) Thus,

$$0110 \quad\quad 1110$$
$$\downarrow \quad\quad\quad\ \downarrow$$
$$6 \quad\quad\quad\ E$$

so that $1101110_2 = 6E_{16}$.   ◀

To change from hexadecimal to binary notation, we reverse the procedure in Example 3, as shown next.

EXAMPLE 4   Convert $9AD_{16}$ to binary notation.

Solution   Using Table 3.4b, we find the binary equivalents of 9, A, and D. Thus,

$$9 \quad\quad\ A \quad\quad\ D$$
$$\downarrow \quad\quad\ \downarrow \quad\quad\ \downarrow$$
$$1001 \quad 1010 \quad 1101$$

so that $9AD_{16} = 100110101101_2$.   ◀

C.   Octal and Hexadecimal Conversion

To convert between octal and hexadecimal notation, we first convert the given numeral to binary notation and then convert this result to the desired notation, as illustrated in the next examples.

EXAMPLE 5 Convert $9AD_{16}$ to octal notation.

Solution From Example 4, we have $9AD_{16} = 100110101101_2$. Converting this binary numeral to octal (Table 3.4a), we obtain

$$
\begin{array}{cccc}
100 & 110 & 101 & 101 \\
\downarrow & \downarrow & \downarrow & \downarrow \\
4 & 6 & 5 & 5
\end{array}
$$

Thus, $100110101101_2 = 4655_8$, so that $9AD_{16} = 4655_8$. ◀

EXAMPLE 6 Convert 357_8 to hexadecimal notation.

Solution **1.** First, convert the given numeral to binary form (Table 3.4a):

$$
\begin{array}{ccc}
3 & 5 & 7 \\
\downarrow & \downarrow & \downarrow \\
011 & 101 & 111
\end{array}
$$

Thus, $357_8 = 11101111_2$.

2. Now, convert 11101111_2 to hexadecimal notation (Table 3.4b):

$$
\begin{array}{cc}
1110 & 1111 \\
\downarrow & \downarrow \\
E & F
\end{array}
$$

Therefore, $357_8 = EF_{16}$ ◀

EXERCISE 3.4

A. In Problems 1–4, convert each numeral to octal notation.

1. 110111_2 **2.** 101101_2 **3.** 1101101_2 **4.** 10111101_2

In Problems 5–8, convert each numeral to binary notation.

5. 65_8 **6.** 57_8 **7.** 306_8 **8.** 472_8

B. In Problems 9–12, convert each numeral to hexadecimal notation.

9. 110111_2 **10.** 101101_2 **11.** 1101101_2 **12.** 10111101_2

In Problems 13–16, convert each numeral to binary notation.

13. 95_{16} **14.** $8B_{16}$ **15.** $7CD_{16}$ **16.** $A9C_{16}$

C. In Problems 17 and 18, convert each numeral to octal notation.

17. 109_{16} **18.** $2BF_{16}$

In Problems 19 and 20, convert each numeral to hexadecimal notation.

19. 537_8 **20.** 6235_8

Since $4 = 2^2$, the ideas presented in this section should apply to conversions back and forth between the binary system and the base 4 system of numeration. The digits of the base 4 system are 0, 1, 2, and 3, and all these can be obtained as sums formed from the numbers 0, 1, and 2. Thus, we can construct a table for base 4 numerals similar to the tables for octal and hexadecimal numerals, as shown in the margin. The table tells us that we can use exactly the same ideas for conversion between the base 4 and the binary system as we did for the octal and hexadecimal systems. For instance, we can convert from binary to base 4 notation as illustrated below, where we change 11110_2 to base 4:

BINARY	BASE 4
00	0
01	1
10	2
11	3

$$
\begin{array}{ccc}
01 & 11 & 10 \\
\downarrow & \downarrow & \downarrow \\
1 & 3 & 2
\end{array}
$$

This shows that $11110_2 = 132_4$.

1. Show how to change 312_4 to binary notation.
2. Convert 203_4 to binary notation.
3. Convert 1101111_2 to base 4 notation.
4. Convert 1010101_2 to base 4 notation.
5. Convert 301_4 to octal notation.
6. Convert 223_4 to hexadecimal notation.

On July 14, 1965, a camera installed on the spacecraft Mariner IV took the first pictures of the planet Mars and sent them by radio signals back to Earth. On Earth, a computer received the "pictures" in the form of binary numerals consisting of six bits. (A **bit** is a binary digit.) The shade of each dot in the final picture was determined by six bits.

The numeral 000000_2 (0 in base 10) indicated a white dot, and the numeral 111111_2 (63 in base 10) indicated a black dot. The 62 numerals between represented various shades of gray between white and black. To make a complete picture, 40,000 dots, each described by six bits, were needed!

1. If one of the numerals received was 110111_2, can you discover the corresponding decimal numeral?
2. Does the dot corresponding to the numeral received in Problem 1 represent a shade of gray closer to white or to black?
3. Can you discover what binary numeral would represent the lightest shade of gray that is not white?
4. Can you discover what binary numeral would represent the darkest shade of gray that is not black?

5. Can you discover what binary numeral would represent the shade numbered 31?

▶ **Calculator Corner 3.4**

You can convert numbers from one base to another provided you have a calculator with a key displaying the desired bases. For instance, to convert 11101110_2 to octal (Example 1), you need `bin` *and* `oct` *(or equivalent) keys. First, set the calculator in the binary mode by pressing* `2nd` `mode` `bin`*. Enter the number 11101110_2 and enter* `2nd` `mode` `oct`*. The answer 356_8 appears on the screen. Conversely, to change 5732_8 to binary (Example 2), we place the calculator in the octal mode (press* `2nd` `mode` `oct`*), enter 5732, and change it to binary by pressing* `2nd` `mode` `bin`*. In this case, an error message appears because the answer (101111011010_2) has twelve digits and the calculator can enter only ten digits. This problem cannot be done on the calculator. Instead, let us change 472_8 to binary. As before, place the calculator in the octal mode by pressing* `2nd` `mode` `oct` *and enter 472_8. We then convert to binary by entering* `2nd` `mode` `bin`*. The answer 100111010_2 appears on the screen.*

1. Use a calculator to check the answers to Problems 1–19 in Exercise 3.4.

▶ **Computer Corner 3.4**

Computers also can be used to change numbers from one base to another. The program on page 743 will change numbers from any base x (less than 11) to any base y (less than 11). Thus, the program can change numbers from binary to octal, and vice versa.

1. Use the program to check Problems 1, 3, 5, and 7 in Exercise 3.4.

We also give a program to change numbers from binary to hexadecimal, and vice versa. This program appears on page 743.

2. Use this program to check Problems 9, 11, 13, 15, and 17 in Exercise 3.4.

The conversion of numbers involving binary, octal, and hexadecimal bases requires the use of Tables 3.4a and 3.4b. We can avoid using these tables if we use the program on page 743. This program converts a number with a given base to a number with a specified base as long as both bases are less than 17.

3.5

BINARY ARITHMETIC

Now that we know how to represent numbers in the binary system, we look at how computations are done in that system. First, we can construct addition and multiplication tables like the ones for base 10 arithmetic. See Tables 3.5a and 3.5b on page 134.

Table 3.5a	Binary Addition	
+	0	1
0	0	1
1	1	10

Table 3.5b	Binary Multiplication	
×	0	1
0	0	0
1	0	1

The only entry that looks peculiar is the 10 in the addition table, but recall that 10_2 means 2_{10}, which is exactly the result of adding $1 + 1$. (Be sure to read 10 as "one zero" *not* as "ten.")

A. Addition

Binary addition is done in the same manner as addition in base 10. We line up the corresponding digits and add column by column.

EXAMPLE 1   Add 1010_2 and 1111_2.

Solution

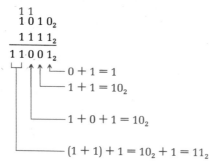

$$\begin{array}{r} 1\,1 \\ 1\,0\,1\,0_2 \\ 1\,1\,1\,1_2 \\ \hline 1\,1\,0\,0\,1_2 \end{array}$$

$0 + 1 = 1$

$1 + 1 = 10_2$ Write the 0 and carry the 1 to the next column.

$1 + 0 + 1 = 10_2$ Write the 0 and carry the 1 to the next column.

$(1 + 1) + 1 = 10_2 + 1 = 11_2$

Thus, $1010_2 + 1111_2 = 11001_2$. ◄

EXAMPLE 2   Perform the addition $1101_2 + 110_2 + 11_2$.

Solution   We shall omit the subscript 2 in the computation, but keep in mind that all numerals are binary.

$$\begin{array}{r} 1\,1\,1 \\ 1\,1\,0\,1 \\ 1\,1\,0 \\ 1\,1 \\ \hline 1\,0\,1\,1\,0 \end{array}$$

$1 + 0 + 1 = 10$ Write 0 and carry 1 to the next column.

$(1 + 1) + 0 + 1 = 10 + 1 = 11$ Write 1 and carry 1 to the next column.

$(1 + 1) + 1 = 10 + 1 = 11$ Write 1 and carry 1 to the next column.

$1 + 1 = 10$

Thus, $1101_2 + 110_2 + 11_2 = 10110_2$. ◄

B. Subtraction

To subtract in the binary system, we line up corresponding digits and subtract column by column, "borrowing" as necessary.

EXAMPLE 3   Perform the subtraction $1111_2 - 110_2$.

Solution   Again, we omit the subscript 2 in the computation:

$$
\begin{array}{r}
1\,1\,1\,1 \\
-\ \ 1\,1\,0 \\
\hline
1\,0\,0\,1
\end{array}
$$

$1 - 0 = 1$

$1 - 1 = 0$

$1 - 1 = 0$

1 is left in this column.

This shows that $1111_2 - 110_2 = 1001_2$. You can check this by adding $1001_2 + 110_2$ to get 1111_2.   ◀

Example 3 did not require "borrowing," but the next example does.

EXAMPLE 4   Subtract 101_2 from 1010_2.

Solution

$$
\begin{array}{r}
0\ 10\ 0\ 10 \\
1\ \ 0\ \ 1\ \ 0 \\
-1\ \ 0\ \ 1 \\
\hline
1\ \ 0\ \ 1
\end{array}
$$

$0 - 1$ requires borrowing; then, $10 - 1 = 1$.

$0 - 0 = 0$

$0 - 1$ requires borrowing; then, $10 - 1 = 1$.

The result is $1010_2 - 101_2 = 101_2$. You can check this answer by adding $101_2 + 101_2$ to get 1010_2.   ◀

C. Multiplication

Multiplication in the binary system also is done in a manner similar to that for base 10, as the following examples illustrate.

EXAMPLE 5   Multiply $101_2 \times 110_2$.

Solution

$$
\begin{array}{r}
1\,1\,0 \\
\times 1\,0\,1 \\
\hline
1\,1\,0 \\
0\,0\,0 \\
1\,1\,0 \\
\hline
1\,1\,1\,1\,0
\end{array}
$$

← Start from the right and multiply by 1.

← Indent and multiply by 0.

← Indent and multiply by the left-hand 1.

← Add the above products.

The result shows that $101_2 \times 110_2 = 11110_2$. Note that we could abbreviate a little by omitting the two leftmost 0's in the second partial product and writing the third product on the same line.   ◀

EXAMPLE 6   Multiply 1110_2 by 110_2.

Solution

$$
\begin{array}{r}
1\,1\,1\,0 \\
\times 1\,1\,0 \\
\hline
\end{array}
$$

1 1 1 0 0   ← Multiply by 0, and write one 0. Then multiply by 1, using the same line.

  1 1 1 0   ← Indent two places and multiply by 1.

1 0 1 0 1 0 0   ← Add the partial products.

The computation shows that $110_2 \times 1110_2 = 1010100_2$.   ◄

D. Division

The procedure used for division in the base 10 system also can be used in the binary system. This is illustrated next.

EXAMPLE 7   Divide 1011_2 by 10_2.

Solution

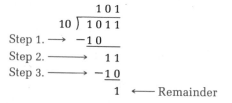

Step 1. ⟶ $-1\,0$

Step 2. ⟶   1 1

Step 3. ⟶ $-1\,0$

1   ←— Remainder

Step 1. $1 \times 10 = 10$. Write 1 in the quotient above the 10. Then subtract.

Step 2. Bring down the next digit. 10 does not go into 1, so write 0 in the quotient and bring down the final digit.

Step 3. $1 \times 10 = 10$. Write 1 in the quotient and then subtract to get the remainder.

The computation shows that $1011_2 \div 10_2 = 101_2$ with remainder 1. You can check this answer by multiplying 101_2 by 10_2 and adding 1 to the product. Thus,

$$(10_2 \times 101_2) + 1 = 1010_2 + 1 = 1011_2$$

showing that the answer is correct.   ◄

EXAMPLE 8   Divide 110111_2 by 101_2.

Solution

$$
\begin{array}{r}
1\,0\,1\,1 \\
101\,)\,1\,1\,0\,1\,1\,1 \\
\end{array}
$$

Step 1. ⟶ $-1\,0\,1$

Step 2. ⟶   1 1 1

Step 3. ⟶ $-1\,0\,1$

Step 4. ⟶   1 0 1

Step 5. ⟶ $-1\,0\,1$

0   ←— Remainder

Step 1. $1 \times 101 = 101$. Write 1 in the quotient above the 110. Then subtract.

Step 2. Bring down the next digit. 101 does not go into 11, so write 0 in the quotient and bring down the next digit.

Step 3. $1 \times 101 = 101$. Write 1 in the quotient. Then subtract.

Step 4. Bring down the final digit.

Step 5. $1 \times 101 = 101$. Write 1 in the quotient and then subtract to get the remainder.

The division here is exact; the quotient is 1011_2 and the remainder is 0. You can check the answer by multiplying the quotient 1011_2 by the divisor 101_2. Thus,

$$
\begin{array}{r}
1\,0\,1\,1 \\
\times\;\;1\,0\,1 \\
\hline
1\,0\,1\,1 \\
1\,0\,1\,1\,0 \\
\hline
1\,1\,0\,1\,1\,1
\end{array}
$$

Therefore, the answer is correct. ◄

EXERCISE 3.5

A. In Problems 1–6, perform the indicated additions.

1. $\begin{array}{r} 111_2 \\ +\;\;10_2 \\ \hline \end{array}$

2. $\begin{array}{r} 111_2 \\ +101_2 \\ \hline \end{array}$

3. $\begin{array}{r} 1101_2 \\ +\;\;110_2 \\ \hline \end{array}$

4. $\begin{array}{r} 1111_2 \\ +1101_2 \\ \hline \end{array}$

5. $\begin{array}{r} 110_2 \\ +101_2 \\ +111_2 \\ \hline \end{array}$

6. $\begin{array}{r} 1101_2 \\ +1110_2 \\ +\;\;101_2 \\ \hline \end{array}$

B. In Problems 7–12, perform the indicated subtractions.

7. $\begin{array}{r} 111_2 \\ -\;\;10_2 \\ \hline \end{array}$

8. $\begin{array}{r} 110_2 \\ -\;\;11_2 \\ \hline \end{array}$

9. $\begin{array}{r} 1000_2 \\ -\;\;111_2 \\ \hline \end{array}$

10. $\begin{array}{r} 1101_2 \\ -\;\;111_2 \\ \hline \end{array}$

11. $\begin{array}{r} 1111_2 \\ -\;\;101_2 \\ \hline \end{array}$

12. $\begin{array}{r} 1010_2 \\ -\;\;101_2 \\ \hline \end{array}$

C. In Problems 13–18, multiply as indicated.

13. $\begin{array}{r} 110_2 \\ \times\;\;11_2 \\ \hline \end{array}$

14. $\begin{array}{r} 101_2 \\ \times\;\;10_2 \\ \hline \end{array}$

15. $\begin{array}{r} 1111_2 \\ \times\;\;11_2 \\ \hline \end{array}$

16. $\begin{array}{r} 1110_2 \\ \times\;\;111_2 \\ \hline \end{array}$

17. $\begin{array}{r} 1011_2 \\ \times\;101_2 \\ \hline \end{array}$

18. $\begin{array}{r} 1011_2 \\ \times\;111_2 \\ \hline \end{array}$

D. In Problems 19–24, divide as indicated.

19. $10_2\overline{)1101_2}$ **20.** $11_2\overline{)1101_2}$ **21.** $11_2\overline{)1110_2}$

22. $11_2\overline{)11011_2}$ **23.** $101_2\overline{)111011_2}$ **24.** $111_2\overline{)1110111_2}$

▶ **Calculator Corner 3.5**   *Some scientific calculators will perform operations in different bases. Thus, to add $1101_2 + 110_2 + 11_2$ (Example 2), we place the calculator in the binary mode by pressing* `2nd` `mode` `bin` *and proceed as in regular addition by entering $1101 + 110 + 11 =$. The answer will appear on the screen as 10110. You can do addition, subtraction, and multiplication in the same manner, but you must be careful with division. Thus, in Example 7 we divided 1011_2 by 10_2. The answer is 101_2 with a remainder of 1. To do this on your calculator, enter* `2nd` `mode` `bin` *(now you are in binary) and $1011 \div 10$. The calculator will give you the quotient 101 but not the remainder. To discover that there is a remainder, you must check your division by multiplying the quotient (101_2) by the divisor (10_2). The result is 1010_2, not 1011_2, so the remainder must be 1. The moral of the story is that when doing division problems in different bases, you must check the problem by multiplication and, during this process, find the remainder.*

1. Use your calculator to find the remainder (if any) in Problems 19, 21, and 23 of Exercise 3.5.

3.6

OCTAL AND HEXADECIMAL ARITHMETIC

Arithmetic in the base 8 and base 16 systems can be done with addition and multiplication tables in much the same way as in the base 10 system. We consider the octal system first and construct the required tables (Tables 3.6a and 3.6b).

Table 3.6a Octal Addition Table

+	0	1	2	3	4	5	6	7
0	0	1	2	3	4	5	6	7
1	1	2	3	4	5	6	7	10
2	2	3	4	5	6	7	10	11
3	3	4	5	6	7	10	11	12
4	4	5	6	7	10	11	12	13
5	5	6	7	10	11	12	13	14
6	6	7	10	11	12	13	14	15
7	7	10	11	12	13	14	15	16

Table 3.6b Octal Multiplication Table

×	0	1	2	3	4	5	6	7
0	0	0	0	0	0	0	0	0
1	0	1	2	3	4	5	6	7
2	0	2	4	6	10	12	14	16
3	0	3	6	11	14	17	22	25
4	0	4	10	14	20	24	30	34
5	0	5	12	17	24	31	36	43
6	0	6	14	22	30	36	44	52
7	0	7	16	25	34	43	52	61

A. Octal Addition

To add octal numbers, we align corresponding digits and add column by column, carrying over from column to column as necessary.

As in the preceding section, we shall omit the subscripts in the computations, but keep in mind the system with which we are working.

EXAMPLE 1   Add: $673_8 + 52_8$

Solution

$$
\begin{array}{r}
1 \\
6\,7\,3 \\
+\ 5\,2 \\
\hline
7\,4\,5
\end{array}
$$

$7\,4\,5 \longleftarrow 3 + 2 = 5$

$ 7 + 5 = 14$ (Table 3.6a)   Write 4 and carry 1. (Read 14 as "one-four" *not* as "fourteen.")

$ 1 + 6 = 7$

Thus, $673_8 + 52_8 = 745_8$.   ◀

EXAMPLE 2   Do the addition: $705_8 + 374_8$

Solution

$$
\begin{array}{r}
1\,1 \\
7\,0\,5 \\
+3\,7\,4 \\
\hline
1\,3\,0\,1
\end{array}
$$

$1\,3\,0\,1 \longleftarrow 5 + 4 = 11$ (Table 3.6a)   Write 1 and carry 1.

$ 1 + 0 + 7 = 10$ (Table 3.6a)   Write 0 and carry 1.

$ 1 + (7 + 3) = 1 + 12 = 13$ (Table 3.6a)

The required sum is 1301_8.   ◀

B. Octal Multiplication

EXAMPLE 3   Multiply 46_8 by 5_8.

Solution

$$
\begin{array}{r}
4\,6 \\
\times\ \ 5 \\
\hline
2\,7\,6
\end{array}
$$

$5 \times 6 = 36$ (Table 3.6b)   Write 6 and carry 3.

$5 \times 4 = 24$ (Table 3.6b)   Add the 3 to get 27.

The answer is 276_8. [*Note:* You can check the answer by converting all the numbers to base 10. Thus, since $46_8 = 38$ and $5_8 = 5$, in base 10, we have $5 \times 38 = 190$. We also find that $276_8 = 190$, so the answer checks.   ◀

EXAMPLE 4   Multiply 237_8 by 14_8.

$$\begin{array}{r} 2\ 3\ 7 \\ \times\ \ 1\ 4 \\ \hline 1\ 1\ 7\ 4 \\ 2\ 3\ 7 \\ \hline 3\ 5\ 6\ 4 \end{array}$$

$4 \times 7 = 34$ (Table 3.6b) Write 4 and carry 3.
$4 \times 3 = 14$ (Table 3.6b); $14 + 3 = 17$ Write 7 and carry 1.
$4 \times 2 = 10$ (Table 3.6b); $10 + 1 = 11$
$1 \times 237 = 237$

Addition of the partial products gives the final answer, 3564_8. The check is left to you. ◄

C. Octal Subtraction and Division

Subtraction and division also can be done by using the tables as in the following examples.

EXAMPLE 5 Subtract 56_8 from 747_8.

Solution

$$\begin{array}{r} {\scriptstyle 6\ 14} \\ 7\ 4\ 7 \\ -\ \ 5\ 6 \\ \hline 6\ 7\ 1 \end{array}$$

7 − 6 = 1 (Table 3.6a shows $6 + 1 = 7$.)

4 − 5 requires borrowing. Change the leftmost 7 to 6 and add 10 to the 4. Then subtract. To get this result, refer to Table 3.6a, go down the left-hand column under + to the 5 and read across to 14. The 7 in the top line above the 14 is the required number.

Bring down the 6.

The final answer is 671_8. You can check this answer by adding $56_8 + 671_8$ to get 747_8. ◄

EXAMPLE 6 Subtract as indicated: $643_8 - 45_8$

Solution

$$\begin{array}{r} {\scriptstyle 5\ 13} \\ {\scriptstyle 3\ 13} \\ 6\ 4\ 3 \\ -\ \ 4\ 5 \\ \hline 5\ 7\ 6 \end{array}$$

3 − 5 requires borrowing. Change the 4 to 3 and add 10 to the first 3 to give 13. Then subtract: $13 - 5 = 6$ (Table 3.6a).

3 − 4 requires borrowing. Change the 6 to 5 and add 10 to the 3 to give 13. Then subtract: $13 - 4 = 7$ (Table 3.6a).

Bring down the 5.

The result shows that $643_8 - 45_8 = 576_8$. This can be checked by addition as in Example 5. ◄

EXAMPLE 7   Divide 765_8 by 24_8.

Solution

$$
\begin{array}{r}
3\,1 \\
24\overline{)7\,6\,5} \\
7\,4 \\
\hline
2\,5 \\
2\,4 \\
\hline
1
\end{array}
$$

← $3 \times 24 = 74$ (Table 3.6b). Write 3 in the quotient above the 76.

← Subtract and bring down the 5.

← $1 \times 24 = 24$

← Subtract to get the remainder.

This computation shows that $765_8 \div 24_8 = 31_8$ with a remainder of 1_8. This can be checked by multiplying 31_8 by 24_8 and adding 1 to the product.   ◄

EXAMPLE 8   Do the indicated division: $4357_8 \div 21_8$

Solution

$$
\begin{array}{r}
2\,0\,6 \\
21\overline{)4\,3\,5\,7} \\
4\,2 \\
\hline
1\,5\,7 \\
1\,4\,6 \\
\hline
1\,1
\end{array}
$$

← $2 \times 21 = 42$. Write 2 in the quotient above the 43.

← Subtract and bring down the 5. 21 does not go into 15, so write 0 in the quotient and bring down the 7.

← $6 \times 21 = 146$ (Table 3.6b). Write 6 in the quotient.

← Subtract to get the remainder.

The computation shows that the quotient is 206_8 and the remainder is 11_8. As before, the answer can be checked by multiplying the quotient by the divisor and adding the remainder. Thus,

$$
\begin{array}{r}
206 \\
\times\ \ 21 \\
\hline
206 \\
414 \\
\hline
4346 \\
+\ \ 11 \\
\hline
4357
\end{array}
$$

which shows the answer is correct.   ◄

D. Hexadecimal Addition and Multiplication

To do arithmetic in the hexadecimal system, we first construct the addition and multiplication tables, as shown in Tables 3.6c and 3.6d. These tables furnish the "number facts" we need to do the computations.

Table 3.6c Hexadecimal Addition Table

+	0	1	2	3	4	5	6	7	8	9	A	B	C	D	E	F
0	0	1	2	3	4	5	6	7	8	9	A	B	C	D	E	F
1	1	2	3	4	5	6	7	8	9	A	B	C	D	E	F	10
2	2	3	4	5	6	7	8	9	A	B	C	D	E	F	10	11
3	3	4	5	6	7	8	9	A	B	C	D	E	F	10	11	12
4	4	5	6	7	8	9	A	B	C	D	E	F	10	11	12	13
5	5	6	7	8	9	A	B	C	D	E	F	10	11	12	13	14
6	6	7	8	9	A	B	C	D	E	F	10	11	12	13	14	15
7	7	8	9	A	B	C	D	E	F	10	11	12	13	14	15	16
8	8	9	A	B	C	D	E	F	10	11	12	13	14	15	16	17
9	9	A	B	C	D	E	F	10	11	12	13	14	15	16	17	18
A	A	B	C	D	E	F	10	11	12	13	14	15	16	17	18	19
B	B	C	D	E	F	10	11	12	13	14	15	16	17	18	19	1A
C	C	D	E	F	10	11	12	13	14	15	16	17	18	19	1A	1B
D	D	E	F	10	11	12	13	14	15	16	17	18	19	1A	1B	1C
E	E	F	10	11	12	13	14	15	16	17	18	19	1A	1B	1C	1D
F	F	10	11	12	13	14	15	16	17	18	19	1A	1B	1C	1D	1E

Table 3.6d Hexadecimal Multiplication Table

×	0	1	2	3	4	5	6	7	8	9	A	B	C	D	E	F
0	0	0	0	0	0	0	0	0	0	0	0	0	0	0	0	0
1	0	1	2	3	4	5	6	7	8	9	A	B	C	D	E	F
2	0	2	4	6	8	A	C	E	10	12	14	16	18	1A	1C	1E
3	0	3	6	9	C	F	12	15	18	1B	1E	21	24	27	2A	2D
4	0	4	8	C	10	14	18	1C	20	24	28	2C	30	34	38	3C
5	0	5	A	F	14	19	1E	23	28	2D	32	37	3C	41	46	4B
6	0	6	C	12	18	1E	24	2A	30	36	3C	42	48	4E	54	5A
7	0	7	E	15	1C	23	2A	31	38	3F	46	4D	54	5B	62	69
8	0	8	10	18	20	28	30	38	40	48	50	58	60	68	70	78
9	0	9	12	1B	24	2D	36	3F	48	51	5A	63	6C	75	7E	87
A	0	A	14	1E	28	32	3C	46	50	5A	64	6E	78	82	8C	96
B	0	B	16	21	2C	37	42	4D	58	63	6E	79	84	8F	9A	A5
C	0	C	18	24	30	3C	48	54	60	6C	78	84	90	9C	A8	B4
D	0	D	1A	27	34	41	4E	5B	68	75	82	8F	9C	A9	B6	C3
E	0	E	1C	2A	38	46	54	62	70	7E	8C	9A	A8	B6	C4	D2
F	0	F	1E	2D	3C	4B	5A	69	78	87	96	A5	B4	C3	D2	E1

EXAMPLE 9   Add: $2B4_{16} + A1_{16}$

Solution

$$\begin{array}{r} 1 \\ 2\,B\,4 \\ +\ \ A\,1 \\ \hline 3\,5\,5 \end{array}$$

$4 + 1 = 5$

$B + A = 15$ (Table 3.6c)   Write 5 and carry 1.

$1 + 2 = 3$

The answer is 355_{16}. (You can check this by converting to base 10.) ◀

EXAMPLE 10   Add: $1AB2_{16} + 2CD3_{16}$

Solution
```
  1 1
  1 A B 2
 +2 C D 3
  4 7 8 5
```

— $2 + 3 = 5$

— $B + D = 18$ (Table 3.6c)   Write 8 and carry 1.

— $(A + C) + 1 = 16 + 1 = 17$ (Table 3.6c)   Write 7 and carry 1.

— $1 + 1 + 2 = 4$

The answer is 4785_{16}. (You can check this by converting to base 10.) ◀

EXAMPLE 11   Multiply $1A2_{16}$ by B_{16}.

Solution
```
  1 A 2
 ×   B
 1 1 F 6
```

— $B \times 2 = 16$ (Table 3.6d)   Write 6 and carry the 1.

— $B \times A = 6E$ (Table 3.6d);
  $6E + 1 = 6F$ (Table 3.6c)   Write F and carry the 6.

— $B \times 1 = B$; $B + 6 = 11$ (Table 3.6c)

The answer is $11F6_{16}$. ◀

EXAMPLE 12   Multiply $2B4_{16}$ by $B1_{16}$.

Solution
```
    2 B 4
  ×   B 1
    2 B 4 ← 1 × 2B4 = 2B4
  1 D B C ←
  1 D E 7 4
```

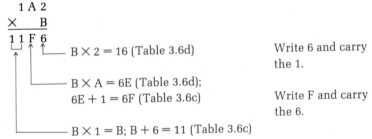

$1 \times 2B4 = 2B4$

$B \times 4 = 2C$ (Table 3.6d); write C and carry the 2. Then, $B \times B = 79$ (Table 3.6d); $79 + 2 = 7B$ (Table 3.6c); write B and carry the 7. Then, $B \times 2 = 16$ (Table 3.6d); $16 + 7 = 1D$ (Table 3.6c).

— $B + C = 17$ (Table 3.6c); write 7 and carry the 1.

— $(2 + B) + 1 = D + 1 = E$ (Table 3.6c)

The answer is $1DE74_{16}$. ◀

A. In Problems 1–4, do the indicated additions.

    1. $531_8 + 47_8$ **2.** $425_8 + 364_8$

    3. $7256_8 + 634_8$ **4.** $5732_8 + 747_8$

B. In Problems 5–8, do the indicated multiplications.

    5. $57_8 \times 6_8$ **6.** $45_8 \times 7_8$

    7. $216_8 \times 32_8$ **8.** $312_8 \times 65_8$

C. In Problems 9–12, do the indicated subtractions.

    9. $534_8 - 25_8$ **10.** $617_8 - 47_8$

    11. $3264_8 - 756_8$ **12.** $4763_8 - 654_8$

In Problems 13–16, do the indicated divisions.

    13. $317_8 \div 7_8$ **14.** $4355_8 \div 5_8$

    15. $4215_8 \div 15_8$ **16.** $7342_8 \div 31_8$

D. In Problems 17–20, do the indicated additions.

    17. $3CB_{16} + 4C_{16}$ **18.** $4FE_{16} + 35_{16}$

    19. $98D_{16} + 2B_{16}$ **20.** $CBD_{16} + AF_{16}$

In Problems 21–24, do the indicated multiplications.

    21. $2C5_{16} \times 3B_{16}$ **22.** $4DE_{16} \times 12_{16}$

    23. $6F3_{16} \times AB_{16}$ **24.** $29A_{16} \times EOF_{16}$

USING YOUR KNOWLEDGE 3.6

You know that we use decimals in the base 10 system with the understanding that the place values to the right of the decimal point are

$$\frac{1}{10^1}, \quad \frac{1}{10^2}, \quad \frac{1}{10^3}, \quad etc.$$

For example, the decimal numeral 23.759 stands for

$$(2 \times 10) + 3 + \frac{7}{10^1} + \frac{5}{10^2} + \frac{9}{10^3}$$

The same idea is used in other number systems. Thus, in the binary system, the numeral

$$11.101_2 = (1 \times 2) + 1 + \frac{1}{2^1} + \frac{1}{2^2} + \frac{1}{2^3}$$

In the octal system, the numeral

$$73.524_8 = (7 \times 8) + 3 + \frac{5}{8^1} + \frac{2}{8^2} + \frac{4}{8^3}$$

To convert numbers from binary or octal form to decimal form, it is convenient to have the values

$$\tfrac{1}{2} = 0.5, \quad \tfrac{1}{4} = 0.25, \quad \tfrac{1}{8} = 0.125, \quad etc.$$

For instance,

$$1.111_2 = 1 + 0.5 + 0.25 + 0.125 = 1.875$$
$$3.5_8 = 3 + (5 \times 0.125) = 3.625$$

Change to decimal form:

1. 10.101_2
2. 11.011_2
3. 10.001_2
4. 21.4_8
5. 72.6_8
6. 31.7_8

▶ **Calculator Corner 3.6** *Some scientific calculators are able to perform arithmetic in different bases. (You need a key displaying the desired base.) Thus, to add $673_8 + 52_8$ (Example 1), we place the calculator in the octal mode by pressing* `2nd` `mode` `oct` *and entering the indicated operation 673 + 52 =, as in regular addition. The answer 745 will appear on the calculator screen. Keep in mind that when doing division the calculator will not provide the remainder. Thus, if you divide 765_8 by 24_8 (Example 7), the calculator gives the quotient as 31. Is there a remainder? If you multiply the quotient 31_8 by the divisor 24_8, the result is 764_8 and not 765_8. Thus, the remainder must be 1.*

What about the hexadecimal system? First, you need a `hex` *key and A, B, C, D, E, F buttons to represent the numbers 10, 11, 12, 13, 14, and 15, respectively. You can then multiply $1A2_{16}$ by B_{16} (Example 11) by entering*

`2nd` `mode` `hex` `1A2` `×` `B` `=`

The calculator gives the answer 11F6.

1. Use your calculator to find the remainder (if any) in Problems 13 and 15, Exercise 3.6.

SUMMARY

Section	Item			Meaning
	Egyptian	Babylonian	Roman	
3.1	|	▼	I	One
3.1	|||||	▼▼▼▼▼	V	Five
3.1	∩	⬎	X	Ten

Section	Item			Meaning	Example
	Egyptian	Babylonian	Roman		
3.1	∩∩∩∩∩	⟨⟨⟨⟨⟨	L	Fifty	
3.1	ꝯ		C	One hundred	
3.1	ꝯꝯꝯꝯꝯ		D	Five hundred	
3.1	𓆼		M	One thousand	
3.1	𓂭		$\overline{\text{X}}$	Ten thousand	
3.1	𓆐		$\overline{\text{C}}$	One hundred thousand	
3.1	𓁨		$\overline{\text{M}}$	One million	
3.2A	a^n			$\underbrace{a \times a \times \cdots \times a}_{n\ a\text{'s}}$	$10^3 = 10 \times 10 \times 10$
3.2A	$(2 \times 10^2) + (4 \times 10^1) + (5 \times 10^0)$			Expanded form of 245	
3.2B	$a^m \times a^n = a^{m+n}$			Law of exponents	$5^3 \times 5^6 = 5^9$
3.2B	$a^m \div a^n = a^{m-n}$			Law of exponents	$4^7 \div 4^2 = 4^5$
3.3A	43_{five}			$(4 \times 5) + (3 \times 5^0)$	
3.3B	10001_{two}			$(1 \times 2^4) + (1 \times 2^0)$	
3.3B	$A_{16}, B_{16}, C_{16}, D_{16}, E_{16}, F_{16}$			10, 11, 12, 13, 14, 15	
3.3B	17_{eight}			$(1 \times 8) + (7 \times 8^0)$	

PRACTICE TEST 3

1. Write in Egyptian numerals:
   a. 63   b. 735
2. Write in decimal notation:
   a. ∩∩||| b. ꝯ∩∩|
3. Write in Babylonian numerals:
   a. 63   b. 735
4. Write in decimal notation:
   a. ▼ ⟨⟨▼▼   b. ▼▼ ⟨▼
5. Do the multiplication 23 × 21 using:
   a. The Egyptian method of successive duplication
   b. The Egyptian method of mediation and duplation
6. Write in Roman numerals:
   a. 53   b. 42   c. 22,000
7. Write in decimal notation:
   a. LXVII   b. $\overline{\text{XLVIII}}$
8. Write in expanded form:
   a. 2507   b. 189

9. Write in decimal notation:

    a. $(3 \times 10^3) + (7 \times 10^2) + (2 \times 10^0)$

    b. $(5 \times 10^4) + (9 \times 10^3) + (4 \times 10)$

10. Do the following computations in the usual way and in expanded form:

    a. $75 + 32$ **b.** $56 - 24$

11. Perform the indicated operations, leaving the answer in exponential form:

    a. $3^4 \times 3^8$ **b.** $2^9 \div 2^3$

12. Do the following computations in the usual way and in expanded form:

    a. 83×21 **b.** $54 \div 7$

13. Change to decimal notation:

    a. 203_4 **b.** 143_5 **c.** 1101_2

14. Change to decimal notation:

    a. 152_8 **b.** $A2C_{16}$

15. Convert the number 33 to:

    a. Base 5   **b.** Base 6

16. Convert to binary notation:

    a. 39   **b.** 527

17. Convert the number 47 to:

    a. Base 8   **b.** Base 16

18. **a.** Convert 1011101_2 to octal notation.

    b. Convert 327_8 to binary notation.

19. **a.** Convert 1011101_2 to hexadecimal notation.

    b. Convert $2BD_{16}$ to binary notation.

20. **a.** Convert $2B_{16}$ to octal notation.

    b. Convert 27_8 to hexadecimal notation.

21. Do the indicated computations in the binary system:

    a. 1101_2 **b.** 1101_2

       $+ \ 101_2$ $- \ 111_2$

22. Do the indicated computations in the binary system:

    a. $1101_2 \times 11_2$ **b.** $10110_2 \div 11_2$

23. Do the following computations in the octal system:

    a. $632_8 + 46_8$ **b.** $37_8 \times 5_8$

24. Do the following computations in the octal system:

    a. $632_8 - 46_8$ **b.** $572_8 \div 6_8$

25. Do the following computations in the hexadecimal system:

    a. $2BC_{16} + 5D_{16}$ **b.** $3C4_{16} \times 2B_{16}$

CHAPTER 4

THE RATIONAL NUMBERS

Leopold Kronecker was born of prosperous Jewish parents on December 7, 1823, in Liegnitz, Prussia. The boy's early education was under a private tutor who was supervised by Leopold's father, a well-educated man whose love of philosophy was handed on to his son.

Throughout his schooling, Kronecker's performance was many-sided and brilliant. He mastered Greek, Latin, Hebrew, philosophy, and mathematics with ease. As a youngster he specialized in no particular area, although it became obvious that his greatest talent was in mathematics. In addition to his formal studies, he took music lessons and became an accomplished pianist and vocalist. A small but compact man, not over 5 ft tall, he was an expert gymnast and swimmer.

In 1841, Kronecker entered the University of Berlin, where he began to specialize in mathematics although he continued his broad education. He was soon doing mathematics research, and his dissertation for the Ph.D. degree was accepted in 1845 when he was 22 years old. His main doctoral work was in the theory of numbers and involved special algebraic problems stemming from the attempt to construct a regular polygon of n sides with only a straightedge and compass.

In the same year in which Kronecker got his degree, a rich uncle died, leaving the young Leopold to manage a large estate and to run a banking business for his cousin (whom he married in 1848). As in everything else, he was an extremely successful businessman, having a genius for making the right friends and investments.

Despite his diversion to business and finance, Kronecker did not neglect his mathematics completely. In fact, he finished an outstanding memoir on the theory of equations, which was published in 1853. Throughout his mathematical work, much of it in number theory and theory of equations, he tried to make concise and expressive formulas tell the whole story so that the development shone forth in a direct and simple fashion. Most of Kronecker's papers have a strong arithmetical flavor; he wanted to explain everything in terms of the whole numbers and in a finite number of steps, something that has had a great influence on modern mathematics.

Until the last decade of his life, Kronecker was a free man with obligations to no employer. However, in 1883, the University of Berlin offered him a professorship, which he accepted and held until his death in December 1891, at the age of 68.

LEOPOLD KRONECKER (1823–1891)

David Eugene Smith Collection, Rare Book and Manuscript Library, Columbia University

In short, Kronecker was an artist who used mathematical formulas as his medium.

E. T. BELL

THE NATURAL, OR COUNTING, NUMBERS

Look at the numbers on the keys of your calculator. They can be used to form a special set of numbers, called the set N of **natural,** or **counting, numbers** {1, 2, 3, . . .}. In this section, we shall discuss some uses and properties of the set of natural numbers.

A. Using the Natural Numbers

One of the earliest uses of the set of natural numbers was for counting. Tribes wanted to know how many members they had, how many sheep were in a flock, or how many days remained before the harvest. This counting was probably done by some simple tally method. For example, to count the number of elements in the set $A = \{a, b, c\}$, we associate the number 1 with a, 2 with b, and 3 with c, and **count** the elements in the set as 1, 2, 3. This **one-to-one-correspondence** used to do the counting is illustrated below:

$$A = \{a, \, b, \, c\}$$
$$\updownarrow \ \updownarrow \ \updownarrow$$
$$1 \ 2 \ 3$$

In mathematics, the number of elements in a set A is called the **cardinal number** of the set and is denoted by $n(A)$. For example, the cardinal number of $B = \{*, \$, @, ¢\}$ is 4, and the cardinal number of $C = \{£, ¢, \square, §, \%\}$ is 5, so that $n(B) = 4$ and $n(C) = 5$. Thus, the natural numbers can be used as cardinal numbers; that is, they can be used for counting.

Numbers can also be used to assign an **order,** or position, to the elements of a set, that is, to indicate which element is **first, second, third,** and so on. We then refer to these numbers as **ordinal numbers.**

As you can see from the Snuffy Smith cartoon, numbers can be used for purposes other than counting and ordering. What Snuffy needed was the correct **identification** (or catalog) number. When a person is assigned the number 123-45-6789 by the Social Security Administration, this number is being used neither as a cardinal nor as an ordinal number but merely as identification. The same is true of your charge card, and savings account numbers; they are used for identification purposes.

EXAMPLE 1   Determine whether the underlined word is used as a cardinal or an ordinal number, or for identification.

    a. If she kissed you once, would she kiss you <u>two</u> times?
    b. My account number is <u>123,456</u>.
    c. This is my <u>first</u> and last warning.

Solution   **a.** Cardinal   **b.** Identification   **c.** Ordinal   ◄

B. Properties of the Natural Numbers

We now explore some of the properties of the set N of natural numbers under the operations of addition (+) and multiplication (×). These two operations are examples of *binary operations*. In general, a **binary operation** is an operation that associates with any two elements of a given set a unique result. If the given set is the set of natural numbers and the operation is addition, the result is another natural number. For example, $2 + 3 = 5$ and $3 + 7 = 10$. Similarly, if the operation is multiplication, $2 \times 3 = 6$ and $3 \times 7 = 21$. The fact that the results of addition and multiplication are again natural numbers is described by saying that **the set of natural numbers is closed under addition and under multiplication.**

►   **Definition 4.1a**

If an operation is defined on a set and the result of this operation is always an element of the set, then the set is said to be **closed** under this operation.

EXAMPLE 2   Show that the set {0, 1} is closed under ordinary multiplication.

Solution   In the table in the margin, we show the operation table for the given set. For any two elements of the set, the result after multiplication is in the set, so the given set is closed under the operation of multiplication. Note that you only need to look at the body of the table. If all the entries are in the original set, then the set is closed under the operation. It is not closed otherwise. (See the next example.)   ◄

×	0	1
0	0	0
1	0	1

EXAMPLE 3   Show that the set {0, 1} is not closed under the operation of addition.

Solution   In the table in the margin, we show the operation table for the given set. The addition of 1 and 1 yields 2, which is not a member of the original set, so the given set is not closed under the operation of addition.   ◄

+	0	1
0	0	1
1	1	2

Notice that the set of natural numbers is not closed under subtraction (because, for example, $3 - 5 = -2$, which is not a natural number) or division (because $3 \div 2$ is not a natural number).

Another property that we can easily observe when adding natural numbers is that the order in which the addition is performed is immaterial to the sum. Thus, if we are asked, "How much is 1 and 3?" we answer, "4." If we are asked, "How much is 3 and 1?" we again answer, "4." Thus, $3 + 1 = 1 + 3$. Similarly, if we multiply two natural numbers together, the order in which we perform the multiplication does not affect the product. For example, $3 \times 4 = 4 \times 3$. These two properties are briefly described by saying that the operations of addition and multiplication have the *commutative* property. In general, if we have a set and a binary operation on the elements of this set such that the result is the same no matter in which order the operation is performed, then we say that the operation has the **commutative property**.

Commutative Property of Addition

If a and b are natural numbers, then

$$a + b = b + a$$

Commutative Property of Multiplication

If a and b are natural numbers, then

$$a \times b = b \times a$$

These two properties can be proved using the ideas of sets as shown in Discovery 4.1.

A third property of importance becomes evident when we try to add more than two numbers, say, $4 + 3 + 5$. Because we usually add two numbers at a time, the question arises, do we add $4 + 3$ and then add that sum to 5, that is, $(4 + 3) + 5 = 7 + 5 = 12$? Or do we add 4 to the sum of 3 and 5, that is, $4 + (3 + 5) = 4 + 8 = 12$? From the results obtained, we see that it makes no difference which numbers are added first; that is, we may group (associate) the numbers in either way and obtain the same answer. This property of addition is called the **associative property**.

Associative Property of Addition

If a, b, and c are natural numbers, then

$$a + (b + c) = (a + b) + c$$

Multiplication also has the associative property, as illustrated by $2 \times (3 \times 5) = (2 \times 3) \times 5$, since in both cases the result is 30.

Associative Property of Multiplication

If a, b, and c are natural numbers, then

$$a \times (b \times c) = (a \times b) \times c$$

WANTED SMALL ENGINE MECHANIC

Do you think they want a small (engine mechanic) or a (small engine) mechanic? These are not the same; English does not have the associative property.

The method of proving these two associative properties is indicated in Discovery 4.1.

There is another property of addition and multiplication that is of great importance. Suppose we are asked to find the product of 4 and $(10 + 6)$, which is denoted by $4(10 + 6)$ or by $4 \cdot (10 + 6)$, where the raised dot is used in place of the times sign. We can find the product by adding 10 and 6 first, getting

$$4 \cdot (10 + 6) = 4 \cdot 16 = 64$$

However, the same result can be obtained by multiplying by 4 first:

$$4 \cdot (10 + 6) = 4 \cdot 10 + 4 \cdot 6 = 40 + 24 = 64$$

In general, if a, b, and c are natural numbers, then

$$a \times (b + c) = (a \times b) + (a \times c)$$

This property is called the **distributive property of multiplication over addition.** Note that $a(b + c)$ means $a \times (b + c)$. This notation and the fact that $a \times b$ is usually written ab will be used throughout this book. Thus, using this notation, we state:

Distributive Property

If a, b, and c are any natural numbers, then

$$a(b + c) = ab + ac$$

Note that the distributive property is used in ordinary multiplication by hand. For example, to multiply 14 by 21, we may write

$$14 \cdot 21 = 14(20 + 1) = 14 \cdot 20 + 14 \cdot 1$$
$$= 280 + 14 = 294$$

The ordinary multiplication would appear as follows:

$$
\begin{array}{r}
14 \\
\times\ 21 \\
\hline
14 \\
28 \\
\hline
294
\end{array}
$$

so that we have actually multiplied 14 by 1 and then by 20 and added the partial products, just as is indicated by the distributive property. (The second partial product is actually $280 = 14 \cdot 20$ since we moved to the left one place.)

The preceding properties of the natural numbers may be summarized as follows:

Properties of Addition

Closure If a and b are natural numbers, then $a + b$ is also a natural number. The set N is closed with respect to addition.

Commutativity If a and b are natural numbers, then

$$a + b = b + a$$

Addition is commutative.

Associativity If a, b, and c are natural numbers, then

$$(a + b) + c = a + (b + c)$$

Addition is associative.

Properties of Multiplication

Closure If a and b are natural numbers, then $a \times b$ is also a natural number. The set N is closed with respect to multiplication.

Commutativity If a and b are natural numbers, then

$$a \times b = b \times a$$

Multiplication is commutative.

Associativity If a, b, and c are natural numbers, then

$$a \times (b \times c) = (a \times b) \times c$$

Multiplication is associative.

Distributivity If a, b, and c are natural numbers, then

$$a \times (b + c) = (a \times b) + (a \times c)$$

Multiplication is distributive over addition.

Table 4.1a

+	1	2	3	4	. . .
1	2	3	4	5	. . .
2	3	4	5	6	. . .
3	4	5	6	7	. . .
4	5	6	7	8	. . .
.	.	.	.	.	. . .
.	.	.	.	.	
.	.	.	.	.	

It is worth noticing that we can check the closure and commutative properties of addition by examining an addition table such as Table 4.1a. Because all the entries in this table are natural numbers (no new numbers are introduced), the set N is closed with respect to addition.

To check commutativity, we draw a diagonal from the upper left-hand corner to the lower right and note that each entry is identical to the entry that is symmetrically placed with respect to this diagonal. Thus, the entry in the first row, fourth column and that in the fourth row, first column are identical. When this happens for every entry in the table, the operation is commutative (assuming that the entries in the top row and the left-hand column are listed in the same order).

EXAMPLE 4

*	a	b	c
a	a	c	b
b	b	a	c
c	c	b	a

Consider the set $A = \{a, b, c\}$ and the operation $*$ as defined by the table in the margin.

a. Is the set closed under $*$? Why?
b. Is the set commutative under $*$? Why?
c. Is the set associative under $*$? Why?

Solution

a. Yes, because all entries are elements of A.
b. No. The table is not symmetric to the diagonal shown. For instance, $a * b = c$, but $b * a = b$.
c. No. For instance, $a * (b * c) = a * c = b$, but $(a * b) * c = c * c = a$. ◀

EXERCISE 4.1

A. In Problems 1–5, identify the underlined items as cardinal numbers, as ordinal numbers, or for identification only.

1. My telephone number is <u>123-7643</u>.
2. This is the <u>second</u> problem in this exercise.
3. It takes <u>two</u> to tango.
4. <u>One</u>, <u>two</u>, <u>three</u>, go!
5. <u>First</u> National Bank is number <u>one</u>.

6. Three numbers are circled on the check shown here. Identify each of these numbers as a cardinal number, an ordinal number, or for identification only.

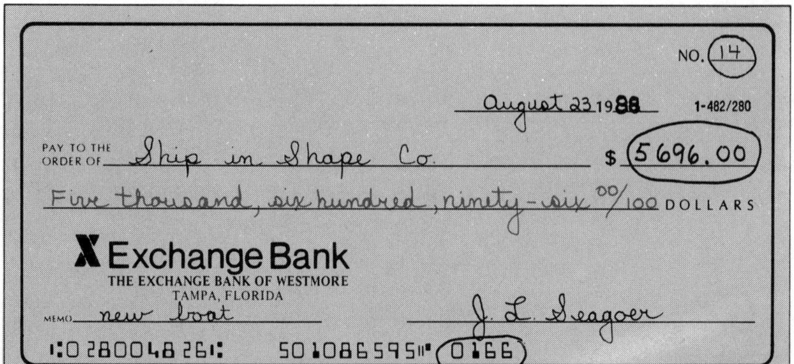

B. **7.** Construct a multiplication table similar to Table 4.1a and answer the following questions on the basis of your table.
  a. Would you say that multiplication is a binary operation? Why?

⊕	0	1	2	3
0	0	1	2	3
1	1	2	3	1
2	2	3	1	2
3	3	1	2	3

*	*a*	*b*	*c*
a	b	c	a
b	c	a	b
c	a	b	c

⊗	0	1	2
0	1	2	0
1	2	0	1
2	0	1	2

b. Would you say that the set of natural numbers is closed under multiplication? Why?

c. Is multiplication of natural numbers commutative? Why?

8. Consider the set $A = \{0, 1, 2, 3\}$ and the operation $\oplus$ as given by the table in the margin.

   a. Is $\oplus$ a binary operation? Why?

   b. Is the set A closed under $\oplus$? Why?

   c. Is $\oplus$ commutative? Why?

9. Use the distributive property to multiply the following:

   a. $4(3 + 8)$ **b.** $5(9 + 6)$ **c.** $8(3 + 8)$ **d.** $9(4 + 9)$

10. Sometimes, the distributive property can be used to simplify multiplication. For example, $8 \cdot 19 = 8(10 + 9) = 8 \cdot 10 + 8 \cdot 9 = 80 + 72 = 152$. Use this technique to compute the following:

   a. $6 \cdot 17$ **b.** $8 \cdot 16$ **c.** $7 \cdot 23$ **d.** $9 \cdot 18$

11. Is the set of even natural numbers, $E = \{2, 4, 6, \ldots\}$, closed under addition? Under multiplication?

12. Is the set of odd numbers, $O = \{1, 3, 5, \ldots\}$, closed under addition? Under multiplication?

13. Consider the set $A = \{a, b, c\}$ and the operation $*$ as given by the table in the margin.

   a. Is $*$ a binary operation? Why?

   b. Is the set A closed under $*$? Why?

   c. Is $*$ commutative? Why?

   d. Is $*$ associative? Why?

14. Consider the set $A = \{0, 1, 2\}$ and the operation $\otimes$ as given by the table in the margin. Show that this set with the operation $\otimes$ has exactly the same properties as the set described in Problem 13; only the notation is different.

In Problems 15–23, state which of the properties discussed in this section is being applied.

15. **a.** $7 + (5 + 3) = 7 + (3 + 5)$ **b.** $7 + (5 + 3) = (7 + 5) + 3$

16. **a.** $3(4 \times 8) = (3 \times 4) \times 8$ **b.** $3(4 \times 3) = 3 \times (3 \times 4)$

17. **a.** $3(4 + 8) = 3 \times 4 + 3 \times 8$ **b.** $3(4 + 8) = 3 \times (8 + 4)$

18. **a.** $5 \cdot (7 \cdot 9) = (7 \cdot 9) \cdot 5$ **b.** $3 \cdot (4 \cdot 7) = (7 \cdot 4) \cdot 3$

19. **a.** $(5 + 9) \cdot 3 = 5 \cdot 3 + 9 \cdot 3$ **b.** $(5 + 9) \times 3 = 3(5 + 9)$

20. **a.** $a(b + c) = (b + c)a$ **b.** $a(b + c) = ab + ac$

21. **a.** $a(b + c) = a(c + b)$ **b.** $(b + c)a = a(b + c)$

22. **a.** $ab + ac = ac + ab$ **b.** $ab + ac = ab + ca$

23. **a.** $a(bc) = (ab)c$ **b.** $a(bc) = (bc)a$

24. After adding a column of numbers such as

$$\begin{array}{r} 17 \\ 3 \\ 9 \\ \underline{8} \end{array}$$

from top to bottom, one can check the result by adding from bottom to top. Which of the properties discussed in this section guarantees the validity of the check?

25. Use the distributive property to fill in each blank with the correct number:
    a. $4(3 + 5) = (4 \cdot 3) + (4 \cdot \underline{\quad})$
    b. $5(6 + 7) = (5 \cdot \underline{\quad}) + (5 \cdot 7)$
    c. $8(4 + 6) = (\underline{\quad} \cdot 4) + (8 \cdot 6)$

26. Use the distributive property to fill in each blank with the correct number:
    a. $\underline{\quad}(2 + 5) = (3 \cdot 2) + (3 \cdot 5)$
    b. $8(9 + \underline{\quad}) = (8 \cdot 9) + (8 \cdot 4)$
    c. $7(\underline{\quad} + 5) = (7 \cdot 9) + (7 \cdot 5)$

27. We already know that multiplication is distributive over addition. Is addition distributive over multiplication?

28. Is multiplication distributive over subtraction? That is, does
    $$a(b - c) = (a \cdot b) - (a \cdot c)?$$

29. Is subtraction distributive over multiplication? That is, does
    $$a - (b \times c) = (a - b) \times (a - c)?$$

30. The Taste-T Noodle Company has 3 secretaries and 6 sales representatives. If each of them makes $280 a week, the weekly payroll can be calculated by multiplying 280 by 9. It could also be calculated by first finding the total payroll for the secretaries and adding this to the total payroll for the sales representatives. Which of the properties discussed in this section guarantees that the final results will be the same?

31. A grocer found that a shelf containing canned beans held 4 layers of cans, and each layer contained 5 rows of 3 cans. He counted the cans in each row (3) and the number of rows (5) and found that he had 15 cans per layer. Then he multiplied the result by 4, and knew he had $(5 \times 3) \times 4 = 60$ cans. His son noticed that there were 4 cans in each pile. There were 3 cans in each row, so each row had $4 \cdot 3 = 12$ cans. He then multiplied by the number of rows and obtained $(4 \times 3) \times 5 = 60$ cans. Which of the properties discussed in this section guarantee that the total, calculated either way, will be the same?

32. Refer to the figure below. Person A claims that there are 3 rows of 5 objects in front of him, while person B says that there are 5 rows of 3 objects in front of her. Which property discussed in this section guarantees that they both see the same total number of objects?

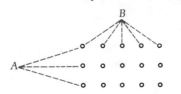

As we have seen, the cardinal number of a set A is the number of elements in A and is denoted by $n(A)$. Thus, if $A = \{@, \#, \%, \&, *\}$ and $B = \{a, b, c\}$, then $n(A) = 5$ and $n(B) = 3$. The addition of 5 and 3 can be defined using the sets A and B and noting that $5 + 3 = n(A) + n(B)$. Since $A \cup B = \{@, \#, \%, \&, *, a, b, c\}$, then $n(A \cup B) = 8$ and $n(A) + n(B) = n(A \cup B)$. In general, we use the following definition:

> If A and B are two sets such that $n(A) = a$, $n(B) = b$, and $A \cap B = \emptyset$, then
>
> $$a + b = n(A \cup B)$$

Note that we require $A \cap B$ to be empty, because, for example, if $A = \{x, y, z\}$ and $B = \{y, z, u, v\}$, $n(A) = 3$, $n(B) = 4$, and $n(A) + n(B) = 4 + 3 = 7$. But $n(A \cup B) = n(\{x, y, z, u, v\}) = 5$; that is, $n(A \cup B) \neq n(A) + n(B)$. Using these ideas, we can prove the commutative property of addition for natural numbers, $a + b = b + a$, as follows:

i. Let A and B be disjoint sets such that $n(A) = a$ and $n(B) = b$; that is, A has a elements and B has b elements.

ii. By the above definition, $a + b = n(A \cup B)$.

iii. By the commutative law for set union, $A \cup B = B \cup A$.

iv. By the above definition, $b + a = n(B \cup A)$.

v. From steps ii, iii, and iv, it follows that $a + b = n(A \cup B) = n(B \cup A) = b + a$.

vi. Therefore, $a + b = b + a$.

1. Can you discover the proof for the associative law of addition?

The multiplication of natural numbers can be defined in terms of sets. Thus, if you are taking three classes, x, y, and z, and there are two exams in each class, the number of exams you will have is $3 \cdot 2 = 6$. If the exams in each of the classes are identified as 1 and 2, the set of all exams will be $\{(x, 1), (x, 2), (y, 1), (y, 2), (z, 1), (z, 2)\}$, where $(x, 1)$ denotes the first exam in class x, $(x, 2)$ the second exam in class x, and so on. The number of elements in this set is 6. If the set of classes is $A = \{x, y, z\}$ and the set of exams is $B = \{1, 2\}$, the set $\{(x, 1), (x, 2), (y, 1), (y, 2), (z, 1), (z, 2)\}$ is called the **Cartesian product** of A and B and is denoted by $A \times B$. As you can see from this example, $2 \cdot 3 = n(A) \cdot n(B) = 6$ and $n(A \times B)$ is also 6.

2. Can you discover a definition for the product of any two natural numbers a and b?

3. Using the definition from Problem 2, can you discover the proof of the commutative law for multiplication?

4. Using the definition from Problem 2, can you discover the proof of the associative law for multiplication?

PRIME NUMBERS

Suppose there are 40 students in your class and you wish to divide them into 5 equal groups. Each group will have 8 students, and no student will be left out. Thus, we say that 40 divided by 5 is exactly 8 with no remainder. In other words, 5 is an **exact divisor** of 40 or, more briefly, 5 is a **divisor** or **factor** of 40. Similarly, 63 divided by 7 is exactly 9, so 7 is a divisor or factor of 63. However, 7 is not a divisor or factor of 60, because 60 is not exactly divisible by 7.

A. Prime and Composite Numbers

If 40 is written as a product of factors, the product—for example, 5×8—is called a **factorization** of 40. The 5 and the 8 are factors (divisors) of 40, and the number 40 is a **multiple** of each of these factors. Note that the 8 has divisors other than 8 and 1 (2 and 4), but the only divisors of 5 are itself and 1. We say that 5 is a *prime number*.

▶ **Definition 4.2a**

A natural number with exactly two distinct divisors (1 and itself) is called a **prime** or a **prime number.**

Any number with more than two distinct divisors is said to be **composite.**

According to Definition 4.2a, 5 is a prime number but 40 and 8 are composite. Notice that the number 1 is neither prime nor composite, because it has only one divisor (itself).

In the third century B.C., the Greek mathematician Eratosthenes devised the following procedure to find prime numbers: He listed all the numbers from 2 up to a given number (say, from 2 through 50 as in Fig. 4.2a). He concluded that 2 is prime, but every multiple of 2 (4, 6, 8, etc.) is composite, because it is divisible by 2. He then circled 2 as a prime and crossed out all multiples of 2. Next, because 3 is prime, he circled it, and crossed out all following multiples of 3. Similarly, he circled 5 (the next

	2	3	~~4~~	5	~~6~~	7	~~8~~	~~9~~	~~10~~
11	~~12~~	13	~~14~~	~~15~~	~~16~~	17	~~18~~	19	~~20~~
~~21~~	~~22~~	23	~~24~~	~~25~~	~~26~~	~~27~~	~~28~~	29	~~30~~
31	~~32~~	~~33~~	~~34~~	~~35~~	~~36~~	37	~~38~~	~~39~~	~~40~~
41	~~42~~	43	~~44~~	~~45~~	~~46~~	47	~~48~~	~~49~~	~~50~~

Figure 4.2a

prime), and crossed out its multiples. He continued this process until he reached 11. We note that all the multiples of 11 (11 · 2, 11 · 3, 11 · 4) were eliminated earlier when the multiples of 2 and 3 were crossed out. Similarly, the multiples of any prime larger than 11 also were eliminated. Hence, all the numbers left in the table are primes. Eratosthenes then circled those numbers. If the table had gone as far as 200, he would have had to continue his stepwise exclusion of composite numbers until he came to 17, the first prime whose square exceeds 200. (Can you explain why?) For obvious reasons, the preceding method is called the *Sieve of Eratosthenes*.

Note that the set of primes is not finite; that is, there is no largest prime. This fact is proved in Problem 9 of Exercise 4.2.

The applications of prime numbers that we shall consider depend on a theorem that we state below without proof.

Fundamental Theorem of Arithmetic

Every composite number can be expressed as a unique product of primes (disregarding the order of the factors).

Thus,

$$180 = 2 \cdot 2 \cdot 3 \cdot 3 \cdot 5 = 2^2 \cdot 3^2 \cdot 5$$

and

$$92 = 2 \cdot 2 \cdot 23 = 2^2 \cdot 23$$

B. Factorization

The prime factorization of a composite number can be found by "dividing out" the prime factors of the number, starting with the smallest factor. Thus,

$$40 = 2 \cdot 20 = 2 \cdot 2 \cdot 10 = 2 \cdot 2 \cdot 2 \cdot 5 = 2^3 \cdot 5$$

and

$$63 = 3 \cdot 21 = 3 \cdot 3 \cdot 7 = 3^2 \cdot 7$$

We can write these computations using repeated division as follows:

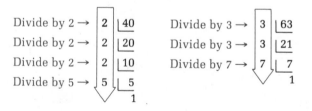

Reading downward, as indicated by the arrows, we get

$$40 = 2 \cdot 2 \cdot 2 \cdot 5 = 2^3 \cdot 5 \qquad \text{and} \qquad 63 = 3 \cdot 3 \cdot 7 = 3^2 \cdot 7$$

as before.

EXAMPLE 1   Write 1440 as a product of primes.

Solution

$$\begin{array}{r|r} 2 & 1440 \\ 2 & 720 \\ 2 & 360 \\ 2 & 180 \\ 2 & 90 \\ 3 & 45 \\ 3 & 15 \\ 5 & 5 \\ & 1 \end{array}$$

Hence, $1440 = 2 \cdot 2 \cdot 2 \cdot 2 \cdot 2 \cdot 3 \cdot 3 \cdot 5 = 2^5 \cdot 3^2 \cdot 5$.   ◄

EXAMPLE 2   Is the number 197 prime or composite?

Solution   To answer this question, we must determine whether 197 has factors other than 1 and itself. Thus, we try dividing by the consecutive prime numbers 2, 3, 5, 7, 11, 13, ... Since 197 is not even, it is not divisible by 2. It is also not divisible by 3 or 5 or 7 or 11 or 13, as you can find by actual trial. Where do we stop?

In trying to divide 197 by 3, 5, 7, 11, and 13, you should notice that the quotient is *greater* than the trial divisor each time (for example, dividing 197 by 3 gives a quotient of 65). The Sieve of Eratosthenes (Fig. 4.2a) gives the next prime as 17. If we try to divide 197 by 17, we get a quotient (11) that is *less* than 17 and a remainder showing that 17 is not a factor of 197. Consequently, we need go no further (Can you see why?), and we have shown that 197 is prime.   ◄

C.   Divisibility Rules

Finding the prime factorization of a number or determining whether the number is prime or composite may require a number of trial divisions. Some of these divisions can be avoided if we know the **divisibility rules** given in Table 4.2a on page 162.

Table 4.2a

DIVISIBLE BY	TEST	EXAMPLE
2	The number ends in 0, 2, 4, 6, or 8.	345,678 ends in 8, so it is divisible by 2.
3	The sum of its digits is divisible by 3.	258 is divisible by 3, since $2 + 5 + 8 = 15$, which is divisible by 3.
5	The number ends in 0 or 5.	365 ends in 5, so it is divisible by 5.

There are other rules for divisibility by larger primes, but most people feel these rules are too complicated to bother with; it is easier to check by direct division.

EXAMPLE 3 Which of the following numbers are divisible by 2, 3, or 5?

a. 8925   **b.** 39,120   **c.** 2553

Solution **a.** 8925 is not divisible by 2.

$8 + 9 + 2 + 5 = 24$, which is divisible by 3; hence, 8925 is divisible by 3.

Since 8925 ends in 5, it is divisible by 5.

b. 39,120 is divisible by 2, 3, and 5. Note that $3 + 9 + 1 + 2 = 15$, which is divisible by 3.

c. 2553 is not divisible by 2 or 5, but is divisible by 3 because the sum of its digits is divisible by 3.   ◄

EXERCISE 4.2

A. **1.** Continue the Sieve of Eratosthenes (Fig. 4.2a), and find all the primes between 50 and 100.

Use Fig. 4.2a and the results of Problem 1 to find how many primes there are between:

2. 1 and 25   **3.** 25 and 50   **4.** 50 and 75   **5.** 75 and 100

6. Refer to Fig. 4.2a and find:
   a. The smallest prime   **b.** An even prime

7. Refer to Fig. 4.2a.
   a. Find a pair of primes that are consecutive counting numbers.
   b. Can there be a second pair of primes that are consecutive counting numbers? Why?

8. Primes that differ by 2 are called **twin primes.** The smallest twin primes are 3 and 5. Refer to Fig. 4.2a and find two other pairs of twin primes.

9. To show that there is no largest prime, Euclid (in about 300 B.C.) gave the following *proof by contradiction:* Assume that there is a largest prime; call it P. Now form a number, say m, by taking the product of all the primes from 2 through P and adding 1 to the result. This gives

$$m = (2 \cdot 3 \cdot 5 \cdot 7 \cdot \cdots \cdot P) + 1$$

  a. m is not divisible by 2. Why?
  b. m is not divisible by 3. Why?
  c. m is not divisible by 5. Why?
  d. m is not divisible by any of the primes from 2 through P. Why?
  e. m is greater than P, so cannot be a prime. Why?
  f. m cannot be a composite number. Why?

  Now we have a contradiction! Since m is a natural number greater than 1, it must be either prime or composite. But if our assumption that P is the largest prime is correct, then m cannot be either prime or composite. Therefore, the assumption is invalid, and there is no largest prime.

B. In Problems 10–15, *find all the factors (divisors) of the given number.*

  10. 28   **11.** 50   **12.** 119
  13. 128   **14.** 1365   **15.** 1001

  In Problems 16–21, *find the prime factorization of the given number. If the number is prime, state so.*

  16. 24   **17.** 41   **18.** 82
  19. 91   **20.** 191   **21.** 148

  In Problems 22–25, *find the natural number whose prime factorization is given.*

  22. $2 \cdot 3^2 \cdot 5^2$ **23.** $2 \cdot 5 \cdot 7^2$
  24. $2 \cdot 3 \cdot 5 \cdot 11$ **25.** $2^4 \cdot 3 \cdot 5^2$

C. In Problems 26 and 27, *determine whether the given numbers are divisible by 2, 3, or 5.*

  26. a. 468   **b.** 580   **c.** 795   **d.** 3942
  27. a. 6345   **b.** 8280   **c.** 11,469,390

28. Mathematicians dream of finding a formula that will yield only primes and infinitely many of these when natural numbers are substituted into the formula. No such formula has ever been found, but there are formulas that give a large number of primes. One such formula is $n^2 - n + 41$. This formula gives primes for all natural numbers n less than 41. For example,

For n = 1, $1^2 - 1 + 41 = 41$, a prime.
For n = 2, $2^2 - 2 + 41 = 4 - 2 + 41 = 43$, a prime.
For n = 3, $3^2 - 3 + 41 = 9 - 3 + 41 = 47$, a prime.

a. What prime do you get for n = 4?
b. What prime do you get for n = 5?
c. What number do you get for n = 41?
d. Is the number obtained in part c prime?

29. There are many **conjectures** (unproved theories) regarding primes. One of these was transmitted to Euler in 1742 by C. Goldbach. Goldbach conjectured that every even natural number, except 2, could be written as the sum of two primes. Thus, $4 = 2 + 2$, $6 = 3 + 3$, $8 = 3 + 5$, $10 = 5 + 5$, and so on.
   a. Write 100 as the sum of two primes.
   b. Write 200 as the sum of two primes. [*Hint:* Look at the examples above.]
30. In 1931, the Russian mathematician Schnirelmann proved that every natural number can be written as the sum of not more than 300,000 primes. Another mathematician, I. M. Vinogradoff, has proved that every sufficiently large natural number can be expressed as the sum of at most four primes!
   a. Try to write 20 as the sum of three primes.
   b. Try to write 43 as the sum of four primes.

USING YOUR KNOWLEDGE 4.2

The validity of the divisibility by 3 rule can be shown by using some of the properties of the natural numbers. For example, consider the number 2853. Since $10 = 9 + 1$, $100 = 99 + 1$, and $1000 = 999 + 1$, we may write

$$2853 = 2(999 + 1) + 8(99 + 1) + 5(9 + 1) + 3$$
$$= 2 \cdot 999 + 2 + 8 \cdot 99 + 8 + 5 \cdot 9 + 5 + 3$$
$$= 2 \cdot 999 + 8 \cdot 99 + 5 \cdot 9 + (2 + 8 + 5 + 3)$$

(We have used the distributive property first and then the commutative and associative properties of addition.)

Now, as you can see, the first three terms of the last expression are all divisible by 3, because 999, 99, and 9 are all divisible by 3. Thus, the number is divisible by 3 if and only if the sum of the numbers in parentheses is divisible by 3. But the numbers in parentheses are exactly the digits 2, 8, 5, and 3 of the number 2853. Since the sum of these digits is 18, which is exactly divisible by 3, the number 2853 is also divisible by 3.

The reasoning we have used for 2853 applies to any natural number. Hence, the divisibility by 3 rule is valid.

1. If some of the digits of a number are 3's, 6's, or 9's, you can omit these in figuring the sum of the digits to check for divisibility by 3. For example, 2,963,396,607 is divisible by 3, because $2 + 7 = 9$ is divisible by 3. Explain this.
2. Is 5,376,906,391 divisible by 3?
3. The way in which we wrote

   $$2853 = 2(999 + 1) + 8(99 + 1) + 5(9 + 1) + 3$$

   shows that 2853 is divisible by 9 if and only if the sum of its digits is divisible by 9. Explain this.

Here are some other simple divisibility rules for the natural numbers:

DIVISIBLE BY	IF AND ONLY IF
4	The last two digits of the number form a number divisible by 4.
6	The number is divisible by both 2 and 3.
8	The last three digits of the number form a number divisible by 8.
9	The sum of the digits of the number is divisible by 9.
10	The number ends in 0.
12	The number is divisible by both 3 and 4.

The rule for divisibility by 8 may be obtained as follows: Consider the number 2,573,649,336. This can be written in the form 2,573,649 × 1000 + 336. Since 1000 is divisible by 8, the entire number is divisible by 8 if and only if the 336 is. Since 336 is divisible by 8, the given number is also divisible by 8. This reasoning applies to the natural numbers in general. Therefore, the rule is valid.

4. Which of the following numbers are divisible by 9? By 6?
   a. 405   b. 676   c. 7488   d. 309,907,452
5. Which of the following numbers are divisible by 4? By 8?
   a. 1436   b. 21,408   c. 347,712   d. 40,924
6. Which of the following numbers are divisible by 10? By 12?
   a. 4920   b. 943   c. 52,341,120   d. 60,210

DISCOVERY 4.2

*A natural number is said to be **perfect** if the number is the sum of its proper divisors. (The **proper divisors** of a number include all its divisors except the number itself.) For example, 6 is a perfect number because the proper divisors of 6 are 1, 2, and 3, and $1 + 2 + 3 = 6$.*

Some historians believe that the Pythagoreans were the first to define perfect numbers. In any event, it is certain that the Pythagoreans knew about these numbers and endowed them with mystical properties. In the ancient Greek numerology, 6 was regarded as the most beautiful of all numbers; it is not only a perfect number, but it is also the product of all its proper divisors: $6 = 1 \times 2 \times 3$.

1. Can you discover why 6 is the smallest perfect number? This fact may be what led St. Augustine in about the year 400 to assert, "God created all things in 6 days because 6 is a perfect number."
2. The next perfect number after 6 is a number between 25 and 30. Can you discover what number this is?
3. The third perfect number is 496. Can you find its proper divisors and so prove that 496 is a perfect number?
4. In about the year 800, Alcuin remarked that the whole human race descended from the 8 souls of Noah's Ark, and he regarded this as imperfect. Can you discover why?

When the sum of the proper divisors of a number is less than the number, the number is called **deficient.** For example, 4 is a deficient number because its proper divisors are 1 and 2, and $1 + 2 = 3$, which is less than 4. Similarly, 7 is a deficient number, because it has only 1 as a proper divisor.

5. If n is any prime number, can you discover whether n is a deficient number?
6. It is known that there are infinitely many prime numbers. Can you use this fact to prove that there are infinitely many deficient numbers?

▶ **Computer Corner 4.2**   *Do you want to see the Sieve of Eratosthenes being constructed by a computer before your very own eyes? If you type and run the program on page 744, the computer will show all the numbers up to 320 first, then eliminate those divisible by 2, those divisible by 3, and so on, until only the primes that are less than 320 are left on the screen. If you want a faster program, the prime searcher program on page 744 will list directly all the primes up to a desired number. We also have a program that will factor a number as a product of primes. This program appears on page 744.*

4.3

HOW PRIME NUMBERS ARE USED

A common fraction is simply an indicated division where only the horizontal bar of the division symbol is used. For example,

$$\frac{1}{2} = 1 \div 2, \quad \frac{3}{8} = 3 \div 8, \quad \frac{15}{35} = 15 \div 35, \quad \text{and so on}$$

A typical arithmetic problem is that of **reducing a fraction to lowest terms,** that is, dividing out all common factors (other than 1) of the numerator and denominator. Thus, to reduce the fraction $\frac{15}{35}$, we divide the numerator and denominator by 5 to obtain $\frac{3}{7}$, a fraction having no common factors (other than 1) in the numerator and denominator.

The New York Stock Exchange reports use the fractions ⅛, ¼, ⅜, ½, ⅝, ¾, and ⅞.

NYSE

TUESDAY COMPOSITE TRADING FOR THE 1,500 MOST ACTIVE NEW YORK STOCK EXCHANGE ISSUES.

52- High	week Low	Stock	Div	Yld.	PE	Sales 100s	High	Low	Last	Chg.
28¼	14¾	Alberto	.30	1.2	16	92	25	24⅜	24⅞	+ ⅛
24	12⅝	AlbCulA	.30	1.6	12	50	19¼	18½	18⅞	− ⅜
34	20¼	Albtsn s	.48	1.7	16	1492	28¼	26⅝	27⅝	+ ¾
37⅞	18	Alcan s	.45i	1.8	10	9537	24⅝	23⅞	24½	+ ½
30	15¼	AlcoS s	.68	3.2	12	137	21¾	21½	21½	...
32	15⅞	AlexAlx	1.00	5.1	14	354	19⅝	19⅜	19½	...
24¾	2⅜	AlgInt		...	...	1315	5¼	4⅝	4⅝	− ¾
20⅛	5¼	Algin pr		...	...	43	8½	7¾	7¾	−1
88½	20	Algl pfC		...	...	27	33	31	31	−2½
34	15⅛	AlgLud n	.20e	.9	10	159	22	21¾	21¾	− ¼
48¼	31⅜	AllgPw	3.00	7.4	10	363	40⅞	40½	40⅝	...
105⅞	53⅛	Allegis		...	12	11577	71	70¼	71	+ ½
19⅜	5½	AllenG		...	...	130	8¾	8⅝	8⅝	− ⅛
24⅞	9¼	Allen pf	1.75	13.3	...	33	13⅛	13	13⅛	...
49¼	26	AldSgnl	1.80	5.6	9	1544	32	30¾	31⅞	+ ½
10⅛	8⅞	AlstMu n	.30e	3.0	...	716	10⅛	10	10⅛	+ ⅛
34⅛	23	ALLTL s	1.52	5.3	9	140	28⅞	28¼	28¾	...
64¾	33¾	Alcoa	1.20	3.0	16	8981	40⅜	38¾	39½	+ ¾
32	14	AmxG n	.04e	.2	...	239	19¾	18⅞	19	− ⅛
29¼	12½	Amax		...	6	8146	16¼	15⅝	16⅛	+ ¼
41⅞	21½	AmHes	.45e	1.7	10	2327	26⅜	25¾	26⅜	+ ⅝
30⅜	9⅞	ABrck s	.05e	...	...	1096	18⅝	18⅜	18½	+ ½
60	36½	AmBrnd	2.20	4.8	10	5410	45⅞	44	45½	+ ⅞

A. Greatest Common Factor

The number 5 is the **greatest common factor** of 15 and 35. We abbreviate greatest common factor by GCF and write GCF(15, 35) = 5. A fraction

$$\frac{a}{b}$$

is said to be in **simplest form** if GCF(a, b) = 1. If GCF(a, b) $\neq$ 1, the fraction is reduced by dividing a and b by their GCF. The most commonly used way of finding the GCF of a set of numbers is as follows:

How to Find the GCF

1. Write the prime factorization of each number.
2. Select *all* the primes that occur in *all* the factorizations, applying to each such prime the smallest exponent to which it occurs.
3. The product of the factors selected in step 2 is the GCF of all the numbers.

(It will be easier for you to see if you write the same primes in a column.)

EXAMPLE 1   Find GCF(216, 234).

Solution   **Step 1.** We write the prime factorization of each number:

Pick the one with the smallest exponent in each column.

$$216 = 2^3 \cdot 3^3$$
$$234 = 2 \ \cdot 3^2 \cdot 13$$

Step 2. We select the *common* prime factors with their *smallest* exponents; these factors are 2 and 3^2.

Step 3. GCF(216, 234) $= 2 \cdot 3^2 = 18$.   ◀

EXAMPLE 2   Reduce the fraction $\frac{216}{234}$ to lowest terms.

Solution   We need to divide out GCF (216, 234), which we found in Example 1 to be 18. Thus, we write

$$\frac{216}{234} = \frac{18 \cdot 12}{18 \cdot 13} = \frac{12}{13}$$

(Notice that you can get the 12 and the 13 from the factorization in Example 1. Just use all the factors except those in the GCF.)   ◀

In many instances, the greatest common factor of two numbers is 1. For example, GCF(3, 7) $= 1$, because 3 and 7 are both prime. Also, GCF(3, 8) $= 1$. When the GCF of two numbers is 1, we say that the numbers are **relatively prime.** Thus, the final numerator and denominator in Example 2 (12 and 13) are relatively prime, as is the case for any fraction in simplest form.

B.   Least Common Multiple

A second use of prime factorization occurs in the addition of fractions, where we need to find a common denominator. For example, to add $\frac{3}{8}$ and $\frac{5}{12}$, we must first select a common denominator for the two fractions. Such a denominator is any natural number that is exactly divisible by both 8 and 12. Thus, we could use $8 \cdot 12 = 96$. However, it is usually most efficient to use the **least** (smallest) **common denominator** possible. In our case, we note that

$$8 = 2^3 \qquad \text{and} \qquad 12 = 2^2 \cdot 3.$$

To have a common multiple of 8 and 12, we must include 2^3 and 3, at the very least. Thus, we see that the LCM (least common multiple) of 8 and 12 is $2^3 \cdot 3 = 24$.

The procedure that we used to find LCM(8, 12) can be generalized to find the LCM of any two or more natural numbers.

> **How to Find the LCM**
>
> 1. Write the prime factorization of each number.
> 2. Select every prime that occurs, to the highest power to which it occurs, in these factorizations.
> 3. The product of the factors selected in step 2 is the LCM.

EXAMPLE 3   Find LCM(18, 21, 28).

Solution   **Step 1.**   Pick the one with the greatest
$$18 = 2 \cdot 3^2 \qquad \text{exponent.}$$
$$21 = \qquad 3 \cdot 7$$
$$28 = 2^2 \quad \cdot \quad 7$$

Step 2. We select every prime factor (not just the common factors) with the greatest exponent to which it occurs, obtaining 2^2, 3^2, and 7.

Step 3. LCM(18, 21, 28) = $2^2 \cdot 3^2 \cdot 7 = 252$.   ◀

EXAMPLE 4   Do the addition: $\dfrac{1}{18} + \dfrac{1}{21} + \dfrac{1}{28}$

Solution   We use the LCM found in Example 3 to replace the given fractions by equivalent fractions with a common denominator. To obtain these equivalents, we refer to the factored forms of the denominators and the LCM. Since the LCM is

$$252 = 2^2 \cdot 3^2 \cdot 7$$

and

$$18 = 2 \cdot 3^2$$

we have to multiply 18 by $2 \cdot 7 = 14$ to get 252. Hence, we multiply the fraction $\frac{1}{18}$ by 1 in the form $\frac{14}{14}$ to get

$$\frac{1 \times 14}{18 \times 14} = \frac{14}{252}$$

The same procedure can be used for the other two fractions.

$$\frac{1}{18} = \frac{14}{252} \qquad \frac{1}{21} = \frac{12}{252} \qquad \frac{1}{28} = \frac{9}{252}$$

Then we add:

$$\frac{1}{18} + \frac{1}{21} + \frac{1}{28} = \frac{14 + 12 + 9}{252} = \frac{35}{252}$$

Since $35 = 5 \times 7$ and $252 = 36 \times 7$, we can reduce the last fraction by dividing out the 7. This gives a final answer of $\frac{5}{36}$.   ◀

EXAMPLE 5   Do the subtraction: $\dfrac{7}{24} - \dfrac{5}{84}$

Solution   To do this subtraction, we have to change the fractions to equivalent fractions with a common denominator. We factor the denominators to get

$$24 = 2^3 \cdot 3$$
$$84 = 2^2 \cdot 3 \cdot 7$$

Thus, the LCM is $2^3 \cdot 3 \cdot 7 = 168$. Consequently, we multiply the first fraction by $\frac{7}{7}$ and the second by $\frac{2}{2}$ to get

$$\frac{7}{24} - \frac{5}{84} = \frac{7 \cdot 7}{24 \cdot 7} - \frac{5 \cdot 2}{84 \cdot 2}$$

$$= \frac{49 - 10}{168} = \frac{39}{168} = \frac{3 \cdot 13}{3 \cdot 56} = \frac{13}{56} \qquad \blacktriangleleft$$

C.   Applications

EXAMPLE 6   A motorist on a 2500-mile trip drove 500 miles the first day, 600 miles the second day, and 750 miles the third day. What (reduced) fraction of the total distance did he drive each of these days?

Solution   First day: $\dfrac{500}{2500} = \dfrac{1 \times 500}{5 \times 500} = \dfrac{1}{5}$

Second day: $\dfrac{600}{2500} = \dfrac{6 \times 100}{25 \times 100} = \dfrac{6}{25}$

Third day: $\dfrac{750}{2500} = \dfrac{3 \times 250}{10 \times 250} = \dfrac{3}{10}$ $\qquad \blacktriangleleft$

EXAMPLE 7   A cake recipe calls for $\frac{3}{4}$ cup of sugar for the cake itself and $\frac{1}{6}$ cup of sugar for the frosting. How much sugar is needed in all?

Solution   We need to add $\frac{3}{4}$ and $\frac{1}{6}$, which requires a common denominator. Since

$$4 = 2^2 \quad \text{and} \quad 6 = 2 \cdot 3$$

the LCM $(4, 6) = 2^2 \cdot 3 = 12$. To add the two fractions, we replace them by equivalent fractions with denominator 12:

$$\frac{3}{4} + \frac{1}{6} = \frac{3 \cdot 3}{4 \cdot 3} + \frac{1 \cdot 2}{6 \cdot 2}$$

$$= \frac{9}{12} + \frac{2}{12} = \frac{11}{12}$$

Thus, the amount of sugar needed is $\frac{11}{12}$ cup. $\qquad \blacktriangleleft$

A. In Problems 1–10, find the GCF of the given numbers. If two numbers are relatively prime, state so.

1. 14 and 210

2. 135 and 351

3. 315 and 350

4. 147 and 260

5. 368 and 80

6. 282 and 329

7. 12, 18, and 30

8. 12, 15, and 20

9. 285, 315, and 588

10. 100, 200, and 320

In Problems 11–18, reduce the given fractions to lowest terms.

11. $\dfrac{80}{92}$ **12.** $\dfrac{62}{88}$ **13.** $\dfrac{140}{280}$ **14.** $\dfrac{88}{96}$

15. $\dfrac{156}{728}$ **16.** $\dfrac{315}{420}$ **17.** $\dfrac{96}{384}$ **18.** $\dfrac{716}{4235}$

B. In Problems 19–28, find the LCM of the given numbers.

19. 15 and 55

20. 17 and 136

21. 32 and 124

22. 124 and 155

23. 180 and 240

24. 284 and 568

25. 12, 18, and 30

26. 12, 15, and 20

27. 285, 315, and 588

28. 100, 200, and 320

In Problems 29–34, do the indicated operations.

29. $\dfrac{1}{7} + \dfrac{1}{8}$ **30.** $\dfrac{1}{4} + \dfrac{5}{18}$ **31.** $\dfrac{11}{12} - \dfrac{7}{18}$

32. $\dfrac{7}{8} - \dfrac{15}{32}$ **33.** $\dfrac{1}{2} + \dfrac{2}{3} - \dfrac{5}{12}$ **34.** $\dfrac{1}{3} + \dfrac{1}{10} - \dfrac{1}{15}$

C. **35.** A recipe for *flan* (an egg custard) calls for $2\frac{1}{2}$ cups of sugar for the mixture itself and $\frac{1}{8}$ cup of sugar for the caramel. How much sugar is needed for this recipe?

36. A Simplicity pattern calls for $\frac{3}{4}$ yd of material for the dress and $\frac{1}{8}$ yd of the same material for the collar. How much material is needed?

37. The baseboards must be installed in a room 10 ft by $9\frac{1}{4}$ ft. If the door uses $3\frac{1}{8}$ ft of space that does not require baseboard, how much baseboard is needed?

38. My house is $36\frac{1}{8}$ ft long. If I increase the length by adding another room of dimensions $10\frac{1}{4}$ ft by $10\frac{1}{4}$ ft, how long will the remodeled house be?

39. My lot is $100\frac{3}{4}$ ft wide. If the city takes an easement $16\frac{1}{8}$ ft wide off one side of the lot, what is the width of the remaining lot?

40. A cyclist traveled only $\frac{1}{6}$ of his route yesterday because of inclement weather. He decides to go $\frac{3}{8}$ of the route today and $\frac{2}{7}$ tomorrow. How much more will he have to go to finish his route the day after tomorrow?

41. A survey by the National Opinion Research Center indicated that $\frac{3}{100}$ of the workers were dissatisfied with their jobs, $\frac{1}{100}$ were a little dissatisfied, and $\frac{9}{25}$ were moderately satisfied. The rest were very satisfied. What fraction of the workers is that?

42. In a recent year, $410 billion was spent on health care in the U.S.A. Of this amount, $\frac{1}{10}$ went for drugs, eyeglasses, and necessities; $\frac{1}{4}$ for professional services; $\frac{2}{25}$ for nursing homes; $\frac{3}{20}$ for miscellaneous; and the rest for hospital costs. What fraction of the money went for hospital costs?

43. A car depreciates an average of $\frac{1}{4}$ of its value the first year, $\frac{3}{20}$ the second, and $\frac{1}{10}$ the third. What fraction of its original value is the car worth after the third year?

44. A book store owner spent $\frac{2}{5}$ of her initial investment on bookcases, $\frac{1}{10}$ on office machines, and the rest on books. What portion of her initial investment was spent on books?

45. The operating ratios for hardware stores in the United States suggest that $\frac{3}{5}$ of their sales go to pay for merchandise, $\frac{1}{4}$ for expenses, and the rest is profit. What portion of sales is profit?

USING YOUR KNOWLEDGE 4.3

There is a second method for finding the GCF and the LCM of given numbers. This method depends on successive division by primes.

> ### To Find the GCF (Second Method)
>
> **1.** Write the numbers in a horizontal line and divide each number by a prime divisor that is common to all the numbers. Start with the smallest such divisor, that is, 2 or 3 or 5 or 7, and so on.
> **2.** Repeat the procedure with the quotients until there is no longer any common divisor.
> **3.** The product of all the divisors used in steps 1 and 2 is the GCF of the given numbers.

For example, to find the GCF of 252, 420, and 588.

1. *We write the 252, 420, and 588 in a horizontal line, and, since these are all even numbers, divide by 2.*

2. *The quotients are all even numbers, so we divide by 2 again. The next quotients are all divisible by 3 and the resulting quotients all can be divided by 7. The final quotients have no common prime factor, so we stop.*

3. *The required GCF is the product of all the divisors, that is, $2 \cdot 2 \cdot 3 \cdot 7 = 84$. (See the arrow in the diagram.)*

2	252	420	588
2	126	210	294
3	63	105	147
7	21	35	49
	3	5	7

$$\text{GCF} = 2 \cdot 2 \cdot 3 \cdot 7$$

This method produces the GCF because it finds all the common prime divisors, each to the highest power possible.

1. Rework Problems 1–10 in Exercise 4.3 by the division method.

To Find the LCM (Second Method)

1. Write the numbers in a horizontal line and divide by a prime factor common to two or more of the numbers. If any of the other numbers is not divisible by this prime, simply circle this number and carry it down to the next line.

2. Repeat step 1 with the quotients and carry-downs until no two numbers have a common prime divisor.

3. The LCM is the product of all the divisors from steps 1 and 2 and the numbers in the final row.

This method gives all the prime factors common to two or more of the given numbers, as well as the final quotients when the numbers are divided by these prime factors. Consequently, the product described in step 3 is the required LCM.

The following diagram shows the method used to find LCM (15, 21, 28):

The LCM is the product of all the divisors and the numbers in the final row (see the arrow). Thus, LCM(15, 21, 28) = 420.

2. Rework Problems 19–28 in Exercise 4.3 by the division method.

▶ **Calculator Corner 4.3** The GCF of two numbers can be found by another method that is attributed to Euclid (about 300 B.C.) and is called ***Euclid's algorithm.*** This method is based on repeated divisions, and is particularly useful for large numbers. To see why the method works, we shall use it to find GCF(54, 42). We do successive divisions as follows, stopping as soon as we get a 0 remainder:

Step 1. Divide 54 by 42 and list the quotient and the remainder:

$$\begin{array}{r} 1 \\ 42\overline{)54} \\ \underline{42} \\ 12 \end{array} \qquad 54 = (1 \times 42) + 12$$

Step 2. Divide the preceding divisor (42) by the remainder (12):

$$\begin{array}{r} 3 \\ 12\overline{)42} \\ \underline{36} \\ 6 \end{array} \qquad 42 = (3 \times 12) + 6$$

Step 3. Divide the preceding divisor (12) by the remainder (6):

$$6\overline{)12} \quad \begin{array}{r} 2 \\ \hline \end{array}$$

$$\begin{array}{r} 12 \\ \hline 0 \end{array} \qquad 12 = (2 \times 6) + 0$$

Stop, because the remainder is 0. The last divisor, 6, is the GCF.

To see why this works, look at the equations we wrote, but in reverse order:

$12 = (2 \times 6) + 0$ shows that 6 divides 12.

$42 = (3 \times 12) + 6$ shows that 6 divides the right-hand side, so must divide 42.

$54 = (1 \times 42) + 12$ shows that 6 divides the right-hand side, so must divide 54.

Thus, 6 is a common divisor of 42 and 54; that it is the **greatest** common divisor follows from the fact that we did not get a 0 remainder until the last step.

A calculator is a useful tool for performing the divisions required by Euclid's algorithm, but we first need an easy way to get the remainders on the calculator. Actually, the calculator gives the quotient with a decimal part if the remainder is not 0. If we multiply this decimal part by the divisor, we get the remainder. For example, if you divide 54 by 42 on your calculator, it shows the quotient 1.2857143 (the number of decimal places will depend on the calculator). If we subtract the whole number part (1 in this case), we get the decimal we want to multiply by the divisor 42. Doing this multiplication on the calculator gives 12, the remainder we found in our long-hand division.

Now let us find GCF(259, 888) on the calculator. Here is what you must key in:

| 8 | 8 | 8 | ÷ | 2 | 5 | 9 | = | *Calculator shows 3.4285714.*

| − | 3 | = | × | 2 | 5 | 9 | = | *Calculator shows 111, the first remainder.*

| 2 | 5 | 9 | ÷ | 1 | 1 | 1 | = | *Calculator shows 2.3333333.*

| − | 2 | = | × | 1 | 1 | 1 | = | *Calculator shows 37, the second remainder.*

| 1 | 1 | 1 | ÷ | 3 | 7 | = | *Calculator shows 3, no remainder.*

The GCF is 37.

Use your calculator to find the GCF of each of the following pairs of numbers:

1. 154 and 286   **2.** 255 and 323

3. 2244 and 2210   **4.** 2002 and 3276

5. 50,625 and 1062   **6.** 48,314 and 41,055

► **Computer Corner 4.3** A process similar to the one discussed in Calculator Corner 4.3 can be used to find the GCF of two numbers with a computer. The program appears on page 745. We also have a program that will reduce fractions (page 745) and another one that finds the LCM of two numbers (page 745).

4.4

THE WHOLE NUMBERS AND THE INTEGERS

The set of natural numbers N and the operations of addition, subtraction, multiplication, and division form a **mathematical system.** We have already discussed many properties of this system, but there is one more property that leads to some interesting and important ideas.

A. Whole Number Properties

© 1973 King Features Syndicate, Inc. World rights reserved.

As you can see from the cartoon, the number 1 has the unique property that multiplication of any natural number a by 1 gives the number a again. Because the **identity** of the number a is preserved under multiplication by 1, the number 1 is the **multiplicative identity,** and this property is the **identity property for multiplication.**

Identity Property for Multiplication

If a is any natural number, then

$$a \cdot 1 = 1 \cdot a = a$$

It can be shown (Problem 49, Exercise 4.4) that 1 is the only element with this property.

Is there an **additive identity?** That is, is there an element z such that $a + z = a$ for every a in N? There is no such element in N. This lack of an additive identity spoils the usefulness of N in many everyday applications of arithmetic. The Babylonians realized this difficulty and simply used a space between digits as a zero place-holder, but it was not until about A.D. 1400 that the Hindu–Arabic system popularized the idea of zero. The

Hindus used the word *sunya*, (meaning "void"), which was later adopted by the Arabs as *sifr*, or "vacant." This word passed into Latin as *zephirum* and became, over the years, *zero*.

The number zero (0) provides us with an **identity for addition.** Therefore, we enlarge the set N by adjoining to it this new element. The set consisting of all the natural numbers and the number 0, that is, the set {0, 1, 2, 3, . . .}, is called the set of **whole numbers** and is denoted by the letter W. It is an important basic assumption that the set W obeys the same fundamental laws of arithmetic as do the natural numbers. Moreover, adding 0 to a whole number does not change the identity of the whole number. Thus, we have the following:

Identity Property for Addition

If a is any whole number, then

$$0 + a = a \quad \text{and} \quad a + 0 = a$$

As with the number 1, we can prove (Problem 50, Exercise 4.4) that the number 0 is unique. It also can be shown that if a is any whole number, then

$$a - 0 = a, \quad 0 \cdot a = 0, \quad \text{and} \quad 0 \div a = \frac{0}{a} = 0$$

Note that $a \div 0$ is **not** defined (Problems 51–53, Exercise 4.4).

The importance of 0 is also evident in algebra. When solving an equation such as

$$(x - 1)(x - 2) = 0$$

we argue that either $x - 1 = 0$ or else $x - 2 = 0$, and conclude that $x = 1$ or $x = 2$. This argument is based on the following theorem:

▶ **Theorem 4.4a** | If $a \cdot b = 0$, then $a = 0$ or $b = 0$.

Proof   Assume that $a \neq 0$. Divide both sides by a:

$$\frac{a \cdot b}{a} = \frac{0}{a} = 0$$

Since

$$\frac{a \cdot b}{a} = b$$

then $b = 0$.

A similar argument shows that if $b \neq 0$, then $a = 0$. Of course, if a and b are both 0, the theorem is obviously true. ◀

B. The Set of Integers

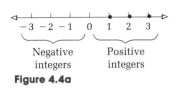

Readings above 0 are positive.
Readings below 0 are negative.

The set W is extremely useful when the idea of "How many?" is involved. However, the whole numbers are inadequate even for some simple everyday problems. For instance, the below-zero temperature on a winter's day cannot be described by a whole number, and the simple equation $x + 3 = 0$ has no solution in W. Because similar problems occur repeatedly in the applications of mathematics, the set of whole numbers is extended to include the **negative numbers,** $-1, -2, -3, \ldots$. This new set, called the set of **integers,** is denoted by the letter I, so that $I = \{\ldots, -3, -2, -1, 0, 1, 2, 3, \ldots\}$. Note that the set I consists of three subsets:

1. The **positive integers:** $1, 2, 3, \ldots$
2. The number **0**
3. The **negative integers:** $-1, -2, -3, \ldots$

We make the important assumption that the set I obeys the same basic laws of arithmetic as do the natural numbers. However, some of the properties of the set I must be examined more closely. First, you must realize that with the set I, an important concept has been added to the idea of a number, the concept of *direction*.

A **number line** offers a good representation for this concept. Thus, we draw a horizontal straight line, choose any point on it, and label that point 0. We then measure successive equal intervals to the right and to the left of 0 (see Fig. 4.4a). The endpoints of the intervals to the right of 0 are labeled with the positive integers in order $1, 2, 3, \ldots$. Those to the left of 0 are labeled with the negative integers in order $-1, -2, -3, \ldots$. The resulting picture is called a number line. Note that the number line continues indefinitely in the positive (right) and negative (left) directions. We **graph** the integers 1, 2, and 3 by adding dots to the number line, as shown in Fig. 4.4a.

We can give a physical interpretation of addition using the number line. For example, $3 + 2$ may be regarded as a move of 3 units from 0 in the positive direction (right) followed by a move of 2 units more in the positive direction (right). The terminal point is 5. Similarly, $3 + (-5)$ may be interpreted as a move of 3 units in the positive direction followed by a move of 5 units in the negative direction (left). The terminal point is -2 (see Fig. 4.4b). Thus,

$$3 + (-5) = -2$$

Using the same idea,

$$3 + (-3) = 0$$

Note that the numbers 3 and -3 are the **same distance** from 0 on the number line, but in **opposite directions.** Thus, 3 and -3 are **opposites.**

Negative   Positive
integers   integers

Figure 4.4a

Figure 4.4b

Similarly, 5 and −5, and −12 and 12 are opposites. This fact can be easily verified, since $5 + (−5) = 0$ and $(−12) + 12 = 0$. Because of this, 5 and −5, and −12 and 12 are also known as **additive inverses.** This discussion can be summarized as follows:

Additive Inverse Property

If n is any integer, then there exists an integer $−n$ such that

$$n + (−n) = 0 \qquad \text{and} \qquad (−n) + n = 0$$

n and −n are said to be **additive inverses (opposites)** of each other.

It can be proved (see Problem 54, Exercise 4.4) that each integer has exactly one additive inverse.

The operation of subtraction can also be represented on the number line. For instance, $3 − 5$ may be thought of as a move of 3 units from 0 in the positive (right) direction, followed by a move of 5 units in the negative (left) direction. The terminal point is −2. Thus, we have $3 − 5 = −2$. Notice that this is the same result we found for $3 + (−5)$ (see Fig. 4.4b); that is,

$$3 − 5 = 3 + (−5)$$

The minus sign is used in two different ways in this last equation: On the left side of the equal sign, it means to subtract 5 from 3; on the right, it means that −5 is the integer 5 units in the negative direction from 0.

EXAMPLE 1   Illustrate the subtraction $−3 − (−4)$ on the number line.

Solution   This subtraction may be considered as a move of 3 units to the left of 0, followed by a move of 4 units to the right (in the opposite direction from that indicated by the −4). The terminal point is 1 unit to the right of 0, as indicated in the figure, so we have $−3 − (−4) = 1$. Notice that this is the same result as $−3 + 4 = 1$.   ◀

This discussion should convince you that subtraction may be viewed as the opposite of addition, and any subtraction problem can be transformed into an addition problem, as shown next.

Subtraction Problem	Equivalent Addition Problem
$3 − 1$	$3 + (−1)$
$10 − 7$	$10 + (−7)$
$−3 − 5$	$−3 + (−5)$
$2 − (−3)$	$2 + 3$

This motivates the following definition:

Definition of Subtraction

If a and b are any integers, the **subtraction** of b from a is defined as

$$a - b = a + (-b)$$

This means that subtracting b from a is the same as adding the inverse (opposite) of b to a.

EXAMPLE 2   Change the subtraction problem $4 - 3$ to an equivalent addition problem and illustrate on the number line.

Solution   By the above definition, $4 - 3 = 4 + (-3)$. On the number line, we think of this as a move of 4 units to the right, followed by a move of 3 units to the left. The result, as shown in the figure, is 1. ◀

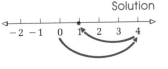

We assume that the reader is familiar with the operations of addition and subtraction of integers. However, the rules for the multiplication of integers usually have been memorized without any attempt to see or understand their development. We now give these results and prove one of them.

1. The product of two positive integers is positive.
2. The product of a positive integer and a negative integer is negative.
3. The product of two negative integers is positive.

Table 4.4a summarizes these results. Note that the product of integers with *like* signs is *positive*; with *unlike* signs the product is *negative*.

Table 4.4a

RULE OF SIGNS IN MULTIPLICATION	EXAMPLE
1. Positive × Positive = Positive	$4 \times 3 = 12$
2. Positive × Negative = Negative	$4 \times (-3) = -12$
3. Negative × Negative = Positive	$(-4) \times (-3) = 12$

The next theorem illustrates the way in which we can prove the rule of signs.

Theorem 4.4b

If a and b are positive, then

$$\underset{\text{Positive}}{(a)} \quad \times \quad \underset{\text{Negative}}{(-b)} \quad = \quad \underset{\text{Negative}}{-(ab)}$$

Before proceeding with the proof, we note that in Theorem 4.4b, a is a positive number, $-b$ is negative (because b is positive), $a \cdot b$ is positive (by property 1), and hence, $-(ab)$ is negative. Thus, Theorem 4.4b states that a positive number (a) times a negative number $(-b)$ yields a negative number $-(ab)$.

Proof   **i.** $b + (-b) = 0$ Definition of inverse

        ii. $a[b + (-b)] = a \cdot 0 = 0$ Multiplying both sides of the equation by a and recalling that $a \cdot 0 = 0$

       iii. $a \cdot b + a(-b) = 0$ Distributive law

       iv. By step iii, the additive inverse of $a \cdot b$ is $a(-b)$, but because $a \cdot b$ is positive, its additive inverse is also $-(ab)$.

       v. However, the additive inverse of a number is unique, so we conclude that $a(-b) = -(ab)$.   ◄

We now summarize the properties of the integers under the operations of addition and multiplication.

Properties of Addition

1. The integers are closed with respect to addition. (The sum of any two integers a and b is always an integer.)
2. The integers are commutative with respect to addition. (For any two integers a and b, $a + b = b + a$.)
3. The integers are associative with respect to addition. [For any three integers a, b, and c, $a + (b + c) = (a + b) + c$.]
4. The integers have a unique identity (the number 0) for addition. ($a + 0 = a$ and $0 + a = a$.)
5. Each integer has a unique additive inverse. [$a + (-a) = 0$.]

Properties of Multiplication

1. The integers are closed with respect to multiplication. (The product of two integers a and b is always an integer.)
2. The integers are commutative with respect to multiplication. (For any two integers a and b, $a \cdot b = b \cdot a$.)
3. The integers are associative with respect to multiplication. [For any three integers a, b, and c, $a \cdot (b \cdot c) = (a \cdot b) \cdot c$.]
4. The integers have a unique identity (the number 1) for multiplication. ($1 \cdot a = a$ and $a \cdot 1 = a$.)
5. Not all the integers have multiplicative inverses. For example, the multiplicative inverse for 3 would be a number b such that $3 \cdot b = 1$. But there is no such integer as b.

Besides these properties, the integers have the distributive property of multiplication over addition. Thus,

$$a(b + c) = ab + ac \qquad \text{and} \qquad (a + b)c = ac + bc$$

C.   Order of Operations

Suppose a calculation involves several operations, as in the evaluation of

You can remember the order of operations if you remember

Please
Excuse
My
Dear
Aunt
Sally

$$7 \times 8 \div 4 \times 10^2 - 3(-5 + 2) \times 10^3$$

In such calculations, we always use the following order of operations:

1. Evaluate quantities inside **p**arentheses.
2. Do all **e**xponentiations.
3. Do all **m**ultiplications and **d**ivisions in order from left to right.
4. Do all **a**dditions and **s**ubtractions in order from left to right.

EXAMPLE 3 Evaluate $7 \times 8 \div 4 \times 10^2 - 3(-5 + 2) \times 10^3$.

Solution $7 \times 8 \div 4 \times 10^2 - 3(-5 + 2) \times 10^3$

Step 1. $= 7 \times 8 \div 4 \times 10^2 - 3(-3) \times 10^3$
Step 2. $= 7 \times 8 \div 4 \times 100 - 3(-3) \times 1000$
Step 3. $= 56 \div 4 \times 100 - (-9) \times 1000$
$= 14 \times 100 - (-9000)$
$= 1400 - (-9000)$
Step 4. $= 1400 + 9000 = 10,400$ ◀

EXERCISE 4.4

A. *In Problems 1–4, graph the indicated set of numbers.*

1. The integers between 4 and 9, inclusive
2. The integers between -5 and 6, inclusive
3. The whole numbers between -2 and 5, inclusive
4. The natural (counting) numbers between -2 and 4, inclusive

In Problems 5–8, find the additive inverse of the given number.

5. 3 **6.** 47 **7.** -8 **8.** -1492

B. *In Problems 9–16, illustrate on the number line the indicated operation.*

9. $4 - 5$ **10.** $-3 - 2$ **11.** $4 - 2$ **12.** $3 + 4$
13. $-5 + 4$ **14.** $4 - 4$ **15.** $3 + (-5)$ **16.** $5 + (-2)$

In Problems 17–28, write each problem as an equivalent addition problem and find the answer.

17. $3 - 8$ **18.** $8 - 3$ **19.** $3 - 4$ **20.** $-3 - 4$
21. $-5 - 2$ **22.** $-3 - 5$ **23.** $5 - (-6)$ **24.** $6 - (-3)$
25. $-3 - (-4)$ **26.** $-5 - (-6)$ **27.** $-5 - (-3)$ **28.** $-10 - (-5)$

In Problems 29–36, find each product.

29. $(-5) \times 3$ **30.** $(-8) \times 9$ **31.** $(-3) \times (-4)$ **32.** $(-9) \times (-3)$
33. $4 \times (-5)$ **34.** $3 \times (-13)$ **35.** $0 \times (-4)$ **36.** $(-0) \times 0$

In Problems 37–46, perform the indicated operations.

37. a. $-3(4 + 5)$ **b.** $-4(4 - 5)$
38. a. $-2(-3 + 1)$ **b.** $-5(-4 + 2)$
39. a. $-5 + (-5 + 1)$ **b.** $-8 + (-2 + 5)$
40. a. $-2(4 - 8) - 9$ **b.** $-3(5 - 7) - 11$
41. $(-2 - 4)(-3) - 8(5 - 4)$
42. $(-3 - 5)(-2) + 8(3 + 4 - 5)$
43. $6 \times 2 \div 3 + 6 \div 2 \times (-3)$
44. $8 \div 2 \times 4 - 8 \times 2 \div 4$
45. $4 \times 9 \div 3 \times 10^3 - 2 \times 10^2$
46. $5 \times (-2) \times 3^2 + 6 \div 3 \times 5 \times 3^2$

Courtesy of PSH

47. The highest point on Earth is Mount Everest, 9 km above sea level. The lowest point is the Marianas Trench, 11 km below sea level. Use signed numbers to find the difference in altitude between these two points.

48. Julius Caesar died in 44 B.C. Nero is said to have set Rome afire in A.D. 64. Use signed numbers to find how many years elapsed between Caesar's death and the burning of Rome.

49. Supply reasons for steps 1, 3, and 4 in the following proof by contradiction that 1 is the unique multiplicative identity: Suppose there is another multiplicative identity, say m, possibly different from 1. Then for every element a of N:

Step 1. $m \cdot a = a$
Step 2. $m \cdot 1 = 1$ Substituting 1 for a
Step 3. $m \cdot 1 = m$
Step 4. Therefore, $m = 1$.

50. Fill in the blanks in the following proof that 0 is the unique additive identity: Assume that there are two additive identity elements, say 0 and z. Then:
   a. $0 + z = z$, because 0 is an additive _____.
   b. $0 + z = 0$, because z is an additive _____.
   c. Therefore, $z = 0$, because they are both equal to _____.
   This shows that the assumed two identity elements are the same; that is, 0 is the unique additive identity.

51. Supply reasons for steps 1, 3, and 5 in the following proof that $a - 0 = a$:

Step 1. $a - 0 = a + (-0)$
Step 2. $\quad = [a + (-0)] + 0$ Because 0 is the additive identity
Step 3. $\quad = a + [(-0) + 0]$
Step 4. $\quad = a + 0$ Because 0 and -0 are additive inverses of each other
Step 5. $\quad = a$

52. Fill in the blanks in the following proof that $0 \cdot a = 0$:

    a. $0 \cdot a = [1 + (-1)] \cdot a$ Because 1 and -1 are _____ inverses

    b. $= 1 \cdot a + (-1) \cdot a$ By the _____ property

    c. $= a + (-a)$ Multiplying and using the rule of signs

    d. $= 0$ Because a and $-a$ are _____

53. Fill in the blanks in the following proof that for $a \neq 0$, $\dfrac{0}{a} = 0$:

    a. Let the quotient $\dfrac{0}{a} = q$. Then by the definition of division, $q \cdot a = 0$.

    b. $q \cdot a + a = 0 + a$ by _____ to both sides.

    c. $q \cdot a + a = a$, because 0 is the additive _____.

    d. $q \cdot 1 + 1 = 1$ by dividing both sides by a.

    e. Therefore, $q + 1 = 1$, because $q \cdot 1 =$ _____.

    f. Consequently, $q = 0$, because the additive _____ is _____.

54. Fill in the blanks in the following proof that each integer has only one additive inverse, $-n$: Suppose that the integer n has another such inverse, say x. Then because $n + (-n) = 0$,

    a. $x = x + [n + (-n)]$ 0 is the _____ identity.

    b. $= (x + n) + (-n)$ _____ law of addition

    c. $= (n + x) + (-n)$ _____ law of addition

    d. $= 0 + (-n)$ x was assumed to be an additive inverse of n.

    e. $= -n$ 0 is the _____.

55. Supply reasons for steps 1, 3, and 5 in the following alternative proof that $a \cdot 0 = 0$:

    Step 1. $0 + 0 = 0$

    Step 2. $a \cdot (0 + 0) = a \cdot 0$ Multiplying both sides by a

    Step 3. The left side,
         $a \cdot (0 + 0) = a \cdot 0 + a \cdot 0$

    Step 4. Therefore,
         $a \cdot 0 + a \cdot 0 = a \cdot 0$ Substituting from step 3 into step 2

    Step 5. Hence, $a \cdot 0 = 0$.

56. Supply reasons for steps 1, 3, 4, and 6 in the following proof that $(-a)(-b) = ab$:

    Step 1. $a + (-a) = 0$

    Step 2. $[a + (-a)](-b) = 0$ Multiplying both sides by $(-b)$ and using $0 \cdot (-b) = 0$

    Step 3. $(a)(-b) + (-a)(-b) = 0$

    Step 4. $(a)(-b) = -ab$

    Step 5. $-ab + (-a)(-b) = 0$ Substituting from step 4 into step 3

    Step 6. Hence, $(-a)(-b)$ is the additive inverse of $-ab$.

    Step 7. Therefore, $(-a)(-b) = ab$.   The additive inverse of a number is unique.

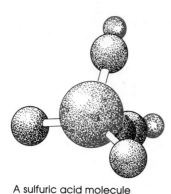

A sulfuric acid molecule

The oxidation number (or valence) of a molecule is found by using the oxidation numbers of the atoms present in the molecule. For example, the oxidation number of hydrogen (H) is $+1$, that of sulfur (S) is $+6$, and that of oxygen (O) is -2. Thus, we can get the oxidation number of sulfuric acid (H_2SO_4) as follows:

$$H_2SO_4$$
$$2(+1) + (+6) + 4(-2) = 2 + 6 - 8 = 0$$

Use this idea to find the oxidation number of:

1. Phosphate, PO_4, if the oxidation number of phosphorus (P) is $+5$ and that of oxygen (O) is -2
2. Sodium dichromate, $Na_2Cr_2O_7$, if the oxidation number of sodium (Na) is $+1$, that of chromium (Cr) is $+6$, and that of oxygen (O) is -2
3. Baking soda, $NaHCO_3$, if the oxidation number of sodium (Na) is $+1$, that of hydrogen (H) is $+1$, that of carbon (C) is $+4$, and that of oxygen (O) is -2

Nomographs are graphs that can be drawn to perform various numerical operations. For example, the nomograph in the figure can be used to do addition and subtraction. To add any two numbers, we locate one of the numbers on the lower scale and the other number on the upper scale. We then connect these by a straight line. The point where the line crosses the middle scale gives the result of the addition. The figure shows the sum $-2 + 4 = 2$.

An addition nomograph

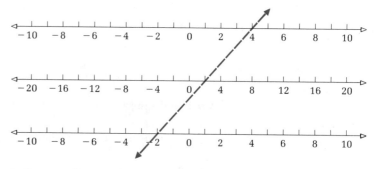

1. Can you discover how to subtract on a nomograph?

▶ **Calculator Corner 4.4**   Most scientific calculators perform operations in **algebraic order,** that is, they perform multiplications and divisions first, followed by additions and subtractions. Thus, if you enter ⊡3⊡+⊡5⊡×⊡2⊡=⊡ in an algebraic calculator, the answer is 13. (If you do not get this answer, you have to multiply 5×2 first and then add 3 to the result.) Of course, if you have an expression such as $-3(-5 + 2)$, remember that the parentheses represent an indicated multiplication. Thus, you must enter this expression as

Practice this concept by evaluating the expression from Example 3 on your calculator:

$$7 \times 8 \div 4 \times 10^2 - 3(-5 + 2) \times 10^3$$

(Recall that $10^2 = 10 \times 10$ and $10^3 = 10 \times 10 \times 10$.)

1. Use your calculator to check the answers to Problems 41, 43, and 45 in Exercise 4.4.

4.5

THE RATIONAL NUMBERS

In Section 4.4 we extended the set of natural numbers and obtained the set of integers. As we saw, in studying multiplication of integers, not all the integers have multiplicative inverses; for instance, there is no integer b such that $3 \cdot b = 1$. Just as the lack of an additive identity and additive inverses impairs the usefulness of the set of natural numbers, so does the lack of multiplicative inverses impair the usefulness of the set of integers.

A. Rational Numbers

The difficulty lies in the fact that the set of integers is not closed under division; division of two integers can produce fractions such as $\frac{3}{4}, \frac{5}{7}, \frac{8}{3}$, and so on. Thus, we extend the system of integers to include such fractions and call the resulting system the set of **rational numbers.** The set of rational numbers is symbolized by the letter Q. (Note that a common fraction is often called a *ratio*; hence the name *rational numbers.*)

Definition 4.5a ▶

> A **rational number** is a number that can be put in the form
>
> $$\frac{a}{b}$$
>
> where a and b are integers and $b \neq 0$

Here are some important facts about rational numbers:

1. Every integer is a rational number, because the integers can be written in the form $\ldots, -\frac{2}{1}, -\frac{1}{1}, \frac{0}{1}, \frac{1}{1}, \frac{2}{1}, \ldots$.

Thou shalt not divide by zero

Cartoon by Paul Kicklighter

2. The integer a in Definition 4.5a is called the **numerator,** and the integer b is called the **denominator.**

3. The restriction $b \neq 0$ is a necessary one. By the definition of division, if $a/b = c$, then $a = bc$. So, if $b = 0$, this means that $a = 0$. Thus, if $a \neq 0$, then the attempted division by 0 leads to a contradiction. If $a = 0$, then the equation $0 = 0 \cdot c$ is true for every number c; that is, the quotient c is not uniquely defined. The only way to avoid this dilemma is to **forbid division by 0.**

4. Definition 4.5a includes the words "a number that *can* be put in the form a/b, where a and b are integers and $b \neq 0$." Why couldn't we say "a number *of* the form a/b, where a and b are integers and $b \neq 0$"? To answer this question, consider the number $0.333 \ldots$. This number is *not* of the form a/b, but because $0.333 \ldots = \frac{1}{3}$, it *can* be put in the form a/b, where $a = 1$ and $b = 3$. Similarly, $1\frac{3}{4}$ is not of the form a/b. But because $1\frac{3}{4} = \frac{7}{4}$, the number *can* be put in the form a/b, where $a = 7$ and $b = 4$. From this discussion we conclude that $0.333 \ldots$ and $1\frac{3}{4}$ are rational numbers.

How do we recognize the fact that $\frac{1}{3}$ and $\frac{3}{9}$ represent the same rational number? To do this we need to define equality of rational numbers.

Definition 4.5b

$$\frac{a}{b} = \frac{c}{d} \qquad \text{if and only if} \qquad ad = bc$$

Thus, $\frac{1}{3} = \frac{3}{9}$ because $1 \cdot 9 = 3 \cdot 3$, and $\frac{1}{8} = \frac{4}{32}$ because $1 \cdot 32 = 8 \cdot 4$. Using Definition 4.5b, we can prove the following useful result:

Theorem 4.5a

$$\frac{a}{b} = \frac{ak}{bk} \qquad k \neq 0$$

Proof

$$\frac{a}{b} = \frac{ak}{bk} \qquad \text{if and only if} \qquad a(bk) = b(ak) \quad \text{By Definition 4.5b}$$

But this equality is true, because we can use the associative and commutative laws of multiplication for the integers to show that $a(bk) = b(ak)$. ◄

Theorem 4.5a enables us to reduce $\frac{4}{6}$ by writing $\frac{4}{6} = (2 \cdot 2)/(3 \cdot 2)$ and "canceling" the 2's to obtain $\frac{4}{6} = \frac{2}{3}$. The theorem also assures us that if we multiply the numerator and denominator of a fraction by the same nonzero number k, the fraction is unchanged in value. Thus,

$$\frac{1}{2} = \frac{1 \cdot 2}{2 \cdot 2} = \frac{2}{4} \qquad \text{and} \qquad \frac{1}{8} = \frac{1 \cdot 3}{8 \cdot 3} = \frac{3}{24}$$

EXAMPLE 1   Use Theorem 4.5a to reduce $\frac{10}{30}$.

Solution $\qquad \dfrac{10}{30} = \dfrac{1 \cdot 10}{3 \cdot 10} = \dfrac{1}{3}$ ◀

EXAMPLE 2 $\quad$ Using Theorem 4.5a, find a rational number with a denominator of 12 and equal to $\frac{5}{6}$.

Solution $\qquad \dfrac{5}{6} = \dfrac{5 \cdot 2}{6 \cdot 2} = \dfrac{10}{12}$ ◀

B. Operations with Rational Numbers

We now define the operations of addition, subtraction, multiplication, and division of rational numbers.

▶ **Definition 4.5c**

> The **product** of two rational numbers a/b and c/d is defined by
>
> $$\frac{a}{b} \cdot \frac{c}{d} = \frac{ac}{bd}$$

Thus,

$$\frac{2}{7} \cdot \frac{3}{5} = \frac{2 \cdot 3}{7 \cdot 5} = \frac{6}{35}$$

▶ **Definition 4.5d**

> The **sum** of two rational numbers a/b and c/d is defined as
>
> $$\frac{a}{b} + \frac{c}{d} = \frac{ad + bc}{bd}$$

For example,

$$\frac{2}{7} + \frac{3}{5} = \frac{(2 \cdot 5) + (7 \cdot 3)}{7 \cdot 5} = \frac{10 + 21}{35} = \frac{31}{35}$$

We can arrive at this definition by noting that

$$\frac{a}{b} + \frac{c}{d} = \frac{ad}{bd} + \frac{bc}{bd} \quad \text{By Theorem 4.5a}$$

$$= \frac{ad + bc}{bd} \quad \text{The fractions have a common denominator.}$$

[Note: If $d = b$, it is easy to show that Definition 4.5d gives

$$\frac{a}{b} + \frac{c}{b} = \frac{a + c}{b}$$

as we should expect.]

► **Definition 4.5e**

> The **difference** of two rational numbers a/b and c/d is defined as
>
> $$\frac{a}{b} - \frac{c}{d} = \frac{a}{b} + \frac{-c}{d} = \frac{ad - bc}{bd}$$

Thus,

$$\frac{1}{3} - \frac{1}{4} = \frac{1}{3} + \frac{-1}{4} = \frac{(1 \cdot 4) - (3 \cdot 1)}{3 \cdot 4} = \frac{4 - 3}{12} = \frac{1}{12}$$

Now suppose that we want to do the addition $\frac{3}{4} + \frac{1}{16}$. Using Definition 4.5d, we proceed as follows:

$$\frac{3}{4} + \frac{1}{16} = \frac{(3 \cdot 16) + (4 \cdot 1)}{4 \cdot 16} = \frac{48 + 4}{64} = \frac{52}{64} = \frac{13}{16}$$

However, it is easier to use the fact that

$$\frac{3}{4} = \frac{3 \cdot 4}{4 \cdot 4} = \frac{12}{16}$$

and thus,

$$\frac{3}{4} + \frac{1}{16} = \frac{12}{16} + \frac{1}{16} = \frac{13}{16}$$

The number 16 is the **least common denominator (LCD)** of $\frac{3}{4}$ and $\frac{1}{16}$. Recall from Section 4.3 that this number is the LCM of the given denominators, so it can be obtained by following the procedure discussed in Section 4.3. The next example illustrates the use of this idea.

EXAMPLE 3   Perform the addition: $\dfrac{3}{8} + \dfrac{7}{36}$.

Solution   We first find the LCM of 8 and 36 by using the procedure given in Section 4.3:

$$8 = 2^3$$
$$\underline{36 = 2^2 \cdot 3^2}$$
$$\text{LCM} = 2^3 \cdot 3^2 = 72$$

As you can see, LCM(8, 36) = 72. Thus,

$$\frac{3}{8} = \frac{3 \cdot 9}{8 \cdot 9} = \frac{27}{72} \quad \text{and} \quad \frac{7}{36} = \frac{7 \cdot 2}{36 \cdot 2} = \frac{14}{72}$$

so that

$$\frac{3}{8} + \frac{7}{36} = \frac{27}{72} + \frac{14}{72} = \frac{41}{72}$$

◄

To define division, we introduce the idea of a *reciprocal*.

▶ **Definition 4.5f**

> The **reciprocal** of a rational number $\dfrac{a}{b}$ is $\dfrac{b}{a}$, where $a \neq 0$ and $b \neq 0$.

Note that

$$\frac{a}{b} \cdot \frac{b}{a} = 1$$

Thus, the product of any nonzero rational number and its reciprocal is 1. For this reason, the reciprocal of a rational number is also called its **multiplicative inverse.**

EXAMPLE 4 Find the reciprocal of:

a. $\frac{1}{3}$ **b.** $-\frac{3}{4}$ **c.** $1\frac{3}{8}$

Solution **a.** The reciprocal of $\frac{1}{3}$ is $\frac{3}{1} = 3$.
b. The reciprocal of $-\frac{3}{4}$ is $-\frac{4}{3}$. Note that $(-\frac{3}{4})(-\frac{4}{3}) = (-1)(-1)(\frac{3}{4})(\frac{4}{3}) = 1$.
c. We first write $1\frac{3}{8}$ in the form a/b. Because $1\frac{3}{8} = \frac{11}{8}$, the reciprocal of $1\frac{3}{8}$ is $\frac{8}{11}$. ◀

Note: The reciprocal of any negative rational number can be handled as in part b of Example 4. This shows that the reciprocal of a negative number is always negative.

▶ **Definition 4.5g**

> The **quotient** of two rational numbers a/b and c/d is defined as
>
> $$\frac{a}{b} \div \frac{c}{d} = \frac{a}{b} \cdot \frac{d}{c} \qquad c \neq 0$$

Briefly, we say, "Invert the divisor and multiply." Thus,

$$\frac{1}{3} \div \frac{2}{7} = \frac{1}{3} \cdot \frac{7}{2} = \frac{7}{6} \quad \text{and} \quad \frac{4}{5} \div \frac{3}{5} = \frac{4}{5} \cdot \frac{5}{3} = \frac{4 \cdot 5}{3 \cdot 5} = \frac{4}{3}$$

We can check Definition 4.5g by multiplying the quotient, $(a/b) \cdot (d/c)$, by the divisor, c/d, to obtain the dividend, a/b. Thus,

$$\left(\frac{a}{b} \cdot \frac{d}{c}\right) \cdot \frac{c}{d} = \frac{a}{b} \cdot \left(\frac{d}{c} \cdot \frac{c}{d}\right) = \frac{a}{b} \cdot 1 = \frac{a}{b}$$

as required.

As a consequence of these definitions of the basic operations, the set of rational numbers under addition and multiplication is closed and has the associative, commutative, and distributive properties. It also has an additive identity (0) and additive inverses. With respect to multiplication, the rational numbers have a multiplicative identity (1), and every rational number, except 0, has a multiplicative inverse (its reciprocal).

It can be shown that the definitions we have given are the only possible ones if we require that the rational numbers obey the same basic laws of arithmetic as do the integers. (Just imagine how unpleasant it would be if $\frac{1}{2} + \frac{1}{4} \neq \frac{1}{4} + \frac{1}{2}$.)

From the preceding definitions and Theorem 4.5a, we can obtain some additional important results. By the definition of the rational number -1, we know that $-1 = \frac{-1}{1}$, and by Theorem 4.5a,

$$\frac{-1}{1} = \frac{(-1)(-1)}{(1)(-1)} = \frac{1}{-1}$$

Therefore,

$$-\frac{a}{b} = -1 \cdot \frac{a}{b} = \frac{-1}{1} \cdot \frac{a}{b} = \frac{-a}{b}$$

Similarly,

$$-\frac{a}{b} = -1 \cdot \frac{a}{b} = \frac{1}{-1} \cdot \frac{a}{b} = \frac{a}{-b}$$

Thus, we have the important conclusion that

$$-\frac{a}{b} = \frac{-a}{b} = \frac{a}{-b}$$

Furthermore, because $1 = \frac{1}{1} \cdot \frac{-1}{-1} = \frac{-1}{-1}$, we see that

$$\frac{a}{b} = \frac{-1}{-1} \cdot \frac{a}{b} = \frac{-a}{-b}$$

If you think of a and b as positive numbers, you can see that the rule of signs in division is exactly the same as that in multiplication: The quotient of two numbers with like signs is positive and of two numbers with unlike signs is negative. For example,

$$\frac{9}{3} = \frac{-9}{-3} = 3 \qquad \text{and} \qquad \frac{-9}{3} = \frac{9}{-3} = -3$$

If the operations under discussion involve **mixed numbers,** first write the mixed numbers as fractions. Thus, to add $2\frac{3}{4} + \frac{5}{6}$, note that $2\frac{3}{4} = 2 + \frac{3}{4} = \frac{8}{4} + \frac{3}{4} = \frac{11}{4}$. A simpler way to do this (in just one step) is

$$2\frac{3}{4} = \frac{2 \cdot 4 + 3}{4} = \frac{11}{4}$$

Same denominator

Thus,

$$2\frac{3}{4} + \frac{5}{6} = \frac{11}{4} + \frac{5}{6} = \frac{33}{12} + \frac{10}{12} = \frac{43}{12} = 3\frac{7}{12}$$

Note that to convert the fraction $\frac{43}{12}$ to a mixed number, we divide 43 by 12 (the answer is 3) and write the remainder 7 as the numerator of the remaining fraction, with the denominator unchanged.

EXAMPLE 5   Perform the indicated operations:

a. $3\frac{1}{4} + 4\frac{1}{6}$ **b.** $5 - 2\frac{1}{7}$ **c.** $3\frac{1}{4} \times (-8)$ **d.** $21 \div (-4\frac{1}{5})$

Solution   **a.** We first change the mixed numbers to fractions:

$$3\frac{1}{4} = \frac{13}{4} \quad \text{and} \quad 4\frac{1}{6} = \frac{25}{6}$$

We then obtain the LCD, which is 12, and change the fractions to equivalent ones with 12 as a denominator. We then have

$$3\frac{1}{4} + 4\frac{1}{6} = \frac{13}{4} + \frac{25}{6} = \frac{39}{12} + \frac{50}{12} = \frac{89}{12} = 7\frac{5}{12}$$

b. Changing the $2\frac{1}{7}$ to $\frac{15}{7}$, and 5 to $\frac{35}{7}$, we write

$$5 - 2\frac{1}{7} = \frac{35}{7} - \frac{15}{7} = \frac{20}{7} = 2\frac{6}{7}$$

c. Write $3\frac{1}{4}$ as $\frac{13}{4}$, and recall that the product of two numbers with unlike signs is negative. We have

$$3\frac{1}{4} \times (-8) = \frac{13}{\underset{1}{4}} \times (-\overset{-2}{8}) = -26$$

d. Change the $-4\frac{1}{5}$ to $-\frac{21}{5}$, invert, and then multiply to get

$$21 \times \left(-\frac{5}{21}\right) = -5$$

(Note that the answer is negative.)   ◀

EXERCISE 4.5

A.   In Problems 1–4, identify the numerator and denominator of the given rational number.

1. $\frac{3}{4}$ **2.** $\frac{4}{5}$ **3.** $\frac{3}{-5}$ **4.** $\frac{-4}{5}$

In Problems 5–7, identify which rational numbers are equal by using Definition 4.5b.

5. a. $\frac{17}{41}$ **b.** $\frac{289}{697}$ **c.** $\frac{714}{1682}$

6. a. $\frac{438}{529}$ **b.** $\frac{19}{23}$ **c.** $\frac{323}{391}$

7. a. $\frac{11}{91}$ **b.** $\frac{111}{911}$ **c.** $\frac{253}{2093}$

In Problems 8–10, reduce the given rational number.

8. $\frac{42}{86}$ **9.** $\frac{21}{48}$ **10.** $\frac{148}{370}$

B. It is possible to add and subtract rational numbers by converting them to equivalent rational numbers with the same denominator. In Problems 11–13, express each sum as a sum of rational numbers with a denominator of 18.

11. $\frac{2}{9} + \frac{1}{6} + \frac{7}{18}$ **12.** $\frac{7}{3} + \frac{7}{9} + \frac{5}{6}$ **13.** $\frac{1}{3} + \frac{1}{6} + \frac{1}{9}$

In Problems 14–55, perform the indicated operations and reduce the answer (if possible).

14. $\frac{1}{7} + \frac{1}{3}$ **15.** $\frac{1}{7} + \frac{1}{9}$ **16.** $\frac{2}{7} + \frac{3}{11}$ **17.** $\frac{3}{4} + \frac{5}{6}$

18. $\frac{1}{12} + \frac{7}{18}$ **19.** $\frac{3}{17} + \frac{7}{19}$ **20.** $\frac{1}{3} - \frac{1}{7}$ **21.** $\frac{1}{7} - \frac{1}{9}$

22. $\frac{2}{7} - \frac{3}{11}$ **23.** $\frac{3}{4} - \frac{5}{6}$ **24.** $\frac{7}{18} - \frac{1}{12}$ **25.** $\frac{7}{19} - \frac{3}{17}$

26. $\frac{3}{4} \times \frac{2}{7}$ **27.** $\frac{2}{5} \times \frac{5}{3}$ **28.** $\frac{7}{9} \times \frac{3}{8}$ **29.** $\frac{3}{4} \div \frac{2}{7}$

30. $\frac{2}{5} \div \frac{5}{3}$ **31.** $\frac{7}{9} \div \frac{3}{8}$ **32.** $\left(\frac{-2}{5}\right) \times \frac{4}{9}$ **33.** $\left(-\frac{6}{7}\right) \times \left(-\frac{3}{11}\right)$

34. $\frac{4}{5} \div \left(\frac{-7}{9}\right)$ **35.** $\left(-\frac{3}{4}\right) \div \left(-\frac{7}{6}\right)$ **36.** $\frac{3}{4} \div \left(-\frac{1}{5}\right)$ **37.** $\frac{1}{8} \div \left(-\frac{3}{4}\right)$

38. $\left(-\frac{1}{4}\right) + \left(-\frac{1}{8}\right)$ **39.** $\left(-\frac{1}{8}\right) + \left(\frac{1}{4}\right)$ **40.** $\left(\frac{1}{3} + \frac{1}{4}\right) + \frac{7}{8}$ **41.** $\frac{3}{8} - \left(\frac{1}{4} - \frac{1}{8}\right)$

42. $\left(\frac{1}{5} \times \frac{1}{4}\right) \times \frac{3}{7}$ **43.** $\frac{1}{2} \times \left(\frac{7}{8} \times \frac{7}{5}\right)$ **44.** $\frac{1}{2} \div \left(\frac{1}{2} \div \frac{1}{4}\right)$ **45.** $\left(\frac{1}{2} \div \frac{1}{2}\right) \div \frac{1}{4}$

46. $\frac{3}{4} + \frac{1}{2}\left(\frac{3}{2} + \frac{1}{4}\right)$ **47.** $\frac{2}{3}\left(\frac{1}{2} + \frac{3}{4}\right) + \frac{2}{3}$ **48.** $\frac{1}{2}\left(\frac{3}{4} - \frac{1}{2}\right) - \frac{1}{12}$ **49.** $\frac{1}{3}\left(\frac{3}{2} - \frac{1}{5}\right) - \frac{1}{30}$

50. $\frac{1}{2}\left(\frac{5}{2} - \frac{1}{3}\right) - \frac{5}{12}$

51. **a.** $1\frac{1}{2} + \frac{1}{7}$ **b.** $5 - 1\frac{1}{3}$
  c. $\frac{1}{4} \times 1\frac{1}{7}$ **d.** $5 \div \left(-2\frac{1}{2}\right)$

52. **a.** $3\frac{1}{4} + \frac{1}{6}$ **b.** $4 - 2\frac{1}{4}$
  c. $\frac{1}{5} \times 2\frac{1}{7}$ **d.** $6 \div \left(-1\frac{1}{5}\right)$

53. **a.** $-3 + 2\frac{1}{4}$ **b.** $-\frac{2}{3} - (-2)$
  c. $(-8) \times 2\frac{1}{4}$ **d.** $7 \div \left(-2\frac{1}{3}\right)$

54. **a.** $-2 + 1\frac{1}{5}$ **b.** $-\frac{3}{4} - (-3)$
  c. $(-9) \times 3\frac{1}{3}$ **d.** $\left(-\frac{1}{6}\right) \div \left(-\frac{5}{7}\right)$

55. **a.** $7\frac{1}{4} + \left(-\frac{1}{8}\right)$ **b.** $-3\frac{1}{8} - (-2)$
  c. $\left(-1\frac{1}{4}\right) \times \left(-2\frac{1}{10}\right)$ **d.** $\left(-1\frac{1}{8}\right) \div \left(-2\frac{1}{4}\right)$

USING YOUR KNOWLEDGE 4.5

Road maps and other maps are drawn to **scale**. The scale on a map shows what distance is represented by a certain measurement on the map. For example, the scale on a certain map is

1 inch = 36 miles

Thus, if the distance on the map is $2\frac{1}{4}$ inches from Indianapolis to Dayton, the actual distance is

$$36 \times 2\frac{1}{4} = 36 \times \frac{9}{4} = 9 \times 4 \times \frac{9}{4} = 81 \text{ miles}$$

1. Find the actual distance from Indianapolis to Cincinnati, a distance of $3\frac{1}{2}$ inches on the map.
2. Find the actual distance from Indianapolis to Terre Haute, a distance of $3\frac{1}{4}$ inches on the map.

3. If the actual distance between two cities is 108 miles, what is the distance on the map?
4. If the actual distance between two cities is 162 miles, what is the distance on the map?

DISCOVERY 4.5

Let us try to discover how many positive rational numbers there are. We arrange the first few fractions as shown below:

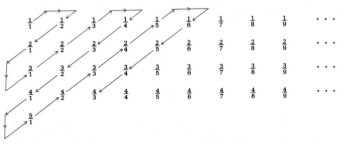

1. Can you discover the next two rows in this array?
2. Notice that the number $\frac{1}{5}$ appears in the first row, fifth column. Can you discover where the number $\frac{37}{43}$ will be?
3. By following the path indicated in the array, we can set up a correspondence between the set of natural numbers N and the set of positive rational numbers. The first few terms look like this:

$$
\begin{array}{ccccccccccc}
1 & 2 & 3 & 4 & 5 & 6 & 7 & 8 & 9 & 10 & 11 \\
\updownarrow & \updownarrow & \updownarrow & \updownarrow & \updownarrow & \updownarrow & \updownarrow & \updownarrow & \updownarrow & \updownarrow & \updownarrow \\
1 & \frac{1}{2} & \frac{2}{1} & \frac{3}{1} & \frac{1}{4} & \frac{1}{3} & \frac{2}{3} & \frac{3}{2} & \frac{4}{1} & \frac{5}{1} & \frac{1}{5}
\end{array} \; \cdots
$$

You will notice that in this pairing we skipped $\frac{2}{2}, \frac{4}{2}, \frac{3}{3}$, and $\frac{2}{4}$. Can you discover why?
4. The correspondence indicated in Problem 3 can be continued indefinitely. If we assign a cardinal number to the set N of natural numbers, say $n(N) = \aleph_0$ (aleph zero), so that we can speak of the "number of natural numbers" as being $\aleph_0$, what would $n(Q^+)$ be? That is, how many positive rational numbers are there?
5. Based on your answer to Problem 4, would you say that there are more or fewer rational numbers than natural numbers?

▶ **Computer Corner 4.5**   *Some calculators will perform operations with fractions; however, most of them will convert fractions to decimals and then perform the operations, giving answers in decimal form. But a computer can be programmed to do calculations involving fractions. The programs on page 746 add fractions. (One program reduces the answer, the other one does not.) Note that the addition is done by using Definition 4.5d without converting to decimals, thus eliminating round-off errors.*

RATIONAL NUMBERS IN DECIMAL FORM

As you are probably aware, the number $\frac{1}{2}$ can be written in decimal form, that is, as 0.5. Because all rational numbers can be written in the form a/b, it is always possible to change a rational number a/b to its decimal form by simply dividing a by b. For example, $\frac{1}{2} = 0.5$, $\frac{1}{8} = 0.125$, And $\frac{1}{4} = 0.25$. Of course, if the denominator of a fraction is already a power of 10 (10, 100, 1000, and so on), then the fraction can easily be written as a decimal. Thus,

$$\frac{7}{10} = 0.7, \qquad \frac{19}{100} = 0.19, \qquad \text{and} \qquad \frac{17}{1000} = 0.017$$

You should keep in mind that the successive places to the right of the decimal point have the place values

$$\frac{1}{10^1}, \quad \frac{1}{10^2}, \quad \frac{1}{10^3}, \quad \frac{1}{10^4}, \quad \cdots$$

The exponent on the 10 is the number of the place. For example, the underlined digit in 2.567<u>9</u>4 has the place value $1/10^4$, because the 9 is in the fourth decimal place.

A. Terminating and Nonterminating Decimals

Numbers such as $\frac{1}{2}$, $\frac{1}{8}$, and $\frac{1}{5}$ are said to have **terminating decimal representations,** because division of the numerator by the denominator terminates (ends). However, some rational numbers—for example, $\frac{1}{3}$—have **infinite repeating decimal representations.** Such a representation is obtained by dividing the numerator of the fraction by its denominator. In the case of $\frac{1}{3}$, we obtain

$$
\begin{array}{r}
0.333\ \ldots \\
3\overline{)1.0} \\
\underline{9} \\
10 \\
\underline{9} \\
10 \\
\underline{9} \\
1 \text{ etc.}
\end{array}
$$

For convenience, we shall write 0.333 . . . as $0.\overline{3}$. The bar over the 3 indicates that the 3 repeats indefinitely. Similarly, $\frac{1}{7} = 0.142857142857\ \ldots = 0.\overline{142857}$.

EXAMPLE 1   Write as decimals and give the value in the second decimal place:

a. $\frac{3}{4}$ **b.** $\frac{2}{3}$

Solution **a.** Dividing 3 by 4, we obtain

$$
\begin{array}{r}
0.75 \\
4\overline{)3.0} \\
\underline{2\ 8} \\
20 \\
\underline{20} \\
0
\end{array}
$$

Thus, $\frac{3}{4} = 0.75$, and the value in the second decimal place is $\frac{5}{100}$.

b. Dividing 2 by 3, we obtain

$$
\begin{array}{r}
0.666\ \ldots \\
3\overline{)2.0} \\
\underline{1\ 8} \\
20 \\
\underline{18} \\
20 \\
\underline{18} \\
2 \text{ etc.}
\end{array}
$$

Thus, $\frac{2}{3} = 0.666\ \ldots = 0.\overline{6}$, and the value in the second decimal place is $\frac{6}{100}$. ◀

You should be able to convince yourself of the truth of the following theorem, which indicates which rational numbers have terminating decimal representations.

▶ **Theorem 4.6a** A rational number a/b (in lowest terms) has a terminating decimal expansion if and only if b has no prime factors other than 2 and 5.

Notice that b does not have to have 2 *and* 5 as factors; it might have only one of them as a factor, or perhaps neither, as in the following illustrations:

$$
\frac{1}{25} = 0.04
$$

Since $25 = 5 \times 5$, $\frac{1}{25}$ has a terminating decimal expansion.

$$
\frac{1}{4} = 0.25
$$

Since $4 = 2 \times 2$, $\frac{1}{4}$ has a terminating decimal expansion.

$$
\frac{8}{1} = 8.0
$$

The denominator is 1, which has no prime factors, so $\frac{8}{1}$ has a terminating decimal expansion. Of course, if the rational number is given as a mixed number—for example, $5\frac{1}{2}$—we could first convert it to $\frac{11}{2}$ and notice that the denominator has only 2 as a factor. Thus, $5\frac{1}{2} = \frac{11}{2}$ has a terminating decimal expansion.

The stock markets use the fractions with denominators 2, 4, 8, and 16 because these fractions have simple terminating decimal forms. The denominators are powers of 2 and

$$\frac{1}{2^n} = (0.5)^n$$

so there will be exactly n decimal places.

It is easy to see that every rational number has an infinite repeating decimal representation. In the case of a terminating decimal, we can agree simply to adjoin an infinite string of 0's. For example, $\frac{3}{4} = 0.75\overline{0}$, $\frac{1}{20} = 0.05\overline{0}$, and so on. If the rational number a/b has no terminating decimal representation, then it must have a repeating decimal representation, as we can see by carrying out the division of a by b. The only possible remainders are $1, 2, 3, \ldots, b - 1$. Therefore, after at most $(b - 1)$ steps of the division, a remainder must occur for the second time. Thereafter, the digits of the quotient must repeat. The following division illustrates the idea:

Of the first twenty counting numbers, only 7, 17, and 19 have reciprocals with the maximum possible number of digits in the repeating part of their decimal representations:

$\frac{1}{7} = 0.\overline{142857}$

$\frac{1}{17} = 0.\overline{0588235294117647}$

$\frac{1}{19} = 0.\overline{052631578947368421}$

$$
\begin{array}{r}
1.692307 \\
13\overline{)22} \\
\underline{13} \\
90 \\
\underline{78} \\
120 \\
\underline{117} \\
30 \\
\underline{26} \\
40 \\
\underline{39} \\
100 \\
\underline{91} \\
9
\end{array}
$$

Notice that the remainder **9** occurs just before the digit 6 appears in the quotient and occurs again just after the digit 7 appears in the quotient. Thus, the repeating part of the decimal must be 692307, and $\frac{22}{13} = 1.\overline{692307}$.

B. Changing Infinite Repeating Decimals to Fractions

The preceding discussion shows that **every rational number can be written as an infinite repeating decimal.** Is the converse of this statement true? That is, does every infinite repeating decimal represent a rational number? If the repeating part is simply a string of 0's so that the decimal is actually terminating, it can be written as a rational number with a power of 10 as the denominator. For example, $0.73 = \frac{73}{100}$, $0.7 = \frac{7}{10}$, and $0.013 =$

$\frac{13}{1000}$. If the decimal is repeating but not terminating, then we can proceed as follows:

Suppose we wish to write $x = 0.\overline{23}$ as a quotient of integers. If $x = 0.232323 \ldots$, then $100x = 23.232323 \ldots$. Thus,

$$100x = 23.232323 \ldots$$
$$x = 0.232323 \ldots$$

Subtracting, we get

$$99x = 23$$

and dividing by 99, we get

$$x = \frac{23}{99}$$

Here is the procedure:

1. Let $x =$ the given decimal.
2. Multiply by a power of 10 to move the decimal point to the right of the first sequence of digits that repeats.
3. If the decimal point is not at the left of the first repeating sequence of digits, multiply by a power of 10 to place it there.
4. Subtract the result of step 3 from that of step 2.
5. Divide by the multiplier of x in the result of step 4 to get the desired fraction.

The idea in steps 2 and 3 is to line up the repeating parts so they drop out in the subtraction in step 4.

EXAMPLE 2   Write $3.5212121 \ldots$ as a quotient of two integers.

Solution
1. Let $x = 3.5212121 \ldots$. ——We want the decimal point here.
2. Since we want the decimal point to the right of the first 21, we multiply by 1000:

$$1000x = 3521.212121 \ldots$$

3. In this step, we want the decimal point to the left of the first 21,

$3.5212121 \ldots$ Here

so we multiply x by 10:

$$10x = 35.212121 \ldots$$

4. $ 1000x = 3521.212121 \ldots$
   $\underline{(-) 10x = 35.212121 \ldots}$
   $ 990x = 3486$ \qquad The decimal parts drop out.

5. We divide by 990 to get

$$x = \frac{3486}{990} = \frac{581 \cdot 6}{165 \cdot 6} = \frac{581}{165}$$ ◀

This discussion can be summarized as follows: **Every rational number has a repeating decimal representation, and every repeating decimal represents a rational number.**

C. Decimals in Expanded Form

You learned in Section 3.2 that a positive integer can always be written in expanded form, as, for instance,

$$361 = (3 \times 10^2) + (6 \times 10^1) + (1 \times 10^0)$$

Can we do something like this for a decimal, say 3.52? You know that

$$0.5 = \frac{5}{10} \quad \text{and} \quad 0.02 = \frac{2}{100}$$

so

$$3.52 = (3 \times 10^0) + \left(5 \times \frac{1}{10^1}\right) + \left(2 \times \frac{1}{10^2}\right)$$

This is somewhat awkward, so to make it more convenient, we define **negative exponents** as follows:

$$10^{-1} = \frac{1}{10^1} = \frac{1}{10}, \qquad 10^{-2} = \frac{1}{10^2} = \frac{1}{100}, \qquad 10^{-3} = \frac{1}{10^3} = \frac{1}{1000}, \; \cdots$$

▶ **Definition 4.6a**

In general, if n is a positive integer and $a \neq 0$.

$$a^{-n} = \frac{1}{a^n}$$

The definition of negative exponents gives a good pattern:

$$10^3 = 1000$$
$$10^2 = 100$$
$$10^1 = 10$$
$$10^0 = 1$$
$$10^{-1} = \frac{1}{10}$$
$$10^{-2} = \frac{1}{100}$$
$$10^{-3} = \frac{1}{1000}$$

And so on.

It can be shown that **negative exponents obey exactly the same laws as positive exponents.**

Using these negative exponents, you can write, for example,

$$3.52 = (3 \times 10^0) + (5 \times 10^{-1}) + (2 \times 10^{-2})$$

and

$$362.754 = (3 \times 10^2) + (6 \times 10^1) + (2 \times 10^0) + (7 \times 10^{-1}) + (5 \times 10^{-2})$$
$$+ (4 \times 10^{-3})$$

which looks like a very natural generalization of the expanded form that we used for positive exponents.

EXAMPLE 3   **a.** Write 25.603 in expanded form.

b. Write $(7 \times 10^2) + (2 \times 10^0) + (6 \times 10^{-1}) + (9 \times 10^{-4})$ in standard decimal form.

Solution   **a.** $25.603 = (2 \times 10^1) + (5 \times 10^0) + (6 \times 10^{-1}) + (0 \times 10^{-2}) + (3 \times 10^{-3})$
$= (2 \times 10^1) + (5 \times 10^0) + (6 \times 10^{-1}) + (3 \times 10^{-3})$

b. Notice that the exponent in 10^{-n} tells you that the first nonzero digit comes in the nth decimal place. With this in mind, we have

$$(7 \times 10^2) + (2 \times 10^0) + (6 \times 10^{-1}) + (9 \times 10^{-4}) = 702.6009 \qquad \blacktriangleleft$$

D. Scientific Notation

In science and in other areas, there frequently occur very large or very small numbers. For example, a red cell of human blood contains 270,000,000 hemoglobin molecules, and the mass of a single carbon atom is 0.000 000 000 000 000 000 000 019 9 gram. Numbers in this form are difficult to write and to work with, so they are written in *scientific notation*.

▶   **Definition 4.6b**

A number is said to be in **scientific notation** if it is written in the form

$$m \times 10^n$$

where m is a number greater than or equal to 1 and less than 10, and n is an integer.

For any given number, the m is obtained by placing the decimal point so that there is exactly one nonzero digit to its left. The n is then the number of places that the decimal point must be moved from its position in m to its original position; it is positive if the point must be moved to the right, and negative if the point must be moved to the left. Thus,

$5.3 = 5.3 \times 10^0$	Decimal point in 5.3 must be moved 0 places.
$87 = 8.7 \times 10^1 = 8.7 \times 10$	Decimal point in 8.7 must be moved 1 place to the right to get 87.
$68,000 = 6.8 \times 10^4$	Decimal point in 6.8 must be moved 4 places to the right to get 68,000.
$0.49 = 4.9 \times 10^{-1}$	Decimal point in 4.9 must be moved 1 place to the left to get 0.49.
$0.072 = 7.2 \times 10^{-2}$	Decimal point in 7.2 must be moved 2 places to the left to get 0.072.
$0.0003875 = 3.875 \times 10^{-4}$	Decimal point in 3.875 must be moved 4 places to the left to get 0.0003875.

EXAMPLE 4 Write in scientific notation:

    a. 270,000,000

    b. 0.000 000 000 000 000 000 000 019 9

Solution   **a.** $270{,}000{,}000 = 2.7 \times 10^8$

    b. $0.000\ 000\ 000\ 000\ 000\ 000\ 000\ 019\ 9 = 1.99 \times 10^{-23}$ ◀

EXAMPLE 5 Write in standard decimal notation:

    a. 2.5×10^{10} **b.** 7.4×10^{-6}

Solution   **a.** $2.5 \times 10^{10} = 25{,}000{,}000{,}000$

    b. $7.4 \times 10^{-6} = 0.0000074$ ◀

We noted earlier in this section that negative exponents obey the same laws as positive exponents. You can verify this quite easily in simple cases. For example,

$$10^4 \times 10^{-2} = 10{,}000 \times \frac{1}{100} = 100 = 10^2 = 10^{4+(-2)}$$

$$10^{-2} \times 10^{-3} = 0.01 \times 0.001 = 0.00001 = 10^{-5} = 10^{-2+(-3)}$$

This knowledge can be used to do calculations with numbers in scientific notation.

EXAMPLE 6 Do the following calculation, and write the answer in scientific notation: $(5 \times 10^4) \times (9 \times 10^{-7})$

Solution $\begin{aligned}
(5 \times 10^4) \times (9 \times 10^{-7}) &= (5 \times 9) \times (10^4 \times 10^{-7}) \\
&= 45 \times 10^{4-7} \\
&= 45 \times 10^{-3} \\
&= 4.5 \times 10^1 \times 10^{-3} \\
&= 4.5 \times 10^{1-3} \\
&= 4.5 \times 10^{-2}
\end{aligned}$ ◀

EXERCISE 4.6

A. In Problems 1–6, write the given number in decimal form.

    1. a. $\frac{9}{10}$ **b.** $\frac{3}{10}$ **c.** $\frac{11}{10}$

    2. a. $\frac{17}{100}$ **b.** $\frac{38}{100}$ **c.** $\frac{121}{100}$

    3. a. $\frac{3}{1000}$ **b.** $\frac{143}{1000}$ **c.** $\frac{1243}{1000}$

    4. a. $\frac{3}{5}$ **b.** $\frac{5}{7}$ **c.** $\frac{5}{4}$

    5. a. $\frac{3}{8}$ **b.** $\frac{7}{6}$ **c.** $\frac{4}{25}$

    6. a. $6\frac{1}{4}$ **b.** $7\frac{1}{7}$ **c.** $3\frac{2}{3}$

In Problems 7 and 8, determine whether the given number has a terminating decimal expansion. If it does, give the expansion.

7. a. $\frac{3}{16}$ **b.** $\frac{3}{14}$ **c.** $\frac{1}{64}$
8. a. $\frac{4}{28}$ **b.** $\frac{31}{3125}$ **c.** $\frac{9}{250}$

B. In Problems 9–15, write the given number as a fraction (a quotient of two integers). Reduce if possible.

9. a. $0.\overline{8}$ **b.** $0.\overline{6}$
10. a. $0.\overline{31}$ **b.** $0.\overline{21}$
11. a. $0.1\overline{14}$ **b.** $0.\overline{102}$
12. a. $2.\overline{31}$ **b.** $5.\overline{672}$
13. a. $1.\overline{234}$ **b.** $0.\overline{017}$
14. a. $1.2\overline{7}$ **b.** $2.4\overline{8}$
15. a. $0.45\overline{75}$ **b.** $0.2\overline{315}$

C. In Problems 16 and 17, write the given numbers in expanded notation.

16. a. 692.087   **b.** 30.2959
17. a. 0.00107   **b.** 4.30008

In Problems 18–20, write the given number in standard decimal notation.

18. $(5 \times 10^3) + (2 \times 10^1) + (3 \times 10^{-1}) + (9 \times 10^{-2})$
19. $(4 \times 10^2) + (5 \times 10^0) + (6 \times 10^{-2}) + (9 \times 10^{-4})$
20. $(4 \times 10^{-3}) + (7 \times 10^{-4}) + (2 \times 10^{-6})$

D. In Problems 21 and 22, write the given numbers in scientific notation.

21. a. 935   **b.** 0.372
22. a. 0.0012   **b.** 3,453,000

In Problems 23 and 24, write the given numbers in standard decimal form.

23. a. 8.64×10^4 **b.** 9.01×10^7
24. a. 6.71×10^{-3} **b.** 4.02×10^{-7}

25. Simplify and express the result in scientific notation:

$$\frac{(2 \times 10^6)(6 \times 10^{-5})}{4 \times 10^3}$$

26. Simplify and express the result in scientific notation:

$$\frac{(8 \times 10^2)(3 \times 10^{-2})}{24 \times 10^{-3}}$$

27. The most plentiful form of sea life known is the nematode sea worm. It is estimated that there are

40,000,000,000,000,000,000,000,000

of these worms in the world's oceans. Write this number in scientific notation.

28. Sir Arthur Eddington claimed that the total number of electrons in the universe is 136×2^{256}. It can be shown that 2^{256} is approximately 1.16×10^{77}. Use this result to write Eddington's number in scientific notation.

In Problems 29–31, write the answer in scientific notation.

29. The width of the asteroid belt is 2.8×10^8 km. The speed of *Pioneer 10* in passing through this belt was 1.4×10^5 km/hr. Thus, *Pioneer 10* took

$$\frac{2.8 \times 10^8}{1.4 \times 10^5} \text{ hr}$$

to go through the belt. How many hours was that?

30. The mass of the Earth is 6×10^{21} tons. The Sun is about 300,000 times as massive. Thus, the mass of the Sun is $(6 \times 10^{21}) \times 300,000$ tons. How many tons is that?

31. The velocity of light can be measured by knowing the distance from the Sun to the Earth (1.47×10^{11} m) and the time it takes for sunlight to reach the Earth (490 sec). Thus, the velocity of light is

$$\frac{1.47 \times 10^{11}}{490} \text{ m/sec}$$

How many meters per second is that?

32. United States oil reserves are estimated to be 3.5×10^{10} barrels. Production amounts to 3.2×10^9 barrels per year. At this rate, how long would U.S. oil reserves last? (Give answer to the nearest year.)

33. The world's oil reserves are estimated to be 6.28×10^{11} barrels. Production is 2.0×10^{10} barrels per year. At this rate, how long would the world's oil reserves last? (Give answer to the nearest year.)

34. Scientists have estimated that the total energy received from the Sun each minute is 1.02×10^{19} calories. Since the area of the Earth is 5.1×10^8 km² (square kilometers), the amount of energy received per square centimeter of Earth's surface per minute (the solar constant) is

$$\frac{1.02 \times 10^{19}}{(5.1 \times 10^8) \times 10^{10}} \quad \text{Note:} \quad 1 \text{ km}^2 = 10^{10} \text{ cm}^2$$

How many calories per square centimeter is that?

DISCOVERY 4.6

In this section we developed a procedure for expressing an infinite repeating decimal as a quotient of two integers. For example,

$$0.\overline{3} = 0.333 \ldots = \tfrac{3}{9}$$
$$0.\overline{6} = 0.666 \ldots = \tfrac{6}{9}$$
$$0.\overline{21} = 0.212121 \ldots = \tfrac{21}{99}$$
$$0.\overline{314} = 0.314314314 \ldots = \tfrac{314}{999}$$

1. From these examples, can you discover how to express $0.\overline{4} = 0.444 \ldots$ as a quotient of integers?
2. Can you express $0.\overline{4321}$ as a quotient of integers?

If we have a repeating decimal with one digit as a **repetend** *(the part that repeats), the number can be written as a fraction by dividing the repetend by 9. For example, $0.\overline{7} = \tfrac{7}{9}$. If the number has two digits in the repetend, the number can be written as a fraction by dividing the repetend by 99. For example, $0.\overline{21} = \tfrac{21}{99}$.*

3. Using these ideas, you can write $0.\overline{9} = 0.999 \ldots = \tfrac{9}{9} = 1$. If you are wondering why $0.\overline{9} = 0.999 \ldots = \tfrac{9}{9} = 1$, here are some arguments that may convince you of this fact.

    a. $\tfrac{1}{3} = 0.333 \ldots$
       $\tfrac{2}{3} = 0.666 \ldots$

    Can you discover what happens when you add the two equations?

    b. $0.\overline{1} = \tfrac{1}{9}, 0.\overline{2} = \tfrac{2}{9}, 0.\overline{3} = \tfrac{3}{9}, 0.\overline{4} = \tfrac{4}{9}, 0.\overline{5} = \tfrac{5}{9}, 0.\overline{6} = \tfrac{6}{9}, 0.\overline{7} = \tfrac{7}{9}, 0.\overline{8} = \tfrac{8}{9}$. To what is $0.\overline{9}$ equal?

    c. $\tfrac{1}{3} = 0.333 \ldots$. Can you discover what result you obtain if you multiply both sides of this equation by 3?

▶ **Calculator Corner 4.6**   *Most operations in this section can be done with a calculator. For example, to change a fraction to a decimal, divide the numerator by the denominator. But can you find the decimal expansion for $\tfrac{3}{14}$? Since the calculator has ten decimal places, when you divide 3 by 14 the answer is given as 0.214285714, and you must know that the repeating part is 142857.*

On the other hand, numbers can be converted automatically to scientific notation provided you have a $\boxed{\text{sci}}$ key or its equivalent. Thus, to write 270,000 in scientific notation, enter $\boxed{2}\ \boxed{7}\ \boxed{0}\ \boxed{0}\ \boxed{0}\ \boxed{0}\ \boxed{=}$ and $\boxed{\text{2nd}}\ \boxed{\text{sci}}$. If you do not have a $\boxed{\text{sci}}$ key, you must know how to enter numbers in scientific notation by using the $\boxed{\text{exp}}$ key. Thus, to enter 270,000, you have to know that

$$270{,}000 = 2.7 \times 10^5$$

and enter $\boxed{2}\ \boxed{.}\ \boxed{7}\ \boxed{\text{exp}}\ \boxed{5}$. The display shows 2.7 05.
To perform operations in scientific notation (without a $\boxed{\text{sci}}$ key), we enter the numbers as discussed. Thus, to find $(5 \times 10^4) \times (9 \times 10^{-7})$ (as we did in Example 6), enter $\boxed{(}\ \boxed{5}\ \boxed{\text{exp}}\ \boxed{4}\ \boxed{)}\ \boxed{\times}\ \boxed{(}\ \boxed{9}\ \boxed{\text{exp}}\ \boxed{7}\ \boxed{\pm}\ \boxed{)}\ \boxed{=}$. The display will give the answer as 4.5 −02,

that is, 4.5×10^{-2}. If you have a $\boxed{\text{sci}}$ *key or if you can place the calculator in scientific mode, then enter* $\boxed{5}$ $\boxed{\text{exp}}$ $\boxed{4}$ $\boxed{\times}$ $\boxed{9}$ $\boxed{\text{exp}}$ $\boxed{7}$ $\boxed{\pm}$ $\boxed{=}$ *(no need to use parentheses), and the same result as before will appear in the display.*

1. Use your calculator to check Problems 15, 27, 29, 31, and 33 in Exercise 4.6.

SUMMARY

Section	Item	Meaning	Example
4.1A	$N = \{1, 2, 3, \ldots\}$	The natural numbers	Any counting numbers such as 10, 27, 38, and so on
4.1A	$n(A)$	The cardinal number of A	If $A = \{a, b\}$, then $n(A) = 2$.
4.1A	1st, 2nd, 3rd, . . .	Ordinal numbers	This is the first one.
4.1A	123-45-6789	Number used for identification	A Social Security number
4.1B	$+, -, \times, \div$	Binary operations that associate with any two elements of a set a unique result	Addition, subtraction, multiplication, and division are binary operations
4.1B	Closed set	A set is closed if when an operation is performed on elements of the set, the result is also an element of the set.	The natural numbers are closed under multiplication.
4.1B	$a + b = b + a$	Commutative property of addition	$3 + 5 = 5 + 3$
4.1B	$a \times b = b \times a$	Commutative property of multiplication	$6 \times 7 = 7 \times 6$
4.1B	$a + (b + c) = (a + b) + c$	Associative property of addition	$4 + (2 + 5) = (4 + 2) + 5$
4.1B	$a \times (b \times c) = (a \times b) \times c$	Associative property of multiplication	$2 \times (4 \times 7) = (2 \times 4) \times 7$
4.1B	$a \cdot (b + c) = a \cdot b + a \cdot c$	Distributive property	$3 \cdot (4 + 7) = 3 \cdot 4 + 3 \cdot 7$
4.2A	Prime number	A number with exactly two divisors, 1 and itself	2, 3, 5, 7, 11, etc.
4.2A	Composite	A number with more than two divisors	4, 33, 50, etc.
4.2B	$12 = 2^2 \cdot 3$	Prime factorization of 12	
4.3A	GCF	Greatest common factor	18 is the GCF of 216 and 234.

Section	Item	Meaning	Example
4.3B	LCM	Least common multiple	252 is the LCM of 18, 21, and 28.
4.4A	$a \cdot 1 = 1 \cdot a = a$	Identity property for multiplication	$1 \cdot 97 = 97$ and $83 \cdot 1 = 83$
4.4A	$W = \{0, 1, 2, \ldots\}$	The set of whole numbers	
4.4A	$0 + a = a + 0 = a$	Identity property for addition	$0 + 13 = 13$ and $84 + 0 = 84$
4.4B	$I = \{\ldots, -1, 0, 1, \ldots\}$	The set of integers	
4.4B	$-2 \quad -1 \quad 0 \quad 1 \quad 2$	The number line	
4.4B	$n + (-n) = 0$	Additive inverse property	$3 + (-3) = 0$ and $(-5) + 5 = 0$
4.4B	$a - b = a + (-b)$	Definition of subtraction	$3 - 7 = 3 + (-7)$
4.5A	$Q = \left\{ r \,\middle|\, r = \dfrac{a}{b}, a, b \in I, b \neq 0 \right\}$	The set of rational numbers	$\frac{3}{5}, -\frac{7}{3}, 3\frac{1}{2}$
4.5B	$\dfrac{b}{a}$	The reciprocal, or multiplicative inverse, of $\dfrac{a}{b}$	$\frac{3}{4}$ and $\frac{4}{3}$ are reciprocals
4.6A	$0.\overline{142857}$	A nonterminating, repeating decimal	$0.\overline{142857} = 0.142857142857 \ldots$
4.6C	$a^{-n} = \dfrac{1}{a^n}$	Definition of negative exponents	$4^{-2} = \dfrac{1}{4^2}$
4.6D	$m \times 10^n$, where m is greater than or equal to 1 and less than 10, and n is an integer	Scientific notation	7.4×10^{-6}

Properties of Addition

1. The rational numbers are closed under addition. (The sum of any two rational numbers a and b is always a rational number.)
2. Addition of rational numbers is commutative. (For any two rational numbers a and b, $a + b = b + a$.)
3. Addition of rational numbers is associative. [For any three rational numbers a, b, and c, $a + (b + c) = (a + b) + c$.]
4. The rational numbers have a unique identity (the number 0) for addition. ($a + 0 = 0 + a = a$.)
5. Each rational number has a unique additive inverse. [$a + (-a) = 0$.]

Properties of Multiplication

1. The rational numbers are closed under multiplication. (The product of two rational numbers a and b is always a rational number.)
2. Multiplication of rational numbers is commutative. (For any two rational numbers a and b, $a \cdot b = b \cdot a$.)
3. Multiplication of rational numbers is associative. [For any three rational numbers a, b, and c, $a \cdot (b \cdot c) = (a \cdot b) \cdot c$.]
4. The rational numbers have a unique identity (the number 1) for multiplication. ($1 \cdot a = a \cdot 1 = a$.)
5. Each rational number, except 0, has a unique multiplicative inverse (its reciprocal). [For any nonzero rational number a/b,

$$\frac{a}{b} \cdot \frac{b}{a} = 1$$

The number 0 does not have a multiplicative inverse.]

PRACTICE TEST 4

1. Tell whether the underlined item is used as a cardinal number, an ordinal number, or for identification.
   a. Sally came in <u>third</u> in the 100-yard dash.
   b. Bill's lottery ticket won <u>two</u> dollars.
   c. Jane's auto license number was <u>270-891</u>.
2. Consider the set $A = \{0, 1, 2\}$ and the operation table in the margin.

*	0	1	2
0	0	1	2
1	1	2	0
2	2	0	1

   a. Is this a binary operation? Why?
   b. Is the set A closed under *? Why?
   c. Is the operation commutative? Why?
3. What properties of the system of natural numbers are used in the following computations?
   a. $8 + 9 + 2 = 8 + 2 + 9$ b. $8 \times 12 = 8(10 + 2) = 80 + 16$
   c. $5 \times 27 \times 4 = 5 \times 4 \times 27$
4. Write 1220 as a product of primes.
5. Is 143 prime or composite?
6. Of the numbers 2345, 436, 387, and 1530, identify those divisible by:
   a. 2   b. 3   c. 5
7. Find the GCF of 18, 54, and 60.
8. Reduce the fraction $\frac{210}{254}$ to lowest terms.
9. Find the LCM of 18, 54, and 60.
10. Perform the indicated operations:
    a. $\frac{1}{18} + \frac{1}{24} + \frac{1}{28}$ b. $\frac{11}{24} - \frac{5}{44}$
11. A father left $\frac{1}{4}$ of his estate to his daughter, $\frac{1}{2}$ to his wife, and $\frac{1}{8}$ to his son. If the rest went for taxes, what fraction of the estate was that?

12. Change to an equivalent addition problem and give the result:
    a. $8 - 19$ **b.** $8 - (-19)$ **c.** $-8 - 19$ **d.** $-8 - (-19)$
13. Evaluate $4 \times 12 \div 3 \times 10^3 - 2(-6 + 4) \times 10^4$.
14. Find a rational number with a denominator of 16 and equal to $\frac{3}{4}$.
15. Find the reciprocal of:
    a. $\frac{2}{3}$ **b.** $-\frac{4}{7}$ **c.** $2\frac{5}{8}$ **d.** -8
16. Perform the indicated operations:
    a. $\frac{7}{8} \times \left(-\frac{5}{16}\right)$ **b.** $-\frac{7}{8} \div \left(-\frac{5}{16}\right)$
17. Write as decimals:
    a. $\frac{3}{4}$ **b.** $\frac{1}{15}$
18. Write as a quotient of two integers:
    a. $0.\overline{12}$ **b.** $2.6555 \ldots$
19. **a.** Write 23.508 in expanded form.
    b. Write $(8 \times 10^2) + (3 \times 10^0) + (4 \times 10^{-2})$ in decimal form.
20. Do the following calculation and write the answer in scientific notation: $(6 \times 10^4) \times (8 \times 10^{-6})$

CHAPTER 5

THE REAL AND THE COMPLEX NUMBERS

Pythagoras was one of the dominant figures of Greek mathematics in the period immediately preceding Euclid. Many legends but very few facts are known about Pythagoras' life. It seems probable that he was born on the island of Samos about 569 B.C., and that as a young man he traveled to Egypt and Babylonia. It is known that he settled in Crotona in southern Italy, and that in about 540 B.C. he formed a secret society (known as the Pythagoreans) with some 300 young aristocrats. Among other things, Pythagoras taught this group to worship numbers; never to eat beans; and always to remain anonymous and sign the name of the Pythagoreans to any writing or discovery.

The motto of the Pythagoreans, "Number rules the universe," expressed their philosophy and indicated their belief in a mystic relationship between number and reality. The esoteric character of the society eventually aroused the suspicions of the people of Crotona, who drove the Pythagoreans out and burned their buildings. One story says that Pythagoras died in the flames; another that he fled, only to be murdered in another town. All we know for sure is that his life ended about 501 B.C.

The Pythagoreans must be credited with two of the greatest contributions in the history of mathematics. They were the first to treat mathematical concepts as abstractions and to employ the process of deductive proof exclusively and systematically. They gave the first clear answer to the question, "What is a mathematical proof?" Pythagoras himself is said to have given the first proof of the famous theorem that bears his name, the Pythagorean theorem: *The square of the hypotenuse of a right triangle equals the sum of the squares of the two sides.* Although the Babylonians and the Egyptians had been aware of the theorem itself for many centuries, there seems to be no evidence that anyone before the Pythagoreans had tried to prove it. No one knows who first discovered the relationship among the sides of a right triangle.

Pythagoras' second major contribution was the discovery that $\sqrt{2}$ is an irrational number—that it cannot be expressed exactly as the ratio of two whole numbers. To the Pythagoreans, this was a dreadful thing, and it is said that they threatened death to any member who revealed the awful secret. Before this, Pythagoras had believed and preached that everything in the universe is built on the whole numbers. "God is number," he said; but one small mathematical discrepancy demolished all his mystic philosophy.

PYTHAGORAS (ca. 569–501 B.C.)

Picture Collection, the Branch Libraries, The New York Public Library

Pythagoras, mystic, mathematician, investigator of nature to the best of his self-hobbled ability, "one-tenth of him genius, nine-tenths sheer fudge."

E. T. BELL

DECIMALS AND THE REAL NUMBERS

In Chapter 4 we presented the usefulness and beauty of the set of integers. The Pythagoreans (see the introduction) were so certain that the entire universe was made up of the whole numbers that they classified them into categories such as "perfect" and "amicable." They labeled the even numbers as "feminine" and the odd numbers as "masculine," except 1, which was the generator of all other numbers. (At that time, the symbol for marriage was the number 5, the sum of the first feminine number, 2, and the first masculine number, 3.) In the midst of these charming fantasies, the discovery of a new type of number was made — a type of number so unexpected that the brotherhood tried to suppress their discovery. They had found the numbers that we call **irrational numbers** today.

"We have reason to believe Bingleman is an irrational number himself."

Courtesy Sydney Harris

Here is a general idea of what happened: Suppose you draw a number line 2 units long (see Fig. 5.1a):

Figure 5.1a

We divide the unit interval into 2 equal parts and graph $\frac{1}{2}$, $\frac{2}{2} = 1$, $\frac{3}{2}$, and $\frac{4}{2} = 2$. We then proceed in the same way and divide the unit interval into three parts, marking $\frac{1}{3}$, $\frac{2}{3}$, $\frac{3}{3} = 1$, $\frac{4}{3}$, $\frac{5}{3}$, and $\frac{6}{3} = 2$ as shown. For any whole number q, we can divide the unit interval into q equal parts and then

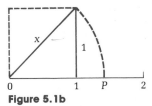

Figure 5.1b

graph $1/q$, $2/q$, and so on. It seems reasonable to assume that this process continued indefinitely would assign a rational number to every point on the line.

Now, suppose we construct a unit square and draw its diagonal as shown in Fig. 5.1b. Turn this diagonal (called the **hypotenuse** of the resulting triangle) clockwise to coincide with the number line extending from 0 to a point marked P. What rational number corresponds to P? None! Why? Because the required number is not obtainable by dividing any whole number by another whole number; it is not a rational number.

Ironically enough, Pythagoras himself had proved the famous theorem bearing his name: *The square of the hypotenuse of a right triangle is equal to the sum of the squares of the other two sides.* If we let x be the length of this hypotenuse (the diagonal of the square in Fig. 5.1b), the theorem says that

$$x^2 = 1^2 + 1^2 = 2$$

so that

$x = \sqrt{2}$ Square root of 2, a positive number
 that when squared yields 2

The Pythagoreans were able to prove that $\sqrt{2}$ is not obtainable by dividing any whole number by another, that is, that $\sqrt{2}$ is not a rational number. To do so, they used a method of proof called *reductio ad absurdum*, meaning "reduction to the absurd" (we now call this **proof by contradiction**). Their proof may have gone as follows:

Proof   Assume that $\sqrt{2}$ is a rational number, say a/b, in lowest terms; that is, a and b have no common factor other than 1. Thus,

1. $$\sqrt{2} = \frac{a}{b}$$

2. Multiply by b: $\sqrt{2}\, b = a$
3. Square both sides: $(\sqrt{2}\, b)^2 = a^2$
4. Simplify: $2b^2 = a^2$
5. Thus, a^2 is an even number, and since only even numbers have squares that are even, a is an even number. Suppose $a = 2c$.
6. Substitute $a = 2c$ in step 4:
$$2b^2 = (2c)^2 = 4c^2$$
7. Divide by 2: $b^2 = 2c^2$
8. This last equation says that b^2 is an even number, so that b is an even number.
9. Now, we have a contradiction: Step 5 says that a is even, and step 8 says that b is even, which means that a and b would have the common factor 2, contrary to our assumption. Hence, our assumption is invalid, and $\sqrt{2}$ is not a rational number.   ◄

Numbers that are not rational are called **irrational.**

A. Square Roots

The proof above can be generalized to show that the only numbers with rational square roots are the **perfect squares:**

$$\sqrt{1} = 1 \quad \text{because} \quad 1^2 = 1$$
$$\sqrt{4} = 2 \quad \text{because} \quad 2^2 = 2$$
$$\sqrt{9} = 3 \quad \text{because} \quad 3^2 = 9$$

and so on. The square roots of all other natural numbers are irrational.

EXAMPLE 1 Classify as rational or irrational:

a. $\sqrt{36}$ **b.** $\sqrt{44}$ **c.** $\sqrt{81}$

Solution **a.** $\sqrt{36} = 6$; it is rational.   **b.** $\sqrt{44}$ is irrational.
c. $\sqrt{81} = 9$; it is rational.   ◀

B. Irrational Numbers and Decimal Numbers

Irrational numbers such as $\sqrt{2}$ can be approximated to any finite number of decimal places. But these decimal numbers can never repeat (as in $\frac{1}{3}$) or terminate (as in $\frac{1}{2}$), because if they did, they would be rational numbers. For example,

$$\sqrt{2} = 1.4142 \ldots \quad \text{or} \quad \sqrt{2} = 1.414213 \ldots$$

We use this idea to define irrational numbers.

▶ **Definition 5.1a** | An **irrational number** is a number that has a **nonterminating, nonrepeating** decimal representation.

For example, 0.909009000 . . . (the successive sets of digits are 90, 900, 9000, etc.) is nonterminating and nonrepeating, and thus is irrational. Another irrational number is the decimal 1.23456789101112 . . . , where we continue writing the digits of the successive counting numbers. Here again, although there is a definite pattern, the decimal is nonrepeating and nonterminating.

The set consisting of all decimals, terminating or nonterminating, repeating (that is, the rational numbers) and nonterminating, nonrepeating (that is, the irrational numbers) is called the set R of **real numbers.** (Keep in mind that the rationals can be written as quotients of two integers and the irrationals cannot.) The set R includes all the numbers we have studied: the natural numbers, the integers, the rational numbers, and the irrational numbers. The rationals and the irrationals completely cover the number line: To each point on the line there corresponds a unique real number, and to each real number there corresponds a unique point.

EXAMPLE 2   Classify the following numbers as rational or irrational:

   a. 0.35626262 . . .   **b.** 0.305300530005 . . .   **c.** 0.12345678
   d. $-\frac{1}{3}$ **e.** $\sqrt{65}$ **f.** $\sqrt{144}$

Solution   **a.** A repeating decimal; therefore, rational
   b. A nonrepeating, nonterminating decimal; therefore, irrational
   c. A terminating decimal; therefore, rational
   d. A fraction; therefore, rational
   e. Irrational
   f. $\sqrt{144} = 12$, which is rational   ◄

Looking at the decimal approximations of $\sqrt{2}$ given earlier, we can see that 1.4 is less than $\sqrt{2}$ but 1.5 is greater; that is, $\sqrt{2}$ is between 1.4 and 1.5. Can we always find an irrational number between any two rational numbers? In order to answer this question, we must first make the ideas of **less than** ($<$) and **greater than** ($>$) more precise, as in the following definition.

► **Definition 5.1b**

> If a and b are real numbers, then:
>
> **1.** $a < b$ if and only if there is a positive number c such that $a + c = b$.
> **2.** $b > a$ if and only if $a < b$.

Thus, $3 < 5$ because $3 + 2 = 5$; $-4 < -1$ because $-4 + 3 = -1$; $5 > 3$ because $3 < 5$; and $\frac{1}{3} = 0.333 \ldots < \frac{1}{2} = 0.5$ because $\frac{1}{3} + \frac{1}{6} = \frac{1}{2}$.

A basic property of the real numbers is given by the following statement.

> **The Trichotomy Law**
>
> If a and b are any real numbers, then exactly one of the following relations must occur:
>
> $a = b$ (1)
> $a < b$ (2)
> $a > b$ (3)

Thus, if $a \not> b$ (a is not greater than b), then $a \le b$ (a is less than or equal to b). And if $a \not< b$ (a is not less than b), then $a \ge b$ (a is greater than or equal to b).

We now return to the question, "Can we always find an irrational number between any two rational numbers?" The answer is affirmative. For example, to find an irrational number between $\frac{1}{3}$ and $\frac{1}{2}$, we first write $\frac{1}{3}$ and $\frac{1}{2}$ as decimals:

$$\tfrac{1}{2} = 0.5$$

$$\tfrac{1}{3} = 0.333 \ldots$$

Obviously, the number 0.4 is between 0.5 and 0.333 . . . ; however, this number is not irrational. We now add a nonrepeating, nonterminating part to this number as shown:

$\frac{1}{2} = 0.5$

    0.4101001000 . . .   Nonterminating, nonrepeating

$\frac{1}{3} = 0.333$ . . .

The number 0.4101001000 . . . is bigger than 0.333 . . . , smaller than 0.5, and irrational. We could have found infinitely many other numbers using a similar technique. Can you find two more?

EXAMPLE 3   Find:

a. A rational number between 0.121 and 0.122
b. An irrational number between 0.121 and 0.122

Solution   **a.** The rational number 0.1215 is between 0.121 and 0.122 as shown:

    0.121
    0.1215
    0.122

b. The irrational number 0.1215101001000 . . . is between 0.121 and 0.122 as shown:

    0.121
    0.1215101001000 . . .   Nonterminating, nonrepeating
    0.122   ◄

In the preceding discussion we have seen that we are able to find an irrational number between any two given rational numbers. We can show that it is also possible to find a rational number between any two given rational numbers. For example, given the rational numbers $\frac{4}{7}$ and $\frac{5}{7}$, we write

$$\frac{4}{7} = \frac{4 \cdot 2}{7 \cdot 2} = \frac{8}{14} \quad \text{and} \quad \frac{5}{7} = \frac{5 \cdot 2}{7 \cdot 2} = \frac{10}{14}$$

We can now see by inspection that one rational number between $\frac{4}{7}$ and $\frac{5}{7}$ is $\frac{9}{14}$. If the two fractions do not have the same denominator, then we can proceed as in the next example.

EXAMPLE 4   Find a rational number between $0.\overline{4}$ and $\frac{6}{13}$.

Solution   Since $0.\overline{4} = \frac{4}{9}$, the problem is equivalent to that of finding a rational number between $\frac{4}{9}$ and $\frac{6}{13}$. We can do this easily by changing to fractions with common denominators. Thus,

$$\frac{4}{9} = \frac{4 \cdot 13}{9 \cdot 13} = \frac{52}{117} \quad \text{and} \quad \frac{6}{13} = \frac{6 \cdot 9}{13 \cdot 9} = \frac{54}{117}$$

Therefore, an obvious choice for the required number is $\frac{53}{117}$. ◄

C. The Number π

Measure the circumference and diameter of a soda can. Divide the circumference by the diameter. What is your answer?

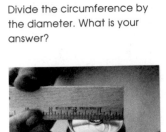

There is one more important irrational number that caused the Pythagoreans further difficulties. The ratio of the circumference of a circle C to its diameter d is itself an irrational number, 3.14159 . . . , which we symbolize by π (the Greek letter pi, probably inspired by the first letter in the Greek word *periphereia*, meaning "periphery"). Thus,

$$\frac{C}{d} = \pi$$

or, solving for the circumference,

$$C = \pi d$$

Actually proving that π is irrational is a very difficult mathematical problem and we shall simply accept the fact. The value of π has been approximated by various people throughout the ages. In the ancient Orient, the value was taken as 3; the Egyptians gave it a value of $(\frac{4}{3})^4$, which is about 3.16. Then about 240 B.C., Archimedes showed that π is between $\frac{223}{71}$ ($=3.1408$) and $\frac{22}{7}$ ($=3.1429$). Since that time, many people have used a variety of methods to calculate the value of π to a large number of decimal places. Now, the value of π is known to millions of decimal places. Here is the value to a mere 32 places:

$$\pi = 3.141\ 592\ 653\ 589\ 793\ 238\ 462\ 643\ 383\ 279\ 50$$

In most applications, however, we use only two decimal places, 3.14, but if greater accuracy is desired, the value 3.1416 is customarily used.

EXAMPLE 5   A manufacturer of cylindrical tanks wants to put a reinforcing steel strap around a 2-foot-diameter tank. The strap is to be 2 inches longer than the circumference to allow for riveting the overlap. How long should the strap be? (Answer to the nearest 0.1 inch.)

Solution   We use the formula $C = \pi d$, taking π as 3.14. Thus,

$$C = 3.14 \times 2 \text{ ft}$$
$$= 3.14 \times 2 \times 12 \text{ inches}$$
$$= 75.4 \text{ inches}$$

Adding the 2-inch overlap, we get the required length to be 77.4 inches. ◄

A. In Problems 1–3, classify the given numbers as rational or irrational.

    1. a. $\sqrt{120}$ **b.** $\sqrt{121}$

    2. a. $\sqrt{125}$ **b.** $\sqrt{169}$

    3. a. $\sqrt{\frac{9}{16}}$ **b.** $\sqrt{\frac{9}{15}}$

B. In Problems 4–10, classify the given numbers as rational or irrational.

    4. a. $\frac{3}{5}$ **b.** $-\frac{22}{7}$

    5. a. $-\frac{5}{3}$ **b.** -0

    6. a. 0.232323 . . .   **b.** 0.023002300023 . . .

    7. a. 0.121231234 . . .   **b.** 0.121231234

    8. a. $6\frac{1}{4}$ **b.** $\sqrt{6\frac{1}{4}}$

    9. a. 0.24681012 . . .   **b.** 0.1122334455

    10. a. 3.1415   **b.** π

In Problems 11–20, insert $<$, $>$, or $=$, as appropriate.

11. 3 _____ 4   **12.** 17 _____ 11

13. 0.333 . . . _____ 0.333444 . . .   **14.** $\frac{1}{5}$ _____ $\frac{1}{4}$

15. $\frac{12}{19}$ _____ $\frac{11}{17}$

16. 0.101001000 _____0.1101001000 . . .

17. 0.999 . . . _____ 1

18. 0.333 . . . $+$ 0.666 . . . _____ 1

19. 3(0.333 . . .) _____ 1

20. 0.112233 _____ 0.111222333 . . .

21. Find a rational number between 0.31 and 0.32.

22. Find a rational number between 0.28 and 0.285.

23. Find an irrational number between 0.31 and 0.32.

24. Find an irrational number between 0.28 and 0.285.

25. Find a rational number between 0.101001000 . . . and 0.102002000 . . . .

26. Find a rational number between 0.303003000 . . . and 0.304004000 . . . .

27. Find an irrational number between 0.101001000 . . . and 0.102002000 . . . .

28. Find an irrational number between 0.303003000 . . . and 0.304004000 . . . .

29. Find a rational number between $\frac{3}{11}$ and $\frac{4}{11}$.

30. Find a rational number between $\frac{7}{9}$ and $\frac{9}{11}$.

31. Find an irrational number between $\frac{4}{9}$ and $\frac{5}{9}$.

32. Find an irrational number between $\frac{2}{11}$ and $\frac{3}{11}$.

33. Find a rational number between $0.\overline{5}$ and $\frac{2}{3}$.

34. Find a rational number between 0.1 and $0.\overline{1}$.

In Problems 35 and 36, list the given numbers in order from smallest to largest.

35. 0.21, 0.212112111 . . . , 0.21211, 0.2121, 0.21212

36. 3.14, 3.1414, 3.141411411 . . . , 3.141, 3.1

In Problems 37–40, use the approximate value 3.14 for π.

37. The largest circular crater in northern Arizona is 5200 feet across. If you were to walk around this crater, how many miles would you walk? (1 mi = 5280 ft.)

38. The Fermi National Accelerator Laboratory has a circular atom smasher that is 6562 feet in diameter. Find the distance in miles that a particle travels in going once around in this accelerator. (1 mi = 5280 ft.)

39. The U.S. Department of Energy is studying the possibility of building a circular superconductivity collider that will be 52 miles in diameter. How far would a particle travel in going once around this collider? Give your answer to the nearest mile.

40. The smallest functional record, a rendition of "God Save the King," is $1\frac{3}{8}$ inches in diameter. To the nearest hundredth of an inch, what is the circumference of this record?

USING YOUR KNOWLEDGE 5.1

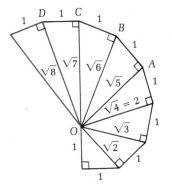

The figure shows an interesting spiral made up of successive right triangles. This spiral can be used to construct lengths corresponding to the square roots of the integers on the number line. For example, the second triangle has sides 1 and $\sqrt{2}$, so if the hypotenuse is of length x, then

$$x^2 = 1^2 + (\sqrt{2})^2$$
$$= 1 + 2 = 3$$

Therefore, $x = \sqrt{3}$.

1. Verify that the ray OB is of length $\sqrt{6}$.
2. Verify that the ray OD is of length $\sqrt{8}$.
3. Find the length of the hypotenuse of the triangle shown in color.
4. If the triangles are numbered 1, 2, 3, . . . , starting with the lowest triangle in the figure, what would be the number of the triangle whose hypotenuse is of length $\sqrt{17}$?

OPERATIONS WITH DECIMALS

We have already studied the operations of addition, subtraction, multiplication, and division with whole numbers, integers, and fractions. We now consider the same operations with decimals.

A. Addition and Subtraction

To add or subtract decimals, we proceed as with whole numbers, making sure that the numbers to be added have the *same* number of decimal places. Thus, to add 5.1 and 2.81, we attach a **0** to 5.1 so that it has two decimal places and then add. To see why this works, we write the numbers in expanded form and add as follows:

$$5.10 = 5 + \tfrac{1}{10} + \tfrac{0}{100}$$
$$\underline{2.81 = 2 + \tfrac{8}{10} + \tfrac{1}{100}}$$
$$7.91 = 7 + \tfrac{9}{10} + \tfrac{1}{100}$$

In practice, we place the numbers to be added in a vertical column *with the decimal points aligned*, and then add or subtract as required. The rules for "carrying" in addition and "borrowing" in subtraction are justified here also, as shown by the expanded forms in the next examples.

EXAMPLE 1   Add 4.81 and 3.7.

Solution   We attach a 0 to the 3.7, align the decimal points, and add.

Short Form	Expanded Form

Short Form

$$4.81$$
$$\underline{+3.70} \leftarrow \text{Attach a 0}$$
$$8.51$$

Expanded Form

$$4 + \tfrac{8}{10} + \tfrac{1}{100}$$
$$\underline{3 + \tfrac{7}{10} + \tfrac{0}{100}}$$
$$7 + \tfrac{15}{10} + \tfrac{1}{100}$$
$$= 7 + \tfrac{10}{10} + \tfrac{5}{10} + \tfrac{1}{100}$$
$$= 8 + \tfrac{5}{10} + \tfrac{1}{100}$$
$$= 8.51$$

◄

EXAMPLE 2   Subtract 6.53 from 8.71.

Solution

Short Form   **Expanded Form**

$$8.71$$
$$-6.53$$
$$\overline{2.18}$$

$$8 + \tfrac{7}{10} + \tfrac{1}{100} = 8 + \tfrac{6}{10} + \tfrac{11}{100}$$
$$6 + \tfrac{5}{10} + \tfrac{3}{100} = 6 + \tfrac{5}{10} + \tfrac{3}{100}$$
$$\overline{\qquad\qquad 2 + \tfrac{1}{10} + \tfrac{8}{100} = 2.18}$$

Note: In the first line of the expanded form, we wrote

$$\tfrac{7}{10} = \tfrac{6}{10} + \tfrac{1}{10} = \tfrac{6}{10} + \tfrac{10}{100}$$

and then combined the hundredths.   ◀

B. **Multiplication and Division**

To understand the rule for multiplying decimals, look at the following example, where the decimals are first replaced by equivalent fractions.

$$\tfrac{1}{100} \times \tfrac{1}{10} = \tfrac{1}{1000}$$

EXAMPLE 3   Multiply 0.37 × 7.2.

Solution $$0.37 \times 7.2 = \frac{37}{100} \times \frac{72}{10} = \frac{37 \times 72}{1000}$$

Must be able ✱ *to justify this*

The numerator 37 × 72 tells us that the 37 and the 72 are to be multiplied as usual for whole numbers. The 1000 in the denominator tells us that the final answer will have three decimal places, the sum of the number of decimal places in the two factors. Thus, we can multiply the 0.37 and the 7.2 just as if they were whole numbers and then point off three decimal places in the product. The short form of the multiplication follows:

$$0.37 \leftarrow 2 \text{ decimal places}$$
$$\times \quad 7.2 \leftarrow 1 \text{ decimal place}$$
$$\overline{\quad 74 \qquad\qquad}$$
$$\underline{\quad 259 \qquad\qquad}$$
$$2.664 \leftarrow 2 + 1 = 3 \text{ decimal places}$$ ◀

If signed numbers are involved, the rules of signs apply. Thus,

$$(-0.37) \times 7.2 = -2.664$$
$$0.37 \times (-7.2) = -2.664$$
$$(-0.37) \times (-7.2) = 2.664$$

A division involving decimals is easier to understand if the divisor is a whole number. If the divisor is not a whole number, we can make it so by moving the decimal point to the right the same number of places in both the divisor and the dividend. This procedure can be justified as in the next example.

EXAMPLE 4   A 6½ ounce can of tuna fish costs 55¢. What is the cost per ounce to the nearest tenth of a cent?

Solution   The cost per ounce can be obtained by dividing the total cost by the number of ounces. Thus, we have to find $55 \div 6.5$, and we can make the divisor a whole number by writing the division in fractional form:

$$\frac{55}{6.5} = \frac{550}{65} \quad \text{Multiply numerator and denominator by 10.}$$

This is equivalent to moving the decimal point one place to the right in both the divisor and the dividend. Now, we divide in the usual way:

```
        8.46
   65) 550.00
        520
         30 0
         26 0
          4 00
          3 90
            10
```

To two decimal places, the cost per ounce is 8.46¢. Therefore, to the nearest tenth of a cent, the answer is 8.5¢.   ◄

Again, if signed numbers are involved, the rules of signs apply. For example,

$$\frac{-14}{2.5} = -5.6 \qquad \frac{14}{-2.5} = -5.6 \qquad \frac{-14}{-2.5} = 5.6$$

C.   Rounding

In Example 4, we **rounded off** the quotient 8.46 to one decimal place to get 8.5. This type of procedure is necessary in many practical problems and is done in the following way:

To Round Off Numbers

1. Underline the digit in the place to which you are rounding.
2. If the first digit to the right of the underlined digit is 5 or more, add 1 to the underlined digit. Otherwise, do not change the underlined digit.
3. Drop all digits to the right of the underlined digit, attaching 0's to fill in the place values if necessary.

For instance, to round off 8.46 to one decimal place, that is, to the tenths place:

1. Underline the 4:

8.4̲6

2. The digit to the right of the underlined digit is 6 (greater than 5). So add 1 to the underlined digit.
3. Drop all digits to the right of the underlined digit. The result is 8.5.

If it is necessary to round off 2632 to the hundreds place, you should underline the 6 to show 26̲32. Since the next digit is 3 (less than 5), the underlined 6 is unchanged and the 32 is replaced by 00 to give 2600.

In many applications, the numbers are obtained by measurement and are usually only approximate numbers correct to the stated number of decimal places. When calculations are performed with approximate numbers, two rules are customarily used to avoid presenting results with a false appearance of accuracy. The rule for addition and subtraction is:

The result of an addition or subtraction of approximate numbers should be given to the number of decimal places possessed by the approximate number with the least number of decimal places.

EXAMPLE 5   The three sides of a triangle are measured to be approximately 26.3, 7.41, and 20.64 cm long, respectively. What is the perimeter of the triangle?

Solution   The perimeter is the total distance around the triangle, so we must add the lengths of the three sides. We get

$$\begin{array}{r} 26.3 \\ 7.41 \\ +20.64 \\ \hline 54.35 \end{array}$$

However, the least precise of the measurements is 26.3, with just one decimal place. Therefore, we round off the answer to 54.4 cm.   ◄

Before giving the rule that is used when multiplying or dividing approximate numbers, we must explain what is meant by the **significant digits** of a number.

1. The digits 1, 2, 3, 4, 5, 6, 7, 8, 9 are always significant.
2. The digit 0 is significant if it is preceded and followed by other significant digits.
3. The digit 0 is significant unless its only purpose is to place the decimal point.

Thus, all the digits in the numbers 2.54 and 1.73205 are significant. However, the number 0.00721 has only three significant digits: 7, 2, and 1;

the 0's in this number do nothing except place the decimal point. There is one ambiguous case left. For example, how many significant digits does the number 73,200 have? We cannot answer this question unless we know more about the final 0's. To avoid this type of difficulty, we write 73,200 in the form

7.32×10^4,   if three digits are significant
7.320×10^4,   if four digits are significant
7.3200×10^4,   if all five digits are significant

EXAMPLE 6   State the number of significant digits in each of the following:

a. 0.50 inch   **b.** 0.05 pound
c. 8.2×10^3 liters   **d.** 8.200×10^3 kilometers

Solution   **a.** 0.50 has two significant digits.
b. 0.05 has one significant digit. The 0 before the 5 only places the decimal point.
c. 8.2×10^3 has two significant digits.
d. 8.200×10^3 has four significant digits.   ◀

We can now state the multiplication and division rule for approximate numbers:

> The result of a multiplication or division of approximate numbers should be given with the same number of significant digits as are possessed by the approximate number with the fewest significant digits.

For instance, if we wish to convert 6 centimeters to inches, we note that 1 cm is approximately 0.394 in. Therefore, we multiply

$6 \times 0.394 = 2.364$

However, 0.394 has only three significant digits, so we must round our answer to three significant digits. Our final result is

6 cm ≈ 2.36 in.   ≈ means "approximately equal to"

Notice carefully that the 6 in 6 cm is an *exact*, not an approximate, number.

EXAMPLE 7   Two adjacent sides of a rectangular lot are measured to be approximately 91.5 feet and 226.6 feet. Find the area of this lot.

Solution   The area of a rectangle of sides a and b is $A = ab$. Thus, we have

$A = 91.5 \times 226.6$
$= 20,733.90$

However, the 91.5 has only three significant digits, so our answer must be rounded to three significant digits. This gives 20,700 square feet for the area. A better form for the answer is 2.07×10^4 square feet. Use of scientific notation frequently helps to display only the significant digits of the answer. ◄

EXAMPLE 8 A rectangular piece of silver approximately 0.5 inch by 6.75 inches is to be formed into a circular bracelet. If the ends are to meet exactly, what will be the diameter of the bracelet? Use the approximate value 3.14 for π.

Solution We know that the circumference, C, of a circle is given by the formula $C = \pi d$, where d is the diameter. Thus, we have, upon dividing by π,

$$d = \frac{C}{\pi} = \frac{6.75}{3.14}$$
$$= 2.14968 \quad \text{By direct division}$$

Since the 6.75 and the value used for π both have only three significant digits, we must round our answer to three significant digits. This gives $d = 2.15$ in. ◄

D. Percent

In many of the daily applications of decimals, information is given in terms of **percents.** The interest rate on a mortgage may be 12% (read, "12 percent"), the Dow Jones stock average may increase by 2%, your savings account may earn interest at 6%, and so on. The word "percent" comes from the Latin words *per* and *centum* and means "by the hundred." Thus, 12% is the same as $\frac{12}{100}$ or the decimal 0.12.

> **To Change a Percent to a Decimal**
>
> Move the decimal point in the number two places to the left and omit the % symbol.

EXAMPLE 9 Write as a decimal:

    a. 18%   **b.** 11.5%   **c.** 0.5%

Solution   **a.** 18% = 0.18   **b.** 11.5% = 0.115   **c.** 0.5% = 0.005   ◄

To change a decimal to a percent, we just reverse the procedure.

> **To Change a Decimal to a Percent**
>
> Move the decimal point two places to the right and affix the percent sign.

EXAMPLE 10   Change to a percent: **a.** 0.25   **b.** 1.989

Solution   **a.** 0.25 = 25%   **b.** 1.989 = 198.9%   ◄

To Change a Fraction to a Percent

Divide the numerator by the denominator, and then convert the resulting decimal to a percent.

EXAMPLE 11   Write as a percent: **a.** $\frac{2}{5}$ **b.** $\frac{3}{7}$ (Give answer to one decimal place.)

Solution   **a.** $\frac{2}{5}$ = 0.40 = 40%   **b.** $\frac{3}{7}$ = 0.42856   By ordinary division
                                             = 0.429   Rounded to three decimal places
                                             = 42.9%   ◄

EXAMPLE 12   When you buy popcorn at the theater, you get popcorn, butter substitute, and a bucket. Which of these is the most expensive? The bucket! Here are the costs to the theater: popcorn, 5¢; butter substitute, 2¢; bucket, 25¢.

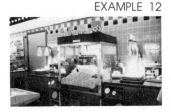

a. Find, to two decimal places, the percent of the total cost for each of these.

b. If the average profit on this popcorn is 86% of the cost, how much is that?

Solution   **a.** Cost of popcorn is $\frac{5}{32}$ = 0.15625, or 15.63%.
            Cost of butter substitute is $\frac{2}{32}$ = 0.0625, or 6.25%.
            Cost of bucket is $\frac{25}{32}$ = 0.78125, or 78.13%.
    b. 86% of 32 = 0.86 × 32 = 27.52, or nearly 28¢.   ◄

EXAMPLE 13   Theater concession stands make 25% of all the popcorn sales in the United States. If these stands make about 250 million sales annually, about how many popcorn sales are made per year in the United States?

Solution   We know that 25% of all sales is about 250 million. If we let s be the total number of sales, then

$$0.25s = 250 \text{ million}$$

Thus, dividing both sides by 0.25, we get

$$s = \frac{250 \text{ million}}{0.25}$$
$$= 1000 \text{ million, or about 1 billion sales per year} \qquad ◄$$

E.   Percent Increase or Decrease

EXAMPLE 14   **a.** In a certain company, male engineers earned $23,000, while female engineers earned $15,500. To the nearest percent, how much more did male engineers earn?

b. Male managers in a certain manufacturing company earned an average of $24,300. Their female counterparts earned about $13,300. To the nearest percent, how much more did the male managers make?

Solution **a.** The difference in salaries is $23,000 − $15,500 = $7500. Thus, the male engineers earned

$$\frac{7500}{15,500} = 0.484$$

or a whopping 48% more. (Note that the female engineers' salary was in the denominator of the fraction, because that salary was the basis for comparison.)

b. Here, the salary difference is $24,300 − $13,300 = $11,000. Thus, the male managers earned

$$\frac{11,000}{13,300} = 0.827$$

or an incredible 83% more. ◀

EXERCISE 5.2

In Problems 1–8, perform the indicated operations.

1. a. $3.81 + 0.93$ **b.** $-3.81 + (-0.93)$

2. a. $18.64 - 0.983$ **b.** $-18.64 - 0.983$

3. a. $2.08 - 6.238$ **b.** $3.07 - 8.934$

4. a. 2.48×2.7 **b.** $(-2.48) \times (-2.7)$

5. a. $(-0.03) \times (-1.5)$ **b.** $(-3.2) \times (-0.04)$

6. a. $10.25 \div 0.05$ **b.** $2.16 \div 0.06$

7. a. $(-0.07) \div 1.4$ **b.** $(-0.09) \div (-4.5)$

8. a. $(-1.8) \div (0.09)$ **b.** $3.6 \div (-0.012)$

9. The revenues (in millions) of City Investing were derived from the following sources: manufacturing, $926.3; printing, $721.21; international, $488.34; housing, $674; food services, $792.45; insurance, $229.58. Find the total revenue.

10. The sides of a triangular lot are measured to be 392.1, 307.25, and 507 feet. Find the perimeter of the lot to the correct number of decimal places.

11. The mileage indicator on a car read 18,327.2 miles at the beginning of a trip. It read 18,719.7 at the end. How far did the car go?

12. A business reported annual sales of $8.5 million. If the expenses were reported as $6.52 million, find the profit to the correct number of decimal places.

13. The average weekly circulation of T.V. Guide is 19,230,000 copies. If each copy is sold for 75¢, how much money (on the average) do sales amount to each week?

14. On October 17, 1965 the *New York Times* contained a total of 946 pages. If the thickness of one sheet of newspaper is 0.0040 inch, find (to the correct number of decimal places) the thickness of this edition of the paper.

15. A light-year is about 5.878×10^{12} miles. The *Great Galaxy* in *Andromeda* is about 2.15 million light-years away. How many miles is the *Great Galaxy* from us? (Write your answer in scientific notation.) By the way, this galaxy is the most remote heavenly body visible to the naked eye.

16. Lionel Harrison drove the 1900 miles from Oxford, England, to Moscow in a Morris Minor fitted with a 62-gallon tank. If he used all the 62 gallons of gas, how many miles per gallon did he get? (Round the answer to the nearest tenth.)

17. Stuart Bladon holds the record for the greatest distance driven without refueling, 1150.3 miles on 19.41 gallons of fuel. To the correct number of decimal places, how many miles per gallon is that?

18. The highest average yardage gain for a football season belongs to Beattie Feathers. He gained 1004 yards in 101 carries. What was his average number of yards per carry? (Answer to the nearest tenth of a yard.)

19. In 1925, a 3600-revolution-per-minute (rpm) motor was attached to a gramaphone with a 46 : 1 gear ratio. The resulting speed was $\frac{3600}{46}$ rpm. What is this ratio to the nearest whole number? This ratio gave birth to a new kind of record, and if you find the answer, you will find what kind of record we mean.

20. The longest and heaviest freight train on record was one about 21,120 feet in length and consisting of 500 coal cars. To the nearest foot, how long was each coal car?

21. Write as a decimal:
   a. 29% **b.** 23.4% **c.** 0.9%

22. Write as a decimal:
   a. 56.9% **b.** 45.69% **c.** 0.008%

23. Write as a decimal:
   a. 34.15% **b.** 93.56% **c.** 0.0234%

24. Write as a percent:
   a. 0.38 **b.** 3.45 **c.** 9.998

25. Write as a percent:
   a. 0.567 **b.** 0.0045 **c.** 9.003

26. Write as a percent:
   a. 0.0004 **b.** 0.0045 **c.** 0.0008

27. Write as a percent:
   a. $\frac{3}{5}$ **b.** $\frac{4}{7}$ (to one decimal place)

28. Write as a percent:
   a. $\frac{5}{6}$ (to one decimal place) **b.** $\frac{7}{8}$

29. In a recent year, about 16,300,000 cars were sold by the U.S. auto industry. Of these, 8,214,671 were made in the United States. What percent is that? (Answer to one decimal place.)

30. The most expensive British standard car is the Rolls-Royce Phantom VI, quoted at $312,954. An armor-plated version was quoted at $560,000. What percent more does the armor plating cost?

31. By weight, the average adult is composed of 43% muscle, 26% skin, 17.5% bone, 7% blood, and 6.5% organs. If a person weighs 150 pounds:

    a. How many pounds of muscle does the person have?
    b. How many pounds of skin does the person have?

32. Referring to Problem 31:

    a. How many pounds of bone does the person have?
    b. How many pounds of organs does the person have?

33. A portable paint compressor is priced at $196.50. If the sales tax rate is 5.5%, what is the tax?

34. The highest recorded shorthand speed was 300 words per minute for 5 minutes with 99.64% accuracy. How many errors were made?

35. In a recent year, the United States had 168,607,000 cars. This represented about 37.9% of the total world stock of cars. How many cars were in the world stock?

36. In a recent year, 41.2 million households with televisions watched the Superbowl. This represented 47.1% of homes with televisions in major cities. How many homes with televisions are there in major cities?

37. Joe Long is a computer programmer. His salary was increased by $8000. If his previous salary was $28,000, what was his percent increase? (Answer to the nearest percent.)

38. Felicia Perez received a $1750 annual raise from the state of Florida. If her salary was $25,000 before the raise, what percent raise did she receive?

39. Joseph Clemons had a salary of $20,000 last year. This year, his salary was increased to $22,000. What percent increase is this?

40. Here are some professions and the average salaries earned by males and females. To the nearest percent, how much more did males make?

	Male	Female
a. Doctors and dentists	$90,000	$72,000
b. Elementary and high school teachers	$21,600	$17,280
c. Sales clerks	$18,000	$9,180

If you are an incurable romantic, read on and weep! If you are not, then read on anyway. You will be glad you do not have to indulge in these costs of loving. For Problems 1–8, find the percent change.

SOME COSTS OF LOVING

	Item	Mid-1950s	Recent Price	Percent Change
1.	French perfume (an ounce)	$35.00	$515.00	
2.	Tips to circling violinists	.50	5.00	
3.	Dom Pérignon champagne	12.00	65.00	
4.	A movie (first run)	1.00	5.00	
5.	Intimate dinner	9.00	40.00	
6.	Beer and cheeseburger	1.00	4.00	
7.	Mai Tai	1.25	4.25	
8.	Soave Bolla	1.55	4.39	

Reprinted from the May 1981 issue of *Money Magazine* by special permission. © 1981, Time, Inc.

▶ **Calculator Corner 5.2** The operations of addition, subtraction, multiplication, and division involving decimals are easy to perform with a calculator. You just press the appropriate numbers and indicated operations and the calculator does the rest! However, you must know how to round off answers using the rules we have discussed. Thus, in Example 5, you enter $26.3 + 7.41 + 20.64$, but you have to know that the answer must be rounded off to 54.4.

Some calculators have a $\boxed{\%}$ key that changes percents to decimals. You can determine whether your calculator does this by entering 11.5 $\boxed{\text{2nd}}$ $\boxed{\%}$. The answer should appear as 0.115 (Example 9b).

1. Use your calculator to check Problems 21 and 23 in Exercise 5.2.

5.3

OPERATIONS WITH RADICALS

Photograph courtesy of Parachutes, Inc.

What is the velocity v (in feet per second) of the man in the picture? It depends on the distance d he has fallen. The formula is

$$v = \sqrt{32d}$$

Thus, after he has fallen 1 foot ($d = 1$), his velocity is $\sqrt{32}$, and after 2 feet ($d = 2$), it is $\sqrt{64}$. The number $\sqrt{32}$ is an irrational number, because 32 is not a perfect square, but 64 is a perfect square ($8^2 = 64$), so that $\sqrt{64} = 8$ is a rational number. Note that $\sqrt{32}$ is positive. In general, if n is positive, $\sqrt{n}$ is positive.

A. Simplifying Radicals

Can we simplify $\sqrt{32}$? We say that $\sqrt{n}$ is in **simplest form** if n has no factor (other than 1) that is a perfect square. Using this definition, we can see that $\sqrt{32}$ is not in simplest form, because the perfect square 16 is a factor of 32. The simplification can be done using the following property:

If a and b are nonnegative real numbers, then

$$\sqrt{a \cdot b} = \sqrt{a} \cdot \sqrt{b}$$

Thus,

$$\sqrt{32} = \sqrt{16 \cdot 2} = \sqrt{16} \cdot \sqrt{2} = 4\sqrt{2}$$

In general, the simplest form of a number involving the radical sign $\sqrt{}$ is obtained by using the perfect squares 1, 4, 9, 16, 25, 36, 49, 64, 81, 100, and so on, as factors under the radical and then using the property given above, as in the next example.

EXAMPLE 1   Simplify if possible:

a. $\sqrt{75}$ **b.** $\sqrt{70}$

Solution   **a.** The largest perfect square dividing 75 is 25. Thus, we write

$$\sqrt{75} = \sqrt{25 \cdot 3} = \sqrt{25} \cdot \sqrt{3} = 5\sqrt{3}$$

b. There is no perfect square (except 1) that divides 70. (Try dividing by 4, 9, 16, 25, 36.) Thus, $\sqrt{70}$ cannot be simplified any further.   ◄

B. Rationalizing the Denominator

The property $\sqrt{a \cdot b} = \sqrt{a} \cdot \sqrt{b}$ can be used to **rationalize** the denominator of certain expressions, that is, to free the denominator of radicals. Thus, if we wish to rationalize the denominator in the expression $6/\sqrt{3}$, we use the fundamental principle of fractions and multiply the numerator and the denominator of the fraction by $\sqrt{3}$, as follows:

$$\frac{6}{\sqrt{3}} = \frac{6 \cdot \sqrt{3}}{\sqrt{3} \cdot \sqrt{3}} = \frac{6 \cdot \sqrt{3}}{\sqrt{9}} = \frac{6 \cdot \sqrt{3}}{3} = 2\sqrt{3}$$

EXAMPLE 2   Rationalize the denominator in the expression $5/\sqrt{10}$.

Solution   We multiply the numerator and the denominator by $\sqrt{10}$ and then simplify. Thus,

$$\frac{5}{\sqrt{10}} = \frac{5 \cdot \sqrt{10}}{\sqrt{10} \cdot \sqrt{10}} = \frac{5 \cdot \sqrt{10}}{\sqrt{100}} = \frac{5 \cdot \sqrt{10}}{10} = \frac{\sqrt{10}}{2}$$ ◄

C. Quotients of Radicals

Can we simplify $\sqrt{\frac{3}{4}}$? (This is one of the two answers you will get if you solve the equation $x^2 = \frac{3}{4}$.) This time, the perfect square 4 appears in the denominator, so to simplify the expression, we use the following property:

If a and b are positive numbers, then

$$\sqrt{\frac{a}{b}} = \frac{\sqrt{a}}{\sqrt{b}}$$

Thus,

$$\sqrt{\frac{3}{4}} = \frac{\sqrt{3}}{\sqrt{4}} = \frac{\sqrt{3}}{2}$$

EXAMPLE 3 Simplify:

a. $\sqrt{\dfrac{32}{25}}$ b. $\sqrt{\dfrac{36}{7}}$

Solution a. $\sqrt{\dfrac{32}{25}} = \dfrac{\sqrt{32}}{\sqrt{25}} = \dfrac{\sqrt{32}}{5} = \dfrac{\sqrt{16 \cdot 2}}{5} = \dfrac{4 \cdot \sqrt{2}}{5}$

b. $\sqrt{\dfrac{36}{7}} = \dfrac{\sqrt{36}}{\sqrt{7}} = \dfrac{6}{\sqrt{7}}$

But now we must rationalize the denominator by multiplying the numerator and the denominator of the fraction $6/\sqrt{7}$ by $\sqrt{7}$, obtaining

$$\frac{6 \cdot \sqrt{7}}{7}$$

as our final answer. An easier way to get this result would be to multiply the numerator and the denominator of the original fraction $\frac{36}{7}$ by 7 first, obtaining

$$\sqrt{\frac{36}{7}} = \sqrt{\frac{36 \cdot 7}{7 \cdot 7}} = \frac{\sqrt{36 \cdot 7}}{\sqrt{7 \cdot 7}} = \frac{6\sqrt{7}}{7}$$

Keep this in mind when working the exercises! ◀

D. Multiplication and Division of Radicals

The two properties we have presented can serve as the definitions for the multiplication and division of radicals. Thus,

$$\sqrt{6} \cdot \sqrt{2} = \sqrt{12} \qquad \text{Using the first property to multiply}$$
$$= \sqrt{4 \cdot 3} = 2\sqrt{3} \qquad \text{Using the first property to simplify}$$

Similarly,

$$\frac{\sqrt{32}}{\sqrt{2}} = \sqrt{\frac{32}{2}} = \sqrt{16} = 4 \quad \text{Using the second property}$$

EXAMPLE 4   Perform the indicated operations and simplify: **a.** $\sqrt{6} \cdot \sqrt{3}$ **b.** $\dfrac{\sqrt{40}}{\sqrt{5}}$

Solution   **a.** $\sqrt{6} \cdot \sqrt{3} = \sqrt{18} = \sqrt{9 \cdot 2} = 3\sqrt{2}$ **b.** $\dfrac{\sqrt{40}}{\sqrt{5}} = \sqrt{\dfrac{40}{5}} = \sqrt{8} = \sqrt{4 \cdot 2} = 2\sqrt{2}$ ◄

E. Addition and Subtraction of Radicals

The addition and subtraction of radicals can be accomplished using the distributive law. Thus, to add $5\sqrt{2} + 3\sqrt{2}$ or subtract $5\sqrt{2} - 3\sqrt{2}$, we write:

$$5\sqrt{2} + 3\sqrt{2} = (5 + 3)\sqrt{2} = 8\sqrt{2}$$

or

$$5\sqrt{2} - 3\sqrt{2} = (5 - 3)\sqrt{2} = 2\sqrt{2}$$

Sometimes, you may have to use the properties we mentioned before the additions or subtractions can be accomplished. Thus, to add $\sqrt{48} + \sqrt{27}$, we use the first property to write $\sqrt{48} = \sqrt{16 \cdot 3} = 4\sqrt{3}$ and $\sqrt{27} = \sqrt{9 \cdot 3} = 3\sqrt{3}$. We then have

$$\sqrt{48} + \sqrt{27} = 4\sqrt{3} + 3\sqrt{3} = 7\sqrt{3}$$

EXAMPLE 5   Perform the indicated operations:

   a. $\sqrt{50} - \sqrt{20}$ **b.** $\sqrt{75} + \sqrt{48} - \sqrt{147}$

Solution   **a.** $\sqrt{50} - \sqrt{20} = \sqrt{25 \cdot 2} - \sqrt{4 \cdot 2}$
$$= 5\sqrt{2} - 2\sqrt{2}$$
$$= 3\sqrt{2}$$
   b. $\sqrt{75} + \sqrt{48} - \sqrt{147} = \sqrt{25 \cdot 3} + \sqrt{16 \cdot 3} - \sqrt{49 \cdot 3}$
$$= 5\sqrt{3} + 4\sqrt{3} - 7\sqrt{3}$$
$$= 2\sqrt{3}$$ ◄

F. Applications

EXAMPLE 6   The greatest speed s (in miles per hour) at which a bicyclist can safely turn a corner of radius r feet is $s = 4\sqrt{r}$. Find the greatest speed at which a bicyclist can safely turn a corner with a 20-foot radius, and write the answer in simplest form.

Solution $s = 4\sqrt{r} = 4\sqrt{20} = 4\sqrt{4 \cdot 5} = 4 \cdot 2\sqrt{5} = 8\sqrt{5}$

This is slightly less than 18 miles per hour.   ◄

A. In Problems 1–10, simplify if possible.

1. $\sqrt{90}$ $3\sqrt{10}$
2. $\sqrt{72}$
3. $\sqrt{122}$
4. $\sqrt{175}$
5. $\sqrt{180}$
6. $\sqrt{162}$
7. $\sqrt{200}$
8. $\sqrt{191}$
9. $\sqrt{384}$
10. $\sqrt{486}$

B. In Problems 11–16, rationalize the denominator.

11. $\dfrac{3}{\sqrt{7}}$

12. $\dfrac{6}{\sqrt{5}}$

13. $\dfrac{-\sqrt{2}}{\sqrt{5}}$

14. $\dfrac{-\sqrt{3}}{\sqrt{7}}$

15. $\dfrac{4}{\sqrt{8}}$

16. $\dfrac{3}{\sqrt{27}}$

C. In Problems 17–24, simplify the given expression.

17. $\sqrt{\frac{3}{49}}$
18. $\sqrt{\frac{7}{16}}$
19. $\sqrt{\frac{4}{3}}$
20. $\sqrt{\frac{25}{11}}$
21. $\sqrt{\frac{8}{49}}$
22. $\sqrt{\frac{18}{25}}$
23. $\sqrt{\frac{18}{50}}$
24. $\sqrt{\frac{24}{75}}$

D. In Problems 25–34, perform the indicated operations and simplify.

25. $\sqrt{7} \cdot \sqrt{8}$
26. $\sqrt{5} \cdot \sqrt{50}$
27. $\sqrt{6} \cdot \sqrt{12}$
28. $\sqrt{10} \cdot \sqrt{5}$

29. $\dfrac{\sqrt{28}}{\sqrt{2}}$

30. $\dfrac{\sqrt{22}}{\sqrt{2}}$

31. $\dfrac{\sqrt{10}}{\sqrt{250}}$

32. $\dfrac{\sqrt{10}}{\sqrt{490}}$

33. $\dfrac{\sqrt{33}}{\sqrt{22}}$

34. $\dfrac{\sqrt{18}}{\sqrt{12}}$

E. In Problems 35–40, perform the indicated operations.

35. $6\sqrt{7} + \sqrt{7} - 2\sqrt{7}$
36. $\sqrt{3} + 11\sqrt{3} - 3\sqrt{3}$
37. $5\sqrt{7} - 3\sqrt{28} - 2\sqrt{63}$
38. $3\sqrt{28} - 6\sqrt{7} - 2\sqrt{175}$
39. $-3\sqrt{45} + \sqrt{20} - \sqrt{5}$
40. $-5\sqrt{27} + \sqrt{12} - 5\sqrt{48}$

F. 41. The time t (in seconds) it takes an object dropped from a certain distance (in feet) to hit the ground is

$$t = \sqrt{\frac{\text{Distance in feet}}{16}}$$

Find the time it takes an object dropped from a height of 50 feet to hit the ground, and write the answer in simplified form.

42. The time t (in seconds) it takes an object dropped from a certain distance (in meters) to hit the ground is

$$t = \sqrt{\frac{\text{Distance in meters}}{5}}$$

Find the time it takes an object dropped from a height of 160 meters to hit the ground, and write the answer in simplified form.

43. The compound interest r that is paid when you borrow $P and pay $A at the end of 2 years is

$$r = \sqrt{\frac{A}{P}} - 1$$

Find the rate when $100 is borrowed and the amount paid at the end of the 2 years is $144.

44. When you are at an altitude of a feet above the Earth, your view V_m (in miles) extends as far as a circle called the *horizon* and is given by

$$V_m = \sqrt{\tfrac{3}{2}\,a}$$

The greatest altitude reached in a manned balloon is 123,800 feet and was attained by Nicholas Piantanida.

a. In simplified form, what was the view in miles from this balloon?
b. If $\sqrt{1857} \approx 43$, what was the view in miles?

USING YOUR KNOWLEDGE 5.3

*At the beginning of this section we mentioned that after the man in the picture had fallen 1 foot, his velocity was $\sqrt{32} = 4\sqrt{2}$ feet per second. Can you estimate what $\sqrt{32}$ is? Mathematicians use a method called **interpolation** to approximate this answer. Since we know that $\sqrt{25} = 5$ and $\sqrt{36} = 6$, $\sqrt{32}$ should be between 5 and 6. If we place $\sqrt{25}$, $\sqrt{32}$, and $\sqrt{36}$ in a column, the interpolation is done as shown in the diagram:*

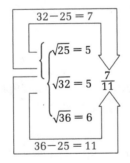

Thus, $\sqrt{32}$ is approximately $5\frac{7}{11}$. If we wish, we can write this answer as a decimal by dividing 7 by 11, obtaining $0.\overline{63}$ and writing the approximation as $5.\overline{63}$. (If you use a calculator to find the square root of 32, the actual answer is 5.6569.)

   Use this knowledge to approximate the following roots. Give each answer as a mixed number and then to two decimal places.

1. $\sqrt{40}$ $6\frac{4}{13} = 6.31$ **2.** $\sqrt{68}$ **3.** $\sqrt{85}$ $9\frac{4}{7}$ 9.21   **4.** $\sqrt{108}$

THE COMPLEX NUMBERS

In Chapter 4, we adjoined the number 0 to the set N of counting numbers to obtain the set W of whole numbers. We then adjoined the negative numbers, the additive inverses of the counting numbers, to construct the set I of integers. By dividing one integer by another nonzero integer, we obtained the set Q of rational numbers. Then, with the help of the Pythagoreans, who solved the equation $x^2 = 2$, the irrational numbers were born. These numbers were viewed as nonrepeating, nonterminating decimals and were adjoined to the rational numbers to form the set R of real numbers. Clearly, the real numbers include all the rationals and irrational numbers; the rational numbers include all the integers (positive, negative, and 0); the integers include all the whole numbers; and, in turn, the whole numbers include the natural numbers.

At first glance, it appears that the set of real numbers is so vast that it contains the solution for any possible equation, but this is not the case! If we consider the equation $x^2 = -1$, we seek a number which, when multiplied by itself equals -1. No real number will satisfy this equation, because every real number multiplied by itself results in a nonnegative product. This means that the set R is not closed with respect to the operation of taking square roots. To avoid this difficulty, sixteenth century mathematicians expanded the set of real numbers and created a new set of numbers called the **complex numbers.** In this system, we are allowed to take the square root of a negative number. In 1545, Girolamo Cardano, an Italian mathematician, used the square root of a negative number to solve the following problem: Find two numbers whose sum is 10 and whose product is 40. The answer?

$$5 + \sqrt{-15} \quad \text{and} \quad 5 - \sqrt{-15}$$

As you can see, if you add these numbers, their sum is 10. But what about their product? If we use ordinary multiplication and the rules we have studied, we obtain

$$
\begin{array}{r}
5 + \sqrt{-15} \\
\times \quad 5 - \sqrt{-15} \\
\hline
25 + 5\sqrt{-15} \\
-5\sqrt{-15} - (-15) \\
\hline
25 \qquad\qquad + 15 = 40
\end{array}
$$

A. Pure Imaginary Numbers

The great mathematician Leonhard Euler introduced the following notation for the square root of a negative number:

$$i = \sqrt{-1} \quad \text{so that} \quad i^2 = -1$$

Then, the square root of any other negative number can be expressed in terms of i. For example,

$$\sqrt{-4} = (\sqrt{-1})(\sqrt{4}) = (i)(2) = 2i$$

and

$$\sqrt{-7} = (\sqrt{-1})(\sqrt{7}) = (i)(\sqrt{7}) = \sqrt{7}\, i$$

Similarly, if c is any positive number, then $\sqrt{-c}$ is expressed in the form $\sqrt{c}\, i$ or $i\sqrt{c}$.

EXAMPLE 1   Express in the form bi, where b is a real number:

     a. $\sqrt{-25}$ **b.** $\sqrt{-19}$

Solution   **a.** $\sqrt{-25} = \sqrt{25}\, i = 5i$ **b.** $\sqrt{-19} = \sqrt{19}\, i$ ◄

B. The Complex Numbers

Any number of the form bi, where b is a nonzero real number, is called a **pure imaginary** number. For example,

$$i, \quad 3i, \quad -5i, \quad \tfrac{4}{7}i, \quad \sqrt{2}\, i, \quad \text{and} \quad \pi i$$

are pure imaginary numbers. We then define a **complex number** as the indicated sum of a real number and a pure imaginary number.

► **Definition 5.4a**

> The set C of **complex numbers** is defined as follows:
>
> $C = \{a + bi \mid a$ and b are real numbers$\}$

Note that if $a = 0$, $a + bi = 0 + bi = bi$ is a pure imaginary number. If $b = 0$, we define $0 \cdot i$ to be 0, so that every real number can be written in the form $a + 0i$. With this convention, every real number may be regarded as a complex number—that is, the real numbers are a subset of the set C.

EXAMPLE 2   Classify the given numbers by making a check mark in the appropriate row:

Set	$5 + 3i$	$7i$	$-\tfrac{3}{4}$	$\sqrt{5}\, i$
Real number				
Pure imaginary number				
Complex number				

Solution   The correct check marks are shown below:

Set	$5 + 3i$	$7i$	$-\tfrac{3}{4}$	$\sqrt{5}\, i$
Real number			✓	
Pure imaginary number		✓		✓
Complex number	✓	✓	✓	✓

◄

You should be warned that the adjective *imaginary* is a most unfortunate choice of words and must not be taken literally. The complex numbers have many *real* practical applications. Electric circuit analysis and mechanical vibration analysis are two of these applications. If you thought imaginary numbers were imaginary in the popular sense of that word, then you would be very surprised (not to say shocked) if you stuck your finger into a 110-volt electric socket with the power on and felt the "imaginary" current!

This completes our discussion of the relationship of the various sets of numbers that we have studied. In particular, we see that

$$N \subset W \subset I \subset Q \subset R \subset C$$

as shown in Fig. 5.4a.

Figure 5.4a

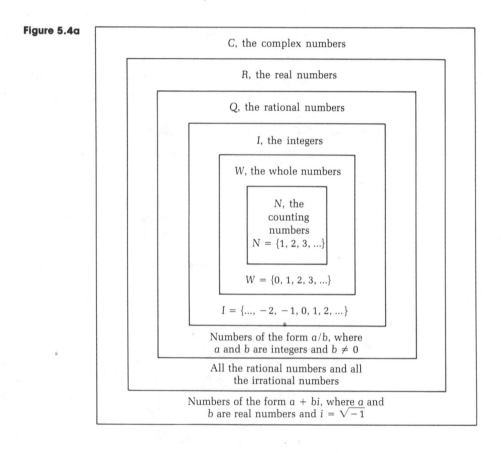

EXAMPLE 3   Classify the given numbers by making a check mark in the appropriate row:

Set	$\sqrt{4}$	$\sqrt{5}\,i$	$\sqrt{2}$	$3 + 5i$
Natural numbers				
Integers				
Rational numbers				
Irrational numbers				
Real numbers				
Complex numbers				

Solution   The correct check marks are shown below:

Set	$\sqrt{4}$	$\sqrt{5}\,i$	$\sqrt{2}$	$3 + 5i$
Natural numbers	✓			
Integers	✓			
Rational numbers	✓			
Irrational numbers			✓	
Real numbers	✓		✓	
Complex numbers	✓	✓	✓	✓

◀

EXERCISE 5.4

A.   In Problems 1–6, express the given number in the form bi, where b is a real number.

1. $\sqrt{-49}$ 2. $\sqrt{-55}$ 3. $\sqrt{-63}$
4. $\sqrt{-64}$ 5. $\sqrt{-100}$ 6. $\sqrt{-144}$

In Problems 7–12, express the given number in the form $i\sqrt{b}$, where $\sqrt{b}$ is in simplified form.

7. $\sqrt{-50}$ 8. $\sqrt{-32}$ 9. $\sqrt{-200}$
10. $\sqrt{-98}$ 11. $\sqrt{-48}$ 12. $\sqrt{-80}$

B.   In Problems 13–17, classify the given numbers by making a check mark in the appropriate row.

Set	13. $9i$	14. $7 - 4i$	15. $-\frac{5}{8}$	16. $\sqrt{18}\,i$	17. $0.8i$
Real numbers					
Pure imaginary numbers					
Complex numbers					

In Problems 18–24, classify the given numbers by making a check mark in the appropriate row.

Set	18. $-\frac{2}{8}$	19. 0	20. $\sqrt{3}$	21. $\sqrt{9}$	22. 5	23. $\sqrt{3}\,i$	24. $-2+i$
Natural numbers							
Whole numbers							
Integers							
Rational numbers							
Irrational numbers							
Real numbers							
Complex numbers							

▶ **Calculator Corner 5.4**   *Some scientific calculators perform complex number calculations. For example, to add $(3 - 8i) + (-5 + 2i)$, you place the calculator in the complex number mode (press* `mode` `cplx`*). Now, recall that $3 - 8i$ is written in the form $a + bi$, so $a = 3$ and $b = -8$. Similarly, in the complex number $-5 + 2i$, $a = -5$ and $b = 2$. Thus, to add these two numbers, enter the sequence*

$$\boxed{3}\,\boxed{a}\,\boxed{8}\,\boxed{\pm}\,\boxed{b}\,\boxed{+}\,\boxed{5}\,\boxed{\pm}\,\boxed{a}\,\boxed{2}\,\boxed{b}\,\boxed{=}$$

The display shows the real part -2. To find the imaginary part, press $\boxed{b}$*. The display shows -6. Thus, the complete answer is $-2 - 6i$.*

5.5

NUMBER SEQUENCES

Photograph courtesy of Parachutes, Inc.

The photograph shows a skydiver plunging to the ground. Do you know how far he will fall in the first 5 seconds? A free-falling body travels about 16 feet in the first second, 48 feet in the next second, 80 feet in the third second, and so on. The number of feet traveled in each successive second is:

16, 48, 80, 112, 144, . . .

This list of numbers is an example of a **number sequence.** In general, a list of numbers having a first number, a second number, a third number, and so on, is called a **sequence;** the numbers in the sequence are called the **terms.** Here are some examples of sequences:

a. The odd positive integers   1, 3, 5, 7, . . .
b. The positive multiples of 3   3, 6, 9, 12, . . .
c. The powers of 10 $10^1, 10^2, 10^3, . . .$
d. The interest on the first three payments on a $8.33, $8.11, $7.89, . . .
   $10,000 car being paid over 3 years at 12%
   annual interest

A. Arithmetic Sequences

The sequences a, b, and d given above are *arithmetic sequences*. An **arithmetic sequence** is a sequence in which each term after the first is obtained by *adding* a quantity called the **common difference** to the preceding term. Thus,

16, 48, 80, 112, 144, . . .

is an arithmetic sequence in which each term is obtained by adding the common difference 32 to the preceding term. This means that the common difference for an arithmetic sequence is just the difference between any two consecutive terms.

EXAMPLE 1   Find the common difference in each sequence.

a. 7, 37, 67, 97, . . .   **b.** 10, 5, 0, −5, . . .

Solution   **a.** The common difference is $37 - 7 = 30$ (or $67 - 37$, or $97 - 67$).
b. The common difference is $5 - 10 = -5$ (or $0 - 5$, or $-5 - 0$). ◄

It is customary to denote the first term of an arithmetic sequence by a_1 (read, "a sub 1"), the common difference by d, and the nth term by a_n. Thus, in the sequence 16, 48, 80, 112, 144, . . . , we have $a_1 = 16$ and $d = 32$. The second term of the sequence, a_2, is

$$a_2 = a_1 + 32 = 16 + 32 = 48$$

Since each term is obtained from the preceding one by adding 32,

$$a_3 = a_2 + 32 = (a_1 + 32) + 32 \quad = a_1 + 2 \cdot 32 = 80$$
$$a_4 = a_3 + 32 = (a_1 + 2 \cdot 32) + 32 = a_1 + 3 \cdot 32 = 112$$
$$a_5 = a_4 + 32 = (a_1 + 3 \cdot 32) + 32 = a_1 + 4 \cdot 32 = 144$$

By following this pattern, we find the **general term** a_n to be

$$a_n = a_1 + (n - 1) \cdot d$$

EXAMPLE 2   Consider the sequence 7, 10, 13, 16, ... Find:

a. a_1, the first term   **b.** d, the common difference
c. a_{11}, the 11th term   **d.** a_n, the nth term

Solution   **a.** The first term a_1 is 7.
b. The common difference d is $10 - 7 = 3$.
c. The 11th term is $a_{11} = 7 + (11 - 1) \cdot 3 = 7 + 10 \cdot 3 = 37$.
d. $a_n = a_1 + (n - 1) \cdot d = 7 + (n - 1) \cdot 3 = 4 + 3n$ ◄

B. Sum of an Arithmetic Sequence

Let us go back to our original problem of finding how far the skydiver falls in 5 seconds. The first five terms of the sequence are 16, 48, 80, 112, and

144; thus, we need to find the sum

$$16 + 48 + 80 + 112 + 144$$

Since successive terms of an arithmetic sequence are obtained by adding the common difference d, the sum S_n of the first n terms is

$$S_n = a_1 + (a_1 + d) + (a_1 + 2d) + (a_1 + 3d) + \cdots + a_n \qquad (1)$$

We can also start with a_n and obtain successive terms by subtracting the common difference d. Thus, with the terms written in reverse order,

$$S_n = a_n + (a_n - d) + (a_n - 2d) + \cdots + a_1 \qquad (2)$$

Adding equations (1) and (2), we find that the d's drop out, and we obtain

$$2S_n = (a_1 + a_n) + (a_1 + a_n) + \cdots + (a_1 + a_n)$$
$$= n(a_1 + a_n)$$

Thus,

$$S_n = \frac{n(a_1 + a_n)}{2}$$

We are now able to determine the sum S_5, the distance the skydiver dropped in 5 seconds. The answer is

$$S_5 = \frac{5(16 + 144)}{2} = 400 \text{ feet}$$

C. Geometric Sequences

The sequence 10, 100, 1000, and so on, is *not* an arithmetic sequence, since there is no common difference. This sequence is obtained by *multiplying* each term by 10 to get the next term. Such sequences are called *geometric sequences.* A **geometric sequence** is a sequence in which each term after the first is obtained by multiplying the preceding term by a number called the **common ratio,** r. Thus, the common ratio can be found by taking the ratio of two successive terms. For example, in the sequence 8, 16, 32, $\ldots$ the first term a_1 is 8 and the common ratio is $\frac{16}{8} = 2$ (or $\frac{32}{16}$). Thus, the first n terms in a geometric sequence are

$$a_1, a_1 r, a_1 r^2, a_1 r^3, \ldots, a_1 r^{n-1}$$

EXAMPLE 3 Consider the sequence 1, $\frac{1}{10}$, $\frac{1}{100}$, $\frac{1}{1000}$, $\ldots$. Find:

a. a_1 **b.** r **c.** a_n

Solution **a.** a_1 is the first term, 1.

b. r is the common ratio of any two successive terms. Thus,

$$r = \frac{\frac{1}{10}}{1} = \frac{1}{10}$$

c. $a_n = a_1 r^{n-1} = 1 \cdot \left(\dfrac{1}{10}\right)^{n-1} = \dfrac{1}{10^{n-1}}$ ◀

D. Sum of a Geometric Sequence

Can we find the sum S_n of the first n terms in a geometric sequence?

By definition: $\quad\quad\quad\quad\quad\quad\quad S_n = a_1 + a_1 r + a_1 r^2 + \cdots + a_1 r^{n-1}$

Multiply by r: $\quad\quad\quad\quad\quad\quad rS_n = a_1 r + a_1 r^2 + a_1 r^3 + \cdots + a_1 r^n$

Subtract: $\quad\quad\quad\quad\quad\quad S_n - rS_n = a_1 - a_1 r^n = a_1(1 - r^n)$

By the distributive law: $\quad S_n(1 - r) = a_1(1 - r^n)$

Divide by $1 - r$: $\quad\quad\quad\quad\quad\quad S_n = \dfrac{a_1(1 - r^n)}{1 - r}$

Thus, the sum of the first three powers of 10, that is, $10 + 10^2 + 10^3$, can be found by noting that $a_1 = 10$, $r = 10^2/10 = 10$, and

$$S_3 = \frac{10(1 - 10^3)}{1 - 10} = \frac{(10)(-999)}{-9} = 1110$$

as expected.

EXAMPLE 4   The first term of a geometric sequence is $\frac{1}{5}$, and $r = \frac{1}{2}$. Find:

a. a_5, the fifth term   **b.** S_5, the sum of the first five terms

Solution   **a.** The nth term in a geometric sequence is $a_1 r^{n-1}$; thus,

$$a_5 = \left(\frac{1}{5}\right)\left(\frac{1}{2}\right)^{5-1} = \left(\frac{1}{5}\right)\left(\frac{1}{2}\right)^4 = \left(\frac{1}{5}\right)\left(\frac{1}{16}\right) = \frac{1}{80}$$

b. The sum of the first n terms of a geometric sequence is $S_n = a_1(1 - r^n)/(1 - r)$, so

$$S_5 = \frac{\left(\frac{1}{5}\right)\left[1 - \left(\frac{1}{2}\right)^5\right]}{1 - \frac{1}{2}} = \frac{\frac{1}{5}\left[\frac{31}{32}\right]}{\frac{1}{2}} = \frac{62}{160} = \frac{31}{80}$$ ◀

E. Infinite Geometric Sequences

Let us now return to the repeating decimals we discussed in Chapter 4. These decimals can be written using an **infinite geometric sequence.** Thus, the decimal $0.333\ldots$ can be written as

$$0.333\ldots = \frac{3}{10} + \frac{3}{100} + \frac{3}{1000} + \cdots$$

where the common ratio is $\frac{1}{10}$. The sum of the first n terms of this sequence is

$$S_n = \frac{a_1(1 - r^n)}{1 - r} = \frac{a_1}{1 - r} \cdot (1 - r^n)$$

where $a_1 = \frac{3}{10}$ and $r = \frac{1}{10}$. If we want to find the sum of *all* the terms, we note that as n increases, $(\frac{1}{10})^n$ becomes very small. Thus, S_n is very close to

$$\frac{a_1}{1 - r} = \frac{\frac{3}{10}}{1 - \frac{1}{10}} = \frac{\frac{3}{10}}{\frac{9}{10}} = \frac{1}{3}$$

that is, $0.333 \ldots = \frac{1}{3}$. We can generalize this discussion to obtain the following result:

> If r is a number between -1 and 1, the sum of the infinite geometric sequence $a_1, a_1r, a_1r^2, \ldots$ is
> $$S = \frac{a_1}{1 - r}$$

EXAMPLE 5   Use the sum of an infinite geometric sequence to write the repeating decimals as fractions.

a. 0.666 . . .   **b.** 0.121212 . . .   **c.** 3.222 . . .

Solution   **a.** $0.666 \ldots = \frac{6}{10} + \frac{6}{100} + \frac{6}{1000} + \cdots$

This is a geometric sequence with first term $a_1 = \frac{6}{10}$ and ratio $r = \frac{1}{10}$. The sum of this sequence is

$$\frac{a_1}{1 - r} = \frac{\frac{6}{10}}{1 - \frac{1}{10}} = \frac{\frac{6}{10}}{\frac{9}{10}} = \frac{6}{9} = \frac{2}{3}$$

Thus, $0.666 \ldots = \frac{2}{3}$.

b. $0.121212 \ldots = \frac{12}{100} + \frac{12}{10,000} + \cdots$

This is a geometric sequence with $a_1 = \frac{12}{100}$, $r = \frac{1}{100}$, and sum

$$\frac{a_1}{1 - r} = \frac{\frac{12}{100}}{1 - \frac{1}{100}} = \frac{\frac{12}{100}}{\frac{99}{100}} = \frac{12}{99} = \frac{4}{33}$$

Thus, $0.121212 \ldots = \frac{4}{33}$.

c. $3.222 \ldots = 3 + \frac{2}{10} + \frac{2}{100} + \frac{2}{1000} + \cdots$

The repeating part, $0.222 \ldots$ is a geometric sequence with $a_1 = \frac{2}{10}$, $r = \frac{1}{10}$, and sum

$$\frac{a_1}{1 - r} = \frac{\frac{2}{10}}{1 - \frac{1}{10}} = \frac{\frac{2}{10}}{\frac{9}{10}} = \frac{2}{9}$$

Thus, $3.222 \ldots = 3\frac{2}{9} = \frac{29}{9}$.   ◀

A. In Problems 1–10, an arithmetic sequence is given. Find:

    a. The first term   **b.** The common difference d
    c. The 10th term   **d.** The nth term

    1. 7, 13, 19, 25, . . .   **2.** 3, 6, 9, 12, . . .
    3. 43, 34, 25, 16, . . .   **4.** 3, -1, -5, -9, . . .
    5. 2, -3, -8, -13, . . .   **6.** $\frac{2}{3}, \frac{5}{6}, 1, \frac{7}{6}, \ldots$
    7. $\frac{-5}{6}, \frac{-1}{3}, \frac{1}{6}, \frac{2}{3}, \ldots$ **8.** $\frac{-1}{4}, \frac{1}{4}, \frac{3}{4}, \frac{5}{4}, \ldots$
    9. 0.6, 0.2, -0.2, -0.6, . . .   **10.** 0.7, 0.2, -0.3, -0.8, . . .

B. In Problems 11–20, find S_{10} and S_n for the sequences given in Problems 1–10.

C. In Problems 21–26, a geometric sequence is given. Find:

    a. The first term   **b.** The common ratio r
    c. The tenth term   **d.** The nth term

    21. 3, 6, 12, 24, . . .   **22.** 5, 15, 45, 135, . . .
    23. $\frac{1}{3}$, 1, 3, 9, . . .   **24.** $\frac{1}{5}$, 1, 5, 25, . . .
    25. 16, -4, 1, $\frac{-1}{4}$, . . .   **26.** 3, -1, $\frac{1}{3}$, $\frac{-1}{9}$, . . .

D. In Problems 27–32, find S_{10} and S_n for the sequences given in Problems 21–26. Give answers in simplified exponential form.

E. In Problems 33–36, find the sum of the infinite geometric sequence.

    33. 6, 3, $\frac{3}{2}, \frac{3}{4}, \ldots$ **34.** 12, 4, $\frac{4}{3}, \frac{4}{9}, \ldots$
    35. -8, -4, -2, -1, . . .   **36.** 9, -3, 1, $\frac{-1}{3}$, . . .

In Problems 37–40, use sequences to write the given repeating decimal as a fraction.

    37. 0.777 . . .   **38.** 1.555 . . .
    39. 2.101010 . . .   **40.** 1.272727 . . .

    41. A property valued at $30,000 will depreciate $1380 the first year, $1340 the second year, $1300 the third year, and so on.
      a. What will be the depreciation the tenth year?
      b. What will be the value of the property at the end of the tenth year?
    42. Strikers at a plant were ordered to return to work and were told they would be fined $100 the first day they failed to do so, $150 the second day, $200 the third day, and so on. If the strikers stayed out for 10 days, what was their fine?
    43. A well driller charges $50 for the first foot; for each succeeding foot, the charge is $5 more than that for the preceding foot. Find:
      a. The charge for the 10th foot
      b. The total charge for a 50-foot well

44. When dropped on a hard surface, a Super Ball takes a series of bounces, each one being about $\frac{9}{10}$ as high as the preceding one. If a Super Ball is dropped from a height of 10 feet, find:

  a. How high it will bounce on the 10th bounce

  b. The approximate distance the ball travels before coming to rest
   [*Hint:*   Draw a picture.]

45. If $100 is deposited at the end of each year in a savings account paying 10% compounded annually, at the end of 5 years the compound amount of each deposit is:

$$100, \quad 100(1.10), \quad 100(1.10)^2, \quad 100(1.10)^3, \quad 100(1.10)^4$$

How much money is in the account right after the last deposit? [*Hint:* $(1.10)^5 = 1.61051$.]

USING YOUR KNOWLEDGE 5.5

Leonardo Fibonacci, one of the greatest mathematicians of the Middle Ages, wrote a book called the Liber Abaci. In this book, Fibonacci proposed the following problem: Let us suppose you have a 1-month-old pair of rabbits, and assume that in the second month, and every month thereafter, they produce a new pair. If each new pair does the same, and none of the rabbits die, can we find out how many pairs of rabbits there will be at the beginning of each month?

The figure illustrates what happens in the first 5 months. The rabbits shown in color indicate newborn pairs of rabbits. As you can see, the number of pairs of rabbits at the beginning of each of the 5 months is: 1, 1, 2, 3, 5. The resulting sequence is called a **Fibonacci sequence.**

1. Is the sequence an arithmetic sequence? A geometric sequence?

2. What is the relationship between the first two terms of the sequence and the third term?

3. Look at the second and third terms of the sequence. What is their relationship to the fourth term?

4. Based on the pattern found in Problems 2 and 3, write the first 10 terms of the Fibonacci sequence.

Section	Item	Meaning	Example
5.1	Irrational number	A number that is not rational	$\sqrt{2}$, $3\sqrt{5}$, π
5.1B	Irrational number	A number that has a nonterminating, nonrepeating decimal representation	$0.101001000\ldots$
5.1B	R	Real numbers	
5.1C	$C = \pi d$	Circumference of a circle	
5.2D	%	Percent sign	
5.3A	$\sqrt{}$	Radical sign	
5.3A	$\sqrt{a \cdot b} = \sqrt{a} \cdot \sqrt{b}$		$\sqrt{32} = \sqrt{16 \cdot 2} = \sqrt{16} \cdot \sqrt{2} = 4\sqrt{2}$
5.3C	$\sqrt{\dfrac{a}{b}} = \dfrac{\sqrt{a}}{\sqrt{b}}$		$\sqrt{\dfrac{2}{7}} = \dfrac{\sqrt{2}}{\sqrt{7}}$
5.4A	i	$\sqrt{-1}$	
5.4B	C	Complex numbers	
5.4B	$a + bi$	A complex number	$3 + 4i$
5.5A	d	The common difference of an arithmetic sequence	In the sequence 4, 9, 14, . . . , $d = 5$.
5.5A	$a_n = a_1 + (n - 1) \cdot d$	The nth term of an arithmetic sequence	In the above sequence, $a_n = 4 + (n - 1) \cdot 5 = 5n - 1$.
5.5B	$S_n = \dfrac{n(a_1 + a_n)}{2}$	The sum of the first n terms of an arithmetic sequence	In the above sequence, $S_5 = \dfrac{5(4 + 24)}{2} = 70$
5.5C	r	The common ratio of a geometric sequence	The common ratio of 5, 10, 20, . . . is $r = 2$.
5.5C	$a_n = a_1 r^{n-1}$	The nth term of a geometric sequence	In the sequence 5, 10, 20, . . . , $a_n = 5 \cdot 2^{n-1}$.
5.5D	$S_n = \dfrac{a_1(1 - r^n)}{1 - r}$	The sum of the first n terms of a geometric sequence	For the sequence 5, 10, 20, . . . , $S_4 = \dfrac{5(1 - 2^4)}{1 - 2} = 75$
5.5E	$\dfrac{a_1}{1 - r}$	The sum of the infinite geometric sequence a_1, $a_1 r$, $a_1 r^2$, . . . , where r is between -1 and 1	The sum of the sequence 2, 1, $\frac{1}{2}$, $\frac{1}{4}$, . . . is $S = \dfrac{2}{1 - \frac{1}{2}} = 4$

1. Classify as rational or irrational:
   a. $\sqrt{49}$ b. $\sqrt{45}$ c. $\sqrt{121}$
   d. 0.41252525 . . .   e. 0.212112111 . . .   f. 0.246810 . . .
2. Find:
   a. A rational number between $0.\overline{2}$ and 0.25
   b. An irrational number between $0.\overline{2}$ and 0.25
3. The diameter of a circular hamburger is 4 inches. To the nearest tenth of an inch, what is the circumference of this hamburger? (Use $\pi = 3.14$.)
4. Perform the indicated operations:
   a. $6.73 + 2.8$ b. $9.34 - 4.71$
   c. 0.29×6.7 d. $17.36 \div 3.1$
5. The three sides of a triangle are measured to be 18.7, 6.25, and 19.63 centimeters long, respectively. What is the perimeter of the triangle?
6. The dimensions of a college basketball court are 93.5 by 50.6 feet. What is the area of this court?
7. The circumference of a basketball is 29.5 inches. Find the diameter. (Use $\pi = 3.14$.)
8. Write as a decimal:
   a. 21%   b. 9.35%   c. 0.26%
9. Write as a percent:
   a. 0.52   b. 2.765
   c. $\frac{3}{5}$ d. $\frac{2}{11}$ (Answer to one decimal place.)
10. A 2-liter bottle of soda sells for 86¢ and costs the store 48¢. Find, to two decimal places, the percent of profit.
11. About 4 million women make between $25,000 and $50,000 annually. This represents 15% of all working women. About how many working women are there?
12. Simplify if possible:
    a. $\sqrt{96}$ b. $\sqrt{58}$
13. Simplify:
    a. $\dfrac{4}{\sqrt{20}}$ b. $\sqrt{\dfrac{48}{49}}$
14. Perform the indicated operations and simplify:
    a. $\sqrt{8} \cdot \sqrt{6}$ b. $\dfrac{\sqrt{56}}{\sqrt{7}}$
15. Perform the indicated operations:
    a. $\sqrt{90} - \sqrt{40}$ b. $\sqrt{32} + \sqrt{18} - \sqrt{50}$
16. Express in the form bi, where b is a real number:
    a. $\sqrt{-36}$ b. $\sqrt{-43}$

17. Classify the given numbers by making a check mark in the appropriate row:

Set	a. $2 + 7i$	b. $-3i$	c. $-\frac{2}{5}$	d. $\sqrt{3}\,i$
Real numbers				
Pure imaginary numbers				
Complex numbers				

18. Classify the given numbers by making a check mark in the appropriate row:

Set	a. $\sqrt{16}$	b. $\sqrt{7}\,i$	c. $\sqrt{5}$	d. $9 - 2i$
Natural numbers				
Integers				
Rational numbers				
Irrational numbers				
Real numbers				
Complex numbers				

19. Classify as an arithmetic or geometric sequence:
   a. 2, 4, 8, 16, . . .   **b.** 5, 8, 11, 14, . . .

20. Find the common difference in each sequence:
   a. 3, 43, 83, 123, . . .   **b.** 12, 6, 0, -6, . . .

21. Consider the sequence 9, 13, 17, 21, ... Find:
   a. The first term   **b.** The common difference
   c. The 10th term   **d.** The nth term

22. Find the sum of the first ten terms of 9, 13, 17, 21, ...

23. Consider the sequence 1, $\frac{1}{2}$, $\frac{1}{4}$, $\frac{1}{8}$, ... Find:
   a. a_1 **b.** r **c.** a_n

24. Find the sum of the first five terms of the sequence 1, $\frac{1}{2}$, $\frac{1}{4}$, $\frac{1}{8}$, ...

25. Use the sum of an infinite geometric sequence to write the following repeating decimals as fractions:
   a. 0.444 . . .   **b.** 0.212121 . . .   **c.** 2.555 . . .

CHAPTER 6

THE METRIC SYSTEM

In his report to the Congress in 1821, John Quincy Adams, sixth president of the United States, said:

> Weights and measures may be ranked among the necessaries of life to every individual of human society. They enter into the economical arrangements and daily concerns of every family. They are necessary to every occupation of human industry; to the distribution and security of every species of property; to every transaction of trade and commerce; to the labors of the husbandman; to the ingenuity of the artificer; to the studies of the philosopher; to the researches of the antiquarian, to the navigation of the mariner, and the marches of the soldier; to all the exchanges of peace; and all the operations of war. The knowledge of them, as in established use, is among the first elements of education, and is often learned by those who learn nothing else, not even to read and write.

Early Babylonian and Egyptian records and the Bible indicate that length was first measured with the forearm, hand, or finger, and time was measured by the periods of the sun, moon, and other heavenly bodies. To measure capacities of containers, they were filled with plant seeds, which were then counted to obtain the volume. As civilization developed, weights and measures became more complex, but the invention of numeration systems made it possible to create systems of weights and measures suited to trade and commerce, land division, taxation, and science.

The measurement system commonly used in the United States today is essentially the one brought by the colonists from England. These measures originated in a variety of cultures, and the ancient units evolved into the *inch*, *foot*, and *yard* through a complicated transformation not yet fully understood.

In 1790, the French Academy of Sciences, at the request of the French government, created a system of weights and measures that was both simple and scientific. Measures for capacity (volume) and mass (weight) were derived from the unit of length, which was selected to be a portion of the circumference of the Earth. Thus, the basic units were related to each other and to nature. Moreover, the larger and smaller versions of these units were obtained by multiplying or dividing by powers of 10, so that a simple shifting of the decimal point avoided all the awkward conversions of the English system, such as dividing by 36 to change inches to yards.

JOHN QUINCY ADAMS (1767–1848)

Picture Collection, the Branch Libraries, The New York Public Library

© 1974 United Feature Syndicate, Inc. Reprinted by permission of UFS, Inc.

6.1

THE METRIC SYSTEM

Weights and measures were among the earliest tools invented by humans. Rudimentary measures for constructing dwellings, fashioning clothing, and bartering food were used in many primitive societies. At first, people naturally used parts of their bodies as measuring instruments. Thus, the length of a human *foot*, or the circumference of a person's waist (*gird* in Saxon, later evolving into *yard*) were used as standards of measurement. Later, when means for weighing objects were invented, seeds and stones served as standard weights. For instance, the *carat*, a unit still used for measuring precious stones such as diamonds, was derived from the carob seed.

As mentioned in the introduction, in an effort to standardize measurements, the National Assembly of France in 1790 requested the French Academy of Sciences to "deduce an invariable standard for all the measures and all the weights." The result was the metric system.

In 1960, the General Conference on Weights and Measures adopted a revised and simplified version of the metric system, the **International System of Units,** which is now called **SI** (after the French name, *Le Système International d'Unités*). The SI is the system that we study in the following pages.

A. Metric Units

The metric system is a decimal system of weights and measures. The basic units in the metric system are:

1. The **meter** (the unit of length, a little longer than a yard)

A meter is a little longer than a yard.

A gram is about the weight of a paper clip.

2. The **liter** (the unit of volume or capacity, a little more than a quart)
3. The **gram** (the unit of weight, about the weight of a regular paper clip)
4. The **second** (the unit of time)

Multiples and subdivisions of these units are given in powers of 10. **Greek** prefixes are used to denote **multiples** of the units, and **Latin** prefixes to denote **subdivisions.** Here are the prefixes, with asterisks indicating the ones that are most commonly used in everyday life. Corresponding abbreviations are in parentheses.

kilo* (k)	one thousand	1000
hecto (h)	one hundred	100
deka (dk)	ten	10
STANDARD UNIT	one	1
deci (d)	one-tenth	0.1
centi* (c)	one-hundredth	0.01
milli* (m)	one-thousandth	0.001

[*Note:* deka is often abbreviated da rather than dk.]

EXAMPLE 1   How many grams are there in a kilogram?

Solution   Since the prefix **kilo** means **1000,** a kilogram is 1000 g.   ◄

EXAMPLE 2   What prefix would you use to indicate

a. One-hundredth of a basic unit?
b. One-thousandth of a basic unit?

Solution   **a.** Since the prefix **centi** means **one-hundredth,** we should use the prefix centi.
b. The prefix corresponding to **one-thousandth** is **milli.**   ◄

B. Which Unit to Use

A liter is a little more than a quart.

Let us see how the basic ideas of the metric system will relate to you. You are probably accustomed to the notion of buying carpet by the *square yard*. If we change to the metric system, you will buy it by the *square meter*. For measuring distances along a road, you now use *miles*; in the metric system, you will use *kilometers*. Smaller dimensions, such as wrench sizes, will not be handled in *quarters of an inch*, but in *millimeters*. See if you can do the following examples. It will help you to remember that:

1 quart (qt)	is about	1 liter (l)
1 inch (in.)	is about	2.5 centimeters (cm)
1 kilogram (kg)	is about	2.2 pounds (lb)
1 ounce (oz)	is about	28 grams (g)

EXAMPLE 3   Determine in what kind of metric units the following products should be sold.

a. 2 oz of cheese   **b.** 1 qt of milk   **c.** 3 lb of meat

Solution   **a.** The ounce is a small unit of weight; hence, the cheese should be sold by the gram.

b. The quart is a unit of volume close to a liter; thus, milk should be sold by the liter.

c. The closest metric unit of weight to the pound is the kilogram. Hence, meat should be sold by the kilogram.   ◄

EXERCISE 6.1

A.   In Problems 1–6, indicate what prefix you should use to express the given multiples or subdivisions of a basic unit.

1. 0.01   **2.** 0.1   **3.** Thousandths
4. Tenths   **5.** 1000   **6.** 10

7. How many liters are there in a kiloliter?
8. How many grams are there in a milligram?
9. How many centimeters are there in a meter?
10. How many meters are there in a kilometer?

B.   In Problems 11–15, indicate the appropriate metric units in which the given items should be sold.

11. A quart of vinegar   **12.** A 100-lb bag of cement
13. $\frac{1}{2}$-in.-wide adhesive tape   **14.** 14-in. shoelaces
15. A 2-oz candy bar

In Problems 16–25, select the answer that is most nearly correct.

16. A high school football quarterback weighs:
    a. 100 kl   **b.** 100 kg   **c.** 100 g
17. The height of a professional basketball player is:
    a. 200 mm   **b.** 200 m   **c.** 200 cm
18. The dimensions of the living room in an ordinary home are:
    a. 4 m by 5 m   **b.** 4 cm by 5 cm
19. The diameter of an aspirin tablet is:
    a. 1 cm   **b.** 1 mm   **c.** 1 m
20. The amount of milk in a carton is:
    a. 1 cl   **b.** 1 kl   **c.** 1 liter
21. The height of the Empire State Building (1250 ft) is:
    a. 400 cm   **b.** 400 m   **c.** 400 mm
22. The length of the 100-yd dash is:
    a. 100 cm   **b.** 100 mm   **c.** 100 m

Is this player 200 mm, 200 m, or 200 cm tall?

Courtesy Golden State Warriors

23. The weight of an average human male is:
  a. 60 kg   **b.** 60 g   **c.** 60 mg

24. The length of an ordinary lead pencil is:
  a. 19 mm   **b.** 19 cm   **c.** 19 m

25. The height of an ordinary door is:
  a. 2 m   **b.** 2 cm   **c.** 2 km

Problems 26 – 29 refer to the photograph showing a can of evaporated milk, a bottle of vitamin tablets, a package of Scotch tape, and a can of tuna fish. Identify which of these items has the given measure.

26. 6.35 m   **27.** 384 ml   **28.** 354 g   **29.** 1000 mg

30. Which is the most nearly correct answer? A 12-oz box of cereal has a net weight of:
  a. 2 kg   **b.** 340 g   **c.** 1000 mg

USING YOUR KNOWLEDGE 6.1

1. Use the fact that an inch is about 2.5 cm to answer the following question: Is the 100-m dash longer than the 100-yd dash by about 5 yd, 10 yd, or 20 yd?

2. A **metric ton** is 1000 kg. About how many pounds more than our customary ton (2000 lb) is this?

3. A liter is the volume of a cube that is 10 cm on each edge. The volume of a cube 1 m on each edge is how many liters?

4. A gram is the weight of 1 cubic centimeter (cm^3) of water (under certain standard conditions). What is the weight of a liter of water? (See Problem 3.)

5. The *humerus* is the bone in a person's upper arm. With this bone as a clue, an anthropologist can tell about how tall a person was. If the bone is that of a female, then the height of the person is about

$(2.75 \times \text{Humerus length}) + 71.48 \text{ cm}$

Suppose the humerus of a female was found to be 31 cm long. About how tall was the person?

DISCOVERY 6.1

In this section we have discussed the basic units in the metric system: the meter, the liter, and the gram. Many unbelievable feats have been recorded in terms of our customary units: miles, pints or quarts, and pounds. Can you discover how to convert the measurements involved into the metric system? It will help you to know that 1 mi is about 1.6 km, 1 qt is about 1 liter, and 1 oz is about 28 g.

1. You have probably heard the expression, "Hang in there!" Well, Rudy Kishazy did just that. He hung onto a glider that took off from Mount Blanc and landed 35 minutes later at Servoz, France, a distance of 15 mi. How many kilometers is that?

2. Romantic young ladies always dream about that knight in shining armor, but do you know what is the longest recorded ride in full armor? It is 146 mi. How many kilometers is that? (By the way, Dick Brown took the ride on June 12–15, 1973.)

3. Do you get thirsty on a hot summer day? Of course, you do! But nothing compares with the unquenchable thirst of Miss Helge Anderson of Sweden. She has been drinking 40 pt of water every day since 1922. About how many liters of water per day is that? (Remember that 2 pt = 1 qt.)

4. Do you know how to knit? If you do, you might try to equal the feat of Mrs. Gwen Mathewman of Yorkshire, England. In 1974, she knitted 836 garments. How much wool do you think she used? An unbelievable 9770 oz! How many grams is that?

5. What did you have for breakfast today? An omelet? We'll bet that you couldn't eat the largest omelet ever made—it weighed 1234 lb. About how many grams is that?

6.2

LINEAR (LENGTH) MEASURES

A. U.S. Units to Metric Units

In the *Tiger* cartoon, Hugo is measuring a board. When we measure an object, we assign to it a number indicating the size of the object in terms of some standard unit. As we saw in the preceding section, the standard unit

A paper clip is about 3 cm long.

of length in the metric system is the meter. The meter (39.37 in.) is a little longer than a yard. The meter was originally defined to be 1 ten-millionth of the distance from the North Pole to the Equator. However, the 1960 conference redefined it in terms of the wavelength of the orange-red line in the spectrum of krypton-86. This definition makes it easy for any scientific laboratory in the world to reproduce the length of the meter.

A yard is divided into 36 equal parts (inches), whereas a meter is divided into 100 equal parts (centimeters). The centimeter is about 0.4 in., and the inch is about 2.5 cm. To give you an idea of the relative lengths of the inch and the centimeter, here are two line segments 1 in. and 1 cm long, respectively:

_____ 1 in.

_____ 1 cm

EXAMPLE 1   A student is 60 in. tall. About how tall is she in centimeters?

Solution   Since an inch is about 2.5 cm, she is about

$$60 \times 2.5 \text{ cm} = 150 \text{ cm tall}$$ ◄

B.   Metric Units to Metric Units

From these comparisons of the inch and the centimeter, you can probably see that in the metric system we measure small distances (the length of a pencil, the width of a book, your height) in centimeters. For long distances, we use the kilometer. The kilometer is about 0.6 mi, and the mile is about 1.6 km. If you recall that **kilo** means **1000,** then you know that a kilometer is 1000 m. Table 6.2a shows the relationships among some metric units of length. Notice that the prefixes are the familiar ones we mentioned in Section 6.1. The abbreviations for the more commonly used units are in boldface type.

Table 6.2a

kilometer	hectometer	dekameter	meter	decimeter	centimeter	millimeter
km	hm	dkm	**m**	dm	**cm**	**mm**
1000 m	100 m	10 m	1 m	0.1 m	0.01 m	0.001 m

As you can see from the table, each metric unit of length is a multiple or a submultiple of the basic unit, the meter. Thus, to change from one unit to another, we simply substitute the correct equivalence. For example, if we wish to know how many centimeters there are in 3 km, we proceed as follows:

1 kilometer = 1000 **m**

= 1000 × **(100 cm)**

= 100,000 cm

Hence, 3 km is 300,000 cm.

Another way to change from one unit to another is to notice that **a shift of one place to the right in Table 6.2a moves the decimal point one place to the right,** and **a shift of one place to the left moves the decimal point one place to the left.** Thus, to change 3 km to centimeters, a shift of five places to the right in the table, we move the decimal point in the 3 five places to the right to obtain 300,000 cm. Similarly, to change 256 mm to meters, a shift of three places to the left in the table, we move the decimal point in the 256 three places to the left to get 256 mm = 0.256 m.

EXAMPLE 2   The height of a basketball player is 205 cm. How many meters is that?

Solution   From Table 6.2a, we have

$$1 \text{ cm} = 0.01 \text{ m}$$

Thus,

$$205 \text{ cm} = 205 \times 0.01 \text{ m}$$
$$= 2.05 \text{ m}$$

The height of the player is 2.05 m. (Again, note that to change from centimeters to meters, two jumps left in the table, we moved the decimal point two places to the left in the number 205.)   ◄

These examples illustrate that a change from one unit of length to another in the metric system simply moves the decimal point.

EXAMPLE 3   The screen on a certain pocket calculator is 65 mm long. How many centimeters is that?

Solution   Since *milli* means 0.001 and *centi* means 0.01, there is 0.1 cm in a millimeter. Thus, 65 **mm** = 65 × **0.1 cm** = 6.5 cm.   ◄

EXERCISE 6.2

A.   In Problems 1–4, convert the given measurement to centimeters by using the approximate equivalence 1 in. = 2.5 cm.

1. 40 in.   **2.** 50 in.   **3.** 3.4 in.   **4.** 12 in.

5. A door is 76 in. high. How many centimeters is that?
6. A car is 120 in. long. How many centimeters is that?
7. A wall panel is 8 ft high. How many centimeters is that?
8. A room is 20 ft long. How many meters is that?

B.   In Problems 9–16, fill in the blanks with the correct numbers.

9. 5 km = _____ m   **10.** 4 m = _____ cm
11. 3409 cm = _____ m   **12.** 49.4 mm = _____ cm

13. 8413 mm = _____ m   **14.** 7.3 m = _____ mm

15. 319 mm = _____ m   **16.** 758 m = _____ km

17. A bed is 210 cm long. How many meters is that?

18. The diameter of a vitamin C tablet is 6 mm. How many centimeters is that?

19. The length of a certain race is 1.5 km. How many meters is that?

20. The depth of a swimming pool is 1.6 m. How many centimeters is that?

An international rule that went into effect June 1, 1975, changed track events from yards to meters. Using the fact that 1 yd is approximately 0.914 m, determine whether the metric events in Problems 21–24 are longer or shorter than the old events and by how much.

Old Event	New Event
21. 100 yd	100 m
22. 220 yd	200 m
23. 440 yd	400 m
24. 880 yd	800 m

25. Dr. James Strange of the University of South Florida wishes to explore Mount Ararat in Turkey searching for Noah's ark. According to the book of Genesis, the dimensions of the ark are as given below. If a cubit is 52.5 cm, give each dimension in meters.

   a. Length, 300 cubits   **b.** Breadth, 50 cubits

   c. Height, 30 cubits

USING YOUR KNOWLEDGE 6.2

Match each item in the first column with an appropriate measure in the second column.

1. A letter-size sheet of paper   **a.** 20 × 25 mm

2. A newspaper   **b.** 54 × 86 mm

3. A credit card   **c.** 80 × 157 mm

4. An American Express check   **d.** 22 × 28 cm

5. A postage stamp   **e.** 38 × 68 cm

6.3

VOLUME

The Coke bottle shown in the margin on the next page holds 2 qt, or 1892 ml. In the metric system, the basic unit of volume is the liter. A liter is the volume of a cube that is 10 cm on each edge; that is, a liter is equivalent to 1000 cm³ (cubic centimeters). As before in the metric system, changing from one volume unit to another is just a matter of moving the decimal point. Table 6.3a shows the relationships among the various units. The abbreviations in boldface type are for the most commonly occurring units.

Table 6.3a

kiloliter	hectoliter	dekaliter	liter	deciliter	centiliter	milliliter
kl	hl	dkl	**l**	dl	cl	**ml**
1000 liters	100 liters	10 liters	1 liter	0.1 liter	0.01 liter	0.001 liter

A. Volume Units

If we wish to know how many liters there are in the Coke bottle that holds 1892 ml, we note that 1 liter = 1000 ml, or 1 ml = $\frac{1}{1000}$ liter. Thus,

$$1892 \text{ ml} = 1892 \times \tfrac{1}{1000} \text{ liter} = 1.892 \text{ liters}$$

You can get a good idea of the relative sizes of the quart and the liter by remembering that a 2-qt bottle of Coke contains 1892 ml, or 1.892 liters. This means that 1 qt is 0.946 liter. To recognize relative sizes, keep in mind that a liter is just a little more than a quart.

EXAMPLE 1   A soft drink bottle contains 0.946 liter. How many milliliters is this?

Solution   We use the same relationship as before. Thus,

$$0.946 \textbf{ liter} = 0.946 \times \textbf{1000 ml} = 946 \text{ ml}$$

Note that changing from liters to milliliters is a shift of three places to the right in Table 6.3a. Thus, you can get the answer by just moving the decimal point three places to the right.   ◀

In the metric system, small volumes are measured in milliliters. For instance, a teaspoonful of oil is about 5 ml of oil, and an 8-oz glass holds about 250 ml.

EXAMPLE 2   A man buys 2 liters of ice cream. How many 250-ml servings of ice cream can he make?

Solution   Since 2 liters = 2000 ml, he can make

$$\tfrac{2000}{250} = 8 \text{ servings}$$ ◀

EXAMPLE 3   Did you know that a cup is one-fourth of a quart? It has been proposed that in the metric system we use a *metricup*, one-fourth of a liter. How many milliliters is that?

Solution   A liter is 1000 ml. Therefore, a metricup is

$$\frac{1000}{4} = 250 \text{ ml}$$ ◄

EXAMPLE 4   Fill in the blanks:

a. 8.9 liters = _____ cl   **b.** 387 ml = _____ liters
c. 7.9 kl = _____ liters

Solution   **a.** Since a centiliter is 0.01 liter, we know that **1 liter = 100 cl.** Thus,

$$8.9 \text{ liters} = 8.9 \times \textbf{100 cl}$$
$$= 890 \text{ cl}$$

b. Similarly,

$$387 \text{ ml} = 387 \times \textbf{0.001 liter} = 0.387 \text{ liter}$$

c. In the same way,

$$7.9 \text{ kl} = 7.9 \times \textbf{1000 liters} = 7900 \text{ liters}$$ ◄

EXERCISE 6.3

A.   *In Problems 1–8, fill in the blanks with the correct numbers.*

1. 6.3 kl = _____ liters   **2.** 9.23 liters = _____ cl
3. 4.2 cl = _____ liters   **4.** 72.3 ml = _____ liters
5. 1.3 liters = _____ ml   **6.** 3.7 cl = _____ ml
7. 49 ml = _____ cl   **8.** 3479 ml = _____ kl

The beakers hold 10, 100, 250, 600, and 1000 ml.

9. A man buys a liter bottle of soda. How many 200-ml cups of soda can he serve?
10. Howard Hues bought a 0.5-liter can of paint. Which do you think he is going to paint, his desk or his whole house?
11. A woman bought a $\frac{1}{2}$-liter pan. Do you think she will use it to heat water for instant coffee or to boil potatoes for her family?
12. A person drank 60 ml of milk. Is that more or less than half a liter of milk?
13. Do you think a 2-liter bottle of wine would properly serve 4, 12, or 20 persons?
14. Do you think an average drinking cup holds 25, 250, or 2500 ml?
15. Seawater contains 3.5 g of salt per liter. How many grams of salt are there in 1000 ml of seawater?
16. Hydrogen weighs about 0.0001 g per milliliter. How much would 1 liter of hydrogen weigh?
17. A liter is equivalent to 1000 cm³. How many liters of liquid will a rectangular container 50 cm long and 20 cm wide hold when filled to a depth of 10 cm?

18. A gallon of gas is about 3.8 liters. An American car takes about 20 gal of gas. How many liters is that?
19. A certain medicine has 20 ml of medication per liter of solution. How many milliliters of solution are needed to obtain 5 ml of medication?
20. A tank truck delivers 10 kl of gasoline to a service station.
   a. How many liters is that?
   b. If 100 liters is about 26.4 gal, how many gallons were delivered?

USING YOUR KNOWLEDGE 6.3

In each of the following, fill in the blank with the most appropriate one of the three given numbers.

1. The gas tank of an average automobile holds about _____ liters of gasoline. (20, 60, 400)
2. It takes about _____ liters of water to fill a swimming pool. (600, 6000, 60,000)
3. It takes about _____ liters of water to make a pitcher of lemonade. (2, 10, 15)
4. A tablespoon holds 3 teaspoonfuls. Therefore, a tablespoon holds _____ ml. (6, 15, 50)
5. A thermos bottle holds about _____ liter(s) of coffee. (1, 5, 10)

DISCOVERY 6.3

1. Here is a problem just for fun: One glass is half full of wine. A second glass, which is twice the size of the first glass, is one-quarter full of wine. Both glasses are filled with water, and the contents are then mixed in a third container. Can you discover what part of the total mixture is wine?
2. Millie's Tavern is a peculiar place indeed. Her wine "liters" contain only 400 ml, and she insists on calling them "Millie liters"! Moreover, she has quite a time measuring her wine, since she has only a very large pitcher and two small glasses that hold 500 ml and 300 ml, respectively. Can you discover how Millie measures her Millie liter (400-ml) drinks using the two glasses and the pitcher?
3. Barely a half block from Millie's Tavern, there is a dairy called the Half-and-Half, not because they sell this product, but because they give you only half of what you pay for, cleverly keeping the other half. In this dairy, a "liter" of milk gets you only 500 ml. Can you discover how they measure their "liters" (500 ml) if they have only a large pitcher and two glasses that hold 300 and 700 ml, respectively?

WEIGHT AND TEMPERATURE

In the *Crock* cartoon, the prisoner is getting three glops of food a day. To measure how much food that is, we should measure the *mass* of the food.

CROCK

by Bill Rechin & Brant Parker

© 1976 Field Enterprises, Inc. Courtesy of Field Newspaper Syndicate.

Many people use the words *mass* and *weight* as though they were synonyms, but the two concepts are entirely different. The **mass** of an object is the *quantity of matter* in the object. The **weight** of the object is the *force* with which the Earth pulls on it. To measure your weight, you can stand on a spring scale; but to measure your mass, you would have to sit on a balance scale and be balanced against some standard masses. Suppose you are on the moon. The gravitational pull of the moon is only one-sixth the gravitational pull of the Earth. Thus, if your Earth weight were 150 lb, your moon weight would be only one-sixth of that, or 25 lb. However, your mass on the moon is exactly the same as it is on the Earth.

A. Units of Weight

A spring scale with 1 kg of apples.

The basic unit of mass is the gram, originally defined to be the mass of 1 cubic centimeter of water. However, for most everyday affairs, the gram is simply regarded as the basic unit weight; it is the weight of a cubic centimeter of water under certain standard conditions on the surface of the Earth. (This does not apply to scientific work, where a separate unit of weight must be defined.) The gram is used to weigh small objects. For example, a candy bar, the contents of a tuna fish can, and your breakfast cereal are suitably weighed in grams. Very small objects, such as pills, are weighed in milligrams. For heavier objects, we use the kilogram. Thus, we would buy meat, vegetables, and coffee in kilograms. A kilogram is about 2.2 lb, and a pound is about 0.45 kg. Table 6.4a gives the relationships among the customary metric units of weight. As before, the abbreviations for the most commonly used units are in boldface type.

Table 6.4a

kilogram **kg** 1000 g	hectogram hg 100 g	dekagram dkg 10 g	gram **g** 1 g	decigram dg 0.1 g	centigram cg 0.01 g	milligram **mg** 0.001 g

EXAMPLE 1   A woman weighs 50 kg. About how many pounds is that?

Solution   Since **1 kg ≈ 2.2 lb,**

$$50 \textbf{ kg} \approx 50 \times \textbf{2.2 lb}$$
$$\approx 110 \text{ lb}$$ ◄

EXAMPLE 2   A man bought a can containing 100 g of tuna fish. How many kilograms is that?

Solution   Since **1 g** = $\frac{1}{1000}$ **kg,**

$$100 \textbf{ g} = 100 \times \frac{1}{1000} \textbf{ kg}$$
$$= 0.1 \text{ kg}$$ ◄

EXAMPLE 3   Fill in the blanks with the appropriate numbers.

a. 0.34 kg = _____ g   **b.** 50 mg = _____ g
c. 7.1 g = _____ cg

Solution   **a.** Since **1 kg = 1000 g,**

0.34 **kg** = 0.34 × **1000 g** = 340 g

b. Because **1 mg** = $\frac{1}{1000}$ **g,**

50 **mg** = 50 × $\frac{1}{1000}$ **g** = 0.05 g

c. 7.1 **g** = 7.1 × **100 cg** = 710 cg   ◄

B.   Celsius and Fahrenheit Temperatures

We have now discussed length, volume, and weight. What about temperature? The temperature scale we normally use was invented by Gabriel Robert Fahrenheit. In the **Fahrenheit** scale, the boiling point of water is labeled 212°F (read, "212 degrees Fahrenheit"), and the freezing point of water is labeled 32°F. The temperature scale was modified by Anders Celsius, who avoided the awkward numbers 32 and 212. In the **Celsius** scale, the freezing point of water is taken to be 0°C (read, "0 degrees Celsius"), and the boiling point is 100°C. Because this scale is based on the number 100, it is sometimes called the **centigrade** scale. Figure 6.4a shows the comparison between the two scales.

Figure 6.4a

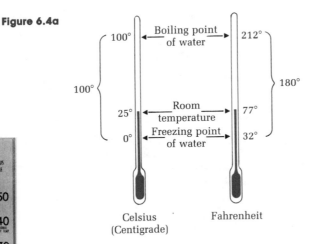

Some thermometers give the conversion formulas we use.

Here are the formulas for converting from one scale to the other. In these formulas, C stands for the Celsius temperature and F for the Fahrenheit temperature.

$$C = \frac{5(F - 32)}{9} \qquad F = \frac{9C}{5} + 32$$

To understand these formulas, note that there are 100 Celsius degrees and 180 Fahrenheit degrees between the freezing and boiling points of water (Fig. 6.4a). Therefore, any piece of the temperature scale has $\frac{100}{180} = \frac{5}{9}$ as many Celsius as Fahrenheit degrees. If we subtract 32 from the Fahrenheit reading to bring the freezing point back to 0, then the Celsius reading must be $\frac{5}{9}(F - 32)$. Similar reasoning leads to the formula $F = \frac{9}{5}C + 32$.

LUTHER   **By Brumsic Brandon, Jr.**

EXAMPLE 4 Desi had the flu. Her temperature was 104°F. What is that on the Celsius scale?

Solution We replace F by 104 in the formula

$$C = \frac{5(F - 32)}{9} \quad \text{to get} \quad C = \frac{5(104 - 32)}{9} = \frac{5(72)}{9} = 40$$

Thus, her temperature was 40°C. ◄

EXAMPLE 5 The melting point of gold is 1000°C. What is that on the Fahrenheit scale?

Solution We substitute 1000 for C in the formula

$$F = \frac{9C}{5} + 32$$

to find

$$F = \frac{9(1000)}{5} + 32$$

$$= 1800 + 32 = 1832$$

Hence, the melting point of gold is 1832°F. ◄

EXERCISE 6.4

A. In Problems 1–8, fill in the blanks with the correct numbers. Recall that 1 kg ≈ 2.2 lb and 1 lb ≈ 0.45 kg.

1. 3 lb = _____ kg
2. 5 lb = _____ kg
3. 8 lb = _____ kg
4. 10 lb = _____ kg
5. 8 kg = _____ lb
6. 3 kg = _____ lb
7. 10 kg = _____ lb
8. 25 kg = _____ lb

In Problems 9–16, fill in the blanks with the correct numbers.

9. 14 kg = _____ g
10. 4.8 kg = _____ g
11. 2.8 g = _____ kg
12. 3.9 g = _____ mg
13. 37 mg = _____ g
14. 49 mg = _____ kg
15. 41 g = _____ kg
16. 3978 g = _____ kg

B. In Problems 17–26, fill in the blanks with the correct numbers.

17. 59°F = _____ °C
18. 113°F = _____ °C
19. 86°F = _____ °C
20. −4°F = _____ °C
21. −22°F = _____ °C
22. 0°F = _____ °C

23. $10°C =$ _____ $°F$

24. $25°C =$ _____ $°F$

25. $-10°C =$ _____ $°F$

26. $-15°C =$ _____ $°F$

27. In an Air Force experiment, heavily clothed men endured temperatures of 500°F. How many degrees Celsius is that?

28. The temperature in Death Valley has been recorded at 131°F. How many degrees Celsius is that?

29. A temperature of 41°C is a dangerously high body temperature for a human being. How many degrees Fahrenheit is that?

30. For a very short time in September 1933, the temperature in Coimbra, Portugal, rose to 70°C. What is that on the Fahrenheit scale?

31. The average normal human body temperature is 98.6°F. What is that on the Celsius scale?

32. Tungsten, which is used for the filament in electric light bulbs, has a melting point of 3410°C. What is that on the Fahrenheit scale?

33. What Celsius temperatures should we use for the following?
    a. Cool to 41°F   b. Boil at 212°F

34. During the second quarter of the 1981 play-offs in Cincinnati, the wind–chill factor reached −58°F. (It got worse later.) What is the equivalent Celsius wind–chill factor?

35. Dry ice changes from a solid to a vapor at −78°C. Express this temperature in degrees Fahrenheit.

USING YOUR KNOWLEDGE 6.4

Match each item in the first column with an appropriate measure in the second column.

1. A man
2. A book
3. A bicycle
4. A small "compact" automobile
5. A common pin
6. A tennis ball
7. A teaspoonful of water
8. An orange

a. 80 mg
b. 5 g
c. 65 g
d. 166 g
e. 1 kg
f. 14 kg
g. 72 kg
h. 1000 kg

DISCOVERY 6.4

Americans are probably the most weight-conscious people in the world. The tables at the top of the next page show the desirable weights for certain heights at age 25 or over. Can you discover the desirable weights in kilograms? (Remember that 1 lb is about 0.45 kg.)

Men

	HEIGHT		WEIGHT	
	Feet	Inches	Pounds	Kilograms
1.	5	2	130	————
2.	5	6	142	————
3.	5	10	158	————
4.	6	2	178	————
5.	6	3	184	————

Women

	HEIGHT		WEIGHT	
	Feet	Inches	Pounds	Kilograms
6.	5	2	120	————
7.	5	6	135	————
8.	5	10	145	————
9.	5	11	155	————

▶ **Computer Corner 6.4** *Converting temperatures from Celsius to Fahrenheit or vice versa is a matter of using the formulas given in the text. A computer program (page 749) can do it for you! You simply enter the temperature followed by a C for Celsius or an F for Fahrenheit and the program does the rest.*

6.5

CONVERT IF YOU MUST

The preceding sections of this chapter have clearly shown the advantages of the metric system. Despite these advantages, many people are still opposed to changing to this system of measurement. Most of these people think that if the metric system is put into effect, then they must continually convert U.S. (formerly called English) units into metric units, and vice versa. Nothing is further from the truth! Once you learn to "think metric," you will usually not have to make conversions between the U.S. and metric systems. However, in case you occasionally must make such conversions, Table 6.5a should prove helpful.

Table 6.5a U.S. Units and Equivalents*

1 in. = 2.54 cm†	1 cm ≈ 0.394 in.
1 yd ≈ 0.914 m	1 m ≈ 1.09 yd
1 mi ≈ 1.61 km	1 km ≈ 0.621 mi
1 lb ≈ 0.454 kg	1 kg ≈ 2.20 lb
1 qt ≈ 0.946 liter	1 liter ≈ 1.06 qt

* The symbol ≈ means "is approximately equal to."
† The inch is legally defined to be *exactly* 2.54 cm.

As you can see from Table 6.5a, with the exception of the 2.54, the numbers on the right sides of the conversion equations are all approximate. Some of these numbers are stated with two decimal places and some with three. The reason is that these numbers are either the results of certain measurements or else are rounded-off approximations to the true values. For example, when we say that a yard is 0.914 m, we are giving the result to the nearest thousandth of a meter; that is, we mean that the actual number is between 0.9135 and 0.9145. Thus, in calculations with the numbers in the table, we must use the round-off rules and the rules for approximate numbers given in Section 5.2.

Let us look at the conversion from centimeters to inches. We have the exact equivalence 1 in. = 2.54 cm. To express centimeters in terms of inches, we must divide by 2.54 to get

$$\frac{1}{2.54} \text{ in.} = \frac{2.54}{2.54} \text{ cm}$$

or

$$\text{cm} = \frac{1}{2.54} \text{ in.}$$

If we carry out this division, we find

$$\frac{1}{2.54} = 0.3937007 \ldots$$

which, rounded off to three decimal places, gives the result in the table, 0.394.

EXAMPLE 1   The record distance reached by a boomerang before it starts to return to the thrower is about 90 yd (to the nearest yard). How many meters is that?

Solution   From Table 6.5a, we see that 1 yd ≈ 0.914 m. Thus, we must multiply 90 by 0.914, giving 82.26. Since the 90 yd is correct to the nearest yard, the 90 has two significant digits and our answer must be rounded to two significant digits. Hence, we see that 90. yd ≈ 82 m. (Note that we have just followed the custom of putting a decimal point after an integer with terminal 0's when all the 0's are significant.)   ◄

EXAMPLE 2   How many grams are there in an ounce?

Solution   From Table 6.5a, we see that 1 lb ≈ 0.454 kg. Since there are 16 oz in a pound and 1000 g in a kilogram, we have

$$16 \text{ oz} \approx 454 \text{ g}$$

$$1 \text{ oz} \approx \frac{454}{16} \text{ g}$$

If we divide 454 by 16, we get 28.37, but this must be rounded to three significant digits because there are only three significant digits in 0.454. Therefore,

$$1 \text{ oz} \approx 28.4 \text{ g}$$ ◄

EXAMPLE 3   The maximum speed limit on some highways is 55 mi/hr. How many kilometers per hour (km/hr) is this? (Assume the 55 to be exact.)

Solution   From Table 6.5a, we have 1 mi ≈ 1.61 km. Thus, we multiply 1.61 by 55 to get 88.55. This result must be rounded to agree with the three significant digits in the 1.61. Therefore,

$$55 \text{ mi/hr} \approx 88.6 \text{ km/hr}$$ ◄

EXAMPLE 4   The top speed of a certain European car is 200 km/hr. How many miles per hour is this? (Assume the 200 to be exact.)

Solution
$$1 \text{ km} \approx 0.621 \text{ mi}$$
$$200 \text{ km} \approx 200 \times 0.621 \text{ mi}$$
$$\approx 124 \text{ mi}$$

Thus, 200 km/hr is equivalent to 124 mi/hr.   ◄

EXAMPLE 5   Mary bought 3 qt of milk. How many liters of milk is this?

Solution
$$1 \text{ qt} \approx 0.946 \text{ liter}$$
$$3 \text{ qt} \approx 3 \times 0.946 \text{ liter}$$
$$\approx 2.84 \text{ liters}$$ ◄

Finally, here is a conversion that you probably see almost every day: gallons to liters, or liters to gallons. As you can see from the brochure shown below, 1 gal is about 3.7854 liters, and 4 liters is a little more than 1 gal. A more precise relationship is:

1 liter ≈ 0.2642172 gal

1 gal ≈ 3.785412 liters

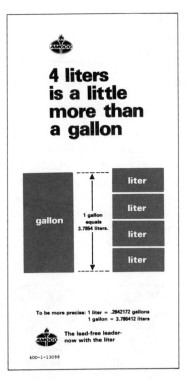

Suppose you fill your tank, and it takes 38 liters. To the nearest tenth of a gallon, how many gallons would that be? We write

38 **liters** ≈ (38)(**0.2642172 gal**)

≈ 10.040254 gal

≈ 10.0 gal

EXAMPLE 6   The tank of a Datsun 210 has a capacity of 13.2 gal. To the nearest liter, how many liters is that?

Solution   We write

13.2 **gal** ≈ (13.2)(**3.785412 liters**)

≈ 49.967438 liters

≈ 50 liters   ◀

In Problems 1–28, fill in the blank with the appropriate number. (Assume the given numbers are all exact.)

1. 8 in. = _____ cm

2. 5.2 in. = _____ cm

3. 12 cm = _____ in.

4. 25 cm = _____ in.

5. 51 yd = _____ m

6. 1.2 yd = _____ m

7. 3.7 m = _____ yd

8. 4.5 m = _____ yd

9. 4 mi = _____ km

10. 6.1 mi = _____ km

11. 3.7 km = _____ mi

12. 14 km = _____ mi

13. 6 lb = _____ kg

14. 8 lb = _____ kg

15. 5 kg = _____ lb

16. 1.2 kg = _____ lb

17. 5 qt = _____ liters

18. 6.1 qt = _____ liters

19. 8.1 liters = _____ qt

20. 11 liters = _____ qt

21. 75 cm = _____ ft

22. 800 m = _____ ft

23. 1 in.^{3} = _____ cm^3
(to the nearest hundredth)

24. 1 cm^3 = _____ in.^{3}
(to the nearest hundredth)

25. 2 yd = _____ cm

26. 3 yd = _____ cm

27. 78 cm = _____ yd

28. 100 cm = _____ yd

29. The speed limit is 40 mi/hr. How many kilometers per hour is that? (Answer to the nearest kilometer per hour.)

30. Find the distance (to the nearest kilometer) from Tampa, Florida, to:
a. Zephyrhills, Florida, 22 mi   **b.** Ocala, Florida, 93 mi

31. The longest field goal in National Football League competition was 63 yd. How many meters is that?

32. The *U.S.S. New Jersey* is the longest battleship, 296 yd long. How many meters is that?

33. Mount Everest is 8848 m high. How many feet is that?

34. The screen of a television set measures 24 in. diagonally. How many centimeters is that?

35. The largest omelet ever made weighed 1234 lb. How many kilograms is that?

36. Miss Helge Anderson of Sweden has been drinking 20 qt of water per day since 1922. How many liters per day is that?

37. The longest street in the world is Figueroa Street in Los Angeles. This street is 30 mi long. How many kilometers is that?

38. A car is traveling 125 km/hr. How many miles per hour is that?

39. The maximum allowable weight for a flyweight wrestler is 52 kg. How many pounds is that?

40. The largest car ever built was the Bugatti Royale, type 41, with an eight-cylinder engine of 12.7-liter capacity. How many quarts is this?

41. Two adjacent sides of a rectangle are measured and found to be 52.3 and 96.84 m long, respectively. How many meters long is the perimeter of the rectangle?

42. A rectangle is measured and found to be 21.5 ft by 32.63 ft. How many meters long is the perimeter?

43. A road sign warns of a bridge with a safe load of 14 tons. A metric ton is 1000 kg (very close to 2200 lb). What is the safe load of the bridge in metric tons? Answer to the nearest tenth.

44. Another road sign warns of a bridge with a safe load of 15 tons. What is this safe load in metric tons? (See Problem 43.)

45. The Pontiac 6000 had a highway mileage estimated to be 40 mi/gal. To the nearest tenth, how many kilometers per liter is that?

46. If your car delivers 25 mi/gal, how many kilometers per liter is that? Answer to the nearest tenth.

USING YOUR KNOWLEDGE 6.5

1. A Boeing 747 plane requires about 1900 m for a takeoff runway. About how many miles is this?

2. A fully loaded Boeing 747 weighs about 320,000 kg. About how many tons is this? (1 ton = 2000 lb)

3. The longest recorded distance for throwing (and catching) a raw hen's egg without breaking it is 316 ft 5¾ in. About how many meters is this?

4. The largest champagne bottle made is called a *Nebuchadnezzar*; it holds 16 liters. How many gallons is this?

5. In 1954, the winner of the Miss World contest had the Junoesque measurements 40–26–38. These measurements are in inches. What would they be in centimeters?

6. An 1878 bottle (1 liter) of Chartreuse (a liqueur) sold for $42. About how much per ounce is this?

Are you puzzled by the material in this chapter? We hope not, for here is a "cross-number" puzzle that will amuse you. See if you can discover all the correct numbers.

Across

1. $3 \text{ g} + 125 \text{ mg} = $ _____ mg
2. $2000 \text{ g} - 0.275 \text{ kg} = $ _____ g
5. $13.2 \text{ g} - 0.65 \text{ g} = $ _____ cg
7. $180 \text{ g} + 200 \text{ dg} = $ _____ dkg
8. $11 \text{ dg} + 13 \text{ cg} = $ _____ cg
9. $100 \text{ dg} = $ _____ g
10. $10.22 \text{ hg} - 0.938 \text{ kg} = $ _____ g
12. $50 \text{ dg} + 120 \text{ cg} - 2075 \text{ mg} = $ _____ mg
13. $2.5 \text{ kg} = $ _____ g

Down

1. $55 \text{ kg} - 20 \text{ kg} = $ _____ hg
3. $60 \text{ dkg} - 0.1 \text{ kg} = $ _____ g
4. $3.5 \text{ g} + 40 \text{ dg} = $ _____ dg
6. $510{,}000 \text{ g} = $ _____ kg
7. $1.133 \text{ g} - 0.915 \text{ g} = $ _____ mg
8. $110{,}500 \text{ mg} + 3500 \text{ mg} = $ _____ g
9. $1200 \text{ cg} = $ _____ g
11. $4.20 \text{ dg} = $ _____ mg

Note: The prefixes hecto (h), deka (da), and deci (d), are not as commonly used as the others in the puzzle, but they are (SI).

Solve like a crossword puzzle except that numbers replace words. Be sure to express answers in the indicated units.

Courtesy of *American Metric Journal* Publishing Co./The American Metric Journal.

Section	Item	Meaning	Example
6.1A	kilo (k)	One thousand	1 kilometer = 1000 meters
6.1A	hecto (h)	One hundred	1 hectogram = 100 grams
6.1A	deka (dk)	Ten	1 dekaliter = 10 liters
6.1A	deci (d)	One-tenth	1 decimeter = 0.1 meter
6.1A	centi (c)	One-hundredth	1 centigram = 0.01 gram
6.1A	milli (m)	One-thousandth	1 millimeter = 0.001 meter
6.2A	kilometer (km)	1000 meters	
6.2A	hectometer (hm)	100 meters	
6.2A	dekameter (dkm)	10 meters	
6.2A	meter (m)	Basic unit of length	
6.2A	decimeter (dm)	0.1 meter	
6.2A	centimeter (cm)	0.01 meter	
6.2A	millimeter (mm)	0.001 meter	
6.3A	kiloliter (kl)	1000 liters	
6.3A	hectoliter (hl)	100 liters	
6.3A	dekaliter (dkl)	10 liters	
6.3A	liter (l)	Basic unit of volume	
6.3A	deciliter (dl)	0.1 liter	
6.3A	centiliter (cl)	0.01 liter	
6.3A	milliliter (ml)	0.001 liter	
6.4A	kilogram (kg)	1000 grams	
6.4A	hectogram (hg)	100 grams	
6.4A	dekagram (dkg)	10 grams	
6.4A	gram (g)	Basic unit of weight	
6.4A	decigram (dg)	0.1 gram	
6.4A	centigram (cg)	0.01 gram	
6.4A	milligram (mg)	0.001 gram	
6.4B	$C = \dfrac{5(F - 32)}{9}$	Formula to convert Fahrenheit to Celsius degrees	
6.4B	$F = \dfrac{9C}{5} + 32$	Formula to convert Celsius to Fahrenheit degrees	

1. Write the power of 10 that corresponds to the following prefixes:
   a. centi **b.** milli **c.** kilo

2. A person is buying some paint to paint a house. In the metric system, the paint would be measured in which of the following?
   a. Grams **b.** Centimeters **c.** Liters

3. A fruit market sells oranges by weight. In the metric system, the oranges should be weighed in which of the following units?
   a. Liters **b.** Kilograms **c.** Meters

4. A steel rod is $\frac{3}{4}$ m long. How many centimeters is that?

5. A racetrack is 0.9 km long. How many meters is that?

6. The sides of a triangle are 3 cm, 4 cm, and 6 cm long, respectively. What are these lengths in millimeters?

7. A piece of a straight line is 0.082 m long. How many millimeters is that?

8. A water tank holds 8.3 kl of water. How many liters is that?

9. A bottle with a capacity of 92 cl holds how many liters?

10. A small container holds 19 ml of liquid. How many centiliters is that?

11. A liter of wine will fill about ten large wine glasses. How many milliliters does one of these glasses hold?

12. A 1-carat diamond weighs 0.2 g. How many milligrams will a 2-carat diamond weigh?

13. A market prices onions at $1 per kilogram. How many cents per gram is that?

14. Convert 275 mg to kilograms.

15. A person bought a 250-g can of tuna fish. How many kilograms is that?

16. A baby weighed 4.1 kg at birth. How many grams is that?

17. The temperature in San Francisco was 77°F. What is that in degrees Celsius?

18. The temperature in New York was 20°C. What is that in degrees Fahrenheit?

19. In some Canadian towns, the posted speed limit is 40 km/hr. How many miles per hour is this?

20. A woman weighs 52 kg. How many pounds is this?

21. A wine bottle contains 750 ml of wine.
    a. How many liters is this? **b.** How many quarts?

22. The sides of a triangle are found to be 152, 178, and 135 m long. How many kilometers long is the perimeter of the triangle?

23. The side of a square is 3.5 in. long. How many centimeters long is the perimeter of the square?

24. The sides of a triangle are measured to be 16.3, 18.9, and 15.46 cm long, respectively. How many centimeters long is the perimeter of the triangle?

25. Which of the following is most nearly correct? Ten inches is about:
    a. 30 cm **b.** 25 cm **c.** 20 cm

CHAPTER 7

EQUATIONS, INEQUALITIES, AND PROBLEM SOLVING

Karl Friedrich Gauss, who has been called the Prince of Mathematicians, was born in Brunswick, Germany, in the spring of 1777. His father was a laborer, a poor but upright man, who did all he could to prevent his son from getting the education suited to his great talents. It was only by accident that Gauss became a mathematician.

Throughout his life, Gauss was noted for his miraculous ability to perform stupendous mental calculations. No one in the history of mathematics can compare with him in this respect. Before he was 3 years old, while watching his father make out a weekly payroll, he noted an error in the long computation and told his father what the answer should be. A check of the account showed that the boy was correct.

When Gauss was 10 years old, his genius attracted the attention of a young mathematician by the name of Bartels, who not only taught the boy some elementary mathematics but was able to bring his young friend to the attention of the Duke of Brunswick. The Duke was so attracted by Gauss that he made the boy his protégé.

Gauss entered the Caroline College in Brunswick at the age of 15, and in a short time began the research into higher arithmetic that made him one of the two or three greatest mathematicians of all time. When he left the College in 1795 at the age of 18, he had already invented the method of least squares. From the college, he entered the University of Göttingen, where he spent three fruitful years essentially completing his *Disquisitiones Arithmeticae (Arithmetical Researches)*. Then, in 1798, he went to the University of Helmstedt, where he was awarded the Ph.D. degree one year later. His doctoral thesis gave the first proof of the fundamental theorem of algebra, that every algebraic equation has at least one root among the complex numbers. Gauss was 21 or 22 when he solved this famous problem.

His *Disquisitiones*, which was published in 1801, is regarded as the basic work in the theory of numbers, but this was only one of his many accomplishments. During his long life (he died in 1855 at the age of 78), he made great contributions to astronomy, geodesy (the measurement of the Earth), geometry (including non-Euclidean geometry), theoretical physics (especially electromagnetism), and complex numbers and functions. Along with his masterful theoretical researches, he was also a well-known inventor; among other things, he invented the electric telegraph in 1833.

KARL FRIEDRICH GAUSS (1777–1855)
The Bettmann Archive

Archimedes, Newton, and Gauss, these three, are in a class by themselves among the great mathematicians, and it is not for ordinary mortals to attempt to range them in the order of merit.

E. T. BELL

OPEN SENTENCES AND STATEMENTS

Elementary algebra was first treated in a systematic fashion by the Arabs during the period before the Renaissance, when Europe was almost at a standstill intellectually. By the early 1600's, algebra had become a fairly well-developed branch of mathematics, and mathematicians were beginning to discover that a marriage of algebra and geometry could be highly beneficial to both subjects.

It has been said that algebra is arithmetic made simple, and it is true that a small amount of elementary algebra enables us to solve many problems that would be quite difficult by purely arithmetic means. In this chapter we shall learn some of the simpler algebraic techniques that are used in problem solving.

We have already made frequent use of various symbols, usually letters of the alphabet, as place-holders for the elements of a set of numbers. For example, we wrote

$$a + b = b + a \qquad a, b \text{ real numbers}$$

as a symbolic way of stating the commutative property of addition. Of course, we mean that a and b may each be replaced by any real number. In this case the set of real numbers is the **replacement set** for a and b. A symbol that may be replaced by any one of a set of numbers is called a **variable.**

Letters of the alphabet as well as symbols such as $\square$ are often used to indicate variables in arithmetic. The study of sentences and expressions involving variables is, however, a part of algebra.

A sentence in arithmetic involves only numbers and operations on numbers as well as an appropriate **verb phrase.** Examples of such sentences are

1. $2 + 3 = 5$
2. $2 - 3 = 7$
3. $3 + 4 < 2 + 4$
4. $2 + 6 > 6 - 2$

Each of these sentences may be recognized as true or false, and hence is a statement. Thus, statements 1 and 4 are true, and statements 2 and 3 are false. The verb phrase in statements 1 and 2 is **is equal to** ($=$); that in statement 3 is **is less than** ($<$); and that in statement 4 is **is greater than** ($>$).

In algebra, as in arithmetic, the commonly used verb phrases are

$=$: is equal to
$\neq$: is not equal to
$>$: is greater than $\geq$: is greater than or equal to
$<$: is less than $\leq$: is less than or equal to

By using these verb phrases along with specific numbers and variables joined by the usual operations from arithmetic, we can form many types of sentences. Some examples of simple algebraic sentences are

1. $x - 1 = 3$
2. $x - 2 \neq 4$
3. $x - 1 \geq 3$
4. $x + 7 < 9$

A. Equations and Inequalities

In the sentences above, x is a variable—that is, a place-holder for the numbers by which it may be replaced. Until x is replaced by a number, none of these sentences is a statement, because it is neither true nor false. For this reason, we call such sentences **open sentences**. Because only one variable is involved, we may refer to the sentences as **open sentences in one variable**. Sentences in which the verb phrase is $=$ are called **equations**; if the verb phrase is any of the others we have listed, then the sentence is called an **inequality**.

In order to study an open sentence in one variable, we obviously must know what is the replacement set for that variable. We are interested in knowing for which of the possible replacements the sentence is a true statement. The set of elements of the replacement set that make the open sentence a true statement is called the **solution set** for the given replacement set.

EXAMPLE 1   Suppose the replacement set for x is $\{2, 4, 6\}$. For each of the following open sentences, find the solution set:

a. $x - 1 = 3$ **b.** $x - 2 \neq 4$ **c.** $x - 1 \geq 3$ **d.** $x + 7 < 9$

Solution   **a.** We substitute the elements of the replacement set into the open sentence $x - 1 = 3$:

For $x = 2$, we get $2 - 1 = 1$, not 3.

For $x = 4$, we get $4 - 1 = 3$, which makes the sentence a true statement.

For $x = 6$, we get $6 - 1 = 5$, not 3.

Thus, $x = 4$ is the only replacement that makes the sentence $x - 1 = 3$ a true statement, so the solution set is $\{4\}$.

b. We make the permissible replacements into the open sentence $x - 2 \neq 4$:

For $x = 2$, we get $2 - 2 = 0$, which is not equal to 4, so the sentence is a true statement.

For $x = 4$, we get $4 - 2 = 2$, which is not equal to 4, so the sentence is a true statement.

For x = **6**, we get **6** − 2 = 4, which does not satisfy the "≠4," so the sentence is a false statement.

Thus, the solution set is {2, 4}.

c. Making the permissible replacements into x − 1 ≥ 3:

For x = **2**, we get **2** − 1 = 1, which is less than 3, not greater than or equal to 3. Hence, the sentence is a false statement.

For x = **4**, we get **4** − 1 = 3, which satisfies the "≥3," so that the sentence is a true statement.

For x = **6**, we get **6** − 1 = 5, which satisfies the "≥3," so that the sentence is a true statement.

The solution set is thus {4, 6}.

d. Making the permissible replacements into x + 7 < 9:

For x = **2**, we get **2** + 7 = 9, which is not less than 9, so the sentence is a false statement.

For x = **4**, we get **4** + 7 = 11, which is not less than 9, so the sentence is a false statement.

For x = **6**, we get **6** + 7 = 13, which is not less than 9, so the sentence is a false statement.

Since none of the replacements make the sentence x + 7 < 9 a true statement, the solution set is ∅. ◀

EXAMPLE 2 Find the solution set for the inequality x + 7 < 9 if x is an integer.

Solution The replacement set for the inequality x + 7 < 9, x an integer, is the set of integers. Of this set, the integer 1 is the largest for which x + 7 < 9, because 1 + 7 = 8, which is less than 9; but 2 + 7 = 9, which is *not* less than 9. The solution set is the set of all integers less than or equal to 1, that is, {. . . , −3, −2, −1, 0, 1}. Test for yourself that the integers less than 1 satisfy the given inequality. ◀

EXAMPLE 3 Find the solution set for 2(y − 2) = −4 + 2y if y is a real number.

Solution We work with the left-hand side of the given sentence. By the distributive property, for any real number y,

$$2(y - 2) = 2y - 4$$

Then by the commutative property of addition, for any real number y,

$$2y - 4 = -4 + 2y$$

which is the same as the right-hand side of the given sentence. Thus, the equation 2(y − 2) = −4 + 2y is true for all real numbers, so the solution set is the set of all real numbers. ◀

Open sentences that are true statements for every number in the replacement set are called **identities.** Thus, the equation given in Example 3 is an identity as is the equation $x + 1 = 1 + x$, x a real number.

B. Applications

In the preceding examples, we learned how to determine whether a given number **satisfies** an equation or an inequality. This idea can be used to do some detective work, as you will see in the next example.

EXAMPLE 4　The relationship between the length f of the human female femur bone and the height H of the female is given (in centimeters) by the formula

$$H = 1.95f + 72.85$$

Suppose the police find a female femur bone 40 cm long. If a missing girl is known to be 120 cm tall, can the bone belong to her?

Solution　If the bone belongs to the missing girl (120 cm tall), its length f must satisfy the equation

$$120 = 1.95f + 72.85$$

But the right side, for $f = $ **40,** is

$$1.95(\textbf{40}) + 72.85 = 150.85$$

Since $120 \neq 150.85$, the bone does not belong to the missing girl. ◀

EXERCISE 7.1

A.　In Problems 1 – 16, take the replacement set to be the set of positive integers. Find the solution set.

1. $x + 2 = 4$　　2. $x + 3 = 7$　　3. $x + 2 \geq 2$　　4. $x + 1 \geq 1$
5. $x + 3 < x$　　6. $x < x + 2$　　7. $x + 4 \leq 7$　　8. $x + 1 \leq 5$

9. $x - 1 < 5$ **10.** $x - 2 < 1$ **11.** $x + 1 \neq 7$ **12.** $3 + x \neq 4$

13. $x + 3 < 3$ **14.** $x - 1 < 1$ **15.** $x - 5 = 18$ **16.** $x - 3 = 2$

In Problems 17–24, take the replacement set to be the set of all integers. Find the solution set.

17. $x + 1 = 5$ **18.** $x + 4 < 0$ **19.** $2 + x \leq x + 2$

20. $x - 5 = \frac{1}{2}$ **21.** $x + 3 = 2$ **22.** $x + 2 = 2$

23. $x + 1 > 3$ **24.** $x + 3 > 1$

In Problems 25–32, take the replacement set to be the set of real numbers. Find the solution set.

25. $x - 1 > 0$ **26.** $x + 1 < 3$ **27.** $x + \frac{1}{2} = 2$ **28.** $x - \frac{1}{2} \neq 0$

29. $x + 2 = 2$ **30.** $x + 2 \neq 2$ **31.** $x - 3 \leq 4$ **32.** $x + 1 \geq 2$

B. **33.** The relationship between the length h of the human male humerus bone and the height H of the male is given (in centimeters) by the formula

$$H = 2.89h + 70.64$$

Can a 36-cm humerus bone belong to a man 174.68 cm tall?

34. The number of chirps, C, that a cricket makes per minute satisfies the equation

$$C = 4(F - 40)$$

where F is the temperature in degrees Fahrenheit. A farmer claimed that a cricket chirped 150 times a minute when the temperature was 80°F. Is this possible?

Courtesy of PSH

35. Referring to Problem 34, if the temperature was 77.5°F, how many chirps per minute would the cricket make?

36. The perimeter, P, of a rectangle of height h and base b is given by the formula

$$P = 2(b + h)$$

A wire of length 20 cm is to be bent into a rectangle of height 4 cm. What will be the length of the base of the rectangle?

USING YOUR KNOWLEDGE 7.1

The ideas presented in this section can be used to solve many problems that involve certain simple formulas. For example, the weekly salary of a salesperson is given by

(Salary) plus (Commission) equals (Total pay)
  S   +   C   =   T

1. If a salesperson made $66 on commissions and her total pay was $176, what was her salary?

2. If a salesperson's salary was $150 and his total pay amounted to $257, what was his commission?

One's bank balance is given by

(Deposits) minus (Withdrawals) equals (Balance)
$$D \quad - \quad W \quad = \quad B$$

3. If a person deposited $304 and her balance was $102, how much money did she withdraw?
4. If a person withdrew $17 and her balance was $39, how much money did she deposit?

The ideal weight W (in pounds) of a man is related to his height H (in inches) by the formula

$$W = 5H - 190$$

5. If a man weighs 200 lb, what should his height be?
6. What should be the weight of a man whose height is 5 ft 10 in.?

DISCOVERY 7.1

Here are some problems just for fun!

1. Sally has 20 coins, all dimes and quarters. She wishes the dimes were quarters and the quarters were dimes because she would then have $1.20 more than she has now. Can you discover how many of each coin she has?
2. A clever little girl who knows her addition and multiplication tables for the integers from 1 through 5 uses finger reckoning to multiply numbers between 5 and 10. On one hand she extends the number of fingers equal to the excess of one of the numbers over 5, and does the same with the second number on her other hand. The sum of the extended fingers gives her the first digit of the product, and the product of the unextended fingers gives her the second digit of the product. Try it! Can you discover why this works?
3. Not very much is known about the Greek algebraist, Diophantus, except how old he was when he died. This fact has been preserved in the following 1500-year-old riddle. Can you discover Diophantus' age at death?

Diophantus' youth lasted $\frac{1}{6}$ of his life.

After $\frac{1}{12}$ more, he grew a beard.

After $\frac{1}{7}$ more of his life, he married, and 5 years later, he had a son.

The son lived exactly $\frac{1}{2}$ as long as his father, and Diophantus died just 4 years after his son.

[*Hint:* Let x be the number of years that Diophantus lived.]

SOLUTION OF FIRST-DEGREE SENTENCES

Suppose you wish to rent an economy car for 1 day. The cost C is $9 per day plus 10¢ per mile. If m is the number of miles traveled in 1 day, the cost can be written as

$$C = 0.10m + 9$$

If at the end of the day you paid $25, how far did you drive? To find the answer, you must solve the equation

$$25 = 0.10m + 9$$

An open sentence such as $25 = 0.10m + 9$, in which the unknown quantity has an exponent of 1, is called a **first-degree** sentence. We shall study first-degree sentences here.

A. Solving Equations

First, we consider what operations may be performed on a sentence to obtain an **equivalent** sentence — that is, one with exactly the same solution set as the original sentence. Such operations are called **elementary operations.**

For the equation

$$a = b$$

the following elementary operations yield equations **equivalent** to the original equation:

1. **Addition** $a + c = b + c$
2. **Subtraction** $a - c = b - c$
3. **Multiplication** $a \times c = b \times c$ $c \neq 0$
4. **Division** $a \div c = b \div c$ $c \neq 0$

Briefly stated, we may add or subtract the same number on both sides, or multiply or divide both sides by the same nonzero number. To solve an equation, we use the elementary operations as needed to obtain an equivalent equation of the form

$$x = n \qquad \text{or} \qquad n = x$$

where the number n is the desired solution. For example to solve the equation

$$25 = 0.10m + 9$$

we must get the m all by itself on one side of the equation. Hence, we proceed as follows:

Subtract 9:	$25 - 9 = 0.10m + 9 - 9$
or	$16 = 0.10m$
Divide by 0.10:	$\dfrac{16}{0.10} = \dfrac{0.10\ m}{0.10}$
or	$160 = m$

Thus, if you paid $25 at the end of the day, you drove 160 miles.

EXAMPLE 1 Solve the equation $x + 2 = 5$.

Solution To solve the equation, we first want to get the variable x by itself on one side. Therefore, we subtract 2 from both sides to eliminate the "+2" on the left-hand side:

$$x + 2 - 2 = 5 - 2$$

which simplifies to

$$x = 3$$

The solution set is {3}. We used elementary operation 2. ◀

EXAMPLE 2 Solve the equation $2x - 1 = x - 5$.

Solution First, we add 1 to both sides to eliminate the "−1" on the left side:

$$2x - 1 + 1 = x - 5 + 1$$

or

$$2x = x - 4$$

Then, to eliminate the x on the right side, we subtract x from both sides to get

$$2x - x = x - 4 - x$$

or

$$x = -4$$

The solution set is {−4}. Here, we have used elementary operations 1 and 2. ◀

Note: In the second operation of the solution to Example 2, we subtracted x from both sides. The only restriction on the addition or subtraction of an expression whose value depends on the value of x is that this expression be a real number for each element of the replacement set of the given equation. It is possible to get into trouble if this is not so. For instance, the equation

$$x + 1 = 2$$

has the solution set {1}. If we add $1/(x - 1)$ to both sides, to get

$$x + 1 + \frac{1}{x-1} = 2 + \frac{1}{x-1}$$

then we have an equation that is not equivalent to the given equation, because neither side of the new equation is defined for $x = 1$. (Why?)

EXAMPLE 3   Solve the equation $2x - 8 = 5x - 6$.

Solution   Because the coefficient of x on the right is greater than that on the left, we subtract **2x** on both sides to get

$$2x - 8 - 2x = 5x - 6 - 2x$$

or

$$-8 = 3x - 6$$

To eliminate the "-6" on the right side, we add **6** to both sides to obtain

$$-8 + 6 = 3x - 6 + 6$$

or

$$-2 = 3x$$

Since the 3 on the right multiplies the x, we divide both sides by **3**.

$$\frac{-2}{3} = \frac{3x}{3}$$

That is,

$$-\frac{2}{3} = x \quad \text{or} \quad x = -\frac{2}{3}$$

The solution set is $\{-\frac{2}{3}\}$, and we have used elementary operations 1, 2, and 4.   ◀

To help you in solving an equation, we suggest the following procedure:

Procedure to Solve an Equation

1. Simplify both sides of the equation, if necessary. (Remove parentheses and combine like terms.)
2. Add or subtract the same expression on both sides so that the variable occurs on one side only.
3. Add or subtract the same numbers on both sides so that only the variable term is left on one side.
4. If the coefficient of the variable is not 1, divide both sides by this coefficient.
5. The resulting equation is of the form $x = a$ (or $a = x$), where the number a is the solution of the equation.
6. Check your answer by substituting it into the original equation. Both sides must simplify to the same number.

EXAMPLE 4   Solve the equation $2(x + 1) = 5(x - 2) + 18$.

Solution   We follow the suggested procedure for the given equation:

$$2(x + 1) = 5(x - 2) + 18$$

1. Simplify both sides:

$$2x + 2 = 5x - 10 + 18$$

or

$$2x + 2 = 5x + 8$$

2. Since the coefficient of x is greater on the right than on the left, we subtract **2x** on both sides to get

$$2x + 2 - 2x = 5x + 8 - 2x$$

or

$$2 = 3x + 8$$

3. To eliminate the 8 on the right-hand side, we subtract **8** on both sides:

$$2 - 8 = 3x + 8 - 8$$

or

$$-6 = 3x$$

4. Divide both sides by **3**:

$$\frac{-6}{3} = \frac{3x}{3}$$

5. Or

$$-2 = x$$

Thus, the solution is $x = -2$.

6. *Check.* For $x = -2$, the left side of the given equation becomes

$$2(-2 + 1) = 2(-1) = -2$$

and the right side becomes

$$5(-2 - 2) + 18 = 5(-4) + 18$$
$$= -20 + 18 = -2$$

Since the two sides agree, the solution checks.   ◀

An expression of the form $ax + b$, where a and b are real numbers and $a \neq 0$, is called a **first-degree (or linear) expression in x,** and an equation of the form

$$ax + b = 0 \qquad a \neq 0$$

is called a **first-degree (or linear) equation in x.** The methods we have used in Examples 1–4 always suffice to solve first-degree equations in x.

B. Solving Inequalities

An **inequality of the first degree in x** is an inequality of the form

$$ax + b < 0 \qquad a \neq 0$$

or of the form

$$ax + b > 0 \qquad a \neq 0$$

We can solve such inequalities by means of elementary operations that produce equivalent inequalities. These operations are as follows:

For the inequality

$$a < b$$

the following elementary operations yield inequalities **equivalent** to the original inequality:

1. Addition	$a + c < b + c$	
2. Subtraction	$a - c < b - c$	
3. Multiplication	$ac < bc$	for $c > 0$
	$ac > bc$	for $c < 0$
4. Division	$\dfrac{a}{c} < \dfrac{b}{c}$	for $c > 0$
	$\dfrac{a}{c} > \dfrac{b}{c}$	for $c < 0$

Briefly stated, we may add or subtract the same number on both sides. The sense of the inequality is **unchanged** if both sides are multiplied or divided by the same *positive* number. The sense of the inequality is **reversed** if both sides are multiplied or divided by the same *negative* number. For instance, if both sides of $-2 < 1$ are multiplied by -3, we get $6 > -3$.

The preceding operations have been stated for the inequality $a < b$, but the same operations are valid for $a > b$. You can convince yourself of the validity of these operations by noting that the geometric equivalent of $a < b$ is *a precedes b on the number line.* (The diagram in Figure 7.2a will clarify this idea.) Since $a \le b$ means $a < b$ **or** $a = b$, the elementary operations may be used for inequalities of the type $a \le b$ and $a \ge b$.

Figure 7.2a

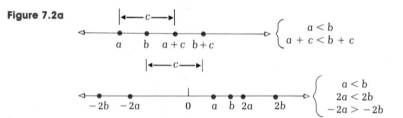

Linear (first-degree) inequalities can be solved by a procedure similar to that used for equations. This is illustrated in the next examples, where x represents a real number.

EXAMPLE 5 Solve the inequality $3x + 2 > x + 6$.

Solution 1. The two sides are already in simplified form.
2. Subtract x from both sides:

$$3x + 2 - x > x + 6 - x$$
$$2x + 2 > 6$$

3. Subtract 2 from both sides:

$$2x + 2 - 2 > 6 - 2$$
$$2x > 4$$

4. Divide both sides by 2:

$$\frac{2x}{2} > \frac{4}{2}$$
$$x > 2$$

5. The solution set is $\{x | x > 2\}$.
6. A partial check can be made by substituting a number from the proposed solution set into the original inequality. For instance, 3 is in the set $\{x | x > 2\}$. For $x = 3$, the left side becomes

$$3(3) + 2 = 11$$

and the right side becomes

$$3 + 6 = 9$$

Since $11 > 9$, $x = 3$ does satisfy the inequality. Because the solution set contains infinitely many numbers, we cannot check by substituting one number at a time. However, if the number selected did *not* check, then something would be wrong, and we could check the work to find the error. ◄

EXAMPLE 6 Solve the inequality $2x - 3 < 5x + 7$.

Solution 1. The two sides are already in simplified form.
2. Subtract $5x$ from both sides:

$$2x - 3 - 5x < 5x + 7 - 5x$$
$$-3x - 3 < 7$$

3. Add 3 to both sides:

$$-3x - 3 + 3 < 7 + 3$$
$$-3x < 10$$

4. Divide both sides by -3:

$$x > -\frac{10}{3}$$

5. The solution set is $\{x | x > -\frac{10}{3}\}$. Be sure to notice that **division by -3 reversed the sense of the inequality.**

6. The check is left for you to do. An easy number to use is 0.

If, in step 2, you were to subtract $2x$ from both sides of the inequality, you would avoid dividing by a negative number later. However, in this case, the answer would be $-\frac{10}{3} < x$. If you are asked what x is, then you must write the equivalent answer $x > -\frac{10}{3}$. Which way should you do it? Whichever way you understand best! ◀

EXERCISE 7.2

A. *Find the solution set for each of the following if the replacement set is the set of real numbers.*

1. $x + 10 = 15$ **2.** $x - 5 = 8$

3. $2x - 1 = 5$ **4.** $3x + 1 = 4$

5. $2x + 2 = x + 4$ **6.** $3x + 1 = x - 3$

7. $3x + 1 = 4x - 8$ **8.** $2x + 3 = 3x - 1$

9. $7 = 3x + 4$ **10.** $22 = 4x + 2$

11. $4 = 3x - 2$ **12.** $1 = 5x - 8$

13. $2(x + 5) = 13$ **14.** $6(x + 2) = 17$

15. $8(x - 1) = x + 2$ **16.** $6(x - 1) = x + 6$

17. $x + 6 = 2(x + 2)$ **18.** $x + 5 = 3(x + 1)$

B. **19.** $x - 3 < 1$ **20.** $x - 2 < 2$

21. $x - 4 > -1$ **22.** $x + 3 > -2$

23. $2x - 1 > x + 2$ **24.** $3x - 3 > 2x + 1$

25. $2x + 3 \leq 9 + 5x$ **26.** $x + 8 \geq 2x - 1$

27. $x + 1 > \frac{1}{2}x - 1$ **28.** $x - 1 < \frac{1}{2}x + 2$

29. $x \geq 4 + 3x$ **30.** $x - 1 \leq 5 + 3x$

31. $\frac{1}{3}x - 2 \geq \frac{2}{3}x + 1$ **32.** $\frac{1}{4}x + 1 \leq \frac{3}{4}x - 1$

33. $2x - 2 > x + 1$ **34.** $3x - 2 > 2x + 2$

35. $x + 3 > \frac{1}{2}x + 1$ **36.** $x + 1 < \frac{1}{2}x + 4$

37. $x \geq 2 + 4x$ **38.** $x - 2 \leq 6 + 3x$

39. $2x + 1 < 2x$ **40.** $5x \leq 5x + 4$

41. $8x + 2 \leq 3(x + 4)$ **42.** $9x + 3 \leq 4(x + 2)$

43. $3(x + 4) > -5x - 4$ **44.** $5(x + 2) < -3x + 2$

45. $-2(x + 1) \geq 3x - 4$ **46.** $-3(2 - x) \geq 5x - 7$

One of the most important ideas in elementary mathematics is that of **percent.** Basically, there are three types of problems involving percent. These may be illustrated as follows:

1. 40% of 60 is what number?
2. What percent of 50 is 10?
3. 20 is 40% of what number?

All these problems (which, incidentally, can be stated in different ways) can easily be solved by using the knowledge we have gained in this section. The basic idea is that "r% of n is p" translates into the equation

$$rn = 100p \quad \text{Recall that } r\% = \frac{r}{100} \tag{1}$$

Each type of percent problem can be solved by substituting the known data into equation (1) and then solving for the unknown as follows:

Problem 1. 40% of 60 is what number? Here, $r = 40$, $n = 60$, and p is unknown. Equation (1) becomes

$$40 \times 60 = 100p \quad \text{or} \quad 2400 = 100p$$

To find p, divide both sides by 100. This gives

$$p = 24$$

Problem 2. What percent of 50 is 10? Here, r is unknown, $n = 50$, and $p = 10$. Equation (1) becomes

$$50r = 100 \times 10 = 1000$$

To find r, divide both sides by 50 to get

$$r = 20$$

So, 10 is 20% of 50.

Problem 3. 20 is 40% of what number? Here, $r = 40$, $p = 20$, and n is unknown. Equation (1) becomes

$$40n = 100 \times 20 = 2000$$

Dividing by 40, we find $n = 50$.

Use these ideas to solve the following problems.

1. 40% of 80 is what number?
2. Find 15% of 60.
3. 315 is what percent of 3150?
4. 8 is what percent of 4?
5. What percent of 40 is 5?

6. 20 is what percent of 30?

7. 30% of what number is 60?

8. 10 is 40% of what number?

9. North America has approximately 7% of the world's oil reserves. If the North American reserves represent 47 billion barrels of oil, what are the world's oil reserves? (Round your answer to the nearest billion.)

10. In a recent year, about 280 million tons of pollutants were released into the air in the United States. If 47% of this amount was carbon monoxide, how many tons was that?

11. On a 60-item test, a student got 40 correct. What percent is that? (Round to the nearest percent.)

12. The price of an article on sale was 90% of the regular price. If the sale price was $18, what was the regular price?

13. Two stores sell an item that they normally price at $140. Store A advertises a sale price of 25% off the regular price, and store B advertises a sale price of $100. Which is the lower price?

14. The ABC Savings & Loan loans the Adams family $40,000 toward the purchase of a $48,000 house. What percent of the purchase price is the loan?

DISCOVERY 7.2

1. A problem that comes from the Rhind papyrus, one of the oldest mathematical documents known, reads like this: "$\frac{2}{3}$ added and then $\frac{1}{3}$ taken away, 10 remains. . . ." The document then goes on to tell how to find the number with which you started. Can you discover what this number is? [*Hint:* If you let x be the unknown number, then the problem intends you to add $\frac{2}{3}x$ and then take away $\frac{1}{3}$ of the result to get a remainder of 10.]

2. Mr. C usually takes the 5 o'clock train from the city and is met at the station by Mrs. C, who then drives him home. One day Mr. C took the 4 o'clock train, and when he reached his station, he started walking home. Mrs. C met him on the way and drove him the rest of the way home. If they reached home 20 min earlier than usual, how long did Mr. C walk? [*Hint:* You do not need any algebra for this problem. Just draw a diagram and see on what portion of the trip Mrs. C saved the 20 min.]

3. There is a peculiar three-digit number. It ends with a 4. If the 4 is moved to the front, the new number is as much greater than 400 as the original number was less than 400. Can you discover what was the original number? [*Hint:* If the digits of a three-digit number are a, b, c, then the number is $100a + 10b + c$. If you simply move the c to the front, the new number is $100c + 10a + b$.]

GRAPHS OF ALGEBRAIC SENTENCES

The solution set of an open sentence in one variable can always be represented by a set of points on the number line. This set of points is often called the **graph** of the equation or the inequality, as the case may be. We shall illustrate various types of graphs in the following examples.

EXAMPLE 1   Graph the solution set of the equation $x + 1 = 0$, where x is an integer.

Solution   Subtracting 1 from both sides, we see that the solution set is the singleton set $\{-1\}$. The graph consists of the single point -1 on the number line. We draw a solid dot to indicate this graph (see Fig. 7.3a).

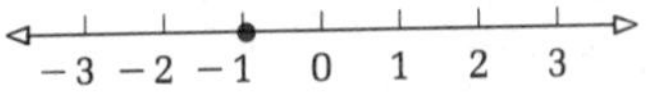

Figure 7.3a
The singleton set $\{-1\}$

EXAMPLE 2   Graph the solution set of the inequality $x + 1 \leq 0$, where x is an integer.

Solution   Subtracting 1 from both sides yields the equivalent inequality $x \leq -1$. Thus, the solution set is the set of all integers that are less than or equal to -1, that is, $\{\ldots, -3, -2, -1\}$. To show this graph we draw dots at the corresponding points of the number line (see Fig. 7.3b).

Figure 7.3b
$\{\ldots, -3, -2, -1\}$

EXAMPLE 3   Graph the solution set of the inequality $x + 1 \leq 0$, where x is a real number.

Solution   Proceeding as in Example 2, we see that the solution set is the set of all real numbers less than or equal to -1, that is, $\{x \mid x \leq -1, x \text{ real}\}$. We display this set by drawing a heavy line starting at -1 on the number line and going to the left. The point at -1 is marked with a solid dot to show that it is included in the set (see Fig. 7.3c).

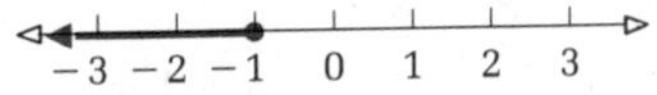

Figure 7.3c
The set $\{x \mid x \leq -1\}$

EXAMPLE 4   Graph the solution set of the inequality $x + 2 \neq 5$, where x is a real number.

Solution   The number 3 is the only replacement for x such that $x + 2 = 5$, so the solution set is all real numbers except 3, that is, $\{x \mid x \neq 3, x \text{ real}\}$. The graph consists of the entire number line except for the point 3. In Fig. 7.3d, the

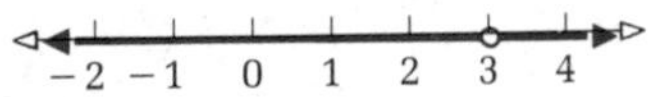

Figure 7.3d
The set $\{x \mid x \neq 3\}$

graph is shown in color and the point 3 is marked with an open circle to indicate its exclusion from the solution set. ◄

EXAMPLE 5   Graph the solution set of the inequality $-2 \leq x < 1$, where x is a real number.

Solution   The solution set consists of all the real numbers between -2 and 1, with the -2 included and the 1 excluded. The graph is shown in color in Fig. 7.3e.

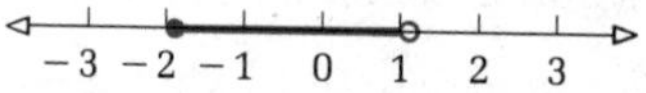

Figure 7.3e
The set $\{x|-2 \leq x < 1\}$

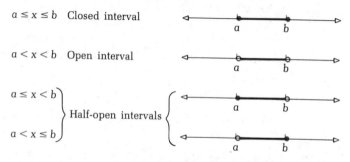

The piece of the number line such as that in Fig. 7.3e is called a **finite interval** (or a **line segment**). The endpoints are -2 and 1. We call the interval **closed** if both endpoints are included, **open** if both endpoints are excluded, and **half-open** if only one of the endpoints is included. The interval in Fig. 7.3e is half-open.

If a and b are real numbers with $a < b$, then the various types of finite intervals are as shown in Fig. 7.3f.

Figure 7.3f
Finite intervals

$a \leq x \leq b$ Closed interval

$a < x < b$ Open interval

$\left. \begin{array}{l} a \leq x < b \\ \\ a < x \leq b \end{array} \right\}$ Half-open intervals

EXERCISE 7.3

In Problems 1–8, take x to be an integer and graph the solution set of the given sentence.

1. $x + 2 = 4$
2. $x - 1 = 3$
3. $x + 1 \geq 2$
4. $x + 2 < 5$
5. $x - 3 \neq 1$
6. $-3 < x \leq 2$
7. $-2 \leq x \leq 4$
8. $-2 < x < 4$

In Problems 9–26, take x to be a real number and graph the solution set of the given sentence.

9. $x < 4$
10. $x \geq 2$
11. $x - 2 \leq 0$
12. $x - 2 \geq 4$
13. $-2 \leq x \leq 4$
14. $x - 3 \neq 1$
15. $x + 2 > 5$
16. $x - 2 = 1$
17. $-1 < x < 2$
18. $x \geq 3$
19. $x + 4 < 5$
20. $x + 5 < 4$

21. $x + 1 < x$ **22.** $x + 1 > x$ **23.** $2x + 3 < x + 1$

24. $-x + 5 \leq 2x + 2$ **25.** $3x - 7 \geq -7$ **26.** $2x + 5 < 5$

DISCOVERY 7.3

Can we discover a base 6, three-digit number that has its digits just reversed when expressed in base 5? Here is how to go about it. Let the digits in base 6 be a, b, c. Then, we want to have

$$abc_6 = cba_5$$

But, in expanded notation, this means

$$(a \times 6^2) + (b \times 6) + c = (c \times 5^2) + (b \times 5) + a$$

that is,

$$36a + 6b + c = 25c + 5b + a$$

By subtracting a, 5b, and c from both sides, we get

$$35a + b = 24c$$

Since a, b, and c are to be digits in base 5 notation, the domain of the equation is the set {0, 1, 2, 3, 4}. Also, to have actual three-digit numbers, the a and the c must not be 0. Now, we use trial and error to solve the equation.

1. Try $c = 1$. Then $35a + b = 24$. If $a \geq 1$, then b would be negative. Thus, there is no solution for $c = 1$.
2. Try $c = 2$. Then $35a + b = 48$. If $a = 1$, b would be 13, which is not in the domain. If $a \geq 2$, b would be negative. Thus, there is no solution for $c = 2$.
3. Try $c = 3$. Then $35a + b = 72$. If $a = 1$, b would be 37, which is not in the domain. If $a = 2$, we get $70 + b = 72$, so that $b = 2$. Therefore, $a = 2$, $b = 2$, $c = 3$ is a solution, and

$$223_6 = 322_5$$

If $a \geq 3$, b would be negative.
4. Try $c = 4$. Then $35a + b = 96$. You can show that there is no solution for b in the domain for any permissible value of a.

The only solution is that found in step 3.

Check. Convert to base 10:

$$223_6 = (2 \times 36) + (2 \times 6) \times 3 = 87$$
$$322_5 = (3 \times 25) + (2 \times 5) + 2 = 87$$

Thus, the answer is correct.

1. Can you discover a base 10, three-digit number that has its digits just reversed when expressed in base 9?
2. Can you discover a base 9, three-digit number that has its digits just reversed when expressed in base 7?
3. Can you discover a base 9, three-digit number that has its digits just reversed when expressed in base 5?
4. Can you discover a base 7, three-digit number that has its digits just reversed when expressed in base 5?

7.4

COMPOUND SENTENCES

In this section we consider compound algebraic sentences consisting of two or more simple sentences of the type that occurred in the preceding sections. We shall be concerned with the connectives **and** and **or** used in exactly the same sense as in Chapter 2.

A. Sentences with "and"

EXAMPLE 1   Consider the sentence $x + 1 < 3$ *and* $x - 1 > -1$. Find its solution set.

Solution   The sentence given here is a compound sentence of the type $p \wedge q$ (p and q), where p is $x + 1 < 3$ and q is $x - 1 > -1$. Such a sentence as $p \wedge q$ is true only when both p and q are true. Consequently, the solution set of the compound sentence is the **intersection** of the solution sets of the two components.

We have

$$x + 1 < 3 \qquad and \qquad x - 1 > -1$$
$$x + 1 - 1 < 3 - 1 \qquad and \qquad x - 1 + 1 > -1 + 1$$
$$x < 2 \qquad and \qquad x > 0$$

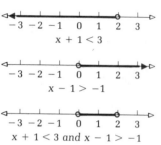

Reversing the second inequality, we see that x must satisfy the conditions

$$0 < x \qquad and \qquad x < 2$$

Thus, the solution set can be written

$$\{x | 0 < x\} \cap \{x | x < 2\}$$

or, more efficiently, $\{x | 0 < x < 2\}$.

This result is most easily seen from the figure in the margin.   ◀

Note: No replacement set was specified in Example 1. Whenever this is the case, we assume that the replacement set consists of all real numbers for which the members of the inequalities are defined.

EXAMPLE 2 Find the solution set of $x - 1 > 4$ and $x + 2 < 5$.

Solution We have

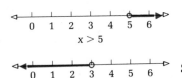

$$x - 1 > 4 \qquad and \qquad x + 2 < 5$$
$$x - 1 + 1 > 4 + 1 \qquad and \qquad x + 2 - 2 < 5 - 2$$
$$x > 5 \qquad and \qquad x < 3$$

Since there are no numbers satisfying both of these conditions (see the figure in the margin), there are no solutions; the solution set is empty. ◄

EXAMPLE 3 Find the solution set of the sentence $x \leq 5$ and $x + 1 \geq 0$ if x is an integer.

Solution If x is an integer, then the solution set of $x \leq 5$ is the set of all integers less than or equal to 5. Similarly, the solution set of $x + 1 \geq 0$ is the set of all integers greater than or equal to -1. The intersection of these two sets is the set $\{-1, 0, 1, 2, 3, 4, 5\}$, which is the desired solution set. ◄

B. Sentences with "or"

EXAMPLE 4 Find the solution set of the sentence $x + 1 < 5$ or $x - 1 > 6$.

Solution The replacement set is the set of all real numbers. (Why?) Because this is a sentence of the type $p \lor q$ (p or q), we know that the solution set is the **union** of the solution sets of the two components. We have

$$x + 1 < 5 \qquad or \qquad x - 1 > 6$$
$$x < 4 \qquad or \qquad x > 7$$

Thus, the required solution set is

$$\{x | x < 4\} \cup \{x | x > 7\}$$

or stated in another way,

$$\{x | x < 4 \text{ or } x > 7\}$$

The graph in the figure illustrates the solution.

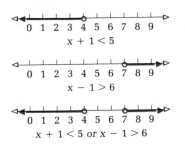

◄

A – B. In Problems 1–8, take the replacement set to be the set of integers. Give the solution set by listing its elements.

1. $x \leq 4$ and $x - 1 \geq -2$
2. $x > 0$ and $x \leq 5$
3. $x + 1 \leq 7$ and $x > 2$
4. $x > -5$ and $x - 1 < 0$
5. $2x - 1 > 1$ and $x + 1 < 4$
6. $x - 1 < 1$ or $3x - 1 > 11$
7. $x < -5$ or $x > 5$
8. $x \geq 0$ or $x < -2$

In Problems 9–26, take the replacement set to be the set of real numbers. Graph the solution set whenever it is not the empty set.

9. $x + 1 \geq 2$ and $x \leq 4$
10. $x \leq 5$ and $x > -1$
11. $x > 2$ and $x < -2$
12. $x + 2 \leq 4$ or $x + 2 \geq 6$
13. $x - 1 > 0$ and $x + 1 < 5$
14. $x \leq 0$ or $x > 3$
15. $x \leq x + 1$ and $x \geq 2$
16. $x + 2 \geq -2$ or $x < 0$
17. $x - 2 \geq 2$ and $x < 0$
18. $x + 3 \leq 0$ or $x - 1 > 0$
19. $x \geq 0$ and $x - 1 \geq 2$
20. $x \geq 0$ and $x - 1 \leq 2$
21. $x < 0$ or $x - 1 < 2$
22. $x < 0$ and $x - 1 > 2$
23. $x + 1 > 2$ and $x - 2 < 3$
24. $x - 1 > 3$ and $x + 1 > 2$
25. $x - 1 > 0$ or $x + 2 < 4$
26. $x + 1 > 2$ or $x - 1 < 2$

DISCOVERY 7.4

In ancient history it is said that King Solomon was the wisest of men. There is a legend that goes like this: One day when he was resting in his palace garden, Solomon, who was so wise that he could even understand the language of all animals and plants, heard gentle voices close to him. Upon further inspection he discovered two snails in a solemn meeting.

The first snail said, "Brother, see'st thou yon straight pole that riseth upright from the ground 30 cubits high?" And the second snail answered, "Yea, even so do I."

"It is my desire," said the first snail, "to climb to the very top of it. How long thinketh thou it will take me?"

"That certainly shall depend on the speed with which thou climbest."

"It is not as simple as that," said the would-be climber, "I can ascend but 3 cubits during the day, but in the evening I fall asleep and slip back 2 cubits so that, in effect, I move up but one cubit every 24 hours."

"Tis plain then," said the second snail, "that thou will take 30 days to reach the top of the pole. Why dost thou plague me with such a simple problem? Prithee be silent and allow me to sleep."

King Solomon smiled. He alone knew whether the second snail was right.

1. What do you think? [Hint: It will obviously take the snail 25 days to reach 25 cubits of height. From then on, draw a graph on the number line in the margin and find the number of days it really took the snail to get to the top.]

SENTENCES INVOLVING ABSOLUTE VALUES

A.   Absolute Value

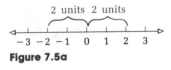

Figure 7.5a

Sometimes we need to solve equations or inequalities that involve *absolute values*. The **absolute value** of a number x is defined to be the distance on the number line from 0 (the origin) to x, and is denoted by |x| (read, "absolute value of x"). For example, the number 2 is 2 units away from 0, so |2| = 2. The number −2 is also 2 units away from 0, so |−2| = 2. (See Fig. 7.5a.) Similarly, we see that |8| = 8, |½| = ½, |−5| = 5, and |−¼| = ¼. In general, if x is any real nonnegative number (x ≥ 0), then |x| is simply x itself. But if x is a negative number (x < 0), then |x| is the corresponding positive number obtained by reversing the sign of x. Thus,

$$|x| = \begin{cases} x & \text{if } x \geq 0 \\ -x & \text{if } x < 0 \end{cases}$$

B.   Equations and Inequalities with Absolute Values

EXAMPLE 1   Find and graph the solution set of |x| = 2.

Solution   We look for all numbers that are 2 units away from the origin (the 0 point). Since 2 and −2 are the only two numbers that satisfy this condition, the solution set of the equation |x| = 2 is {2, −2}. The graph of this set is shown in the figure in the margin. Note that this solution set can be described by the compound sentence x = 2 or x = −2.   ◄

EXAMPLE 2   Find and graph the solution set of the inequality |x| < 2.

Solution   Here we look for all real numbers x that are less than 2 units away from 0. Since 2 and −2 are each exactly 2 units away from 0, we need all points between −2 and 2, that is, all numbers that satisfy the inequality

−2 < x < 2

The solution set is thus {x|−2 < x < 2}. The graph appears in the figure in the margin.   ◄

In general, if a is any positive number, then

|x| < a   is equivalent to   −a < x < a

EXAMPLE 3   Graph the solution set of $|x| \leq 4$.

Solution   Since $|x| \leq 4$ is equivalent to $-4 \leq x \leq 4$, the graph of the solution set is as shown in the figure.

$$\xleftarrow{\quad} \underset{-5}{\bullet} \quad \underset{-4}{\mid} \quad \underset{-3}{\mid} \quad \underset{-2}{\mid} \quad \underset{-1}{\mid} \quad \underset{0}{\mid} \quad \underset{1}{\mid} \quad \underset{2}{\mid} \quad \underset{3}{\mid} \quad \underset{4}{\bullet} \quad \underset{5}{\mid} \xrightarrow{\quad}$$ ◄

EXAMPLE 4   Find the solution set of the inequality $|x - 1| < 4$.

Solution   Since $|x| < a$ is equivalent to $-a < x < a$,

$$|x - 1| < 4 \quad \text{is equivalent to} \quad -4 < x - 1 < 4$$

As in the case of equations, we still want to have x alone; so in the middle member of the inequality, we need to add 1. Of course, we must do the same to all the members. Thus, we get

$$-4 + 1 < x - 1 + 1 < 4 + 1$$

or

$$-3 < \quad x \quad < 5$$

Thus, the solution set is $\{x | -3 < x < 5\}$. ◄

EXAMPLE 5   Find and graph the solution set of $|x| \geq 3$.

Solution   The solution set of $|x| = 3$ consists of all points that are exactly 3 units away from 0. Hence, the solution set of $|x| \geq 3$ consists of all points that are 3 or *more* units away from 0. As you can see from the figure, these points can be described by the compound sentence.

$$x \geq 3 \quad \text{or} \quad x \leq -3$$

$$\xleftarrow{\quad} \underset{-5}{\blacktriangleleft} \quad \underset{-4}{\mid} \quad \underset{-3}{\bullet} \quad \underset{-2}{\mid} \quad \underset{-1}{\mid} \quad \underset{0}{\mid} \quad \underset{1}{\mid} \quad \underset{2}{\mid} \quad \underset{3}{\bullet} \quad \underset{4}{\mid} \quad \underset{5}{\mid} \xrightarrow{\quad}$$

We may therefore write the solution set of $|x| \geq 3$ in the form

$$\{x | x \geq 3\} \cup \{x | x \leq -3\}$$

Note that the answer may also be written in the form $\{x | x \geq 3 \text{ or } x \leq -3\}$. ◄

The idea in Example 5 can be generalized as follows:

If $a \geq 0$, then $|x| > a$ is equivalent to the compound sentence

$$x > a \quad \text{or} \quad x < -a$$

Note: Be sure to remember to reverse the inequality sign for the $-a$, as in the next examples.

EXAMPLE 6   Graph the solution set of $|x| > 2$.

Solution |x| > 2 is equivalent to the compound sentence x > 2 or x < −2. The graph of the solution set appears in the figure.

EXAMPLE 7 Graph the solution set of |x + 1| ≥ 2.

Solution |x| ≥ a is equivalent to x ≥ a or x ≤ −a. Therefore, |x + 1| ≥ 2 is equivalent to x + 1 ≥ 2 or x + 1 ≤ −2; that is, x ≥ 1 or x ≤ −3. The graph of the solution set is shown in the figure.

EXERCISE 7.5

A. In Problems 1–10, evaluate the given expression.

1. |−10|
2. |15|
3. |−⅛|
4. |¾|
5. |5 − 8|
6. |8 − 5|
7. |0| + |−2|
8. |−2| − |−3|
9. −|8|
10. −|3| + |−4|

B. In Problems 11–16, find the set of integers for which the given sentence is true.

11. |x| < 1
12. |x| > −2
13. |x| = 5
14. |x| ≤ 3
15. |x| ≥ 1
16. |x| < 4

In Problems 17–34, graph the solution set of the given sentence (if possible).

17. |x| = 1
18. |x| = 2.5
19. |x| ≤ 4
20. |x| > 1
21. |x + 1| < 3
22. |x − 2| < 1
23. |x| ≥ 1
24. |x| > −1
25. |x − 1| > 2
26. |x − 3| ≥ 1
27. |2x| < 4
28. |3x| ≤ 9
29. |3x| ≥ 6
30. |2x| > 5
31. |2x − 3| ≤ 3
32. |3x + 1| ≤ 8
33. |2x − 3| > 3
34. |3x + 1| > 8

7.6

QUADRATIC EQUATIONS

In the preceding sections we have discussed the solution of first-degree sentences. We now turn our attention to a certain type of **second-degree** sentence that is called a *quadratic equation*. A **quadratic equation** is a second-degree sentence whose standard form is

$$ax^2 + bx + c = 0$$

where a, b, c are real numbers and $a \neq 0$.

We first consider quadratic equations where the second-degree expression $ax^2 + bx + c$ can be written as a product of two first-degree expressions. Suppose that we have the product $(x + p)(x + q)$. Then, by the distributive property $a(b + c) = ab + ac$, with $a = x + p$ and $b + c = x + q$, we get

$$(x + p)(x + q) = (x + p)x + (x + p)q$$

Then, using the distributive property $(b + c)a = ba + ca$, we get

$$(x + p)(x + q) = x^2 + px + qx + pq$$
$$= x^2 + (p + q)x + pq$$

A good way to remember this multiplication is to remember the **FOIL** idea. To multiply $(x - 1)(x + 2)$, we write:　$x^2 + 2x - x - 2$

1. Multiply **F** irst terms:　$(x - 1)(x + 2) = x^2 \ldots$
2. Multiply **O** utside terms: $(x - 1)(x + 2) = x^2 + 2x \ldots$
3. Multiply **I** nside terms:　$(x - 1)(x + 2) = x^2 + 2x - x \ldots$
4. Multiply **L** ast terms:　$(x - 1)(x + 2) = x^2 + 2x - x - 2$

Thus,

$$(x - 1)(x + 2) = x^2 + x - 2$$

A. Factoring Quadratics

These ideas can sometimes be "reversed" to write a quadratic expression in the product (factored) form. For example, in order to write $x^2 + 5x + 6$ in factored form, we try to find two numbers p and q so that $pq = 6$ and $p + q = 5$. By inspection, we see that $p = 2$ and $q = 3$ will do the trick. Therefore,

$$x^2 + 5x + 6 = (x + 2)(x + 3)$$

To **factor** an expression means to write it as a product of lower-degree expressions as in the above illustration.

EXAMPLE 1　Factor:

　　　a. $x^2 + x - 2$　　**b.** $x^2 - 2x - 8$

Solution　**a.** We must find two numbers whose product is -2 and whose sum is 1. These numbers are 2 and -1. Thus,

$$x^2 + x - 2 = (x + 2)(x - 1)$$

Note that $2x - x = x$.

b. Here we need two numbers whose product is -8 and whose sum is -2. These numbers are -4 and 2. Thus,

$$x^2 - 2x - 8 = (x - 4)(x + 2)$$

Note that $-4x + 2x = -2x$. ◀

B. Solving Quadratics by Factoring

The next examples show how factoring can sometimes be used to solve quadratic equations.

EXAMPLE 2   Solve the equation $x^2 + 5x + 6 = 0$.

Solution   By the preceding discussion, we see that an equivalent equation is

$$(x + 2)(x + 3) = 0$$

With 0 on one side of the equation, we can make use of the property of the real number system that says that a product of two real numbers is 0 if and only if at least one of them is 0. Thus, the preceding equation is true if and only if

$$x + 2 = 0 \quad \text{or} \quad x + 3 = 0$$
$$x = -2 \quad \text{or} \quad x = -3$$

The solution set of the given equation is $\{-3, -2\}$.

Check   By substitution in the left side of the given equation, we find for $x = -3$, $x^2 + 5x + 6 = (-3)^2 + 5(-3) + 6 = 9 - 15 + 6 = 0$. This checks the solution $x = -3$. For $x = -2$, we get $x^2 + 5x + 6 = (-2)^2 + 5(-2) + 6 = 4 - 10 + 6 = 0$, which checks the solution $x = -2$. ◀

EXAMPLE 3   Solve the equation $x^2 - 4x - 12 = 0$

Solution   We try to write the left side in the product form by finding p and q so that $pq = -12$ and $p + q = -4$. A little trial and error using factors of 12 shows that $p = -6$, $q = 2$ will work. Thus, an equivalent equation is

$$(x - 6)(x + 2) = 0$$

If $x - 6 = 0$, then $x = 6$. If $x + 2 = 0$, then $x = -2$. The solution set is $\{-2, 6\}$. (The reader may check this solution set as in Example 2.) ◀

EXAMPLE 4   Solve the equation $x^2 - 9 = 0$.

Solution   Adding 9 to both sides, we obtain

$$x^2 = 9$$

which has the solution set $\{-3, 3\}$, because $(-3)^2 = 9$ and $(3)^2 = 9$.

The equation $x^2 - 9$ can be solved in another way since the difference of two squares, say $x^2 - p^2$, can always be written as the product $(x - p)(x + p)$. [Checking this multiplication, we get $(x - p)(x + p) = x^2 - \underbrace{px + px}_{0} - p^2 = x^2 - p^2$.] Thus,

$$x^2 - 9 = 0$$

becomes

$$(x - 3)(x + 3) = 0 \quad \text{Since } (x - 3)(x + 3) = x^2 - 9$$

which gives the solution $\{-3, 3\}$, as before. ◀

EXAMPLE 5   Solve the equation $3x^2 + 2 = 50$.

Solution   We first subtract 2 from both sides, obtaining

$$3x^2 = 48$$

We then divide both sides by 3 to get

$$x^2 = 16$$

As in Example 4, we find the solution set by taking square roots of both sides. This gives the solution set $\{-4, 4\}$. ◀

C. The Quadratic Formula

The general quadratic equation

$$ax^2 + bx + c = 0 \qquad a \neq 0$$

can be solved by the **quadratic formula**,

$$x = \frac{-b \pm \sqrt{b^2 - 4ac}}{2a}$$

The derivation of this formula is given in the exercises (see Problem 51). The symbol $\pm$ in the formula means that there are two solutions, one with the plus sign and the other with the minus sign. If the quantity under the radical sign, $b^2 - 4ac$, is positive, there are two real number solutions. If $b^2 - 4ac$ is 0, the two solutions are the same, so that there is actually just one solution, $-b/2a$. If the quantity $b^2 - 4ac$ is negative, then the two solutions are both imaginary numbers.

EXAMPLE 6   Use the quadratic formula to solve $x^2 - 4x - 12 = 0$. (Compare Example 3.)

Solution   To obtain the correct values of a, b, and c, we rewrite the equation in standard quadratic form as

$$1x^2 + (-4)x + (-12) = 0$$

which we compare with

$$ax^2 + \quad bx + \quad c = 0$$

Now, we see that $a = 1$, $b = -4$, and $c = -12$. Thus,

$$x = \frac{-(-4) \pm \sqrt{(-4)^2 - (4)(1)(-12)}}{(2)(1)}$$

$$= \frac{4 \pm \sqrt{16 + 48}}{2} = \frac{4 \pm \sqrt{64}}{2} = \frac{4 \pm 8}{2}$$

so that

$$x = \frac{4 + 8}{2} = \frac{12}{2} = 6 \quad \text{or} \quad x = \frac{4 - 8}{2} = \frac{-4}{2} = -2$$

Hence, the solution set is $\{-2, 6\}$, which agrees with our previous result. ◀

EXAMPLE 7   Solve the equation $3x^2 + x - 5 = 0$.

Solution   We compare

$$ax^2 + bx + c = 0 \quad \text{and} \quad 3x^2 + x - 5 = 0$$

to see that $a = 3$, $b = 1$, and $c = -5$. Hence,

$$x = \frac{-1 \pm \sqrt{1^2 - 4(3)(-5)}}{2(3)} = \frac{-1 \pm \sqrt{61}}{6}$$

The solution set is

$$\left\{ \frac{-1 - \sqrt{61}}{6}, \frac{-1 + \sqrt{61}}{6} \right\}$$

These numbers cannot be expressed exactly in any simpler form. By using the table of square roots in the back of the book (or a calculator), we find that $\sqrt{61} \approx 7.81$. This gives the approximate solutions -1.47 and 1.14. ◀

EXAMPLE 8   Solve the equation $2x^2 - 2x = -1$.

Solution   In order to use the quadratic formula, we must first write the equation in the standard quadratic form. We can do this by adding 1 to both sides of the given equation to obtain

$$2x^2 - 2x + 1 = 0$$

which we compare with

$$ax^2 + bx + c = 0$$

Thus, $a = 2$, $b = -2$, and $c = 1$. Now, we can substitute into the quadratic formula to find

$$x = \frac{-(-2) \pm \sqrt{(-2)^2 - 4(2)(1)}}{2(2)} = \frac{2 \pm \sqrt{-4}}{4}$$

Recall that $\sqrt{-4} = 2i$ (Section 5.4), so the solutions are

$$x = \frac{2 \pm 2i}{4} = \frac{2(1 \pm i)}{4} = \frac{1 \pm i}{2}$$

The solution set is

$$\left\{ \frac{1+i}{2}, \frac{1-i}{2} \right\}$$
◄

Examples 6–8 display the tremendous advantage of the quadratic formula over other methods of solving quadratic equations. If is not necessary to attempt to write the left side as a product of first-degree expressions. You need only recognize the values of a, b, c and make a direct substitution into the quadratic formula.

EXERCISE 7.6

A. In Problems 1–10, factor the given expression.

1. $x^2 + 6x + 8$
2. $x^2 + 7x + 10$
3. $x^2 - x - 12$
4. $x^2 - 3x - 10$
5. $x^2 + 7x - 18$
6. $x^2 - 12x + 11$
7. $x^2 - 10x + 25$
8. $x^2 - 8x + 16$
9. $x^2 + 10x + 25$
10. $x^2 + 16x + 64$

B. In Problems 11–26, solve the given equation.

11. $(x - 2)(x - 4) = 0$
12. $(x + 2)(x + 3) = 0$
13. $(x + 2)(x - 3) = 0$
14. $(x + 5)(x - 6) = 0$
15. $x(x - 1)(x + 1) = 0$
16. $(x + 1)(x + 2)(x - 3) = 0$
17. $(2x - 1)(x + 2) = 0$
18. $(3x + 5)(4x + 7) = 0$
19. $x^2 - 16 = 0$
20. $3x^2 - 27 = 0$
21. $5x^2 = 125$
22. $4x^2 + 1 = 65$
23. $(3x - 6)(2x + 3)(5x - 8) = 0$
24. $2x(x + 7)(2x - 3) = 0$
25. $6x^2 - 1 = 215$
26. $4x^2 + 1 = 50$

C. In Problems 27–34, rewrite the given equation with the left side factored. Then use your result to solve the equation.

27. $x^2 - 12x + 27 = 0$
28. $x^2 - 6x + 8 = 0$
29. $x^2 - 8x - 20 = 0$
30. $x^2 - 9x - 36 = 0$
31. $x^2 + 19x - 20 = 0$
32. $x^2 + 5x - 24 = 0$
33. $x^2 - x - 12 = 0$
34. $x^2 - x - 30 = 0$

In Problems 35–50, solve by using the quadratic formula.

35. $2x^2 + 3x - 5 = 0$
36. $3x^2 - 7x + 2 = 0$

37. $2x^2 + 5x - 7 = 0$
38. $4x^2 - 7x - 15 = 0$
39. $x^2 + 5x + 3 = 0$
40. $2x^2 + 7x - 4 = 0$
41. $5x^2 - 8x + 2 = 0$
42. $3x^2 + 5x + 1 = 0$
43. $7x^2 - 6x = -1$
44. $7x^2 - 12x = -5$
45. $9x^2 - 6x + 2 = 0$
46. $4x^2 - 8x + 5 = 0$
47. $2x^2 + 2x = -1$
48. $2x^2 - 6x = -5$
49. $4x^2 = -8x - 5$
50. $2x^2 = 2x - 5$

*51. The following procedure can be used to obtain the quadratic formula. Suppose that the quadratic equation is given in the standard form

$$ax^2 + bx + c = 0 \qquad a \neq 0$$

Then,

$$ax^2 + bx = -c \qquad \text{Why?}$$

Now multiply both sides by $4a$ to get

$$4a^2x^2 + 4abx = -4ac$$

By adding b^2 to both sides, we get

$$4a^2x^2 + 4abx + b^2 = b^2 - 4ac$$

The left side of this equation is the square of $(2ax + b)$. (Verify this!) Thus, we have

$$(2ax + b)^2 = b^2 - 4ac$$

Next, take the square roots of both sides to get

$$2ax + b = \pm\sqrt{b^2 - 4ac}$$

From this equation, we get

$$2ax = -b \pm \sqrt{b^2 - 4ac} \quad \text{Explain}$$

and

$$x = \frac{-b \pm \sqrt{b^2 - 4ac}}{2a} \quad \text{Explain}$$

(These solutions can be checked in the original equation.)

USING YOUR KNOWLEDGE 7.6

The distance h (in feet) traveled in t seconds by an object dropped from a point above the surface of the Earth is given by the formula

$$h = 16t^2$$

* Optional.

1. An object is dropped from the top of a 64-ft building. How long does it take the object to hit the ground?
2. The man in the picture has jumped from a height of 28 ft. How long does it take him to hit the water?
3. An object is dropped from a height of 144 ft. How long does it take for the object to hit the ground?

Henry LaMothe, at age 70, dove from 28 ft up, alongside the Flatiron Building in New York City, into a $12\frac{1}{2}$-in. deep plastic wading pool of water.

From the *Guinness Book of World Records*, © 1987 by Sterling Publishing Co., Inc.

People have been interested in right triangles for thousands of years. The right triangle relationship $a^2 + b^2 = c^2$ seems to have been known to the Babylonians and the ancient Egyptians. Among the interesting problems about right triangles is this one: Find the right triangles with integer sides such that the hypotenuse is 1 unit longer than one of the legs. We can solve this problem by letting the legs be x and y units long and the hypotenuse be $y + 1$ units long. Then, we have

$$x^2 + y^2 = (y + 1)^2 \qquad or \qquad \begin{aligned} x^2 &= (y + 1)^2 - y^2 \\ &= y^2 + 2y + 1 - y^2 = 2y + 1 \end{aligned}$$

Since x and y are to be positive integers, $2y + 1$ is an odd positive integer. Thus, x^2 is an odd integer, so x must be an odd integer. You can see that x cannot be 1 because this would make y be 0. However, if x is an odd integer greater than 1, then there is a triangle having the required relationship between the sides. If you choose $x = 3$, then $x^2 = 9 = 2y + 1$, so $y = 4$. This gives the well-known 3-4-5 right triangle. For $x = 5$, you get $x^2 = 25 = 2y + 1$, and $y = 12$. This gives the 5-12-13 right triangle. You see that it is easy to make a table of these triangles.

4. Make a table of the next three of these triangles.

DISCOVERY 7.6

PEANUTS

"DEAR FRIEND, THIS IS A CHAIN LETTER"

"COPY THIS LETTER SIX TIMES AND SEND IT TO SIX OF YOUR FRIENDS"

5-16

DEAR FRIEND, THIS IS A CHAIN LETTER.

SIX TIMES?!!

SCHULZ

Charlie Brown received a chain letter. Several days later, after receiving more letters, he found that the number he had received was a perfect square. (The numbers 1^2, 2^2, 3^2, and so on, are perfect squares.) Charlie decided to throw away some of the letters, and being very superstitious, he threw away 13^2 of them. To his surprise, he found that the number he had left was still a perfect square.

1. What is the maximum number of letters Charlie could have received before throwing any away? [*Hint:* Let x^2 be the initial number of letters and let y^2 be the number he had left. Then $x^2 = y^2 + 13^2$. Remember that x and y are integers, and you will want to make y as large as possible relative to x.]

▶ **Calculator Corner 7.6**   *Your calculator can be extremely helpful in finding the roots of a quadratic equation by using the quadratic formula. Of course, the roots you obtain are being approximated by decimals. It is most convenient to start with the radical part in the solution of the quadratic equation and then store this value so you can evaluate both roots without having to backtrack or copy down any intermediate steps. Let us look at the following equation:*

$2x^2 + 7x - 4 = 0$

Using the quadratic formula, the solution is obtained by following these key strokes:

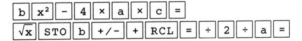

The display will show 0.5. To obtain the other root, key in

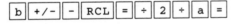

which yields −4. In general, to solve the equation $ax^2 + bx + c = 0$ using your calculator, key in the following:

| b | x^2 | − | 4 | × | a | × | c | = |
| √x | STO | b | +/− | + | RCL | = | ÷ | 2 | ÷ | a | = |

and

| b | +/− | − | RCL | = | ÷ | 2 | ÷ | a | = |

▶ **Computer Corner 7.6**   *In this section, we learned how to solve quadratic equations by using the* **quadratic formula.** *To do this, we must first write the equation in the form*

$ax^2 + bx + c = 0$

If you write the equation in this form, we have a program (page 751) that will do the rest! You need only enter the coefficients a, b, and c. One word of caution: Make sure you enter a 1 as the coefficient of x^2 when you have an equation such as $x^2 + 2x + 1 = 0$.

PROBLEM SOLVING

Courtesy of NASA

In the preceding sections, we learned how to solve certain kinds of equations. Now, we are ready to apply this knowledge to solve problems. These problems will be stated in words, and are consequently called **word,** or **story, problems.** Word problems frighten many students, but do not panic. We have a surefire method for tackling such problems.

Let us start with a problem that you might have heard about. Do you know the name of the heaviest glider in the world? It is the *Columbia!* When fully loaded, the glider and its payload weigh 215,000 lb. The glider itself weighs 85,000 lb more than the payload. What is the weight of each? There you have it, a word problem. Here is the way to solve such problems.

Procedure for Solving Word Problems

1. **R**ead the problem carefully, and decide what it asks for (the unknown).
2. **S**elect a variable to represent this unknown.
3. **T**ranslate the problem into the language of algebra.
4. **U**se the rules of algebra to solve for the unknown.
5. **V**erify the solution.

How can you remember these five steps? It's easy — just look at the first letter in each sentence. We call this method the **RSTUV** method.

Here is how we use this method to solve the glider problem.

1. **R**ead the problem slowly, not once but two or three times.

2. **S**elect the variable p to be the number of pounds in the weight of the payload. Since the glider weighs 85,000 lb more than the payload, the glider weighs $p + 85,000$ lb.

3. **T**ranslate the problem into the language of algebra.

   The glider and its payload weigh 215,000 lb.
   $(p + 85,000)\ +\quad p\qquad =\quad 215,000$

4. **U**se algebra to solve the equation.

Given:	$(p + 85,000) + p = 215,000$
Remove parentheses:	$p + 85,000 + p = 215,000$
Combine the p's:	$2p + 85,000 = 215,000$
Subtract 85,000:	$2p = 215,000 - 85,000$
	$2p = 130,000$
Divide by 2:	$p = 65,000$

Thus, the payload weighs 65,000 lb and the glider weighs a total of 65,000 + 85,000 = 150,000 lb.

5. Verifying the answer here consists of making sure that the conditions of the problem are satisfied. In our solution, we made sure that the glider weighs 85,000 lb more than the payload. Now, we check the total weight: 150,000 + 65,000 = 215,000, which agrees with the weight given for glider and payload. Thus, our result is correct.

EXAMPLE 1

Angie bought a 6-month, $10,000 certificate of deposit. At the end of the 6 months, she received $650 simple interest. What rate of interest did the certificate pay?

Solution

1. Read the problem. It asks for the rate of simple interest.
2. Select the variable r to represent this rate.
3. Translate the problem. Here, we need to know that the formula for simple interest is

$$I = Prt$$

where I is the amount of interest, P is the principal, r is the interest rate, and t is the time in years. For our problem, $I = \$650$, $P = \$10,000$, r is unknown, and $t = \frac{1}{2}$ year. Thus, we have

$$650 = (10,000)(r)(\tfrac{1}{2})$$

or

$$650 = 5000r$$

4. Use algebra to solve the equation:

$$650 = 5000r$$

Divide by 5000: $\frac{650}{5000} = r$

Express decimally: $r = 0.13 = 13\%$

Hence, the certificate paid 13% simple interest.

5. Verify the answer. Is the interest earned on a $10,000, 6-month certificate at a 13% rate $650? Evaluating Prt, we have

$$(10,000)(0.13)(\tfrac{1}{2}) = 650$$

Since the answer is yes, 13% is correct. ◀

The next problem may save you some money when you rent a car. As the accompanying chart shows, you can rent a car and use either the mileage rate or the flat rate. For example, if you wish to rent an intermediate sedan for 1 day, you can pay $12 plus 12¢ for each mile traveled, or a $19 flat rate. Which is the better deal? That depends on how far you plan to go. Let's be more specific. How many miles could you go for $19 if you used the mileage rate? We shall find the answer in the next example.

TYPE	MILEAGE RATE			FLAT RATE	
	DAY	WEEK	MILE	DAY	WEEK
ECONOMY	$ 9.00	$49.00	.10	$15.00	$ 79.00
SUBCOMPACT	10.00	55.00	.10	17.00	89.00
COMPACT	11.00	60.00	.11	17.00	89.00
INTERMEDIATE SEDAN	12.00	65.00	.12	19.00	99.00
SEDANS	13.00	70.00	.13	21.00	109.00
PREMIUM SEDANS	15.00	85.00	.15	25.00	125.00
STATION WAGON	18.00	90.00	.18	35.00	135.00

EXAMPLE 2   Jim Jones rented an intermediate sedan at $12 per day plus 12¢ per mile. How many miles can Jim travel for $19?

Solution   Again, we proceed by steps:

1. Read the problem carefully. We are looking for the number of miles Jim can travel for $19.
2. Select m to represent this number of miles.
3. Translate the problem into an equation. To do this, you must realize that Jim is paying 12¢ for each mile plus $12 for the day. Thus, if Jim travels

1 mi,   the cost is $0.12(1) + 12$
2 mi,   the cost is $0.12(2) + 12$
m mi,   the cost is $0.12m + 12$

Because we want to know how many miles Jim can go for $19, we must put the cost for m miles equal to $19, which gives the equation

$$0.12m + 12 = 19$$

4. Use algebra to solve the equation

$$0.12m + 12 = 19$$

Subtract 12: $0.12m = 7$

Multiply by 100: $12m = 700$ This gets rid of the decimal.

Divide by 12: $m = \frac{700}{12} = 58\frac{1}{3}$

Thus, Jim can travel $58\frac{1}{3}$ mi for $19.

5. Verify that $(0.12)(58\frac{1}{3}) + 12 = 19$. (We leave this to you.)   ◀

The car rental agency's mileage rate contract undoubtedly reads "$0.12 per mile or fraction thereof." Thus, you would have to modify the answer to Example 2 to 58 mi. (If Jim drove $58\frac{1}{3}$ mi, he would be charged for 59 mi.) From this information, you can deduce that if you want to rent an intermediate sedan for 1 day and plan to drive over 58 mi, then the flat rate is the better of the two options.

Talking about distances, we think that the next example could save some lives. Have you seen the Highway Patrol booklet that indicates the **braking distance** b (in feet) that it takes to stop a car after the brakes are applied? This information is usually given in a chart, but there is a formula that gives close estimates under normal driving conditions. The braking distance formula is

$$b = 0.06v^2$$

where v is the speed of the car (in miles per hour) when the brakes are applied. Thus, if you are traveling 20 mi/hr, you will go

$$b = 0.06(20^2) = 0.06(400) = 24 \text{ ft}$$

after you apply the brakes.

EXAMPLE 3   A car traveled 150 ft *after* the brakes were applied. (It might have left a skid mark that long.) How fast was the car going when the brakes were applied?

Solution   1. Read the problem carefully.
2. Select the variable v to represent the velocity.
3. Translate: The braking distance $b = 150$, and the braking distance formula reads $b = 0.06v^2$. Thus, we have the equation

$$0.06v^2 = 150$$

4. Use algebra to solve the last equation:

Multiply by 100:	$6v^2 = 15{,}000$	This gets rid of the decimal.
Divide by 6:	$v^2 = \frac{15{,}000}{6} = 2500$	
Take square roots:	$v = 50$ or -50	

Since the -50 makes no sense in this problem, we discard it. Thus, the car was going 50 mi/hr when the brakes were applied.

5. Verify the answer by substituting 50 for v in the braking distance formula.   ◀

You have probably noticed the frequent occurrence of certain words in the statements of word problems. Because these words are used frequently, we give a small mathematics dictionary in Table 7.7a to help you translate them properly.

Table 7.7a   Mathematics Dictionary

WORDS	TRANSLATION	EXAMPLE	TRANSLATION
Add, more than, sum, increased by, added to	$+$	Add n to 7 7 more than n The sum of n and 7 n increased by 7 7 added to n	$n + 7$
Subtract, less than, minus, difference, decreased by, subtracted from	$-$	Subtract 9 from x 9 less than x x minus 9 Difference of x and 9 x decreased by 9 9 subtracted from x	$x - 9$
Of, the product, times, multiply by	$\times$	$\frac{1}{2}$ of a number x The product of $\frac{1}{2}$ and x $\frac{1}{2}$ times a number x Multiply $\frac{1}{2}$ by x	$\frac{1}{2}x$
Divide, divided by, the quotient	$\div$	Divide 10 by x 10 divided by x The quotient of 10 and x	$\frac{10}{x}$

Of course, most of these words are used in conjunction with the word "equals." Here are some words that mean equals:

The same, yields, gives, is $=$

The next example shows how some of these words are used.

EXAMPLE 4   If 7 is added to twice the square of a number n, the result is 9 times the number. Find n.

Solution   **1.** Read the problem.

2. Select the unknown: n in this case.

3. Translate the problem:

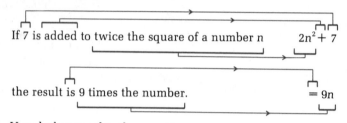

4. Use algebra to solve the equation:

$2n^2 + 7 = 9n$

Since this equation has both a squared term and a first-degree term, we solve it as a quadratic equation, first putting it in the standard form $ax^2 + bx + c = 0$. Hence, we subtract 9n from both sides to get

$2n^2 - 9n + 7 = 0$

Then, we use the quadratic formula with $a = 2, b = -9,$ and $c = 7$ to get the value of n:

$$n = \frac{-(-9) \pm \sqrt{(-9)^2 - 4(2)(7)}}{2(2)}$$

$$= \frac{9 \pm \sqrt{81 - 56}}{4} = \frac{9 \pm \sqrt{25}}{4}$$

$$= \frac{9 \pm 5}{4}$$

$$= \frac{14}{4} \quad \text{or} \quad \frac{4}{4}$$

$$= \frac{7}{2} \quad \text{or} \quad 1$$

5. Verification for $n = \frac{7}{2}$:

$$2\left(\frac{7}{2}\right)^2 + 7 = 2 \times \frac{7}{2} \times \frac{7}{2} + 7 = \frac{49}{2} + 7 = \frac{63}{2}$$

and

$$9 \times \frac{7}{2} = \frac{63}{2}$$

Thus, we see that if 7 is added to twice the square of $\frac{7}{2}$, the result is 9 times $\frac{7}{2}$. The verification for $n = 1$ is left for you to do. We have now shown that the number n can be either $\frac{7}{2}$ or 1. ◄

EXAMPLE 5   The Better Business Bank has two types of checking accounts. Type 1 has a monthly service charge of $3 plus 6¢ for each check cashed. Type 2 has a monthly service charge of $5 plus 3¢ for each check cashed. What is the greatest number of checks that can be cashed before the type 1 account becomes the more expensive of the two?

Solution   Let x be the number of checks cashed. Then the cost of the type 1 account is

   $3 + 0.06x$ dollars

and the cost of the type 2 account is

   $5 + 0.03x$ dollars

Hence, we need to find the greatest value of x such that

   $3 + 0.06x < 5 + 0.03x$

This inequality can be solved as follows:

$3 + 0.03x < 5$	Subtract 0.03x from both sides.
$0.03x < 2$	Subtract 3 from both sides.
$x < \dfrac{2}{0.03}$	Divide both sides by 0.03.
$x < 66\frac{2}{3}$	

Since x must be a whole number, it follows that 66 is the required answer. ◄

EXERCISE 7.7

In Problems 1–10, write the given statement as an equation, and then solve it.

1. If 4 times a number is increased by 5, the result is 29. Find the number.
2. Eleven more than twice a number is 19. Find the number.
3. The sum of 3 times a number and 8 is 29. Find the number.
4. If 6 is added to 7 times a number, the result is 69. Find the number.

5. If the product of 3 and a number is decreased by 2, the result is 16. Find the number.

6. Five times a certain number is 9 less than twice the number. What is the number?

7. Two times the square of a certain number is the same as twice the number increased by 12. What is the number?

8. If 5 is subtracted from half the square of a number, the result is 1 less than the number itself. Find the number.

9. One-third the square of a number decreased by 2 yields 10. Find the number.

10. One-fifth the square of a certain number plus 2 times the number is 15. What is the number?

In Problems 11–32, use the RSTUV method to obtain the solution.

Courtesy of NASA

11. The space shuttle *Columbia* consists of the orbiter, the external tank, two solid-fuel boosters, and fuel. At the time of lift-off, the weight of all these components was 4.16 million pounds. The tank and the boosters weighed 1.26 million pounds less than the orbiter and fuel. What was the weight of the orbiter and fuel?

12. The external tank and the boosters of the *Columbia* weighed 2.9 million pounds, and the boosters weighed 0.34 million pounds more than the external tank. Find the weight of the tank and of the boosters.

13. Russia and Japan have the greatest number of merchant ships in the world. The combined total is 15,426 ships. If Japan has 2276 ships more than Russia, how many ships does each have?

14. Mary is 12 years old and her brother Joey is 2 years old. In how many years will Mary be just twice as old as Joey?

15. The cost of renting a car is $18 per day plus 20¢ per mile traveled. Margie rented a car and paid $44 at the end of the day. How many miles did Margie travel?

16. In baseball, the slugging average of a player is obtained by dividing his total bases (1 for a single, 2 for a double, 3 for a triple, and 4 for a home run) by his official number of at bats. Dickey Fantle has 2 home runs, 1 triple, 2 doubles, and 9 singles. His slugging average is 1.2. How many times has he been at bat?

17. Peter buys a $10,000, 3-month certificate of deposit. At the end of the 3 months, he receives interest of $350. What was the rate of interest on the certificate?

18. A loan company charges $588 for a 2-year loan of $1400. What interest rate is the loan company charging?

19. Referring to the car-rental chart for Example 2, you can see that the cost of renting a sedan for 1 day is $(0.13m + 13)$ dollars, where m is the number of miles traveled.

   a. How many miles could you go for $21? (Answer to the nearest mile.)

b. If you plan to make a 60-mi round trip, should you use the flat rate or the mileage rate for the sedan?

20. The Greens need to rent a station wagon to move some furniture. If the round trip distance is 50 mi, should the Greens use the flat rate or the mileage rate? (Use the car-rental chart for Example 2.)

21. If P dollars is invested at r percent compounded annually, then at the end of 2 years, the amount will have grown to

$$A = P(1 + r)^2$$

At what rate of interest will $1000 grow to $1210 in 2 years?

22. Use the formula in Problem 21 to find at what rate of interest $1000 will grow to $1440 in 2 years.

23. As given in the text, the braking distance b (in feet) for a car traveling v miles per hour is

$$b = 0.06v^2$$

After the driver applied the brakes, a car traveled 54 ft. How fast was the car going when the brakes were applied?

Courtesy of PSH

24. The **reaction distance** r (in feet) is the distance a car travels while the driver is moving his or her foot (reacting) to apply the brakes. The formula for this distance is

$$r = 1.5tv$$

where t is the driver's reaction time (in seconds) and v is the speed of the car (in miles per hour). A car going 30 mi/hr travels 22.5 ft while the driver is reacting to apply the brakes. What is the driver's reaction time?

*Use the following information in Problems 25–30: The **stopping distance** d (in feet) for a car traveling v miles per hour when the driver has a reaction time of t seconds is given by*

$$d = 1.5tv + 0.06v^2$$

25. A car is traveling 30 mi/hr. If the driver's reaction time is 0.5 sec, what is the stopping distance?

26. The reaction time of a driver is 0.5 sec. If the car is going 50 mi/hr, what is the stopping distance?

27. An automobile going 20 mi/hr is 42 ft away from an intersection when the traffic light turns red. If the automobile stops right at the intersection, what is the driver's reaction time?

28. You may have heard about the reflecting collars that can save the life of a dog or a cat. If your car's headlights will illuminate objects up to 200 ft away and you are driving 50 mi/hr when you see a dog on the

road at the edge of the illuminated distance, what must be your reaction time if you can stop just short of hitting the dog?

29. Loren, who has a reaction time of $\frac{2}{3}$ sec, was taking a driving test. When the examiner signaled for a stop, she stopped her car in 44 ft. How fast was she going at the instant of the stop signal?

30. Pedro reacts very quickly. In fact, his reaction time is 0.4 sec. When driving on a highway, Pedro saw a danger signal ahead and tried to stop. If his car traveled 120 ft before stopping, how fast was he going when he saw the signal?

31. Two consecutive integers are such that 6 times the smaller is less than 5 times the larger. Find the largest integers for which this is true.

32. A wallet contains 20 bills, all $1 and $5 bills. If the total value is less than $80, what is the greatest possible number of $5 bills? What is the largest total value possible?

USING YOUR KNOWLEDGE 7.7

Have you ever heard of **Chamberlain's formula,** which purports to tell you how many years you should drive your present car before you buy a new one? If y is this number of years, then Chamberlain's formula reads

$$y = \frac{GMC}{(G - M)DP}$$

where G is the new car's gas mileage, M is your present car's gas mileage, C is the cost in dollars of the new car, D is the number of miles you drive in a year, and P is the dollar price of gasoline per gallon.

For instance, suppose that the new car's gas mileage is 24 mi/gal, the old car's mileage is 12 mi/gal, the price of the new car is $6400, you drive 12,000 mi/yr, and the cost of gasoline is $1.40/gal. This means that G = 24, M = 12, C = 6400, D = 12,000, and P = 1.40, so

$$y = \frac{(24)(12)(6400)}{(12)(12,000)(1.4)} = \frac{64}{7} \approx 9.1$$

Thus, according to the formula, your old car should have been driven about 9.1 years before the purchase is justified.

Now try these:

1. Suppose the new car's gas mileage is 32 mi/gal, the old car's gas mileage is 10 mi/gal, the price of the new car is $7000, you drive 15,000 mi/year, and the cost of gasoline is $1.30/gal. How many years should your old car have been driven to justify buying the new one?

2. Suppose that G, M, D, and P have the same values as in Problem 1, and you have driven your present car for 5 years. What price (to the nearest $10) would you be justified in paying for the new car?

RATIO, PROPORTION, AND VARIATION

You have probably heard the term "junk food," but do you know what foods fall into this category? The U.S. Department of Agriculture suggests looking at the *ratio* of nutrients to calories and multiplying the result by 100. The actual procedure totals the RDA's (recommended daily allowances) for the first eight nutrients listed on the label, divides by the number of calories in one serving, and multiplies by 100. For instance, the sum of the first eight nutrients for the cereal label shown is 139, and the number of calories in one serving is 100. Thus, the ratio of nutrients to calories is

$$\frac{139}{100} \times 100 = 139$$

If the answer is less than 32, the nutritional value of the food is in question.

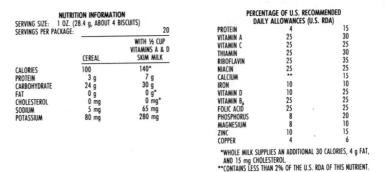

NUTRITION INFORMATION
SERVING SIZE: 1 OZ. (28.4 g, ABOUT 4 BISCUITS)
SERVINGS PER PACKAGE: 20

	CEREAL	WITH ½ CUP VITAMINS A & D SKIM MILK
CALORIES	100	140*
PROTEIN	3 g	7 g
CARBOHYDRATE	24 g	30 g
FAT	0 g	0 g*
CHOLESTEROL	0 mg	0 mg*
SODIUM	5 mg	65 mg
POTASSIUM	80 mg	280 mg

PERCENTAGE OF U.S. RECOMMENDED
DAILY ALLOWANCES (U.S. RDA)

PROTEIN	4	15
VITAMIN A	25	30
VITAMIN C	25	25
THIAMIN	25	30
RIBOFLAVIN	25	35
NIACIN	25	25
CALCIUM	**	15
IRON	10	10
VITAMIN D	10	25
VITAMIN B₆	25	25
FOLIC ACID	25	25
PHOSPHORUS	8	20
MAGNESIUM	8	10
ZINC	10	15
COPPER	4	6

*WHOLE MILK SUPPLIES AN ADDITIONAL 30 CALORIES, 4 g FAT, AND 15 mg CHOLESTEROL.
**CONTAINS LESS THAN 2% OF THE U.S. RDA OF THIS NUTRIENT.

A. Ratios

Quantities are often compared by using ratios.

▶ **Definition 7.8a** | A **ratio** is a quotient of two numbers.

The ratio of a number a to another number b can be written as:

$$a \text{ to } b \quad \text{or} \quad a:b \quad \text{or} \quad \frac{a}{b}$$

The last form is used most often, but is frequently written as a/b.

EXAMPLE 1 Write the ratio of nutrients (139) to calories (100) in three different ways.

Solution   139 to 100   139:100 $\dfrac{139}{100}$ ◀

Of course, if the ratio in Example 1 had been 140 to 100, you could write it in reduced form as

$$7 \text{ to } 5 \qquad \text{or} \qquad 7:5 \qquad \text{or} \qquad \frac{7}{5}$$

You encounter many ratios in everyday life. For example, the expression "miles per gallon" is actually a ratio, the ratio of the number of miles traveled to the number of gallons of gas used. Thus, if your car travels 294 miles on 12 gallons of gas, your miles per gallon ratio is $\frac{294}{12} = \frac{49}{2} = 24.5$.

Ratios can also be used to compare prices. For example, most people have the misconception that the more you buy of an item, the cheaper it is. Is this always true? Not necessarily. The picture shows two cans of Hunt's tomato sauce, bought in the same store. The 15-oz can cost 59¢, while the 8-oz can cost 28¢. Which can is the better buy? To compare these prices, we need to find the price of 1 ounce of tomato sauce; that is, we need the **unit price** (the price per ounce). This unit price is given by the ratio

$$\frac{\text{Price}}{\text{Number of ounces}}$$

For the 15-oz can,   For the 8-oz can,

$$\frac{59}{15} = 3.9\overline{3} \qquad\qquad \frac{28}{8} = 3.5$$

Thus, the 15-oz can is more expensive. (Note that you could buy 16 oz, two 8-oz cans, for only 56¢, instead of 59¢ for the 15-oz can!)

EXAMPLE 2   The 4-oz can of mushrooms in the picture costs 50¢, while the 8-oz can costs $1.09. Find the cost per ounce for:

a. The 4-oz can   **b.** The 8-oz can

(Round answers to the nearest tenth of a cent.)

Solution   **a.** For the 4-oz can, the cost per ounce is

$$\frac{50}{4} = 12.5 \text{ cents}$$

b. For the 8-oz can, the cost per ounce is

$$\frac{109}{8} = 13.6 \text{ cents} \qquad\qquad\blacktriangleleft$$

Note that if we wish to find which of the two cans in Example 2 is the better buy, we do not have to look at the unit prices. Two 4-oz cans would give us 8 oz for $1.00 instead of $1.09.

B. Proportion

Let us go back to the car that traveled 294 miles on 12 gallons of gas. How many miles would this car go on 10 gallons of gas? The ratio of miles per gallon for the car is $\frac{294}{12}$ or $\frac{49}{2}$. If we let m be the number of miles the car would travel on 10 gallons of gas, the ratio of miles per gallon would be $m/10$. Since the two ratios must be equal,

$$\frac{49}{2} = \frac{m}{10}$$

This equation, which is an equality between two ratios, is called a **proportion.**

Definition 7.8b | A **proportion** is an equality between two ratios.

To solve the proportion

$$\frac{49}{2} = \frac{m}{10}$$

which is simply an equation involving fractions, we proceed as before.

1. Multiply both sides by **10** so the m is by itself on the right. $\qquad \frac{49}{2} \cdot 10 = \frac{m}{10} \cdot 10$

3. Simplify. $\qquad\qquad\qquad\qquad\qquad\qquad\qquad 245 = m$

3. Thus, the car can go 245 miles on 10 gallons of gas.

Proportions can often be solved by a shortcut method that depends on the definition of equality of fractions:

If $\qquad \dfrac{a}{b} = \dfrac{c}{d} \qquad$ then $\qquad ad = bc$

Thus, to solve the proportion

$$\frac{3}{4} = \frac{15}{x}$$

We use the cross-products, and write

$$3 \cdot x = 15 \cdot 4$$
$$3x = 60$$
$$x = 20$$

EXAMPLE 3 The ratio of your foot length to your height (in inches) is 1 to 6.7. In 1951, Eric Shipton published photographs of what he thought were the "Abominable Snowman's" footprints. Each footprint was 23 inches long. How tall is the Abominable Snowman?

Solution   We use the five-step procedure outlined in Section 7.7.

1. Read the problem carefully.
2. Selct a variable to represent the unknown. Here, we let h be the height of the Abominable Snowman.
3. Translate the problem: Since the original ratio is $1/6.7$, the ratio for the Snowman is $23/h$. But these ratios must be equal, so

$$\frac{1}{6.7} = \frac{23}{h}$$

4. Use cross-products to solve the equation:

$$1 \cdot h = 6.7 \cdot 23$$
$$h = 154.1 \text{ inches}$$

5. Verify your answer: If we substitute $h = 154.1$ in the original proportion, we obtain

$$\frac{1}{6.7} = \frac{23}{154.1}$$

which is a true statement since $1 \cdot 154.1 = 6.7 \cdot 23$. ◀

C. Variation

Sometimes, we say that a variable y is **proportional to** or **varies directly as** another variable x. For example, the number m of miles you drive a car is proportional to or varies directly as the number g of gallons of gas used. This means that the ratio m/g is a constant, the miles per gallon the car attains. In general, we have the following definition:

▶   **Definition 7.8c**

> y **varies directly as** x if there is a constant k such that
>
> $y = kx$

EXAMPLE 4   The length L of a moustache varies directly as the time t.

a. Write an equation of variation.
b. The longest moustache on record was owned by Masuriya Din. His moustache grew 56 inches (on each side) over a 14-year period. Find k and explain what it represents.

Solution   **a.** Since the length L varies directly as the time t,

$$L = kt$$

b. We know that when $L = 56$, $t = 14$. Thus,

$$56 = k \cdot 14$$
$$4 = k$$

This means that the moustache grew 4 inches each year. ◄

Sometimes, as a quantity increases, a related quantity decreases proportionately. For example, the more time you spend practicing a task, the less time it will take you to do the task. In such cases, we say that the quantities **vary inversely** as each other.

Definition 7.8d

> y **varies inversely as** x if there is a constant k such that
>
> $$y = \frac{k}{x}$$

EXAMPLE 5 The speed s that a car travels is inversely proportional to the time t it takes to travel a given distance.

a. Write the equation of variation.
b. If a car travels at 60 miles per hour for 3 hours, what is k, and what does it represent?

Solution **a.** The equation is

$$s = \frac{k}{t}$$

b. We know that $s = 60$ when $t = 3$. Thus,

$$60 = \frac{k}{3}$$

$$k = 180$$

In this case, k represents the distance traveled. ◄

EXAMPLE 6 Have you ever heard one of those loud "boom" boxes or a car sound system that makes your stomach tremble? The loudness L of sound is inversely proportional to the square of the distance d that you are from the source.

a. Write an equation of variation.
b. The loudness of rock music coming from a boom box 5 feet away is 100 decibels. Find k.
c. If you move to 10 feet away from the boom box, how loud is the sound?

Solution **a.** The equation is

$$L = \frac{k}{d^2}$$

b. We know that $L = 100$ for $d = 5$, so that

$$100 = \frac{k}{5^2} = \frac{k}{25}$$

Multiplying both sides by 25, we find that $k = 2500$.

c. Since $k = 2500$,

$$L = \frac{2500}{d^2}$$

When $d = 10$,

$$L = \frac{2500}{10^2} = 25 \text{ decibels}$$ ◀

EXERCISE 7.8

A.

1. The *Voyager* was the first plane to fly nonstop around the world without refueling. At the beginning of the trip, the fuel in the 2000-pound plane weighed 7000 pounds. Write the ratio of fuel to plane weight in three different ways.

2. During the last 6 years of his life, Vincent Van Gogh produced 700 drawings and 800 oil paintings. Write the ratio of drawings to oil paintings in three different ways.

3. The first suspension bridge built in England has a 70-foot span. The world's longest suspension bridge, the Verrazano Narrows Bridge, is 4260 feet long. Write the ratio of the span of the first suspension bridge to that of the Verrazano Narrows Bridge in three different ways.

4. A woman has a 28-inch waist and 34-inch hips. Write the waist-to-hip ratio in three different ways. (If the waist-to-hip ratio is over 1.0 for men or 0.8 for women, the risk of heart attack or stroke is five to ten times greater than for persons with a lower ratio.)

5. The transmission ratio in an automobile is the ratio of engine speed to drive-shaft speed. Find the transmission ratio for an engine running at 2000 revolutions per minute when the drive-shaft speed is 600 revolutions per minute.

6. Most job seekers can expect about 6 job leads and/or interviews for every 100 resumes they mail out. What is the ratio of job leads and/or interviews to resumes?

7. The average American car is driven 12,000 miles a year and burns 700 gallons of gas. How many miles per gallon does the average American car get? (Give the answer to the nearest whole number.)

8. Are generic products always cheaper than name brands? Not necessarily! It depends on where you buy. The photograph shows two cans of mushrooms. The name brand 8-oz can costs $1.09 and the generic 4-oz can costs 53¢.

   a. To the nearest tenth of a cent, what is the cost per ounce of the name brand 8-oz can?

b. To the nearest tenth of a cent, what is the cost per ounce of the generic 4-oz can?

c. Which can is the better buy?

d. In Example 2, the 4-oz can cost 12.5¢ per ounce. Which is the better buy, the generic can here or the 4-oz can of Example 2?

9. As a consumer, you probably believe that cheaper is always better. But be careful. Here is a situation that should give you food for thought!

a. Dermassage dishwashing liquid costs $1.31 for 22 oz. To the nearest cent, what is the cost per ounce? 6¢

b. White Magic dishwashing liquid costs $1.75 for 32 oz. To the nearest cent, what is the cost per ounce? 5¢

c. Based on price alone, which is the better buy, Dermassage or White Magic? *White Magic*

But how much do you use per wash? *Consumer Reports* estimated that it costs 10¢ for 10 washes with Dermassage and 18¢ for the same number of washes with White Magic. Thus, Dermassage is more economical.

10. Is cheaper still better? Here is another problem. A&P Wool Washing Liquid costs 79¢ for 16 oz. Ivory Liquid is $1.25 for 22 oz.

a. To the nearest cent, what is the price per ounce of the A&P Wool Washing Liquid?

b. To the nearest cent, what is the price per ounce of the Ivory Liquid?

c. Based on price alone, which is the better buy?

But wait. How much do you have to use? According to *Consumer Reports*, it costs 17¢ for 10 washes with the A&P Wool Washing Liquid, while Ivory Liquid is only 12¢ for 10 washes!

B. In Problems 11–16, solve the given proportion.

11. $\dfrac{x}{9} = \dfrac{4}{3}$

12. $\dfrac{x}{6} = \dfrac{5}{12}$

13. $\dfrac{8}{x} = \dfrac{4}{3}$

14. $\dfrac{6}{x} = \dfrac{18}{7}$

15. $\dfrac{3}{8} = \dfrac{9}{x}$

16. $\dfrac{3}{5} = \dfrac{9}{x}$

17. When flying a hot-air balloon, you get $\frac{1}{2}$ hour of flight time for each 20-pound tank of propane gas. How many tanks of gas do you need for a 3-hour flight?

18. When serving shrimp, you need $\frac{1}{2}$ pound of cooked shrimp without the shell to make 3 servings. How many pounds of cooked shrimp do you need for 90 servings?

19. The official ratio of width to length for the U.S. flag is 10 to 19. If a flag is 35 inches wide, how long should it be?

20. Do you like tortillas? Tom Nall does! As a matter of fact, he ate 74 tortillas in 30 minutes. How many could he eat in 45 minutes at that rate?

21. A certain pitcher allowed 10 runs in the last 32 innings he pitched. At this rate, how may runs would he allow in 9 innings? (The answer is called the ERA, or earned run average, for the pitcher and is usually given to two decimal places.)

22. Do you know what a xerus is? It is a small rodent that looks like a cross between a squirrel and a chipmunk. The ratio of tail to body length in one of these animals is 4 to 5. If the body of a xerus is 10 inches long, how long is its tail?

23. A zoologist tagged and released 250 fish from a lake. A few days later, 53 fish were taken from the lake and 5 of them were found to be tagged. Approximately how many fish are there in the lake?

C. 24. The amount of annual interest I received on a savings account is directly proportional to the amount of money m you have in the account.
  a. Write an equation of variation.
  b. If $480 produced $26.40 in interest, what is k?
  c. How much annual interest would you receive if the account had $750?

25. The number of revolutions R a record makes as it is being played varies directly with the time t that it is on the turntable.
  a. Write an equation of variation.
  b. A record that lasted $2\frac{1}{2}$ minutes made 112.5 revolutions. What is k?
  c. If a record makes 108 revolutions, how long does it take to play it?

26. The distance d an automobile travels after the brakes have been applied varies directly as the square of its speed s.
  a. Write an equation of variation.
  b. If the stopping distance for a car going 30 miles per hour is 54 feet, what is k?
  c. What is the stopping distance for a car going 60 miles per hour?

27. The weight of a person varies directly as the cube of the person's height h (in inches). The **threshhold weight** T (in pounds) for a person is defined as "the crucial weight, above which the mortality (risk) for the patient rises astronomically."
  a. Write an equation of variation relating T and h.
  b. If $T = 196$ when $h = 70$, find k.
  c. To the nearest pound, what is the threshhold weight T for a person 75 inches tall?

28. The number S of new songs a rock band needs each year is inversely proportional to the number y of years the band has been in the business.
  a. Write an equation of variation.
  b. If, after 3 years in the business, the band needs 50 new songs, how many songs will it need after 5 years?

29. When set at infinity, the f number on a camera lens varies inversely with the diameter d of the aperture (opening).
   a. Write an equation of variation.
   b. If the f number on a camera is 8 when the aperture is $\frac{1}{2}$ inch, what is k?
   c. Find the f number when the aperture is $\frac{1}{4}$ inch.
30. The weight W of an object varies inversely as the square of its distance d from the center of the Earth.
   a. Write an equation of variation.
   b. An astronaut weighs 121 pounds on the surface of the Earth. If the radius of the Earth is 3960 miles, find the value of k for this astronaut. (Do not multiply out your answer.)
   c. What will this astronaut weigh when she is 880 miles above the surface of the Earth?

SUMMARY

Section	Item	Meaning	Example
7.1	Variable	A symbol that may be replaced by any one of a set of numbers	x, y, z
7.1A	Open sentence	A sentence in which the variable can be replaced by a number	$x + 3 = 5; x - 1 < 7$
7.1A	Equation	Sentences in which the verb phrase is "="	$x + 7 = 9$
7.1A	Inequality	Sentences in which the verb is $>, <, \neq, \geq,$ or $\leq$	$x + 7 < 9; x > 8; x \neq 9$
7.1A	Solution set	The set of elements of the replacement set that make the open sentence a true statement	$\{3\}$ is the solution set of $x + 2 = 5$ when the replacement set is the set of whole numbers.
7.1A	Identities	Open sentences that are true for every number in the replacement set	$x + 0 = x; x + 2 = 2 + x;$ $a(b + c) = ab + ac$
7.2	First-degree sentence	An open sentence in which the unknown quantity has an exponent of 1 only	$x + 7 = 8 - 2x; 20 = 3m$
7.2A	Elementary operations	Operations that may be performed on a sentence to obtain an equivalent sentence	Addition or subtraction of a number to both sides of an equation

Section	Item	Meaning	Example
7.3	Finite intervals	$a \leq x \leq b$ Closed interval	
		$a < x < b$ Open interval	
		$\left.\begin{array}{l} a \leq x < b \\ \\ a < x \leq b \end{array}\right\}$ Half-open intervals $\left\{\begin{array}{l}\end{array}\right.$	
7.5A	Absolute value	The distance on the number line from 0 to the number	$\lvert 3 \rvert = 3; \lvert -7 \rvert = 7; \lvert \frac{-2}{3} \rvert = \frac{2}{3}$
7.6	Quadratic equation	A second-degree sentence that can be written in the form $ax^2 + bx + c = 0, a \neq 0$	$x^2 - 7x = 6; 8x^2 - 3x - 4 = 0$
7.6C	Quadratic formula	The solutions of the equation $ax^2 + bx + c = 0$ are $$x = \frac{-b \pm \sqrt{b^2 - 4ac}}{2a}$$	The solutions of the equation $x^2 - 4x - 12 = 0$ are -2 and 6.
7.7		Procedure for Solving Word Problems	

1. **R**ead the problem carefully, and decide what it asks for (the unknown).
2. **S**elect a variable to represent this unknown.
3. **T**ranslate the problem into the language of algebra.
4. **U**se the rules of algebra to solve for the unknown.
5. **V**erify the solution.

Section	Item	Meaning	Example
7.8A	Ratio	A quotient of two numbers	The ratio of 5 to 7 is $\frac{5}{7}$.
7.8B	Proportion	An equality between two ratios	$\frac{5}{7} = \frac{10}{14}$
7.8C	Varies directly	y varies directly as x if $y = kx$	
7.8C	Varies inversely	y varies inversely as x if $y = \dfrac{k}{x}$	

PRACTICE TEST 7

1. If the replacement set is the set of integers, solve the following equations:

   a. $x + 7 = 2$ **b.** $x - 4 = 9$

2. If the replacement set is the set of integers, find the solution set for each of the following inequalities:

   a. $x + 5 > 4$ **b.** $2 + x \geq -x - 1$

3. Solve: $2x + 2 = 3x - 2$
4. Solve: $2x + 8 \geq -x - 1$
5. Graph the solution set of each of the following:
   a. $x - 3 \leq 0$ b. $-2x + 4 > x + 1$
6. Graph the solution set (if it is not empty) of each of the following:
   a. $x + 2 \geq 3$ and $x \leq 4$ b. $x - 3 \geq 1$ and $x \leq 0$
7. Graph the solution set of each of the following:
   a. $x < 0$ or $x - 2 < 1$ b. $x + 2 < 3$ or $x - 1 > 2$
8. Solve the equation $|x| = 3$.
9. Graph the solution set of $|x| < 2$.
10. Graph the solution set of $|x + 2| < 1$.
11. Graph the solution set of $1 < |x - 2| < 2$.
12. Graph the solution set of $|x| > 2$.
13. Graph the solution set of $|x - 2| > 3$.
14. Factor:
    a. $x^2 + 3x + 2$ b. $x^2 - 3x - 4$
15. Solve:
    a. $(x - 1)(x + 2) = 0$ b. $x(x - 1) = 0$
16. Solve by factoring: $x^2 + 7x + 10 = 0$
17. Solve by factoring: $x^2 - 3x - 10 = 0$
18. Use the quadratic formula to solve: $2x^2 + 3x - 5 = 0$
19. Use the quadratic formula to solve: $3x^2 + 5x - 2 = 0$
20. Solve:
    a. $9x^2 - 16 = 0$ b. $25x^2 - 4 = 0$
21. Suppose you rent a car for 1 day at the rate of $11 per day plus 11¢ per mile. How many miles could you drive for a rental charge of $33?
22. Sally has two rings, one that is 9 years old and one that is 35 years old. In how many years will the older ring be twice as old as the newer one?
23. Three times the sum of two consecutive integers is 45. What are the integers?
24. A piggy bank contains 20 coins, all dimes and nickels. If the total value of the coins is more than $1.20, what is the least possible number of dimes in the bank?
25. A pair of consecutive integers is such that 12 times the smaller is more than 9 times the larger. What is the least pair of integers for which this is true?
26. On a certain day, the New York Stock Exchange reported that 688 stocks went up, 801 went down, and 501 were unchanged.
    a. Write the ratio of gainers to losers in three different ways.
    b. What is the ratio of losers to the total number of stocks?
27. A supermarket is selling a certain kind of cracker for 50¢ for an 8-oz box and 76¢ for a 12-oz box.
    a. Find the unit price for each box.
    b. Which is the better buy?

28. We know that corresponding sides of similar rectangles are proportional. One rectangle is 5 ft by 8 ft, and the short side of a similar rectangle is 9 ft long.

   a. Write a proportion for the length, x ft, of the long side of the second rectangle.

   b. Find the missing length.

29. The cost C of fuel per hour for running an airplane is directly proportional to the square of the speeds.

   a. Write an equation of variation.

   b. If the cost is $100 per hour for a speed of 150 miles per hour, find the value of k.

   c. Find the cost per hour for a speed of 180 miles per hour.

30. The time t of exposure needed to photograph an object at a fixed distance from the camera is inversely proportional to the intensity I of the illumination.

   a. Write an equation of variation.

   b. If the correct exposure is $\frac{1}{30}$ sec when I is 300 units, find the value of k.

   c. If I is increased to 600 units, what is the correct exposure time?

CHAPTER 8

FUNCTIONS AND GRAPHS

René Descartes, soldier, philosopher, and mathematician, was born March 31, 1596, near Tours, France, at a time when Europe was in the throes of religious and political upheaval and in the midst of one of the great intellectual periods in the history of civilization.

René was not a particularly precocious child, and his frail health caused his formal education to be delayed until he was 8. His father enrolled him in a Jesuit school, where the rector noticed that the boy needed more than normal rest and advised him to stay in bed as long as he liked in the morning. Descartes followed this advice, which had a salutary effect on his health, and he formed a lifelong habit of staying late in bed whenever he could.

The boy was trained to be a gentleman, educated in Latin, Greek, and rhetoric. However, he soon developed a healthy skepticism toward all he was taught. He left school at 17, and by 18 was so disenchanted with his studies that he took an interlude of pleasure, settling in Paris. This did not last long; he retired to the suburb of St. Germain, where he worked on mathematics for 2 years.

At the age of 32, Descartes, still not having published any of his work, was persuaded to prepare his researches for publication. At 38, he compiled all that he had done into what was to be one grand treatise on the world, but fear of ecclesiastic displeasure with his conclusions caused him to refrain from having it printed.

Finally, when he was 41, his friends persuaded him to allow the printing of his masterpiece. Briefly known as *The Method*, it was published in 1637. Near the end of this work was an essay on geometry, an essay that is probably the single most important thing that Descartes ever did. This was *analytic geometry*; it revolutionized the entire realm of geometry and made much of modern mathematics possible.

In the spring of 1649, Queen Christine of Sweden succeeded in getting Descartes to visit her. He arrived in Sweden to discover that this amazonian queen expected him to teach her philosophy every day at five o'clock in the morning and to do this in the ice-cold library of her palace. This was too much for Descartes! He soon caught "inflammation of the lungs," from which he died on February 11, 1650, at the age of 54.

RENÉ DESCARTES (1596–1650)
Picture Collection, the Branch Libraries,
The New York Public Library

But there are other men who attain greatness because they embody the potentiality of their own day and magically reflect the future. They express the thoughts that will be everybody's two or three centuries after them. Such a one was Descartes.

THOMAS HUXLEY

RELATIONS AND FUNCTIONS

In Chapter 7 we studied first- and second-degree equations with *one* variable. In this chapter, we shall study similar equations with *two* variables. The main feature here, however, is an introduction to some simple ideas that belong to the area that is called **analytic geometry,** a blend of algebra and geometry in which algebra is used to study geometry and geometry is used to study algebra. The key to this combination is a workable system of associating **points in the plane** with **ordered pairs of numbers.**

Let us first consider an example of an ordered pair of numbers, $(4, -3)$. We say "ordered pair," because there are two numbers and the order of these numbers is important. Thus, the first number is 4 and the second is -3, and we distinguish $(4, -3)$ from $(-3, 4)$. That is,

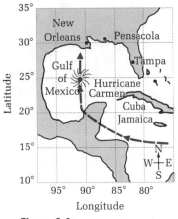

Figure 8.1a

$$(4, -3) \neq (-3, 4)$$

because the order of the two numbers is different in these pairs. You should not confuse this with the idea of equality of sets of numbers, where the order is not considered and we write $\{4, -3\} = \{-3, 4\}$. The use of parentheses for ordered pairs and braces for sets should serve to warn you that they are not the same.

The idea of an ordered pair is widely used in many areas. For example, the map in Fig. 8.1a shows the position of hurricane Carmen. As you can see, the storm center was near longitude 90° and latitude 25°. If we agree to write this information as an ordered pair of numbers showing the longitude (east–west distance) first and the latitude (north–south distance) second, then the location of the storm is (90, 25). These numbers are called the **coordinates** of the hurricane. On the same map, the coordinates of New Orleans are (90, 30).

In mathematics, we often call the first number of an ordered pair the **x value** and the second number the **y value.** The pair is symbolized by (x, y), and the equation

$$(x, y) = (2, -5)$$

means that $x = 2$ and $y = -5$. In general,

$$(x, y) = (a, b) \quad \text{if and only if} \quad x = a, y = b$$

There are many problems in which we need to study sets of ordered pairs. For convenience, we call such a set a *relation.*

▶ **Definition 8.1a**

| A **relation** is a set of ordered pairs. |

For instance, the set $R = \{(4, -3), (2, -5), (-3, 4)\}$ is a relation in which all the pairs have been specifically listed. Notice that the set of first members is $\{-3, 2, 4\}$ and the set of second members is $\{-5, -3, 4\}$. The listing in the relation R shows how the elements of the first set are associated with the elements of the second set to form the ordered pairs of the given relation. Different ways of associating the elements of the two sets will, of course, result in different relations. Thus,

$$S = \{(-3, -5), (2, -3), (4, 4)\}$$

is an example of a relation different from R but formed from the same two sets of numbers.

Relations as simple as the preceding two are usually not of interest. In most cases, we shall be concerned with a set of pairs $\{(x, y)\}$, where a **rule** is given for finding the y value for a given x value. For example,

$$Q = \{(x, y) | y = 4 - 2x, x \text{ an integer}\}$$

is a relation in which the ordered pairs are (x, y) such that x is an integer and $y = 4 - 2x$. For instance, the pair $(1, 2)$ is an element of Q, because if $x = 1$, then $y = 4 - 2 = 2$. Thus, a y value of 2 goes with an x value of 1. Other elements of Q are $(0, 4)$, $(-1, 6)$, and $(3, -2)$. In each case, we replace x by an integer and calculate y from the formula (or rule) $y = 4 - 2x$ to construct an element of Q. Clearly, Q has infinitely many elements.

A. Domain and Range

In the description of the relation Q, the replacement set for the variable x was prescribed; it is the set of integers. The set of all possible x values is called the **domain** of the relation. The domain of Q is the set of integers. Corresponding to any x value from the domain, the y value can be computed by using the rule given for this relation, $y = 4 - 2x$, as we have seen. The set of all possible y values is called the **range** of the relation. The range of the relation Q is the set of all even integers. (Can you see why?) The relations R and S above both have the domain $\{-3, 2, 4\}$ and the range $\{-5, -3, 4\}$.

Unless otherwise specified, the domain of a relation is taken to be the largest set of real numbers that can be substituted for x and that result in real numbers for y. The range is then determined by the rule of the relation.

For example, if

$$Q = \{(x, y) | y = \sqrt{x}\}$$

then we may substitute any *nonnegative* real number for x and obtain a real number for y. But x cannot be replaced by a negative number. (Why?) Thus, the domain is the set of all nonnegative real numbers, and the range, in this case, is the same set. Why?

EXAMPLE 1   Find the domain and the range of the relation

$$R = \{(x, y)|y = 2x\}$$

Solution   The variable x can be replaced by any real value, because 2 times any real number is a real number. Hence, the domain is $\{x|x$ a real number$\}$. The range is also the set of real numbers, because every real number is 2 times another real number. Thus, the range is $\{y|y$ a real number$\}$. ◀

EXAMPLE 2   Find the domain and the range of the relation

$$S = \{(x, y)|y = x^2 + 1\}$$

Solution   The variable x can be replaced by any real value, because the result of squaring a real number and adding 1 is again a real number. Thus, the domain of S is $\{x|x$ a real number$\}$. The square of a real number is never negative, $x^2 \geq 0$, so $x^2 + 1 \geq 1$. Hence, the rule $y = x^2 + 1$ implies $y \geq 1$. Consequently, the range of S is $\{y|y \geq 1\}$. ◀

EXAMPLE 3   Find the domain and the range of the relation

$$R = \{(x, y)|y \leq x, x \text{ and } y \text{ positive integers less than 5}\}$$

Solution   The domain as described is $\{1, 2, 3, 4\}$. For $x = 1$, $y = 1$; for $x = 2$, $y = 1$ or 2; for $x = 3$, $y = 1, 2,$ or 3; for $x = 4$, $y = 1, 2, 3,$ or 4. Thus, the range is also $\{1, 2, 3, 4\}$. We can list the ordered pairs that are elements of R:

$$\{(1, 1), (2, 1), (2, 2), (3, 1), (3, 2), (3, 3), (4, 1), (4, 2), (4, 3), (4, 4)\}$$ ◀

B.   Functions

In many areas of mathematics and its applications, the most important kind of relation is one such that to each domain value there corresponds one and only one range value. The relation in Example 1, $R = \{(x, y)|y = 2x\}$, is an illustration of such a relation. It is clear here that for each x value there corresponds exactly one y value, because the rule is $y = 2x$. On the other hand, the relation given in Example 3 is not this type of relation, because to each of the x values 2, 3, and 4 there corresponds more than one y value.

▶   **Definition 8.1b**

> A **function** is a relation such that to each domain value there corresponds exactly one range value.

All of us bump into functions every day: the association between the weight of a letter and the amount of postage you pay, the association between the cost of a piece of meat and the number of pounds it weighs, the association between the number of miles per gallon that you get and the speed at which you drive. These are all simple examples of functions. You can undoubtedly think of many more.

EXAMPLE 4   Is the relation $\{(x, y)|y = 2x + 3,$ x a real number$\}$ a function? Explain.

Solution   If x is a real number, then y is the unique real number $2x + 3$. For instance, if $x = 2$, then $y = 2(2) + 3 = 7$; if $x = -\frac{1}{2}$, then $y = 2(-\frac{1}{2}) + 3 = 2$. Clearly, for each real x value, the expression $2x + 3$ gives one and only one y value. Thus, the given relation has exactly one range value corresponding to each domain value and is therefore a function.   ◄

EXAMPLE 5   Is the relation $\{(x, y)|x = y^2 + 1\}$ a function? Explain.

Solution   For this relation the domain is $\{x|x \geq 1\}$. (Why?) If we take $x = 5$, then the rule $x = y^2 + 1$ gives $5 = y^2 + 1$, or $y^2 = 4$. Thus, $y = 2$ or -2, because $2^2 = (-2)^2 = 4$. So the pairs $(5, 2)$ and $(5, -2)$ both are elements of this relation. The fact that there are two range values (2 and -2) for the domain value 5 shows that the given relation is not a function.   ◄

C.   Function Notation

We often use letters such as $f, F, g, G, h,$ and H to designate functions. Thus, it is appropriate to write

$$f = \{(x, y)|y = 2x + 3\}$$

because we know this relation to be a function. A very commonly used notation to denote the range value that corresponds to a given domain value x is $f(x)$. (This is usually read, "f of x.")

The $f(x)$ notation, called **function notation,** is quite convenient, because it denotes the value of the function for the given value of x. For example, if

$$f(x) = 2x + 3$$

then

$$f(1) = 2(1) + 3 = 5$$
$$f(0) = 2(0) + 3 = 3$$
$$f(-6) = 2(-6) + 3 = -9$$
$$f(4) = 2(4) + 3 = 11$$
$$f(a) = 2(a) + 3 = 2a + 3$$
$$f(w + 2) = 2(w + 2) + 3 = 2w + 7$$

and so on. Whatever appears between the parentheses in $f(\;\;)$ is to be substituted for x in the rule that defines $f(x)$.

Instead of describing a function in set notation, we frequently say "the function defined by $f(x) = \ldots$" where the three dots are to be replaced by the expression for the value of the function. For instance, "the function defined by $f(x) = 2x + 3$" has the same meaning as "the function $f = \{(x, y)|y = 2x + 3\}$."

EXAMPLE 6   Let $f(x) = 3x + 5$. Find: **a.** $f(4)$ **b.** $f(2)$ **c.** $f(2) + f(4)$ **d.** $f(x + 1)$

Solution   **a.** Since $f(x) = 3x + 5$,

$$f(4) = 3 \cdot 4 + 5 = 12 + 5 = 17$$

b. $f(2) = 3 \cdot 2 + 5 = 6 + 5 = 11$

c. Since $f(2) = 11$ and $f(4) = 17$,

$$f(2) + f(4) = 11 + 17 = 28$$

d. $f(x + 1) = 3(x + 1) + 5 = 3x + 8$ ◀

D. Applications

In recent years, aerobic exercises such as jogging, swimming, and bicycling have been taken up by millions of Americans. To see if you are exercising too hard (or not hard enough), you should stop from time to time and take your pulse to determine your heart rate. The idea is to keep your rate within a range known as the **target zone,** which is determined by your age. The next example explains how to find the **lower limit** of your target zone.

EXAMPLE 7   The lower limit L (heartbeats per minute) of your target zone is a function of your age a (in years) and is given by

$$L(a) = -\tfrac{2}{3}a + 150$$

Find the value of L for a person who is:

a. 30 years old   **b.** 45 years old

Solution   **a.** We need to find $L(30)$, and because

$$L(a) = -\tfrac{2}{3}a + 150$$
$$L(30) = -\tfrac{2}{3}(30) + 150$$
$$= -20 + 150 = 130$$

This result means that a 30-year-old person should try to attain at least 130 heartbeats per minute while exercising.

b. Here, we want to find $L(45)$. Proceeding as before, we obtain

$$L(45) = -\tfrac{2}{3}(45) + 150$$
$$= -30 + 150 = 120$$

(Find the value of L for your own age.)   ◀

EXERCISE 8.1

A.   In Problems 1–14, find the domain and the range of the given relation. [Hint: Remember that you cannot divide by 0.]

1. $\{(1, 2), (2, 3), (3, 4)\}$
2. $\{(3, 1), (2, 1), (1, 1)\}$
3. $\{(1, 1), (2, 2), (3, 3)\}$
4. $\{(4, 1), (5, 2), (6, 1)\}$
5. $\{(x, y)|y = 3x\}$
6. $\{(x, y)|y = 2x + 1\}$
7. $\{(x, y)|y = x + 1\}$
8. $\{(x, y)|y = 1 - 2x\}$
9. $\{(x, y)|y = x^2\}$
10. $\{(x, y)|y = 2 + x^2\}$
11. $\{(x, y)|y^2 = x\}$
12. $\{(x, y)|x = 1 + y^2\}$
13. $\left\{(x, y)|y = \dfrac{1}{x}\right\}$
14. $\left\{(x, y)|y = \dfrac{1}{x - 2}\right\}$

Find the domain and the range of the relation given in each of Problems 15–22. List the ordered pairs in each relation.

15. $\{(x, y)|y = 2x,$ x an integer between -1 and 2, inclusive$\}$
16. $\{(x, y)|y = 2x - 1,$ x a counting number not greater than 5$\}$
17. $\{(x, y)|y = 2x - 3,$ x an integer between 0 and 4, inclusive$\}$
18. $\left\{(x, y)|y = \dfrac{1}{x}, \text{ x an integer between 1 and 5, inclusive}\right\}$

19. $\{(x, y)|y = \sqrt{x}, x = 0, 1, 4, 9, 16, \text{ or } 25\}$
20. $\{(x, y)|y \le x + 1,$ x and y positive integers less than 4$\}$
21. $\{(x, y)|y > x,$ x and y positive integers less than 5$\}$
22. $\{(x, y)|0 < x + y < 5,$ x and y positive integers less than 4$\}$

B. In Problems 23–30, decide whether the given relation is a function. State the reason for your answer in each case.

23. $\{(x, y)|y = 5x + 6\}$
24. $\{(x, y)|y = 3 - 2x\}$
25. $\{(x, y)|x = y^2\}$
26. $\{(x, y)|x + 1 = y^2\}$
27. $\{(x, y)|y = \sqrt{x}, x \ge 0\}$
28. $\{(x, y)|x = \sqrt{y}, y \ge 0\}$
29. $\{(x, y)|x = y^3\}$
30. $\{(x, y)|y = x^3\}$

C. 31. A function f is defined by $f(x) = 3x + 1$. Find:

    a. $f(0)$ **b.** $f(2)$ **c.** $f(-2)$

  32. A function g is defined by $g(x) = -2x + 1$. Find:

    a. $g(0)$ **b.** $g(1)$ **c.** $g(-1)$

  33. A function F is defined by $F(x) = \sqrt{x - 1}$. Find:

    a. $F(1)$ **b.** $F(5)$ **c.** $F(26)$

  34. A function G is defined by $G(x) = x^2 + 2x - 1$. Find:

    a. $G(0)$ **b.** $G(2)$ **c.** $G(-2)$

  35. A function f is defined by $f(x) = 3x + 1$. Find:

    a. $f(x + h)$ **b.** $f(x + h) - f(x)$ **c.** $\dfrac{f(x + h) - f(x)}{h}, \quad h \ne 0$

D. 36. The Fahrenheit temperature reading F is a function of the Celsius temperature reading C. This function is given by

$$F(C) = \tfrac{9}{5}C + 32$$

    a. If the temperature is 15°C, what is the Fahrenheit temperature?

    b. Water boils at 100°C. What is the corresponding Fahrenheit temperature?

c. The freezing point of water is 0°C or 32°F. How many Fahrenheit degrees below freezing is a temperature of −10°C?

d. The lowest temperature attainable is −273°C; this is the zero point on the absolute temperature scale. What is the corresponding Fahrenheit temperature?

37. Refer to Example 7. The **upper limit** U of your target zone when exercising is also a function of your age a (in years), and is given by

$$U(a) = -a + 190$$

Find the highest safe heart rate for a person who is:

a. 50 years old   **b.** 60 years old

38. Refer to Example 7 and Problem 37. The target zone for a person a years old consists of all the heart rates between $L(a)$ and $U(a)$, inclusive. Thus, if a person's heart rate is R, that person's target zone is described by $L(a) \le R \le U(a)$. Find the target zone for a person who is:

a. 30 years old   **b.** 45 years old

39. The ideal weight w (in pounds) of a man is a function of his height h (in inches). This function is defined by

$$w(h) = 5h - 190$$

a. If a man is 70 inches tall, what should his weight be?

b. If a man weighs 200 pounds, what should his height be?

40. The cost C in dollars of renting a car for 1 day is a function of the number m of miles traveled. For a car renting for $20 per day and 20¢ per mile, this function is given by

$$C(m) = 0.20m + 20$$

a. Find the cost of renting a car for 1 day and driving 290 miles.

b. If an executive paid $60.60 after renting a car for 1 day, how many miles did she drive?

41. The pressure P (in pounds per square foot) at a depth of d feet below the surface of the ocean is a function of the depth. This function is given by

$$P(d) = 63.9d$$

What is the pressure on a submarine at a depth of:

a. 10 ft?   **b.** 100 ft?

42. If a ball is dropped from a point above the surface of the Earth, the distance s (in meters) that the ball falls in t seconds is a function of t. This function is given by

$$s(t) = 4.9t^2$$

Find the distance that the ball falls in:

a. 2 sec   **b.** 5 sec

Courtesy PSH

1. There are many interesting functions that can be defined using the ideas of this section. For example, did you know that the frequency with which a cricket chirps is a function of the temperature? The table below shows the number of chirps per minute and the temperature in degrees Fahrenheit. Can you find a function that relates the number (c) of chirps per minute and the temperature x?

TEMPERATURE (°F)	40	41	42	43	44
CHIRPS PER MINUTE	0	4	8	12	16

2. The function relating the number of chirps per minute of the cricket and the temperature is given by $f(x) = 4(x - 40)$. If the temperature is 80°F, how many chirps per minute will you hear from your friendly house cricket?

3. An interesting function in physics was discovered by Galileo Galilei. This function relates the distance an object (dropped from a given height) travels and the time elapsed. The table in the margin shows the time (in seconds) and the distance (in feet) traveled by a rock dropped from a tall building. Can you find the relationship between the number of seconds elapsed, t, and the distance traveled, $f(t)$?

TIME ELAPSED (sec)	DISTANCE (ft)
1	$16 = 16 \times 1$
2	$64 = 16 \times 4$
3	$144 = 16 \times 9$
4	$256 = 16 \times 16$
5	$400 = 16 \times 25$
6	$576 = 16 \times 36$

4. Assume that a rock took 10 sec to reach the ground when dropped from the top of a building. Using the results of Problem 3, can you find the height of the building?

DISCOVERY 8.1

A special kind of relation that is important in mathematics is called an **equivalence relation.** *A relation R is an equivalence relation if it has the following three properties:*

a. **Reflexive property.** *If a is an element of the domain of R, then (a, a) is an element of R.*

b. **Symmetric property.** *If (a, b) is an element of R, then (b, a) is an element of R.*

c. **Transitive property.** *If (a, b) and (b, c) are both elements of R, then (a, c) is an element of R.*

A very simple example of an equivalence relation is

$$R = \{(x, y)|y = x, x \text{ an integer}\}$$

To show that R is an equivalence relation, we check the above three properties:

*a. **Reflexive.*** If a is an integer, then a = a, so (a, a) is an element of R.

*b. **Symmetric.*** Suppose that (a, b) belongs to R. Then, by the definition of R, a and b are integers and b = a. But if b = a, then a = b, so (b, a) also belongs to R.

*c. **Transitive.*** Suppose that (a, b) and (b, c) both belong to R. Then b = a and c = b, so c = a. Hence, (a, c) also belongs to R.

Because R has all three properties, it is an equivalence relation.

The pairs in a relation do not have to be numbers, and some interesting relations occur outside the field of numbers. For example,

$$R = \{(x, y) | y \text{ is a member of the same family as } x, x \text{ is a person}\}$$

is a relation. Is R an equivalence relation?

We check the three properties as before:

*a. **Reflexive.*** Given a person A, is (A, A) an element of R? Yes, A is obviously a member of the same family as A.

*b. **Symmetric.*** Suppose that (A, B) is an element of R. Then B is a member of the same family as A. But then A is a member of the same family as B, so (B, A) is an element of R.

*c. **Transitive.*** Suppose that (A, B) and (B, C) both belong to R. Then A, B, C are all members of the same family. Thus, (A, C) is an element of R.

Again, we see that R has all three properties, so it is an equivalence relation.

Can you discover which of the following are equivalence relations?

1. $R = \{(x, y) | x \text{ and } y \text{ are triangles and } y \text{ is similar to } x\}$
   [Note: Here, is similar to means has the same shape as.]
2. $R = \{(x, y) | x \text{ and } y \text{ are integers and } y > x\}$
3. $R = \{(x, y) | x \text{ and } y \text{ are positive integers and } y \text{ has the same parity as } x\}$; that is, y is odd if x is odd and y is even if x is even
4. $R = \{(x, y) | x \text{ and } y \text{ are boys and } y \text{ is the brother of } x\}$
5. $R = \{(x, y) | x \text{ and } y \text{ are positive integers and when } x \text{ and } y \text{ are divided by } 3, y \text{ leaves the same remainder as } x\}$
6. $R = \{(x, y) | x \text{ is a fraction } a/b, \text{ where } a \text{ and } b \text{ are integers } (b \neq 0), \text{ and } y \text{ is an equivalent fraction } ma/mb, \text{ where } m \text{ is a nonzero integer}\}$

8.2

GRAPHING RELATIONS AND FUNCTIONS

We shall now study a method for drawing pictures of relations and functions. Figure 8.2a shows two number lines drawn perpendicular to each

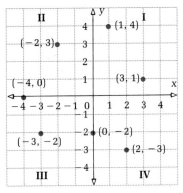

Figure 8.2a
Cartesian coordinate system

other. The horizontal line is labeled x and is called the **x axis.** The vertical line is labeled y and is called the **y axis.** The intersection of the two axes is the **origin.** We mark a number scale with the 0 point at the origin on each of the axes. The four regions into which the plane is divided by these axes are called **quadrants** and are numbered I, II, III, IV, as shown in Fig. 8.2a. This diagram forms a **Cartesian coordinate system** (named after René Descartes). We can make pictures of relations on such a coordinate system.

Figure 8.2a shows the usual way in which the positive directions along the axes are chosen. On the x axis, to the right is positive and to the left is negative. On the y axis, up is positive and down is negative. To locate a point (x, y), we go x units horizontally along the x axis and then y units in the vertical direction. For instance, to locate (−2, 3), we go 2 units horizontally in the negative direction along the x axis and then 3 units up, parallel to the y axis. Several points are located in Fig. 8.2a.

A. Graphs of Relations

The Cartesian coordinate system furnishes us with a one-to-one correspondence between the points in the plane and the set of all ordered pairs of numbers; that is, corresponding to a given ordered pair there is exactly one point, and corresponding to a given point there is exactly one ordered pair. The **graph** of a relation is the set of points corresponding to the ordered pairs of the relation.

EXAMPLE 1   Find the graph of the relation

$$R = \{(x, y) | y = 2x, \text{ x an integer between } -1 \text{ and } 2, \text{ inclusive}\}$$

Solution   The domain of R is {−1, 0, 1, 2}. By using the rule of R, y = 2x, we can find the ordered pairs (−1, −2), (0, 0), (1, 2), and (2, 4) that belong to R. Note that since y = 2x, the y coordinate is always *twice* the x coordinate. The graph of the relation is shown in the figure below.

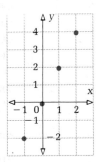

EXAMPLE 2   Graph the relation

$$T = \{(x, y)|y^2 = x, y \text{ an integer between } -3 \text{ and } 3, \text{ inclusive}\}$$

Solution   We are given the range $\{y|y$ an integer between -3 and 3, inclusive$\}$, and by squaring each of these integers, we find the domain, $\{0, 1, 4, 9\}$. The set of pairs $\{(0, 0), (1, 1), (1, -1), (4, 2), (4, -2), (9, 3), (9, -3)\}$ is thus the relation T. The graph of the relation consists of the seven dots shown in the figure below. If the domain of this relation had been $x \geq 0$, the graph would be a curve called a **parabola** (shown dashed in the figure).

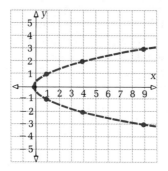

B.   Graphs of Functions

Since a function is simply a special kind of relation, a function can also be graphed on a Cartesian coordinate system. It is customary to represent the x values along the horizontal axis and the $f(x)$, or y, values along the vertical axis.

Because a function has just one value for each value of x in the domain, the graph of the function cannot be cut in more than one point by any vertical line. Thus, we have a simple **vertical line test** for a function. If the graph of a relation is known, we can tell by inspection whether the relation is a function. For example, the parabola in the figure for Example 2 is the graph of a relation, but that relation is *not* a function because any vertical line to the right of the y axis cuts the graph in two points. On the other hand, if the graph were only the lower (or only the upper) portion of the parabola, then the relation would be a function.

EXAMPLE 3   Graph the function f defined by $f(x) = 2 - 3x$ if the domain is the set $\{-1, 0, 1, 2\}$.

Solution   We calculate the values of the function using the rule $f(x) = 2 - 3x$ to obtain the values in the table. The figure shows the graph.

x	y = f(x)	ORDERED PAIR
−1	5	(−1, 5)
0	2	(0, 2)
1	−1	(1, −1)
2	−4	(2, −4)

There is something remarkable about the set of points graphed in this figure. Can you see what it is? The points all lie on a straight line! ◄

EXAMPLE 4   Graph the function g defined by $g(x) = 2 - x^2$ if the domain is $\{-3, -2, -1, 0, 1, 2, 3\}$.

Solution   The rule $g(x) = 2 - x^2$ tells us that to find the range value corresponding to an x value, we must square the x value and subtract the result from 2. Using this procedure, we obtain the values in the table. The graph appears in the figure. Notice that the points all lie on a parabola, as in Example 2.

x	y = g(x)	ORDERED PAIR
−3	−7	(−3, −7)
−2	−2	(−2, −2)
−1	1	(−1, 1)
0	2	(0, 2)
1	1	(1, 1)
2	−2	(2, −2)
3	−7	(3, −7)

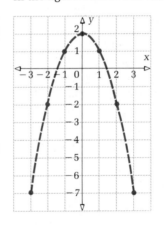

◄

C. Applications

EXAMPLE 5

Mon.–Fri.	8 A.M.–5 P.M.
Initial 1 minute	Each additional minute
$0.36	$0.28

Long-distance telephone costs are functions of time. The dial-direct rate for a Tampa to Los Angeles call made during business hours is given in the margin. Of course, if you talk for a fraction of a minute, you will be charged for the whole minute, so the words "or fraction thereof" should be understood to be included in each column. Find the cost of a call lasting:

a. 1 minute or less
b. More than 1 minute but not more than 2 minutes
c. More than 2 minutes but not more than 3 minutes
d. More than 3 minutes but not more than 4 minutes
e. Draw a graph showing the results of parts a, b, c, and d.

Solution We make a table of values, as in Examples 3 and 4. This table gives the answers for parts a, b, c, and d:

	TIME, t	COST, C
a.	$0 < t \le 1$	\$0.36
b.	$1 < t \le 2$	$\$0.36 + (1)(\$0.28) = \$0.64$
c.	$2 < t \le 3$	$\$0.36 + (2)(\$0.28) = \$0.92$
d.	$3 < t \le 4$	$\$0.36 + (3)(\$0.28) = \$1.20$

e. The graph is shown in the figure. The open circles in the figure indicate that the left-hand endpoints of the line segments are not included in the graph. The solid dots mean that the right-hand endpoints are included. ◀

EXAMPLE 6 The **greatest integer function** is often denoted by the symbol $[\![x]\!]$ and is defined by the rule: $[\![x]\!] =$ the greatest integer that is less than or equal to x.

For example, $[\![2.56]\!] = 2$, $[\![3]\!] = 3$, $[\![0.623]\!] = 0$, $[\![-2.5]\!] = -3$, and so on. Graph this function.

Solution The domain of this function is the set of all real numbers. The definition tells us that if x is between two consecutive integers, then $[\![x]\!]$ is the lesser of the two integers. Thus, if n is an integer, and $n \le x < n + 1$, then $[\![x]\!] = n$. We can list some values as in the table. The graph of these values appears in the figure. The open dots in the figure indicate that the right-hand endpoints of the line segments are not included in the graph; the solid dots indicate that the left-hand endpoints are included.

x	$y = [\![x]\!]$
$-3 \le x < -2$	-3
$-2 \le x < -1$	-2
$-1 \le x < 0$	-1
$0 \le x < 1$	0
$1 \le x < 2$	1
$2 \le x < 3$	2
$3 \le x < 4$	3

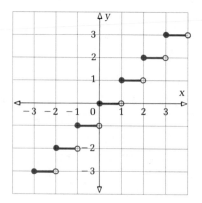

The greatest integer function

If you are curious about a possible connection between the functions in Examples 5 and 6, you should look at Using Your Knowledge 8.2, Problem 5.

EXERCISE 8.2

A. *In Problems 1–10, graph the given relation.*

1. $\{(x, y)|y = x, x$ an integer between -1 and 4, inclusive$\}$
2. $\{(x, y)|y = -x, x$ an integer between -1 and 4, inclusive$\}$
3. $\{(x, y)|y = 2x + 1, x$ an integer between 0 and 5, inclusive$\}$
4. $\{(x, y)|x + 2y = 3, x$ an odd integer between 0 and 10$\}$
5. $\{(x, y)|2x - y = 4, x$ an integer between -2 and 2, inclusive$\}$
6. $\{(x, y)|y = x^2, x$ an integer between -3 and 3, inclusive$\}$
7. $\{(x, y)|y = \sqrt{x}, x = 0, 1, 4, 9, 16, 25,$ or 36$\}$
8. $\{(x, y)|x = \sqrt{y}, x$ an integer between 0 and 3, inclusive$\}$
9. $\{(x, y)|x + y < 5, x, y$ nonnegative integers$\}$
10. $\{(x, y)|y > x, x$ and y positive integers less than 4$\}$

B. *In Problems 11–18, graph the given function.*

11. $f(x) = x + 1, x$ an integer between -3 and 3, inclusive
12. $f(x) = 3x - 1, x$ an integer between -1 and 3, inclusive
13. $g(x) = x^2 + 1, x$ an integer between -3 and 3, inclusive
14. $h(x) = -x^2, x$ an integer between -3 and 3, inclusive
15. $F(x) = x^2 - 2, x$ an integer between -3 and 3, inclusive
16. $f(x) = [\![x]\!] + 1, -3 \le x \le 3$
17. $g(x) = [\![x]\!] - 1, -3 \le x \le 3$
18. $h(x) = -[\![-x]\!], -3 \le x \le 3$

C. 19. Anthropologists can determine a person's height in life by using the person's unearthed bones as a clue. For example, the height (in centimeters) of a man with a humerus bone of length x cm can be obtained by multiplying 2.89 by x and adding 70.64 to the result.
   a. Find a function $h(x)$ that gives the height of a man whose humerus bone is x cm long.
   b. Use your function h from part a to predict the height of a man whose humerus bone is 34 cm long.

20. The daily cost of renting a car at $10 per day plus 10¢ per mile can be expressed as a function $c(x)$, where x is the number of miles traveled.
   a. Find $c(x)$.
   b. Find the total cost of driving 10, 20, and 30 miles.
   c. Graph the points obtained in part b.

21. A finance company will lend you $1000 for a finance charge of $20 plus simple interest at 1% per month. This means that you will pay

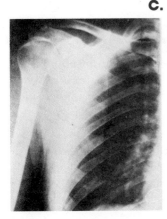

Courtesy PSH

interest of $\frac{1}{100}$ of $1000, that is, $10 per month in addition to the $20 finance charge. The total cost of the loan can be expressed as a function $F(x)$, where x is the number of months before you repay the loan.

a. Find $F(x)$.

b. Find the total cost of borrowing the $1000 for 8 months.

22. A bank charges 2% per month on a high-risk loan. If this is simple interest, describe the function that gives the cost $C(x)$ of borrowing $5000 for x months. (Compare Problem 21.)

23. One of the depreciation (loss of value) methods approved by the IRS is the **straight-line method.** Under this method, if a $10,000 truck is to be fully depreciated in 5 years, the yearly depreciation will be $\frac{1}{5}$ of $10,000, that is, $2000.

a. Write an equation that defines the depreciated value V of the truck as a function of the time t in years.

b. Draw a graph of this function.

24. The rates for a Tampa to Atlanta dial-direct telephone call during business hours are: 33¢ for the first minute or fraction thereof, and 27¢ for each additional minute or fraction thereof. Find the cost C of a call lasting:

a. 1 minute or less

b. More than 1 minute but not more than 2 minutes

c. More than 2 minutes but not more than 3 minutes

d. Draw a graph of the results of parts a, b, and c

25. The cost of a Tampa to Omaha dial-direct telephone call during business hours is given for positive integer values of t by

$$C(t) = 0.33 + 0.27(t - 1)$$

a. Draw a graph of $C(t)$ for $0 < t \le 5$.

b. Santiago made a call from Tampa to his friend Jessica in Omaha. If the call cost Santiago $3.57, how long did they talk?

USING YOUR KNOWLEDGE 8.2

*U.S. first-class postage charges can be described by a **step function** similar to the greatest integer function of Example 6. If the weight of a letter is x oz, where $0 < x \le 12$, then the postage is 25¢ for the first ounce or fraction thereof plus 20¢ for each additional ounce or fraction thereof.*

WEIGHT x (oz)	POSTAGE p (cents)
$\frac{1}{2}$	
$\frac{3}{4}$	
1	
$1\frac{1}{2}$	
2	
$2\frac{1}{4}$	

1. Complete the table in the margin.

2. To describe the first-class postage function by a formula, we first need to look at the function $f(x) = [\![1 - x]\!]$. Compare Example 6 and make a table for values of x between 0 and 12, using unit intervals $0 < x \le 1$, $1 < x \le 2$, etc. You should get $f(x) = 0$ for $0 < x \le 1$, $f(x) = -1$ for $1 < x \le 2$, etc.

3. The function that describes the first-class postage is

$$p(x) = 25 - 20[\![1 - x]\!] \qquad 0 < x \leq 12$$

Use the results of Problem 2 to graph the function p.

4. Show that $p(x) = 25 + 20[\![x - 1]\!]$ is not correct for the postage function. (Try it for $x = 1.5$.)
5. Show that the telephone call cost function of Example 5 is described by

$$C(t) = 0.36 - 0.28[\![1 - t]\!]$$

DISCOVERY 8.2

Have you ever seen a pendulum clock? An Italian scientist named Galileo Galilei made various discoveries about swinging weights, and these discoveries led to the invention of the pendulum clock. Galileo discovered that there was a relationship between the time of the swing of a pendulum and its length. The table shows corresponding values of these two quantities. (The unit length is about 25 cm.)

TIME OF SWING $f(x)$	LENGTH OF PENDULUM (x)
1 sec	1 unit
2 sec	4 units
3 sec	9 units
4 sec	16 units
5 sec	25 units

1. Judging from the table, what do you think is the rule connecting the time, $f(x)$, of the swing and the length, x, of the pendulum?
2. From the pattern given in the table, can you find the length of a pendulum that takes 6 sec for a swing?
3. Can you find the length of a pendulum that takes 100 sec for a swing?
4. The University of South Florida has a Foucault pendulum in its physics building. This pendulum takes 7 sec for a swing. Can you predict the length of the pendulum?

A view of the Foucault pendulum in the main lobby of the General Assembly building at United Nations headquarters.

Courtesy of the United Nations

LINEAR FUNCTIONS AND RELATIONS

A relation of the form

$$\{(x, y)|y = ax + b\}$$

where a and b are real numbers, always defines a function. (Why?) A function of this special form is called a **linear function,** because its graph is a straight line. We shall learn here how to draw the graph of a linear function.

A. Graphs of Linear Functions and Relations

Let us look first at the case in which $a = 0$. Then the rule of the function is $y = b$ for *all real values of* x. Note that x is unrestricted (any real number) but for any value of x, $y = b$. This means that the graph consists of all points such as $(0, b), (-1, b), (\sqrt{2}, b), (10.26, b)$, and so on. Thus, the graph is a straight line parallel to the x axis and b units from this axis. If $b > 0$, the line will be above the x axis; and if $b < 0$, the line will be below the x axis. (What if $b = 0$?) Figure 8.3a shows the graphs of $\{(x, y)|y = b\}$ for $b = -3$ and for $b = 2$.

Figure 8.3a

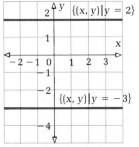

The next example will show the general procedure that can be used to draw the graph of a linear function. Note that the function can be defined in the form $f = \{(x, y)|y = ax + b\}$ or by just a statement of the rule $f(x) = ax + b$.

EXAMPLE 1   Graph the function defined by $f(x) = 3x - 6$.

Solution   Because f is a linear function, we know that its graph is a straight line. Hence, any two points on the line determine the entire line. One point that is always easy to find is the one for which $x = 0$. In this case, $f(x) = 3x - 6$, so $f(0) = 3(0) - 6 = -6$. The point $(0, -6)$ is the point where the graph crosses the y axis. Another point that is usually quite easy to find is the point where the line crosses the x axis. At this point $f(x) = 0$ (that is, $y = 0$).

x	f(x)	ORDERED PAIR
0	-6	(0, -6)
2	0	(2, 0)
3	3	(3, 3)

Thus, we need to solve $3x - 6 = 0$, and we find $x = 2$. The point is (2, 0). Let us find a third point to check our graph. If we put $x = 3$, we get $f(3) = 3(3) - 6 = 3$. The point is (3, 3). The table displays the points we have found, and the graph appears in the figure.

$f(x) = 3x - 6$

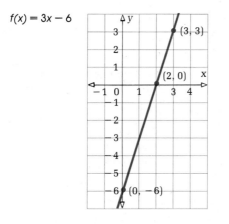

The x coordinate of the point where the line crosses the x axis is called the **x intercept** of the line. Similarly, the y coordinate of the point where the line crosses the y axis is called the **y intercept**. In Example 1, the x intercept is 2 and the y intercept is -6.

Notice that the ordered pairs (x, y) determined by the rule $f(x) = 3x - 6$ in Example 1 all satisfy the equation $y = 3x - 6$. Thus, the graph of the function may also be regarded as the graph of the equation. It makes no difference whether we graph the function f, where $f(x) = ax + b$, or we graph the equation $y = ax + b$.

A relation such as $\{(x, y)|x = c\}$, where c is a real number, is not a function, because y may have any real value. For instance, if $R = \{(x, y)|x = 2\}$, then $(2, -1)$, $(2, 0)$, $(2, 1.75)$, and so on, are all ordered pairs belonging to R. Because x has the fixed value 2, all these points are on the line parallel to and 2 units to the right of the y axis. In general, the equation $x = c$ (or the relation $\{(x, y)|x = c\}$) has for its graph a vertical line, that is, a line parallel to and c units from the y axis.

EXAMPLE 2   Graph the relation $R = \{(x, y)|x = 2\}$.

Solution   As indicated in the above discussion, the graph of this relation is a straight line parallel to the y axis and 2 units to the right of this axis. The graph appears in the figure.   ◄

B. Graphs of Linear Equations

We have already learned that the graph of a relation of the form $\{(x, y)|y = ax + b$, a and b real numbers$\}$ is a straight line. Are there any other relations whose graphs are straight lines? It is easy to see that the relation described by $\{(x, y)|ax + by = c\}$, where a and b are not both 0, is always a linear relation. If $b \neq 0$, we can solve the equation $ax + by = c$ for y, obtaining

$$by = -ax + c$$

$$y = -\frac{a}{b}x + \frac{c}{b}$$

Because a, b, and c are all real numbers, this equation is the rule for a linear function. If $b = 0$, then

$$0 = -ax + c$$

so that

$$x = +\frac{c}{a}$$

which is the rule for a linear relation (not a function), corresponding to a vertical line. Because of these facts, the equation $ax + by = c$, with a and b not both 0, is called a **linear equation;** its graph is always a straight line.

EXAMPLE 3   Graph the equation $2x + 3y = 6$.

Solution   Because the equation $2x + 3y = 6$ is a linear equation, we know that its graph is a straight line. Thus, any two points on the line will determine the line. For $x = 0$, we have $3y = 6$, or $y = 2$, so the point $(0, 2)$ is on the line. For $y = 0$, we get $x = 3$, so $(3, 0)$ is a second point on the line. We graph these two points and draw a straight line through them to get the graph shown in the figure.   ◄

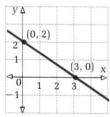

C. Applications

In order to make sensible decisions in certain consumer cost problems, you should be able to make a quick sketch of a straight line, the graph of a linear function. Such a situation is illustrated in the next example.

EXAMPLE 4   Suppose the cost of renting a car from company A is \$15 a day plus 20¢ per mile traveled. If m is the number of miles traveled and $C_1(m)$ is the corresponding cost in dollars, then

$$C_1(m) = 0.20m + 15$$

Suppose that for company B, the cost is \$20 a day plus 15¢ per mile. If $C_2(m)$ is the corresponding cost, then

$$C_2(m) = 0.15m + 20$$

a. Graph C_1 and C_2 on the same set of axes.

b. For what distance is the cost the same for both companies?

c. If cost is the only consideration, which company would you rent from if you planned to drive more than 100 miles?

Solution

a. Since C_1 and C_2 are both linear functions of m, their graphs are straight lines. For $m = 0, C_1(0) = 15$, so $(0, 15)$ is on the graph of C_1. For $m = 100$, $C_1(100) = 35$, so $(100, 35)$ is also on this graph. We draw a line through these two points as shown in the figure. Similarly, for $m = 0, C_2(0) = 20$, and for $m = 100, C_2(100) = 35$. Thus, the two points $(0, 20)$ and $(100, 35)$ are on the graph of C_2. Again, we draw a line through these two points as shown in the figure.

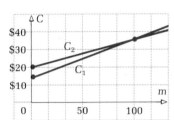

b. Since both lines pass through the point $(100, 35)$, it is clear that the cost is the same for both companies for a distance of 100 miles.

c. You can see from the graphs that the car from company B will cost less if you drive more than 100 miles. ◀

D. Distance Between Two Points

In this section we have learned that the line $y = b$ is a line parallel to the x axis and b units from it (see Fig. 8.3a). Similarly, the line $x = a$ is a line parallel to and a units from the y axis (see Example 2). It is not difficult to find the distance between two points on a horizontal or on a vertical straight line, that is, on a line $y = b$ or on a line $x = a$.

Suppose that we have two points, say A and B, on the same horizontal line, the line $y = b$ in Fig. 8.3b. Then their coordinates would be (x_1, b) and (x_2, b), as in the figure. Because x_1 and x_2 are the **directed distances** of the respective points from the y axis, the length of AB is $|x_2 - x_1|$. Denoting this length by $|AB|$, we have the formula

$$|AB| = |x_2 - x_1| \tag{1}$$

Similarly, if the two points C and D are on the same vertical line, their coordinates are (a, y_1) and (a, y_2), and the length of CD is

$$|CD| = |y_2 - y_1| \tag{2}$$

To obtain a formula for the distance between *any* two points, we use the **Pythagorean theorem:**

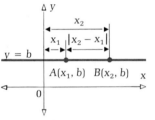

Figure 8.3b

If a triangle is a right triangle, then the square of the hypotenuse equals the sum of the squares of the other two sides. (See Fig. 8.3c.)

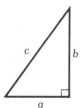

$a^2 + b^2 = c^2$

Figure 8.3c

The Pythagorean theorem

In Fig. 8.3d (page 350), $A(x_1, y_1)$ and $B(x_2, y_2)$ represent two general points. The line BC is drawn parallel to the y axis and the line AC is drawn parallel to the x axis, so ABC is a right triangle with right angle at C.

Figure 8.3d

$|AB| = \sqrt{(x_2 - x_1)^2 + (y_2 - y_1)^2}$

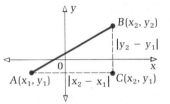

Because A and C are on the same horizontal line, they must have the same y coordinates. Likewise, B and C are on the same vertical line, so they have the same x coordinates. Thus, the coordinates of C must be (x_2, y_1). By the Pythagorean theorem, the length of AB is given by $|AB| = \sqrt{|AC|^2 + |BC|^2}$. Hence, by using equations (1) and (2), which give $|AC| = |x_2 - x_1|$ and $|BC| = |y_2 - y_1|$, we find the distance formula

$$|AB| = \sqrt{(x_2 - x_1)^2 + (y_2 - y_1)^2} \tag{3}$$

EXAMPLE 5   Find the distance between the points $A(2, -3)$ and $B(8, 5)$.

Solution   By equation (3),

$$|AB| = \sqrt{(8 - 2)^2 + [5 - (-3)]^2}$$
$$= \sqrt{6^2 + 8^2} = \sqrt{36 + 64} = \sqrt{100} = 10$$

The distance between A and B is 10 units.   ◀

EXAMPLE 6   Find the distance between the points $A(-1, -4)$ and $B(-3, 5)$.

Solution   As in the preceding example, we find

$$|AB| = \sqrt{[-3 - (-1)]^2 + [5 - (-4)]^2}$$
$$= \sqrt{(-2)^2 + 9^2} = \sqrt{4 + 81} = \sqrt{85}$$

With the aid of a table of square roots (or a calculator), we could express this result in decimal form. However, the indicated root form, $\sqrt{85}$ units, is adequate for our purposes.   ◀

EXAMPLE 7   Use the distance formula to show that $A(-3, -6)$, $B(5, 0)$, and $C(1, 2)$ are the vertices of a right triangle.

Solution   We know that a triangle is a right triangle if the square of one side equals the sum of the squares of the other two sides. This suggests that we find the squares of the sides of the triangle and then see whether two of the squares add up to the third. Using the distance formula, we find

$$|AB|^2 = [5 - (-3)]^2 + [0 - (-6)]^2 = 8^2 + 6^2 = 100$$
$$|BC|^2 = (1 - 5)^2 + (2 - 0)^2 = (-4)^2 + 2^2 = 20$$
$$|AC|^2 = [1 - (-3)]^2 + [2 - (-6)]^2 = 4^2 + 8^2 = 80$$

Thus, $|AB|^2 = |BC|^2 + |AC|^2$. So ABC is a right triangle with AB as its hypotenuse.   ◀

EXAMPLE 8   Determine if the triangle with vertices at $A(2, 7)$, $B(6, 0)$, and $C(12, 3)$ is a right triangle (see the figure).

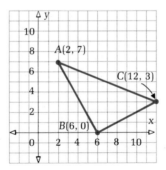

Solution $|AB|^2 = (6 - 2)^2 + (0 - 7)^2 = 16 + 49 = 65$

$|BC|^2 = (12 - 6)^2 + (3 - 0)^2 = 36 + 9 = 45$

$|AC|^2 = (12 - 2)^2 + (3 - 7)^2 = 100 + 16 = 116$

There is no side whose square is the sum of the squares of the other two sides. Therefore, the triangle is not a right triangle. ◄

EXERCISE 8.3

A.–B.   In Problems 1–16, graph the given linear function or equation.

1. $f(x) = 3x + 6$ **2.** $f(x) = 2x + 5$
3. $f(x) = 3$ **4.** $f(x) = -2$
5. $x = -1$ **6.** $x = 4$
7. $f(x) = -x + 2$ **8.** $f(x) = -2x - 4$
9. $g(x) = -3x - 6$ **10.** $g(x) = -2x + 6$
11. $3x + 2y = 6$ **12.** $4x + 3y = 12$
13. $-2x + 3y = 6$ **14.** $-3x + 2y = 12$
15. $4x - 3y = 12$ **16.** $3x - 5y = 15$

D.   In Problems 17–26, find the distance between the given points.

17. $(2, 4)$ and $(-1, 0)$ **18.** $(3, -2)$ and $(8, 10)$
19. $(-4, -5)$ and $(-1, 3)$ **20.** $(5, 7)$ and $(-2, 3)$
21. $(4, -8)$ and $(1, -1)$ **22.** $(-2, -2)$ and $(6, -4)$
23. $(3, 0)$ and $(3, -2)$ **24.** $(4, -1)$ and $(6, -1)$
25. $(-2, 3)$ and $(-2, 7)$ **26.** $(1, -5)$ and $(8, -5)$

In Problems 27–30, take the three given points as the vertices of a triangle. Determine whether the triangle is a right triangle, an isosceles (two sides *equal) triangle, or a scalene (no sides equal) triangle. Note that a triangle can be a right triangle and isosceles or scalene.*

27. $(2, 2)$, $(0, 5)$, $(-20, -12)$ **28.** $(2, 2)$, $(0, 5)$, $(-19, -12)$

29. $(2, 2)$, $(-4, -14)$, $(-20, -8)$ **30.** $(0, 0)$, $(6, 0)$, $(3, 3)$

C.

31. The daily cost C_1 (in dollars) of renting a car from company A is given by

$$C_1(m) = 0.13m + 8.50$$

where m is the number of miles traveled. For company B, the daily cost C_2 is given by

$$C_2(m) = 0.20m + 5$$

     a. Graph C_1 and C_2 on the same set of axes.

     b. Find the number of miles for which the rental cost is the same for both companies.

32. The costs (in dollars) of renting a car from companies A and B are given, respectively, by

$$C_1(m) = 0.10m + 10 \quad \text{and} \quad C_2(m) = 0.05m + 11$$

     a. Graph C_1 and C_2 on the same set of axes.

     b. For how many miles is the rental cost the same for both companies?

33. Economy telephone calls to Spain from Tampa have to be made between 6 P.M. and 7 A.M. The rates for calls dialed direct during those hours are $1.16 for the first minute or fraction thereof and 65¢ for each additional minute or fraction thereof.

     a. Let t (in minutes) be the duration of a call, and make a table of the cost $C(t)$ for $0 < t \leq 5$.

     b. Draw a graph of $C(t)$ for $0 < t \leq 5$.

34. Use the rates described in Problem 33.

     a. José wanted to call his dad in Spain during the economy hours, but José had only $5. How long a call could he make without exceeding his $5?

     b. Maria called her mother in Spain during the economy hours and was charged $7.01. How long was Maria's call?

DISCOVERY 8.3

*The ideas developed in this section for graphing lines can be summarized by means of a flowchart. A **flowchart** is a pictorial representation describing the logical steps that have to be taken in order to perform a task.*

The basic component of a flowchart is a box that contains a command. For example, $\boxed{x \to 0}$ instructs one to take x, a previously given quantity, and let it be equal to 0. In some cases, this instruction is given as

$\boxed{\text{Let } x = 0}$

The figure shows a flowchart that can be used to find the graph of any line not passing through the origin. An example of how it works is given at the right for the line $y = 2x + 6$.

Flowchart for graphing a line

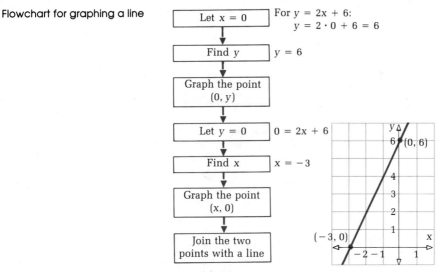

Let $x = 0$ For $y = 2x + 6$:
$\qquad y = 2 \cdot 0 + 6 = 6$

Find y $y = 6$

Graph the point $(0, y)$

Let $y = 0$ $0 = 2x + 6$

Find x $x = -3$

Graph the point $(x, 0)$

Join the two points with a line

Use the flowchart technique to graph the linear relations described by the following problems in Exercise 8.3.

1. Problem 1   2. Problem 2   3. Problem 7   4. Problem 8
5. Problem 9   6. Problem 10   7. Problem 11   8. Problem 12
9. Problem 13   10. Problem 14   11. Problem 15   12. Problem 16

▶ **Calculator Corner 8.3** You can use a calculator to find the distance between two points. For example, to find the distance between $(2, -3)$ and $(8, 5)$, you must use equation (3). Here are the steps you need. Press:

Note that in this case the answer appears as 10. If you work Example 6 using a calculator, your answer will be 9.219544457, an approximation for $\sqrt{85}$.

1. Try Problems 17, 19, 21, 23, and 25 with your calculator.

▶ **Computer Corner 8.3** A computer can be programmed to find the distance between two points when the points are entered (see page 749). Following the notation in equation (3), you must enter x_1, y_1, x_2, and y_2. The program does the rest. Notice that the answer to Example 6 in the text would appear as sqrt(85) = 9.219544. The program gives answers to six decimal places.

1. Try Problems 17, 19, 21, 23, and 25 on the computer.

EQUATIONS OF A LINE

Figure 8.4a

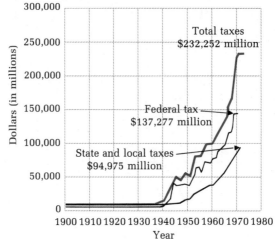

The graph (in color) in Figure 8.4a shows the total tax (in millions of dollars) for various years. This graph indicates that the total taxes paid in the years between 1940 and 1970 increased faster than in the years between 1910 and 1940. This can be seen easily, because if a line segment were drawn from the year 1940 to 1970, it would be steeper than a segment drawn from 1910 to 1940.

A. Slope

In mathematics, an important feature of a straight line is its *steepness.* We can measure the steepness of a nonvertical line by means of the ratio of the **vertical rise (or fall)** to the corresponding **horizontal run.** We call this ratio the **slope.** For example, a staircase that rises 3 ft in a horizontal distance of 4 ft is said to have a slope of $\frac{3}{4}$. (See Fig. 8.4b.)

Figure 8.4b

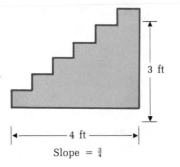

Slope $= \frac{3}{4}$

In general, we use the following definition.

Definition 8.4a

A line going through two points (x_1, y_1) and (x_2, y_2), where $x_1 \neq x_2$, has **slope** m, where

$$m = \frac{y_2 - y_1}{x_2 - x_1}$$

Figure 8.4c shows the horizontal run $x_2 - x_1$ and the vertical rise $y_2 - y_1$ used in calculating the slope. We do not define slope for a vertical line. The slope of a horizontal line is obviously 0, because all points on such a line have the same y values.

Figure 8.4c

The slope of a line:

$$m = \frac{y_2 - y_1}{x_2 - x_1}, \quad x_1 \neq x_2$$

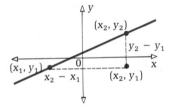

EXAMPLE 1   Find the slope of the line that goes through the points $(0, -6)$ and $(3, 3)$.

Solution   The two given points are shown in the figure. Suppose we choose $(x_1, y_1) = (0, -6)$ and $(x_2, y_2) = (3, 3)$. Then we get

$$m = \frac{3 - (-6)}{3 - 0} = \frac{9}{3} = 3$$

If we choose $(x_1, y_1) = (3, 3)$ and $(x_2, y_2) = (0, -6)$, then

$$m = \frac{-6 - 3}{0 - 3} = \frac{-9}{-3} = 3$$

Line with positive slope

As you can see, it makes no difference which point is labeled (x_1, y_1) and which is labeled (x_2, y_2). Since an interchange of the two points simply changes the sign of both the numerator and the denominator in the slope formula, the result is the same in both cases. ◄

EXAMPLE 2 Find the slope of the line that goes through the two points $(3, -4)$ and $(-2, 3)$. See the figure.

Line with negative slope

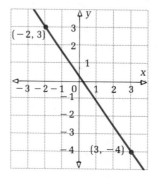

Solution We take $(x_1, y_1) = (-2, 3)$ so that $(x_2, y_2) = (3, -4)$. Then

$$m = \frac{-4 - 3}{3 - (-2)} = -\frac{7}{5}$$

◀

Examples 1 and 2 are illustrations of the fact that a line that rises from left to right has a **positive slope** and one that falls from left to right has a **negative slope**.

B. Equations of Lines

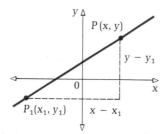

Figure 8.4d

The slope of a line can be used to obtain an equation of the line. For example, suppose the line goes through a point $P_1(x_1, y_1)$ and has slope m. See Fig. 8.4d. We let $P(x, y)$ be any second point (distinct from P_1) on the line. Then, by Definition 8.4a, the slope of the line in terms of these two points is

$$\frac{y - y_1}{x - x_1} = m$$

Multiplying both sides by $(x - x_1)$, we get

$$y - y_1 = m(x - x_1) \tag{1}$$

This equation, which must be satisfied by the coordinates of every point on the line, is known as the **point–slope equation** of the line.

EXAMPLE 3 Find an equation of the line that goes through the point $(2, -3)$ and has slope $m = -4$.

Solution Using the point–slope equation (1), we get

$$y - (-3) = -4(x - 2)$$
$$y + 3 = -4x + 8$$
$$y = -4x + 5$$

◀

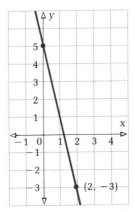

Figure 8.4e
$y = -4x + 5$

An important special case of equation (1) is that in which the given point is the point where the line intersects the y axis. Let this point be denoted by $(0, b)$. Then b is called the **y intercept** of the line. Using equation (1), we obtain

$$y - b = m(x - 0)$$

and, by adding b to both sides, we get

$$y = mx + b \tag{2}$$

This equation, in which m is the slope and b is the y intercept, is called the **slope–intercept form** of the equation of the line. Notice that the answer to Example 3 was given in the slope–intercept form. This form is convenient for reading off the slope and the y intercept of the line. Thus, the answer to Example 3 immediately tells us that the slope of the line is -4 and the y intercept is 5 (see Fig. 8.4e).

EXAMPLE 4   Find the slope and the y intercept of the line with equation $6x + 3y = 5$.

Solution   By equation (2), the slope–intercept form of the equation of a line is $y = mx + b$, where m is the slope and b is the y intercept. We can solve the given equation for y by subtracting $6x$ from both sides and then dividing by 3. This procedure gives

$$y = -2x + \tfrac{5}{3}$$

an equation in the slope–intercept form. Thus, the slope m of the given line is -2 and the y intercept is $\tfrac{5}{3}$.   ◀

The procedure of Example 4 can be followed for any equation of the form $Ax + By = C$, where $B \neq 0$, to obtain an equation in the slope–intercept form. If $B = 0$, then we can divide by A to get $x = C/A$, an equation of a line parallel to the y axis. Thus, we see that every equation of the form

$$Ax + By = C \tag{3}$$

where A, B, and C are real numbers and A and B are not both 0, is an equation of a straight line.

On the other hand, we can show that *every* straight line has an equation of the form $Ax + By = C$. Any line that is not parallel to the y axis has a slope–intercept equation $y = mx + b$, which can be written in the form $mx - y = -b$ by subtracting b and y from both sides. This last equation is of the form $Ax + By = C$, with $A = m$, $B = -1$, and $C = -b$. For example, the equation $y = 2x + 5$ can be written in the form $2x - y = -5$ by subtracting 5 and y from both sides. This equation is the special case of $Ax + By = C$ with $A = 2$, $B = -1$, and $C = -5$. A line parallel to the y axis has an equation $x = a$, which is already of the form $Ax + By = C$, with $A = 1$, $B = 0$, and $C = a$. Thus, we see that every straight line has an equation of the form $Ax + By = C$, which is consequently often called the **general form** of the equation of a straight line.

EXAMPLE 5   Find the general form of the equation of the line that goes through (6, 2) and (3, −2).

Solution   We first find the slope of the line:

$$m = \frac{-2 - 2}{3 - 6} = \frac{4}{3}$$

Then, using the point–slope form, we find the equation

$$y - 2 = \tfrac{4}{3}(x - 6)$$
$$3y - 6 = 4x - 24$$
$$4x - 3y = 18$$

[We can check this answer by seeing that (6, 2) and (3, −2) are both solutions of the equation.]   ◄

The ideas of Example 5 can often be used to obtain a simple formula that summarizes a group of data in a convenient form. For example, Table 8.4a shows the desirable weight range corresponding to a given height for men and for women.

Table 8.4a

HEIGHT (in.)	MEN'S WEIGHT (lb)	WOMEN'S WEIGHT (lb)
62	108–134	98–123
63	112–139	102–128
64	116–144	106–133
65	120–149	110–138
66	124–154	114–143
67	128–159	118–148
68	132–164	122–153
69	136–169	126–158
70	140–174	130–163
71	144–179	134–168
72	148–184	138–173
73	152–189	
74	156–194	
75	160–199	
76	164–204	

Can we find an equation giving the relationship between height and weight? First, we must realize two things:

1. Men are heavier, so there will be one equation for men and another equation for women.

2. Even for men, the table gives only a weight range. For instance, a 76-in. (6-ft 4-in.) man should weigh between 164 and 204 lb.

Hence, we must be more specific, as in the next example.

EXAMPLE 6   Find an equation that gives the relationship between a man's height h (in inches) and his weight w (in pounds) using the lower weights in Table 8.4a as the desirable ones.

Solution   If we examine the heights and the corresponding weights in the table, we see that the heights increase by 1 in. from one entry to the next and the corresponding weights increase by 4 lb. This means that all the points (62, 108), (63, 112), . . . , (76, 164) lie on one straight line, because the slope of a line connecting any consecutive pair of these points is the same as the slope of the line connecting any other consecutive pair. Thus, the equation of this line must be the desired equation. The slope of the line is easily obtained by using the first pair of points (62, 108) and (63, 112):

$$m = \frac{112 - 108}{63 - 62} = 4$$

Then, we can use the point–slope form of the equation to obtain

$$w - 108 = 4(h - 62)$$
$$w - 108 = 4h - 248$$
$$w = 4h - 140$$

You can check this against entries in the table. For example, for a 72-in. (6-ft) man, the equation gives

$$w = 4(72) - 140$$
$$= 288 - 140 = 148$$

which agrees with the table entry.   ◄

C. Parallel Lines

Since the slope of a line determines its direction, it is obvious that **two lines with the same slope and different y intercepts are parallel lines.** The next example makes use of this idea.

EXAMPLE 7   Show that $3y = x + 2$ and $2x - 6y = 7$ describe parallel lines.

Solution   We solve each equation for y to obtain

$$y = \tfrac{1}{3}x + \tfrac{2}{3} \quad \text{and} \quad y = \tfrac{1}{3}x - \tfrac{7}{6}$$

These equations show that both lines have slope $\tfrac{1}{3}$. The y intercepts are different, so the lines are parallel.   ◄

At this point, many students ask, "Which formula should we use in the problems?" Table 8.4b tells you which formula to use, depending on what information is given. Study this table before you attempt the problems in Exercise 8.4.

Table 8.4b

TO FIND THE EQUATION OF A LINE, GIVEN:	USE
Two points (x_1, y_1) and (x_2, y_2), $x_1 \neq x_2$	Point–slope form: $y - y_1 = m(x - x_1)$, where $m = (y_2 - y_1)/(x_2 - x_1)$
A point (x_1, y_1) and the slope m	Point–slope form: $y - y_1 = m(x - x_1)$
The slope m and the y intercept b	Slope–intercept form: $y = mx + b$

The resulting equation can always be written in the general form: $Ax + By = C$.

EXERCISE 8.4

A. In Problems 1–10, find the slope of the line that goes through the two given points.

   1. (1, 2) and (3, 4)
   2. (1, −2) and (−3, −4)
   3. (0, 5) and (5, 0)
   4. (3, −6) and (5, −6)
   5. (−1, −3) and (7, −4)
   6. (−2, −5) and (−1, −6)
   7. (0, 0) and (12, 3)
   8. (−1, −1) and (−10, −10)
   9. (3, 5) and (−2, 5)
   10. (4, −3) and (2, −3)

B. In Problems 11–16, find the slope–intercept form (if possible) of the equation of the line that has the given properties (m is the slope).

   11. Goes through (1, 2); $m = \frac{1}{2}$
   12. Goes through (−1, −2); $m = -2$
   13. Goes through (2, 4); $m = -1$
   14. Goes through (−3, 1); $m = \frac{3}{2}$
   15. Goes through (4, 5); $m = 0$
   16. Goes through (3, 2);   slope is not defined (does not exist)

In Problems 17–26, find:

a. The slope   **b.** The y intercept

of the graph of the given equation.

   17. $y = x + 2$
   18. $2x + y = 3$
   19. $3y = 4x$
   20. $2y = x + 4$
   21. $x + y = 14$
   22. $y - 4x = 8$
   23. $y = 6$
   24. $2y = 16$
   25. $x = 3$
   26. $3x = -6y + 9$

In Problems 27–32, find the general form of the equation of the straight line through the two given points.

27. $(1, -1)$ and $(2, 2)$ **28.** $(-3, -4)$ and $(-2, 0)$

29. $(3, 2)$ and $(2, 3)$ **30.** $(3, 0)$ and $(0, 5)$

31. $(0, 0)$ and $(1, 10)$ **32.** $(-4, -1)$ and $(-4, 3)$

33. Use Table 8.4a to find a formula relating the height h of a man and his ideal weight w given by the *second* number in the weight column. See Example 6.

34. Use Table 8.4a to find a formula relating the height h of a woman and her weight w as given by the *first* number in the weight column. See Example 6.

35. Repeat Problem 34 using the *second* number in the weight column.

36. Based on your answer to Problem 35, if a woman weighs 183 pounds, how tall should she be? (This is not given in Table 8.4a.)

C. In Problems 37–42, determine whether the given lines are parallel.

37. $y = 2x + 5; \quad 4x - 2y = 7$

38. $y = 4 - 5x; \quad 15x + 3y = 3$

39. $2x + 5y = 8; \quad 5x - 2y = -9$

40. $3x + 4y = 4; \quad 2x - 6y = 7$

41. $x + 7y = 7; \quad 2x + 14y = 21$

42. $y = 5x - 12; \quad y = 3x - 8$

43. Find an equation of the line that goes through the point $(1, -2)$ and is parallel to the line $4x - y = 7$.

44. Find an equation of the line that goes through the point $(2, 0)$ and is parallel to the line $3x + 2y = 5$.

*45. It is shown in analytic geometry that two lines, $a_1x + b_1y = c_1$ and $a_2x + b_2y = c_2$, are perpendicular if and only if

$$a_1a_2 + b_1b_2 = 0$$

that is, **two lines are perpendicular if and only if the sum of the products of the corresponding coefficients of x and y in the general form of the equations is 0.** This leads to a very easy way of writing an equation of a line that is perpendicular to a given line. For instance, if the given line is $3x + 5y = 8$, then $5x - 3y = c$ for each real number c is a line perpendicular to the given line. This is obvious because $(3)(5) + (5)(-3) = 0$. Notice that all we need to do to form the left side of the second equation is interchange the coefficients of x and y in the first equation and change the sign of one of these coefficients. If we wish to have the second line go through a specified point, we select the value of c so that this happens. Thus, if the second line is to go through $(3, 2)$, then c has to be selected so that $(3, 2)$ is in the solution set of the equation. Hence,

$$5(3) - 3(2) = c \quad \text{or} \quad c = 9$$

The line $5x - 3y = 9$ goes through $(3, 2)$ and is perpendicular to the line $3x + 5y = 8$. In each of the following problems, find an equation of the line that is perpendicular to the given line and goes through the given point.

a. $2x + 5y = 7$; $(2, 0)$ b. $y = 2x - 3$; $(1, 1)$
c. $x - 2y = 3$; $(2, -2)$ d. $4x + 5y = 9$; $(1, 1)$

*46. If the lines $y = m_1 x + b_1$, $m_1 \neq 0$, and $y = m_2 x + b_2$ are perpendicular, then $m_2 = -1/m_1$. Show this by referring to Problem 45. This leads to the simple statement that **the slopes of perpendicular lines are negative reciprocals of each other.** Why do we need the condition $m_1 \neq 0$?

*47. A line goes through the two points $(0, 0)$ and $(100, 200)$. A second line goes through the two points $(0, 10)$ and $(790, -405)$. Can you see how to determine whether these are perpendicular lines? [Hint: Look at Problem 46.]

USING YOUR KNOWLEDGE 8.4

1. The idea of slope is used in economics in the **consumption possibility line.** For simplicity, we assume that a man has a total of $10 per day to spend. He is confronted with fixed prices for each food and clothing unit. The cost of food is $2 per unit and that of clothing is $1 per unit. If it is known that the man will spend the entire $10 on some combination of food and clothing, make a graph depicting this situation. [Hint: Use the x axis to represent the number of units of food and the y axis to represent the number of units of clothing.]

2. In Problem 1, what is the interpretation of the point $(0, 10)$?

3. In Problem 1, what is the interpretation of the point $(5, 0)$?

4. What does the slope of the line obtained in Problem 1 tell about the ratio of the cost of food units to the cost of clothing units?

5. If the total cost y of producing x units of a product is assumed to be linear, then it can be written as

$$y = mx + b$$

where m is the cost of producing 1 unit (the **marginal cost**) and b is the **fixed cost,** which does not depend on the number of units produced.

a. Find the total cost y of a product whose production cost is $2 per unit and whose fixed cost is $2000.

b. If the total cost y changes from $1200 to $1350 as the number x increases from 400 to 700, find the fixed cost b and the marginal cost m. [Hint: The line must go through the two points $(400, 1200)$ and $(700, 1350)$.]

							3rd Ave.North
3rd St.West	2nd St.West	1st St.West	Main St.	1st St.East	2nd St.East	3rd St.East	2nd Ave.North
							1st Ave.North
			Central Avenue				
			1st Ave.South		4th St.East	5th St.East	6th St.East
			2nd Ave.South				

SQUARESVILLE

The Fun and Games Club of Squaresville is having a treasure hunt. The figure shows a map of a portion of the town. The blocks are all square (of course!) and all of the same size. The instructions for the hunt are as follows: Draw two lines, one line going through the intersection of Central Avenue and Main Street and through the intersection of 20th Street East and 10th Avenue North, the second line going through the intersection of Central Avenue and 35th Street East and through the intersection of 40th Street East and 15th Avenue North. The treasure is in the red brick house at the intersection of these two lines.

Unfortunately, the treasure is in a new part of the town and there are no maps covering the area. Furthermore, no one on the treasure hunt has thought to bring any graph paper.

1. Can you locate the treasure without doing any drawing?

▶ **Computer Corner 8.4**   You learned in this section that you can find the equation of a line if:

a. Two points are given
b. A slope and a point are given

We have a program (page 751) that combines these two features; that is, you can provide two points, or a slope and a point, and the program will give you the equation of the corresponding line. To do Example 3, simply enter the points $(2, -3)$ and the slope -4. As before, the answer is given in slope–intercept form.

1. Work Problems 11, 13, and 15 using this program.

8.5

TWO LINEAR EQUATIONS IN TWO VARIABLES

There are many applications of algebra that require the solution of a **system of linear equations.** We shall consider only the simple case of two equations in two variables. Applications that require the solution of such systems appear in Section 8.7.

As we have seen, a linear equation $Ax + By = C$ has a straight line for its graph. Hence, two such equations will graph into two straight lines in the plane. Two distinct lines in the plane can either **intersect** at a point or else be **parallel** (have no intersection). If the lines intersect, then the coordinates of the point of intersection are called the **solution of the system of equations.** If the lines are parallel, then, of course, the system has no solution.

A. Solution by Graphing

One way to solve a system of two equations in two variables is to graph the two lines and read the coordinates of the point of intersection from the graph. The next two examples illustrate this graphical method.

EXAMPLE 1   Use the graphical method to find the solution of the system

$$2x - y = 4$$
$$x + y = 5$$

Solution   We graph the two equations as shown in the figure. The point of intersection of the two lines appears to be (3, 2). We check this set of coordinates in the given system: For $x = 3$ and $y = 2$,

$$2x - y = 2 \cdot 3 - 2 = 4$$
$$x + y = 3 + 2 = 5$$

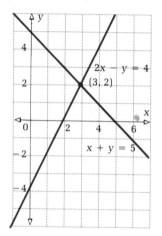

Thus, the two equations are both satisfied, and the desired solution is $x = 3$, $y = 2$, or (3, 2).   ◀

EXAMPLE 2   If you are about to build a house, you might consider two types of heating:

1. Solar heating, which requires a $10,000 initial investment and a $200 annual cost
2. Oil heating, with a $2000 initial investment and an annual cost of $1000

   a. Write an equation giving the total cost y (in dollars) of solar heating over x years.
   b. Repeat part a for the oil heating.
   c. Graph the equations obtained in parts a and b for a period of 12 years.
   d. When will the total costs for the two systems be equal?

Solution   **a.** The cost y of solar heating is given by $y = 10,000 + 200x$.
   b. The cost y of oil heating is given by $y = 2000 + 1000x$.
   c. Note that the x coordinate represents the number of years and the y

coordinate represents the number of dollars in the corresponding total cost. For the solar heating equation, we find the two points (0, 10,000) and (5, 11,000). These two points and the line through them are shown in color in the figure. For the oil heating equation, we find the two points (0, 2000) and (5, 7000). These two points and the line through them are also shown in the figure.

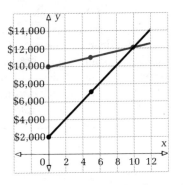

d. As you can see from the figure, the lines intersect at the point where $x = 10$ and $y = 12,000$. Thus, the total costs are equal at the end of 10 years. (After 10 years, the total cost for solar heating is much less than that for oil heating.) ◀

B. Solution by Algebraic Methods

If the solution of the system is a pair of simple numbers that can be read exactly from the graph, then the graphical method is quite satisfactory. Naturally, you should always check a proposed solution in the given equations. Unfortunately, most systems will have solutions that are not easy to read exactly from the graph, and for such systems an algebraic method of solving is needed. There is a quite simple method that is best explained by means of examples, as follows.

EXAMPLE 3   Solve the system of Example 1 by algebraic means.

Solution   The point (x, y) where the lines $2x - y = 4$ and $x + y = 5$ intersect is the intersection of the solution sets of the two equations. This means that the number pair consisting of the x value and the y value of this point must satisfy both equations. Hence, we may use the following procedure: Since the y terms in the two equations are the same except for sign, we add the two equations term by term to eliminate the y terms and get an equation in one variable:

$$2x - y = 4$$
$$\underline{x + y = 5}$$
$$3x \qquad = 9$$

Thus, $x = 3$. By substituting this x value into the second equation, we get

$$3 + y = 5 \quad \text{or} \quad y = 2$$

We have now found the same solution as before, (3, 2). ◄

If both variables occur in both equations, we can proceed as follows:

> **1.** Eliminate one of the variables.
>   **a.** Make this variable have the same coefficients but opposite in sign (by multiplication if necessary).
>   **b.** Add corresponding terms of the two equations.
> **2.** Solve the resulting linear equation for the remaining variable.
> **3.** Substitute the solution back into one of the original equations to find the value of the second variable.

This procedure is illustrated in the next example.

EXAMPLE 4 Find the point of intersection of the two lines $3x - 2y = 19$ and $2x + 5y = -19$.

Solution We multiply the first equation by 5 and the second by 2 to get

$$15x - 10y = 95$$
$$4x + 10y = -38$$

We then add the two equations to get

$$19x = 57$$

This gives $x = 3$, which we substitute into the first of the given equations to obtain

$$9 - 2y = 19$$
$$-2y = 10$$
$$y = -5$$

The point of intersection is thus $(3, -5)$. ◄

EXAMPLE 5 Solve the following system (if possible):

$$2x - y = 5$$
$$4x - 2y = 7$$

Solution We multiply the first equation by -2 to obtain the system

$$-4x + 2y = -10$$
$$4x - 2y = 7$$

Upon adding these two equations, we get the *impossible* result

$$0 = -3$$

Thus, the system has no solution. The lines represented by the two equations are parallel. ◀

Notice that the method of addition of the equations (used in Examples 4 and 5) will detect parallel lines as in Example 5 by arriving at an *impossible* result: $0 = $ a nonzero number.

EXAMPLE 6   Solve the following system (if possible):

$$2x - y = 5$$
$$4x - 2y = 10$$

Solution   We multiply the first equation by 2 to get

$$4x - 2y = 10$$

which is exactly the second equation. Thus, we see that the two equations represent the same line, and every solution of one of these equations satisfies the other. If we put $x = a$ in the first equation and then solve for y to get $y = 2a - 5$, we can write the general solution of the system as $(a, 2a - 5)$, where a is any real number. ◀

EXERCISE 8.5

A.   In Problems 1–6, find the solution of the given system by the graphical method.

1. $x + y = 3$; $2x - y = 0$
2. $x + y = 5$; $x - 4y = 0$
3. $2x - y = 10$; $3x + 2y = 1$
4. $2x - 3y = 1$; $x + 2y = 4$
5. $3x + 4y = 4$; $2x - 6y = 7$
6. $y = 5x - 12$; $y = 3x - 8$

B.   In Problems 7–30, find the solution of the given system by the method of addition that was used in Examples 4 and 5. If there is no solution, state so.

7. $x + y = 3$; $2x - y = 0$
8. $x + y = 5$, $x - 4y = 0$
9. $x + y = 6$; $3x - 2y = 8$
10. $2x - y = 5$; $5x + 3y = 18$
11. $2x - y = 10$; $3x + 2y = 1$
12. $2x - 3y = 1$; $x + 2y = 4$
13. $5x + y = 4$; $15x + 3y = 8$
14. $2x - y = -5$; $4x - 2y = -10$
15. $2x + 5y = 12$; $5x - 3y = -1$
16. $2x + 3y = 9$; $11x + 7y = 2$
17. $3x + 4y = 4$; $2x - 6y = 7$
18. $y = 5x - 12$; $y = 3x - 8$
19. $11x + 3 = -3y$; $5x + 2y = 5$
20. $10x + 6y = 1$; $5x = 9 - 3y$

21. $x = 2y - 3$; $x = -2y - 1$ 22. $y = 4x - 2$; $4x = 2y + 3$
23. $2x + 3y + 11 = 0$; $5x + 6y + 20 = 0$
24. $3x + y = 4$; $2x = 4y - 9$ 25. $3x - 12y = -8$; $2x + 2y = 3$
26. $4x + 8y = 7$; $3x + 4y = 6$ 27. $r - 4s = -10$; $2r - 8s = 13$
28. $3r + 4s = 15$; $4r - s = 20$ 29. $6u - 2v = -27$; $4u + 3v = 8$
30. $8w - 13z = 3z + 4$; $12w - 3 = 18z$

31. The daily cost y (in dollars) of renting a car from company A is given by

$$y = 0.15x + 10$$

where x is the number of miles traveled. For company B, the daily cost y is given by

$$y = 0.20x + 5$$

   a. Find the number of miles for which the two costs are equal. Use the method of addition to solve the system of equations.
   b. Graph the two equations on the same set of axes from $x = 0$ to $x = 200$.
   c. Which of the two companies is cheaper if you have to drive farther than the distance you found in part a?

32. A solar hot-water heating system requires an initial investment of $5400 and an annual cost of $50. An electric hot-water heating system requires an initial investment of $1000 and an annual cost of $600. Let y dollars be the total cost at the end of x years.
   a. Write an equation giving the total cost of the solar heating system.
   b. Write an equation giving the total cost of the electric heating system.
   c. Graph the two equations you found in parts a and b on the same set of axes from $x = 0$ to $x = 10$.
   d. Use the method of addition to solve the system of equations.
   e. How long will it take for the total cost of the solar heat to become less than the total cost of the electric heat?

USING YOUR KNOWLEDGE 8.5

Linear systems can sometimes be used to solve interesting problems. Here is an illustration: Suzie has 15 coins, all nickels and dimes, in a peculiar combination. If the nickels were quarters and the dimes were nickels, the amount of money that Suzie has would be unchanged. How many of each coin does she have?

If we let x be the number of nickels and y be the number of dimes that Suzie has, then we see that $x + y = 15$, the total number of coins. Furthermore, the amount of money she has is $(5x + 10y)$ cents. Now, if the nickels were quarters and the dimes were nickels, the amount would be $(25x + 5y)$

cents. The problem says that the amount would be unchanged, so we must have

$$25x + 5y = 5x + 10y$$

or

$$20x - 5y = 0 \quad \text{By subtracting } 5x \text{ and } 10y \text{ from both sides}$$

and

$$4x - y = 0 \quad \text{By dividing both sides by 5}$$

Thus, we have the system

$$x + y = 15$$
$$4x - y = 0$$

By adding the equations, term by term, we obtain

$$5x = 15 \quad \text{or} \quad x = 3$$

If $x = 3$ is substituted into the first equation, it gives $3 + y = 15$, or $y = 12$. Hence, Suzie has 3 nickels and 12 dimes. (You can check that 3 quarters and 12 nickels will give the same amount of money.)

Try to use these ideas to solve the following problems.

1. Ronnie has $2.95 in nickels and dimes. If the nickels were dimes and the dimes were nickels, Ronnie would have exactly $1.00 more. How many of each coin does he have?
2. Josie has invested $10,000 in two types of bonds. Type A pays a 7% dividend and type B pays a 6% dividend. If the total dividend amounts to $640, how much has Josie invested in each type of bond?
3. An auto rental company charges $10 per day plus 10¢ per mile. A second company charges $20 per day plus 6¢ per mile. Find the number of miles for which the cost of renting a car is the same in both companies.
4. Tickets for a dance were sold to men for $3 and to women for $2. If 130 tickets were sold for $340, how many of each were sold?
5. Admission to a certain movie was $2.50 for adults and $1.50 for children. If 300 tickets were sold for a total of $550, how many adult tickets and how many children's tickets were sold?

▶ **Computer Corner 8.5**   We have a program (page 751) that will solve a system of two or three linear equations when you enter the **coefficients** of the variables and the numbers to the right of the equal signs. Note that in Example 3, you must enter a 1 for the coefficients of x and y in $x + y = 5$. If the system has **infinitely many solutions** or **no solution,** the computer replies, "I can't solve this one."

1. Try the odd-numbered Problems from 7 to 29 in Exercise 8.5.

LINEAR INEQUALITIES

In this section we want to discover how to graph a relation in which the rule is a linear inequality, that is, an inequality of the type $ax + by + c \geq 0$. For short, we speak of graphing a linear inequality.

A. Graphing Linear Inequalities

Let us look at the relation $\{(x, y)|2x + 3y - 6 \geq 0\}$. We already know how to graph the straight line $2x + 3y = 6$, and this is a big start on our present problem. Figure 8.6a shows the graph of the straight line and several points not on the line. We know that any point on the line has coordinates (x, y) such that $2x + 3y - 6 = 0$. In Table 8.6a we have evaluated the linear expression $2x + 3y - 6$ and compared it with 0 at each of the points in the figure. Notice that for all the points on one side of the line, $2x + 3y - 6$ is less than 0 (negative), and for all points on the other side of the line, $2x + 3y - 6$ is greater than 0 (positive). These results lead us to guess that $2x + 3y - 6 < 0$ for all points below the line and that $2x + 3y - 6 > 0$ for all points above the line. This guess is correct; it is a fact that all the points above the line do satisfy the inequality $2x + 3y - 6 > 0$, and no points below the line satisfy this inequality. We shall not prove this. The shaded region in Fig. 8.6a plus the line $2x + 3y = 6$ is the graph of the relation $\{(x, y)|2x + 3y - 6 \geq 0\}$.

Figure 8.6a
$\{(x, y)|2x + 3y \geq 6\}$

Table 8.6a

POINT	$2x + 3y - 6$	
(1, 3)	5	>0
(3, 3)	9	>0
(4, 1)	5	>0
(−3, 1)	−9	<0
(0, 0)	−6	<0
(−2, −2)	−16	<0
(3, −1)	−3	<0

We know that a straight line in a plane divides the plane into three sets of points, the line itself and the two **half-planes,** one half-plane on one side and the other half-plane on the other side of the line. We regard the line as belonging to neither half-plane. The two half-planes are disjoint sets; no point of one set is shared by the other set. Geometrically speaking, we cannot connect a point in one half-plane and a point in the other half-plane by a continuous line (or curve) in the plane without crossing the line that separates one half-plane from the other. It is proved in more advanced mathematics that this intuitive geometric fact has an important algebraic counterpart, namely: **If an equation of the line is $ax + by = c$, then for every point in one half-plane, the linear expression $ax + by - c$ is positive (>0), and for every point in the other half-plane, $ax + by - c$ is negative (<0).**

This result makes it quite easy to graph a linear inequality, as the next example shows.

EXAMPLE 1   Graph the linear inequality $x + 2 < y$.

Solution   We follow Julie Ashmore's amusing verse.

How to Graph a Linear Inequality, by Julie Ashmore*

Look here, my children, and you shall see
How to graph an inequality.
Here's a simple inequality to try:
x plus 2 is less than y. $x + 2 < y$
First, make the "less than" "equal to";
So now y equals x plus 2. $y = x + 2$
Then pick a point for x: say, 10;
Now plug that single constant in.
Add 10 plus 2 and you'll get y; $x = 10, y = 10 + 2$
See if this pair will satisfy!
x: 10, y: 12; you'll find it's right, $(10, 12)$
So graph this point to expedite.
Now find a second ordered pair   See the figure.
That fits in your equation there.
x: 3, y: 5 will do quite well. $(3, 5)$
And it's correct, as you can tell.
Plot this point, and then you've got
To draw a line from dot to dot.   See the figure.
Make it neat and make it straight;
A ruler's edge I'd advocate.
The next step's hard! You've got to choose
Which side of this line you must use.
Change "equal to" back to "less than," $x + 2 < y$
Just as it was when you began.
Pick a point on one side! I
Use 3 for x and 1 for y. $(3, 1)$
Is 1 greater than 3 plus 2?
No! This side will never do! $3 + 2 \not< 1$
On the other side, let's try
1 for x and 4 for y. $(1, 4)$
1 plus 2 (which equals 3)
Is less than 4, as you can see. $1 + 2 = 3 < 4$
Shade in the side that dot is on;
We've one more step to come upon.   See the figure.
Do the points upon your line
Fit the equation I assigned?
Use 1 and 3 for this last test; $(1, 3)$
They're "equal to," but they're not "less."
Make your line dotted to show this is true. $1 + 2 = 3$
And that is *all* you have to do!   See the figure. ◄

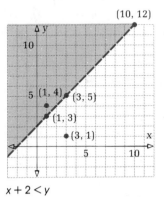

$x + 2 < y$

* Julie Ashmore was a young student in the John Burroughs School in St. Louis, Missouri, when she wrote this verse.

The next example illustrates the solution of an inequality that involves the less than or equal to sign. Notice carefully that the line corresponding to the equals sign is a part of the solution and is drawn solid in the graph.

EXAMPLE 2   Graph the linear inequality $3x - 4y \le 12$.

Solution   We first draw the graph of the line $3x - 4y = 12$ (see the figure). We know that our answer requires one of the half-planes determined by this line. To find which of the half-planes is required, we may select any point not on the line and check whether it satisfies the original inequality. If the origin is not on the line, then $(0, 0)$ is a good choice, because it is so easy to check. Does $(0, 0)$ satisfy the inequality $3x - 4y \le 12$? Yes, because $3(0) - 4(0) = 0 < 12$. Thus, the half-plane containing $(0, 0)$ is the one we need. This half-plane is shaded in the figure.   ◄

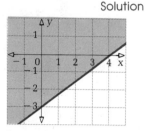

$3x - 4y \le 12$

B. Solving Systems of Inequalities by Graphing

The solution set of a system of linear inequalities in two variables can often be found as in the next example, where the individual inequalities are first solved separately and then the final shading shows the intersection of these solution sets. This intersection is the solution set of the system.

EXAMPLE 3   Graph the solution set of the system of inequalities

$$x + 2y \le 5$$
$$x - \ y < 2$$

Solution   We first graph the lines $x + 2y = 5$ and $x - y = 2$, as shown in the figure. The inequality $x + 2y \le 5$ is satisfied by the points *on or below* the line $x + 2y = 5$, as indicated by the arrows attached to the line. The inequality $x - y < 2$ is satisfied by the points *above* the line $x - y = 2$, as indicated by the arrows attached to the line. This line is drawn dashed to indicate that the points on it do *not* satisfy the inequality $x - y < 2$. The solution set of the system is shown in the figure by the shaded region and the portion of the solid line forming one boundary of this region. The point of intersection of the two lines is *not* in the solution set.   ◄

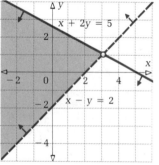

We can also graph the solution set of a system consisting of more than two inequalities, as illustrated by the next example.

EXAMPLE 4   Graph the solution set of the system of inequalities.

$$x + 2y \le 6$$
$$3x + 2y < 10$$
$$x \ge 0$$
$$y \ge 0$$

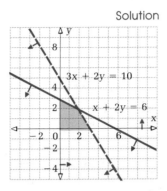

Solution   We first graph the lines $x + 2y = 6$, $3x + 2y = 10$, $x = 0$, $y = 0$, as in the figure. The inequality $x + 2y \leq 6$ is satisfied by the set of points on or below the line $x + 2y = 6$, as indicated by the arrows attached to the line. The inequality $3x + 2y < 10$ is satisfied by the set of points below the line $3x + 2y = 10$. Notice that this line is drawn dashed to show that it does not satisfy the given inequality. The set of inequalities $x \geq 0$ and $y \geq 0$ is satisfied by points in the first quadrant or points on the portions of the axes bounding the first quadrant. The solution set of the system can easily be identified; it is the shaded region plus the solid portions of the boundary lines.   ◄

C.   Applications

In Section 8.1, we mentioned the target zone used to gauge your effort when performing aerobic exercises. This target zone is determined by your pulse rate p and your age a and is found in the next example.

EXAMPLE 5   The target zone for aerobic exercise is defined by the following inequalities, in which a is the age in years and p is the pulse rate:

$$10 \leq a \leq 70$$

$$p \geq -\tfrac{2}{3}a + 150 \quad \text{Lower limit}$$

$$p \leq -a + 190 \quad \text{Upper limit}$$

Graph these inequalities and label the resulting target zone.

Solution   Since a is between 10 and 70, inclusive, it is convenient to label the horizontal axis starting at 10, using 10-unit intervals, and ending at 70. The vertical axis is used for the pulse rate p, and we start at 70 (the normal pulse rate) and go up to 200, as shown in the figure. We then graph the line $p = -\tfrac{2}{3}a + 150$ after finding the two points (30, 130) and (60, 110). Notice that because of the $-\tfrac{2}{3}a$, values of a that are divisible by 3 are most convenient to use. The inequality $p \geq -\tfrac{2}{3}a + 150$ is satisfied for all points that are on or above this line. Next, we graph the line $p = -a + 190$. Two points easy to find on this line are (10, 180) and (70, 120). The inequality $p \leq -a + 190$ is satisfied by all points that are on or below this line. The target zone is shaded in the figure.   ◄

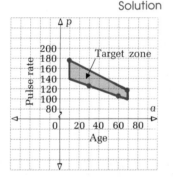

EXERCISE 8.6

A.   *Graph the linear inequalities in Problems 1–10.*

1. $x + 2y \geq 2$ 2. $x - 2y > 0$ 3. $x \leq 4$
4. $y \leq 3$ 5. $3x - y < 6$ 6. $3x + 4y \geq 12$
7. $2x + y \leq 4$ 8. $2x - 3y < 0$ 9. $4x + y > 8$
10. $x - 4y \leq 4$

B. *Graph the solution set of the system of inequalities in Problems 11–22.*

11. $x - y \geq 2$; $x + y \leq 6$ **12.** $x + 2y \leq 3$; $x \leq y$

13. $2x - 3y \leq 6$; $4x - 3y \geq 12$ **14.** $2x - 5y \leq 10$; $3x + 2y \leq 6$

15. $2x - 3y \leq 5$; $x \geq y$; $y \geq 0$ **16.** $x \leq 2y$; $2x \geq y$; $x + y < 4$

17. $x + 3y \leq 6$; $x \geq 0$; $y \geq 0$ **18.** $2x - y \leq 2$; $y \geq 1$; $x \geq \frac{1}{2}$

19. $x \geq 1$; $y \geq 1$; $x - y \leq 1$; $3y - x < 3$

20. $x - y \geq -2$; $x + y \leq 6$; $x \geq 1$; $y \geq 1$

21. $x + y \geq 1$; $x \leq 2$; $y \geq 0$; $y \leq 1$

22. $1 < x + y < 8$; $x < 5$; $y < 5$

C.   **23.** The desirable weight range corresponding to a given height for a man is shown in Table 8.4a. We found the equation for the lower weights in terms of the height to be $w = 4h - 140$. The equation for the upper weights is $w = 5h - 176$. Thus, the desirable weights for men from 62 to 76 in. in height satisfy the system of inequalities

$$62 \leq h \leq 76$$

$$w \geq 4h - 140$$

$$w \leq 5h - 176$$

Graph this system of inequalities and shade the region corresponding to the range of desirable weights. Be sure to take the horizontal axis as the h axis.

24. The desirable weight range corresponding to a given height for a woman is also shown in Table 8.4a. The equation for the lower weights in terms of the height h is $w = 4h - 150$ and for the upper weights is $w = 5h - 187$. Thus, the desirable weights for women from 62 to 72 in. in height satisfy the system of inequalities

$$62 \leq h \leq 72$$

$$w \geq 4h - 150$$

$$w \leq 5h - 187$$

Graph this system of inequalities and shade the region corresponding to the range of desirable weights. Be sure to take the horizontal axis as the h axis.

USING YOUR KNOWLEDGE 8.6

Suppose you want to find two integers x and y with y > 1 and such that the sum of the two integers is less than 10 and the difference 3x − 2y is greater than 6. Here is an easy way to find the possible pairs (x, y):

1. Graph the corresponding system of inequalities:

$$y > 1$$

$$x + y < 10$$

$$3x - 2y > 6$$

You should end up with a shaded region in the plane. Mark each integer point (point with integer coordinates) within this region, and then just read the coordinates. These are the possible pairs of integers.

2. Find all the pairs of integers (x, y) such that x is greater than 1, the sum of the two integers is less than 12, and the difference $2y - x$ is at least 8.

DISCOVERY 8.6

1. Here is a problem just for fun. A square is 2 in. on a side. Five points are marked inside or on the perimeter of the square. Can you show that the distance between at least two of the five points is less than or equal to $\sqrt{2}$? [*Hint:* Cut the square up into four equal squares by connecting the midpoints of the opposite sides.]
2. The figure below is a flowchart for graphing the inequality $ax + by > c$, $abc \neq 0$. Can you explain why this works? Try it for some of the problems in Exercise 8.6.

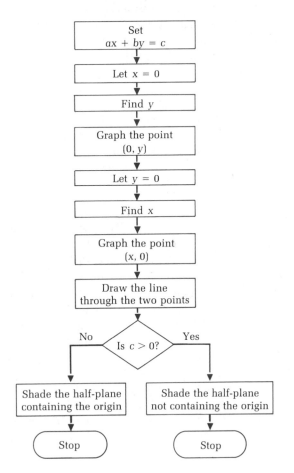

3. Can you discover what changes to make in the flowchart if the inequality is $ax + by < c$, $c \neq 0$?

4. Can you discover what changes to make in the flowchart if $c = 0$?

LINEAR PROGRAMMING

There are many problems that require us to find better ways of accomplishing a given objective. For example, there are many ways of conserving energy, but some of these ways are more costly than others. Consequently, we may be interested in minimizing the cost of our energy requirements. If the equations and inequalities involved in such a problem are linear, then we have a **linear programming problem.**

Most practical linear programming problems involve a large number of variables and are solved on a digital computer using a scheme known as the **simplex algorithm.** The simplex method of solving such problems was developed by George B. Dantzig in the late 1940's as a tool for the solution of some complex problems in allocating supplies for the U.S. Air Force. Linear programming problems are of great importance today in business and in the social sciences.

Although we shall not consider the simplex method, we shall solve some simple problems that illustrate the variety of applications of linear programming methods. The concepts we employ are basic even for the larger problems that are beyond the scope of this book.

As an example of a linear programming problem, let us assume that Sew & Sew, Inc. manufactures pants and vests. The profit on each pair of pants is $6 and on each vest is $5. The pants use 2 yd of material and the vests use 1.5 yd of material each. Because of production limitations, Sew & Sew can manufacture not more than 10 of these garments per day, and can use not more than 18 yd of material per day. If Sew & Sew can sell all the pants and vests they make, find the number of each garment they should produce per day to **maximize their profit.**

One way of presenting a linear programming problem so that it is easier for our minds to grasp is to put the data into tabular form. For the Sew & Sew problem, we tabulate the data and make some minor calculations as shown in the table in the margin. We can now see that the total profit, say P, is

	PANTS	VESTS
Number produced	x	y
Yards used	$2x$	$1.5y$
Profit (dollars)	$6x$	$5y$

$$P = 6x + 5y \text{ dollars}$$

Thus, P is a linear functon of x and y, and we wish to determine x and y so that P has its maximum value.

Next, we must express the restrictions in the problem in terms of x and y. The first restriction is that Sew & Sew produce not more than 10 garments per day. Since they produce x pants and y vests, the total number of garments is $x + y$. Thus,

$$x + y \leq 10$$

A second restriction is that they use not more than 18 yd of material per day. This means that

$$2x + 1.5y \leq 18$$

Another restriction is that x and y cannot be negative. This gives the so-called **positivity conditions**

$$x \geq 0 \qquad y \geq 0$$

We can summarize our problem as follows: We want to maximize the linear function P given by

$$P = 6x + 5y$$

subject to the **constraints** (that is, the restrictions)

$$x + y \leq 10 \qquad\qquad\qquad\qquad (1)$$
$$2x + 1.5y \leq 18 \qquad\qquad\qquad (2)$$
$$x \geq 0 \qquad y \geq 0 \qquad\qquad\qquad (3)$$

To make a start on the solution of this problem, we graph the solution set of the system of inequalities (1), (2), and (3), just as we did in the preceding section. This gives the shaded region in Fig. 8.7a. Each point (x, y) in this region represents a combination of garments that satisfies all the constraints. For this reason, the region is called the **feasible region.**

We need to select a point from the feasible region that will maximize the expression $6x + 5y$. In order to do this, let us examine the equation $6x + 5y = P$. For a given value of P, this is an equation of a straight line, which in slope–intercept form is

$$y = -\frac{6}{5}x + \frac{P}{5}$$

This equation shows that the slope is $-\frac{6}{5}$ regardless of the value of P. Hence, for a set of values of P, we get a set of parallel straight lines. Furthermore, for positive increasing values of P, the lines recede from the origin. (This follows because the y intercept is P/5.) The dashed lines in Fig. 8.7a are the graphs of $6x + 5y = P$ for the values $P = 12, 24, 36$, and 48. The smaller the value of P, the closer is the line to the origin.

These considerations make plausible the following basic theorem for feasible regions that are **convex** (nonreentrant) polygons, that is, regions such that the points of the line segment joining any two points on the boundary lie entirely inside the region or else on the boundary. See Figure 8.7b.

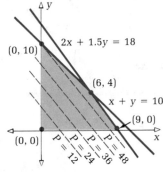

Figure 8.7a

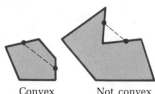

Convex   Not convex

Figure 8.7b

► **Theorem 8.7a**

> If the feasible region is a convex (nonreentrant) polygon, then the desired maximum (or minimum) value of a linear function occurs at a corner point (vertex) of the region.

VERTEX	$P = 6x + 5y$
(0, 0)	0
(9, 0)	54
(6, 4)	56 ← Maximum
(0, 10)	50

To make use of this theorem, we need only check the values of P at the vertices of the polygon. These vertices are indicated in Fig. 8.7a and can, of course, be found by solving the appropriate pairs of linear equations. By direct calculation, we find the values in the table in the margin. Thus, Sew & Sew should produce 6 pants and 4 vests per day to maximize their profit.

EXAMPLE 1   Little Abner raises ducks and geese. He is too lazy to take care of more than 30 birds all together, but wants to make as much profit as possible (naturally). It costs him $1 to raise a duck and $1.50 to raise a goose, and he has only $40 to cover this cost. If Little Abner makes a profit of $1.50 on each duck and $2 on each goose, what is his maximum profit?

Solution   Letting x and y be the number of ducks and geese, respectively, that Little Abner should raise, we tabulate the given information as follows:

	DUCKS	GEESE
Number	x	y
Cost	$1.00 each	$1.50 each
Profit	$1.50 each	$2.00 each
Total cost	$x	$1.5y
Total profit	$1.5x	$2y

It appears that Little Abner's total profit from both ducks and geese is P, where

$$P = 1.5x + 2y$$

The constraints are (in the order stated in the problem):

$x + \quad y \le 30$ Too lazy to raise more than 30 birds   (4)

$x + 1.5y \le 40$ Has only $40 to cover his costs   (5)

Although it is not stated, we must also obey the positivity conditions

$x \ge 0 \qquad y \ge 0$ (6)

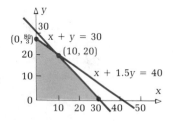

We proceed as before to find the feasible region by graphing the system of inequalities (4), (5), and (6), as shown in the figure. We then find the vertices and check them in the profit function:

VERTEX	$P = 1.5x + 2y$	
(0, 0)	0	
(30, 0)	45	
(10, 20)	55	← Maximum
$(0, \frac{80}{3})$	$53\frac{1}{3}$	

By raising 10 ducks and 20 geese, Little Abner will make the maximum possible profit, $55. ◄

If you have been thinking while studying the preceding problems, you have undoubtedly wondered what to do if the desired maximum (or minimum) occurs at a vertex with noninteger coordinates. In the case of two-variable problems, if the solution must be in integers, we can try the integer points inside the feasible region that are nearest to this vertex and select the point that gives the desired maximum (or minimum). For many-variable problems, more complicated techniques, which are beyond the scope of this book, must be used.

EXERCISE 8.7

1. Find the minimum value of $C = 2x + y$ subject to the constraints
$$x \geq 1$$
$$x \leq 4$$
$$y \leq 4$$
$$x - 3y \leq -2$$

2. Find the maximum value of $P = x + 4y$ subject to the constraints
$$y - x \leq 0$$
$$x \leq 4$$
$$y \geq 0$$
$$x + 2y \leq 6$$

3. Find the minimum value of $W = 4x + y$ subject to the constraints
$$x + y \geq 1$$
$$2y - x \leq 1$$
$$x \leq 1$$

4. Find the minimum value of $C = 2x + 3y$ subject to the constraints
$$2x + y \geq 18$$
$$x + y \geq 12$$
$$3x + 2y \leq 34$$

5. Find the minimum value of $C = x + 2y$ subject to the constraints
$$8 \leq 3x + y \leq 10$$
$$x \geq 1$$
$$y \geq 2$$

6. Find the maximum value of $P = 2x + 3y$ subject to the constraints
$$y - x \leq 2$$
$$x + y \leq 4$$
$$0 \leq x \leq 3$$
$$y \geq 0$$

7. Find the maximum value of $P = x + 2y$ subject to the constraints
$$2x + y \geq 6$$
$$0 \leq y \leq 4$$
$$0 \leq x \leq 2$$

8. Find the maximum value of $P = 4x + 5y$ subject to the constraints
$$y - x \leq 2$$
$$x - y \leq 2$$
$$x + y \leq 6$$
$$x \geq 0$$
$$y \geq 0$$

9. The E-Z-Park storage lot can hold at most 100 cars and trucks. A car occupies 100 ft² and a truck 200 ft², and the lot has a usable area of 12,000 ft². The storage charge for a car is $20 per month and for a truck is $35 per month. How many of each should be stored to bring E-Z-Park the maximum revenue?

10. The Zig-Zag Manufacturing Company produces two products, zigs and zags. Each of these products has to be processed through all three machines, as shown in the table. If Zig-Zag makes $12 profit on each zig and $8 profit on each zag, find the number of each that the company should make in order to maximize its profit.

MACHINE	HOURS AVAILABLE	HOURS/PIECE Zigs	Zags
I	Up to 100	4	12
II	Up to 120	8	8
III	Up to 84	6	0

11. The Kwik-Pep Vitamin Company wishes to prepare two types of vitamin tablets. The first type of tablet contains

1 milligram (mg) of vitamin B_1
1 mg of vitamin B_2

while the second type of tablet contains

1 mg of vitamin B_1
2 mg of vitamin B_2

The profit on the two types of tablets is as follows:

2¢ for each tablet of the first type
3¢ for each tablet of the second type

In manufacturing two bottles of tablets, one of each type, but with the same number of tablets, Kwik-Pep wants to use no more than 100 mg of vitamin B_1 and 150 mg of vitamin B_2. How many tablets should be packed in each bottle to obtain the largest profit?

12. A nutritionist is designing a meal for one of her patients. The meal must include two vegetables, A and B, but not more than 100 g of each. Suppose that each 10-g portion of A contains 2 units of iron and 2 units of vitamin B_{12}, and each 10-g portion of B contains 1 unit of iron and 5 units of vitamin B_{12}. The number of calories in each 10-g portion of these vegetables is 5 for A and 3 for B. If the patient needs at least 20 units of iron and 36 units of vitamin B_{12} in the meal, how many grams of each vegetable should the nutritionist include to satisfy the iron and vitamin requirements while minimizing the number of calories in the meal?

13. The Jeri Tonic Company wishes to manufacture Jeri Tonic so that each bottle contains at least 32 units of vitamin A, 10 units of vitamin B, and 40 units of vitamin C. To supply the vitamins, the company uses additive X, which costs 20¢ per ounce and contains 16 units of vitamin A, 2 units of B, and 4 of C; and additive Y, which costs 40¢ per ounce and contains 4 units of vitamin A, 2 units of B, and 14 of C. If the total amount of additives is not to exceed 10 oz, how many ounces of each additive should the company put into each bottle to minimize its cost?

14. The Write-Right Paper Company operates two factories that manufacture three different grades of paper. There is a demand for each grade, and the company has contracts to supply 16 tons of low-grade, 5 tons of medium-grade, and 20 tons of high-grade paper, all in not more than 8 working days. It costs $1000 per day to operate the first factory and $2000 per day to operate the second factory. In one day's operation, factory number 1 produces 8 tons of low-grade, 1 ton of medium-grade, and 2 tons of high-grade paper, while factory number 2 produces 2 tons of low-grade, 1 ton of medium-grade, and 7 tons of high-grade paper. For how many days should Write-Right operate each factory in order to minimize its cost of filling these contracts?

15. Two oil refineries produce three grades of gasoline, A, B, and C. The refineries operate so that the various grades they produce are in a fixed proportion. Refinery I produces 1 unit of A, 3 units of B, and 1 unit of C per batch, and refinery II produces 1 unit of A, 4 units of B, and 5 units of C per batch. The price per batch is $300 from refinery I and $500 from refinery II. A dealer needs 100 units of A, 340 units of B, and 150 units of C. If the maximum number of batches he can get from either refinery is 100, how should he place his orders to minimize his cost?

16. A local television station is faced with a problem. It found that program A with 20 min of music and 1 min of commercials draws 30,000 viewers, while program B with 10 min of music and 1 min of commercials draws 10,000 viewers. The sponsor insists that at least 6 min per week be devoted to his commercials, and the station can afford no more than 80 min of music per week. How many times per week should each program be run to obtain the maximum number of viewers?

17. A fruit dealer ships her fruit north on a truck that holds 800 boxes of fruit. She must ship at least 200 boxes of oranges, which net her 20¢

profit per box, at least 100 boxes of grapefruit, which net her 10¢ profit per box, and at most 200 boxes of tangerines, which net her 30¢ profit per box. How should she load the truck for maximum profit? [*Hint:* If she ships x boxes of oranges and y boxes of grapefruit, then she ships $800 - x - y$ boxes of tangerines.]

18. Mr. Jones has a maximum of $15,000 to invest in two types of bonds. Bond A returns 8% and bond B returns 10% per year. Because bond B is not as safe as bond A, Jones decides that his investment in bond B will not exceed 40% of his investment in bond A by more than $1000. How much should he invest at each rate to obtain the maximum number of dollars in interest per year?

19. Growfast Nursery is adding imported fruit trees and oriental shrubs to its existing line of nursery products. The trees yield a profit of $6 each, and the shrubs a profit of $7 each. The trees require 2 ft² of display space per tree, and the shrubs require 3 ft² per shrub. In addition, it takes 2 min to prepare a tree for display, and 1 min to prepare a shrub. The space and time constraints are as follows:

a. At most 12 ft² of display space is available.

b. At most 8 min of preparation time is available.

If Growfast can sell all the trees and shrubs it displays, how many trees and how many shrubs should Growfast display each day to maximize its profit? (Assume that it is possible to arrange a display only once per day.)

20. The Excelsior Mining Company operates two mines, EMC no. 1 and EMC no. 2. Mine EMC no. 1 produces 20 tons of lead ore and 30 tons of low-grade silver ore per day of operation. EMC no. 2 produces 15 tons of lead ore and 35 tons of low-grade silver ore per day of operation. Lead ore sells for $14 per ton and low-grade silver ore sells for $34 per ton. The company can sell at most 630 tons of the low-grade silver ore per month, but it can sell all the lead ore it produces. However, there is no space available for stockpiling any silver ore. The company employs one crew and operates only one of the mines at a time. Furthermore, union regulations stipulate that the crew not be worked in excess of 20 days per month. How many days per month should Excelsior schedule for each mine so that the income from the sale of the ore is a maximum?

21. The ABC Fruit Juice Company wants to make an orange–grapefruit drink and is concerned with the vitamin content. The company plans to use orange juice that has 2 units of vitamin A, 3 units of vitamin C, and 1 unit of vitamin D per ounce, and grapefruit juice that has 3 units of vitamin A, 2 units of vitamin C, and 1 unit of vitamin D per ounce. Each can of the orange–grapefruit drink is to contain not more than 15 oz and is to have at least 26 units of vitamin A, 30 units of vitamin C, and 12 units of vitamin D. If the per ounce cost of the orange juice is 4¢ and of the grapefruit juice is 3¢:

a. How many ounces of each should be put into a can if the total cost of the can is as low as possible?

b. What is the minimum cost per can?

c. What is the vitamin content per can?

22. Joey likes a mixture of Grape-nuts, Product 19, and Raisin Bran for his breakfast. Here is some information about these cereals (each quantity in the table is per ounce):

	GRAPE-NUTS	PRODUCT 19	RAISIN BRAN
Calories	100	110	90
Fat	1 g	0	1 g
Sodium	195 mg	325 mg	170 mg

Joey is on a low-sodium diet and tries to make up 12 oz of the mixture so that the number of calories is at least 1200 but not over 1500, the total amount of fat is not more than 10 g, and the sodium content is minimized. Can he do it? If so, what are the per ounce quantities of calories, fat, and sodium in his mixture? [*Hint:* If he uses x oz of Grape-nuts and y oz of Product 19, then he must use $12 - x - y$ oz of Raisin Bran.]

DISCOVERY 8.7

1. Suppose there is a championship prize fight. Gary the Gambler is tired of losing money and wants to hedge his bets so as to win at least $100 on the fight. Gary finds two gambling establishments: *A*, where the odds are 5 to 3 in favor of the champion, and *B*, where the odds are 2 to 1 in favor of the champion. What is the least total amount of money that Gary can bet, and how should he place it to be sure of winning at least $100? You can do this as a linear programming problem. First, let x dollars be placed on the champion to win with *A*, and y dollars on the challenger to win with *B*. Then, verify the following:

a. If the champion wins, Gary wins $\frac{3}{5}x$ dollars and loses y dollars, for a net gain of $(\frac{3}{5}x - y)$ dollars.

b. If the challenger wins, then Gary loses x dollars and wins 2y dollars, for a net gain of $(-x + 2y)$ dollars.

c. The problem now is to minimize $x + y$ subject to the constraints

$$\frac{3}{5}x - \ y \geq 100$$
$$-x + 2y \geq 100$$
$$x \geq 0$$
$$y \geq 0$$

Solve this problem.

2. Show that there is no feasible region if Gary reverses his bets and places x dollars on the challenger to win with A, and y dollars on the champion to win with B.

SUMMARY

Section	Item	Meaning	Example
8.1	Relation	A set of ordered pairs	$S = \{(3, 2), (5, 2), (7, 4)\}$
8.1A	Domain	The set of all possible x values of a relation	The domain of S is $\{3, 5, 7\}$.
8.1A	Range	The set of all possible y values of a relation	The range of S is $\{2, 4\}$.
8.1B	Function	A relation such that to each domain value there corresponds exactly one range value	$\{(x, y)\|y = 2x\}$
8.1C	$f(x)$	Function notation	$f(x) = 3x + 2$
8.2A	Graph of a relation	The set of points corresponding to the ordered pairs of a relation	
8.2C	$[\![x]\!]$	The greatest integer $\leq x$	$[\![2.59]\!] = 2$; $[\![-2.59]\!] = -3$
8.3A	x intercept	x coordinate of the point where the line crosses the x axis $(y = 0)$	The x intercept of $y = 2x - 4$ is 2.
8.3A	y intercept	y coordinate of the point where the line crosses the y axis $(x = 0)$	The y intercept of $y = 2x - 4$ is -4.
8.3B	Linear equation	An equation that can be written in the form $ax + by = c$	$3x + 5y = -2$; $3y = 2x - 1$
8.3D	Distance formula	$d = \sqrt{(x_2 - x_1)^2 + (y_2 - y_1)^2}$	The distance between $(3, 5)$ and $(5, 12)$ is $\sqrt{53}$.
8.4A	Slope of a line	$m = \dfrac{y_2 - y_1}{x_2 - x_1}$	The slope of the line through $(3, 5)$ and $(5, 12)$ is $\frac{7}{2}$.
8.4B	Point–slope equation	$y - y_1 = m(x - x_1)$	$y - 5 = \frac{7}{2}(x - 3)$ is the point–slope equation of the line described above.
8.4B	Slope–intercept equation	$y = mx + b$ (m is the slope, b is the y intercept)	
8.4B	General equation of a line	$Ax + By = C$	
8.4C	Parallel lines	Two lines with the same slope and different y intercepts	$y = 2x + 5$ and $y = 2x - 3$ are equations of parallel lines.
8.7	Convex polygon	A nonreentrant polygon	

1. Find the domain and range of the relation

   $R = \{(5, 3), (3, -1), (2, 2), (0, 4)\}$

2. Find the domain and range of the relation

   $R = \{(x, y)|y = -3x\}$

3. Find the domain and range of the relation

   $R = \{(x, y)|y \geq x, x \text{ and } y \text{ are positive integers less than 5}\}$

4. Which of the following relations are functions?
   a. $\{(x, y)|y^2 = x\}$ **b.** $\{(3, 1), (4, 1), (6, 1)\}$
   c. $\{(x, y)|y = x^2\}$

5. A function is defined by $f(x) = x^2 - x$. Find:
   a. $f(0)$ **b.** $f(1)$ **c.** $f(-2)$

6. For a car renting for $15 per day plus 10¢ per mile, the cost for 1 day is

   $C(m) = 15 + 0.10m$ dollars

   where m is the number of miles driven. If a person paid $35.30 for 1 day's rental, how far did the person drive?

7. Graph the relation

   $R = \{(x, y)|y = 3x, x \text{ is an integer between } -1 \text{ and } 3, \text{ inclusive}\}$

8. Graph the relation

   $Q = \{(x, y)|x + y < 3, x \text{ and } y \text{ are nonnegative integers}\}$

9. Graph the function defined by $g(x) = 2x^2 - 1$, x is an integer and $-2 \leq x \leq 2$.

10. Graph the function defined by $f(x) = 2x - 6$.

11. Graph the equation $3x - 2y = 5$.

12. Find the distance between the two given points:
    a. $(4, 7)$ and $(7, 3)$ **b.** $(-3, 8)$ and $(-3, -2)$

13. Determine whether the triangle with vertices at $A(2, 0)$, $B(4, 4)$, and $C(1, 3)$ is a right triangle.

14. **a.** Find the slope of the line that goes through the two points $(-1, -3)$ and $(9, -2)$.
    b. Find the general equation of the line in part a.

15. **a.** Find the slope–intercept form of the equation of the line that goes through the point $(3, -1)$ and has slope of -2.
    b. Find the slope–intercept form of the equation $2y = 4 - 8x$. What is the slope and what is the y intercept of the line?

16. Determine whether the two given lines are parallel. If they are not parallel, find the coordinates of the point of intersection.
    a. $2x + y = 1$; $12x + 3y = 4$ **b.** $y = 2x - 5$; $4x - 2y = 7$

17. Find the general equation of the line that passes through the point $(1, -2)$ and is parallel to the line $2x - 3y = -5$.

18. Find the point of intersection of the lines

$$3x + 2y = 9 \quad \text{and} \quad 2x - 3y = 19$$

19. Graph the solution set of the inequality $4x - 3y \leq 12$.

20. Graph the solution set of the system of inequalities

$$x + 3y \leq 6 \quad \text{and} \quad x - y \geq 2$$

21. Graph the solution set of the system of inequalities

$$x + 2y \leq 3$$
$$x \leq y$$
$$x \geq 0$$

22. Solve the following system if possible. If not possible, explain why.

$$y = 2x - 3$$
$$6x - 3y = 9$$

23. Find the maximum value of $C = x + 2y$ subject to the constraints

$$3x + y \leq 8$$
$$x \leq 1$$
$$y \geq 2$$
$$x \geq 0$$

24. Find the minimum value (if possible) of $P = 3y - 2x$ subject to the constraints

$$y - x \leq 2$$
$$x + y \leq 4$$
$$x \leq 3$$
$$x \geq 0$$
$$y \geq 0$$

25. Two machines produce the same item. Machine A can produce 10 items per hour and machine B can produce 12 items per hour. At least 420 of the items must be produced each 40-hr week, but the machines cannot be operated at the same time. If it costs $20/hr to operate A and $25/hr to operate B, determine how many hours per week to operate each machine in order to meet the production requirement at minimum machine cost.

CHAPTER 9

GEOMETRY

One of the most famous mathematicians of all time is Euclid, who taught in about 300 B.C. at the university in Alexandria, the main Egyptian seaport. Unfortunately, very little is known about Euclid personally; even his birthdate and birthplace are unknown. However, two stories about him seem to have survived. One concerns the Emperor Ptolemy, who asked if there was no easy way to learn geometry and received Euclid's reply, "There is no royal road to geometry." The other story is about a student who studied geometry under Euclid and, when he had mastered the first theorem, asked, "But what shall I get by learning these things?" Euclid called a slave and said, "Give him a penny, since he must make gain from what he learns."

Geometry evolved from the more or less rudimentary ideas of the ancient Egyptians (about 1500 B.C.), who were concerned with practical problems involving measurement of areas and volumes. The Egyptians were satisfied with the geometry that was needed to construct buildings and pyramids; they cared little about mathematical derivations or proofs of formulas.

The geometry of the Greeks, which is said to have begun with Thales about 600 B.C., was very different from that of the Egyptians. The Greek geometers tried to apply the principles of Greek logic to the study of geometry and to prove theorems by a sequence of logical steps that proceeded from certain basic assumptions to a conclusion.

Euclid's greatest contribution was his collection and systematization of most of the Greek mathematics of his time. His reputation rests mainly on his work titled *Elements,* which contains geometry, number theory, and some algebra. Most American textbooks on plane and solid geometry contain essentially the material in the geometry portions of Euclid's *Elements.* No work, except the *Bible,* has been so widely used or studied, and probably no work has influenced scientific thinking more than this one. Over a thousand editions of *Elements* have been published since the first one appeared in 1482, and for more than 2000 years, this work has dominated the teaching of geometry.

EUCLID
The Bettmann Archive

Euclid alone has looked on Beauty bare.

EDNA ST. VINCENT MILLAY

POINTS, LINES, AND PLANES

The basic elements of Euclidean geometry are **points, lines,** and **planes.** These three words cannot be precisely defined, because this would require other words, which are also undefined. For example, we can say that a line is a set of points. But, what is a point? A point is that which has no dimension. But, again, what is dimension? However, since all other geometric terms are defined on the basis of these three words, we shall try to give you an idea of their meaning.

A. Points and Lines

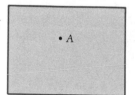

Figure 9.1a

A **point** may be regarded as a location in space. A point has no breadth, no width, and no length. We can picture a point as a small dot, such as A in Fig. 9.1a. (The sharper the pencil, the better the picture.)

A **line** is a set of points, and each point of the set is said to be on the line. You may think of a line as the path of a point moving in a fixed direction in space. A line extends without end in both its directions. Here are some properties that may help to clarify what is meant by the word *line*.

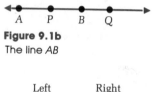

Figure 9.1b
The line *AB*

1. Two distinct points, A and B, determine a line AB (Fig. 9.1b). In other words, **one and only one line can be drawn through the two points.** If P and Q are points on the line AB, then the line PQ is the same as the line AB. Thus, a line may be named by any two of its points. We designate the line AB by the symbol $\overleftrightarrow{AB}$ (read, "the line AB"). Points on the same line are said to be **collinear.** If A, B, P, and Q are collinear, then $\overleftrightarrow{AB} = \overleftrightarrow{PQ}$.

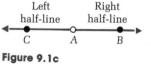

Figure 9.1c

2. **Any point A on a line separates the line into three sets: the point A itself and two half-lines, one on each side of A.** The half-lines do not include the point A, although A is regarded as an endpoint of both (see Fig. 9.1c). The two half-lines may be termed half-line AB and half-line AC, respectively, and are designated by $\overset{\circ}{\overrightarrow{AB}}$ and $\overset{\circ}{\overrightarrow{AC}}$. The open circle at the end of the arrow in the symbol $\overset{\circ}{\overrightarrow{AB}}$ indicates that the half-line does *not* include the point A.

It is sometimes convenient to consider the set of points consisting of a half-line and its endpoint. Such a set is called a **ray.** The ray consisting of $\overset{\circ}{\overrightarrow{AB}}$ and the point A will be designated by $\overrightarrow{AB}$. The ray consisting of the half-line $\overset{\circ}{\overrightarrow{BA}}$ and the point B will be denoted by $\overrightarrow{BA}$. Note that $\overrightarrow{AB} \neq \overrightarrow{BA}$ because rays are named using the endpoint first.

A **line segment AB** consists of the points A and B and that portion of the line AB that lies between A and B. We designate this segment by $\overline{AB}$. Note that $\overline{AB} = \overline{BA}$.

Figure 9.1d shows the figures and notations we have described.

Figure 9.1d

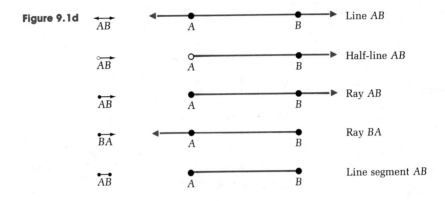

$\overleftrightarrow{AB}$ — Line AB

$\overset{\circ\;\longrightarrow}{AB}$ — Half-line AB

$\overrightarrow{AB}$ — Ray AB

$\overrightarrow{BA}$ — Ray BA

$\overline{AB}$ — Line segment AB

EXAMPLE 1 Refer to the figure in the margin, and state what each of the following describes:

a. $\overrightarrow{AC} \cap \overrightarrow{CA}$ **b.** $\overline{AB} \cap \overrightarrow{BD}$ **c.** $\overline{AC} \cup \overrightarrow{BD}$

d. $\overleftrightarrow{AB} \cap \overleftrightarrow{CD}$ **e.** $\overleftrightarrow{BC} \cap \overrightarrow{BA}$ **f.** $\overrightarrow{AB} \cup \overrightarrow{BD}$

Solution   **a.** The figure below shows the two rays $\overrightarrow{AC}$ and $\overrightarrow{CA}$. The set of points they have in common is the segment $\overline{AC}$. Thus, $\overrightarrow{AC} \cap \overrightarrow{CA} = \overline{AC}$.

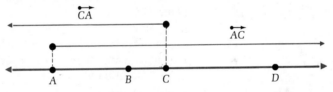

b. You can see from the figure below that the segment $\overline{AB}$ and the ray $\overrightarrow{BD}$ have only the point B in common. Therefore, $\overline{AB} \cap \overrightarrow{BD} = \{B\}$.

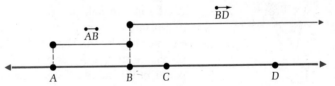

c. The figure below shows that the union of the segment $\overline{AC}$ and the ray $\overrightarrow{BD}$ is the ray $\overrightarrow{AD}$. Thus, $\overline{AC} \cup \overrightarrow{BD} = \overrightarrow{AD}$.

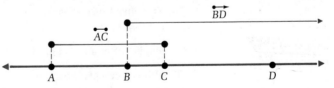

d. The figure below shows that the segments $\overleftrightarrow{AB}$ and $\overleftrightarrow{CD}$ have no points in common. Hence, $\overleftrightarrow{AB} \cap \overleftrightarrow{CD} = \varnothing$.

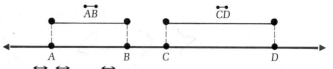

e. Since $\overleftrightarrow{BC}$, $\overleftrightarrow{BA}$, and $\overleftrightarrow{AD}$ are all symbols for the same line, we have

$$\overleftrightarrow{BC} \cap \overleftrightarrow{BA} = \overleftrightarrow{AD}$$

f. As we can see in the lines shown above, the union of the segment $\overleftrightarrow{AB}$ and the segment $\overleftrightarrow{BD}$ is the segment $\overleftrightarrow{AD}$, so we may write

$$\overleftrightarrow{AB} \cup \overleftrightarrow{BD} = \overleftrightarrow{AD} \qquad \blacktriangleleft$$

B. Planes

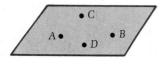

Figure 9.1e

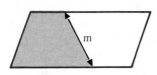

Figure 9.1f
Two half-planes

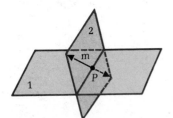

Figure 9.1g
Intersecting planes

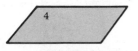

Figure 9.1h
Parallel planes

As with the terms "point" and "line," we give no formal definition of a **plane.** To help you visualize a plane, you can think of the surface of a very large flat floor or of a straight wall extending indefinitely in all directions. Here are some basic properties of planes:

1. **Any three noncollinear (not on the same line) points determine one and only one plane.** We often name planes by giving three noncollinear points in the plane—for example, ABC in Fig. 9.1e. A set of points all in the same plane are called **coplanar.** In Fig. 9.1e, the point D is in the plane ABC, so the points A, B, C, and D are coplanar.

2. **Every pair of distinct points in a given plane determines a line that lies entirely in the plane.** For example, if a line were drawn through the points A and B in Fig. 9.1e, every point of the line would lie in the plane ABC. (A builder uses this property to test the flatness of a floor or a wall by placing a straightedge on the surface and determining whether every point of the straightedge touches the surface.)

3. **Any line m in a plane separates the plane into three parts: the line m itself and two half-planes.** The points of m do not belong to either half-plane, although the line is often called the **edge** of both half-planes. As indicated by the shading in Fig. 9.1f, we regard the points of the plane that are on one side of m as forming one of the half-planes and the points of the plane on the other side of m as forming the other half-plane.

4. **Two distinct planes either have a line in common or else have no point in common at all.** In other words, two distinct planes either intersect in a line or do not intersect at all. See Figs. 9.1g and 9.1h. Two planes that have no common point are **parallel.**

   If a line is not in a given plane, then there are two possibilities:

   a. The line **intersects** the plane in exactly one point. For instance, line m in plane 2 of Fig. 9.1g intersects plane 1 in the point P.

b. The line is **parallel** to the plane. In Fig. 9.1h any line such as n in plane 3 will be parallel to plane 4.

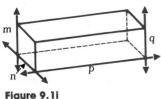

m

q

n

p

Figure 9.1i
A rectangular box

If two distinct lines in space are given, then there may or may not be a plane that contains both lines. If the lines are parallel or if they intersect, then there is exactly one plane that contains both lines. If the lines are neither parallel nor intersecting, so that no plane can contain both lines, then they are called **skew** lines. A simple example of skew lines is the line of intersection of the ceiling and the front wall of an ordinary rectangular classroom, and the line of intersection of the floor and one of the side walls. Figure 9.1i shows a rectangular box. The edges determine various straight lines. For instance, lines m and n intersect at a vertex (corner) of the box; m and p are skew lines; m and q are parallel lines.

EXERCISE 9.1

A. In Problems 1 and 2, draw a line or a portion of a line that corresponds to the given symbol.

1. **a.** $\overrightarrow{PQ}$ **b.** $\overrightarrow{QP}$ **c.** $\overleftrightarrow{QP}$
2. **a.** $\overrightarrow{PQ}$ **b.** $\overrightarrow{QP}$ **c.** $\overleftrightarrow{PQ}$

A   B   C   D

In Problems 3–14, use the figure in the margin and determine what each union or intersection describes.

3. $\overleftrightarrow{AB} \cap \overleftrightarrow{BC}$ 4. $\overleftrightarrow{AC} \cap \overleftrightarrow{BC}$ 5. $\overleftrightarrow{AC} \cup \overleftrightarrow{BC}$
6. $\overrightarrow{AD} \cup \overrightarrow{CB}$ 7. $\overleftrightarrow{AC} \cap \overrightarrow{DA}$ 8. $\overrightarrow{BD} \cap \overrightarrow{DC}$
9. $\overleftrightarrow{AC} \cup \overrightarrow{DC}$ 10. $\overleftrightarrow{AC} \cup \overrightarrow{DB}$ 11. $\overrightarrow{BA} \cap \overleftrightarrow{CD}$
12. $\overleftrightarrow{CB} \cap \overleftrightarrow{CD}$ 13. $\overleftrightarrow{AC} \cap \overleftrightarrow{DC}$ 14. $\overleftrightarrow{AB} \cap \overleftrightarrow{DB}$

15. Let A, B, C, and D be four points, no three of which are collinear. How many straight lines do these points determine? Name each of these lines.
16. Suppose that A, B, C, D, and E are five points, no three of which are collinear. How many straight lines do these points determine? Name each of these lines.
17. If the four points in Problem 15 are not coplanar (do not lie in the same plane), how many planes do the four points determine? Name each of these planes.
18. If no four of the five points in Problem 16 are coplanar, how many planes do the five points determine? Name each of these planes.
19. The figure in the margin represents a triangular pyramid.
    a. Name all the edges.
    b. State which pairs of edges determine skew lines.
    c. Do any of the edges determine parallel lines?

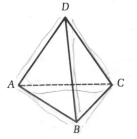

D

A   C

B

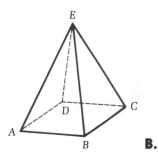

20. The figure represents a pyramid with a square base *ABCD*.
  a. Name all the edges of the pyramid.
  b. Which pairs of edges determine parallel lines?
  c. Which determine skew lines?
  d. Which lines are intersecting?

B. *Determine whether each of the following statements seems to be true or false in Euclidean geometry. You will find it helpful to use pencils, pieces of cardboard, walls, floors, and so on, to represent lines and planes.*

21. Given any plane *ABC* and any point *P* not on *ABC*, there is exactly one plane that contains *P* and is parallel to plane *ABC*.

22. Given any line *m* and any point *P* not on *m*, there is exactly one plane that contains *P* and is parallel to *m*.

23. Given any line *m* and any point *P* not on *m*, there is exactly one line that contains *P* and is parallel to *m*.

24. Given any plane *ABC* and any point *P* not on *ABC*, there are any number of lines containing *P* and parallel to *ABC*.

25. Given any line *m* and any point *P* not on *m*, there are any number of lines that contain *P* and are skew to *m*.

26. Given any plane *ABC* and any line *m* parallel to *ABC*, there is exactly one plane that contains *m* and intersects *ABC*.

27. Two nonparallel lines always determine a plane.

28. Given any line *m* and any point *P* not on *m*, there is exactly one plane that contains both *m* and *P*.

29. Given a plane *ABC* and a line *m* that intersects *ABC*, there is a plane that contains *m* and that does not intersect *ABC*.

30. If a plane intersects two parallel planes, the lines of intersection are parallel.

Problems 31–38 refer to the figure in the margin, which represents a triangular pyramid ABCD with its base on the plane ABC.

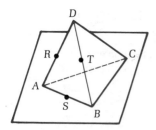

31. Are the points *A*, *R*, and *D* collinear or noncollinear?

32. Are the points *R*, *S*, and *T* collinear or noncollinear?

33. Are the points *B*, *C*, *D*, and *T* coplanar or noncoplanar?

34. Are the points *A*, *C*, *D*, and *T* coplanar or noncoplanar?

35. Are lines *AB* and *CD* skew lines?

36. Name a line that is skew to the line *AD*.

37. Name the line that is coplanar with the lines *BC* and *CD*.

38. If a line is drawn through the points *R* and *T*, what three lines would be skew to the line *RT*?

USING YOUR KNOWLEDGE 9.1

Suppose you have a large number of points, with no three of them being collinear. Can you calculate how many distinct lines these points determine? Let us say there are 10 points in all. Then you can select any one of

A striking example of modern geometric design is the atrium of the Hyatt Regency Hotel, San Francisco.

Courtesy of Hyatt Regency, San Francisco

them as a first point, which means there are 10 such choices. Having made a choice, you can select a second point from the remaining 9 points, so that there are 9 choices possible for the second point. Since for each of the 10 first choices, there are 9 second choices, there are 10×9, or 90, ways of choosing a first point and then a second point. Will there be 90 distinct lines? No, because if you select a point A for the first point and a point B for the second point, the line is the same line as if you had chosen B first and A second. This means that of the 90 lines, half are duplicates of the other half. Thus, the number of distinct lines is $(10 \times 9)/2$, or 45.

Use this idea to answer the following questions. Assume in each case that no three of the points are collinear.

1. If the number of points is 20, how many distinct lines are determined?
2. If the number of points is 50, how many distinct lines are determined?
3. If the number of points is n, how many distinct lines are determined?

Now suppose that you have a large number of points with no four of them being coplanar. Can you calculate how many distinct planes these points determine? Try it for 10 points. As you just saw, 10 points determine $(10 \times 9)/2$, or 45, distinct lines. It takes 2 of the 10 points to determine one of these lines, and you can use any one of the remaining 8 points to pair with the line to determine a plane. Thus, for each of the 45 lines, 8 planes are determined; but these planes are not all distinct. You can check that only one-third are distinct by noting that if a line AB and a point C are selected, this gives the same plane as the line AC and the point B or the line BC and the point A. Hence, the number of distinct planes is $(10 \times 9 \times 8)/(2 \times 3)$, or 120.

Use these ideas to answer the following questions. Assume in each case that no four of the points are coplanar.

4. If the number of points is 20, how many distinct planes are determined?
5. If the number of points is n, how many distinct planes are determined?

DISCOVERY 9.1

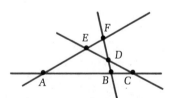

A complete quadrangle

1. The figure in the margin is called a **complete quadrangle.** It is a model for which the following three statements are true:

   i. There are exactly four lines.
   ii. Each pair of lines has exactly one point in common.
   iii. Every point is on exactly two lines.

   Can you discover how to show that any model of lines and points satisfying properties i, ii, and iii also satisfies statements a, b, and c below?

   a. There are exactly six points (A, B, C, D, E, F in the figure).

b. There are exactly three of the points on each line.

c. Corresponding to each point there is exactly one other point not on the same line with it. (For instance, the point D in the figure is the only point not on the same line with A.)

2. Can you write the statements that correspond to statements i, ii, and iii of Problem 1 if the words "line" and "point" are interchanged throughout? Can you discover a simple figure for which the three new statements are true? What do statements a, b, and c become with this interchange of "line" and "point"?

3. Draw a pyramid with a triangular base. Can you discover with what words to replace "line" and "point" in Problem 1 so that the pyramid is an appropriate model? Check to see whether the statements that correspond to statements a, b, and c are true for the pyramid.

4. In Problem 1, replace the word "line" by "club" and the word "point" by "member." If statements i, ii, and iii are true, will statements a, b, and c also be true? Justify your answer.

9.2

ANGLES

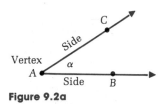

Figure 9.2a

One of the most important concepts in mathematics is that of a *plane angle*. In elementary geometry, we think of a **plane angle** as the figure formed by two rays with a common endpoint, as in Fig. 9.2a. The common endpoint (A in Fig. 9.2a) is called the **vertex** of the angle, and the two rays ($\overrightarrow{AB}$ and $\overrightarrow{AC}$ in the figure) are called the **sides** of the angle. We often use the symbol $\angle$ (read, "angle") in naming angles. The angle in Fig. 9.2a can be named in three ways:

1. By using a letter or a number inside the angle. Thus, we would name the angle in Fig. 9.2a $\angle\alpha$ (read, "angle alpha").

2. By using the vertex letter only, such as $\angle A$ in Fig. 9.2a.

3. By using three letters with the vertex letter in the middle. The angle in Fig. 9.2a would be named $\angle BAC$ or $\angle CAB$.

EXAMPLE 1   Consider the angle in the figure in the margin.

a. Name the angle in three different ways.
b. Name the vertex of the angle.
c. Name the sides of the angle.

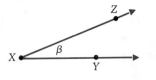

Solution   **a.** The angle can be named $\angle\beta$ (Greek letter beta), $\angle X$, or $\angle YXZ$ (or $\angle ZXY$).

b. The vertex is the point X.

c. The sides are the rays $\overrightarrow{XZ}$ and $\overrightarrow{XY}$.   ◄

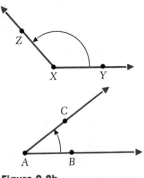

Figure 9.2b

For most practical purposes, we need to have a way of measuring angles. We first consider the amount of **rotation** needed to turn one side of an angle so that it *coincides with* (falls exactly on top of) the other side. Figure 9.2b shows two angles, $\angle CAB$ and $\angle ZXY$, with curved arrows to indicate the rotation needed to turn the rays $\overrightarrow{AB}$ and $\overrightarrow{XY}$ so that they coincide with the rays $\overrightarrow{AC}$ and $\overrightarrow{XZ}$, respectively. Clearly, the amount needed for $\angle ZXY$ is greater than that for $\angle CAB$. To find how much greater, we have to measure the amounts of rotation, and we look at this idea next.

The most common unit of measure for an angle is the *degree*. We can trace the degree system back to the ancient Babylonians, who were responsible for the base 60 system of numeration. The Babylonians considered a *complete revolution* of a ray as indicated in Fig. 9.2c, and divided that into 360 equal parts. Each part is **one degree,** denoted by **1°**. Thus, a complete revolution is equal to 360°. One-half of a complete revolution is 180° and gives us an angle that is called a **straight angle** (see Fig. 9.2d). One-quarter of a complete revolution is 90°, giving a **right angle** (Fig. 9.2e). Notice the small square at Y to denote that this is a right angle.

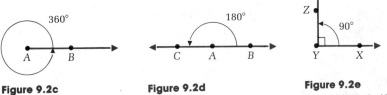

Figure 9.2c
A complete revolution

Figure 9.2d
The straight angle *CAB*

Figure 9.2e
The right angle *XYZ*

The next example, which involves some familiar notions, should help to clarify the preceding ideas.

EXAMPLE 2   Through how many degrees does the hour hand of a clock move in going from:

a. 1 to 2 o'clock?   **b.** 1 to 4 o'clock?

c. 12 to 5 o'clock?   **d.** 12 to 9 o'clock?

Solution   **a.** One complete revolution is 360°, and the face of the clock is divided into 12 equal parts. Thus, the hour hand moves through

$$\frac{360°}{12} = 30°$$

in going from 1 to 2 o'clock.

b. From 1 to 4 o'clock is 3 hr. Since a 1-hr move corresponds to 30°, a 3-hr move corresponds to $3(30°) = 90°$. (Thus, the hour hand moves through one right angle.)

c. From 12 to 5 o'clock is 5 hr. Hence, the hour hand moves through $5(30°) = 150°$.

d. From 12 to 9 o'clock is 9 hr, so the hour hand moves through $9(30°) = 270°$. ◄

In practice, the size of an angle is measured with a protractor (see Fig. 9.2f). The protractor is placed with its center at the vertex of the angle and the straight side of the protractor along one side of the angle as in Fig. 9.2g. The measure of ∠BAC is then read as 70° (because it is obviously less than 90°) and the measure of ∠DAC is read as 110°. Surveying and navigational instruments, such as a sextant, use the idea of a protractor to measure angles very precisely.

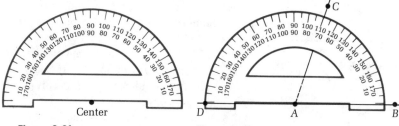

Figure 9.2f A protractor   **Figure 9.2g** Measuring an angle

We have already named two angles: a *straight angle* (180°) and a *right angle* (90°). Certain other angles are classified as follows:

1. An **acute angle** is an angle of measure *greater* than 0° and *less* than 90°.
2. An **obtuse angle** is an angle of measure *greater* than 90° and *less* than 180°.

In Fig. 9.2g, ∠BAC is an acute angle and ∠DAC is an obtuse angle.

Geometric figures frequently appear in highway signs. Do you know what the sign in Fig. 9.2h means? It is an advance warning for a railroad crossing. The angles B and D that are marked in Fig. 9.2i are called *vertical angles*. In general, when two lines intersect, the opposite angles so formed are called **vertical angles.** Two pairs of vertical angles are shown in Fig. 9.2j. As you can see from this figure, the opposite angles are of equal size. Thus, in Fig. 9.2j, the measure of angle A, denoted by m∠A, is the same as that of angle C. This fact is simply written as m∠A = m∠C. Similarly, m∠B = m∠D.

In Fig. 9.2j, angles A and B together form a straight angle, so the sum of their measures must be 180°. For this reason, A and B are called *supplementary angles*. In general, any two angles whose measures add up to 180° are called **supplementary angles.** Other pairs of supplementary angles in Fig. 9.2j are B and C, C and D, and A and D. Figure 9.2j illustrates the obvious fact that *supplements of the same angle are equal.* For example, angles A and C are both supplements of ∠B.

Figure 9.2h

Figure 9.2i

EXAMPLE 3   Refer to Fig. 9.2j.

a. If the measure of ∠A is 25°, what are the measures of the other three angles?
b. If the two lines are to be drawn so that ∠B is twice the size of ∠A, what should the measure of ∠A be?

Figure 9.2j

Solution   **a.** Angles A and B are supplementary, so their measures add to 180°. Hence, the measure of $\angle B$ is 180° minus the measure of $\angle A$, that is, $180° - 25° = 155°$. Since $m\angle D = m\angle B$, the measure of $\angle D$ is also 155°. Also, $m\angle C = m\angle A$, so the measure of $\angle C$ is 25°.

b. We let the measure of angle A be $x°$. Then, the measure of angle B is $2x°$. Because angles A and B are supplementary, we must have

$$x + 2x = 180$$
$$3x = 180$$
$$x = 60$$

Thus, if we make $\angle A$ a 60° angle, $\angle B$ will be a 120° angle, twice the size of $\angle A$.   ◀

Figure 9.2k

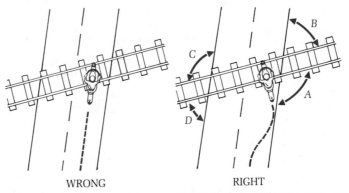

WRONG   RIGHT

If you drive a motorcycle, you should look closely at Fig. 9.2k; it tells you to cross the railroad tracks at right angles (because there is less danger of a wheel catching in the tracks). Now, look at the angles A, B, C, and D that we have marked on the right side of Fig. 9.2k. What can you say about the angles B and D? They are of equal measure, of course (and so are angles A and C). As you can see, the railroad track crosses the two parallel black lines in the figure. In geometry, a line that crosses two or more other lines is called a **transversal.** Thus, each railroad track is a transversal of the pair of parallel black lines. If a transversal crosses a pair of parallel lines, some of the resulting angles are of equal measure. See Fig. 9.2ℓ. The exact relationships are:

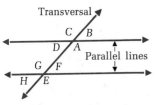

Figure 9.2ℓ

Corresponding Angles Are of Equal Measure:

✓ $m\angle A = m\angle E$ ✓ $m\angle B = m\angle F$

✓ $m\angle C = m\angle G$ ✓ $m\angle D = m\angle H$

Alternate Interior Angles Are of Equal Measure:

✓ $m\angle A = m\angle G$ ✓ $m\angle D = m\angle F$

Alternate Exterior Angles Are of Equal Measure:

✓ $m\angle B = m\angle H$ ✓ $m\angle C = m\angle E$

In Fig. 9.2ℓ, angles A and B form a straight angle and are thus supplementary. Because $m \angle B = m \angle F$, it follows that angles A and F are also supplementary. The same idea applies to angles D and G as well as to angles B and E and angles C and H. We can summarize these facts by saying: **Interior angles on the same side of the transversal are supplementary, and exterior angles on the same side of the transversal are also supplementary.**

The next example will help to clarify and fix this information in your mind.

EXAMPLE 4   In the figure below, find the measure of the angle:

a. Y   **b.** Z   **c.** X   **d.** R   **e.** S   **f.** T   **g.** U

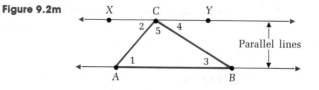

Solution   **a.** Since Y and A are vertical angles, $m \angle Y = m \angle A = 50°$.
b. Angles A and Z are supplementary, so $m \angle Z = 180° - 50° = 130°$.
c. X and Z are vertical angles. Thus, $m \angle X = m \angle Z = 130°$.
d. R and A are alternate exterior angles. Hence, $m \angle R = m \angle A = 50°$.
e. S and Y are interior angles on the same side of the transversal and so are supplementary. Therefore, $m \angle S = 180° - m \angle Y = 130°$.
f. T and A are corresponding angles. Thus, $m \angle T = m \angle A = 50°$.
g. U and A are exterior angles on the same side of the transversal. Thus, $m \angle U = 180° - m \angle A = 130°$.   ◀

Parallel lines and the associated angles allow us to obtain quite easily one of the most important results in the geometry of triangles. (Of course, you know what a triangle is and we give no formal definition here.) In Fig. 9.2m, ABC represents any triangle. The line XY has been drawn through the point C parallel to the side AB of the triangle. Note that $\angle 1 = \angle 2$ and $\angle 3 = \angle 4$, because they are respective pairs of alternate interior angles.

Figure 9.2m

Furthermore, angles 2, 5, and 4 form a straight angle, so

$$m \angle 2 + m \angle 5 + m \angle 4 = 180°$$

By substituting $\angle 1$ for $\angle 2$ and $\angle 3$ for $\angle 4$, we obtain

$$m\angle 1 + m\angle 5 + m\angle 3 = 180°$$

Thus, we have shown that **the sum of the measures of the angles of any triangle is 180°.**

EXAMPLE 5
a. In a triangle ABC, $m\angle A = 47°$ and $m\angle B = 59°$. Find the measure of $\angle C$.

b. Is it possible for a triangle ABC to be such that $\angle A$ is twice the size of $\angle B$, and $\angle C$ is three times the size of $\angle B$?

Solution
a. Because $m\angle A + m\angle B + m\angle C = 180°$, we have

$$47° + 59° + m\angle C = 180°$$
$$m\angle C = 180° - 47° - 59°$$
$$= 180° - 106° = 74°$$

b. To answer this question, let $m\angle B = x°$, so $m\angle A = 2x°$ and $m\angle C = 3x°$. Then, since the sum of the angles is 180°,

$$x + 2x + 3x = 180$$
$$6x = 180$$
$$x = 30, \quad 2x = 60, \quad 3x = 90$$

This means that there is such a triangle, and $m\angle A = 60°$, $m\angle B = 30°$, and $m\angle C = 90°$. (Note that this is a right triangle because one of the angles is a right angle.) ◄

In part b of Example 5, we found angles A and B to be of measure 60° and 30°, respectively, so $m\angle A + m\angle B = 90°$. Two angles whose sum is 90° are called **complementary angles,** and each angle is called the **complement** of the other.

EXAMPLE 6
a. Find the complement of a 38° angle.

b. Can two complementary angles be such that one is three times the size of the other?

Solution
a. For an angle to be the complement of a 38° angle, its measure must be $90° - 38° = 52°$.

b. Let the smaller angle be of measure $x°$, so the larger is of measure $3x°$. Since the angles are to be complementary,

$$x + 3x = 90$$
$$4x = 90$$
$$x = \tfrac{90}{4} = 22\tfrac{1}{2}, \quad 3x = 67\tfrac{1}{2}$$

The answer is yes, and the angles would be of measure $22\tfrac{1}{2}°$ and $67\tfrac{1}{2}°$. ◄

Two lines that intersect at right angles are said to be **perpendicular** to each other. The lines are called **perpendicular lines.** For example, two adjacent outside edges of a page of this book are perpendicular to each other.

Problems 1–14 refer to the following figure:

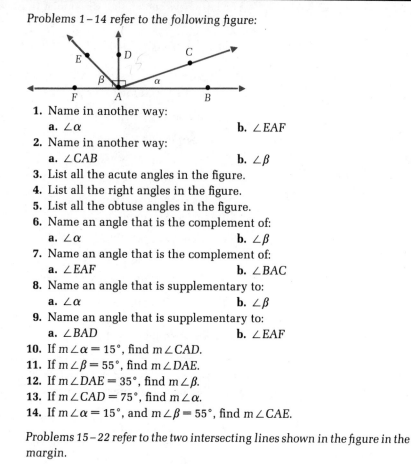

1. Name in another way:

   a. $\angle\alpha$ **b.** $\angle EAF$

2. Name in another way:

   a. $\angle CAB$ **b.** $\angle\beta$

3. List all the acute angles in the figure.

4. List all the right angles in the figure.

5. List all the obtuse angles in the figure.

6. Name an angle that is the complement of:

   a. $\angle\alpha$ **b.** $\angle\beta$

7. Name an angle that is the complement of:

   a. $\angle EAF$ **b.** $\angle BAC$

8. Name an angle that is supplementary to:

   a. $\angle\alpha$ **b.** $\angle\beta$

9. Name an angle that is supplementary to:

   a. $\angle BAD$ **b.** $\angle EAF$

10. If $m\angle\alpha = 15°$, find $m\angle CAD$.

11. If $m\angle\beta = 55°$, find $m\angle DAE$.

12. If $m\angle DAE = 35°$, find $m\angle\beta$.

13. If $m\angle CAD = 75°$, find $m\angle\alpha$.

14. If $m\angle\alpha = 15°$, and $m\angle\beta = 55°$, find $m\angle CAE$.

Problems 15–22 refer to the two intersecting lines shown in the figure in the margin.

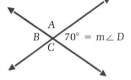

15. Name the angle that is vertical to the 70° angle.

16. Name the two angles that are each supplementary to the 70° angle.

17. Find $m\angle A$.

18. Find $m\angle B$.

19. What is the measure of an angle complementary to the 70° angle?

20. Find the sum of the measures of angles A, B, and C.

21. Find the sum of the measures of angles A and C.

22. If $m\angle D = x°$ (instead of 70°), write an expression for the measure of $\angle A$.

Problems 23–25 refer to the two parallel lines and the transversal shown in the figure in the margin.

23. Find:

   a. $m\angle A$ **b.** $m\angle B$ **c.** $m\angle C$

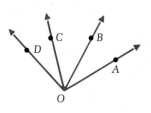

24. Find:
  a. $m\angle D$ **b.** $m\angle E$ **c.** $m\angle F$
25. Name all the angles that are supplementary to $\angle B$.

26. Refer to the angles shown in the figure in the margin.
  a. If $m\angle AOB = 30°$ and $m\angle AOC = 70°$, find $m\angle BOC$.
  b. If $m\angle AOB = m\angle COD$, $m\angle AOD = 100°$, and $m\angle BOC = 2x°$, find $m\angle COD$ in terms of x.

27. If $m\angle A = 41°$, find $m\angle B$ if:
  a. The two angles are complementary
  b. The two angles are supplementary

28. If $m\angle A = 19°$, find $m\angle B$ if:
  a. The two angles are complementary
  b. The two angles are supplementary

29. Given that $m\angle A = (3x + 15)°$, $m\angle B = (2x - 5)°$, and the two angles are complementary, find x.

30. Rework Problem 29 if the two angles are supplementary.

In Problems 31–34, the figures show the number of degrees in each angle in terms of x. Use algebra to find x and the measure of each angle.

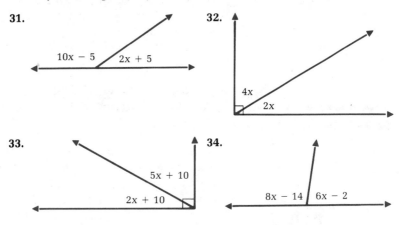

31. $10x - 5$ $2x + 5$

32. $4x$ $2x$

33. $5x + 10$ $2x + 10$

34. $8x - 14$ $6x - 2$

35. Through how many degrees does the hour hand of a clock move in going from:
  a. 11 o'clock to 12 o'clock?   **b.** 11 o'clock to 5 o'clock?

36. Through how many degrees does the hour hand of a clock move in going from:
  a. 12 o'clock to 7 o'clock?   **b.** 12 o'clock back to 12 o'clock?

37. In a triangle ABC, $m\angle A = 37°$ and $m\angle C = 53°$. Find $m\angle B$.

38. In a triangle ABC, $m\angle B = 67°$ and $m\angle C = 105°$. Find $m\angle A$.

39. In a triangle ABC, $m\angle A = (x + 10)°$, $m\angle B = (2x + 10)°$, and $m\angle C = (3x + 10)°$. Find x.

40. In a triangle ABC, $m\angle A$ is 10° less than $m\angle B$, and $m\angle C$ is 40° greater than $m\angle B$. Find the measure of each angle.

Angles are extremely important in surveying and in navigation. With the knowledge you have gained in this section, you should be able to do the following problems.

1. A surveyor measured a triangular plot of ground and reported the three angles as 48.2°, 75.9°, and 56.1°. How much of an error did the surveyor make?

In land surveying, angles are measured with respect to due north and due south. For example, a direction of N 30° W means a direction that is 30° west of due north, and S 60° E means a direction that is 60° east of due south. Surveyors use acute angles only. In aerial navigation, angles are always measured clockwise from due north. A navigator's bearing of 90° corresponds to due east, 180° to due south, and 270° to due west. Thus, a navigator's bearing of 225° would correspond to a surveyor's bearing of S 45° W.

2. Write a navigator's bearing of 135° in the surveyor's terminology.
3. Write a navigator's bearing of 310° in the surveyor's terminology.
4. Write the direction S 40° W as a navigator's bearing.
5. Write the direction N 40° W as a navigator's bearing.

9.3

TRIANGLES AND OTHER POLYGONS

A popular children's puzzle consists of joining in order a set of numbered dots by straight line segments to form a path from the first to the last point. An example of such a puzzle and its solution is shown in Fig. 9.3a. (What is pictured, an antelope or a bird?)

Figure 9.3a

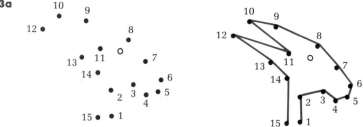

A. Broken Lines and Polygons

A **path** (such as the one shown in the puzzle) consisting of a sequence of connected straight line segments is called a **broken line.** Such a path can

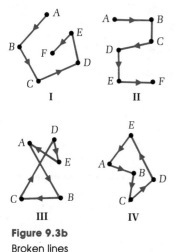

Figure 9.3b
Broken lines

be traced without lifting the pencil from the paper. Certain broken line paths have two characteristics that are of interest to us in this section:

1. A **simple path** is a path that does *not* cross itself.
2. A **closed path** is one that starts and ends at the same point.

The path in the solution of the puzzle in Fig. 9.3a is simple but not closed. Figure 9.3b shows four broken lines. The paths in I, II, and IV are simple, but path III is not simple. Moreover, paths III and IV are closed; paths I and II are not closed.

Any path that consists of a **simple closed broken line** is called a **polygon.** Path IV, *ABCDE*, in Fig. 9.3b is an example of a polygon. The line segments of the path are the **sides** of the polygon, and the endpoints of the sides are the **vertices.** A polygon is said to be **convex** if no line segment *XY* joining any two points on the path ever extends outside the polygon (see Fig. 9.3c). The points *X* and *Y* may be any two points not on the same side of the polygon. Except for its endpoints, the line segment *XY* lies entirely inside the polygon. (Of course, if *X* and *Y* were on the same side, then the segment would lie on that side.)

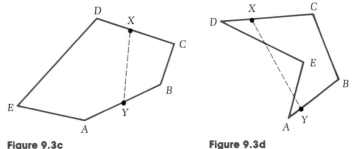

Figure 9.3c
Convex polygon

Figure 9.3d
Concave polygon

A polygon that is not convex is called a **concave,** or **reentrant,** polygon (see Fig. 9.3d). Here, a portion of the line segment *XY* lies outside the polygon—something that never occurs for a convex polygon.

EXAMPLE 1   Consider the polygon *ABCDE* in the margin.
a. Name its sides.   **b.** Name its vertices.
c. Is this a concave or a convex polygon?

Solution   **a.** The sides are *AB*, *BC*, *CD*, *DE*, and *EA*.

b. The vertices are *A*, *B*, *C*, *D*, and *E*.

c. This is a reentrant, or concave, polygon.   ◀

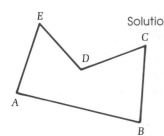

Plane polygons are customarily named according to the number of sides. Here are some of the usual names:

Sides	Name	Sides	Name
3	Triangle	8	Octagon
4	Quadrilateral	9	Nonagon
5	Pentagon	10	Decagon
6	Hexagon	12	Dodecagon
7	Heptagon		

As you probably know, traffic signs are most often in the shape of a polygon. For example, the stop sign is in the shape of an octagon. In fact, it is in the shape of a *regular* octagon. A **regular polygon** is a polygon with all its sides of equal length and all its angles of equal size. A regular triangle has three equal sides and three 60° angles; it is called an **equilateral triangle.** A regular quadrilateral has four equal sides and four 90° angles. You undoubtedly know that it is called a **square.** No special names are given to other regular polygons.

EXAMPLE 2   Some standard traffic signs are shown below.

| School | Stop | Yield | Railroad warning | Warning | Regulatory or information | Railroad crossing | No passing |

a. Which ones are regular polygons?
b. Name the shape of the school sign.
c. Name the shape of the yield sign.
d. Name the shape of the warning sign.

Solution   **a.** The stop sign, the yield sign, and the warning sign are all in the shape of regular polygons.
b. The school sign has five sides, so it is in the shape of a pentagon (but not a regular pentagon).
c. The yield sign has three sides; it is in the shape of an equilateral triangle.
d. The warning sign has four sides; it is in the shape of a regular quadrilateral, a square.   ◀

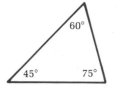

B. Triangles and Quadrilaterals

Triangles are often classified according to their angles:

1. An acute triangle: All three angles are acute, as shown in the margin.

2. A right triangle: One of the angles is a right angle.

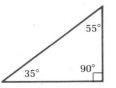

3. An obtuse triangle: One of the angles is an obtuse angle.

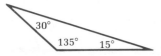

Figure 9.3e
Can you see the triangular region? It is an optical illusion. The triangle exists only in your mind.

Triangles are also classified according to the number of equal sides:

1. A scalene triangle: No equal sides
2. An isosceles triangle: Two equal sides
3. An equilateral triangle: Three equal sides

An equilateral triangle is also **equiangular;** it has three 60° angles. The triangle that you see in the optical illusion in Fig. 9.3e is an equilateral triangle.

EXAMPLE 3   Classify the given triangles according to their angles and their sides.

a.   **b.**

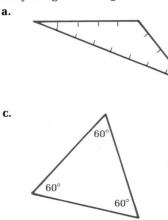

c.

Solution   **a.** The triangle has an obtuse angle and no equal sides; it is an obtuse, scalene triangle.
b. The triangle has two equal sides and a right angle; it is an isosceles right triangle.
c. The triangle has three 60° angles; it is an equilateral triangle, which is also equiangular.   ◄

Certain special quadrilaterals (four-sided polygons) are named as follows:

1. A quadrilateral with two parallel and two nonparallel sides is called a **trapezoid.**
2. A quadrilateral with both pairs of opposite sides parallel is a **parallelogram.**
3. A parallelogram with four equal sides is a **rhombus.**
4. A parallelogram whose angles are right angles is a **rectangle.**
5. A rectangle with four equal sides is a **square.**

All of these polygons are shown in Fig. 9.3f.

Figure 9.3f

Trapezoid   Parallelogram   Rhombus   Rectangle   Square

Many of the applications of geometry involve finding the length of a polygonal path. For instance, fencing a field, laying tile around a rectangular pool, or finding the amount of baseboard needed for a room all involve measuring around polygons. The distance around a plane figure is generally called the **perimeter** of the figure. In the case of a polygon, the perimeter is just the sum of the lengths of the sides. Table 9.3a gives the formulas for the perimeters in terms of the sides for some of the polygons we have discussed.

How many kinds of polygons do you see in this tower?

Table 9.3a

NAME	GEOMETRIC SHAPES	PERIMETER
Triangle		$P = s_1 + s_2 + b$
Trapezoid		$P = s_1 + s_2 + b_1 + b_2$
Parallelogram		$P = 2L + 2W$
Rectangle		$P = 2L + 2W$
Square		$P = 4s$

EXAMPLE 4 The *Mona Lisa* by Leonardo da Vinci was assessed at $100 million for insurance purposes. The picture measures 30.5 by 20.9 inches. Find the length of the frame around this picture.

Courtesy of PSH

Solution Because the picture is rectangular, its perimeter is given by

$$P = 2L + 2W$$

where $L = 30.5$ and $W = 20.9$. Thus,

$$P = 2(30.5) + 2(20.9)$$
$$= 102.8 \text{ in.} \qquad \blacktriangleleft$$

The ideas we have been studying here can be combined with the algebra you know to solve certain kinds of problems. Here is an interesting problem.

EXAMPLE 5 The largest recorded poster was a rectangular greeting card 166 ft long and with a perimeter of 458.50 ft. How wide was this poster?

Solution The perimeter of a rectangle is $P = 2L + 2W$ and we know that $L = 166$ and $P = 458.50$. Thus, we can write

$$2(166) + 2W = 458.50$$
$$332 + 2W = 458.50$$
$$2W = 126.50$$
$$W = 63.25$$

Thus, the poster was 63.25 ft wide. $\qquad \blacktriangleleft$

EXAMPLE 6 John and Emily Gardener want to fence in a small rectangular plot for Emily's kitchen garden. John has 60 ft of fencing and decides that the length of the plot should be one and one-half times its width. What will be the dimensions of Emily's garden plot?

Solution We let x ft be the width of the plot. Then the length must be $\frac{3}{2}x$ ft. We know the perimeter is to be 60 ft, so

$$2L + 2W = P$$
$$2(\tfrac{3}{2}x) + 2x = 60$$
$$3x + 2x = 60$$
$$5x = 60$$
$$x = 12, \quad \tfrac{3}{2}x = 18$$

Emily's plot will be 18 ft long and 12 ft wide. $\qquad \blacktriangleleft$

C. Angles of a Polygon

In Section 9.2, we learned that the sum of the measures of the three angles of any triangle is 180°. This knowledge can be used to find the correspond-

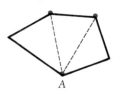

Figure 9.3g

ing sum S for any convex polygon. For example, suppose the polygon is a pentagon as shown in Fig. 9.3g. Select any vertex, say A, and draw lines from A to each nonadjacent vertex, as shown by the dashed lines in the figure. The resulting number of triangles is three (two less than the five sides of the polygon), and the number of degrees in the angles of these three triangles is $3 \times 180° = 540°$.

This result can be generalized to any convex polygon, because if the polygon has n sides, there will be $n - 2$ (instead of three) triangles. The total number of degrees in the angles of these triangles is

$$S = (n - 2)180°$$

Furthermore, if the polygon is a regular polygon, then the angles are all equal. In this case, the measure of a single angle of the polygon is just the preceding result divided by n. This idea is illustrated in the next example.

EXAMPLE 7   Find the measure of an angle of a regular heptagon.

Solution   Since a heptagon has seven sides, the formula gives

$$S = (7 - 2)180° = 5 \times 180° = 900°$$

Because the polygon is regular, each angle has measure

$$\frac{900°}{7} = 128\frac{4}{7}°$$ ◄

EXERCISE 9.3

A.   *In Problems 1 and 2, sketch a broken line path as described.*

1. a. Closed but not simple   **b.** Simple but not closed
2. a. Both simple and closed   **b.** Neither simple nor closed

In Problems 3–7, use the alphabet as printed here.

ABCDEFGHIJKLMNOPQRSTUVWXYZ

3. Which letters form a path that is:
  a. Simple?   **b.** Closed?
4. Which letters form a path that is:
  a. Closed but not simple?   **b.** Simple but not closed?
5. Which letters form a path that is:
  a. Simple and closed?   **b.** Neither simple nor closed?
6. If the lowest points of the legs of the letter M are joined by a straight line segment, will the resulting polygon be concave or convex?
7. If the highest points on the sides of the letter V are joined by a straight line segment, will the resulting polygon be concave or convex?

8. Refer to the traffic signs in Example 2, and name the shape of:

    a. The information sign   **b.** The no passing sign

B.   In Problems 9–16, name the given quadrilaterals.

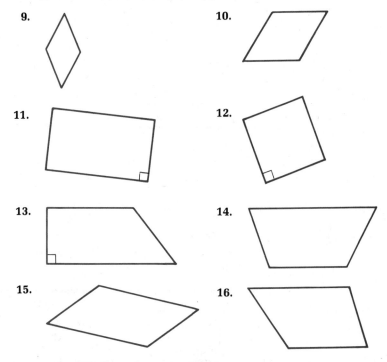

9.

10.

11.

12.

13.

14.

15.

16.

In Problems 17–24, classify the given triangles as scalene, isosceles, or equilateral, and as acute, right, or obtuse.

17.

18.

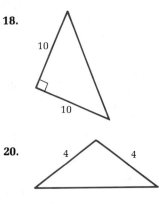

10

10

19.

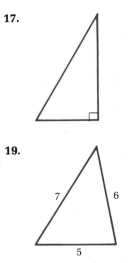

7   6

5

20.

4   4

21.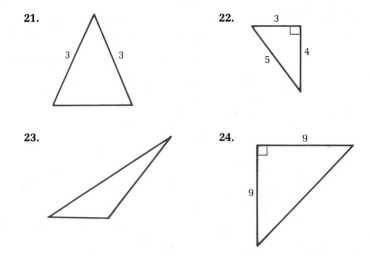

22.

23.

24.

In Problems 25–28, use the following table, which lists all possible triangle classifications. For example, IR stands for an isosceles right triangle, SO stands for a scalene obtuse triangle, and so on. The problems ask you to draw certain types of triangles. In any case where you think no such triangle exists, write "impossible."

	ACUTE	RIGHT	OBTUSE	
SCALENE	SA	SR	SO	
ISOSCELES	IA	IR	IO	
EQUILATERAL	EA	ER	EO	

25. Draw an example of type:
    a. SA   **b.** IA   **c.** EA

26. Draw an example of type:
    a. SR   **b.** IR   **c.** ER

27. Draw an example of type:
    a. SO   **b.** IO   **c.** EO

28. Which of the triangles in the table are impossible?

In Problems 29–36, find the perimeter of the given polygon.

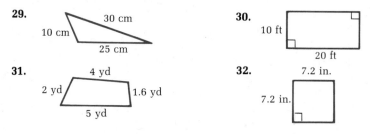

29.

30.

31.

32.

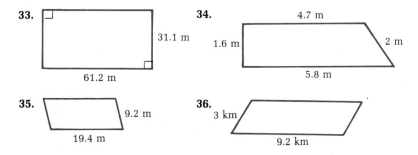

33. 31.1 m · 61.2 m

34. 4.7 m · 1.6 m · 2 m · 5.8 m

35. 9.2 m · 19.4 m

36. 3 km · 9.2 km

37. If one side of a regular pentagon is 6 cm long, find the perimeter of the pentagon.

38. If one side of an octagonal stop sign is 6 in. long, what is the perimeter of the stop sign?

39. The largest rectangular omelet ever cooked was 30 ft long and had an 80-ft perimeter. How wide was it?

40. Do you have a large pool? If you were to walk around the largest pool in the world, in Casablanca, Morocco, you would go more than 1 km. To be exact, you would go 1110 m. If the pool is 480 m long, how wide is it?

41. A baseball diamond is actually a square. A batter who hits a home run must run 360 ft around the bases. What is the distance to first base?

42. The playing surface of a football field is 120 yd long. A player jogging around the perimeter of this surface covers 346 yd. How wide is the playing surface?

43. Have you seen the largest scientific building in the world? It is in Cape Canaveral, Florida. If you were to walk around the perimeter of this building, you would cover 2468 ft. If this rectangular building is 198 ft longer than it is wide, what are its dimensions?

44. The largest regular hexagon that can be cut from a circular sheet of cardboard has each side equal to the radius of the circle. If you cut such a hexagon from a sheet of radius 5 in., how much shorter is the perimeter of the hexagon than the circumference of the circle?

C. *In Problems 45–50, find the measure of one angle of the indicated polygon.*

45. A regular pentagon
46. A regular hexagon
47. A regular octagon
48. A regular nonagon
49. A regular decagon
50. A regular dodecagon

USING YOUR KNOWLEDGE 9.3

In geometry, figures of the same shape and size are called **congruent**, symbolized by ≅. If two figures are congruent, they can be placed one on the other to coincide exactly. In the case of two polygons, congruence means that the corresponding sides of the polygons are equal and the corresponding angles are equal. If the polygons are triangles, there are three useful statements about congruence:

1. **The SSS statement:** If the three sides of one triangle are equal to the corresponding sides of a second triangle, the triangles are congruent.

   This statement says that if $\overrightarrow{AB} = \overrightarrow{DE}$, $\overrightarrow{BC} = \overrightarrow{EF}$, and $\overrightarrow{AC} = \overrightarrow{DF}$, then triangle $ABC \cong$ triangle DEF. (Corresponding parts of the triangles are marked in the following figures.)

2. **The SAS statement:** If two sides and the included angle of one triangle are equal to the corresponding two sides and the included angle of a second triangle, the triangles are congruent.

   This statement says that if $\overrightarrow{AB} = \overrightarrow{DE}$, $\overrightarrow{AC} = \overrightarrow{DF}$, and $m\angle BAC = m\angle EDF$, then triangle $ABC \cong$ triangle DEF.

   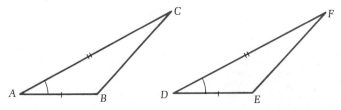

3. **The ASA statement:** If two angles and the included side of one triangle are equal to the corresponding two angles and the included side of a second triangle, the triangles are congruent.

   This statement says that if $m\angle A = m\angle D$, $m\angle B = m\angle E$, and $\overrightarrow{AB} = \overrightarrow{DE}$, then triangle $ABC \cong$ triangle DEF.

   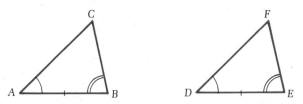

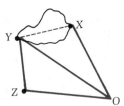

These three statements can supply useful knowledge in surveying and in construction problems. For example, to measure the distance across a swamp (see the figure in the margin), a surveyor was able to sight the points X and Y from a point O, and he could measure the distance from O to X. He then used his transit to lay out $\angle YOZ = \angle YOX$ and laid out $\overrightarrow{OZ} = \overrightarrow{OX}$. Finally, he measured the length of $\overrightarrow{ZY}$, which gave him the distance across the swamp. Why? Since $\overleftrightarrow{OX} = \overleftrightarrow{OZ}$, $\overleftrightarrow{OY} = \overleftrightarrow{OY}$, and $m \angle YOZ = m \angle YOX$, the SAS statement applies and triangle $OXY \cong$ triangle OZY. Therefore, the corresponding sides $\overleftrightarrow{ZY}$ and $\overleftrightarrow{XY}$ are equal.

Try to use these ideas to solve the following problems:

1. A parallelogram $ABCD$ is cut in two along the diagonal AC. Which of the SSS, SAS, ASA statements guarantees that the two triangles so formed are congruent?

2. A carpenter needs a pair of shelf braces. He cuts a wooden rectangle in half along one of the diagonals. Explain why this gives him two congruent right triangles for his braces.

3. Sonya wants to find the distance across a river. She finds a point A on one bank where she can sight a tree T directly across the river on the other bank. (See the diagram in the margin.) She walks 40 yd along the stream in a direction perpendicular to the line $\overleftrightarrow{AT}$, and arrives at a point B, where she can sight the tree T again. Using a surveyor's transit, and with the help of an assistant, she locates a point C on the line $\overleftrightarrow{AT}$ and such that $m \angle ABC = m \angle ABT$. She then finds the length of $\overleftrightarrow{AC}$ to be 67 yd and announces this to be the distance across the river. Justify Sonya's answer.

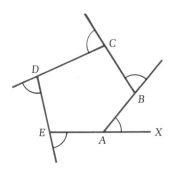

DISCOVERY 9.3

If you extend each side of a convex polygon in one direction, as shown in the figure in the margin, you form a set of exterior angles of the polygon. An **exterior angle** is an angle between the extension of one side and the next side of the polygon. For instance, $\angle XAB$ is one such angle in the figure.

Imagine yourself walking around the perimeter of the polygon. Every time you come to a vertex you must turn through the exterior angle at that vertex to stay on the perimeter of the polygon. When you finish your little trip and arrive back at your starting point, you will have made a complete revolution. Therefore, it follows that the sum of the measures of one set of exterior angles is 360°.

If the polygon is regular, then all the exterior angles are equal. Thus, if we know the measure of one exterior angle, we can discover the number of sides of the polygon. For instance, if the measure of one exterior angle of a regular polygon is 30°, we let n be the number of sides. Then we know that n × 30° = 360°. To discover how many sides the polygon has, we just have to solve the equation

$$30n = 360$$

This gives n = 12, so that the polygon is a dodecagon. Use this idea to find the number of sides of a regular convex polygon, if one exterior angle has measure:

1. 60°   **2.** 24°   **3.** 72°   **4.** 15°

5. It is not possible for an exterior angle of a regular convex polygon to have measure 25°. Can you discover why?

9.4

SIMILAR TRIANGLES

Figure 9.4a

In everyday life, we learn to recognize objects such as buildings and automobiles more on account of their shape than their size. Geometric figures with exactly the same shape, but not necessarily the same size, are called **similar figures.** The six-pointed stars in Fig. 9.4a are similar polygons.

Look at the two similar triangles in Fig. 9.4b. Because they have the same shape, the corresponding angles are equal. In the figure, $m\angle A = m\angle D$, $m\angle B = m\angle E$, and $m\angle C = m\angle F$. The corresponding sides of the two triangles are $\overleftrightarrow{AB}$ and $\overleftrightarrow{DE}$, $\overleftrightarrow{BC}$ and $\overleftrightarrow{EF}$, and $\overleftrightarrow{AC}$ and $\overleftrightarrow{DF}$. Notice that the length of each side of the smaller triangle is one-half the length of the corresponding side of the larger triangle.

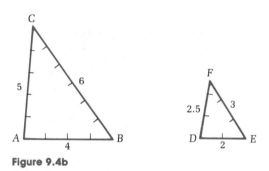

Figure 9.4b

▶   **Definition 9.4a**

> Two triangles are **similar** if and only if they have equal corresponding angles and the corresponding sides are proportional.

Recall that the ratio of two numbers, a and b, is the fraction a/b. For Fig. 9.4b, we have

$$\frac{DE}{AB} = \frac{EF}{BC} = \frac{DF}{AC} = \frac{1}{2}$$

as the ratio of corresponding sides.

EXAMPLE 1   Two similar triangles are shown at the top of the next page. Find d and f for the triangle on the right.

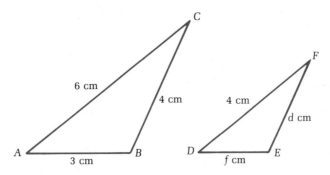

Solution — Since the triangles are similar, the corresponding sides must be proportional. Thus,

$$\frac{DE}{AB} = \frac{EF}{BC} = \frac{DF}{AC}$$

so that

$$\frac{f}{3} = \frac{d}{4} = \frac{4}{6}$$

Since $\frac{4}{6} = \frac{2}{3}$, we have

$$\frac{d}{4} = \frac{2}{3} \quad \text{and} \quad \frac{f}{3} = \frac{2}{3}$$

Solving these equations for d and f, we get

$$d = \frac{8}{3} = 2\frac{2}{3} \quad \text{and} \quad f = \frac{6}{3} = 2 \qquad \blacktriangleleft$$

For similar triangles, we have statements corresponding to the SSS, SAS, and ASA statements for congruent triangles (see Using Your Knowledge 9.3).

1. **The AA statement:** If two angles of a triangle are equal to the corresponding two angles of a second triangle, the triangles are similar.

Note that if there are two pairs of equal angles, then the remaining pair must also be equal. (Remember that the sum of the measures of the angles of any triangle is 180°.) Thus, the two triangles would have the same shape.

2. **The SAS similarity statement:** If an angle of one triangle is equal to an angle of a second triangle, and if the sides including this angle are proportional, then the triangles are similar.

The proof of this statement is given in Discovery 9.4. The usefulness of the statement is illustrated in the next example.

EXAMPLE 2

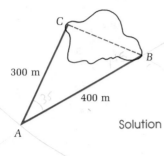

300 m

400 m

A

Wally wanted to find the distance across a lake, so he located a point A from which he could sight points B and C on opposite ends of the lake. See the diagram in the margin. With a surveyor's sextant, Wally found that $m\angle A = 35°$. He was able to measure $\overleftrightarrow{AB}$ and $\overleftrightarrow{AC}$ as 400 m and 300 m, respectively. He took this data to his office and drew a triangle which he labeled $A_1 B_1 C_1$ with $m\angle A_1 = 35°$, $\overleftrightarrow{A_1 B_1} = 40$ cm, and $\overleftrightarrow{A_1 C_1} = 30$ cm. He then measured $\overleftrightarrow{B_1 C_1}$ and found it to be 23.1 cm long. How could Wally use all this information to find the distance across the lake?

Solution

Since Wally made $m\angle A_1 = m\angle A$, and he used the same scale (10 meters per centimeter) for $\overleftrightarrow{A_1 B_1}$ and $\overleftrightarrow{A_1 C_1}$, his triangle $A_1 B_1 C_1$ is similar to triangle ABC. Therefore, if the length of $\overleftrightarrow{BC}$ is x meters, then

$$\frac{400}{40} = \frac{300}{30} = \frac{x}{23.1}$$

so that

$$\frac{x}{23.1} = 10 \quad \text{and} \quad x = 231$$

Thus, Wally found the distance across the lake to be 231 m. ◄

3. **The SSS similarity statement:** If the ratios of the lengths of the three sides of one triangle to the lengths of the corresponding sides of a second triangle are all the same, the triangles are similar.

The proof of this statement is also left for Discovery 9.4.

EXAMPLE 3

Polly, who worked in an architect's office, had to make a scale drawing of a panel in the shape of a parallelogram. The base of the panel was 10 ft long, the adjacent side was 4 ft long, and the longer diagonal was 12 ft long. Polly drew a parallelogram, $ABCD$, with the base $AB = 5$ cm, the adjacent side $BC = 2$ cm, and the longer diagonal $AC = 6$ cm (see the figure). Why is her parallelogram similar to the panel parallelogram?

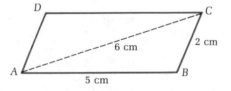

Solution

Let A_1, B_1, C_1, D_1 be the vertices of the panel corresponding to the points A, B, C, D in Polly's drawing. Since Polly used the scale 1 cm per 2 ft for her drawing, the ratios of the corresponding lines are all the same. Thus, triangle $A_1 B_1 C_1$ is similar to triangle ABC, and triangle $A_1 D_1 C_1$ is similar to triangle ADC. Therefore, the corresponding angles of these triangles are equal, so that the corresponding angles of the two parallelograms are equal. Because corresponding sides have the same ratio and corresponding angles are equal, the two parallelograms are similar. ◄

1. Which (if any) of the rectangles below are similar?

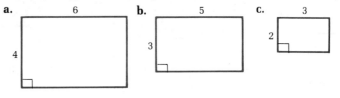

a. 6 b. 5 c. 3

2. The marked angles in the parallelograms below are all equal. Which (if any) of the parallelograms are similar?

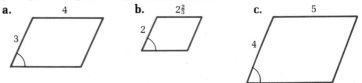

a. 4 b. $2\frac{2}{3}$ c. 5

3. Which (if any) of the following triangles are similar?

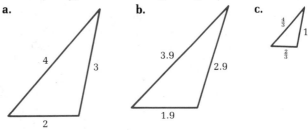

a. b. c.

4. The marked angles in the following triangles are all equal. Which (if any) of the triangles are similar?

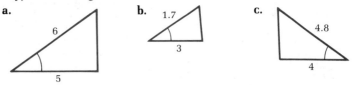

a. b. c.

5. The following triangles are similar. Find the lengths marked x and y.

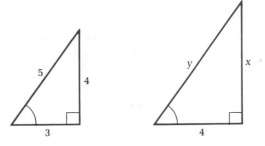

6. The following parallelograms are similar. Find the length of the diagonal PR.

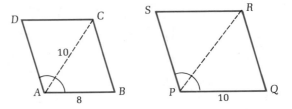

Problems 7–10 refer to the figure in the margin. In this figure, PQ is parallel to AB. In each problem, certain lengths are given. Find the indicated missing length.

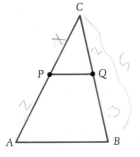

	AP	PC	BQ	BC
7.	3	4	6	?
8.	5	4	?	6
9.	2	?	3	5
10.	?	4	4	8

11. The sides of a triangle measure 6, 9, and 12 cm, respectively. The longest side of a similar triangle measures 7 cm. Find the lengths of the other two sides of the smaller triangle.

12. In Problem 11, if the length of the shortest side of the smaller triangle is 5 cm, find the lengths of the other two sides.

13. The sides of a triangle measure 2, 3, and 4 in., respectively. The perimeter of a similar triangle is 36 in. long. Find the length of each side of the second triangle.

14. Jackie has a piece of wire 18 in. long. She wants to bend this into a triangle similar to a triangle whose sides are 3 in., 4 in., and 5 in. long, respectively. What must be the dimensions of her triangle?

15. A telephone pole casts a shadow 30 ft long at the same time that a 5 ft fence post casts a shadow 8 ft long. How tall is the telephone pole?

16. Betty wants to measure the height of a flagpole. She puts up a vertical post 8 ft tall and moves back in line with the flagpole and the post until her line of sight hits the tops of both post and pole. (See the figure.) If the known distances are as shown in the figure, what height does Betty find for the flagpole?

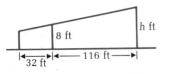

17. Ronny wants to find the height of a tree standing on level ground. He places a 5-ft stake vertically so that the sun throws a shadow of the tree and the stake as shown in the diagram. He then measures the lengths of the shadows, with the results shown in the diagram. What height did Ronny find for the tree?

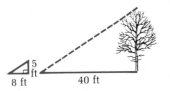

18. Gail wants to measure the distance AB across a small lake. (See the diagram.) She walks 240 m away in a direction perpendicular to the line AB to a point C from which she can sight the point B. She then

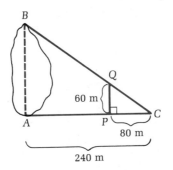

60 m

80 m

240 m

walks back 80 m along *CA* to a point *P*, and then walks in a direction perpendicular to *CA* to a point *Q* in line with *B* and *C*. She finds that *PQ* = 60 m. With this data, what does Gail find for the distance *AB*?

19. Sonny, the surveyor, needs to find the length of a tunnel to be bored through a small hilly area. He makes the marked angles equal and finds the measurements shown in the diagram. What length should Sonny find for the tunnel?

20. To find the distance between two points *P* and *Q* separated by an inaccessible area, Andy makes the measurements shown in the diagram. He returns to his office and draws a triangle *ABC* with $m \angle A = m \angle O$, $m \angle B = m \angle P$, and $\overleftrightarrow{AB} = 10$ cm. He then measures $\overleftrightarrow{BC}$, finding it to be 18 cm. What length would Andy find for *PQ*?

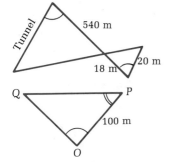

Tunnel 540 m

18 m 20 m

Q P

100 m

O

USING YOUR KNOWLEDGE 9.4

You can use what you know about similar triangles to prove the following important result:

> A line parallel to one side of a triangle and cutting the other two sides, divides these sides proportionally.

In the figure in the margin, if *PQ* is parallel to *AB*, then

$$\frac{AC}{PC} = \frac{BC}{QC}$$

This follows because triangle *ABC* is similar to triangle *PQC*. But $AC = AP + PC$ and $BC = BQ + QC$, so by substituting for *AC* and *BC*, we get

$$\frac{AP + PC}{PC} = \frac{BQ + QC}{QC}$$

or

$$\frac{AP}{PC} + \frac{PC}{PC} = \frac{BQ}{QC} + \frac{QC}{QC}$$

which is the same as

$$\frac{AP}{PC} + 1 = \frac{BQ}{QC} + 1$$

Therefore,

$$\frac{AP}{PC} = \frac{BQ}{QC}$$

This result can be used to show that:

> On any two transversals, parallel lines cut off segments whose lengths are proportional.

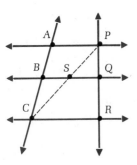

In the figure in the margin, AC and PR are two transversals, cut by the three parallel lines AP, BQ, and CR. The dashed line is drawn joining P and C and intersecting BQ at the point labeled S. In the triangle CPR, the segment SQ is parallel to CR, so that

$$\frac{PS}{SC} = \frac{PQ}{QR}$$

In the triangle ACP, the line segment BS is parallel to AP, so that

$$\frac{PS}{SC} = \frac{AB}{BC}$$

Since AB/BC and PQ/QR are both equal to PS/SC, the desired proportion follows; that is,

$$\frac{AB}{BC} = \frac{PQ}{QR}$$

Here is an application of this result. In the figure below, the lengths of LM, MN, and RS are as shown. The length x of ST must be such that

$$\frac{x}{5} = \frac{3}{2}$$

Therefore,

$$x = \frac{15}{2} = 7.5$$

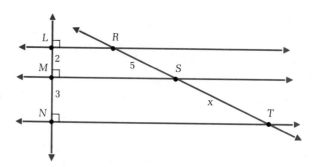

Use these ideas to solve the following problems.

1. The marked angles in the figure below are all equal. Find the value of x and the value of y.

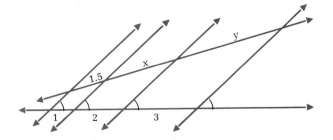

2. Bobby marks points *A*, *B*, and *C* on one transversal so that $AB = 1$ cm and $BC = 3$ cm. He then marks corresponding points *D*, *E*, and *F* on a second transversal so that $DE = 3$ cm and $EF = 5$ cm. If Bobby draws the lines *AD*, *BE*, and *CF*, will these lines be parallel? Explain.

3. Mr. Smith and Ms. Brown own the lots shown in the diagram in the margin. A survey of Mr. Smith's lot showed a frontage of 100 ft on A Street and 96 feet on 1st Avenue. Ms. Brown measured her 1st Avenue frontage to be 120 ft. How much frontage does Brown have on A Street?

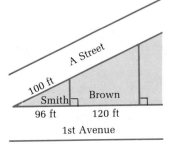

4. The owners of the four lots in the diagram below are the Allens (*A*), the Bakers (*B*), the Cooks (*C*), and the Danes (*D*). The county agrees to pave L Street between 1st Avenue and 2nd Avenue if the property owners pay for the cost of the materials, with each owner to pay a share proportional to the owner's frontage on L Street. If the frontage figures are as shown in the diagram, find what fraction of the cost of materials each owner will have to pay.

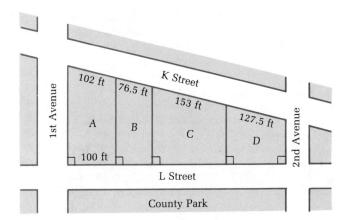

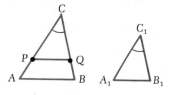

In order to prove the SAS similarity statement, let the two triangles be ABC and $A_1B_1C_1$ as shown in the figure in the margin. Suppose that

$$m \angle C = m \angle C_1 \quad \text{and} \quad \frac{AC}{A_1C_1} = \frac{BC}{B_1C_1}$$

We want to prove that triangles ABC and $A_1B_1C_1$ are similar.

Suppose triangle ABC is the larger of the two triangles. Mark a point P on AC so that $PC = A_1C_1$. Then draw a line PQ parallel to AB and meeting BC at Q.

1. $m \angle CPQ = m \angle A$ and $m \angle CQP = m \angle B$. Why?
2. Therefore, triangle PQC is similar to triangle ABC. Can you discover the reason for this?
3. Now, $\dfrac{AC}{PC} = \dfrac{BC}{QC}$ Why?
4. Since $PC = A_1C_1$, it follows that

$$\frac{AC}{A_1C_1} = \frac{BC}{B_1C_1} = \frac{BC}{QC} \qquad \text{Remember that the equality of the first two fractions was given.}$$

5. From the equality of the second two fractions, we see that $B_1C_1 = QC$.
6. The SAS congruence statement tells us that triangles PQC and $A_1B_1C_1$ are congruent. Can you discover why?
7. It now follows that triangles ABC and $A_1B_1C_1$ are similar. Why?
8. The SSS similarity statement can be proved in about the same way as the preceding proof. Can you discover how to put the proof together?

9.5

THE CIRCLE

You are probably familiar with the geometric figure that we call a *circle*. The circle has been of great interest ever since prehistoric times; it always has been used for many decorative and practical purposes. Here is a modern definition:

▶ **Definition 9.5a**

A **circle** is the set of all coplanar points at a given fixed distance (the **radius**) from a given point (the **center**).

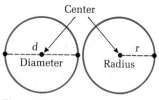

Figure 9.5a

Figure 9.5a shows two circles with the radius and the diameter indicated. Note that the **diameter** consists of two collinear radii, so that $d = 2r$; that is, the length of the diameter is twice the length of the radius.

The perimeter of a circle is known as the **circumference**. We noted earlier that the length of the circumference of a circle is given by the formula

$$C = \pi d = 2\pi r$$

The irrational number $\pi \approx 3.14159$ was discussed in Section 5.1. Unless otherwise noted, you may use the approximate value 3.14 for the problems in this book.

EXAMPLE 1   Andy has a circular swimming pool with a diameter of 25 ft. To keep his little boy from falling into the water, Andy wants to put a low wire fence around the circumference of the pool. How much fencing does he need?

Solution   We use $C = \pi d$ to get

$$C \approx (3.14)(25) = 78.5 \text{ ft}$$

Since 3.14 is a little less than π, Andy should play safe and get 79 feet of fencing.   ◄

EXAMPLE 2   According to the *Guinness Book of World Records*, one of the largest beef hamburgers had a circumference of 27.50 ft. Find the diameter of this hamburger to the nearest hundredth of a foot. Use $\pi \approx 3.14$.

Solution   Since the circumference of a circle is $C = \pi d$, we have

$$27.50 \approx 3.14 d$$

Dividing by 3.14, we find

$$d = 8.76 \qquad \text{to the nearest hundredth}$$

Thus, the diameter of this mammoth beef hamburger was about 8.76 ft. (It also weighed 2859 lb.) [*Note:*   If you used a more accurate value of π, you would find the diameter to be 8.75 ft to the nearest hundredth of a foot.]

Largest hamburger
From the *Guinness Book of World Records*
© 1987 by Sterling Publishing Co., Inc.

◄

EXERCISE 9.5

Use $\pi \approx 3.14$.

1. The diameter of a bicycle tire is 61 cm. Through what distance does the bicycle go when the wheel makes one complete turn? Give your answer to the nearest centimeter.

2. The lid on a garbage can has a diameter of 17 in. Find the length of the circumference of this lid. Give your answer to the nearest tenth of an inch.

3. The minute hand of a clock is 8 cm long. How far does the tip of the hand move in 1 hr? Give your answer to the nearest tenth of a centimeter.

4. The dial on a telephone has a radius of 4.5 in. How far does a point on the rim of this dial travel when the dial makes three-fourths of a complete revolution?

5. A long-playing record has a radius of 6 in. How far does a point on the rim move when the record goes around once? Give your answer to the nearest tenth of an inch.

6. The circumference of a circle is 15π cm. Find the diameter and the radius of this circle.

7. A thin metal rod 8 ft long is to be bent into a circular hoop. Find the radius of this hoop to the nearest tenth of an inch.

8. To make a wedding band for a man who wears a size 12 ring, a strip of gold 7 cm long is needed. Find the diameter of this ring.

9. The largest pizza ever made had a circumference of 251.2 ft. What was its diameter?

10. You already know from Problem 5 that a long-playing record has a 6 in. radius. However, do you know the circumference of the smallest functional record? It is an amazing $4\frac{1}{8}$ in.! Find the diameter of this tiny record to the nearest hundredth of an inch.

DISCOVERY 9.5

A pencil compass and a straightedge are the traditional tools for making geometric constructions. Do you know how to draw an equilateral triangle? Just decide how long you want the side to be. Draw a line segment, say AB, of this length (see Fig. 9.5b). Open a pencil compass to this radius. Then, putting the point at one end, say A, draw an arc. Do the same, putting the point at the other end. Where the two arcs intersect is the third vertex of the equilateral triangle.

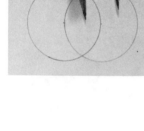

Figure 9.5b
An equilateral triangle

It is also quite easy to construct a regular hexagon. Again, decide on the length of the side. Draw a circle with this length as radius. Then use this

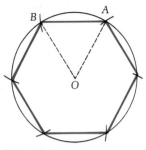

Figure 9.5c
A regular hexagon

Figure 9.5d

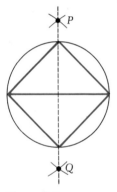

Figure 9.5e

radius to mark off six equidistant points on the circle. These points are the vertices of the hexagon (see Fig. 9.5c). Can you see why this works? [Hint: Look at triangle OAB.]

If you start with a regular polygon with its vertices on a circle as in Fig. 9.5c, it is easy to construct a regular polygon with double the number of sides. All you need to do is bisect the circle arc between two consecutive vertices of the given polygon. Look at Fig. 9.5d, which shows one side AB of a given polygon. Open your compass to any convenient radius greater than one-half of AB. Draw two intersecting arcs, one with center at A and the other with center at B. Call the point of intersection of the arcs P (see Fig. 9.5d). Now draw a straight line through P and the center of the circle O. The point C where the line OP cuts the circle is one vertex of the desired polygon. The line segments BC and AC are two sides of that polygon.

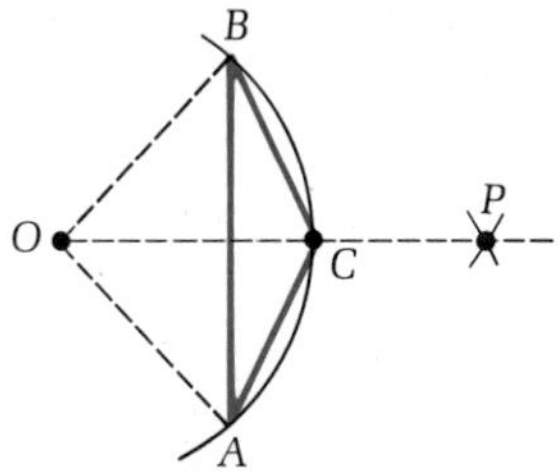

Can you discover how to use this idea to construct a square with its vertices on a given circle? You can start by drawing any diameter of the circle. The ends of this diameter will be two vertices of the square. Use the method of construction shown in Fig. 9.5d to find two points, each equidistant from the ends of the diameter and one on each half of the circle. These points will be the other two vertices of the square. See Fig. 9.5e.

There is a rather complicated way to construct a regular pentagon and a regular decagon with straightedge and compass. We shall not consider this here. But it is interesting to know that mathematicians have proved that it is impossible to construct a regular heptagon or a regular nonagon with straightedge and compass alone.

1. Use the preceding ideas to construct a regular octagon with its vertices on a given circle.
2. Use the preceding ideas to construct a regular dodecagon with its vertices on a given circle.

▶ **Calculator Corner 9.5**

Many calculators come equipped with a $\boxed{\pi}$ key (to access it, you may have to press $\boxed{2nd}$ $\boxed{\pi}$) that gives a nine-decimal-place approximation for π. If you use this approximation to solve Example 2, the answer obtained when dividing 27.50 by π comes out to be 8.75352187, or 8.75 to the nearest hundredth of a foot, as mentioned in the text.

1. Rework Problems 1, 2, 5, 7, and 9 of Exercise 9.5 on your calculator.

AREA MEASURE AND THE PYTHAGOREAN THEOREM

Consider the question, "How much glass was used in the construction of the building in the picture?" If you think a moment, you will realize that the meaning of the question is not clear. Does it ask for the number of pieces of glass, the number of pounds of glass, or what? Let us be more specific and say, "What *area* was covered by the glass in this building?" In order to answer questions like this, we must have a good understanding of what we mean by *area*.

A. Area of a Plane Region

Figure 9.6a
The unit of area measure

We start by choosing the **unit region** to be that of a **square,** each of whose sides is **1 unit in length,** and we say that this region has an **area of 1 square unit** (see Fig. 9.6a). The side of the unit square may be 1 in., 1 ft, 1 mi, 1 cm, 1 m, and so on. The corresponding units of area are the square inch (in.²), the square foot (ft²), the square mile (mi²), the square centimeter (cm²), the square meter (m²), and so on.

The area of a plane region is simply the extent of that region relative to the extent of the unit region. In everyday language, the area measure of a region is the **number of these unit regions contained in the given region.** Thus, suppose that we have a square with side 2 units long. Then, as shown in Fig. 9.6b, we can draw lines joining the midpoints of the opposite sides and dividing the square into $2 \times 2 = 4$ unit squares. We say that the area of the square is 4 square units. Similarly, for a rectangle that is 3 units by 4 units, we can draw lines parallel to the sides that divide the rectangle into $3 \times 4 = 12$ unit squares (see Fig. 9.6c). We say that the area of the rectangle is 12 square units.

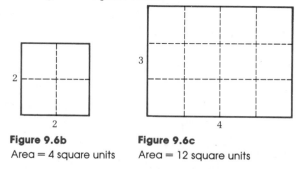

Figure 9.6b
Area = 4 square units

Figure 9.6c
Area = 12 square units

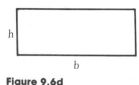

Figure 9.6d
Area = *bh*

The above illustrations show how the area measure can be found if the sides of the rectangle are whole numbers. In order to avoid a complicated mathematical argument, we define the area of a rectangle of sides *b* units long and *h* units long to be *bh* square units (see Fig. 9.6d). Written as a formula, with *A* standing for area, this means

$A = bh$ Area of a rectangle

Notice that we are following the usual custom of saying "area of a rectangle" to mean "area of a rectangular region." Similarly, we shall say "area of a polygon" to mean "area of a polygonal region."

Knowing the area of a rectangle, we can find the area of a parallelogram. The idea is to construct a rectangle with the same area as the parallelogram. As Fig. 9.6e shows, we simply cut the right triangle *ADE* from one end of the parallelogram and attach it to the other end. This forms a rectangle *CDEF* with the same area as the parallelogram. Since the base and height are the same for both figures, it follows that the area of a parallelogram of base *b* and height *h* is

$A = bh$ Area of a parallelogram

(Be sure to note that *h* is the perpendicular height, not just a side of the parallelogram.)

Figure 9.6e

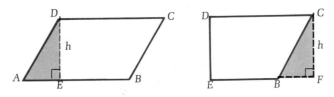

It is now an easy matter to find the area of any triangle. Suppose that triangle *ABC* in Fig. 9.6f is given. We draw a line through *C* parallel to *AB* and a line through *A* parallel to *BC*. These lines meet at a point *D*, and the quadrilateral *ABCD* is a parallelogram. Clearly, the line *AC* is a diagonal of the parallelogram and divides the parallelogram into two equal pieces. Because the area of parallelogram *ABCD* is *bh*, the area of the triangle *ABC* is $\frac{1}{2}bh$. Thus, we have a formula for the area of any triangle with base *b* and height *h*:

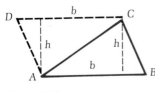

Figure 9.6f

$A = \frac{1}{2}bh$ Area of a triangle

Note: If the angle *B* in Fig. 9.6f is a right angle, the parallelogram obtained is a rectangle. This does not change the final result.

EXAMPLE 1   Find the area of the triangle shown in the margin.

Solution   This triangle has a base of 4 in. and a height of 1.5 in. Thus,

$$A = \frac{1}{2}(4)(1.5) = 3 \text{ in.}^2 \qquad \blacktriangleleft$$

The formulas we have developed can be used for finding the areas of many polygonal regions. This is done by subdividing these regions into nonoverlapping rectangular and/or triangular regions, finding the areas of these subdivisions, and adding the results. This procedure is illustrated in the next example.

EXAMPLE 2   Find the area of the region given in the figure below.

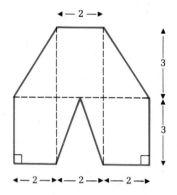

Solution   We subdivide the region as shown:

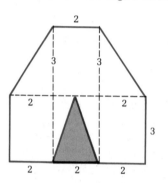

The area of the lower rectangle is $6 \times 3 = 18$. The area of the shaded triangle is $\frac{1}{2}(2 \times 3) = 3$. The upper two triangles both have area 3, and the upper rectangle has area $2 \times 3 = 6$. Thus, the required area is

$$18 - 3 + 3 + 3 + 6 = 27 \text{ square units} \qquad \blacktriangleleft$$

Thus far, we have been concerned entirely with the areas of polygonal regions. How about the circle? Can we find the area of a circle by using one of the preceding formulas? Interestingly enough, the answer is yes. The required area can be found by using the formula for the area of a rectangle. Here is how you can go about it. Look at Fig. 9.6g. Cut the lower half of the

Figure 9.6g

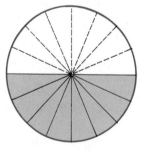
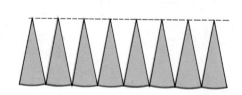

circular region into small equal slices and arrange them as shown in the figure. Then cut the remaining half of the circle into the same number of slices and arrange them as shown in Fig. 9.6h. The result is approximately a parallelogram whose longer side is of length πr (half the circumference of the circle) and whose shorter side is r (the radius of the circle). The more pieces you cut the circle into, the more accurate this approximation becomes. You should also observe that the more pieces you use, the more nearly the parallelogram becomes a rectangle of length πr and height r. Mathematicians have proved that the area of the circle is actually the same as the area of this rectangle, that is, $(\pi r)(r)$, or πr^2. Thus, we arrive at the formula for the area of a circle of radius r:

$$A = \pi r^2 \quad \text{Area of a circle}$$

Figure 9.6h

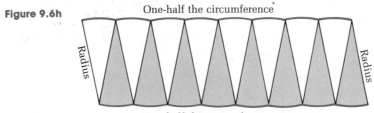

One-half the circumference

Radius

Radius

One-half the circumference

EXAMPLE 3   The circular dish for the Arecibo radiotelescope has a radius of 500 ft. What area does the dish cover? (Use $\pi \approx 3.14$.)

Solution   Here, r = 500, so

$$A = \pi(500)^2$$
$$= \pi(250,000) \approx 785,000 \text{ ft}^2 \blacktriangleleft$$

Your knowledge of algebra can be used in conjunction with the geometric formulas you have just studied. This is illustrated in the next example.

EXAMPLE 4   The Fermi National Accelerator Laboratory has the atom smasher shown in the photo. The smasher covers an area of 1.1304 mi^2. What is the diameter of this atom smasher? Use $\pi \approx 3.14$, and give the answer to the nearest hundredth of a mile.

Solution   The formula for the area of a circle of radius r is

$$A = \pi r^2$$

Here, A = 1.1304 and $\pi \approx 3.14$. Thus,

$$1.1304 \approx 3.14r^2$$

$$r^2 \approx \frac{1.1304}{3.14} = 0.36$$

$$r \approx 0.60 \text{ mi}$$

Thus, the diameter is about 1.20 mi. If your calculator has a square root key, you can do this calculation by keying in

$$\boxed{1.1304}\div\boxed{3.14}\boxed{=}\boxed{\sqrt{x}}$$ ◀

B. The Pythagorean Theorem

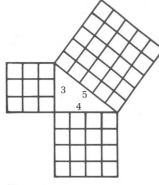

Figure 9.6i
$5^2 = 3^2 + 4^2$

The solution of Example 4 required taking the square root of both sides of an equation. The same technique is used in many problems involving the sides of a right triangle. One of the most famous and important theorems of all time is the **Pythagorean theorem,** which says, **"The square on the hypotenuse (the longest side) of a right triangle is equal to the sum of the squares on the other two sides."** Figure 9.6i illustrates the theorem for a 3-4-5 right triangle.

The perpendicular sides of a right triangle are often called **legs.** In Fig. 9.6j, the legs are labeled a and b, and the hypotenuse is labeled c. In terms of these lengths, the Pythagorean theorem states that

$$c^2 = a^2 + b^2$$

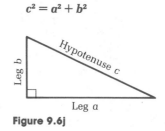

Figure 9.6j

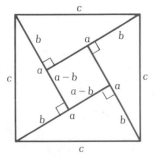

Figure 9.6k

The early Egyptians, and even the Babylonians, knew some special cases of this result; in particular, they knew the result illustrated in Fig. 9.6i. However, the ancient Greeks seem to have been the first to prove the general theorem. There are many different proofs of the theorem, some of which depend on areas in a simple fashion. Here is one of these proofs.

The area of the large square in Fig. 9.6k is c^2. The area of the small square is $(a - b)^2 = a^2 - 2ab + b^2$, and the area of each of the four right triangles is $\frac{1}{2}ab$. Thus,

$$c^2 = a^2 - 2ab + b^2 + (4)(\tfrac{1}{2}ab)$$
$$= a^2 - 2ab + b^2 + 2ab$$
$$= a^2 + b^2$$

EXAMPLE 5   Find the length of the hypotenuse of a right triangle whose legs are 5 and 12 units long.

Solution   We use the Pythagorean theorem to get

$$c^2 = a^2 + b^2$$
$$= 5^2 + 12^2 = 25 + 144 = 169$$

Therefore, $c = \sqrt{169} = 13$. ◀

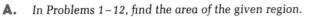

A. *In Problems 1–12, find the area of the given region.*

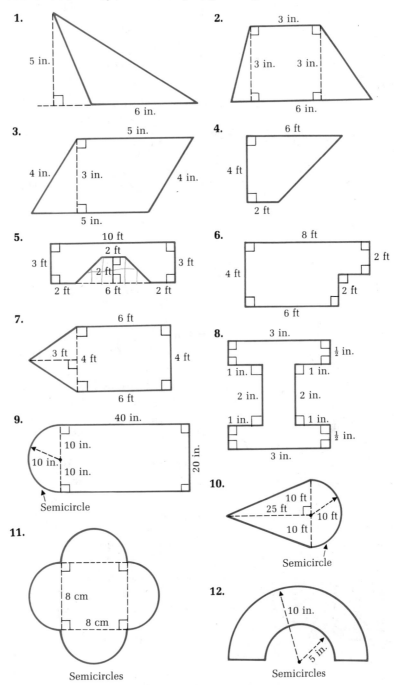

1.

2. 3 in. / 3 in. / 3 in. / 6 in.

3. 5 in. / 4 in. / 3 in. / 4 in. / 5 in.

4. 6 ft / 4 ft / 2 ft

5. 10 ft / 2 ft / 3 ft / 3 ft / 2 ft / 2 ft / 6 ft / 2 ft

6. 8 ft / 2 ft / 4 ft / 2 ft / 6 ft

7. 6 ft / 3 ft / 4 ft / 4 ft / 6 ft

8. 3 in. / ½ in. / 1 in. / 1 in. / 2 in. / 2 in. / 1 in. / 1 in. / ½ in. / 3 in.

9. 40 in. / 10 in. / 10 in. / 10 in. / 20 in. / Semicircle

10. 10 ft / 25 ft / 10 ft / 10 ft / Semicircle

11. 8 cm / 8 cm / Semicircles

12. 10 in. / 5 in. / Semicircles

In Problems 13–16, find the shaded area.

13.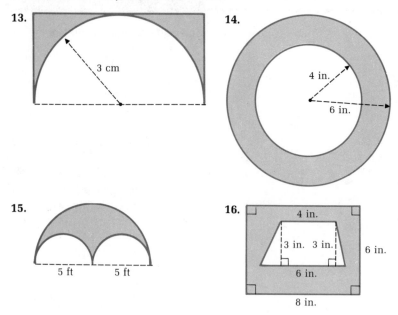

3 cm

14.

4 in.

6 in.

15.

5 ft 5 ft

16.

4 in.

3 in. 3 in.

6 in.

6 in.

8 in.

B. **17.** The diagonal of a rectangle is 17 ft long and one of the sides is 15 ft long. How long is the other side?

18. A rectangle is 24 cm long and 7 cm wide. How long is its diagonal?

19. The base of a parallelogram is 12 ft long, the height is 8 ft, and the shorter diagonal is 10 ft long. How long are the other sides of the parallelogram?

20. The base of a parallelogram is 10 cm long, the height is 5 cm, and the longer diagonal is 13 cm long. How long is the shorter diagonal?

21. The playing surface of a football field is 120 yd long and $53\frac{1}{3}$ yd wide. How many square yards of artificial turf are needed to cover this surface?

22. The floors of three rooms in a certain house measure 9 by 10 ft, 12 by 12 ft, and 15 by 15 ft, respectively.

　a. How many square yards of carpet are needed to cover these three floors?

　b. If the price of the carpet is $14/yd², how much would it cost to cover these floors?

23. The Louisiana Superdome covers an area of 363,000 ft². Find the diameter of this round arena to the nearest foot. Use $\pi \approx 3.14$.

24. The largest cinema screen in the world is in the Pictorium Theater in Santa Clara, California; it covers 6720 ft². If this rectangular screen is 70 ft tall, how wide is it?

25. The area of the biggest pizza was about 5024 ft². What was its diameter?

26. Find the height of a triangle of area 70 in.² if its base is 20 in. long.

27. Glass for picture frames costs $3/ft². If the cost of the glass for a rectangular frame is $4.50 and the frame is one and one-half times as long as it is wide, find the dimensions in inches.

28. Diazinon is a toxic chemical used for insect control in grass. Each ounce of this chemical, diluted in 3 gal of water, covers 125 ft². The transportation department used 32 oz of Diazinon to spray the grass in the median strip of a highway. If the strip was 16 ft wide, how long was it?

29. John Carpenter has a tabletop that is 4 ft wide by 5 ft long. He wants to cut down the length and the width both by the same amount so as to decrease the area by $4\frac{1}{4}$ ft². What would be the new dimensions?

30. In Problem 29, suppose John wants to decrease the 4-ft side and increase the 5-ft side by the same amount so as to decrease the area by $\frac{3}{4}$ ft². What would be the new dimensions?

31. Show that the area of an equilateral (all sides equal) triangle of side s is $s^2 \cdot \sqrt{3}/4$. [*Hint:* A perpendicular from one vertex to the opposite side bisects that side.]

32. What happens to the area of a triangle:
    a. If the length of its base is doubled?
    b. If both base and height are doubled in length?

33. What happens to the area of a rectangle if both its dimensions are:
    a. Doubled?   **b.** Tripled?
    c. Multiplied by a constant k?

34. Use the result of Problem 31 to find the area of a regular hexagon whose side is 4 in. long.

35. The circumference of a circle and the perimeter of a square are both 20 cm long. Which has the greater area, and by approximately how much?

36. In the figure in the margin, the two circles have their centers on the diagonal AC of the square $ABCD$. The circles just touch each other and the sides of the square, as shown in the figure. If the side of the square is of length s, find the total area of the two circles in terms of s.

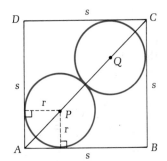

USING YOUR KNOWLEDGE 9.6

You can use what you learned in this section to help you solve some commonly occurring problems. Do the following problems to see how.

1. A gallon of Lucite wall paint costs $14 and covers 450 ft². Three rooms in a house measure 10 by 12 ft, 14 by 15 ft, and 12 by 12 ft, and the ceiling is 8 ft high.
    a. How many gallons of paint are needed to cover the walls of these rooms if you make no allowance for doors and windows?

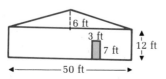

b. What will be the cost of the paint? (The paint is sold by the gallon only.)

2. The diagram in the margin shows the front of a house. House paint costs $17/gal and covers 400 ft².

   a. What is the minimum number of gallons of paint needed to cover the front of the house? (The paint is sold by the gallon only.)

   b. How much will the paint for the front of the house cost?

3. My house and lot are shown in the diagram below. The entire lot, except for the buildings and the drive, is lawn. A bag of lawn fertilizer costs $4 and covers 1200 ft² of grass.

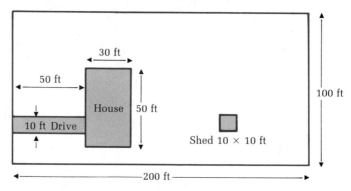

   a. What is the minimum number of bags of fertilizer needed for this lawn?

   b. What will be the cost of the fertilizer?

4. A small pizza (11-in. diameter) costs $5 and a large pizza (15-in. diameter) costs $8. Use $\pi \approx 3.14$ and find, to the nearest square inch:

   a. The area of the small pizza

   b. The area of the large pizza

   c. Which is the better deal, two small pizzas or one large pizza?

5. A frozen apple pie of 8-in. diameter sells for $1.25. The 10-in. diameter size sells for $1.85.

   a. What is the unit price (price per square inch), to the nearest hundredth of a cent, of the 8-in. pie?

   b. What is the unit price of the 10-in. pie?

   c. Which pie gives you the most for your money?

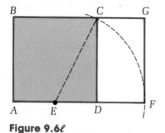

Figure 9.6ℓ

DISCOVERY 9.6

*Artists and architects, past and present, have used the **Golden Ratio** in their art and their architecture. Do you know what the Golden Ratio is? Begin with a square (shaded in Fig. 9.6ℓ). Let E be the midpoint of the base AD. Put the point of your pencil compass at E and the pencil point at C and draw the circular arc as shown. Extend AD to meet this arc and call the* point of intersection F. Now draw a line perpendicular to AF at F and extend BC to meet this perpendicular at G. The rectangle ABGF is known as the **Golden Rectangle.** Its proportions are supposed to be particularly pleasing to the eye. The ratio of the longer to the shorter side of this rectangle is the Golden Ratio.

1. Can you discover the numerical value of the Golden Ratio? [*Hint:* Let the side of the square ABCD be 2a. Then EC is the hypotenuse of a right triangle with legs of length a and 2a, respectively.]

Artists are interested in the areas of their paintings and sometimes meet with the problem of drawing a rectangle of height h that will have the same area as that of a given rectangle. Figure 9.6m shows the problem. The given rectangle is of length L and width W, and the artist wants a rectangle of height h that will have the same area. Of course, you could find the area LW and divide by h to get the second dimension of the desired rectangle. But the artist can do the job very quickly without any arithmetic at all! Here is how: Look at Fig. 9.6n. Draw a line across the given rectangle at height h (line BCD in the figure). Next, draw a line through A and C and extend the top line of the rectangle to meet line AC at point F. Then drop a perpendicular from F to the extended base of the given rectangle. The rectangle ABDE is the desired rectangle. Can you discover why? Look at Fig. 9.6n again. The triangles ABC and FDC are similar (have exactly the same shape). Corresponding sides of similar triangles are always in the same ratio. Thus,

$$\frac{y}{W} = \frac{x}{h}$$

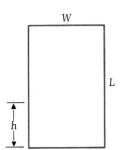

Figure 9.6m

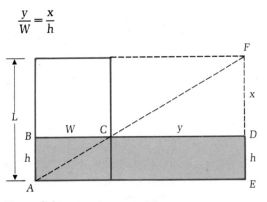

Figure 9.6n

2. Can you discover how to use this result to show that the area of rectangle ABDE is equal to the area of the original rectangle? [*Hint:* What is the area of the rectangle taken away from the given rectangle? What is the area of the rectangle added on?]

3. Draw a careful diagram of a rectangle 4 cm wide and 6 cm high. Use the construction described above to find a rectangle 5 cm high and with the same area. If you do this very carefully, the result will help convince you that this is a neat construction.

THREE-DIMENSIONAL FIGURES AND VOLUMES

The picture shows one of the famous Egyptian pyramids. These pyramids were built over 4000 years ago. Because the Egyptian pyramids have square bases, they are called *square pyramids;* they are examples of solids bounded by polygons.

A. Three-Dimensional Figures

Courtesy of PSH

A solid bounded by plane polygons is called a **polyhedron.** The polygons are the **faces** of the polyhedron, the sides of the polygons are the **edges,** and the vertices of the polygons are the **vertices** of the polyhedron. The Egyptian pyramids are polyhedrons with five faces, one of which is a square (the base) and the others are triangles. A square pyramid has eight edges and five vertices. The Transamerica Pyramid is a striking example of the use of a square pyramid in the design of a modern building.

A **convex polyhedron** is one that lies entirely on one side of the plane of each of its faces. A polyhedron that is not convex is called **concave,** or **reentrant.** The polyhedrons shown in Fig. 9.7a are a **cube,** a **rectangular parallelepiped,** a six-sided polyhedron with triangular faces (part III), and a seven-sided polyhedron (part IV). The first three of these are convex, and the fourth is concave (reentrant).

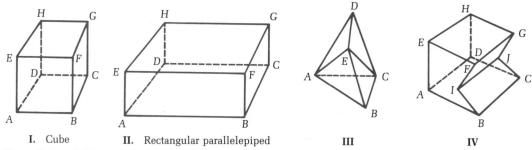

    I. Cube   **II.** Rectangular parallelepiped   **III**   **IV**

Figure 9.7a Polyhedrons

If two faces of a polyhedron lie in parallel planes and if the edges that are not in these planes are all parallel to each other, then the polyhedron is called a **prism.** In Fig. 9.7a, parts I, II, and IV all illustrate prisms. The faces of a prism that are in the two parallel planes are called the **bases.** The parallel lines joining the bases are the **lateral edges.** The two bases of a prism are congruent polygons.

Figure 9.7b shows a **triangular prism.** The mineral crystal shown in the photograph is in the shape of a triangular prism.

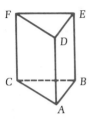

Figure 9.7b
Triangular prism

A crocoite crystal
Courtesy of PSH

If all but one of the vertices of a polyhedron lie in one plane, then the polyhedron is a **pyramid.** The face that lies in this one plane is the **base,** and the remaining vertex is the **vertex of the pyramid.** Figure 9.7c shows a pentagonal pyramid; the base is a pentagon. Prisms and pyramids are named by the shapes of their bases.

EXAMPLE 1   Name the edges and the vertices of the pyramid in Fig. 9.7c.

Figure 9.7c
Pentagonal pyramid

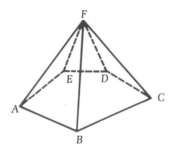

Solution   The edges are the line segments AB, BC, CD, DE, EA, FA, FB, FC, FD, and FE. The vertices are the points A, B, C, D, E, and F.   ◄

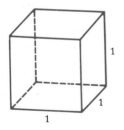

Figure 9.7d
Unit cube:
Volume = 1 cubic unit

B. Volume

The **volume** of a three-dimensional region is measured in terms of a **unit volume,** just as area is measured in terms of a unit area. For the unit volume, we choose the region enclosed by a **unit cube,** as shown in Fig. 9.7d. The unit volume may be the cubic inch (in.³), cubic foot (ft³), cubic centimeter (cm³), cubic meter (m³), and so on, according to the unit of length used.

Volumes may be considered in a manner similar to that used for areas. If a rectangular parallelepiped (box) is such that the lengths of its edges are all whole numbers, then the region can be cut up by planes parallel to the

Figure 9.7e

$V = 4 \times 3 \times 2 = 24$ cubic units

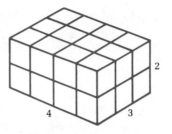

faces as in Fig. 9.7e. In general, we define the volume of a rectangular box of length a, width b, and height c to be abc. Thus,

$$V = abc = \text{Length} \times \text{Width} \times \text{Height} \qquad \text{Volume of a rectangular box}$$

EXAMPLE 2   Each of the little cubes in Rubik's Cube is $\frac{3}{4}$ in. on a side. What is the volume of Rubik's Cube?

Solution   Because there are three little cubes in a row, the edge of the large cube is

$$(3)(\tfrac{3}{4}) = \tfrac{9}{4} \text{ in.}$$

Since this is a cube, we take $a = b = c$ in the above formula to get

$$V = a^3 = (\tfrac{9}{4})^3 = \tfrac{729}{64} \text{ in.}^3$$

If you prefer to express this answer decimally, you will get 11.390625, or about 11.4 in.^{3} ◄

Let us denote the area measure of the base of a prism or a pyramid by B and the height by h. It is shown in solid geometry that the formulas for the volumes of these figures are

$$V = Bh \qquad \text{Volume of a prism}$$
$$V = \tfrac{1}{3}Bh \qquad \text{Volume of a pyramid}$$

EXAMPLE 3   If the crocoite crystal shown earlier (page 437) has a triangular base of height 3 cm and base length 2.6 cm, and if the crystal is 30 cm long, what is its volume?

Solution   First we find B, the area of the base of the prism. Since the base is a triangle of base length 2.6 cm and height 3 cm,

$$B = \tfrac{1}{2}(2.6)(3) = 3.9 \text{ cm}^2$$

Then we use the formula for the volume of a prism to obtain

$$V = Bh = (3.9)(30) = 117 \text{ cm}^3 \qquad ◄$$

EXAMPLE 4   The Transamerica Pyramid is 260 m high, and its square base has a perimeter of 140 m. Find the volume.

Solution   We first find the length of one side of the base. Since this is a square of perimeter 140 m, one side is of length $\frac{140}{4} = 35$ m. The area of this square is $B = (35)^2 = 1225$ m. Using the formula for the volume of a pyramid, we get

$$V = \tfrac{1}{3}Bh = \tfrac{1}{3}(1225)(260)$$
$$= \tfrac{318,500}{3} = 106,166\tfrac{2}{3} \text{ m}^3$$

Note: If we assume that the dimensions are correct just to the nearest meter, then an answer of 106,000 m³ is sufficiently accurate.   ◀

EXAMPLE 5   The figure in the margin shows a polyhedron that consists of a rectangular box surmounted by a pyramid with the top of the box for its base. The base of the box is 2 ft by 3 ft, and the height of the box is 5 ft. The pyramid is 2 ft in height. What is the volume of this figure?

Solution   The volume of the rectangular box portion of the figure is

$$V_1 = abc = (2)(3)(5) = 30 \text{ ft}^3$$

The volume of the pyramid is

$$V_2 = \tfrac{1}{3}Bh = \tfrac{1}{3}(6)(2) = 4 \text{ ft}^3$$

Thus, the entire volume is 34 ft³.   ◀

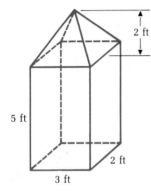

Mathematicians have proved that the formulas $V = Bh$ and $V = \tfrac{1}{3}Bh$ also apply to cylinders and cones, respectively. Thus, for a circular cylinder of radius r and height h (see Fig. 9.7f), we have $B = \pi r^2$, so that

$$V = \pi r^2 h \qquad \text{Volume of a circular cylinder}$$

Similarly, for a circular cone of radius r and height h (see Fig. 9.7g),

$$V = \tfrac{1}{3}\pi r^2 h \qquad \text{Volume of a circular cone}$$

The volume of a sphere of radius r (see Fig. 9.7h) has been shown to be

$$V = \tfrac{4}{3}\pi r^3 \qquad \text{Volume of a sphere}$$

Figure 9.7f   Figure 9.7g   Figure 9.7h

EXAMPLE 6   The water tank for a small town is in the shape of a cone (vertex down) surmounted by a cylinder, as shown in the figure in the margin on the next page. If a cubic foot of water is about 7.5 gal, what is the capacity of the tank in gallons? Use $\pi \approx 3.14$.

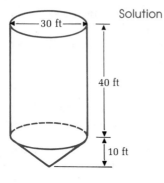

Solution For the cylindrical portion of the tank, $r = 15$ and $h = 40$, so that

$$V = \pi(15)^2(40) = 9000\pi \text{ ft}^3$$

For the conical portion, $r = 15$ and $h = 10$. Thus,

$$V = \tfrac{1}{3}\pi(15)^2(10) = 750\pi \text{ ft}^3$$

The total volume is the sum of these—that is, $9750\pi \text{ ft}^3$. Using 7.5 gal/ft³ and 3.14 for π, we get

$$9750\pi \text{ ft}^3 \approx (9750)(3.14)(7.5) \text{ gal}$$
$$\approx 230,000 \text{ gal} \qquad \blacktriangleleft$$

EXAMPLE 7   A solid metal sphere of radius 3 m just fits inside a cubical tank. If the tank is full of water and the sphere is slowly lowered into the tank until it touches bottom, how much water is left in the tank?

Solution The amount of water left in the tank is the difference between the volume of the tank and the volume of the sphere. Thus, since the edge of the cube equals the diameter of the sphere, the required volume is

$$V = 6^3 - \tfrac{4}{3}\pi(3^3)$$
$$= 216 - 36\pi$$
$$\approx 103 \text{ m}^3 \qquad \blacktriangleleft$$

EXERCISE 9.7

A.   1. Refer to Fig. 9.7a, part III. Name:
    a. The vertices   **b.** The edges
  2. Refer to Fig. 9.7b and repeat Problem 1.
  3. Refer to Fig. 9.7a, part IV, and name the bottom face.
  4. Refer to Fig. 9.7a, part IV, and name the left-hand back face.

In each of Problems 5–8, make a sketch of the figure.

  5. A triangular pyramid
  6. A triangular prism surmounted by a triangular pyramid with the top base of the prism as the base of the pyramid
  7. A six-sided polyhedron that is convex and is not a parallelepiped
  8. An eight-sided polyhedron with triangular faces

  9. **a.** If the edges of a cube are doubled in length, what happens to the volume?
    b. What if the lengths are tripled?

B.   10. A pyramid has a rectangular base. Suppose that the edges of the base and the height of the pyramid are all doubled in length. What happens to the volume?

11. A pentagonal prism has a base whose area is 10 in.². If the prism is 5 in. high, what is its volume?

12. The base of a prism is a triangle whose base is 3 ft and whose height is 4 ft. If the prism is 5 ft high, what is its volume?

13. The edge of the base of a square pyramid is 4 in. long, and the pyramid is 6 in. high. Find the volume of the pyramid.

14. A convex polyhedron consists of two pyramids with a common base that is an equilateral triangle 3 in. on a side. The height of one of the pyramids is 2 in. and the height of the other is 4 in. What is the volume of the polyhedron?

15. A container consists of a cube 10 cm on an edge surmounted by a pyramid of height 15 cm and with the top face of the cube as its base. How many liters does this container hold?

16. The Great Pyramid of Egypt (the Pyramid of Cheops) is really huge. It is 148 m high, and its square base has a perimeter of 930 m. What is the volume of this pyramid?

In Problems 17–20, find the volume of:
a. *The circular cylinder of given radius and height*
b. *The circular cone of given radius and height*
Use the approximate value 3.14 for π.

17. Radius 5 in., height 9 in.

18. Radius 10 cm, height 6 cm

19. Radius 3 ft, height 4 ft

20. Radius 6 cm, height 12 cm

21. Find the volume of a sphere of radius 6 in.

22. Find the volume of a sphere of radius 12 cm.

23. The Peachtree Plaza Hotel tower in Atlanta, Georgia, is 70 stories high. If the height of this cylinder is 754 ft and its diameter is 116 ft, what is the volume?

24. No, it is not SLIME; it is S1-LIME. Each of these silos consists of a cone, vertex down, surmounted by a cylinder. If the diameter of the cylinder is 10 ft, the cylinder is 30 ft high, and the cone is 10 ft high, how many cubic yards of lime does one silo hold?

25. A pile of salt is in the shape of a cone 12 m high and 32 m in diameter. How many cubic meters of salt are in the pile?

26. The fuel tanks on some ships are spheres of which only the top halves are above deck. If one of these tanks is 120 ft in diameter, how many gallons of fuel does it hold? Use 1 ft³ ≈ 7.5 gal.

27. A popular-sized can in American supermarkets is 3 in. in diameter and 4 in. high (inside dimensions). About how many grams of water will one of these cans hold? (Recall that 1 cm³ of water weighs 1 g, and 1 in. = 2.54 cm.)

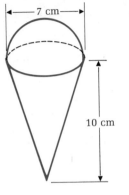

28. An ice cream cone is 7 cm in diameter and 10 cm deep. The inside of the cone is packed with ice cream and a hemisphere of ice cream is put on top. If ice cream weighs $\frac{1}{2}$ g/cm³, how many grams of ice cream are there in all?

1. A **regular polyhedron** is one whose faces are all congruent regular polygons, that is, regular polygons of exactly the same shape and size. The appearance of such a polyhedron at any vertex is identical with its appearance at any other vertex; the same is true at the edges. The early Greeks discovered that only five regular polyhedrons are possible. Can you repeat this discovery? Consider the regular polygons one at a time: equilateral triangles, squares, regular pentagons, and so on. In case you use squares, how many squares can you put together at a vertex to form a polyhedron? Look at a cube. This is the only regular polyhedron with squares for its faces. The regular polyhedrons are as follows:

   a. A tetrahedron with four equilateral triangles for faces
   b. A cube with six squares for faces
   c. An octahedron with eight equilateral triangles for faces
   d. A dodecahedron with 12 pentagons for faces
   e. An icosahedron with 20 equilateral triangles for faces

Dodecahedron-shaped dice

2. Copy the pattern of equilateral triangles in the figure below on a piece of stiff cardboard. Cut around the outside edges and fold on the heavy lines. You can build an icosahedron by holding the cut edges together with transparent tape. There are five triangles at each vertex.

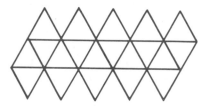

3. Count the number of faces (F), vertices (V), and edges (E) for each of the figures listed in the accompanying table. Compare the value of E with the value of $F + V$ and see whether you can discover Euler's famous formula for polyhedrons.

FIGURE	F	V	E
9.7a, II			
9.7a, III			
9.7a, IV			
9.7b			
9.7c			

4. Can you check the formula you got in Problem 3 by using the diagram in Problem 2?

When higher exponents are present in an expression, calculators with a $\boxed{y^x}$ *key are especially helpful. As the notation indicates, this key raises a number y to a power x. Thus, to perform the calculations in Example 7, we enter*

$$\boxed{6}\ \boxed{y^x}\ \boxed{3}\ \boxed{-}\ \boxed{4}\ \boxed{\times}\ \boxed{\pi}\ \boxed{\times}\ \boxed{3}\ \boxed{y^x}\ \boxed{3}\ \boxed{\div}\ \boxed{3}\ \boxed{=}$$

The result is given as 102.9026645, or about 103 m³.

1. Rework Problem 21 in Exercise 9.7 using your calculator.

---

*9.8

NETWORKS

Courtesy of PSH

Figure 9.8a
Simple networks

Any connected set of line segments or curves is called a **network.** (For the purposes of this section, your intuitive notion of what a curve is will be sufficient.) If the network can be drawn by tracing each line segment or curve exactly once without lifting the pencil from the paper, the network is said to be **traversable.** Any simple network (one that does not cross itself) is traversable. If the network is both simple and closed, then you may choose any point of the network as the starting point, and this point will also be the terminal point of the drawing. If the network is simple but not closed, then you must start at one of the endpoints and finish at the other. See Fig. 9.8a.

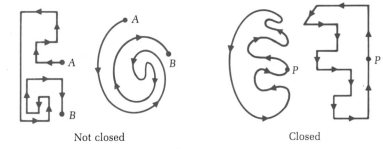

Not closed   Closed

A famous puzzle problem known as the **bridges of Königsberg** probably started the study of the traversability of networks. There was a river flowing through the city, and in the river were two islands connected to each other and to the city by seven bridges, as shown in Fig. 9.8b. The people of the city loved a Sunday walk, and thought it would be fun to follow a route that would take them across each of the seven bridges exactly once. But they found that no matter where they started or what path they took, they could not cross each bridge exactly once. Use Fig. 9.8b

Figure 9.8b
The bridges of Königsberg

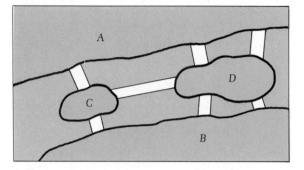

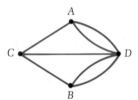

Figure 9.8c
Network for the bridges of
Königsberg

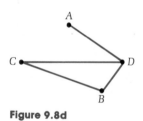

Figure 9.8d

and play with this puzzle for a little while. Perhaps you will see what is involved.

The great Swiss mathematician Leonhard Euler observed that the problem is basically concerned with the network of paths joining the two sides of the river A, B and the two islands C, D, as indicated in Fig. 9.8b. Thus, the problem may be represented by a network, as shown in Fig. 9.8c, and the question is that of traversability of this network.

Euler found this problem interesting, and was able to devise a theory of traversable networks. He regarded the points where segments of the network came together as vertices (A, B, C, D in Fig. 9.8c). Then he classified each vertex as odd or even according to whether it is an endpoint of an odd or an even number of segments. In Fig. 9.8c, all four of the vertices A, B, C, D are odd. In Fig. 9.8d, A and D are odd vertices; B and C are even vertices. Every segment has two endpoints, so it is impossible for a network to have an odd number of odd vertices. A network may have 0, 2, 4, 6, . . . , any even number of odd vertices. Notice that there were four odd vertices in the bridge problem and two odd vertices in Fig. 9.8d.

In his theory, Euler showed that any network that has only even vertices is traversable. Furthermore, the route may start at any vertex and terminate at the same vertex. If a network has exactly two odd vertices, then it is again traversable; but the route must start at one of the odd vertices and will terminate at the other odd vertex. If a network has more than two odd vertices, then it is not traversable.

Euler's theory disposed of the Königsberg bridge problem by showing that the network is not traversable; it has four odd vertices. The network in Fig. 9.8d is traversable. One way of traversing it is to go from A to D to B to C and then to D. Can you find some other routes?

Network theory has many practical applications. It is of great importance in computer science and technology; it is also used to solve problems in the design of city streets, to analyze traffic patterns, to find the most efficient routes for garbage collection, and so on. Networks are also used in connection with PERT (Program Evaluation and Review Technique) diagrams in planning complicated projects. These diagrams help to determine how long a project will take and when to schedule different phases of the project.

In Problems 1–10, find:
a. The number of even vertices
b. The number of odd vertices
c. Whether the network is traversable and which vertices are possible starting points if the network is traversable

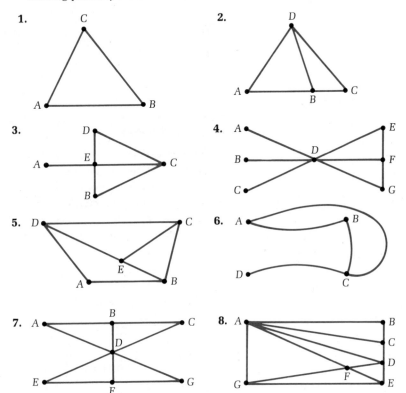

1.

2.

3.

4.

5.

6.

7.

8.

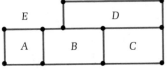

Figure 9.8e

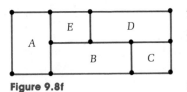

Figure 9.8f

9. The network formed by the edges of a square pyramid
10. The network formed by the edges of a rectangular box

11. Use a network to find whether it is possible to draw a simple connected broken line that crosses each line segment of Fig. 9.8e exactly once. The line segments are those that join successive dots.
12. Repeat Problem 11 for Fig. 9.8f.

13. Use a network to find whether it is possible to take a walk through the building with the floor plan in Fig. 9.8g and pass through each doorway exactly once. Is it possible if you must start and end outside?

Figure 9.8g

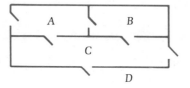

14. Repeat Problem 13 for the floor plan in Fig. 9.8h.

Figure 9.8h

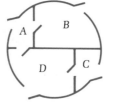

DISCOVERY 9.8

*A plane curve that does not cross itself is called **simple**, just as in the case of a broken line. A **closed curve** is one that starts and ends at the same point. **A simple closed curve divides the plane into two parts, the region interior and the region exterior to the curve.** This important statement is the **Jordan curve theorem**, a very deep theorem and one that is very difficult to prove in spite of the fact that it seems so obvious.*

Which dot is inside?

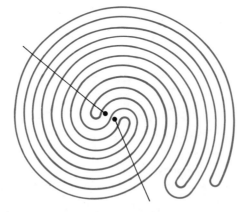

A closed curve that is not simple divides the plane into three or more parts. We call the points where the curve crosses itself **vertices.** Any other points on the curve may also be designated as vertices. We mark the ver-

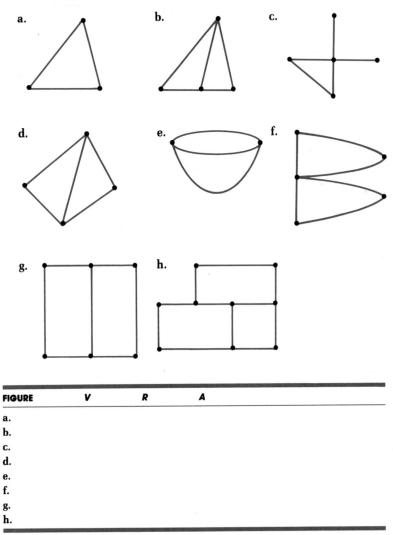

Figure 9.8i
A nonsimple closed curve

tices with black dots as in Fig. 9.8i, where five vertices are indicated. The regions into which the plane is divided are numbered 1, 2, 3, 4 in the figure. The simple curves with vertices as endpoints and containing no other endpoints are called **arcs**. In the figure, there are seven arcs. Can you count them?

1. For the networks below, fill in the table with the number of vertices (V), regions (R) into which the plane is divided, and arcs (A) for each figure. (Line segments connecting vertices are also called arcs.) See if you can discover a formula for A in terms of V and R. This formula is one form of the **Euler formula for networks.**

a. b. c.

d. e. f.

g. h.

FIGURE	V	R	A
a.			
b.			
c.			
d.			
e.			
f.			
g.			
h.			

Section	Item	Meaning	Example
9.1A	Collinear	On the same line	
9.1A	$\overleftrightarrow{AB}$	The line AB	
9.1A	Ray, $\overrightarrow{AB}$	A half-line and its endpoint	The ray AB:
9.1A	Line segment AB, $\overline{AB}$	The points A and B and the part of the line between A and B	The line segment AB:
9.1B	Coplanar	On the same plane	
9.1B	Skew lines	Lines that are not coplanar	
9.2	Plane angle	The figure formed by two rays with a common endpoint	
9.2	Vertex of an angle	The common point of the two rays	
9.2	Sides of the angles	The rays forming the angle	
9.2	Degree	$\frac{1}{360}$th of a complete revolution	
9.2	Straight angle	One-half of a complete revolution	
9.2	Right angle	One-quarter of a complete revolution	
9.2	Vertical angles	The opposite angles formed by two intersecting lines	
9.2	Supplementary angles	Angles whose measures add to 180°	
9.2	Transversal	A line that crosses two or more other lines	
9.2	Complementary angles	Angles whose measures add to 90°	

Section	Item	Meaning	Example
9.2	Perpendicular lines	Lines that intersect at right angles	
9.3A	Broken line	A sequence of connected straight line segments	
9.3A	Simple path	A path that does not cross itself	
9.3A	Closed path	A path that starts and ends at the same point	
9.3A	Polygon	A simple, closed, broken line	
9.3A	Sides (of polygon)	The line segments of the path	
9.3A	Vertices (of polygon)	The endpoints of the sides of the polygon	
9.3A	Convex polygon	A polygon that is not reentrant	
9.3A	Concave polygon	A polygon that is reentrant	
9.3A	Regular polygon	A polygon with all sides of equal length and all angles of equal size	
9.3A	Equilateral triangle	A triangle with three equal sides	
9.3B	Acute triangle	All three angles are acute	
9.3B	Right triangle	One of the angles is a right angle	
9.3B	Obtuse triangle	One of the angles is an obtuse angle	
9.3B	Scalene triangle	No equal sides	

Section	Item	Meaning	Example
9.3B	Isosceles triangle	Two equal sides	
9.3B	Trapezoid	A quadrilateral with two parallel and two nonparallel sides	
9.3B	Parallelogram	A quadrilateral with both pairs of opposite sides parallel	
9.3B	Rhombus	A parallelogram with four equal sides	
9.3B	Rectangle	A parallelogram whose angles are right angles	
9.3B	Square	A rectangle with four equal sides	
9.3B	Perimeter	Distance around a polygon	
9.3C	$S = (n - 2) \cdot 180°$	Sum of the measures of the angles of a polygon of n sides	
9.4	Similar figures	Figures with exactly the same shape	
9.5	Circle	The set of all coplanar points at a given fixed distance (the radius) from a point (the center)	
9.5	Circumference, $C = \pi d = 2\pi r$	The perimeter of a circle of diameter d (radius r)	
9.6	$A = bh$	The area of a rectangle of base b and height h	
9.6	$A = bh$	The area of a parallelogram of base b and height h	
9.6	$A = \frac{1}{2}bh$	The area of a triangle of base b and height h	

Section	Item	Meaning	Example
9.6	$A = \pi r^2$	The area of a circle of radius r	
9.6	Pythagorean theorem, $c^2 = a^2 + b^2$	The square of the hypotenuse c of a right triangle equals the sum of the squares of the other two sides, a and b	
9.7A	Polyhedron	A solid bounded by plane polygons	
9.7A	Convex polyhedron	One that lies entirely on one side of the plane of each of its faces	
9.7A	Concave polyhedron	One that is not convex	
9.7A	Prism	A polyhedron, two of whose faces are parallel and whose edges not on these faces are all parallel	
9.7A	Pyramid	A polyhedron with all but one of its vertices in one plane	
9.7B	$V = abc$	The volume of a rectangular box of length a, width b, and height c	
9.7B	$V = Bh$	The volume of a prism with base area B and height h	
9.7B	$V = \frac{1}{3}Bh$	The volume of a pyramid with base area B and height h	
9.7B	$V = \pi r^2 h$	The volume of a circular cylinder of radius r and height h	
9.7B	$V = \frac{1}{3}\pi r^2 h$	The volume of a circular cone of radius r and height h	
9.7B	$V = \frac{4}{3}\pi r^3$	The volume of a sphere of radius r	
9.8	Network	A connected set of line segments or curves	
9.8	Traversable network	One that can be drawn by tracing each line segment or curve exactly once without lifting the pencil from the paper	

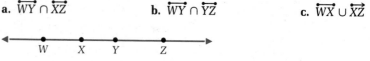

1. Refer to the line shown below and state what each of the following describes:

   a. $\overleftrightarrow{WY} \cap \overleftrightarrow{XZ}$ **b.** $\overleftrightarrow{WY} \cap \overrightarrow{YZ}$ **c.** $\overrightarrow{WX} \cup \overrightarrow{XZ}$

   W   X   Y   Z

2. Let A, B, C, and D be four points, no three of which are collinear. How many straight lines do these points determine? Name each of these lines. Would you change your answers if the four points were not coplanar?

3. Sketch a triangular prism. Label the vertices and then name the edges that are:

   a. Parallel lines   **b.** Skew lines   **c.** Intersecting lines

4. Is it true that two skew lines always determine a plane? Explain.

5. Sketch a broken line that is:

   a. Simple but not closed   **b.** Closed but not simple

6. Through how many degrees does the hour hand of a clock turn in going from:

   a. 12 to 4 o'clock?   **b.** 3 to 5 o'clock?

7. If $\angle A$ and $\angle B$ are supplementary angles, and $m\angle A = 3 \times m\angle B$, find the measures of the two angles.

8. In the figure below, lines PQ and RS are parallel. Find:

   a. $m\angle C$ **b.** $m\angle E$ **c.** $m\angle D$

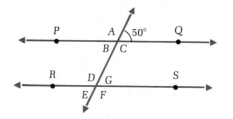

9. **a.** In a triangle ABC, $m\angle A = 38°$ and $m\angle B = 43°$. Find $m\angle C$.
   b. In a triangle ABC, $m\angle A = m\angle B = 2 \times m\angle C$. Find the measures of the three angles.

10. Angles A and B are the acute angles of a right triangle, and $m\angle A = 4 \times m\angle B$. Find $m\angle A$ and $m\angle B$.

11. What is the measure of one of the interior angles of a regular polygon with:

    a. Nine sides?   **b.** Ten sides?

12. A rectangular plot of ground is to be enclosed with 120 yd of fencing. If the plot is to be twice as long as it is wide, what must be its dimensions?

13. Triangles ABC and XYZ are similar, with $m\angle A = m\angle X$ and $m\angle B = m\angle Y$. If AB, BC, and AC are 2 in., 3 in., and 4 in. long, respectively, and XY is 3 in. long, find the lengths of YZ and XZ.

14. The center of a circle is also the center of a square of side 2 cm. The circle passes through the four vertices of the square. Find the circumference of the circle. (Leave your answer in terms of π.)

15. In Problem 14, find the area of the region that is inside the circle and outside the square. (Leave your answer in terms of π.)

16. If you rolled up an $8\frac{1}{2}$ by 11-in. sheet of paper into the largest possible cylinder $8\frac{1}{2}$ in. high, what would be the diameter of the cylinder? Use $\frac{22}{7}$ as the approximate value of π.

17. A window is in the shape of a rectangle surmounted by an isosceles triangle. The window is 3 ft wide, 6 ft high at the center, and 4 ft high on the sides. Find the total area of the window.

18. A rectangle is 84 ft long and 13 ft wide. Find the length of its diagonal.

19. A window is in the shape of a rectangle surmounted by a semicircle. The width of the window is 3 ft, which is also the diameter of the circle, and the height of the rectangular part is 4 ft. Find the total area of the window.

20. A circle of diameter 2 in. has its center at the center of a square of side 2 in. Find the area of the region that is inside the square and outside the circle.

21. A solid consists of a cone and a hemisphere mounted base to base. If the radius of the common base is 2 in. and the volume of the cone is equal to the volume of the hemisphere, what is the height of the cone?

22. The area of a regular hexagon is $(3\sqrt{3}/2)s^2$, where s is the length of the side. A circular cylinder of radius s and a regular hexagonal prism of side s have the same height. What is the ratio of the volume of the prism to the volume of the cylinder?

23. The base of a prism is a triangle whose base is 5 ft and whose height is 3 ft. If the prism is 4 ft high, what is its volume?

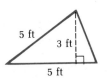

24. The triangle shown in the margin is the base of a pyramid that is 4 ft high. What is the volume of this pyramid?

25. Draw two networks, each with five vertices, such that one of them is traversable and the other is not traversable.

5 ft · 3 ft · 5 ft

CHAPTER 10

MATRICES

Arthur Cayley was born August 16, 1821, at Richmond, Surrey, England. When he was 8 years old, his father sent him to a private school, and at 14, the boy entered King's College School in London. All his teachers recognized his mathematical genius and encouraged him even at an early age to become a mathematician.

At the age of 17, Arthur was enrolled in Trinity College, where he was again quickly recognized, and by his third year was rated in a class by himself, "above the first." At 21, he placed first in the Mathematical Tripos and in the even more difficult Smith's Prize examination.

Elected a Fellow of Trinity College and an assistant tutor for 3 years, Cayley had all the time he wanted for his research. Before reaching the age of 25, he published 25 papers, which set the pattern of his work for the rest of his life.

With all this prolific creativity, Cayley was not a "greasy grind." He loved to read; he enjoyed tramping through the countryside, mountaineering, and watercolor sketching. He took delightful vacations in Switzerland and Italy.

Cayley left Cambridge in 1846 because he could not persuade himself to take "holy orders," a prerequisite for an additional appointment. He turned to the study of law and was admitted to the bar in 1849. Though successful, he did not enjoy the profession, and worked at it only enough to enable him to do his mathematics.

In 1863, Cambridge University established a new chair in mathematics (the Sadlerian Professorship) and offered the appointment to Cayley, who promptly accepted. During the same year, at the age of 42, he married Susan Moline. Enjoying his university duties and a happy home life, Cayley continued his mathematical research until his death in January 1895. He was truly a noble person, generous with his help to students, admired and respected by all who knew him.

Cayley's collected works comprise over 960 papers containing much that has become part of the mainstream of mathematics. One of his great inventions was that of matrices and matrix theory, the ideas for which originated in one of his papers published in 1858. Sixty-seven years later, Heisenberg recognized matrix algebra as exactly the tool he needed for his revolutionary and far-reaching work in quantum mechanics. This is one of the many illustrations that much applicable mathematics was invented long before anyone even imagined the application.

ARTHUR CAYLEY (1821–1895)
Picture Collection, the Branch Libraries, The New York Public Library

But, as for everything else, so for a mathematical theory — beauty can be perceived but not explained.

ARTHUR CAYLEY

MATRIX OPERATIONS

In this chapter we shall study a mathematical system involving *matrices*. Since about 1948, matrices have become an important tool in business and the social sciences, just as they have been in the physical sciences and technology for some decades before that. Matrices can be used to store information. For example, suppose the ABC Toy Company manufactures two types of toys, regular and deluxe. Each of these toys requires bolts, clamps, and screws according to Table 10.1a. If we remember that the rows are labeled bolts, clamps, and screws, in that order, then the rectangular array of numbers

Table 10.1a

	REGULAR	DELUXE
BOLTS	5	3
CLAMPS	2	4
SCREWS	7	10

$$R = \begin{bmatrix} 5 & 3 \\ 2 & 4 \\ 7 & 10 \end{bmatrix}$$

gives all the necessary information regarding the assembly requirements of each toy.

▶ **Definition 10.1a**

> A rectangular array of numbers enclosed in square brackets (or, sometimes, parentheses) is called a **matrix.** (The plural is **matrices.**) The entries in a matrix are called its **elements.**

The matrix

$$\begin{bmatrix} a_{11} & a_{12} \\ a_{21} & a_{22} \\ a_{31} & a_{32} \end{bmatrix}$$

where the subscripts give the **row** and **column** of each element, has 3 rows and 2 columns. We speak of it as a 3 by 2 (3×2) matrix. The number of rows and the number of columns in a matrix are called the **dimensions** of the matrix. Thus, a matrix with m rows and n columns is an $m \times n$ matrix.

For convenience, we often denote matrices by single capital letters, such as A, B, C. We might write

$$A = \begin{bmatrix} 1 & 2 \\ -3 & 0 \end{bmatrix} \qquad B = \begin{bmatrix} 1 \\ 3 \\ 5 \end{bmatrix} \qquad C = \begin{bmatrix} 1 & 0 & -2 & 4 \end{bmatrix}$$

Of these, A is a 2×2 matrix, B is a 3×1 matrix, and C is a 1×4 matrix. A matrix such as B, consisting of a single column, is called a **column matrix.** Likewise, a matrix such as C, consisting of a single row, is called a **row matrix.** A matrix such as A, having the same number of columns as rows, is called a **square matrix.**

Definition 10.1b | Two matrices A and B are said to be **equal** ($A = B$) if and only if the corresponding elements are equal. (This means that they have the same dimensions, and that each element of A equals the element of B in the same row and column.)

For example, if

$$A = \begin{bmatrix} 1 & 2 \\ -3 & 0 \end{bmatrix} \qquad B = \begin{bmatrix} 1 & 4 \\ -3 & 0 \end{bmatrix} \qquad C = \begin{bmatrix} 1 & 2 \\ -3 & 0 \end{bmatrix}$$

then $A \neq B$, because the element in the first row, second column of A is 2 and that in B is 4. On the other hand, $A = C$, because each element of A is the same as the corresponding element of C.

A. Addition, Subtraction, and Multiplication by a Number

Suppose that the ABC Toy Company makes 5 of each of its toys. Of course, it would need 5 times as many bolts, clamps, and screws as for just one of each. It is easily seen that the matrix of parts requirements for 5 of each toy is obtained by multiplying each element of the original matrix by 5 to get

$$\begin{bmatrix} 5 \times 5 & 5 \times 3 \\ 5 \times 2 & 5 \times 4 \\ 5 \times 7 & 5 \times 10 \end{bmatrix} = \begin{bmatrix} 25 & 15 \\ 10 & 20 \\ 35 & 50 \end{bmatrix}$$

If we write R for the original parts requirement matrix, then it is natural to write $5R$ for the new matrix. Thus,

$$5R = 5 \begin{bmatrix} 5 & 3 \\ 2 & 4 \\ 7 & 10 \end{bmatrix} = \begin{bmatrix} 25 & 15 \\ 10 & 20 \\ 35 & 50 \end{bmatrix}$$

This idea motivates us in defining the multiplication of a matrix by a real number.

▶

Definition 10.1c | If A is a matrix and k is a real number, then the **multiplication of A by k,** symbolized by kA or Ak, is the operation that multiplies every element of A by k.

Thus, if

$$A = \begin{bmatrix} 1 & 2 \\ -3 & 0 \end{bmatrix}$$

then

$$-2A = \begin{bmatrix} (-2) \times 1 & (-2) \times 2 \\ (-2) \times (-3) & (-2) \times 0 \end{bmatrix} = \begin{bmatrix} -2 & -4 \\ 6 & 0 \end{bmatrix}$$

EXAMPLE 1 If

$$A = \begin{bmatrix} 1 & x \\ 0 & y \end{bmatrix} \quad \text{and} \quad B = \begin{bmatrix} 2 & 6 \\ 0 & -10 \end{bmatrix}$$

find x and y to satisfy the equation $2A = B$.

Solution   Substituting for A and B in the equation $2A = B$, we get

$$2 \begin{bmatrix} 1 & x \\ 0 & y \end{bmatrix} = \begin{bmatrix} 2 & 6 \\ 0 & -10 \end{bmatrix}$$

Thus,

$$\begin{bmatrix} 2 & 2x \\ 0 & 2y \end{bmatrix} = \begin{bmatrix} 2 & 6 \\ 0 & -10 \end{bmatrix}$$

By Definition 10.1b, then,

$$2x = 6 \quad \text{and} \quad 2y = -10$$

Hence, $x = 3$ and $y = -5$.   ◀

Suppose that a workman in the ABC Toy Company assembles 6 of each of their toys during the first half of the day and 5 of each during the second half. (He gets a little bit tired of regulars and deluxes before the day is over.) What is the matrix of parts requirements for the day's work?

We see that the matrix $6R$ will be the appropriate matrix for his first half-day, and that $5R$ will be the matrix for the second half-day. Thus,

$$6R = \begin{bmatrix} 30 & 18 \\ 12 & 24 \\ 42 & 60 \end{bmatrix} \begin{matrix} \text{Bolts} \\ \text{Clamps} \\ \text{Screws} \end{matrix} \quad \text{and} \quad 5R = \begin{bmatrix} 25 & 15 \\ 10 & 20 \\ 35 & 50 \end{bmatrix} \begin{matrix} \text{Bolts} \\ \text{Clamps} \\ \text{Screws} \end{matrix}$$

are the matrices for the parts requirements for the two half-days. It is easy to see that the total requirements can be found by adding the corresponding elements as follows:

$$\begin{bmatrix} 30 + 25 & 18 + 15 \\ 12 + 10 & 24 + 20 \\ 42 + 35 & 60 + 50 \end{bmatrix} = \begin{bmatrix} 55 & 33 \\ 22 & 44 \\ 77 & 110 \end{bmatrix}$$

It seems perfectly natural to regard this as $6R + 5R$, the "sum" of the two matrices. These and similar considerations motivate the following definition of addition of matrices.

▶   **Definition 10.1d**

The **sum** of two matrices A and B *of the same dimensions* is written $A + B$ and is the matrix obtained by adding corresponding elements of A and B.

EXAMPLE 2   If

$$A = \begin{bmatrix} 2 & 1 \\ 3 & 4 \end{bmatrix} \quad \text{and} \quad B = \begin{bmatrix} 1 & 5 \\ 4 & 2 \end{bmatrix}$$

find:

a. $A + B$ **b.** $2A + 3B$

Solution   **a.** $A + B = \begin{bmatrix} 2+1 & 1+5 \\ 3+4 & 4+2 \end{bmatrix} = \begin{bmatrix} 3 & 6 \\ 7 & 6 \end{bmatrix}$

b. $2A + 3B = \begin{bmatrix} 4 & 2 \\ 6 & 8 \end{bmatrix} + \begin{bmatrix} 3 & 15 \\ 12 & 6 \end{bmatrix} = \begin{bmatrix} 7 & 17 \\ 18 & 14 \end{bmatrix}$ ◀

▶   **Definition 10.1e**

> If A and B are two matrices of the same dimensions, then the **difference** $A - B$ is given by $A - B = A + (-1)B$, which is the same as the matrix formed by subtracting each element of B from the corresponding element of A.

EXAMPLE 3   If A and B are as in Example 2, find $2A - 3B$.

Solution   With $2A$ and $3B$ as in Example 2, we have

$$\begin{aligned} 2A - 3B &= \begin{bmatrix} 4 & 2 \\ 6 & 8 \end{bmatrix} - \begin{bmatrix} 3 & 15 \\ 12 & 6 \end{bmatrix} \\ &= \begin{bmatrix} 4-3 & 2-15 \\ 6-12 & 8-6 \end{bmatrix} = \begin{bmatrix} 1 & -13 \\ -6 & 2 \end{bmatrix} \end{aligned}$$ ◀

B.   Matrix Multiplication

Let us return to the ABC Toy company and recall that the assembly requirement matrix was

$$\begin{array}{c} \\ \text{Bolts} \\ R = \text{Clamps} \\ \text{Screws} \end{array} \begin{array}{cc} \text{Regular} & \text{Deluxe} \\ \begin{bmatrix} 5 & 3 \\ 2 & 4 \\ 7 & 10 \end{bmatrix} \end{array}$$

Suppose now that Jane assembles regulars and Jim assembles deluxes. Jane turns out 18 regulars, and Jim turns out 16 deluxes per day. We can specify the total output by means of the column matrix

$$C = \begin{bmatrix} 18 \\ 16 \end{bmatrix} \begin{array}{l} \text{Regular} \\ \text{Deluxe} \end{array}$$

Can we use these matrices to calculate how many bolts, clamps, and screws are needed for Jane's and Jim's daily output? Yes! We proceed as follows: To calculate the number of bolts, we use a "row–column" type of multiplication, as indicated by the following scheme:

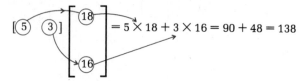

$$= 5 \times 18 + 3 \times 16 = 90 + 48 = 138$$

Similarly, we find the number of clamps to be

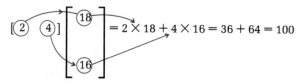

$$= 2 \times 18 + 4 \times 16 = 36 + 64 = 100$$

and the number of screws to be

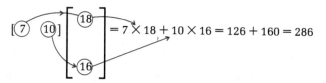

$$= 7 \times 18 + 10 \times 16 = 126 + 160 = 286$$

Next, let us write the result of our computation in the briefer form

$$
\begin{matrix} \text{Bolts} \\ \text{Clamps} \\ \text{Screws} \end{matrix}
\begin{bmatrix} 5 & 3 \\ 2 & 4 \\ 7 & 10 \end{bmatrix}
\begin{bmatrix} 18 \\ 16 \end{bmatrix}
=
\begin{bmatrix} 138 \\ 100 \\ 286 \end{bmatrix}
\begin{matrix} \text{Bolts} \\ \text{Clamps} \\ \text{Screws} \end{matrix}
$$

If we call the daily parts requirement matrix D, then we can write

$$RC = D$$

and regard the RC as a type of product. Notice that this product is a column matrix in which the first element is obtained from a row–column multiplication using the first row of R. Similarly, the second element is obtained in the same way using the second row of R, and likewise for the third element.

Now, suppose the ABC Toy Company needs to produce 18 regular and 16 deluxe toys today and 10 regular and 20 deluxe toys tomorrow. To represent the production matrix P for the 2 days, we write

$$
P =
\begin{matrix} \text{Today} & \text{Tomorrow} \end{matrix}
\begin{bmatrix} 18 & 10 \\ 16 & 20 \end{bmatrix}
\begin{matrix} \text{Regular} \\ \text{Deluxe} \end{matrix}
$$

How many bolts, clamps, and screws are needed in the next 2 days? We need to make the computation for tomorrow only, since we have already done it for today. Thus,

$$\begin{bmatrix} 5 & 3 \\ 2 & 4 \\ 7 & 10 \end{bmatrix} \begin{bmatrix} 10 \\ 20 \end{bmatrix} = \begin{bmatrix} 5 \times 10 + 3 \times 20 \\ 2 \times 10 + 4 \times 20 \\ 7 \times 10 + 10 \times 20 \end{bmatrix} = \begin{bmatrix} 110 \\ 100 \\ 270 \end{bmatrix}$$

If we denote the final requirement matrix by W, then it is natural to write

$$\begin{bmatrix} 5 & 3 \\ 2 & 4 \\ 7 & 10 \end{bmatrix} \begin{bmatrix} 18 & 10 \\ 16 & 20 \end{bmatrix} = \begin{bmatrix} 138 & 110 \\ 100 & 100 \\ 286 & 270 \end{bmatrix}$$

or

$$RP = W$$

where each element in W is calculated by the following rule: **The element of W in the ith row and jth column is the row–column product of the ith row of R and the jth column of P.**

For example, the element of W in the first row, second column position is the row–column product of the first row of R and the second column of P; that is,

$$[5 \quad 3] \begin{bmatrix} 10 \\ 20 \end{bmatrix} = 5 \times 10 + 3 \times 20 = 110$$

You should check all the elements of W to be sure you understand the rule.

Definition 10.1f

> The matrix $W = RP$ is called the **product** of the matrices R and P in that order, and the calculation of W according to the rule given above is called **matrix multiplication.**

The preceding discussion can easily be generalized to the multiplication of two matrices A and B. Suppose that A has as many columns as B has rows. Then we can do a row–column multiplication with any row of A and any column of B. Thus, we have the following definition:

Definition 10.1g

> The matrices A and B are said to be **conformable** with respect to multiplication in the order AB if the number of columns in A is the same as the number of rows in B. In this case, each element of the product AB is formed as follows: The element in row i and column j is the row–column product of the ith row of A and the jth column of B.

If two matrices are not conformable, then we do not attempt to define their product. Notice that A and B might be conformable for multiplication in the order AB and not in the order BA. Thus, if A is $m \times n$ and B is $n \times k$, then A and B are conformable for multiplication in the order AB, but not in the order BA unless $m = k$. Note that AB will be an $m \times k$

matrix. For instance, if A is 2×3 and B is 3×4, then we can form AB, and it will be a 2×4 matrix; but we cannot form BA. Even when the matrices are conformable for both orders of multiplication, there is no reason why the results need to be equal. In fact, it is generally true that $AB \neq BA$.

This failure of matrix multiplication to be commutative was one of the exciting properties of matrices to the mathematicians who first studied them.

EXAMPLE 4   Let

$$A = \begin{bmatrix} 2 & 1 & 0 \\ 3 & 2 & 4 \end{bmatrix} \quad \text{and} \quad B = \begin{bmatrix} 1 & 4 \\ 3 & 1 \\ 2 & 3 \end{bmatrix}$$

Calculate the product AB.

Solution   Because A is 2×3 and B is 3×2, we may proceed according to Definition 10.1g to get the 2×2 product matrix

$$
\begin{aligned}
AB &= \begin{bmatrix} 2 & 1 & 0 \\ 3 & 2 & 4 \end{bmatrix} \begin{bmatrix} 1 & 4 \\ 3 & 1 \\ 2 & 3 \end{bmatrix} \\
&= \begin{bmatrix} 2 \times 1 + 1 \times 3 + 0 \times 2 & 2 \times 4 + 1 \times 1 + 0 \times 3 \\ 3 \times 1 + 2 \times 3 + 4 \times 2 & 3 \times 4 + 2 \times 1 + 4 \times 3 \end{bmatrix} \\
&= \begin{bmatrix} 5 & 9 \\ 17 & 26 \end{bmatrix}
\end{aligned}
$$
◄

EXAMPLE 5   Using the same matrices as in Example 4, form the product BA (if possible).

Solution   We first check that B has as many columns as A has rows (two in each case). This shows that the matrices are conformable for multiplication in the order BA. Thus, by Definition 10.1g, we get the 3×3 matrix

$$
\begin{aligned}
BA &= \begin{bmatrix} 1 & 4 \\ 3 & 1 \\ 2 & 3 \end{bmatrix} \begin{bmatrix} 2 & 1 & 0 \\ 3 & 2 & 4 \end{bmatrix} \\
&= \begin{bmatrix} 1 \times 2 + 4 \times 3 & 1 \times 1 + 4 \times 2 & 1 \times 0 + 4 \times 4 \\ 3 \times 2 + 1 \times 3 & 3 \times 1 + 1 \times 2 & 3 \times 0 + 1 \times 4 \\ 2 \times 2 + 3 \times 3 & 2 \times 1 + 3 \times 2 & 2 \times 0 + 3 \times 4 \end{bmatrix} \\
&= \begin{bmatrix} 14 & 9 & 16 \\ 9 & 5 & 4 \\ 13 & 8 & 12 \end{bmatrix}
\end{aligned}
$$
◄

Examples 4 and 5 offer unassailable evidence that in general $AB \neq BA$. If A and B were both square matrices of the same size, do you think that $AB = BA$ always? Try it for some simple 2×2 matrices!

C. The Identity Matrix

Is there an identity for matrix multiplication? The answer is yes. For example, the multiplication of any 2×2 matrix A by

$$I = \begin{bmatrix} 1 & 0 \\ 0 & 1 \end{bmatrix}$$

gives the matrix A back again. We find that

$$IA = AI = A \quad \text{Try it for } \begin{bmatrix} a & b \\ c & d \end{bmatrix}.$$

It can be shown that I is the only matrix that has this property for *all* 2×2 matrices. Because of this uniqueness, I is called the **multiplicative iden- tity** for 2×2 matrices.

Similarly for 3×3 matrices, the multiplicative identity is the matrix

$$I = \begin{bmatrix} 1 & 0 & 0 \\ 0 & 1 & 0 \\ 0 & 0 & 1 \end{bmatrix}$$

The **main diagonal** of a square matrix is the diagonal set of numbers running from the upper left corner to the lower right corner of the matrix. The two identity matrices, one for 2×2 and the other for 3×3 matrices, are special instances of the general result that the $n \times n$ identity matrix is the $n \times n$ matrix with 1's on the main diagonal and 0's for all other ele- ments.

Addition properties of matrices are discussed in Discovery 10.1.

EXERCISE 10.1

A. For Problems 1–6, suppose that

$$A = \begin{bmatrix} 2 & 1 \\ 0 & -1 \end{bmatrix} \qquad B = \begin{bmatrix} -2 & 4 \\ 3 & 1 \end{bmatrix} \qquad C = \begin{bmatrix} 3 & 5 \\ 2 & 0 \end{bmatrix}$$

1. Find:
   a. $4A$ **b.** $-3B$ **c.** $C - B$

2. Find:
   a. $2A + B$ **b.** $4A + 3B$ **c.** $2A - C$

3. Find:
   a. $A + B + C$ **b.** $A + B - C$

4. Find:
   a. $A - B + 2C$ **b.** $2A + 2B - C$

5. Find:
   a. $7A + 4B - 2C$ **b.** $2A - 2B - 3C$

6. Find:
   a. $5A + 4B - 4C$ **b.** $4A - 2B - 4C$

For Problems 7-12, suppose that

$$A = \begin{bmatrix} 1 & -1 & 2 \\ 3 & 0 & -2 \\ 4 & 2 & 1 \end{bmatrix} \qquad B = \begin{bmatrix} -1 & 2 & 1 \\ 4 & 3 & -1 \\ 0 & 1 & -1 \end{bmatrix} \qquad C = \begin{bmatrix} 0 & -1 & 3 \\ 1 & -2 & 4 \\ 3 & -3 & 0 \end{bmatrix}$$

7. Find:
   a. $A + B$
   b. $A - B$
   c. $A - C$
8. Find:
   a. $2A$
   b. $-2B$
   c. $-3C$
9. Find:
   a. $2A + 3B$
   b. $-2A + 3B$
10. Find:
    a. $4A + 4B$
    b. $4A - 4B$
11. Find:
    a. $3A - 2C$
    b. $B + C$
12. Find:
    a. $3A + B - 4C$
    b. $5A + 2B + 3C$

B. In Problems 13-20, use

$$A = \begin{bmatrix} 1 & -2 & 1 \\ 2 & 0 & 2 \\ -1 & 1 & 3 \end{bmatrix} \qquad B = \begin{bmatrix} 3 & 2 & 0 \\ 1 & 1 & -1 \\ 2 & 0 & 1 \end{bmatrix} \qquad C = \begin{bmatrix} 1 & 0 & 2 \\ 3 & 2 & 1 \\ 2 & 0 & 1 \end{bmatrix}$$

and evaluate the given expression.

13. AB
14. AC
15. BA
16. CA
17. $(A - B)(A + B)$
18. $(A - C)(A + C)$
19. $A^2 - B^2$
20. $A^2 - C^2$

C.
21. If I is the 2×2 identity matrix, find I^2.
22. If n is a positive integer, is it true that $I^n = I$? Why?

In Problems 23-27, verify that $AB = I$, and calculate BA.

23. $A = \begin{bmatrix} 2 & 1 \\ 1 & 1 \end{bmatrix}$; $B = \begin{bmatrix} 1 & -1 \\ -1 & 2 \end{bmatrix}$

24. $A = \begin{bmatrix} -2 & 3 \\ 1 & -1 \end{bmatrix}$; $B = \begin{bmatrix} 1 & 3 \\ 1 & 2 \end{bmatrix}$

25. $A = \begin{bmatrix} 2 & 5 \\ 1 & 3 \end{bmatrix}$; $B = \begin{bmatrix} 3 & -5 \\ -1 & 2 \end{bmatrix}$

26. $A = \begin{bmatrix} 1 & 0 & 1 \\ 0 & 1 & 1 \\ 1 & 0 & 0 \end{bmatrix}$; $B = \begin{bmatrix} 0 & 0 & 1 \\ -1 & 1 & 1 \\ 1 & 0 & -1 \end{bmatrix}$

27. $A = \begin{bmatrix} 1 & 0 & 1 \\ 0 & 2 & -1 \\ 2 & 1 & 2 \end{bmatrix}$; $B = \begin{bmatrix} 5 & 1 & -2 \\ -2 & 0 & 1 \\ -4 & -1 & 2 \end{bmatrix}$

28. Let R be the row matrix [2 1]. Is it possible for there to be a 2×2 matrix, say J, such that $JR = RJ = R$? Explain.

29. If A is a nonsquare matrix of dimensions $m \times n$, do you think there could be a square matrix J such that $AJ = JA = A$? Explain.

30. Find all possible 2×2 matrices A such that $A = 3A$.

31. A square matrix A is called **idempotent** if $A^2 = A$. All identity matrices and all zero matrices (elements all 0's) are idempotent. Show that the following matrices are also idempotent:

a. $\begin{bmatrix} 1 & 2 \\ 0 & 0 \end{bmatrix}$ **b.** $\begin{bmatrix} \frac{1}{2} & \frac{1}{2} \\ \frac{1}{2} & \frac{1}{2} \end{bmatrix}$

32. A square matrix A is called **nilpotent** if there is a positive integer n such that A^n is a zero matrix (elements all 0's). Show that the following matrices are nilpotent:

a. $\begin{bmatrix} 2 & 1 \\ -4 & -2 \end{bmatrix}$ **b.** $\begin{bmatrix} 0 & 0 & 1 \\ 2 & 0 & 3 \\ 0 & 0 & 0 \end{bmatrix}$

33. The E-Z Rest Furniture Company makes armchairs and rocking chairs in three models each: E, an economy model; M, a medium-priced model; L, a luxury model. Each month, the company turns out 20 model E armchairs, 15 model M armchairs, 10 model L armchairs, 12 model E rockers, 8 model M rockers, and 5 model L rockers.

a. Write this information as a 2×3 matrix.

b. Use your answer in part a to obtain a matrix showing the total production for 6 months.

34. Suppose the costs of materials for E-Z Rest's armchairs (see Problem 33) are $30 for model E, $35 for model M, and $45 for model L; and for the rockers, $35 for model E, $40 for model M, and $60 for model L.

a. Write this information as a 3×2 matrix.

b. Suppose that costs increase by 20%. Write the new matrix by multiplying your answer in part a by 1.2.

35. At the end of 6 months, the E-Z Rest Furniture Company (see Problem 33) has sold armchairs as follows: 90 model E, 75 model M, and 50 model L. They also sold rockers as follows: 60 model E, 20 model M, and 30 model L. Use matrix methods to obtain a matrix showing how many of each item are left at the end of the 6 months. Use your answer to Problem 33 and assume that there was no unsold stock at the beginning of the period.

36. The ABC Toy Company figures that the bolts, clamps, and screws used in assembling its toys cost, respectively, 5¢, 10¢, and 2¢ each. Show that the cost of these assembly materials for each toy is obtained by carrying out the multiplication CR, where C is the **cost matrix** [5 10 2] and R is the assembly requirement matrix given at the beginning of this section.

37. The ABC Toy Company has the projected production schedule shown in the table on page 466 for the 5 months preceding December.

	JULY	AUG.	SEPT.	OCT.	NOV.
REGULAR	100	200	300	400	300
DELUXE	50	100	200	200	300

Let M be the matrix with these numbers as the elements, and let R be the same as in Problem 36. Use these matrices to find the schedule of assembly requirements for the 5 months so you can fill in the missing items in the following table:

	JULY	AUG.	SEPT.	OCT.	NOV.
BOLTS	650				
CLAMPS		800			
SCREWS			4100		

38. Let A be the matrix of assembly requirements you calculated in Problem 37, and let C be the cost matrix of Problem 36. Calculate CA to find the cost of assembly materials for each of the 5 months.

USING YOUR KNOWLEDGE 10.1

Here are six special 2×2 matrices:

$$r_1 = \begin{bmatrix} 1 & 0 \\ 0 & -1 \end{bmatrix} \qquad r_2 = \begin{bmatrix} -1 & 0 \\ 0 & 1 \end{bmatrix} \qquad r_3 = \begin{bmatrix} 0 & 1 \\ 1 & 0 \end{bmatrix}$$

$$R_1 = \begin{bmatrix} 0 & -1 \\ 1 & 0 \end{bmatrix} \qquad R_2 = \begin{bmatrix} -1 & 0 \\ 0 & -1 \end{bmatrix} \qquad R_3 = \begin{bmatrix} 0 & 1 \\ -1 & 0 \end{bmatrix}$$

It is interesting to see what effect these matrices have when they multiply a column matrix,

$$P = \begin{bmatrix} a \\ b \end{bmatrix}$$

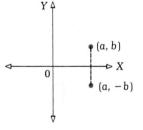

that represents a point (a, b) in the XY plane. For example,

$$r_1 \times P = \begin{bmatrix} 1 & 0 \\ 0 & -1 \end{bmatrix}\begin{bmatrix} a \\ b \end{bmatrix} = \begin{bmatrix} a \\ -b \end{bmatrix}$$

which represents the point $(a, -b)$, the reflection of (a, b) across the X axis. (See the figure in the margin.) Because the matrices r_1, r_2, and r_3 all correspond to reflections across certain lines in the plane, we may think of these three matrices as **reflectors.**
Similarly,

$$R_1 \times P = \begin{bmatrix} 0 & -1 \\ 1 & 0 \end{bmatrix}\begin{bmatrix} a \\ b \end{bmatrix} = \begin{bmatrix} -b \\ a \end{bmatrix}$$

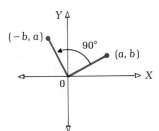

represents the point $(-b, a)$, which can be obtained from the point (a, b) by a rotation of $90°$ around the origin (see the figure in the margin). Because all three of the matrices R_1, R_2, and R_3 correspond to rotations in the plane, we may think of them as **rotators.**

What is accomplished by each of the following?

1. $r_2 \times P$ **2.** $r_3 \times P$ **3.** $R_2 \times P$ **4.** $R_3 \times P$

5. A triangle with its vertices at $(2, 1)$, $(3, 1)$, and $(3, 2)$ can be represented by the matrix

$$T = \begin{bmatrix} 2 & 3 & 3 \\ 1 & 1 & 2 \end{bmatrix}$$

Form the products $r_1 \times T$, $r_2 \times T$, and so on. Then draw graphs to show the effect of each multiplication.

6. Consider another special matrix

$$K = \begin{bmatrix} k & 0 \\ 0 & k \end{bmatrix}$$

What happens to the triangle in Problem 5 if you form the product $K \times T$? Try this for $k = 2$.

DISCOVERY 10.1

We want to discover here in what ways the set S of 2×2 matrices behaves like the set I of integers under addition and multiplication. To do this, we consider the following questions:

1. Is S closed with respect to addition? [Hint: Is the sum of two 2×2 matrices a 2×2 matrix?]
2. Is S associative with respect to addition?
3. Can you find an additive identity for the set S?
   Hint: If

$$A = \begin{bmatrix} 1 & 3 \\ 4 & 6 \end{bmatrix}$$

find a matrix B so that

$$\begin{bmatrix} 1 & 3 \\ 4 & 6 \end{bmatrix} + \begin{bmatrix} & \\ & \end{bmatrix} = \begin{bmatrix} 1 & 3 \\ 4 & 6 \end{bmatrix}$$

4. Does every matrix in S have an additive inverse? For example, if

$$A = \begin{bmatrix} 1 & 2 \\ -3 & 4 \end{bmatrix}$$

try to find a matrix B so that

$$\begin{bmatrix} 1 & 2 \\ -3 & 4 \end{bmatrix} + \begin{bmatrix} \quad & \quad \\ \quad & \quad \end{bmatrix} = \begin{bmatrix} 0 & 0 \\ 0 & 0 \end{bmatrix}$$

5. Is $A + B = B + A$ for any two 2×2 matrices in S?
6. Consider the set S of 2×2 matrices.
   a. Is S closed with respect to multiplication?
   b. Is S associative with respect to multiplication?
   c. Can you find a multiplicative identity for the set S?
7. The inverse of a 2×2 matrix A is defined to be a matrix B such that $AB = BA = I$, where I is the 2×2 identity matrix. Can you show that the matrix given below has no inverse?

$$A = \begin{bmatrix} 1 & 2 \\ 0 & 0 \end{bmatrix}$$

8. Can you discover whether the matrix given below has an inverse?

$$A = \begin{bmatrix} 2 & 4 \\ 1 & 2 \end{bmatrix}$$

▶ **Computer Corner 10.1** *We have supplied a program that adds or subtracts matrices and another one that multiplies matrices, provided you enter the numbers in the rows and columns (see pages 747 and 748). Work some of the problems in this section using these programs.*

10.2

SYSTEMS OF LINEAR EQUATIONS

In Chapter 8 we considered systems of two linear equations in two unknowns and discovered how to solve such systems. There are many practical applications that require the consideration of m linear equations in n unknowns. We shall not try to be so general here; instead, we shall look in detail only into the case of three equations in three unknowns. The techniques we shall use are applicable to all the more general cases.

Let the unknowns be x, y, z, and let the equations be

$$\begin{aligned} a_1x + b_1y + c_1z &= d_1 \\ a_2x + b_2y + c_2z &= d_2 \\ a_3x + b_3y + c_3z &= d_3 \end{aligned} \tag{1}$$

The a's, b's, and c's are usually called the **coefficients;** they, along with the d's, are assumed to be specific known numbers.

Define three matrices A, X, and D as follows:

$$A = \begin{bmatrix} a_1 & b_1 & c_1 \\ a_2 & b_2 & c_2 \\ a_3 & b_3 & c_3 \end{bmatrix} \qquad X = \begin{bmatrix} x \\ y \\ z \end{bmatrix} \qquad D = \begin{bmatrix} d_1 \\ d_2 \\ d_3 \end{bmatrix}$$

Then, system (1) can be written in the brief form

$$AX = D \tag{2}$$

where on the left we mean that the 3×3 matrix A multiplies the column matrix X. This operation gives a column matrix whose elements are the respective left-hand sides of the given system. Because the equation $AX = D$ implies that corresponding elements are equal, this equation signifies system (1) exactly. We can go back and forth, as it suits us, from one form to the other.

For example, let the system of equations be

$$\begin{aligned} 2x - y + z &= 3 \\ x + y &= -1 \\ 3x - y - 2z &= 7 \end{aligned} \tag{3}$$

Then, by writing

$$A = \begin{bmatrix} 2 & -1 & 1 \\ 1 & 1 & 0 \\ 3 & -1 & -2 \end{bmatrix} \qquad X = \begin{bmatrix} x \\ y \\ z \end{bmatrix} \qquad D = \begin{bmatrix} 3 \\ -1 \\ 7 \end{bmatrix}$$

we can symbolize the given set of equations by

$$AX = D \tag{4}$$

(Be sure to verify this!)

The matrix A is called the **matrix of the system** (or sometimes, the matrix of coefficients). The matrix formed by adjoining the column matrix D to the matrix A is denoted by the symbol $[A|D]$ and is called the **augmented matrix.** For system (3), the augmented matrix is

$$[A|D] = \begin{bmatrix} 2 & -1 & 1 & 3 \\ 1 & 1 & 0 & -1 \\ 3 & -1 & -2 & 7 \end{bmatrix}$$

We may regard this symbol as a listing of the detached coefficients of the system. The vertical line takes the place of the equals signs and separates the coefficient matrix from the column of right-hand terms.

Suppose we wish to solve system (3). This means that we wish to find all sets of values of (x, y, z) that make all three equations true. We first need to consider what changes can be made in the system to yield an **equivalent system,** a system that has exactly the same solutions as the original system. We need to consider only three simple operations:

1. **The order of the equations may be changed.** This clearly cannot affect the solutions.
2. **Any of the equations may be multiplied by any nonzero real number.** If (m, n, p) is a solution of $ax + by + cz = d$, then, for $k \neq 0$, it is also a solution of $kax + kby + kcz = kd$, and conversely. (Why?)
3. **Any equation of the system may be replaced by the sum (member by member) of itself and any other equation of the system.** (You can show this by doing Problem 11 of Exercise 10.2.)

These three operations are called **elementary operations.** Let us use these to solve system (3).

Step 1. In the second equation of the system, x and y have unit coefficients, so we interchange the first two equations to get the following more convenient arrangement:

$$\begin{aligned} x + y \quad\ &= -1 \\ 2x - y + z &= \quad 3 \\ 3x - y - 2z &= \quad 7 \end{aligned}$$

Step 2. To make the coefficients of x in the first two equations the same in absolute value but opposite in sign, multiply both sides of the first equation by -2:

$$-2x - 2y = +2$$

Step 3. To eliminate x between the first and second equations, add the equation obtained in step 2 to the second equation to get

$$-3y + z = 5$$

and restore the first equation by dividing out the -2. The system now reads

$$\begin{aligned} x + \ y \quad\ &= -1 \\ -3y + \ z &= \quad 5 \\ 3x - \ y - 2z &= \quad 7 \end{aligned} \tag{5}$$

We proceed next in a similar way to eliminate x between the first and third equations.

Step 4. Multiply the first equation of system (5) by -3:

$$-3x - 3y = 3$$

Step 5. Add this last equation to the third equation to get

$$-4y - 2z = 10$$

and restore the first equation by dividing out the -3. The system now reads

$$x + y \quad\quad = -1$$
$$-3y + z = \quad 5 \tag{6}$$
$$-4y - 2z = \quad 10$$

We now eliminate y between the second and third equations in the following way:

Step 6. Multiply the second equation of system (6) by -4 and the third equation by 3 to get

$$12y - 4z = -20$$
$$-12y - 6z = \quad 30$$

Step 7. Add the last two equations to get

$$-10z = 10$$

and restore the second equation by dividing out the -4. The system is now

$$x + \quad y \quad\quad = -1$$
$$-3y + \quad z = \quad 5 \tag{7}$$
$$-10z = \quad 10$$

Step 8. It is very easy to solve system (7): The third equation immediately gives $z = -1$. Then, by substitution into the second equation, we get

$$-3y - 1 = \quad 5$$
$$-3y = \quad 6$$
$$y = -2$$

By substituting $y = -2$ into the first equation in (7), we find

$$x - 2 = -1$$
$$x = \quad 1$$

Thus, the solution of the system is $x = 1$, $y = -2$, $z = -1$. This is easily checked in the given set of equations.

A general system of the same form as system (7) may be written

$$ax + by + cz = d$$
$$ey + fz = g \tag{8}$$
$$hz = k$$

A. Solution by Matrices

A system such as system (8), in which the first unknown, x, is missing from the second and third equations, and the second unknown, y, is missing from the third equation, is said to be in **echelon form.** A system in echelon form is quite easy to solve. The third equation immediately yields the value of z. Back-substitution of this value into the second equation yields the value of y. Finally, back-substitution of the values of y and z into the first equation yields the value of x. Briefly, we say that we solve the system by **back-substitution.**

Every system of linear equations can be brought into echelon form by the use of the elementary operations. It remains only to organize and make the procedure more efficient by employing the augmented matrix.

Let us compare the augmented matrices of systems (3) and (7):

$$\left[\begin{array}{ccc|c} 2 & -1 & 1 & 3 \\ 1 & 1 & 0 & -1 \\ 3 & -1 & -2 & 7 \end{array}\right] \tag{3}$$

$$\left[\begin{array}{ccc|c} 1 & 1 & 0 & -1 \\ 0 & -3 & 1 & 5 \\ 0 & 0 & -10 & 10 \end{array}\right] \tag{7}$$

The augmented matrix for system (7) shows that the system is in echelon form because it has only 0's below the main diagonal of the coefficient matrix. We should be able to obtain the second matrix from the first by performing operations corresponding to the elementary operations on the equations. These operations are called **elementary row operations** and always yield matrices of equivalent systems. Such matrices are called **row-equivalent.** If two matrices A and B are row-equivalent, we shall write $A \sim B$. The elementary row operations are:

1. **A change of order of the rows**
2. **Multiplication of all the elements of a row by any nonzero number**
3. **Replacement of any row by the element-by-element sum of itself and any other row**

We illustrate the procedure by showing the transition from the matrix of system (3) to that of system (7). To symbolize the steps, we use the notation R_1, R_2, and R_3 for the respective rows of the matrix, along with the following typical abbreviations:

Notation	Meaning
1. $R_1 \leftrightarrow R_2$	Interchange R_1 and R_2.
2. $2 \times R_1$	Multiply each element of R_1 by 2.
3. $2 \times R_1 + R_2 \rightarrow R_2$	Replace R_2 by $2 \times R_1 + R_2$.

Thus, we write

$$\left[\begin{array}{ccc|c} 2 & -1 & 1 & 3 \\ 1 & 1 & 0 & -1 \\ 3 & -1 & -2 & 7 \end{array}\right] \sim \left[\begin{array}{ccc|c} 1 & 1 & 0 & -1 \\ 2 & -1 & 1 & 3 \\ 3 & -1 & -2 & 7 \end{array}\right]$$

$$R_1 \leftrightarrow R_2$$

Next, we proceed to get 0's in the second and third rows of the first column:

$$\begin{bmatrix} 1 & 1 & 0 & | & -1 \\ 2 & -1 & 1 & | & 3 \\ 3 & -1 & -2 & | & 7 \end{bmatrix} \sim \begin{bmatrix} 1 & 1 & 0 & | & -1 \\ 0 & -3 & 1 & | & 5 \\ 0 & -4 & -2 & | & 10 \end{bmatrix}$$

$$-2 \times R_1 + R_2 \rightarrow R_2$$
$$-3 \times R_1 + R_3 \rightarrow R_3$$

To complete the procedure, we get a 0 in the third row of the second column:

$$\begin{bmatrix} 1 & 1 & 0 & | & -1 \\ 0 & -3 & 1 & | & 5 \\ 0 & -4 & -2 & | & 10 \end{bmatrix} \sim \begin{bmatrix} 1 & 1 & 0 & | & -1 \\ 0 & -3 & 1 & | & 5 \\ 0 & 0 & -10 & | & 10 \end{bmatrix}$$

$$-4 \times R_2 + 3 \times R_3 \rightarrow R_3$$

Verify the result, which is the augmented matrix of system (7). Of course, once we obtain this last matrix, we can solve the system by back-substitution, as before.

We shall now give additional illustrations of this procedure, but these will help you only if you take a pencil and paper and carry out the detailed row operations as they are indicated.

EXAMPLE 1   Solve the system

$$2x - y + 2z = 3$$
$$2x + 2y - z = 0$$
$$-x + 2y + 2z = -12$$

Solution   The augmented matrix is

$$\begin{bmatrix} 2 & -1 & 2 & | & 3 \\ 2 & 2 & -1 & | & 0 \\ -1 & 2 & 2 & | & -12 \end{bmatrix} \sim \begin{bmatrix} 2 & -1 & 2 & | & 3 \\ 0 & 3 & -3 & | & -3 \\ 0 & 3 & 6 & | & -21 \end{bmatrix}$$

$$-R_1 + R_2 \rightarrow R_2$$
$$R_1 + 2 \times R_3 \rightarrow R_3$$

$$\sim \begin{bmatrix} 2 & -1 & 2 & | & 3 \\ 0 & 1 & -1 & | & -1 \\ 0 & 0 & 9 & | & -18 \end{bmatrix}$$

$$-R_2 + R_3 \rightarrow R_3$$
$$\tfrac{1}{3} \times R_2 \rightarrow R_2$$

This last matrix is in echelon form and corresponds to the system

$$2x - y + 2z = 3$$
$$y - z = -1$$
$$9z = -18$$

We solve this system by back-substitution. The last equation immediately yields $z = -2$. The second equation then becomes

$$y + 2 = -1$$

so that

$$y = -3$$

The first equation then becomes

$$2x + 3 - 4 = 3$$

so that

$$x = 2$$

The final answer, $x = 2$, $y = -3$, $z = -2$, can be checked in the given system. ◀

EXAMPLE 2 Solve the system

$$2x - y + 2z = 3$$
$$2x + 2y - z = 0$$
$$4x + y + z = 5$$

Solution

$$\begin{bmatrix} 2 & -1 & 2 & \bigm| & 3 \\ 2 & 2 & -1 & \bigm| & 0 \\ 4 & 1 & 1 & \bigm| & 5 \end{bmatrix} \sim \begin{bmatrix} 2 & -1 & 2 & \bigm| & 3 \\ 0 & 3 & -3 & \bigm| & -3 \\ 0 & 3 & -3 & \bigm| & -1 \end{bmatrix}$$

$$-R_1 + R_2 \rightarrow R_2$$
$$-2 \times R_1 + R_3 \rightarrow R_3$$

$$\sim \begin{bmatrix} 2 & -1 & 2 & \bigm| & 3 \\ 0 & 3 & -3 & \bigm| & -3 \\ 0 & 0 & 0 & \bigm| & 2 \end{bmatrix}$$

$$-R_2 + R_3 \rightarrow R_3$$

The final matrix is in echelon form, and the last line corresponds to the equation

$$0x + 0y + 0z = 2$$

which is false for all values of x, y, z. Hence, the given system has *no solution*. ◀

It is important to notice that if reduction to echelon form introduces any row with all 0's to the left and a nonzero number to the right of the vertical line, then the system has no solution.

EXAMPLE 3 Solve the system

$$2x - y + 2z = 3$$
$$2x + 2y - z = 0$$
$$4x + y + z = 3$$

Solution

$$\begin{bmatrix} 2 & -1 & 2 & | & 3 \\ 2 & 2 & -1 & | & 0 \\ 4 & 1 & 1 & | & 3 \end{bmatrix} \sim \begin{bmatrix} 2 & -1 & 2 & | & 3 \\ 0 & 3 & -3 & | & -3 \\ 0 & 3 & -3 & | & -3 \end{bmatrix}$$

$$-R_1 + R_2 \rightarrow R_2$$
$$-2 \times R_1 + R_3 \rightarrow R_3$$

$$\sim \begin{bmatrix} 2 & -1 & 2 & | & 3 \\ 0 & 3 & -3 & | & -3 \\ 0 & 0 & 0 & | & 0 \end{bmatrix}$$

$$-R_2 + R_3 \rightarrow R_3$$

The last matrix is in echelon form, and the last line corresponds to the equation

$$0x + 0y + 0z = 0$$

which is true for all values of x, y, z. Thus, any solution of the first two equations will be a solution of the system. The first two equations are

$$2x - y + 2z = 3$$
$$3y - 3z = -3$$

This system is equivalent to

$$2x - y = 3 - 2z$$
$$y = -1 + z$$

Suppose we put $z = k$, where k is any real number. Then the last equation gives $y = -1 + k$. Substitution into the first equation results in

$$2x + 1 - k = 3 - 2k$$

so that

$$2x = 2 - k$$
$$x = 1 - \tfrac{1}{2}k$$

Thus, if k is any real number, then $x = 1 - \tfrac{1}{2}k$, $y = k - 1$, $z = k$ is a solution of the system. You should verify this by substitution into the original system. We see that the system in this example has infinitely many solutions, because the value of k may be quite arbitrarily chosen. For instance, if $k = 2$, then the solution is $x = 0$, $y = 1$, $z = 2$; if $k = 5$, then $x = -\frac{3}{2}$, $y = 4$, $z = 5$; if $k = -4$, then $x = 3$, $y = -5$, $z = -4$; and so on. ◀

If in the final echelon form there is no row with all 0's to the left and a nonzero number to the right of the vertical line, but there is a row with all 0's both to the left and to the right, then the system has infinitely many solutions.

Examples 1, 2, and 3 illustrate the three possibilities for three linear equations in three unknowns. The system may have **a unique solution** as in Example 1; the system may have **no solution** as in Example 2; the system may have **infinitely many solutions** as in Example 3. The final echelon form of the matrix always shows which case is at hand.

B. Applications

EXAMPLE 4   Tom Jones, who was building himself a workshop, went to the hardware store and bought 1 pound each of three kinds of nails: small, medium, and large. After doing part of the work, Tom found that he had underestimated the number of small and large nails he needed. So he bought another pound of the small nails and 2 pounds more of the large nails. After doing some more of the construction, he again ran short of nails, and had to buy another pound of each of the small and the medium nails. Upon looking over his bills, he found that the hardware store had charged him $2.10 for nails the first time, $2.30 the second time, and $1.20 the third time. The prices for the various sizes of nails were not listed. Find these prices.

Solution   We let x cents per pound, y cents per pound, and z cents per pound be the prices for the small, medium, and large nails, respectively. Then, we know that

$$x + y + z = 210$$
$$x + 2z = 230$$
$$x + y = 120$$

We solve this system as follows:

$$\begin{bmatrix} 1 & 1 & 1 & | & 210 \\ 1 & 0 & 2 & | & 230 \\ 1 & 1 & 0 & | & 120 \end{bmatrix} \sim \begin{bmatrix} 1 & 1 & 1 & | & 210 \\ 0 & 1 & -1 & | & -20 \\ 0 & 0 & 1 & | & 90 \end{bmatrix}$$

$$R_1 - R_2 \rightarrow R_2$$
$$R_1 - R_3 \rightarrow R_3$$

Size	Price per Pound
Small	50¢
Medium	70¢
Large	90¢

The second matrix is in echelon form, and the solution of the system is easily found by back-substitution to be $x = 50$, $y = 70$, $z = 90$, giving the schedule of prices shown in the margin. ◀

A. *In Problems 1–10, find all the solutions (if there are any).*

1. $\begin{aligned} x + y - z &= 3 \\ x - 2y + z &= -3 \\ 2x + y + z &= 4 \end{aligned}$

2. $\begin{aligned} x + 2y - z &= 5 \\ 2x + y + z &= 1 \\ x - y + z &= -1 \end{aligned}$

3. $\begin{aligned} 2x - y + 2z &= 5 \\ 2x + y - z &= -6 \\ 3x + 2z &= 3 \end{aligned}$

4. $\begin{aligned} x + 2y - z &= 0 \\ 2x + 3y &= 3 \\ 2y + z &= -1 \end{aligned}$

5. $\begin{aligned} 3x + 2y + z &= -5 \\ 2x - y - z &= -6 \\ 2x + y + 3z &= 4 \end{aligned}$

6. $\begin{aligned} 4x + 3y - z &= 12 \\ 2x - 3y - z &= -10 \\ x + y - 2z &= -5 \end{aligned}$

7. $\begin{aligned} x + y + z &= 3 \\ x - 2y + z &= -3 \\ 3x + 3z &= 5 \end{aligned}$

8. $\begin{aligned} x + y + z &= 3 \\ x - 2y + 3z &= 5 \\ 5x - 4y + 11z &= 20 \end{aligned}$

9. $\begin{aligned} x + y + z &= 3 \\ x - 2y + z &= -3 \\ x + z &= 1 \end{aligned}$

10. $\begin{aligned} x - y - 2z &= -1 \\ x + 2y + z &= 5 \\ 5x + 4y - z &= 13 \end{aligned}$

*11. Show that elementary operation 3 yields an equivalent system. [*Hint:* Consider the first two equations of system (1) in the text. Show that if (m, n, p) satisfies both these equations, then it satisfies the system consisting of the first equation and the sum of the first two equations, and conversely.]

B. 12. The sum of $8.50 is made up of nickels, dimes, and quarters. The number of dimes is equal to the number of quarters plus twice the number of nickels. The value of the dimes exceeds the combined value of the nickels and the quarters by $1.50. How many of each coin are there?

13. The Mechano Distributing Company has three types of vending machines, which dispense snacks as listed in the table. Mechano fills all the machines once a day and finds them all sold out before the next day. The total daily sales are: candy, 760; peanuts, 380; sandwiches, 660. How many of each type of machine does Mechano have?

SNACK	VENDING MACHINE TYPE		
	I	II	III
Candy	20	24	30
Peanuts	10	18	10
Sandwiches	0	30	30

14. Suppose the total daily income from the various types of machines in Problem 13 is as follows: type I, $32.00; type II, $159.00; type III, $192.00. What is the selling price for each type of snack? (Use the answers for Problem 13.)

CHEMICAL	TYPE OF FERTILIZER		
	I	II	III
A	6%	8%	12%
B	6%	12%	8%
C	8%	4%	12%

15. Gro-Kwik Garden Supply has three types of fertilizer, which contain chemicals A, B, C in the percentages shown in the table in the margin. In what proportions must Gro-Kwik mix these three types to get an 8-8-8 fertilizer (one that has 8% of each of the three chemicals)?

16. Three water supply valves, A, B, C, are connected to a tank. If all three valves are opened, the tank is filled in 8 hr. The tank can also be filled by opening A for 8 hr and B for 12 hr, while keeping C closed, or by opening B for 10 hr and C for 28 hr, while keeping A closed. Find the time needed by each valve to fill the tank by itself. [*Hint:* Let x, y, z, respectively, be the fractions of the tank that valves A, B, C can fill alone in 1 hr.]

USING YOUR KNOWLEDGE 10.2

Instead of stopping with the echelon form of the matrix and then using back-substitution to solve a system of equations, many people prefer to transform the matrix of coefficients into the identity matrix and then read off the solution by inspection. If the final augmented matrix reads

$$\begin{bmatrix} 1 & 0 & 0 & | & a \\ 0 & 1 & 0 & | & b \\ 0 & 0 & 1 & | & c \end{bmatrix}$$

then the solution of the system is $x = a, y = b, z = c$. You need only read the column to the right of the vertical line.

Suppose, for example, that we have reduced the augmented matrix to the form

$$\begin{bmatrix} 2 & -1 & 2 & | & 3 \\ 0 & 3 & -3 & | & -3 \\ 0 & 0 & 2 & | & 5 \end{bmatrix} \qquad (1)$$

We can now divide R_2 by 3 and R_3 by 2 to obtain

$$\begin{bmatrix} 2 & -1 & 2 & | & 3 \\ 0 & 1 & -1 & | & -1 \\ 0 & 0 & 1 & | & \frac{5}{2} \end{bmatrix} \qquad (2)$$

To get 0's in the off-diagonal places of the matrix to the left of the vertical line, we first add R_2 to R_1:

$$\begin{bmatrix} 2 & 0 & 1 & | & 2 \\ 0 & 1 & -1 & | & -1 \\ 0 & 0 & 1 & | & \frac{5}{2} \end{bmatrix} \qquad (3)$$

Next, we subtract R_3 from R_1 and add R_3 to R_2, with the result

$$\begin{bmatrix} 2 & 0 & 0 & | & -\frac{1}{2} \\ 0 & 1 & 0 & | & \frac{3}{2} \\ 0 & 0 & 1 & | & \frac{5}{2} \end{bmatrix} \qquad (4)$$

Finally, we divide R_1 by 2 to obtain

$$\begin{bmatrix} 1 & 0 & 0 & | & -\frac{1}{4} \\ 0 & 1 & 0 & | & \frac{3}{2} \\ 0 & 0 & 1 & | & \frac{5}{2} \end{bmatrix} \qquad (5)$$

from which we can see by inspection that the solution of the system is $x = -\frac{1}{4}$, $y = \frac{3}{2}$, $z = \frac{5}{2}$.

Keep in mind that after you bring the augmented matrix into echelon form, you perform additional elementary operations to get 0's in all the off-diagonal places of the coefficient matrix and 1's on the diagonal. Of course, we assume that the system has a unique solution. If this is not the case, you will see what the situation is when you obtain the echelon form and you will also see that it is impossible to transform the coefficient matrix into the identity matrix if the system does not have a unique solution. Use these ideas to solve Problems 1–10 in Exercise 10.2.

▶ **Computer Corner 10.2**   It is rather tedious to use the methods described in this section to solve a system of equations. We have provided a program (see page 750) that will give you practice solving equations using a computer. The program employs the three elementary row operations we have mentioned in the text, but you must learn how to translate the notation in the text to that used in the program. For example, when working Example 1, the second step is $R_1 + 2 \times R_3 \rightarrow R_3$. In the program, you must multiply $2 \times R_3$ first, and then add $R_1 + R_3$. It will save a lot of time and writing to do the problems using this program.

10.3

THE INVERSE OF A MATRIX

In Section 10.1, we defined the multiplicative identity matrix I, which behaves in matrix multiplication like the integer 1 does in ordinary arithmetic multiplication. In arithmetic, we defined the multiplicative inverse of a number $a \neq 0$ to be a second number a^{-1} such that

Note that $a^{-1} = 1/a$ for $a \neq 0$.

$$a \cdot a^{-1} = a^{-1} \cdot a = 1$$

This suggests a corresponding definition for matrices.

> **Definition 10.3a** | Let A be a square matrix. If there exists a second matrix A^{-1} such that
>
> $$AA^{-1} = A^{-1}A = I$$
>
> then A^{-1} is called the **inverse** of A.

Notice that the superscript $^{-1}$ is not an exponent. The symbol A^{-1} simply stands for the inverse of A. It is clear that if A^{-1} exists, then A and A^{-1} are inverses of each other.

The practical importance of the inverse matrix lies in the fact that a system of n linear equations in n unknowns, symbolized as in Section 10.2 by

$$AX = D$$

can easily be solved if the matrix A^{-1} is known. Thus, by premultiplying both sides by A^{-1}, we get

$$A^{-1}AX = A^{-1}D$$

Because $A^{-1}A = I$ and $IX = X$, we can see that

$$X = A^{-1}D$$

Thus, to get the solution, we only need to premultiply the column matrix of the right-hand terms by the inverse (if it exists) of the coefficient matrix.

Let us now consider the problem of finding A^{-1} for a specific 2×2 matrix, say

$$A = \begin{bmatrix} 2 & 3 \\ 1 & 2 \end{bmatrix}$$

Suppose that

$$A^{-1} = \begin{bmatrix} x & z \\ y & w \end{bmatrix}$$

Then by the definition of A^{-1}, it follows that $AA^{-1} = I$, or

$$\begin{bmatrix} 2 & 3 \\ 1 & 2 \end{bmatrix}\begin{bmatrix} x & z \\ y & w \end{bmatrix} = \begin{bmatrix} 1 & 0 \\ 0 & 1 \end{bmatrix}$$

By multiplying out the matrices on the left and equating corresponding elements on both sides of the equation, we get the following two systems:

$$\left.\begin{array}{l} 2x + 3y = 1 \\ x + 2y = 0 \end{array}\right\} \tag{1}$$

$$\left.\begin{array}{l} 2z + 3w = 0 \\ z + 2w = 1 \end{array}\right\} \tag{2}$$

We see at once that the left sides of the two systems are the same except for the unknowns (x, y) in system (1) being replaced by (z, w) in system (2).

This suggests that we can use matrices to solve the two systems in one set of operations. We write the augmented matrices together as follows:

$$\left[\begin{array}{cc|cc} 2 & 3 & 1 & 0 \\ 1 & 2 & 0 & 1 \end{array}\right]$$

where the first column after the vertical bar is the column of right-hand members of system (1) and the second column is that for system (2). Together, these two columns constitute the identity matrix.

We first reduce the matrix on the left to echelon form:

$$\left[\begin{array}{cc|cc} 2 & 3 & 1 & 0 \\ 1 & 2 & 0 & 1 \end{array}\right] \sim \left[\begin{array}{cc|cc} 2 & 3 & 1 & 0 \\ 0 & -1 & 1 & -2 \end{array}\right]$$
$$R_1 - 2 \times R_2 \rightarrow R_2$$

We then try to reduce the matrix on the left to the identity matrix, because this will give the values of x, y, z, w by inspection. Thus,

$$\left[\begin{array}{cc|cc} 2 & 3 & 1 & 0 \\ 0 & -1 & 1 & -2 \end{array}\right] \sim \left[\begin{array}{cc|cc} 2 & 0 & 4 & -6 \\ 0 & -1 & 1 & -2 \end{array}\right]$$
$$3 \times R_2 + R_1 \rightarrow R_1$$
$$\sim \left[\begin{array}{cc|cc} 1 & 0 & 2 & -3 \\ 0 & 1 & -1 & 2 \end{array}\right]$$
$$\tfrac{1}{2} \times R_1 \rightarrow R_1$$
$$-R_2 \rightarrow R_2$$

This last matrix tells us that $x = 2$, $y = -1$, $z = -3$, and $w = 2$. Since

$$A^{-1} = \left[\begin{array}{cc} x & z \\ y & w \end{array}\right]$$

we have

$$A^{-1} = \left[\begin{array}{cc} 2 & -3 \\ -1 & 2 \end{array}\right]$$

We can check this by multiplying A by A^{-1}:

$$A^{-1}A = \left[\begin{array}{cc} 2 & -3 \\ -1 & 2 \end{array}\right]\left[\begin{array}{cc} 2 & 3 \\ 1 & 2 \end{array}\right]$$
$$= \left[\begin{array}{cc} 2 \times 2 - 3 \times 1 & 2 \times 3 - 3 \times 2 \\ -1 \times 2 + 2 \times 1 & -1 \times 3 + 2 \times 2 \end{array}\right]$$
$$= \left[\begin{array}{cc} 1 & 0 \\ 0 & 1 \end{array}\right] = I$$

You should verify that AA^{-1} also is I.

The preceding method is applicable not only to 2×2 matrices, but in general to $n \times n$ matrices. Of course, there is more work for the higher-order matrices. In practical applications, where the matrices are quite large, the procedure is carried out on a digital computer. The method itself is quite simply described:

1. **Write an augmented matrix consisting of the given square matrix on the left of the vertical bar and the identity matrix on the right of the bar.**
2. **By means of elementary row operations, reduce (if possible) the given matrix to the identity matrix. When this is accomplished, the matrix that appears on the right of the bar is the desired inverse.**

This procedure is always possible if the given matrix has an inverse. It is, of course, not possible otherwise. For instance, if

$$A = \begin{bmatrix} 1 & 1 \\ 2 & 2 \end{bmatrix}$$

we would find

$$\left[\begin{array}{cc|cc} 1 & 1 & 1 & 0 \\ 2 & 2 & 0 & 1 \end{array}\right] \sim \left[\begin{array}{cc|cc} 1 & 1 & 1 & 0 \\ 0 & 0 & -2 & 1 \end{array}\right]$$
$$-2 \times R_1 + R_2 \rightarrow R_2$$

Obviously, it is not possible to reduce the left-hand matrix to the identity matrix! This means that the given matrix has no inverse.

EXAMPLE 1   Find the inverse (if it exists) for the matrix

$$A = \begin{bmatrix} 1 & -2 \\ 2 & 3 \end{bmatrix}$$

Solution   We write

$$\left[\begin{array}{cc|cc} 1 & -2 & 1 & 0 \\ 2 & 3 & 0 & 1 \end{array}\right] \sim \left[\begin{array}{cc|cc} 1 & -2 & 1 & 0 \\ 0 & 7 & -2 & 1 \end{array}\right]$$
$$-2 \times R_1 + R_2 \rightarrow R_2$$

$$\sim \left[\begin{array}{cc|cc} 1 & -2 & 1 & 0 \\ 0 & 1 & -\frac{2}{7} & \frac{1}{7} \end{array}\right]$$
$$\tfrac{1}{7} \times R_2 \rightarrow R_2$$

$$\sim \left[\begin{array}{cc|cc} 1 & 0 & \frac{3}{7} & \frac{2}{7} \\ 0 & 1 & -\frac{2}{7} & \frac{1}{7} \end{array}\right]$$
$$2 \times R_2 + R_1 \rightarrow R_1$$

Hence,

$$A^{-1} = \begin{bmatrix} \frac{3}{7} & \frac{2}{7} \\ -\frac{2}{7} & \frac{1}{7} \end{bmatrix}$$ ◀

EXAMPLE 2　Use the result of Example 1 to solve the system

$$x - 2y = -6$$
$$2x + 3y = 37$$

Solution　We can symbolize this system by

$$AX = D$$

where

$$A = \begin{bmatrix} 1 & -2 \\ 2 & 3 \end{bmatrix} \qquad X = \begin{bmatrix} x \\ y \end{bmatrix} \qquad D = \begin{bmatrix} -6 \\ 37 \end{bmatrix}$$

As you will recall, the solution is given by

$$X = A^{-1}D = \begin{bmatrix} \frac{3}{7} & \frac{2}{7} \\ -\frac{2}{7} & \frac{1}{7} \end{bmatrix} \begin{bmatrix} -6 \\ 37 \end{bmatrix}$$
$$= \frac{1}{7} \begin{bmatrix} 3 & 2 \\ -2 & 1 \end{bmatrix} \begin{bmatrix} -6 \\ 37 \end{bmatrix}$$
$$= \frac{1}{7} \begin{bmatrix} 56 \\ 49 \end{bmatrix} = \begin{bmatrix} 8 \\ 7 \end{bmatrix}$$

Thus, $x = 8$ and $y = 7$. (You do not have to take our word for this. You can check this solution in the given equations!)　◀

EXAMPLE 3　Find the inverse (if it exists) for the matrix

$$A = \begin{bmatrix} 1 & 0 & 1 \\ 1 & 1 & 3 \\ 0 & 1 & 3 \end{bmatrix}$$

Solution　We write

$$\left[\begin{array}{ccc|ccc} 1 & 0 & 1 & 1 & 0 & 0 \\ 1 & 1 & 3 & 0 & 1 & 0 \\ 0 & 1 & 3 & 0 & 0 & 1 \end{array}\right] \sim \left[\begin{array}{ccc|ccc} 1 & 0 & 1 & 1 & 0 & 0 \\ 0 & 1 & 2 & -1 & 1 & 0 \\ 0 & 1 & 3 & 0 & 0 & 1 \end{array}\right]$$
$$-R_1 + R_2 \rightarrow R_2$$
$$\sim \left[\begin{array}{ccc|ccc} 1 & 0 & 1 & 1 & 0 & 0 \\ 0 & 1 & 2 & -1 & 1 & 0 \\ 0 & 0 & 1 & 1 & -1 & 1 \end{array}\right]$$
$$-R_2 + R_3 \rightarrow R_3$$
$$\sim \left[\begin{array}{ccc|ccc} 1 & 0 & 0 & 0 & 1 & -1 \\ 0 & 1 & 0 & -3 & 3 & -2 \\ 0 & 0 & 1 & 1 & -1 & 1 \end{array}\right]$$
$$-R_3 + R_1 \rightarrow R_1$$
$$-2 \times R_3 + R_2 \rightarrow R_2$$

Since the matrix on the left has been reduced to the identity matrix, the matrix on the right is the desired inverse. Thus,

$$A^{-1} = \begin{bmatrix} 0 & 1 & -1 \\ -3 & 3 & -2 \\ 1 & -1 & 1 \end{bmatrix}$$

You can check this result by showing that $AA^{-1} = I$. ◀

EXAMPLE 4   Use the result of Example 3 to solve the system

$$x \quad\;\; + z = 5$$
$$x + y + 3z = 7$$
$$\quad\;\; y + 3z = 4$$

Solution   If we symbolize the system by $AX = D$, where A is the matrix given in Example 3, then the solution is given by

$$X = A^{-1}D = \begin{bmatrix} 0 & 1 & -1 \\ -3 & 3 & -2 \\ 1 & -1 & 1 \end{bmatrix}\begin{bmatrix} 5 \\ 7 \\ 4 \end{bmatrix} = \begin{bmatrix} 3 \\ -2 \\ 2 \end{bmatrix}$$

Thus, the solution is $x = 3$, $y = -2$, $z = 2$. You can check this solution in the given system of equations. ◀

EXERCISE 10.3

In Problems 1–8, find the inverse (if it exists) of the given matrix.

1. $\begin{bmatrix} 2 & 4 \\ 3 & 5 \end{bmatrix}$ 2. $\begin{bmatrix} -1 & 0 \\ 3 & 1 \end{bmatrix}$ 3. $\begin{bmatrix} 5 & -1 \\ 8 & 2 \end{bmatrix}$ 4. $\begin{bmatrix} 3 & -5 \\ 2 & -3 \end{bmatrix}$

5. $\begin{bmatrix} 2 & 3 \\ 4 & 6 \end{bmatrix}$ 6. $\begin{bmatrix} 3 & 4 \\ -1 & 2 \end{bmatrix}$ 7. $\begin{bmatrix} -1 & 0 \\ 0 & -1 \end{bmatrix}$ 8. $\begin{bmatrix} 2 & 2 \\ 4 & 4 \end{bmatrix}$

9. Use your answer to Problem 1 to solve the system

$$2x + 4y = 14$$
$$3x + 5y = 16$$

10. Use your answer to Problem 3 to solve the system

$$5x - \;\; y = 2$$
$$8x + 2y = 3$$

11. Use your answer to Problem 4 to solve the system

$$3x - 5y = -1$$
$$2x - 3y = \;\; 0$$

12. Use your answer to Problem 6 to solve the system

$$3x + 4y = -3$$
$$-x + 2y = \;\; 11$$

In Problems 13–16, decide whether the two given matrices are inverses of each other.

13. $\begin{bmatrix} 1 & 2 & 3 \\ 0 & 1 & 2 \\ 0 & 1 & 3 \end{bmatrix}$ and $\begin{bmatrix} 1 & -3 & 1 \\ 0 & 3 & -2 \\ 0 & -1 & 1 \end{bmatrix}$

14. $\begin{bmatrix} 1 & 3 & 0 \\ 0 & 1 & 1 \\ 1 & 4 & 2 \end{bmatrix}$ and $\begin{bmatrix} -2 & -6 & 3 \\ 1 & 2 & -1 \\ -1 & -1 & 1 \end{bmatrix}$

15. $\begin{bmatrix} 1 & 1 & 1 \\ -1 & -1 & -2 \\ 2 & 3 & 3 \end{bmatrix}$ and $\begin{bmatrix} 3 & 0 & -1 \\ -1 & 1 & 1 \\ -1 & -1 & 0 \end{bmatrix}$

16. $\begin{bmatrix} 1 & 0 & 1 \\ 1 & 1 & 0 \\ 0 & 1 & 1 \end{bmatrix}$ and $\begin{bmatrix} 1 & 1 & 0 \\ 0 & 1 & 1 \\ 1 & 0 & 1 \end{bmatrix}$

17. Find the inverse of the matrix

$$\begin{bmatrix} 1 & 2 & -4 \\ 0 & 1 & -1 \\ 1 & -1 & 0 \end{bmatrix}$$

18. Find the inverse of the matrix

$$\begin{bmatrix} 1 & 0 & -2 \\ 1 & 1 & 0 \\ 0 & 1 & 3 \end{bmatrix}$$

19. Find the inverse of the matrix

$$\begin{bmatrix} 1 & 3 & -1 \\ 0 & 2 & 1 \\ -1 & 1 & 2 \end{bmatrix}$$

20. Find the inverse of the matrix

$$\begin{bmatrix} 2 & -1 & 0 \\ 1 & 2 & -1 \\ -1 & -1 & 2 \end{bmatrix}$$

21. Show that the following matrix has no inverse:

$$\begin{bmatrix} 1 & 2 & 1 \\ -1 & 0 & 1 \\ 1 & 4 & 3 \end{bmatrix}$$

22. Show that the following matrix has no inverse:

$$\begin{bmatrix} 1 & 2 & 1 \\ 1 & 1 & 1 \\ 1 & 0 & 1 \end{bmatrix}$$

23. Use your answer to Problem 17 to solve the system

$$x + 2y - 4z = 16$$
$$y - z = 4$$
$$x - y = 1$$

24. Use your answer to Problem 18 to solve the system

$$x - 2z = 0$$
$$x + y = 5$$
$$y + 3z = 7$$

25. Use your answer to Problem 19 to solve the system

$$x + 3y - z = 3$$
$$2y + z = 11$$
$$-x + y + 2z = 14$$

26. Use your answer to Problem 20 to solve the system

$$2x - y = 4$$
$$x + 2y - z = 3$$
$$-x - y + 2z = 3$$

USING YOUR KNOWLEDGE 10.3

If you are like most people, at this point you are probably not greatly impressed with the idea of solving a system of linear equations by using the inverse of the coefficient matrix. Why go to the bother of finding the inverse matrix, when you can solve the system directly with less work?

Let us recall Problem 15 of Exercise 10.2: Gro-Kwik Garden Supply has three types of fertilizer, which contain chemicals A, B, and C in the percentages shown in the table in the margin. In what proportions must Gro-Kwik mix these three types to get an 8-8-8 fertilizer (one that has 8% of each of the three chemicals)?

For convenience, let us assume that the mixture is to be put up in 100-lb sacks. If x, y, z are the number of pounds of type I, II, and III, respectively, required per sack, then

$$0.06x + 0.08y + 0.12z = 8$$
$$0.06x + 0.12y + 0.08z = 8$$
$$0.08x + 0.04y + 0.12z = 8$$

		TYPE OF FERTILIZER	
CHEMICAL	I	II	III
A	6%	8%	12%
B	6%	12%	8%
C	8%	4%	12%

No matter what the required mix is, the left sides of these equations will be the same; only the right-hand numbers will change. So if we have the inverse of the coefficient matrix, then we can find the values of x, y, z for any possible mix without having to go through the process of solving the equations each time.

This problem is a very much simplified illustration of an important practical application of the inverse of a matrix. In actual practice, the number of components of the mix may be quite large, so hand calculation would be out of the question. The inverse of the coefficient matrix may be stored on tape or on cards. When the requirements of the desired mix are known, the information can be put into a computer, which will multiply the matrix of right-hand numbers by the inverse of the coefficient matrix, and quickly turn out the desired answers.

1. If A denotes the coefficient matrix of the above system of equations, verify that

$$A^{-1} = \frac{25}{8} \begin{bmatrix} -14 & 6 & 10 \\ 1 & 3 & -3 \\ 9 & -5 & -3 \end{bmatrix}$$

2. Suppose that an 8-10-6 mix of the fertilizer is required. Can you see how to use the matrix A^{-1} in Problem 1 to find how many pounds of each type to put in per 100 lb of mix? Be sure to check your answers.

► **Computer Corner 10.3** *We have provided a program (see page 750) that finds the inverse of a 2 × 2 or 3 × 3 matrix, provided you enter the original matrix. Thus, to work Example 1, just enter the numbers in the original matrix. If the matrix has no inverse, the program tells us, "No inverse." Use the program to verify the examples in this section.*

*10.4

DETERMINANTS AND CRAMER'S RULE

Associated with the square matrix

$$A = \begin{bmatrix} a_1 & b_1 \\ a_2 & b_2 \end{bmatrix}$$

is a number called the **determinant of A** and written **det A**. The number det A is symbolized by using vertical bars in place of the brackets in the matrix A. Thus,

$$\det A = \begin{vmatrix} a_1 & b_1 \\ a_2 & b_2 \end{vmatrix}$$

A. Evaluating Determinants

The **value of det A** is defined to be

$$a_1 b_2 - a_2 b_1$$

which can be obtained by multiplying down the diagonals, as indicated by the diagram in the following definition:

► **Definition 10.4a**

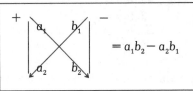

For example,

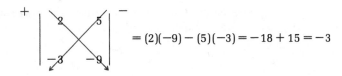

EXAMPLE 1   Evaluate:

a. $\begin{vmatrix} -3 & 7 \\ -5 & 4 \end{vmatrix}$ b. $\begin{vmatrix} -3 & 6 \\ -5 & 10 \end{vmatrix}$

Solution   a. $\begin{vmatrix} -3 & 7 \\ -5 & 4 \end{vmatrix} = (-3)(4) - (7)(-5) = -12 - (-35) = 23$

b. $\begin{vmatrix} -3 & 6 \\ -5 & 10 \end{vmatrix} = (-3)(10) - (6)(-5) = -30 - (-30) = 0$ ◀

B. Cramer's Rule

Notice in Example 1b that the second-column elements of the determinant are both the same multiple of the corresponding first-column elements:

$$6 = (-2)(-3) \quad \text{and} \quad 10 = (-2)(-5)$$

In general, if the elements of one row are just some constant k times the elements of the other row (or if this is true of the columns), then the value of the determinant is 0. This is very easy to see as follows:

$$\begin{vmatrix} a & b \\ ka & kb \end{vmatrix} = kab - kab = 0$$

This result will be of use to us in Using Your Knowledge 10.4.

One of the important applications of determinants is to the solution of a system of linear equations. Let us look at the simplest case, two equations in two unknowns. Such a system can be written

$$a_1 x + b_1 y = d_1$$
$$a_2 x + b_2 y = d_2$$

We can eliminate y from this system by multiplying the first equation by b_2 and the second equation by b_1, and then subtracting as follows:

$$a_1 b_2 x + b_1 b_2 y = d_1 b_2$$
$$\underline{(-)a_2 b_1 x + b_1 b_2 y = d_2 b_1}$$
$$a_1 b_2 x - a_2 b_1 x = d_1 b_2 - d_2 b_1$$
$$(a_1 b_2 - a_2 b_1)x = d_1 b_2 - d_2 b_1$$

Now, if the quantity $a_1 b_2 - a_2 b_1 \neq 0$, we can divide by this quantity to get

$$x = \frac{d_1 b_2 - d_2 b_1}{a_1 b_2 - a_2 b_1}$$

Notice that the denominator, which we shall denote by D, can be written as the determinant:

$$D = \begin{vmatrix} a_1 & b_1 \\ a_2 & b_2 \end{vmatrix}$$

This determinant is naturally called the **determinant of the coefficients.** Notice also that the numerator can be obtained from the denominator by simply replacing the coefficients of x (the *a*'s) by the corresponding constant terms, the *d*'s. We denote the numerator of x by D_x. Thus,

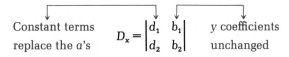

$$\text{Constant terms replace the } a\text{'s} \qquad D_x = \begin{vmatrix} d_1 & b_1 \\ d_2 & b_2 \end{vmatrix} \qquad y \text{ coefficients unchanged}$$

We can now write the solution for x in the form

$$x = \frac{D_x}{D}$$

A similar procedure shows that the solution for y is

$$y = \frac{a_1 d_2 - a_2 d_1}{a_1 b_2 - a_2 b_1}$$

Notice that the denominator is again the determinant D. The numerator, which we shall denote by D_y, can be formed from D by replacing the coefficients of y (the *b*'s) by the corresponding *d*'s. Hence,

$$\text{x coefficients unchanged} \qquad D_y = \begin{vmatrix} a_1 & d_1 \\ a_2 & d_2 \end{vmatrix} \qquad \text{Constant terms replace the } b\text{'s}$$

and we can write the solution for y in the form

$$y = \frac{D_y}{D}$$

We summarize these results as follows:

Cramer's Rule for a System of Two Linear Equations in Two Unknowns

$$a_1 x + b_1 y = d_1$$
$$a_2 x + b_2 y = d_2$$

1. Let

$$D = \begin{vmatrix} a_1 & b_1 \\ a_2 & b_2 \end{vmatrix} \qquad D_x = \begin{vmatrix} d_1 & b_1 \\ d_2 & b_2 \end{vmatrix} \qquad D_y = \begin{vmatrix} a_1 & d_1 \\ a_2 & d_2 \end{vmatrix}$$

2. If $D \neq 0$, then the solution of the system is

$$x = \frac{D_x}{D} \qquad y = \frac{D_y}{D}$$

EXAMPLE 2   Use Cramer's rule to solve the system $2x + 3y =\ \ 7$

$5x + 9y = 11$

Solution $D = \begin{vmatrix} 2 & 3 \\ 5 & 9 \end{vmatrix} = 18 - 15 = 3$

$D_x = \begin{vmatrix} 7 & 3 \\ 11 & 9 \end{vmatrix} = 63 - 33 = 30$

$D_y = \begin{vmatrix} 2 & 7 \\ 5 & 11 \end{vmatrix} = 22 - 35 = -13$

Therefore,

$$x = \frac{D_x}{D} = \frac{30}{3} = 10 \qquad y = \frac{D_y}{D} = \frac{-13}{3} = -\frac{13}{3}$$

Check   Substituting $x = 10$, $y = -\frac{13}{3}$ in the left sides of the two equations, we obtain

$(2)(10) + (3)(-\frac{13}{3}) = 20 - 13 = 7$

$(5)(10) + (9)(-\frac{13}{3}) = 50 - 39 = 11$

Thus, the answers do satisfy the given equations, and the solution is $x = 10$, $y = -\frac{13}{3}$, or $(10, -\frac{13}{3})$.   ◄

We can also solve a system of three linear equations in three unknowns by using determinants. First, we have to define the value of a 3×3 determinant, that is, a determinant with three rows and three columns. We write

$$\det A = \begin{vmatrix} a_1 & b_1 & c_1 \\ a_2 & b_2 & c_2 \\ a_3 & b_3 & c_3 \end{vmatrix}$$

and we define the value of this determinant in terms of the 2×2 determinants that we can form from the second and third rows. The 2×2 determinant that remains when we cross out the row and column of any element is called the **minor** of that element. For example, deleting the first row and the second column of

$$\begin{vmatrix} a_1 & b_1 & c_1 \\ a_2 & b_2 & c_2 \\ a_3 & b_3 & c_3 \end{vmatrix}$$

leaves the 2×2 determinant

$$\begin{vmatrix} a_2 & c_2 \\ a_3 & c_3 \end{vmatrix}$$

which is the minor of b_1. The following definition gives the value of the 3×3 determinant in terms of the minors of the first row.

Definition 10.4b

$$\begin{array}{ccc} + & - & + \end{array}$$
$$\begin{vmatrix} a_1 & b_1 & c_1 \\ a_2 & b_2 & c_2 \\ a_3 & b_3 & c_3 \end{vmatrix} = a_1 \begin{vmatrix} b_2 & c_2 \\ b_3 & c_3 \end{vmatrix} - b_1 \begin{vmatrix} a_2 & c_2 \\ a_3 & c_3 \end{vmatrix} + c_1 \begin{vmatrix} a_2 & b_2 \\ a_3 & b_3 \end{vmatrix}$$

Note that we have written the plus and minus signs above the first row to help you remember this definition.

EXAMPLE 3   Evaluate the determinant

$$\begin{vmatrix} 2 & -4 & 3 \\ 3 & 0 & 5 \\ 1 & -1 & 2 \end{vmatrix}$$

Solution   By Definition 10.4b,

$$\begin{array}{ccc} + & - & + \end{array}$$
$$\begin{vmatrix} 2 & -4 & 3 \\ 3 & 0 & 5 \\ 1 & -1 & 2 \end{vmatrix} = (2)\begin{vmatrix} 0 & 5 \\ -1 & 2 \end{vmatrix} - (-4)\begin{vmatrix} 3 & 5 \\ 1 & 2 \end{vmatrix} + (3)\begin{vmatrix} 3 & 0 \\ 1 & -1 \end{vmatrix}$$
$$= (2)(5) + (4)(1) + (3)(-3) = 5 \qquad \blacktriangleleft$$

We can solve the system

$$a_1 x + b_1 y + c_1 z = d_1$$
$$a_2 x + b_2 y + c_2 z = d_2 \qquad\qquad (1)$$
$$a_3 x + b_3 y + c_3 z = d_3$$

in exactly the same manner as we solved the system of equations with two unknowns. The details are not particularly interesting, and we shall just state the result.

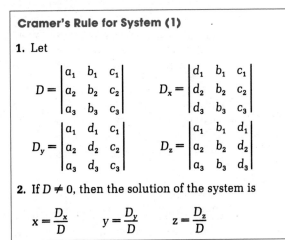

Cramer's Rule for System (1)

1. Let

$$D = \begin{vmatrix} a_1 & b_1 & c_1 \\ a_2 & b_2 & c_2 \\ a_3 & b_3 & c_3 \end{vmatrix} \qquad D_x = \begin{vmatrix} d_1 & b_1 & c_1 \\ d_2 & b_2 & c_2 \\ d_3 & b_3 & c_3 \end{vmatrix}$$

$$D_y = \begin{vmatrix} a_1 & d_1 & c_1 \\ a_2 & d_2 & c_2 \\ a_3 & d_3 & c_3 \end{vmatrix} \qquad D_z = \begin{vmatrix} a_1 & b_1 & d_1 \\ a_2 & b_2 & d_2 \\ a_3 & b_3 & d_3 \end{vmatrix}$$

2. If $D \neq 0$, then the solution of the system is

$$x = \frac{D_x}{D} \qquad y = \frac{D_y}{D} \qquad z = \frac{D_z}{D}$$

Note carefully that D is the determinant of the coefficients; D_x is formed from D by replacing the coefficients of x (the a's) by the corresponding d's; D_y is formed from D by replacing the coefficients of y (the b's) by the corresponding d's, and similarly for D_z.

EXAMPLE 4   Use Cramer's rule to solve the system

$$x + y + 2z = 7$$
$$x - y - 3z = -6$$
$$2x + 3y + z = 4$$

Solution   To use Cramer's rule, we first have to evaluate D, the determinant of the coefficients. If D is not 0, then we calculate the other three determinants, D_x, D_y, and D_z. For the given system,

$$D = \begin{vmatrix} 1 & 1 & 2 \\ 1 & -1 & -3 \\ 2 & 3 & 1 \end{vmatrix} = (1)(-1+9) - (1)(1+6) + (2)(3+2) = 8 - 7 + 10 = 11$$

$$D_x = \begin{vmatrix} 7 & 1 & 2 \\ -6 & -1 & -3 \\ 4 & 3 & 1 \end{vmatrix} = (7)(-1+9) - (1)(-6+12) + (2)(-18+4) = 56 - 6 - 28 = 22$$

$$D_y = \begin{vmatrix} 1 & 7 & 2 \\ 1 & -6 & -3 \\ 2 & 4 & 1 \end{vmatrix} = (1)(-6+12) - (7)(1+6) + (2)(4+12) = 6 - 49 + 32 = -11$$

$$D_z = \begin{vmatrix} 1 & 1 & 7 \\ 1 & -1 & -6 \\ 2 & 3 & 4 \end{vmatrix} = (1)(-4+18) - (1)(4+12) + (7)(3+2) = 14 - 16 + 35 = 33$$

Thus, by Cramer's rule,

$$x = \frac{D_x}{D} = \frac{22}{11} = 2 \qquad y = \frac{D_y}{D} = \frac{-11}{11} = -1 \qquad z = \frac{D_z}{D} = \frac{33}{11} = 3$$

You can check the solution $(2, -1, 3)$ by substituting into the given equations. ◄

EXAMPLE 5   A florist has three types of orchids, which she classifies as large, medium, and small. She will make up a spray of one large, two medium, and four small orchids for $17, or a spray of two large, one medium, and six small orchids for $22, or a spray of two large, three medium, and four small orchids for $25. What price is she placing on each of the three types of orchid?

Solution   We let x, y, and z dollars be the price of a large, a medium, and a small orchid, respectively. Then, for the $17 spray, we have

$$x + 2y + 4z = 17$$

for the $22 spray,

$$2x + y + 6z = 22$$

and for the $25 spray,

$$2x + 3y + 4z = 25$$

Hence, we have to solve the system

$$x + 2y + 4z = 17$$
$$2x + y + 6z = 22$$
$$2x + 3y + 4z = 25$$

For this system,

$$D = \begin{vmatrix} 1 & 2 & 4 \\ 2 & 1 & 6 \\ 2 & 3 & 4 \end{vmatrix} = (1)(4 - 18) - (2)(8 - 12) + (4)(6 - 2) = 10$$

$$D_x = \begin{vmatrix} 17 & 2 & 4 \\ 22 & 1 & 6 \\ 25 & 3 & 4 \end{vmatrix} = (17)(4 - 18) - (2)(88 - 150) + (4)(66 - 25) = 50$$

$$D_y = \begin{vmatrix} 1 & 17 & 4 \\ 2 & 22 & 6 \\ 2 & 25 & 4 \end{vmatrix} = (1)(88 - 150) - (17)(8 - 12) + 4(50 - 44) = 30$$

$$D_z = \begin{vmatrix} 1 & 2 & 17 \\ 2 & 1 & 22 \\ 2 & 3 & 25 \end{vmatrix} = (1)(25 - 66) - (2)(50 - 44) + (17)(6 - 2) = 15$$

By Cramer's rule,

$$x = \frac{50}{10} = 5 \qquad y = \frac{30}{10} = 3 \qquad z = \frac{15}{10} = 1.5$$

Hence, the large orchids were priced at $5.00, the medium orchids at $3.00, and the small orchids at $1.50. You should check that these give the correct prices for the various sprays. ◀

Note that Cramer's rule is not efficient for systems larger than 3×3 because of the amount of work in the evaluation of the determinants.

EXERCISE 10.4

A. In Problems 1–12, evaluate the given determinant.

1. $\begin{vmatrix} 3 & 1 \\ 5 & 2 \end{vmatrix}$

2. $\begin{vmatrix} 5 & 1 \\ 6 & 3 \end{vmatrix}$

3. $\begin{vmatrix} \frac{1}{8} & \frac{1}{6} \\ \frac{1}{4} & \frac{2}{3} \end{vmatrix}$

4. $\begin{vmatrix} -2 & -\frac{1}{2} \\ 3 & \frac{7}{4} \end{vmatrix}$

5. $\begin{vmatrix} a & 2a \\ b & 4b \end{vmatrix}$

6. $\begin{vmatrix} x & -y \\ y & x \end{vmatrix}$

7. $\begin{vmatrix} 2 & 1 & 3 \\ 1 & 2 & -1 \\ 3 & 1 & 5 \end{vmatrix}$

8. $\begin{vmatrix} 1 & 3 & 5 \\ 2 & 0 & 10 \\ -3 & 1 & -15 \end{vmatrix}$

9. $\begin{vmatrix} 2 & 1 & 3 \\ 3 & 4 & 5 \\ 1 & 7 & 2 \end{vmatrix}$

10. $\begin{vmatrix} 8 & 7 & 6 \\ -2 & 3 & -1 \\ 3 & 0 & 4 \end{vmatrix}$

11. $\begin{vmatrix} x & y & 1 \\ 2 & 5 & 1 \\ 3 & -4 & 1 \end{vmatrix}$

12. $\begin{vmatrix} 1 & -1 & a \\ -1 & a & -1 \\ a & -1 & -1 \end{vmatrix}$

B. In Problems 13–24, use Cramer's rule to solve the given system.

13. $2x - 3y = 16$
$x - y = 7$

14. $16x - 9y = -5$
$10x + 18y = -11$

15. $4x + 5y = 5$
$-10x - 4y = -7$

16. $6x + 10y = -1$
$3x + 6y = 1$

17. $3(x + 2) = 2y$
$2(y + 5) = 7x$

18. $4(y - 4x) = 2(x + 5)$
$10(y - x) = 11y - 15x$

19. $x + y - 2z = 13$
$x - 3y - z = -3$
$x - y + 4z = -17$

20. $x + y + z = 13$
$x - 2y + 4z = 10$
$3x + y - 3z = 5$

21. $x + y + z = -1$
$3x - y - 5z = 13$
$5x + 3y + 2z = 1$

22. $x + y + z = 6$
$x - y + 2z = 5$
$x - y - 3z = 10$

23. $y + z = x - 7$
$x + 2z = 1 - 2y$
$x - 7 = 3y + 2z$

24. $x + z = 6 - y$
$x + 3y = 8$
$2x = 2y + 5 - z$

In Problems 25–28, set up the system of equations to solve the problem and then use Cramer's rule to obtain the solution.

25. Exercise 10.2, Problem 13

26. Exercise 10.2, Problem 14

27. Exercise 10.2, Problem 15

28. Exercise 10.2, Problem 16

USING YOUR KNOWLEDGE 10.4

Determinants provide a very convenient and neat way of writing the equation of a line through two points. This is one of the problems that we studied in Section 8.4. Suppose the line is to pass through the points (x_1, y_1) and (x_2, y_2). Then, an equation of the line can be written in the form

$$\begin{vmatrix} x & y & 1 \\ x_1 & y_1 & 1 \\ x_2 & y_2 & 1 \end{vmatrix} = 0$$

It is quite easy to verify this fact. First, think of expanding the determinant by minors along the first row. The coefficients of x and y will be constants, so the equation is linear. Next, you can see that (x_1, y_1) and (x_2, y_2) both satisfy the equation, because if you substitute either of these pairs for (x, y) in the first row of the determinant, the result is a determinant with two identical rows and this you know is 0. Thus, the equation is that of the desired line.

As an illustration, let us find an equation of the line through (1, 3) and $(-5, -2)$. The determinant form of this equation is

$$\begin{vmatrix} x & y & 1 \\ 1 & 3 & 1 \\ -5 & -2 & 1 \end{vmatrix} = 0$$

or

$$[3 - (-2)]x - [1 - (-5)]y + [-2 - (-15)] = 0$$

where the quantities in brackets are the expanded minors of the first-row elements. The final equation is

$$5x - 6y + 13 = 0$$

Use the determinant method to find the equation of the line through the two given points.

1. (2, 7) and (0, 3) **2.** (10, 12) and $(-7, 1)$
3. $(-1, 4)$ and (8, 2) **4.** (5, 0) and $(0, -3)$
5. $(a, 0)$ and $(0, b)$, $ab \neq 0$ **6.** (a, b) and (0, 0), $(a, b) \neq (0, 0)$

The next section describes a very intriguing (no pun intended) use of matrices and their inverses in some cloak-and-dagger work.

▶ **Computer Corner 10.4** The program on page 748 will evaluate the determinant of any 2×2 or 3×3 matrix, provided you enter the original matrix. Verify the answers to Examples 1 and 3 using this program.

*10.5

MATRIX CODES

If you have read Edgar Allan Poe's short story, "The Gold Bug," or Arthur Conan Doyle's Sherlock Holmes story, "The Adventure of the Dancing Men," you will recall the major role played by the mysterious coded messages. In both stories, the codes were broken by statistical methods,

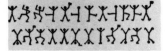

The Dancing Men
The Bettmann Archive

methods that frequently work with *simple substitution codes*. A simple substitution code is one that replaces each letter of the alphabet by a single symbol and uses this same symbol for the same letter each time. If the message is reasonably long, then a study of the frequency with which the different symbols occur may suffice to decode the message.

This idea depends on the fact that in any reasonably long sample of English writing, the percent frequencies of the letters are as follows:

E	13%	S, H	6%	W, G, B	1.5%
T	9%	D	4%	V	1%
A, O	8%	L	3.5%	K, X, J	0.5%
N	7%	C, U, M	3%	Q, Z	0.2%
I, R	6.5%	F, P, Y	2%		

You might have fun by taking a paragraph from a book you are reading and checking the frequencies of the letters against this list. Of course, it is not too difficult to find short sentences for which these frequencies do not hold. An example is, "All boys in Yokohama play with yo-yos." If this sentence were written in a simple substitution code, can you see why it might be difficult to decode by a study of the letter frequencies?

You might guess that simple substitution codes are really too simple for any government to use for important secret messages. Matrices have an interesting application in the science of cryptography (theory of codes) and offer a means of constructing codes that are virtually indecipherable without the key. The following procedure will illustrate the idea.

For simplicity, let us number the letters of the alphabet from 1 to 26 in the usual order, and let us indicate a space between words by the number 27.

1. a	**8.** h	**15.** o	**22.** v
2. b	**9.** i	**16.** p	**23.** w
3. c	**10.** j	**17.** q	**24.** x
4. d	**11.** k	**18.** r	**25.** y
5. e	**12.** l	**19.** s	**26.** z
6. f	**13.** m	**20.** t	**27.** space
7. g	**14.** n	**21.** u	

Suppose we want to encode the message:

Meet me tonight.

We first write the message, replacing letters and spaces by the assigned numbers and disregarding the punctuation, to get

13.5.5.20.27.13.5.27.20.15.14.9.7.8.20.27.

Next we choose a 2 × 2 matrix that has an inverse, say

$$M = \begin{bmatrix} 2 & 3 \\ 1 & 2 \end{bmatrix}$$

We found

$$M^{-1} = \begin{bmatrix} 2 & -3 \\ -1 & 2 \end{bmatrix}$$

in Section 10.3. Because we have chosen a 2×2 matrix, we break the message up into successive pairs of numbers, which we write as column matrices:

$$\begin{bmatrix} 13 \\ 5 \end{bmatrix} \begin{bmatrix} 5 \\ 20 \end{bmatrix} \begin{bmatrix} 27 \\ 13 \end{bmatrix} \begin{bmatrix} 5 \\ 27 \end{bmatrix} \begin{bmatrix} 20 \\ 15 \end{bmatrix} \begin{bmatrix} 14 \\ 9 \end{bmatrix} \begin{bmatrix} 7 \\ 8 \end{bmatrix} \begin{bmatrix} 20 \\ 27 \end{bmatrix}$$

We then multiply each column matrix by M to get

$$\begin{bmatrix} 41 \\ 23 \end{bmatrix} \begin{bmatrix} 70 \\ 45 \end{bmatrix} \begin{bmatrix} 93 \\ 53 \end{bmatrix} \begin{bmatrix} 91 \\ 59 \end{bmatrix} \begin{bmatrix} 85 \\ 50 \end{bmatrix} \begin{bmatrix} 55 \\ 32 \end{bmatrix} \begin{bmatrix} 38 \\ 23 \end{bmatrix} \begin{bmatrix} 121 \\ 74 \end{bmatrix}$$

The coded message is simply the list of numbers in these matrices:

41.23.70.45.93.53.91.59.85.50.55.32.38.23.121.74

The intended receiver of this message has the magic key, M and M^{-1}. Because M is 2×2, the receiver knows that he or she has to separate the message into 2×1 column matrices and multiply each matrix by M^{-1}. This converts the message into its original numbered form, which the receiver can easily read.

Although this code is not very complicated, it is also not very easy to break. Codes of this type can be made as complex as is desired by using higher-order matrices. For instance, the message could have been coded with an 8×8 matrix. In actual practice, the coding and decoding would be done by a digital computer.

EXERCISE 10.5

1. Use the matrix M^{-1} to decode the message in this section.
2. Use the matrix M to encode the message:

   Math can be fun.

3. Use the matrix M^{-1} to decode the message:

   133.80.75.47.66.35.93.57.40.26.91.59.23.15.52.33.121.74.

4. You are given the magic key

   $$C = \begin{bmatrix} 1 & -3 & 1 \\ 0 & 3 & -2 \\ 0 & -1 & 1 \end{bmatrix} \quad \text{and} \quad C^{-1} = \begin{bmatrix} 1 & 2 & 3 \\ 0 & 1 & 2 \\ 0 & 1 & 3 \end{bmatrix}$$

   Decode the following message:

   2.−9.12.−31.18.0.19.−11.10.24.−18.13.24.−25.13.

5. Use the matrix C in Problem 4 to encode the message in Problem 2.

6. Use the key in Problem 4 to decode the message:

$-32.32.-9.-35.21.2.-39.20.0.3.-13.8.-61.79.-26.$
$-57.73.-23.-25.35.-8.$

7. Use the key in Problem 4 to decode the message:

$-36.37.-8.5.-21.13.-61.79.-26.30.-39.22.-12.13.$
$-4.-30.42.-11.-45.39.-9.$

8. Use the key in Problem 4 to decode the message:

$38.-41.21.-75.79.-26.-16.38.-11.-3.-8.11.13.13.$
$-4.3.-12.13.-64.49.-11.-23.14.0.-29.27.0.$

SUMMARY

Section	Item	Meaning	Example
10.1	Matrix	A rectangular array of numbers enclosed in square brackets (or parentheses)	$\begin{bmatrix} 3 & 4 \\ 5 & 6 \end{bmatrix}$
10.1	$A = B$	The corresponding elements of matrices A and B are equal.	$\begin{bmatrix} 2 & 4 \\ 5 & 6 \end{bmatrix} = \begin{bmatrix} \frac{6}{3} & 2 \cdot 2 \\ 3+2 & 6 \end{bmatrix}$
10.1A	kA	Multiplication of each element of the matrix A by the number k	$3\begin{bmatrix} 1 & 2 \\ 3 & 4 \end{bmatrix} = \begin{bmatrix} 3 & 6 \\ 9 & 12 \end{bmatrix}$
10.1A	$A + B$	The sum of the matrices A and B	$\begin{bmatrix} 1 & 2 \\ 3 & 4 \end{bmatrix} + \begin{bmatrix} 4 & 5 \\ 3 & 2 \end{bmatrix} = \begin{bmatrix} 5 & 7 \\ 6 & 6 \end{bmatrix}$
10.1B	AB	The product of the matrices A and B	$[3 \quad 4]\begin{bmatrix} 1 & 2 \\ 0 & 5 \end{bmatrix} = [3 \quad 26]$
10.1C	I	The identity matrix	For a 2×2 matrix, $I = \begin{bmatrix} 1 & 0 \\ 0 & 1 \end{bmatrix}$
10.3	A^{-1}	The inverse of the matrix A	If $A = \begin{bmatrix} 2 & 3 \\ 1 & 2 \end{bmatrix}$, then $A^{-1} = \begin{bmatrix} 2 & -3 \\ -1 & 2 \end{bmatrix}$
10.4A	$\det A = \begin{vmatrix} a_1 & b_1 \\ a_2 & b_2 \end{vmatrix}$	$a_1 b_2 - a_2 b_1$	

Section	Item	Meaning
10.4B	Cramer's rule	For the system

$$a_1x + b_1y = d_1$$
$$a_2x + b_2y = d_2$$

$$x = \frac{D_x}{D} \qquad y = \frac{D_y}{D} \qquad D \neq 0$$

where

$$D = \begin{vmatrix} a_1 & b_1 \\ a_2 & b_2 \end{vmatrix} \qquad D_x = \begin{vmatrix} d_1 & b_1 \\ d_2 & b_2 \end{vmatrix} \qquad D_y = \begin{vmatrix} a_1 & d_1 \\ a_2 & d_2 \end{vmatrix}$$

PRACTICE TEST 10

1. Given the three matrices below, find x and y so that $2A + B = C$.

$$A = \begin{bmatrix} 2 & x \\ 3 & y \end{bmatrix} \qquad B = \begin{bmatrix} 2 & -1 \\ 3 & 2 \end{bmatrix} \qquad C = \begin{bmatrix} 6 & 5 \\ 9 & 10 \end{bmatrix}$$

2. If

$$A = \begin{bmatrix} 4 & -5 \\ 1 & -1 \end{bmatrix} \quad \text{and} \quad B = \begin{bmatrix} -1 & 5 \\ -1 & 4 \end{bmatrix}$$

calculate AB and BA.

3. To build three types of zig-zag toys, the Zee-Zee Toy Company requires materials as listed below. Suppose that Zee-Zee decides to build 20 type I, 25 type II, and 10 type III zig-zag toys. Use matrices to find the number of units of frames, wheels, chains, and paint needed for the job.

Type	Frames	Wheels	Chains	Paint
I	1	2	1	1
II	1	3	2	1
III	4	4	2	2

4. Zee-Zee finds that frames cost $2 each, wheels $1 each, chains 75¢ each, and paint 50¢ per unit. Use matrices to find the total cost of these items for each type of zig-zag toy in Problem 3.

For Problems 5-9, suppose that

$$A = \begin{bmatrix} 2 & 0 & 1 \\ 2 & -1 & 3 \\ 4 & 1 & 2 \end{bmatrix} \quad and \quad B = \begin{bmatrix} -2 & 1 & 0 \\ 4 & 3 & -2 \\ 1 & 2 & -1 \end{bmatrix}$$

5. Find $2A - 3B$.   **6.** Find $A + B$.

7. Find AB and BA.   **8.** Find $(A + B)^2$.

9. Verify that

$$A^{-1} = \frac{1}{4} \begin{bmatrix} 5 & -1 & -1 \\ -8 & 0 & 4 \\ -6 & 2 & 2 \end{bmatrix}$$

10. Find M^{-1} if

$$M = \begin{bmatrix} 3 & 1 \\ 2 & -1 \end{bmatrix}$$

11. Use the method of reducing the augmented matrix to echelon form to solve the system $AX = D$, where

$$A = \begin{bmatrix} 2 & 0 & 1 \\ 2 & -1 & 3 \\ 4 & 1 & 2 \end{bmatrix} \quad X = \begin{bmatrix} x \\ y \\ z \end{bmatrix} \quad D = \begin{bmatrix} 1 \\ 9 \\ 0 \end{bmatrix}$$

12. Use the matrix A^{-1} of Problem 9 to solve the system of Problem 11.

13. Use matrices to solve the system

$$x + 3y = 19$$
$$y + 3z = 10$$
$$z + 3x = -5$$

14. Find the inverse of the matrix

$$\begin{bmatrix} -1 & 1 & 0 \\ 4 & 3 & -2 \\ 2 & 2 & -1 \end{bmatrix}$$

15. Use your answer to Problem 14 to solve the system

$$-x + y \quad\quad = 5$$
$$4x + 3y - 2z = 7$$
$$2x + 2y - z = 5$$

16. Show that the following matrix has no inverse:

$$\begin{bmatrix} 3 & 3 & 3 \\ 4 & 2 & 0 \\ 3 & 3 & 3 \end{bmatrix}$$

17. On counting the money in her piggy bank, Sally found that she had 122 coins, all nickels, dimes, and quarters. If the total value of the coins was $15, and the total value of the quarters was four times the total value of the nickels, how many of each kind of coin did Sally have?

18. The augmented matrix of a system of three equations in the three unknowns x, y, z is reduced to the following form:

$$\left[\begin{array}{ccc|c} 2 & 1 & 1 & 1 \\ 0 & 3 & 2 & 4 \\ 0 & 0 & 1 & 2 \end{array}\right]$$

Find the solution of the system.

19. Suppose the reduced matrix in Problem 18 read as follows:

$$\left[\begin{array}{ccc|c} 2 & 1 & 1 & 1 \\ 0 & 3 & 2 & 4 \\ 0 & 0 & 0 & 2 \end{array}\right]$$

What can be said about the system of equations?

20. Suppose the reduced matrix in Problem 18 read as follows:

$$\left[\begin{array}{ccc|c} 2 & 1 & 1 & 1 \\ 0 & 3 & 2 & 4 \\ 0 & 0 & 0 & 0 \end{array}\right]$$

Find the solution of the system of equations.

21. Evaluate the determinant

$$\begin{vmatrix} 3 & 5 \\ -5 & 8 \end{vmatrix}$$

22. Evaluate the determinant of the matrix in Problem 14.

23. Evaluate the determinant of the matrix in Problem 16.

24. Find the values of x for which

$$\begin{vmatrix} 2x & 3 \\ 4 & x \end{vmatrix} = \begin{vmatrix} x & -2 \\ 2 & 4 \end{vmatrix}$$

25. Solve the system in Problem 15 by Cramer's rule.

CHAPTER 11

MATHEMATICAL SYSTEMS

One of the greatest scientists that Switzerland has ever produced, Leonhard Euler (pronounced, "oiler"), was born in Basel in 1707. His father, a good mathematician himself, taught the boy mathematics but insisted on his attending the University of Basel to study Hebrew and theology. There Leonhard was befriended by the famous Bernoulli family of mathematicians, who persuaded his father that the boy was destined to become a great mathematician rather than a Calvinist pastor. In a short time, the young Euler was on his way. At 19 he wrote his first original paper, which barely missed getting the Paris Academy prize. At 20, through the good offices of the Bernoullis, he received an appointment to the Academy of Sciences at St. Petersburg, Russia.

After deciding to settle down in St. Petersburg, Euler married Catharina, daughter of the Swiss painter Gsell, who had been brought to Russia by Peter the Great.

In about 1735, Euler set himself to win the Paris prize for a problem in astronomy that many leading mathematicians estimated would take several months to solve. Euler solved the problem in 3 days of unremitting effort, but the strain led to an illness from which he lost the sight of his right eye.

In 1740, at the invitation of Frederick the Great, Euler left Russia to join the Berlin Academy. He stayed in Berlin for a productive 26 years, returning to St. Petersburg at the behest of Catherine the Great in 1766. Shortly thereafter, he began to lose the sight in his other eye; but as his vision was fading, he taught himself to write his formulas with chalk on a large slate. When he became totally blind, instead of diminishing, his mathematical productivity increased.

Euler's powerful mind remained clear to the moment of his death. On the afternoon of September 18, 1783, he had outlined the calculation of the orbit of the newly discovered planet Uranus. Later, while playing with one of his grandsons, he suffered a stroke and died.

One of the most prolific mathematicians who ever lived, Euler contributed to every area of mathematics and applied mathematics of his time. Many of these areas he extended far beyond the state in which he found them, and he had an almost unequaled genius for devising algorithms to solve the most difficult problems.

LEONHARD EULER (1707–1783)

Picture Collection, the Branch Libraries, The New York Public Library

Euler calculated without effort, as men breathe, or as eagles sustain themselves in the wind.

FRANCOIS ARAGO

MATHEMATICAL SYSTEMS AND CLOCK ARITHMETIC

"NOW WITH THE NEW MATH..."

Courtesy Sidney Harris

In earlier chapters of this book, we have made frequent reference to *mathematical systems*. For example, in Chapter 4 we defined the set of natural numbers together with the operations of addition and multiplication. We then discussed certain properties of the set of natural numbers with respect to these operations. The set of natural numbers, together with the operations of addition and multiplication, constitutes a mathematical system.

In general, a **mathematical system** consists of the following items:

1. A **set of elements**
2. One or more **operations**
3. One or more **relations** that enable us to compare the elements in the set
4. Some **rules, axioms, or laws** that the elements in the set satisfy

For example, when we refer to the system of integers, we have

1. *Elements.* The elements of this system are the integers in the set $I = \{. \ . \ . \ , -2, -1, 0, 1, 2, \ . \ . \ .\}$.
2. *Operations.* Within the system of integers, we can always perform the operations of addition, subtraction, and multiplication. We can sometimes divide, but most often the result of dividing one integer by another is not an integer. The set of integers is not closed under division.
3. *Relations.* We have three possible relations between any two integers a and b:

$$a < b \qquad a > b \qquad a = b$$

4. *Rules or laws.* Addition and multiplication in the system of integers satisfy the commutative and the associative laws, as well as the distributive law.

Most of the mathematical systems we have discussed involve an *infinite* number of elements. In this section, we wish to acquaint you with a mathematical system that contains only a *finite* number of elements. The idea here is that if you become thoroughly familiar with the workings of a finite system, you will be able to generalize this knowledge and apply it to other systems.

As you are aware, the numbers on the face of a clock are used to tell the time of the day or night. The set of numbers used for this purpose is $S = \{1, 2, 3, 4, 5, 6, 7, 8, 9, 10, 11, 12\}$. We shall define addition on this set by means of an addition table, but we first present some examples to justify the entries in the table.

If it is now, say, 11 A.M., and you have to go to class in 3 hr, it is obvious that you have to go to class at 2 P.M. For this reason, we shall define $11 \oplus 3 = 2$, where $\oplus$ (read, "circle plus") is the operation of clock addi-

tion. If your class were to meet in 5 hr, and it was now 10 A.M., you would have to go to class at 3 P.M., so $10 \oplus 5 = 3$.

EXAMPLE 1 Find the following sums in clock arithmetic:

a. $8 \oplus 3$ **b.** $8 \oplus 7$ **c.** $11 \oplus 12$

Solution **a.** $8 \oplus 3$ means 3 hr after 8 o'clock, so $8 \oplus 3 = 11$. (See Fig. 11.1a.)
b. $8 \oplus 7$ means 7 hr after 8 o'clock. Thus, $8 \oplus 7 = 3$. (See Fig. 11.1b.)
c. $11 \oplus 12$ means 12 hr after 11 o'clock. Thus, $11 \oplus 12 = 11$. ◀

We can now construct a table for the addition facts in clock arithmetic, as shown in Table 11.1a. You should verify the entries in this table before proceeding further.

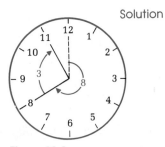

Figure 11.1a

Table 11.1a Addition in Clock Arithmetic

⊕	1	2	3	4	5	6	7	8	9	10	11	12
1	2	3	4	5	6	7	8	9	10	11	12	1
2	3	4	5	6	7	8	9	10	11	12	1	2
3	4	5	6	7	8	9	10	11	12	1	2	3
4	5	6	7	8	9	10	11	12	1	2	3	4
5	6	7	8	9	10	11	12	1	2	3	4	5
6	7	8	9	10	11	12	1	2	3	4	5	6
7	8	9	10	11	12	1	2	3	4	5	6	7
8	9	10	11	12	1	2	3	4	5	6	7	8
9	10	11	12	1	2	3	4	5	6	7	8	9
10	11	12	1	2	3	4	5	6	7	8	9	10
11	12	1	2	3	4	5	6	7	8	9	10	11
12	1	2	3	4	5	6	7	8	9	10	11	12

Figure 11.1b

Of course, other operations can be defined on this set. For example, $3 \ominus 4$* is a number n with the property that $3 = 4 \oplus n$. By looking at Table 11.1a, we can see that the number 11 satisfies this equation, because $3 = 4 \oplus 11$. Accordingly, we state the following definition:

▶ **Definition 11.1a**

$$a \ominus b = n \quad \text{if and only if} \quad a = b \oplus n$$

With this definition, $4 \ominus 6 = 10$, because $4 = 6 \oplus 10$.

If a is a positive number, we can define $-a$ in clock arithmetic as the number obtained by going counterclockwise a hours. With this convention, it follows that

$$-a = 12 - a$$

Figure 11.1c

For example, $-4 = 12 - 4 = 8$, which agrees with Fig. 11.1c.

* The symbol $\ominus$ is read as "circle minus."

With this definition of $-a$, it also follows that

$$a \ominus b = a \oplus (-b)$$

which may be regarded as an alternate way to define circle minus.

EXAMPLE 2 Find a number n so that:

a. $n \oplus 5 = 3$ **b.** $2 \ominus 3 = n$

Solution **a.** We inspect the fifth row in Table 11.1a. We have to find a number n so that when we add n to 5 we obtain 3. The answer is 10, because $10 \oplus 5 = 3$.

b. $2 \ominus 3 = 2 \oplus (-3) = 2 \oplus (12 - 3) = 2 \oplus 9 = 11$. (You can check in the table that $3 \oplus 11 = 2$.) ◀

As usual, multiplication can be defined in terms of addition. Thus, $3 \otimes 5 = (5 \oplus 5) \oplus 5 = 10 \oplus 5 = 3$, so $3 \otimes 5 = 3$.

▶ **Definition 11.1b**

$$a \otimes b = \underbrace{b \oplus b \oplus b \oplus \cdots \oplus b}_{a \text{ times}}$$

(The symbol $\otimes$ is read, "circle times.")

With this definition, $4 \otimes 3 = 3 \oplus 3 \oplus 3 \oplus 3 = 12$ and $2 \otimes 8 = 8 \oplus 8 = 4$.

Notice that a quicker way to obtain the answer is to multiply the two numbers by ordinary arithmetic, divide by 12, and take the remainder as the answer. (If the remainder is 0, the answer is 12, the zero point on a clock.) For example, $6 \otimes 5 = 6$, because $6 \times 5 = 30$ and 30 divided by 12 equals 2 with a remainder of 6. Similarly, $6 \otimes 4 = 12$, because $6 \times 4 = 24$, and 24 divided by 12 is 2 with a 0 remainder.

EXAMPLE 3 Find the following products in clock arithmetic:

a. $3 \otimes 5$ **b.** $4 \otimes 8$ **c.** $9 \otimes 9$

Solution **a.** $3 \otimes 5 = 3$, because $3 \times 5 = 15$, which divided by 12 is 1 with a remainder of 3.

b. $4 \otimes 8 = 8$, because $4 \times 8 = 32$ and 32 divided by 12 is 2 with a remainder of 8.

c. $9 \otimes 9 = 9$, because $9 \times 9 = 81$, which divided by 12 is 6 with a remainder of 9. ◀

Division is defined in clock arithmetic in terms of multiplication. For example, $\frac{4}{8} = 2$ because $4 = 8 \otimes 2$, and $\frac{3}{5} = 3$ because $3 = 5 \otimes 3$.

▶ **Definition 11.1c**

$$\frac{a}{b} = n \qquad \text{if and only if} \qquad a = b \otimes n$$

In discussing multiplication in clock arithmetic, we noted that 12 in this arithmetic corresponds to 0 in ordinary arithmetic. To show further that 12 behaves like 0 in ordinary arithmetic, let us try to find a number n such that

$$\frac{1}{12} = n$$

This equation is true only if $1 = 12 \otimes n$. But there is no n in clock arithmetic such that $12 \otimes n = 1$. So, in clock arithmetic, we cannot divide by 12, just as in ordinary arithmetic we cannot divide by 0.

EXAMPLE 4   Find the following quotients in clock arithmetic:

   a. $\frac{2}{7}$ **b.** $\frac{8}{8}$ **c.** $\frac{3}{4}$

Solution   **a.** $\frac{2}{7} = n$ if and only if $2 = 7 \otimes n$. Thus, we wish to find an n such that $7 \otimes n = 2$. Because $n = 2$ satisfies the given equation, $\frac{2}{7} = 2$. Note that the answer $n = 2$ can be found by trial and error; that is, we let $n = 1$ and see if $7 \otimes 1 = 2$. We then let $n = 2$ and check if $7 \otimes 2 = 2$. Because $7 \otimes 2 = 2$, the desired number has been obtained.

   b. $\frac{8}{8} = n$ if and only if $8 = 8 \otimes n$, an equation that is satisfied by $n = 1$, $n = 4$, $n = 7$, and $n = 10$. Note that unlike ordinary division problems, in which the answer is unique, in clock arithmetic a division problem may have many solutions.

   c. $\frac{3}{4} = n$ if and only if $3 = 4 \otimes n$. There is no n such that $3 = 4 \otimes n$, so the problem $\frac{3}{4} = n$ has no solution.   ◀

EXERCISE 11.1

In Problems 1–8, find the sum in clock arithmetic.

1. $9 \oplus 7$ **2.** $2 \oplus 8$ **3.** $8 \oplus 3$ **4.** $5 \oplus 7$
5. $7 \oplus 8$ **6.** $9 \oplus 9$ **7.** $8 \oplus 11$ **8.** $12 \oplus 3$

In Problems 9–14, find the difference in clock arithmetic.

9. $8 \ominus 3$ **10.** $5 \ominus 8$ **11.** $9 \ominus 12$ **12.** $6 \ominus 9$
13. $8 \ominus 7$ **14.** $1 \ominus 12$

In Problems 15–22, find all n satisfying the given equation in clock arithmetic.

15. $n \oplus 7 = 9$ **16.** $n \oplus 8 = 2$ **17.** $2 \oplus n = 1$
18. $7 \oplus n = 3$ **19.** $3 \ominus 5 = n$ **20.** $2 \ominus 4 = n$
21. $1 \ominus n = 12$ **22.** $3 \ominus 7 = n$

In Problems 23–28, find the indicated products in clock arithmetic.

23. $4 \otimes 3$ **24.** $3 \otimes 8$ **25.** $9 \otimes 2$ **26.** $3 \otimes 9$

27. $2 \otimes 8$ **28.** $12 \otimes 3$

29. Make a table of multiplication facts in clock arithmetic.

In Problems 30–34, find the indicated quotients in clock arithmetic.

30. $\frac{9}{7}$ **31.** $\frac{3}{5}$ **32.** $\frac{3}{9} = n$

                                                     [*Hint:* There are

33. $\frac{1}{11}$ **34.** $\frac{1}{12}$ three answers.]

In Problems 35–40, find all n satisfying the given equations in clock arithmetic.

35. $\dfrac{n}{5} = 8$ **36.** $\dfrac{n}{2} = 4$ **37.** $\dfrac{n}{2} \oplus 4 = 8$

38. $\dfrac{n}{7} = 9$ **39.** $\dfrac{2}{n} = 3$ **40.** $\dfrac{12}{12} = n$

DISCOVERY 11.1

Clock arithmetic can also be defined on the 5-hr clock in the figure in the margin. We agree that on this clock $1 \oplus 4 = 0$ (not 5).

1. Can you discover the addition table for clock arithmetic?
2. Is the set $S = \{0, 1, 2, 3, 4\}$ closed with respect to clock arithmetic?
3. Is the operation $\oplus$ commutative?
4. Is there any element of S that does not have an inverse with respect to $\oplus$?

11.2

MODULAR ARITHMETIC

In Section 11.1, the four fundamental operations were defined on the 12-hr clock. These operations can be generalized to a kind of mathematical system called **modular arithmetic.** This system can be defined on the set of integers as follows:

▶ **Definition 11.2a**

> Two integers a and b are said to be **congruent modulo m,** denoted by $a \equiv b \pmod{m}$, if $a - b$ (or $b - a$) is a multiple of m (m an integer).

Thus, $14 \equiv 4 \pmod 5$, because $14 - 4 = 10$, which is a multiple of 5; and $2 \equiv 18 \pmod 4$, because $18 - 2 = 16$, which is a multiple of 4.

EXAMPLE 1 Are the following statements true or false?

a. $4 \equiv 1 \pmod 3$ **b.** $7 \equiv 2 \pmod 4$ **c.** $1 \equiv 6 \pmod 5$

Solution **a.** True, because $4 - 1 = 3$, a multiple of 3.
b. False, because $7 - 2$ is *not* a multiple of 4.
c. True, because $6 - 1 = 5$, a multiple of 5. ◀

The addition of two numbers in modular arithmetic is very simple. For example, in arithmetic modulo 5, $3 + 3 = 1$, because $3 + 3 = 6 \equiv 1 \pmod 5$. Thus, to add nonnegative numbers in a system modulo m, we proceed as follows:

1. Add the numbers in the ordinary way.
2. If the sum is less than m, the answer is the sum obtained.
3. If the sum is greater than or equal to m, the answer is the remainder obtained upon dividing the sum by m.

For example,

$3 + 4 \equiv 7 \pmod 8$, because $3 + 4 = 7$ is less than 8.

$3 + 5 \equiv 0 \pmod 8$, because $3 + 5 = 8$, which divided by 8 leaves a 0 remainder.

$3 + 7 \equiv 2 \pmod 8$, because $3 + 7 = 10$, which divided by 8 is 1 with a remainder of 2.

We are now in a position to define an addition table for modular arithmetic. Table 11.2a defines addition modulo 5.

When adding numbers in modular arithmetic, we may want to use the following theorem:

Table 11.2a Addition Modulo 5

+	0	1	2	3	4
0	0	1	2	3	4
1	1	2	3	4	0
2	2	3	4	0	1
3	3	4	0	1	2
4	4	0	1	2	3

▶ **Theorem 11.2a**

If $a \equiv b \pmod m$ and $c \equiv d \pmod m$, then

$a + c \equiv b + d \pmod m$

For example, to add 18 and 31 (mod 5), we can use the facts that $18 \equiv 3 \pmod 5$ and $31 \equiv 1 \pmod 5$ to obtain

$18 + 31 \equiv 4 \pmod 5$

Proof The theorem can be proved by using the meanings of the two congruences. Thus,

$a \equiv b \pmod m$ means $a - b = km$, k an integer

$c \equiv d \pmod m$ means $c - d = pm$, p an integer

Hence,

$(a - b) + (c - d) = km + pm = (k + p)m$

Also,

$$(a - b) + (c - d) = (a + c) - (b + d)$$

so that

$$(a + c) - (b + d) = (k + p)m$$

That is, $(a + c) - (b + d)$ is a multiple of m, which means that $a + c \equiv b + d \pmod{m}$.

To multiply two numbers in arithmetic modulo 5, we multiply the numbers in the ordinary way. If the product is less than 5, the answer is this product. If the product is greater than 5, the answer is the remainder when the product is divided by 5. For example, $4 \times 4 \equiv 1$, because the product $4 \times 4 = 16$, which leaves a remainder 1 when divided by 5. The table for multiplication modulo 5 is shown in Table 11.2b.

Subtraction in modulo 5 arithmetic is defined as follows:

$$a - b \equiv c \qquad \text{if and only if} \qquad a \equiv b + c$$

Thus, $3 - 4 \equiv 4 \pmod 5$, because $3 \equiv (4 + 4) \pmod 5$; and $2 - 4 \equiv 3$, because $2 \equiv (4 + 3) \pmod 5$.

Table 11.2b Multiplication Modulo 5

×	0	1	2	3	4
0	0	0	0	0	0
1	0	1	2	3	4
2	0	2	4	1	3
3	0	3	1	4	2
4	0	4	3	2	1

EXAMPLE 2   Find:

a. $(4 - 1) \pmod 5$ **b.** $(2 - 3) \pmod 5$ **c.** $(1 - 4) \pmod 5$

Solution

a. $4 - 1 \equiv 3 \pmod 5$

b. $2 - 3 \equiv n \pmod 5$ if and only if $2 \equiv (3 + n) \pmod 5$. From Table 11.2a, $3 + 4 \equiv 2 \pmod 5$. Thus, $2 - 3 \equiv 4 \pmod 5$.

c. $1 - 4 \equiv n \pmod 5$ if and only if $1 \equiv 4 + n \pmod 5$. From Table 11.2a, we see that $1 \equiv 4 + 2 \pmod 5$. Thus, $1 - 4 \equiv 2 \pmod 5$. ◄

We now examine Table 11.2c to find what properties the set $S = \{0, 1, 2, 3, 4\}$ has under the operation of addition modulo 5.

Table 11.2c Addition Modulo 5

+	0	1	2	3	4
0	0	1	2	3	4
1	1	2	3	4	0
2	2	3	4	0	1
3	3	4	0	1	2
4	4	0	1	2	3

1. All entries in the table are elements of S, so S is **closed** with respect to addition (mod 5).
2. The operation of addition (mod 5) is **associative.** (This fact is a consequence of the properties of ordinary addition, because division by 5 and taking the remainder may be done after the ordinary addition is done.)
3. The operation of addition (mod 5) is **commutative;** that is, if x and y are any elements of S, then $x + y \equiv y + x \pmod 5$. This can easily be checked by inspecting Table 11.2c. As you can see, the results of the operations $2 + 4$ and $4 + 2$ appear in positions that are **symmetric** with respect to a diagonal line drawn from top left to bottom right of the table. If this type of symmetry is present for any operation table, that is, if the top half of the table is the reflection of the bottom half across the diagonal, then the operation is commutative.
4. The **identity** for addition modulo 5 is 0, because for any x in S, $x + 0 = x$; that is, $0 + 0 = 0$, $1 + 0 = 1$, $2 + 0 = 2$, $3 + 0 = 3$, and $4 + 0 = 4$.

5. Every element has an additive **inverse** (mod 5). The inverse of 0 is 0, because $0 + 0 \equiv 0$ (mod 5). The inverse of 1 is 4, because $1 + 4 \equiv 0$ (mod 5). The inverse of 2 is 3, because $2 + 3 \equiv 0$ (mod 5). The inverse of 3 is 2, because $3 + 2 \equiv 0$ (mod 5). The inverse of 4 is 1, because $4 + 1 \equiv 0$ (mod 5).

As we have seen in the preceding discussion, modular arithmetic is a mathematical system in which two operations are defined on the same set. For the set $S = \{0, 1, 2, 3, 4\}$, we defined addition and multiplication modulo 5 as shown in Tables 11.2a and 11.2b. Are there any properties that involve both operations? The answer is yes. For example, to find $3 \times (2 + 4)$ (mod 5), we can proceed in one of two ways:

1. $3 \times (2 + 4) \equiv 3 \times (1) \equiv 3$ (mod 5)
2. $3 \times (2 + 4) \equiv (3 \times 2) + (3 \times 4) \equiv 1 + 2 \equiv 3$ (mod 5)

To get the answer in the second way, we used a property that involves both addition and multiplication. This property is called the **distributive** property; it is true in general that if a, b, and c are elements of S, then

$$a \times (b + c) \equiv (a \times b) + (a \times c) \text{ (mod } m)$$

because $a \times (b + c) - (a \times b) - (a \times c) = 0$ in ordinary arithmetic.

EXAMPLE 3   Find a replacement for n that will make the given sentence true in arithmetic modulo 5.

     a. $4 \times (3 + 1) \equiv (4 \times n) + (4 \times 1)$
     b. $n \times (1 + 3) \equiv (2 \times 1) + (2 \times 3)$

Solution   **a.** By the distributive property, $4 \times (3 + 1) \equiv (4 \times 3) + (4 \times 1)$. Thus, $n = 3$ will make the sentence $4 \times (3 + 1) \equiv (4 \times n) + (4 \times 1)$ true.
     b. By the distributive property, $2 \times (1 + 3) \equiv (2 \times 1) + (2 \times 3)$. Thus, $n = 2$ will make the sentence $n \times (1 + 3) \equiv (2 \times 1) + (2 \times 3)$ true. ◄

EXAMPLE 4   Find a replacement for n that will make the given sentence true.

     a. $3 + n \equiv 2$ (mod 5)   **b.** $\frac{3}{4} \equiv n$ (mod 5)

Solution   **a.** By inspection of Table 11.2a (or 11.2c), we see that $3 + 4 \equiv 2$ (mod 5). Thus, $n = 4$ will make the sentence $3 + n \equiv 2$ (mod 5) true.
     b. $\frac{3}{4} \equiv n$ (mod 5) is true if and only if $3 \equiv 4 \times n$ (mod 5). By inspection of Table 11.2b, we see that $4 \times 2 \equiv 3$ (mod 5). Hence, $n = 2$ will make the sentence $\frac{3}{4} \equiv n$ (mod 5) true. ◄

In Examples 3 and 4, we have given answers in the set $\{0, 1, 2, 3, 4\}$. If we allow n to be any integer that makes the given congruence true, then each answer can be modified by adding or subtracting any desired multiple of 5. This is justified by the addition theorem (Theorem 11.2a) proved earlier. For instance, in Example 4a, all possible integer answers would be given by $n \equiv 4 \pmod 5$. This means that n could have any of the values $\ldots, -6, -1, 4, 9, 14, \ldots$; each of these would make the congruence $3 + n \equiv 2 \pmod 5$ true.

EXERCISE 11.2

In Problems 1–6, classify the given statement as true or false.

1. $2 \equiv 4 \pmod 3$ **2.** $5 \equiv 2 \pmod 3$
3. $6 \equiv 7 \pmod 5$ **4.** $5 \equiv 3 \pmod 2$
5. $8 \equiv 9 \pmod{10}$ **6.** $12 \equiv 8 \pmod 4$

In Problems 7–10, find the indicated sums.

7. $(3 + 4) \pmod 5$ **8.** $(2 + 9) \pmod{10}$
9. $(3 + 1) \pmod 5$ **10.** $(3 + 6) \pmod 7$

In Problems 11–14, find the indicated products.

11. $(4 \times 2) \pmod 5$ **12.** $(4 \times 3) \pmod 5$
13. $(2 \times 3) \pmod 5$ **14.** $(3 \times 3) \pmod 5$

In Problems 15–18, find the indicated differences.

15. $(2 - 4) \pmod 5$ **16.** $(3 - 4) \pmod 5$
17. $(1 - 3) \pmod 5$ **18.** $(0 - 2) \pmod 5$

In Problems 19–34, find a value of n in the set {0, 1, 2, 3, 4} that will make the given congruence true.

19. $4 \times (3 + 0) \equiv (4 \times 3) + (4 \times n) \pmod 5$
20. $2 \times (1 + 3) \equiv (2 \times 1) + (n \times 3) \pmod 5$
21. $2 \times (0 + 3) \equiv (2 \times n) + (2 \times 3) \pmod 5$
22. $4 \times (1 + n) \equiv (4 \times 1) + (4 \times 2) \pmod 5$
23. $2 + n \equiv 3 \pmod 5$ **24.** $n + 3 \equiv 1 \pmod 5$
25. $2 \times n \equiv 4 \pmod 5$ **26.** $3 \equiv 2 \times n \pmod 5$
27. $n - 3 \equiv 4 \pmod 5$ **28.** $2 \equiv n - 1 \pmod 5$
29. $3 \equiv n - 4 \pmod 5$ **30.** $n - 2 \equiv 1 \pmod 5$

31. $\dfrac{n}{2} \equiv 4 \pmod 5$ **32.** $\dfrac{n}{3} \equiv 2 \pmod 5$

33. $\frac{3}{4} \equiv n \pmod 5$ **34.** $\frac{1}{2} \equiv n \pmod 5$

In Problems 35–38, use Table 11.2b and multiplication modulo 5.

35. Is the set $S = \{0, 1, 2, 3, 4\}$ closed with respect to $\times$?
36. Is the set operation $\times$ commutative?
37. Is there an identity for the operation $\times$? If so, what is this identity?
38. Find the inverse (if possible) of:
    a. 0   **b.** 1   **c.** 2   **d.** 3   **e.** 4

A modulo 7 arithmetic is suggested by the fact that a week has 7 days. If one wants to know what day of the week it will be 29 days from Wednesday, we can follow the sequence Wednesday, Thursday, Friday, etc. However, we know that in 28 days it will be Wednesday again, so in 29 days it will be Thursday.

In Problems 39–44, state what day of the week it will be:

39. 30 days from a Monday **40.** 140 days from a Tuesday
41. 150 days from a Wednesday **42.** 80 days from a Thursday
43. 350 days from a Friday **44.** 440 days from a Saturday

USING YOUR KNOWLEDGE 11.2

Look at the back cover of this book. What did you find? The ISBN (International Standard Book Number) is a ten-digit number that encodes certain information about the book. (Some books show the ISBN only on the copyright page.) A certain book bore the ISBN 0-06-040953-3. The following diagram indicates the reference of each part of the ISBN:

┌─The book

English-speaking country ⟶ 0-06-040953-3

└─Publisher

But what about the 3 at the right end? This digit is a check digit that is used to verify orders. The check number is obtained from the other digits as follows: Write the numbers 10, 9, 8, 7, 6, 5, 4, 3, 2 above the first nine digits of the ISBN, and then multiply each of these digits by the number above it and add the results.

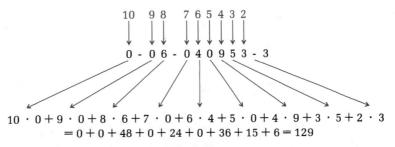

$$10 \cdot 0 + 9 \cdot 0 + 8 \cdot 6 + 7 \cdot 0 + 6 \cdot 4 + 5 \cdot 0 + 4 \cdot 9 + 3 \cdot 5 + 2 \cdot 3$$
$$= 0 + 0 + 48 + 0 + 24 + 0 + 36 + 15 + 6 = 129$$

The check digit (3 in our number) is the number between 0 and 10, inclusive, which when added to 129 gives a multiple of 11. (Note that $129 + 3 = 132$, a multiple of 11.) If the sum of the products obtained in this way is called x, then the check number c is the smallest nonnegative solution of the congruence

$$x + c \equiv 0 \ (mod \ 11)$$

This suggests that we can simplify the arithmetic a great deal by "casting out 11's." Thus, out of the 48, we can throw out 44 and keep 4; out of the 24, throw out 22 and keep 2; out of the 36, throw out 33 and keep 3; out of the 15, throw out 11 and keep 4; keep the 6. Now we just have to add $4 + 2 + 3 + 4 + 6 = 19$. The required solution of

$$19 + c \equiv 0 (mod \ 11)$$

is 3 $(19 + 3 = 22$, a multiple of 11).

Use this knowledge to solve the following problems. Note that if the check number is 10, it is written as the Roman numeral X.

1. If the first nine digits of the ISBN are 0-06-040613, find the check number.
2. If the first nine digits of the ISBN are 0-517-53052, find the check number.
3. Find the check number for ISBN 0-312-87867.
4. The last digit of the book number in the ISBN 0-060-4098▮-3 was blurred as indicated. What must this digit be?
5. Find what the blurred digit must be in the ISBN 0-03-0589▮4-2.
6. Find what the blurred digit must be in the ISBN 0-716▮-0456-0.

DISCOVERY 11.2

The idea of modular arithmetic can be used to create modular designs. To construct a modulo 5 design, we first choose a multiplier (say 2) so that we will get what is called a (5, 2) design. Then we proceed as follows:

a. *Write the multiplication table for mod 5 arithmetic. (See Table 11.2b.)*
b. *Draw a circle and divide the circumference into four equal parts, labeling the points 1, 2, 3, 4 as shown in the figure in the margin.*
c. *From Table 11.2b, we read*

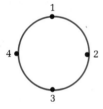

$$2 \times 1 \equiv 2 \ (mod \ 5)$$
$$2 \times 2 \equiv 4 \ (mod \ 5)$$
$$2 \times 3 \equiv 1 \ (mod \ 5)$$
$$2 \times 4 \equiv 3 \ (mod \ 5)$$

d. *We now connect 1 and 2, 2 and 4, 3 and 1, and 4 and 3, as in the figure in the margin.*

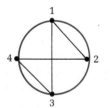

1. Can you discover how the design (5, 3) looks?
2. Can you discover why (5, 3) and (5, 2) are identical? [Hint: Look at the "multipliers" (mod 5).]

3. The designs (19, 9) and (21, 10) are shown in the figure below. Can you draw the design (19, 17)?

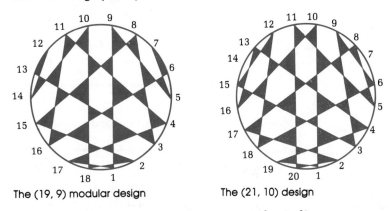

The (19, 9) modular design   The (21, 10) design

4. Can you discover why (19, 9) and (19, 17) are identical?
5. Draw the (13, 3) design. Can you discover for what value of m the (13, m) design is the same as the (13, 3) design?

11.3

ABSTRACT MATHEMATICAL SYSTEMS

·In this section we shall concentrate on the rules and laws that are obeyed by abstract mathematical systems. To illustrate the ideas involved, we shall define a set $A = \{a, b, c, d, e\}$ and an operation $*$, which we shall call *star*. The operations on the set A can be defined by a table similar to the one used to define addition of natural numbers, or addition in base 5. Suppose that the operation $*$ is defined by Table 11.3a.

Table 11.3a

$*$	a	b	c	d	e
a	b	c	d	e	a
b	c	d	e	a	b
c	d	e	a	b	c
d	e	a	b	c	d
e	a	b	c	d	e

To perform the operation star on any pair of elements, say b and c, we find the element that is in the b row and the c column (rows are horizontal and columns are vertical), as shown in Table 11.3b on page 516. From this table we can see that $b * c = e$. Similarly, $a * b = c$ and $c * d = b$.

Table 11.3b

*	a	b	c	d	e
a	b	c	d	e	a
b	c	d	e	a	b
c	d	e	a	b	c
d	e	a	b	c	d
e	a	b	c	d	e

EXAMPLE 1   Use Table 11.3a to find the result of each of the following operations:

a. $b * e$ **b.** $d * b$ **c.** $(a * b) * c$

Solution   **a.** $b * e$ corresponds to the entry in the b row, e column; that is, $b * e = b$.
b. $d * b$ corresponds to the entry in the d row, b column; that is, $d * b = a$.
c. We wrote $(a * b)$ in parentheses, so we must find $(a * b)$ first. Thus,
$(a * b) * c = c * c = a$.   ◄

By proceeding as in Example 1, we can find the result $x * y$ for all elements x and y that are in S, and it is evident that the result is also in S. The name that we give to this property is *closure*.

► **Definition 11.3a**

> A set A is **closed** under the operation $*$ if, for all a and b in A, $a * b$ is also in A.

Intuitively, we say that the set A is closed under the operation $*$ if the operation is always possible, and if no new elements are introduced in the table defining the operation. For example, the set of natural numbers is closed under the operation of addition, because for any two natural numbers a and b, $a + b$ is also a natural number. On the other hand, the set of natural numbers is not closed under subtraction since, for example, $5 - 7 = -2$, which is not a natural number.

EXAMPLE 2   Consider the sets $A = \{0, 1\}$ and $B = \{1, 2\}$. Are these sets closed under ordinary multiplication?

Solution   The set A is closed under multiplication, because all possible products $0 \times 0 = 0, 0 \times 1 = 0, 1 \times 0 = 0$, and $1 \times 1 = 1$ are in A. On the other hand, the set B is not closed under multiplication, because $2 \times 2 = 4$, which is not in B.   ◄

Another important property previously discussed in connection with the natural numbers is *associativity*.

► **Definition 11.3b**

> An operation $*$ defined on a set A is **associative** if, for all a, b, and c in A,
>
> $$(a * b) * c = a * (b * c)$$

For example, the intersection of sets that we studied in Chapter 1 is associative, because for any three sets A, B, C, we have $A \cap (B \cap C) = (A \cap B) \cap C$.

EXAMPLE 3   In Table 11.3a, check to see if:

     a. $(a * b) * d = a * (b * d)$ **b.** $(c * a) * e = c * (a * e)$

Solution   **a.** Table 11.3a gives $a * b = c$, so $a * b$ may be replaced by c to get $(a * b) * d = c * d$. Then, because the figure gives $c * d = b$, we see that $(a * b) * d = b$. Similarly, we find that $a * (b * d) = a * a = b$. The result in both cases is b, so $(a * b) * d = a * (b * d)$.

     b. Again, by using Table 11.3a, we find that $(c * a) * e = d * e = d$ and $c * (a * e) = c * a = d$. Because the result is d in both cases, we have $(c * a) * e = c * (a * e)$.   ◀

Can we conclude from Example 3 that the operation $*$ is associative? The answer is no, because we have not checked all the possibilities. Try some other possibilities and state whether or not you think $*$ is associative.

The next property we shall discuss is the *commutative property*.

Definition 11.3c

> An operation $*$ defined on a set A is **commutative** if, for every a and b in A,
>
> $$a * b = b * a$$

For example, the intersection of sets that we studied in Chapter 1 is commutative, because for any two sets A and B, $A \cap B = B \cap A$.

EXAMPLE 4   In Table 11.3a, check to see if:

     a. $b * d = d * b$ **b.** $e * c = c * e$

Solution   **a.** $b * d = a$ and $d * b = a$; thus, $b * d = d * b$.
     b. $e * c = c$ and $c * e = c$; thus, $e * c = c * e$.   ◀

Can we conclude from Example 4 that the operation $*$ is commutative? The answer is no, because we have not checked all the possibilities. However, since the top half of the table is the reflection of the bottom half across the diagonal, the operation is commutative.

As we learned in Chapter 4, the set of integers has an additive identity (0), with the property that for any integer a, $a + 0 = a = 0 + a$. Similarly, 1 is the identity for multiplication, because $a \cdot 1 = a = 1 \cdot a$. The idea of an *identity* can be generalized by means of the following definition:

Definition 11.3d

> An element e in a set A is said to be an **identity** for the operation $*$ if, for each element x in A,
>
> $$x * e = x = e * x$$

For example, for the operator $*$ defined on the set $A = \{a, b, c, d, e\}$ in Table 11.3a, the identity element is e, as you can easily check. Notice that the column directly under the identity element e is identical with the column at the far left, and the row opposite the element e is identical with the row across the top of the table. This appearance of the operation table is characteristic for any set that has an identity element under the operation.

EXAMPLE 5

Let $A = \{a, b, c\}$, and let $*$ be an operation defined on A. If c is the identity element, what do you know about the table that defines the operation $*$?

Solution

From the preceding discussion, we must have the table shown in the margin. The column at the far right is identical to the column at the far left; the row across the bottom is identical to the row across the top. ◀

$*$	a	b	c
a			a
b			b
c	a	b	c

Closely related to the idea of an identity is the idea of an *inverse*. For example, the additive inverse of 3 is -3, because $3 + (-3) = 0$ (the additive identity). Similarly, the multiplicative inverse of 3 is $\frac{1}{3}$, because $3 \times \frac{1}{3} = 1$ (the multiplicative identity). In order to find the inverse of a number a under addition, we need a number b so that $a + b = 0$; similarly, to find the inverse of a number a under multiplication, we need a number b so that $a \times b = 1$. These ideas can be summarized by the following definition:

▶ **Definition 11.3e**

> If a and b are in A, we say that **a is the inverse of b** under the operation $*$, if $a * b = e = b * a$, where e is the identity.

EXAMPLE 6

Consider the table in the margin, defining the operation $\#$. Find:

a. The identity   **b.** The inverse of a **c.** The inverse of b

Solution

$\#$	a	b
a	b	a
b	a	b

a. The identity is b, because the column under b is identical to the column at the far left and the row opposite b is identical to the row across the top.

b. We have to find an element to place in the box so that $a \# \square = b$. From the table we can see that this element is a, because $a \# a = b$.

c. We have to find an element to place in the box so that $b \# \square = b$. From the table we can see that this element is b, because $b \# b = b$. ◀

EXAMPLE 7

Consider the operation $\#$ defined by the table in the margin.

a. Is there an identity element?

b. Do any of the elements have inverses?

c. Is the operation commutative?

$\#$	a	b	c
a	c	a	b
b	b	c	a
c	a	b	c

Note that this table of operations corresponds to rotations in the plane as shown in the figure at the top of the next page.

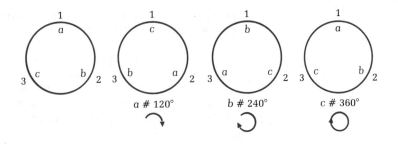

$a \# 120°$ $b \# 240°$ $c \# 360°$

Solution **a.** There is no column identical to the column under #, so there is no identity element.

b. Since there is no identity element, there are no inverses.

c. The operation is not commutative. For example,

$a \# b = a$ but $b \# a = b$ ◄

As in the case of modular arithmetic, there is a property that involves two operations. This property is called the *distributive property* and is defined as follows:

▶ **Definition 11.3f**

> The operation ∗ defined on a set A is said to be **distributive** over the operation # if, for all, a, b, and c in A, we have
>
> $a ∗ (b \# c) = (a ∗ b) \# (a ∗ c)$

For example, the operation of multiplication defined on the set of real numbers is distributive over addition because, for any real numbers a, b, and c,

$a \times (b + c) = (a \times b) + (a \times c)$

On the other hand, addition is *not* distributive over multiplication because

$a + (b \times c) \neq (a + b) \times (a + c)$

as shown by the example

$3 + (4 \times 5) = 3 + 20 = 23$

$(3 + 4) \times (3 + 5) = 7 \times 8 = 56$

Thus,

$3 + (4 \times 5) = 23 \neq (3 + 4) \times (3 + 5) = 56$

Now, let us consider two operations, F and S, defined on the set of natural numbers as follows: If a and b are any two natural numbers, then:

1. a F b means to select the *first* of the two numbers.

2. a S b means to select the *smaller* of the two numbers. (If $a = b$, then a S a is defined to be a.)

Thus, $3 \text{ F } 5 = 3$, $5 \text{ F } 3 = 5$, and $4 \text{ F } 1 = 4$. Similarly, $3 \text{ S } 5 = 3$, $5 \text{ S } 3 = 3$, $4 \text{ S } 1 = 1$, and $2 \text{ S } 2 = 2$.

EXAMPLE 8   Is the operation F distributive over S?

Solution   Recall that $a \text{ F } b$ means to select the first of the numbers and that $a \text{ S } b$ means to select the smaller of the two numbers a and b (if the numbers are the same, select the number). To show that F distributes over S, we have to show that

$$a \text{ F } (b \text{ S } c) = (a \text{ F } b) \text{ S } (a \text{ F } c)$$

We know that $a \text{ F } (b \text{ S } c) = a$ (because a is the first number). Also, $a \text{ F } b = a$ and $a \text{ F } c = a$, so

$$(a \text{ F } b) \text{ S } (a \text{ F } c) = a \text{ S } a = a$$

Thus,

$$a \text{ F } (b \text{ S } c) = (a \text{ F } b) \text{ S } (a \text{ F } c)$$

which shows that F is distributive over S.   ◀

EXERCISE 11.3

@	a	b	c
a	c	a	b
b	a	b	c
c	b	c	a

Consider the set $S = \{a, b, c\}$ and the operation @ defined by the table in the margin. Problems 1–7 refer to this table.

1. Find:
   a. $a @ b$ b. $b @ c$ c. $c @ a$

2. Find:
   a. $a @ (b @ c)$ b. $(a @ b) @ c$
   c. Are the results in parts a and b identical?

3. Find:
   a. $b @ (a @ b)$ b. $(b @ a) @ b$
   c. Are the results in parts a and b identical?

4. Find:
   a. $(a @ b) @ a$ b. $a @ (b @ a)$
   c. Are the results in parts a and b identical?

5. Find:
   a. $b @ c$ b. $c @ b$
   c. Are the results in parts a and b identical?

6. Is the operation @ a commutative operation? Explain.

7. Is the set S closed with respect to the operation @? Explain.

8. Suppose $a \text{ F } b$ means to select the first of two numbers a and b, as in Example 7 of this section. Let $A = \{1, 2, 3\}$.

a. Make a table that will define the operation F on the set A.

b. Is A closed under the operation F?

c. Is the operation F commutative?

d. For any three natural numbers a, b, and c, show that

$$a \text{ F } (b \text{ F } c) = (a \text{ F } b) \text{ F } c$$

9. Let the operation F be defined as in Problem 8, and let N be the set of natural numbers.

a. Is the set N closed under the operation F? Explain.

b. Is the operation F associative? Explain.

c. Is the operation F commutative? Explain.

10. Let S be the set of all multiples of 5 (0, 5, 10, etc.).

a. Is the set S closed with respect to ordinary multiplication?

b. Is the operation of ordinary multiplication commutative on S? Explain.

c. Is the operation of ordinary multiplication associative on S? Explain.

11. Are the following sets closed under the given operation?

a. The odd numbers under addition

b. The odd numbers under multiplication

c. The even numbers under addition

d. The even numbers under multiplication

12. Give an example of an operation under which the set $\{0, 1\}$ is:

a. Not closed   **b.** Not associative   **c.** Not commutative

Let $S = \{\varnothing, \{a\}, \{b\}, \{a, b\}\}$. *The table at the left will be used in Problems 13–18. The entries in the table represent the set intersection of the elements in the corresponding rows and columns.*

13. Supply the missing entries in the table.

14. Find:

a. $(\{a\} \cap \{b\}) \cap \{a, b\}$ **b.** $\{a\} \cap (\{b\} \cap \{a, b\})$

c. Are the results in parts a and b identical?

15. Find:

a. $(\{b\} \cap \{a, b\}) \cap \{a\}$ **b.** $\{b\} \cap (\{a, b\} \cap \{a\})$

c. Are the results in parts a and b identical?

16. In Chapter 1 we learned that for any three sets A, B, and C, $A \cap (B \cap C) = (A \cap B) \cap C$. On the basis of this result, would you say that the operation $\cap$ defined in the table is associative?

17. Is the set S closed with respect to the operation $\cap$? Explain.

18. Is the operation $\cap$ commutative? How can you tell?

19. Suppose that a L b means to select the larger of the two numbers a and b (if the numbers are the same, select the number).

a. Complete the table in the margin.

b. If there is an identity element, what is it?

Table for Problems 13–18

$\cap$	$\varnothing$	$\{a\}$	$\{b\}$	$\{a, b\}$
$\varnothing$	$\varnothing$	$\varnothing$	$\varnothing$	$\varnothing$
$\{a\}$	$\varnothing$	$\{a\}$	$\varnothing$	$\{a\}$
$\{b\}$	$\varnothing$			
$\{a, b\}$	$\varnothing$			

Table for Problem 19

L	1	2	3	4
1				4
2			3	
3		3		
4	4			4

S	1	2	3	4
1	1	1	1	1
2	1	2	2	2
3	1	2	3	3
4	1	2	3	4

Table for Problems 22–24

×	−1	0	1
−1			
0			
1			

Table for Problems 26–28

@	0	1	2	3
0	0	1	2	3
1	1	2	3	0
2	2	3	0	1
3	3	0	1	2

The table at the left will be used in Problems 20 and 21.

20. Does the set $A = \{1, 2, 3, 4\}$ have an identity? If so, what is this identity?
21. Find the inverses of:
    a. 1   **b.** 2   **c.** 3   **d.** 4
22. Consider the set $S = \{-1, 0, 1\}$ and the operation of ordinary multiplication. Complete the table in the margin.
23. Does the set $S = \{-1, 0, 1\}$ have an identity under multiplication? If so, what is this identity?
24. For the set $S = \{-1, 0, 1\}$ under the operation of multiplication, find the inverse (if possible) of:
    a. 1   **b.** −1   **c.** 0
25. If $\mathcal{U}$ is the set of all subsets of a set A:
    a. Find the identity element for the operation of set intersection (∩).
    b. Can you find more than one identity?

The table at the left will be used in Problems 26–28.

26. Is the set $S = \{0, 1, 2, 3\}$ closed under the operation @?
27. Does the set S have an identity with respect to the operation @? If so, what is this identity?
28. Find the inverse (if it exists) of:
    a. 0   **b.** 1   **c.** 2   **d.** 3

In Problems 29–32, let F be defined as in Problem 8 and let a L b mean to select the larger of the two numbers a and b (if a = b, assume a L b = a) as in Problem 19.

29. Find:
    a. 3 F (4 L 5)   **b.** 4 F (5 L 6)
30. Find:
    a. 4 L (4 F 5)   **b.** 5 L (6 F 7)
31. Does the distributive property a F $(b$ L $c) = (a$ F $b)$ L $(a$ F $c)$ hold for all real numbers a, b, c? Explain.
32. Does the distributive property a L $(b$ F $c) = (a$ L $b)$ F $(a$ L $c)$ hold for all real numbers a, b, c? Explain.

33. In ordinary arithmetic, is multiplication distributive over subtraction?
34. In ordinary arithmetic, is division distributive over subtraction? [Hint: Look at two forms: $a \div (b - c) = (a \div b) - (a \div c)$ and $(a - b) \div c = (a \div c) - (b \div c)$.]
35. In the arithmetic of fractions, is multiplication distributive over addition?
36. In the arithmetic of fractions, is addition distributive over multiplication?

The distributive property can be used to shorten the labor in certain multiplication problems. For instance, the product of 6 and 999 can easily be found by writing

$$6 \times 999 = 6 \times (1000 - 1) = 6000 - 6 = 5994$$

Use this idea to calculate the following products:

1. 6×9999 **2.** 8×99 **3.** 7×59
4. 8×999 **5.** 4×9995 **6.** 3×9998

The distributive property can be used in an interesting way in number puzzles. Have you ever seen a magician ask a person in the audience to think of a number and do several things with it? Then, without knowing the original number, the magician knows the number with which the person ended up! Here is one of these puzzles:

Think of a number.
Add 3 to it.
Triple the result.
Subtract 9.
Divide by the number with which you started.
The result is 3.

Here are the calculations of four persons who selected different numbers.

	First	Second	Third	Fourth
Think of a number:	4	6	8	10
Add 3 to it:	7	9	11	13
Triple the result:	21	27	33	39
Subtract 9:	12	18	24	30
Divide by the number with which you started:	3	3	3	3

The result is always 3!

1. Can you discover why the puzzle works? [Hint: Let x be the number you select and work through the puzzle.]
2. Can you discover the result of the following puzzle?

Think of a number.
Add 2 to it. Continued on next page

Double the result.

Subtract 4.

Divide by the number with which you started.

The result is _____ .

GROUPS AND FIELDS

Interest in mathematical systems such as those we studied in Section 11.3 started in the early nineteenth century with the study of groups.

▶ **Definition 11.4a**

A **group** is a mathematical system consisting of a set S and an operation * with the following four properties:

1. **Closure** property
2. **Associative** property
3. There is an **identity** element.
4. Each element of S has an **inverse** in S.

If, in addition, the system has the **commutative** property, then the group is called a **commutative** (or **abelian**) **group**.

For example, the set of integers with the operation of addition has the following properties:

Closure	The sum of two integers is always an integer.
Associative	If a, b, c are integers, then $a + (b + c) = (a + b) + c$.
Identity	The integer 0 is the identity element such that $0 + a = a + 0 = a$ for every integer a.
Inverse	If a is an integer, then $-a$ is an integer such that $a + (-a) = 0$.
Commutative	If a and b are integers, then $a + b = b + a$.

Thus, the set of integers is a commutative group under addition.

Although group theory was developed by many mathematicians, the French mathematician Evariste Galois is usually considered the pioneer in this field. However, the origin of group theory can be traced back to the efforts of the Babylonians in solving equations of degree greater than 2. At the time of the Renaissance, the Italian mathematicians Girolamo Cardano and Nicolo de Brescia, commonly referred to as *Tartaglia* ("the stammerer"), made the first successful attempts to solve equations of third and fourth degree. After their discoveries, it seemed natural to pursue methods to solve equations of degree 5 and higher. At this point, Galois investigated the general properties of the equations involved, as well as the properties of their solutions. These ideas led to the theory of groups.

EXAMPLE 1   Does the set $S = \{0, 1, 2, 3, 4\}$, together with the operation of addition modulo 5, form a commutative group?

Solution   The system has the five properties (closure, associative, identity, inverse, and commutative) as shown following Example 2 in Section 11.2, so the set S together with the operation of addition modulo 5 forms a commutative group.   ◀

EXAMPLE 2   Does the set $S = \{0, 1, 2, 3, 4\}$, together with the operation of multiplication modulo 5, form a commutative group?

Solution   The multiplication table for this system is shown in Table 11.4a. We check the five properties:

Table 11.4a Multiplication Modulo 5

×	0	1	2	3	4
0	0	0	0	0	0
1	0	1	2	3	4
2	0	2	4	1	3
3	0	3	1	4	2
4	0	4	3	2	1

1. The set S is **closed,** because all the entries in the table are elements of S.
2. The operation is **commutative,** because the reflection of the bottom half of the table along the diagonal is identical to the top part.
3. The operation is **associative.**
4. The **identity** for this system is 1.
5. All the elements, except 0, have **inverses.** Zero does not have an inverse because there is no number a in S such that $0 \times a \equiv 1$.

   The inverse of 1 is 1 ($1 \times 1 \equiv 1$).
   The inverse of 2 is 3 ($2 \times 3 \equiv 1$).
   The inverse of 3 is 2 ($3 \times 2 \equiv 1$).
   The inverse of 4 is 4 ($4 \times 4 \equiv 1$).

   The system does not have an inverse for 0, so the system is not a group under the operation of multiplication modulo 5.   ◀

EXAMPLE 3   If the number 0 is omitted from the set in Example 2, is the resulting system a commutative group?

Solution   The multiplication table for this system will be Table 11.4a with the 0 row and the 0 column crossed out. From the discussion in Example 2, we see that this system has all the required properties and is thus a commutative group.   ◀

The distributive law discussed in Section 11.3 is the only property we have studied that involves two operations. We now turn our attention to a kind of mathematical system, called a *field*, which consists of a set S and two operations.

Definition 11.4b

A **field** is a mathematical system consisting of a set S and two operations, say ∗ and #, defined on S and having the following properties:

1. **Closure** property
2. **Associative** property
3. **Identity** property
4. **Inverse** property (except that there is no inverse for the identity with respect to ∗)
5. **Commutative** property
6. **Distributive** property of # with respect to ∗

From Examples 1 and 2, we see that the system consisting of the set $S = \{0, 1, 2, 3, 4\}$ and the operations of addition and multiplication modulo 5 have all the properties required by Definition 11.4b. Thus, this system is a field. In general, by proceeding as we did for modulo 5, it can be shown that if p is any prime number, then the integers modulo p form a field under addition and multiplication.

EXAMPLE 4   Let $S = \{Odd, Even\}$, and let $+$ and $\times$ be two operations defined by Tables 11.4b and 11.4c. You may recognize that these operations correspond to adding or multiplying odd and even numbers. For instance, an odd number added to an even number gives an odd number; thus, Odd + Even = Odd. Similarly, Even × Even = Even. Does the set $S = \{Odd, Even\}$ form a field under the $+$ and $\times$ operations?

Solution

1. **Closure:** The tables show that S is closed under the two operations.
2. **Associative:** Both operations are associative. For example, Odd + (Odd + Even) = Odd + Odd = Even, and (Odd + Odd) + Even = Even + Even = Even, so that Odd + (Odd + Even) = (Odd + Odd) + Even (You can check the other cases in the same way.)
3. **Commutative:** The two tables show that both operations are commutative.
4. **Identity:** The identity for + is Even, because the column under Even in Table 11.4b is identical to the column at the far left, and the row adjacent to Even is identical to the top row. You can check in the same way that the identity for × is Odd.
5. **Inverse:** The inverse of Even under + is Even because Even + Even = Even. The inverse of Odd under + is Odd, because Odd + Odd = Even. The inverse of Odd under × is Odd because Odd × Odd = Odd. There is no inverse of Even under ×.
6. **Distributive:** The distributive property of × over + holds because it holds for multiplication over addition for the real numbers.

Thus, the set $S = \{Odd, Even\}$ with the operations $+$ and $\times$ satisfies all the requirements of Definition 11.4b, so the system is a field. ◀

Table 11.4b

+	Odd	Even
Odd	Even	Odd
Even	Odd	Even

Table 11.4c

×	Odd	Even
Odd	Odd	Even
Even	Even	Even

Table for Problem 1

*	a	b	c
a	b	c	a
b	c	a	b
c	a	b	c

1. Let $S = \{a, b, c\}$, and let $*$ be defined by the table at the left. Is S a group with respect to the operation $*$?

In Problems 2–13, determine whether the given set under the given operation forms a group. For each that is not a group, give one specific example of a condition that is not satisfied.

2. The odd integers under the operation of addition
3. The odd integers under the operation of multiplication
4. The even integers under the operation of addition
5. The even integers under the operation of multiplication
6. The positive integers under the operation of addition
7. The positive integers under the operation of multiplication
8. The integers under the operation of addition
9. The integers under the operation of multiplication
10. The real numbers under the operation of multiplication
11. The real numbers under the operation of addition
12. The set $\{-1, 0, 1\}$ under the operation of addition
13. The set $\{-1, 0, 1\}$ under the operation of multiplication

Table for Problem 14

#	a	b	c
a			b
b		b	c
c			

14. Complete the table in the margin so that the result will be a group under the given operation.

15. Let $S = \{a, b, c, d, e\}$, and let $\#$ be defined on S by the table in the margin. Is the set S under the operation $\#$ a group? If not, give one specific example of a condition that is not satisfied.

Table for Problem 15

#	a	b	c	d	e
a	a	b	c	d	e
b	b	e	a	c	d
c	d	c	e	a	b
d	c	d	b	e	a
e	e	a	d	b	c

In Problems 16–20, determine whether the given sets form a field under the operations of addition and multiplication.

16. The set of positive odd integers
17. The set of positive even integers
18. The set of integral multiples of 5
19. The set of integral multiples of 2
20. The set of all real numbers

*USING YOUR KNOWLEDGE 11.4

If you studied Chapter 10, then you can use the knowledge you gained there to answer the following questions. First, however, we define a nonsingular matrix.

> A square matrix A such that det $A \neq 0$ is said to be a **nonsingular matrix**.

The importance of this definition lies in the fact that every nonsingular matrix has a unique multiplicative inverse. You can see why this is so, because to find the inverse of a matrix A, you must solve a system of linear equations of which A is the matrix of coefficients. This system has a unique solution if and only if det A ≠ 0.

1. Consider the set of all 2×2 matrices under the operation of matrix addition. Is this system a group? If so, is it a commutative group? Explain.

2. Consider the set of all nonsingular 2×2 matrices under the operation of matrix multiplication. Is this system a group? If so, is it a commutative group? Explain.

3. If you adjoined the matrix

$$\begin{bmatrix} 0 & 0 \\ 0 & 0 \end{bmatrix}$$

to the set of matrices in Problem 2, and considered the operations of matrix addition and multiplication, would the system be a field? Explain.

4. Consider the set of all matrices of the special form

$$\begin{bmatrix} a & b \\ -b & a \end{bmatrix}$$

where a and b are real numbers. Is this set a field under matrix addition and multiplication? [Hint: You must determine whether the inverse of a matrix in this set is also in the set, and do not forget to check the commutative property of multiplication.]

DISCOVERY 11.4

There is an interesting problem in group theory that is related to the orientations of a triangle in a plane. The figure in the margin on the facing page shows the six different orientations that can be obtained by the use of one or more of the following operations.

Operation 1. Leave the triangle as is.
Operation 2. Turn the triangle over using the a–a axis.
Operation 3. Rotate the triangle 120° counterclockwise in its own plane.
Operation 4. Turn the triangle over using the b–b axis.
Operation 5. Rotate the triangle 120° clockwise in its own plane.
Operation 6. Turn the triangle over using the c–c axis.

You can check what we are going to do very easily by cutting an equilateral triangle out of a piece of paper and identifying the small triangles as shown in the figure. Then draw three axes a–a, b–b, c–c on another sheet

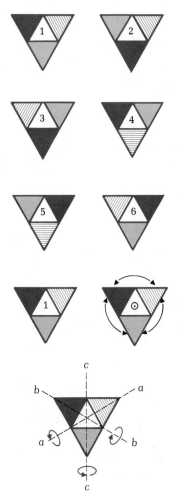

Rotations of a triangle

of paper, which is to be kept in a fixed position. As you can verify, starting with orientation 1, each operation will give the correspondingly numbered new orientation. For instance, operation 5 changes orientation 1 into orientation 5; operation 4 changes orientation 1 into orientation 4.

Let us define an operator C that corresponds to changing from one orientation to another by means of the preceding operations. We could use a symbol such as 3 C 2 to mean that we apply operation 3 to orientation 2. Then the statement 3 C 2 = 6 means that operation 3 applied to orientation 2 gives orientation 6. You should verify that it does. We can also make a table to display the results as shown below, where the numbers in the body of the table give the final orientation in each case. From the table, we can read 2 C 4 = 3, 3 C 3 = 5, and so on.

		STARTING ORIENTATION					
	C	1	2	3	4	5	6
OPERATION	1	1	2	3	4	5	6
	2	2	1	4	3	6	5
	3	3	6	5	2	1	4
	4						
	5						
	6						

1. Can you fill in the rest of the table?
2. Can you discover the identity for the operator C?
3. You can see from the table that C does not have the commutative property. For instance, 3 C 2 ≠ 2 C 3; the first of these gives 6, and the second gives 4. However, as seems plausible from the physical interpretation of rotating the triangle in various ways, C does have the associative property. You need not prove this; but do verify it in a few cases. For example, (3 C 4) C 6 = 2 C 6 = 5 and 3 C (4 C 6) = 3 C 3 = 5. Can you now discover if the set {1, 2, 3, 4, 5, 6} is a group with respect to the operation C? (If it is, then it is an illustration of a noncommutative group.)

SUMMARY

Section	Item	Meaning	Example
11.2	$a \equiv b \pmod{m}$	$a - b$ is a multiple of m	$3 \equiv 8 \pmod 5$
11.3	Closure property	A set S is closed under an operation $*$ if, for all a and b in A, $a * b$ is in A.	
11.3	Associative property	An operation $*$ is associative if, for all a, b, c in A, $(a * b) * c = a * (b * c)$.	

Section	Item	Meaning
11.3	Commutative property	An operation $*$ is commutative if, for all a and b in A, $a * b = b * a$.
11.3	Identity	An element e is an identity for $*$ if, for every a in A, $a * e = a = e * a$.
11.3	Inverse	An element b is the inverse of a if $a * b = e = b * a$ (e is the identity for $*$).
11.3	Distributive property	The operation $*$ is distributive over the operation $\#$ if, for every a, b, c in A, $a * (b \# c) = (a * b) \# (a * c)$.
11.4	Group	A mathematical system consisting of a set S and an operation $*$ that has the closure, associative, identity, and inverse properties
11.4	Commutative group	A group with the commutative property
11.4	Field	A set S and the operations $*$ and $\#$ with the closure, associative, commutative, and identity properties; the distributive property of $\#$ with respect to $*$; and the inverse property (except that there is no inverse for the identity with respect to $*$)

PRACTICE TEST 11

In Problems 1–4, find the answer in clock arithmetic.

1. a. $3 \oplus 11$ b. $8 \oplus 9$
2. a. $3 \ominus 9$ b. $5 \ominus 12$
3. a. $3 \otimes 5$ b. $6 \otimes 8$
4. a. $\frac{3}{11}$ b. $\frac{5}{6}$

5. Are the following statements true or false?
   a. $5 \equiv 2 \pmod 3$ b. $9 \equiv 5 \pmod 4$
   c. $9 \equiv 2 \pmod 6$

In Problems 6–9, find the value of n.

6. $3 + 2 \equiv n \pmod 5$ 7. $4 \times 3 \equiv n \pmod 5$
8. $2 - 4 \equiv n \pmod 5$ 9. $\frac{3}{2} \equiv n \pmod 5$

In Problems 10 and 11, find all possible replacements for n for which the given congruence is true.

10. a. $6 + n \equiv 1 \pmod 7$ b. $3 - n \equiv 4 \pmod 7$

11. a. $2n \equiv 1 \pmod 3$ b. $\dfrac{n}{2} \equiv 2 \pmod 3$

*	#	$	%	¢
#	#	$	%	¢
$	$	%	¢	#
%	%	¢	#	$
¢	¢	#	$	%

Problems 12–16 refer to the operation ∗ defined by the table in the margin.

12. Is the set $S = \{\#, \$, \%, ¢\}$ closed with respect to the operation ∗? Explain.

13. Use the table to find the result of each of the following operations:
    a. ($ ∗ ¢) ∗ #   **b.** $ ∗ (¢ ∗ #)
    c. ($ ∗ #) ∗ (% ∗ ¢)

14. Is the operation ∗ commutative? Explain.

15. Find the identity element for the operation ∗.

16. Find the inverse of:
    a. #   **b.** $ **c.** %   **d.** ¢

17. The table in the margin defines the operation @.
    a. Find the identity element if there is one.
    b. Does any element have an inverse?
    c. Is the operation commutative?

@	$	#	&
$	&	$	#
#	#	&	$
&	$	#	&

In Problems 18 and 19, suppose that a S b means to select the second of the two numbers a and b, and a L b means to select the lesser of the two numbers (if the numbers are equal, select the number).

18. Is S distributive over L? Explain.

19. Is L distributive over S? Explain.

20. Is the set {0, 1, 2} together with addition modulo 3 a commutative group? Explain.

21. Is the set {1, 2} together with multiplication modulo 3 a commutative group? Explain.

22. Is the set of all rational numbers along with the ordinary operations of addition and multiplication a field? Explain.

Problems 23–25 refer to the set {0, 1} and the operations ⊕ and ⊗ as defined by the tables in the margin.

⊕	0	1
0	0	1
1	1	0

⊗	0	1
0	0	0
1	0	1

23. Is the set {0, 1} together with the operation ⊕ a group?

24. Is the set {0, 1} together with the operation ⊗ a group?

25. Is the set {0, 1} together with the two operations ⊕ and ⊗ a field?

CHAPTER 12

COUNTING TECHNIQUES

Gottfried Wilhelm Leibniz, the all-around genius of his time, was born in Leipzig, Germany, in 1646. As a child, he had taught himself to read Latin and Greek, and at an early age, he had mastered the then current textbook knowledge of mathematics, philosophy, theology, and law. By the time he was 20, he had already begun to have ideas for a kind of *universal mathematics*, which later developed into the symbolic logic of George Boole.

When, supposedly because of his youth, he was refused the doctor of laws degree at the University of Leipzig, he moved to Nuremberg. There, a brilliant dissertation on a historical method of teaching law earned him the doctorate at the University of Altdorf in 1666. This led to a commission for the recodification of certain statutes for the Elector of Mainz, in whose service Leibniz remained until 1676. The rest of his life was spent in the diplomatic service for the estate of the Duke of Brunswick at Hanover.

In 1672, while on a diplomatic visit to Paris, he met the noted physicist, Christian Huygens, whom he persuaded to teach him some mathematics. Then, in 1673, on a visit to London, he became acquainted with some of the British mathematicians and he exhibited

his calculating machine (the first mechanical device that could do multiplication). This and some of his earlier work earned him a foreign membership in the Royal Society.

Leibniz's appointment at Hanover allowed him ample time to devote to his favorite studies, and he produced an enormous number of papers on a variety of subjects. He was a good linguist and established a reputation as a scholar of Sanskrit. His writings on philosophy were highly regarded, and he made contributions to law, religion, history, literature, and logic.

In 1682, he helped establish a journal, the *Acta Eruditorm*, of which he became editor-in-chief. Most of his mathematical papers appeared in this journal in the years from 1682 to 1692 and had a wide circulation in continental Europe.

Among other things, Leibniz advocated the use of the binary system, which we discussed in Chapter 3. However, his outstanding achievements in mathematics were his discovery of the *calculus* (independently of Isaac Newton) and his work on *combinatorial analysis*, which involves counting techniques, the subject of this chapter.

GOTTFRIED WILHELM LEIBNIZ (1646–1716)
Courtesy of IBM

In his dream of a "universal characteristic" Leibniz was well over two centuries ahead of his age.

E. T. BELL

THE SEQUENTIAL COUNTING PRINCIPLE (SCP)

FUNKY WINKERBEAN by Tom
Batiuk.

© 1973 Field Enterprises, Inc. Courtesy of
Field Newspaper Syndicate.

Can you count? It is not always as easy as 1, 2, 3. For example, can you tell in how many different ways Funky Winkerbean can answer the questions on a true/false test? There are many situations in which the answer to the question, "How many?" is the first step in the solution of a problem. In Section 1.2, we counted the number of subsets of a given set, and in Section 1.5, we used Venn diagrams to count the number of elements of various sets. In this chapter, we shall consider a few counting techniques that are important in many applications and that we shall use in Chapter 13 when we study probability.

A. Tree Diagrams

Let us return to Funky Winkerbean. He is still taking a true/false test and guessing at the answers. If we assume that there are just two questions and that Funky answers both, in how many different ways can he respond?

In order to answer this question, we have to find all the possible ways in which the two questions can be answered. We do this by constructing a **tree diagram,** as shown in Fig. 12.1a. In the figure, the first set of branches of the tree shows the two ways in which the first question can be answered (T for true, F for false), while the second set of branches shows the ways in which the second question can be answered. By tracing each path from left to right, we find that there are four end results, which correspond to the 2 × 2 = 4 ways in which the two questions can be answered. The four possibilities are TT, TF, FT, and FF.

The tree diagram technique is used again in the next example.

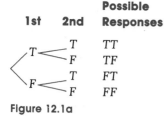

| | | Possible |
1st	2nd	Responses
T	T	TT
	F	TF
F	T	FT
	F	FF

Figure 12.1a

EXAMPLE 1   A woman desires to purchase a car. She has a choice of two body styles (convertible or hardtop) and three colors (red, blue, or green). Make a tree diagram and find how many choices she has.

Solution   We make a tree diagram as in the Funky Winkerbean problem. As shown in the figure on page 535, for each of the body styles, convertible (c) or hardtop (h), the woman has three choices of color, red (r), blue (b), or green (g). Thus, she has 2 × 3 = 6 choices.

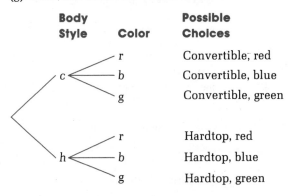

Body Style	Color	Possible Choices
c	r	Convertible, red
	b	Convertible, blue
	g	Convertible, green
h	r	Hardtop, red
	b	Hardtop, blue
	g	Hardtop, green

◄

B. Sequential Counting Principle (SCP)

In Example 1, we found that if there are 2 ways to do a thing (select a body style) and 3 ways to do a second thing (select a color), then there are 2 × 3 = 6 ways of doing the two things in succession in the stated order. This example illustrates a basic principle:

Sequential Counting Principle (SCP)

If one thing can occur in m ways and a second thing can then occur in n ways, then the sequence of two things can occur in $m \times n$ ways.

(The SCP is sometimes called the **FCP, fundamental counting principle**.)

Note that it is assumed that the second thing can occur in n ways for *each* of the ways that the first thing can occur.

For example, the number of ways in which two cards can be drawn in succession from a pack of 52 cards is 52 × 51 = 2652, because the first card can be drawn in 52 ways, while the second card can be drawn in only 51 ways.

EXAMPLE 2   Johnny's Homestyle Restaurant has 12 different meals and 5 different desserts on the menu. How many choices of a meal followed by a dessert does a customer have?

Solution   There are 12 choices for the meal and 5 choices for the dessert. Thus, by the SCP, there are 12 × 5 = 60 choices in all.   ◄

The SCP can be extended to cases in which 3, 4, or more things occur in succession. Thus, if the customer in Example 2 also has the choice of selecting a cookie (vanilla, chocolate, or almond) to go with the dessert, then the number of choices the customer has is

$$12 \times 5 \times 3 = 180$$

We now state a more general sequential counting principle.

Sequential Counting Principle (SCP)

If one thing can occur in m ways and a second thing can then occur in n ways, and a third thing can then occur in r ways, and so on, then the sequence of things can occur in $m \times n \times r \times \cdots$ ways.

EXAMPLE 3   In the cartoon on page 306, Charlie Brown has just received a chain letter. If he sends this letter to 6 of his friends, and these 6 friends send letters to 6 of their friends, and all these people send letters to 6 of their friends, how many letters will be sent?

Solution   Charlie Brown sends 6 letters, and each of the people receiving one of these sends 6 letters. Thus, by the SCP, these 6 people send a total of $6 \times 6 = 36$ letters. Then, each of the 36 people receiving one of these sends 6 letters. Again by the SCP, these 36 people send a total of $6 \times 6 \times 6 = 216$ letters. Therefore, the total number of letters is

$$6 + (6 \times 6) + (6 \times 6 \times 6) = 258 \qquad \blacktriangleleft$$

In some cases it is advantageous to use a diagram to represent the individual events in a sequence of events. We shall use this technique in the following examples.

EXAMPLE 4   A game called Passion Cubes consists of 4 cubes, each with 6 different words or phrases inscribed, one on each face. If it is assumed that the cubes are arranged with a pronoun coming first, then an auxiliary verb, then a verb, and finally an adverb, find how many different phrases can be formed.

Solution   We make 4 boxes representing the 4 events:

There are 6 choices for each of the boxes (each cube has 6 sides), so we enter a 6 in each box:

The number of possibilities, by the SCP, is $6 \times 6 \times 6 \times 6 = 6^4 = 1296.$ $\blacktriangleleft$

EXAMPLE 5   A slot machine has three dials, each having 20 symbols, as listed in the table on page 537. If the symbols are regarded as all different:

a. How many symbol combinations are possible on the three dials?

b. In how many ways can we get 3 bars? (The biggest payoff.)

SYMBOL	DIAL 1	DIAL 2	DIAL 3
Bar	1	3	1 = 3
Bell	1	3	3
Cherry	7	7	0
Lemon	3	0	4
Orange	3	6	7
Plum	5	1	5

Solution

a. We make 3 boxes representing the 3 dials:

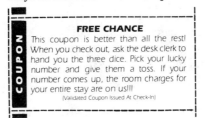

There are 20 choices for each of the boxes (each dial has 20 symbols), so we enter a 20 in each box:

| 20 | 20 | 20 |

The number of possibilities is $20 \times 20 \times 20 = 8000$.

b. The number of different ways we can get 3 bars is $1 \times 3 \times 1 = 3$, because we have one bar on the first dial, three on the second, and one on the third. ◀

Frederick Lewis, Inc.

The Clearwater Hilton Inn (in Clearwater, Florida) has a Free Chance coupon that works like this: When you check out, ask the desk clerk to hand you the three dice. Pick your lucky number and give them a toss. If your number comes up, the room charges are canceled. Does this sound easy? Look at the next example.

COUPON

FREE CHANCE
This coupon is better than all the rest!
When you check out, ask the desk clerk to
hand you the three dice. Pick your lucky
number and give them a toss. If your
number comes up, the room charges for
your entire stay are on us!!!
(Validated Coupon Issued At Check-In)

EXAMPLE 6   Suppose that one of three dice is black, one is green, and one is red.

a. If we distinguish the result of a toss by both numbers and colors, how many different results are possible?

b. If you picked 4 as your lucky number, in how many ways could you get a sum of 4?

Solution   **a.** We make three boxes, one for each die (singular of dice), to represent the possible outcomes:

Black   Green   Red

There are six choices for each box because a die can come up with any number from 1 to 6. So we enter a 6 in each box:

Black   Green   Red

By the SCP, there are $6 \times 6 \times 6 = 216$ outcomes possible.

b. The simplest way to solve this part of the problem is to reason that in order to get a sum of 4, one of the three dice must come up 2 and the other two dice must come up 1. Thus, the only choice we have is which die is to come up 2. This means that there are only three ways to get a sum of 4 out of the 216 possible outcomes. (See the tree diagram in the margin.) ◄

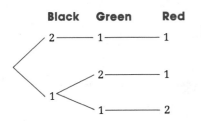

EXAMPLE 7   At the present time, zip codes consist of five digits.

    a. How many different zip codes are possible?

    b. How many are possible if 0 is not to be used as the first digit?

Solution   **a.** We make five boxes, each to be filled with the number of choices for that digit:

        ☐☐☐☐☐

    Since there are 10 digits (0, 1, 2, 3, 4, 5, 6, 7, 8, 9), there are 10 choices for each box:

    | 10 | 10 | 10 | 10 | 10 |

    Thus, by the SCP, there are $10 \times 10 \times 10 \times 10 \times 10$, that is,

    $10^5 = 100{,}000$

    different zip codes possible.

    b. If we cannot use a 0 for the first digit, we will have only 9 choices for the first box, but the others will still have 10 choices. Again, by the SCP, there are $9 \times 10 \times 10 \times 10 \times 10$, that is,

    $9 \times 10^4 = 90{,}000$

    different possible zip codes that do not start with a 0. ◄

EXAMPLE 8   Two cards are drawn in succession and without replacement from a deck of 52 cards. Find:

    a. The number of ways in which we can obtain the ace of spades and the king of hearts, in that order.

    b. The total number of ways in which two cards can be dealt.

Solution   **a.** There is one way of selecting the ace of spades and one way of selecting the king of hearts. Thus, there is $1 \times 1 = 1$ way of selecting the ace of spades and the king of hearts, in that order.

    b. There are 52 ways of selecting the first card and 51 ways of selecting the second card. Thus, there are $52 \times 51 = 2652$ ways in which the two cards can be dealt. ◄

A.

1. A man has 2 suits and 4 shirts. Use a tree diagram to find how many different outfits he can wear.

2. At the end of a meal in a restaurant, Elsie wants to have pie a la mode (pie and ice cream) for dessert. There are 5 flavors of ice cream — chocolate, vanilla, strawberry, peach, and coffee — and there are 2 kinds of pie — apple and cherry. Make a tree diagram to find how many choices Elsie has.

B.

3. Research Associates selects 4 people and asks their preferences regarding 2 different styles of blue jeans. If it is important to know which person prefers which style, how many different outcomes are possible?

4. In 1935, a chain letter fad started in Denver, Colorado. The scheme works like this: You receive a letter with a list of 5 names, send a dime to the person named at the top of the list, cross that name out, and add your name at the bottom. Suppose you receive one of these letters and send it to 5 other persons, each of whom sends it to 5 other persons, who in turn each sends it to 5 others, and so on.
   a. How many letters will have your name on the list?
   b. How much money will you receive if the chain is not broken?

5. Refer to the slot machine in Example 5, and determine in how many ways you can get:
   a. 3 bells   **b.** 3 oranges   **c.** 3 plums

6. An ordinary deck of playing cards contains 52 cards, 26 red and 26 black. If a card is dealt to each of 2 players, find in how many different ways this can be done if:
   a. Both cards are red   **b.** Both cards are black
   c. One card is black and the other is red

7. In poker, a pair of aces with any other pair is a good hand.
   a. In how many ways can you get a pair of aces when 2 cards are dealt from the deck?
   b. If the two pairs are aces and eights, the hand is considered to be unlucky! In how many ways can you get a pair of aces and then a pair of eights, in that order, when 4 cards are dealt from the deck? (This superstition dates back to 1876 when "Wild Bill" Hickok was shot dead by Jack McCall during a poker game. What hand was "Wild Bill" holding when he fell dead? A pair of aces and a pair of eights!)

8. Mr. C. Nile and Mr. D. Mented agreed to meet at 8:00 P.M. in one of the Spanish restaurants in Ybor City. They were both punctual, and they both remembered the date agreed upon. Unfortunately, they forgot to specify the name of the restaurant. If there are 5 Spanish restaurants in Ybor City, and each man goes to one of these, find:

a. The number of ways in which they could miss each other

b. The number of ways in which they could meet

9. In how many ways can 1 man and 1 woman, in that order, be selected from 5 men and 6 women?

10. How many different sets of 2 initials can be constructed from the letters of the English alphabet?

11. How many different sets of 3 initials can be constructed from the letters of the English alphabet?

12. A man wants to buy a ring. Suppose that he has 2 choices of metals (gold and silver) and 3 choices of stones (diamond, emerald, and ruby). How many choices does he have?

13. The Good Taste Restaurant has 7 entrees, 6 vegetables, and 9 desserts on its menu. If you want to order 1 entree, 1 vegetable, and 1 dessert, how many choices do you have?

14. Piedmont Airlines recently introduced their "Newest Hub." There are 2 flights from Tampa to Dayton and 2 flights from Dayton to Lansing. In how many ways can you fly from Tampa to Lansing, via Dayton?

15. Piedmont has 8 flights from Miami to Washington and 2 flights from Washington to Dayton. In how many ways can you fly from Miami to Dayton, via Washington?

16. In Florida, auto license plates carry 3 letters followed by 3 digits.

a. How many arrangements are possible for the 3 letters?

b. How many arrangements are possible for the 3 numbers?

c. How many different license plates can be made using 3 letters followed by 3 numbers?

17. How many two-digit numbers are there in the set of natural numbers? [*Hint:* 10 is a two-digit number, but 01 is not.]

18. How many three-digit numbers are there in the set of natural numbers?

19. Social Security numbers consist of nine digits. If the first digit cannot be 0, how many Social Security numbers are possible?

20. Telephone numbers consist of seven digits. For local calls, the first digit cannot be a 0 or a 1. How many local telephone numbers are possible?

21. A combination lock has 40 numbers on its face. To open this lock, you move right to a certain number, then left to another number, and finally right again to a third number. If no number is used twice, what is the total number of combinations?

22. Romano's Restaurant has 6 items that you can add to your pizza. The dessert menu lists 5 different desserts. If you want a pizza with one of the 6 items added and a dessert, how many choices do you have?

Problems 23–30 refer to the menu below.

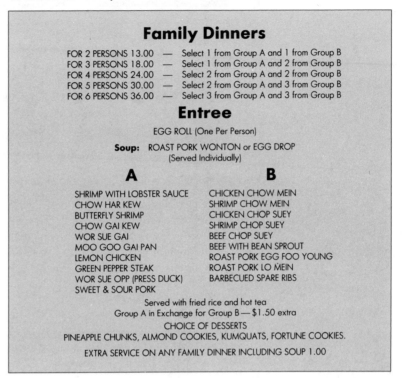

Family Dinners

FOR 2 PERSONS 13.00 — Select 1 from Group A and 1 from Group B
FOR 3 PERSONS 18.00 — Select 1 from Group A and 2 from Group B
FOR 4 PERSONS 24.00 — Select 2 from Group A and 2 from Group B
FOR 5 PERSONS 30.00 — Select 2 from Group A and 3 from Group B
FOR 6 PERSONS 36.00 — Select 3 from Group A and 3 from Group B

Entree

EGG ROLL (One Per Person)

Soup: ROAST PORK WONTON or EGG DROP
(Served Individually)

A

SHRIMP WITH LOBSTER SAUCE
CHOW HAR KEW
BUTTERFLY SHRIMP
CHOW GAI KEW
WOR SUE GAI
MOO GOO GAI PAN
LEMON CHICKEN
GREEN PEPPER STEAK
WOR SUE OPP (PRESS DUCK)
SWEET & SOUR PORK

B

CHICKEN CHOW MEIN
SHRIMP CHOW MEIN
CHICKEN CHOP SUEY
SHRIMP CHOP SUEY
BEEF CHOP SUEY
BEEF WITH BEAN SPROUT
ROAST PORK EGG FOO YOUNG
ROAST PORK LO MEIN
BARBECUED SPARE RIBS

Served with fried rice and hot tea
Group A in Exchange for Group B — $1.50 extra
CHOICE OF DESSERTS
PINEAPPLE CHUNKS, ALMOND COOKIES, KUMQUATS, FORTUNE COOKIES.
EXTRA SERVICE ON ANY FAMILY DINNER INCLUDING SOUP 1.00

23. If Billy decides to get an item from group A, a soup, and a dessert, how many choices does he have?

24. If Sue decides to get an item from group B, a soup, and a dessert, how many choices does she have?

25. If Pedro decides to get an item from group A or B, a soup, and a dessert, how many choices does he have?

26. If Bob and Sue decide to have the family dinner, which includes soup and dessert, one item from group A, and one item from group B, how many choices do they have?

27. If Sam and Sally are having the same type family dinner as Bob and Sue in Problem 26, and Sally decides to get an item from group A, so Sam must get an item from group B, how many choices do they have?

28. In Problem 27, if Sam does not want to eat shrimp, how many choices do they have?

29. In Problem 27, if Sam wants to avoid all the "sueys," how many choices do they have?

30. In Problem 27, if Sam does not want the roast pork, how many choices do they have?

In the cartoon, Tater is juggling 4 blocks.

1. If a word is any arrangement of 4 letters, how many 4-letter words can be formed from the letters B, O, N, K?
2. How many words are possible if only 3 blocks are used?
3. In Problem 11 of Exercise 12.1 we found the number of different sets of 3 initials that are possible. If a town has 27,000 inhabitants, each with exactly 3 initials, can you show that at least two of the inhabitants have the same initials?

DISCOVERY 12.1

In a recent trial in Sweden, the owner of a car was charged with overtime parking. The policeman who accused the man had noted the position of the air valves on the front and rear tires on the curb side of the car, and ascertained that one valve pointed to the place occupied by 12 o'clock on a clock (directly upward) and the other one to 3 o'clock. (In both cases the closest hour was selected.) After the allowed time had elapsed, the car was still there, with the valves pointing to 12 and 3 o'clock. In court, however, the man claimed that he had moved the car and returned later. The valves just happened to come to rest in the same position as before! At this time an expert was called to compute the probability of such an event happening.

If you were this expert and you assumed that the two wheels move independently of each other, could you use the SCP to find the following?

1. The number of ways in which the front air valve could come to rest
2. The number of ways in which the front and rear air valves could come to rest

The defendant, by the way, was acquitted! The judge remarked that if all four wheels had been checked (assuming that they moved independently) and found in the same position as before, he would have rejected the claim as too improbable and convicted the man.

3. Again, assume you are the expert and find the number of ways in which the four air valves could come to rest.
4. The claim that the two wheels on an automobile move independently is not completely warranted. For example, if the front air valve points to 12 and the rear (on the same side) points to 3, after a complete revolution of the front wheel, where will the rear air valve be? (Assume no slipping.)
5. Based on your answer to Problem 4, how many positions were possible for the two wheels on the curb side of the car if the owner did move it and did return later?

12.2

PERMUTATIONS

In Section 12.1 we used the sequential counting principle (SCP) to determine the number of ways in which a sequence of events could happen. A special case of this principle occurs when we want to count the possible *arrangements* of a given set of elements.

EXAMPLE 1   In how many different orders can we write the letters in the set {a, b, c} if no letter is repeated in any one arrangement?

Solution   We have three choices for the first letter, two for the second, and one for the third. By the SCP, the number of arrangements is $3 \times 2 \times 1 = 6$. ◀

A. Permutations

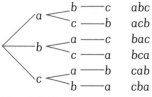

Figure 12.2a

If we are asked to display the arrangements in Example 1, we can draw the tree diagram shown in Fig. 12.2a, in which each path corresponds to one such arrangement. There are six paths, so the total number of arrangements is 6. Notice that in this example *abc* and *acb* are treated as different arrangements, because the order of the letters is not the same in the two arrangements. This type of arrangement, in which the *order is important*, is called a *permutation*.

▶   **Definition 12.2a**

An ordered arrangement of n distinguishable objects, taken r at a time and with no repetitions, is called a **permutation** of the objects. The **number** of such permutations is denoted by **P(n, r)**.

In Example 1, we saw that the number of permutations of the letters in the set {a, b, c} is $3 \times 2 \times 1 = 6$. Thus, P(3, 3) = 6. A notation that is convenient to represent $3 \times 2 \times 1$ is 3! (read, "3 factorial").

Definition 12.2b

> The symbol **$n!$** (read, "**n factorial**"), where n is a positive integer, represents the product of n and each positive integer less than n,
>
> $$n! = n \times (n-1) \times (n-2) \times \cdots \times 3 \times 2 \times 1$$

Thus,

$$4! = 4 \times 3 \times 2 \times 1 = 24$$

and

$$5! = 5 \times 4 \times 3 \times 2 \times 1 = 120$$

EXAMPLE 2   Compute:

a. $6!$ **b.** $7!$ **c.** $\dfrac{6!}{3!}$

Solution   By Definition 12.2b,

a. $6! = 6 \times 5 \times 4 \times 3 \times 2 \times 1 = 720$
b. $7! = 7 \times 6 \times 5 \times 4 \times 3 \times 2 \times 1 = 5040$
c. $\dfrac{6!}{3!} = \dfrac{6 \times 5 \times 4 \times \cancel{3} \times \cancel{2} \times \cancel{1}}{\cancel{3} \times \cancel{2} \times \cancel{1}} = 120$ Note that $\dfrac{6!}{3!} \neq 2$.   ◄

EXAMPLE 3   Wreck-U Car Club organizes a race in which five automobiles, $A, B, C, D,$ and E, are entered. If there are no ties:

a. In how many ways can the race finish?
b. In how many ways can the first three positions come in?

Solution   **a.** The number of ways in which the race can finish if there are no ties is the number of permutations of 5 things taken 5 at a time. By Definition 12.2a, this number is $P(5, 5)$, so, by the SCP.

$$P(5, 5) = 5 \cdot 4 \cdot 3 \cdot 2 \cdot 1 = 5! = 120$$

b. Here we need the number of ordered arrangements of 3 out of the 5 cars, that is, the number of permutations of 5 things taken 3 at a time. Again, by Definition 12.2a, this number is $P(5, 3)$, and by the SCP,

$$P(5, 3) = 5 \cdot 4 \cdot 3 = 60$$ ◄

By Definition 12.2a, $P(n, r)$ is the number of permutations of n objects using r of these objects at a time. We can now obtain formulas to compute these numbers.

1. If we use all n of the objects, we find $P(n, n)$. We think of n boxes to be filled with n objects, as shown in Fig. 12.2b. We have n choices for the first box, $(n-1)$ choices for the second box, $(n-2)$ choices for the third box, and so on, until we come to the last box, when there will be only 1 object left. By the SCP, we have:

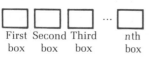

First   Second   Third nth
box   box   box   box

Figure 12.2b

The number of permutations of n objects, n at a time:
$$P(n, n) = n(n - 1)(n - 2) \cdots (3)(2)(1) = n!$$

2. The procedure used to evaluate $P(n, n)$ can be applied to $P(n, r)$. We think of r boxes to be filled by r of the n objects. There are n choices for the first box, $n - 1$ choices for the second box, $n - 2$ choices for the third box, and so on, until there are $n - r + 1$ choices for the rth box. Thus, by the SCP:

The number of permutations of n objects, r at a time:
$$P(n, r) = n(n - 1)(n - 2) \cdots (n - r + 1)$$

You should keep in mind that the n in $P(n, r)$ is the number of objects available and the r is the number of spaces to be filled. (Notice that if $r = n$, then the preceding two formulas agree exactly.)

The symbols $_nP_r$, $P_{n,r}$, and P_r^n are sometimes used to represent the number of permutations of n things, r at a time.

EXAMPLE 4   Compute:

    a. $P(6, 6)$ **b.** $P(7, 3)$ **c.** $P(6, 2)$

Solution   **a.** Here we use the formula for $P(n, n)$ with $n = 6$:

    $P(6, 6) = 6 \cdot 5 \cdot 4 \cdot 3 \cdot 2 \cdot 1 = 6! = 720$

    b. Here we are finding the number of permutations of 7 objects, taken 3 at a time. We use the formula for $P(n, r)$ with $n = 7$ and $r = 3$:

    $P(7, 3) = 7 \cdot 6 \cdot 5 = 210$

    c. We proceed as in part b, but with $n = 6$ and $r = 2$:

    $P(6, 2) = 6 \cdot 5 = 30$ ◀

By using the definitions of n! and $(n - r)!$, we can obtain the useful formula

$$P(n, r) = \frac{n!}{(n - r)!} \qquad r < n$$

We can verify this formula as follows:

$$\frac{n!}{(n - r)!} = \frac{n(n - 1)(n - 2) \cdots (n - r + 1)(n - r) \cdots (3)(2)(1)}{(n - r) \cdots (3)(2)(1)}$$
$$= n(n - 1)(n - 2) \cdots (n - r + 1) = P(n, r)$$

as the formula states. Notice that for $r < n$,

$$P(n, n) = n(n - 1)(n - 2) \cdot \cdots \cdot (n - r + 1)[(n - r)!]$$
$$= P(n, r)[(n - r)!]$$

We want this formula to hold also in the case $n = r$, that is, we want

$$P(n, n) = P(n, n)(0!)$$

For this reason, 0! is defined to be 1:

$$0! = 1$$

EXAMPLE 5 Use the above formula for $P(n, r)$ to compute the answer to part b of Example 4.

Solution
$$P(7, 3) = \frac{7!}{(7 - 3)!} = \frac{7!}{4!}$$
$$= \frac{7 \times 6 \times 5 \times \cancel{4} \times \cancel{3} \times \cancel{2} \times \cancel{1}}{\cancel{4} \times \cancel{3} \times \cancel{2} \times \cancel{1}}$$
$$= 7 \times 6 \times 5 = 210 \qquad \blacktriangleleft$$

B. The Complementary Counting Principle

The number of elements in a set A can sometimes be calculated more easily by an indirect rather than a direct method. If $\mathcal{U}$ is the universal set, the number of elements in A can be obtained by subtracting the number of elements in A' from the number in $\mathcal{U}$. This gives us the **complementary counting principle,**

$$n(A) = n(\mathcal{U}) - n(A')$$

EXAMPLE 6 Out of four dogs, how many ways are there to have *at least* one male?

Solution The only alternative to having at least one male is having no males, that is, the four dogs are all females, which is just one of all the possible cases. Since there are four places to fill, with two choices for each place (male or female), the total number of possible arrangements is

$$n(\mathcal{U}) = 2 \times 2 \times 2 \times 2 = 2^4 = 16$$

Thus,

$$n(\text{At least one male}) = 16 - n(\text{No males})$$
$$= 16 - 1 = 15 \qquad \blacktriangleleft$$

C. The Additive Counting Principle

Another useful counting principle is the **additive counting principle,** which we obtained in Section 1.5. If A and B are two sets, then

$$n(A \cup B) = n(A) + n(B) - n(A \cap B)$$

The use of this formula is illustrated in the next example.

EXAMPLE 7   How many two-digit numbers are divisible by 2 or by 5?

Solution   Let A be the set of two-digit numbers divisible by 2, and let B be the set of two-digit numbers divisible by 5. For two-digit numbers divisible by 2, the first digit can be any digit from 1 to 9 (9 choices), and the second digit can be 0, 2, 4, 6, or 8 (5 choices). Thus,

$$n(A) = 9 \times 5 = 45$$

For two-digit numbers divisible by 5, the first digit can be any digit from 1 to 9 (9 choices), and the second digit can be 0 or 5 (2 choices). Thus,

$$n(B) = 9 \times 2 = 18$$

Since $A \cap B$ is the set of numbers divisible by both 2 and 5, the first digit can still be any digit from 1 to 9 (9 choices), and the second digit can only be 0 (1 choice). Thus,

$$n(A \cap B) = 9 \times 1 = 9$$

and the desired number is

$$n(A \cup B) = n(A) + n(B) - n(A \cap B)$$
$$= 45 + 18 - 9 = 54$$

That is, the number of two-digit numbers divisible by either 2 or 5 is 54.

◀

EXERCISE 12.2

A.   **1.** In how many different orders can the letters in the set $\{a, b, c, d\}$ be written?

2. In how many different ways can 4 people be seated in a row?

3. If 6 horses are entered in a race and they all finish with no ties, in how many ways can they come in?

4. In how many different ways can 7 people be lined up at the checkout counter in a supermarket?

5. An insurance agent has a list of 5 prospects. In how many different orders can the agent phone these 5 prospects?

6. If the agent in Problem 5 decides to phone 3 of the prospects today and the other 2 prospects tomorrow, in how many ways can the agent do this?

In Problems 7–12, compute the given number.

7. 8!

8. $\dfrac{10!}{7!}$

9. $\dfrac{P(5, 2)}{2!}$

10. $\dfrac{P(6, 3)}{4!}$

11. $P(9, 4)$

12. $P(10, 2)$

13. A student is taking 5 classes, each of which uses 1 book. In how many ways can she stack the 5 books she must carry?

14. Suppose 10 people are entered in a race. If there are no ties, in how many ways can the first 3 places come out?

15. A basketball coach must choose 4 players to play in a particular game. (The team already has a center.) In how many ways can the remaining 4 positions be filled if the coach has 10 players who can play any position?

16. Rework Problem 15 if the coach does not have a center already and must fill all 5 positions from the 10 players.

17. In how many ways can 3 hearts be drawn from a standard deck of 52 cards?

18. In how many ways can 2 kings be drawn from a standard deck of 52 cards?

19. In how many ways can 2 red cards be drawn from a standard deck of 52 cards?

20. In how many ways can 4 diamonds be drawn from a standard deck of 52 cards?

21. How many three-digit numbers can be formed from the digits 1, 3, 5, 7, and 9 with no repetitions allowed?

22. How many even three-digit numbers can be formed from the digits 2, 4, 5, 7, and 9 with no repetitions allowed? [*Hint:* Try filling the units place first.]

23. A red die and a green die are tossed. In how many ways is it possible for both dice to come up even numbers? (Distinguish between the two dice.)

24. In Problem 23, in how many ways is it possible for one of the dice to come up an odd number and the other to come up an even number?

B. **25.** Out of 5 children, in how many ways can a family have *at least* 1 boy?

26. If two dice are tossed, in how many ways can at least one of the dice come up a 6? [*Hint:* There are five ways in which a single die can come up not a 6.]

C. **27.** How many of the first 100 natural numbers are multiples of 2 or multiples of 5?

28. How many of the first 100 natural numbers are multiples of 2 or multiples of 3?

If you are interested in horse racing, here are some problems for you.

1. Five horses are entered in a race. If there are no ties, in how many ways can the race end?

2. In Problem 1, if we know that 2 horses are going to be tied for first place (say A and B), in how many ways can the race end?

3. It seems unlikely that if 5 horses are entered in a race, 3 of them will be tied for first place. However, this event actually happened! In the Astley Stakes, at Lewes, England, in August 1880, Mazurka, Wandering Nun, and Scobell triple dead-heated for first place. If it is known that these 3 horses tied for first place, in how many ways could the rest of the horses finish?

4. You probably answered 2 to the question in Problem 3, because it is unlikely that there will be a tie for fourth place. However, the other 2 horses, Cumberland and Thora, did tie for fourth place. If ties are allowed, in how many different ways could Cumberland and Thora have finished the race in the preceding problem?

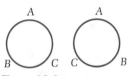

Figure 12.2c

In this section we learned that the number of permutations of n distinct objects is n!. Thus, if we wish to seat 3 people across the table from you, the number of possible arrangements is 3!. However, if 3 persons were to be seated at a circular table, the number of possible arrangements is only $2! = 2$. If the persons are labeled A, B, and C, the two arrangements look like those in Fig. 12.2c.

At first glance it may seem that there should be $3! = 6$ different arrangements, like those in Fig. 12.2d. However, a closer look will reveal that arrangements (1), (4), and (5) are identical; in all of them B is to the right and C to the left of A. Similarly, (2), (3), and (6) are identical, because in every case C is to the right and B is to the left of A. To avoid this difficulty, if we have, say, 4 persons to be seated at a circular table, we seat one of them and use this person as a reference. The rest of the people can be seated in 3! ways. We have the following relationship:

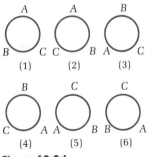

Figure 12.2d

Number of Persons	Number of Different Ways They Can Be Seated Around a Circular Table
3	2!
4	3!

1. From this discussion, can you discover in how many ways n persons can be seated around a circular table?

2. In how many ways can 4 people (including A and B) be seated at a circular table so that A and B are facing each other?

3. In Problem 2, find the number of ways in which the people can be seated so that A and B are *not* facing each other.

4. In Problem 2, find the number of ways in which the people can be seated so that A and B are next to each other.

▶ **Calculator Corner 12.2**   *Many calculators have a factorial* $\boxed{x!}$ *or* $\boxed{n!}$ *key. Thus, to find 6!, you first enter the number 6; then activate the factorial key. The steps are:*

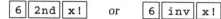

$$\boxed{6}\ \boxed{2\text{nd}}\ \boxed{x!}\qquad or\qquad \boxed{6}\ \boxed{\text{inv}}\ \boxed{x!}$$

In addition, some calculators even have a key that will calculate P(n, r), the $\boxed{{}_nP_r}$ *key. To enter the n and the r, you must use other special keys. If the* $\boxed{a}$ *and* $\boxed{b}$ *keys are those special keys on your calculator and you wish to find P(7, 3), as in Example 4b, you enter*

$\boxed{7}\ \boxed{a}\ \boxed{3}\ \boxed{b}\ \boxed{2\text{nd}}\ \boxed{{}_nP_r}$ *The answer is 210.*

1. Use your calculator to check the answers to Problems 7 and 11.

▶ **Computer Corner 12.2**   *We have several programs that can lessen the labor when working with factorials and permutations. For example, do you know what 30! is? The program on page 747 will actually tell you in 5 seconds (using an AT&T 6300). If you are the impatient kind, the second program on page 747 (factorials less than 30!) will tell you instantly (but you have to know how to read scientific notation because the answer comes out as 2.652529E32, which means 2.652529×10^{32}. Finally, we have a program on page 748 that will compute P(n, r), provided you enter n and r. Again, the answer is in scientific notation.*

1. Use these programs to check the answers to Problems 7 and 11.

12.3

COMBINATIONS

In Section 12.2, we found the number of ordered arrangements that are possible with n distinguishable objects. Sometimes we may wish to count the number of subsets of these objects that can be selected if we *disregard the order* in which the objects are selected. Such subsets are called **combinations.** We use the symbol C(n, r) to denote the number of combinations of r objects that can be formed from a set of n objects. The symbols ${}_nC_r$, $C_{n,r}$, C_n^r, and $\binom{n}{r}$ are also used to represent C(n, r).

EXAMPLE 1   How many different sums of money can be made from a set of coins consisting of a penny, a nickel, and a dime if exactly 2 coins are selected?

Solution   Because the order in which we select the coins is not important, the question asked is equivalent to finding $C(3, 2)$, the number of combinations of 2 things that can be formed using a set of 3 things. One of the sums is 6¢ (it makes no difference whether the penny is selected first and then the nickel, or vice versa), the second sum is 11¢, and the third sum is 15¢. Hence, $C(3, 2)$, the number of combinations of 3 objects taken 2 at a time, is 3. ◄

EXAMPLE 2   Consider the set $S = \{a, b, c, d\}$.

     a. How many combinations of 2 elements are possible using elements of the set S?
     b. How many permutations of 2 elements are possible using elements of the set S?
     c. How many subsets of 2 elements does the set S have?

Solution   **a.** The 6 possible combinations are shown in the table. Hence, $C(4, 2) = 6$.
     b. $P(4, 2) = 4 \times 3 = 12$. The 12 permutations are shown in the table.
     c. This problem is equivalent to finding the number of combinations that can be made from 4 objects using 2 at a time; hence, the answer is 6, as in part a. ◄

COMBINATIONS	PERMUTATIONS
ab	ab, ba
ac	ac, ca
ad	ad, da
bc	bc, cb
bd	bd, db
cd	cd, dc

We can see from the table in Example 2 that every combination determines 2! permutations, so $P(4, 2) = 2! \cdot C(4, 2)$. We use a similar argument to solve Example 3.

EXAMPLE 3   How many sets of 3 letters can be made from the English alphabet?

Solution   Here we want to find $C(26, 3)$. One of the possible combinations is, for example, $\{A, B, C\}$. This choice determines $3! = 6$ permutations (ABC, ACB, BAC, BCA, CAB, CBA). If we were to make a table similar to the one in Example 2, we would see that to each combination there corresponds $3! = 6$ permutations. Hence, there are 6 times as many permutations as there are combinations. That is, $P(26, 3) = 6 \cdot C(26, 3)$; but $P(26, 3) = 26 \times 25 \times 24 = 15{,}600$, so $15{,}600 = 6 \times C(26, 3)$ or $C(26, 3) = 2600$. ◄

The number of ways in which we can select a combination of r objects from a set of n objects is $C(n, r)$. The r objects in any one of these combinations can be arranged (permuted) in $r!$ ways. By the SCP, the total number of permutations is $r! \times C(n, r)$; but this number is $P(n, r)$. Hence,

$$P(n, r) = (r!)C(n, r)$$

so:

The number of combinations of n objects, r at a time:

$$C(n, r) = \frac{P(n, r)}{r!} \qquad 1 \le r \le n$$

The following useful form of the formula for $C(n, r)$ is obtained in the next example:

$$C(n, r) = \frac{n!}{r!(n - r)!}$$

EXAMPLE 4   Show that $C(n, r)$ has the form given in the box above.

Solution   We know that

$$C(n, r) = \frac{P(n, r)}{r!} \quad \text{and} \quad P(n, r) = \frac{n!}{(n - r)!}$$

Thus, by substituting for $P(n, r)$, we obtain

$$C(n, r) = \frac{P(n, r)}{r!} = \frac{n!}{(n - r)!} \div r! = \frac{n!}{r!(n - r)!}$$

as given in the box.   ◄

The meaning of $C(n, r)$ can also be stated in terms of a set of n elements: $C(n, r)$ is the number of subsets of r elements each that can be formed from a set of n elements. Note that this is just a repetition of the statement made at the beginning of this section.

EXAMPLE 5   How many subsets of 3 elements can be formed from a set of 4 elements?

Solution   Using the preceding statement, we find that the number of subsets of 3 elements that can be formed from a set of 4 elements is

$$C(4, 3) = \frac{4!}{3!1!} = 4$$ ◄

EXAMPLE 6   Compute:

a. $C(26, 3)$ **b.** $C(8, 2)$

Solution   **a.** From the second formula for $C(n, r)$, we get

$$C(26, 3) = \frac{26!}{3!23!} = \frac{26 \times 25 \times 24 \times 23!}{3!23!} = 2600$$

b. Similarly,

$$C(8, 2) = \frac{8!}{2!6!} = \frac{8 \times 7 \times 6!}{2!6!} = 28$$ ◄

You can make your work easier by noting possible cancellations, as in the solution for Example 6.

EXAMPLE 7   How many different 2-card hands can be obtained from an ordinary deck of 52 cards?

Solution   The question asked is equivalent to, "How many combinations are there of 52 elements, 2 at a time?" Using the first formula for $C(n, r)$, we find

$$C(52, 2) = \frac{P(52, 2)}{2!} = \frac{52 \cdot 51}{2 \cdot 1} = 1326 \quad\blacktriangleleft$$

Suppose you are asked to find the number of combinations of 10 objects, 8 at a time. You can see that if you take away any combination of 8 of the objects, a combination of 2 of the objects is left. This shows that $C(10, 8) = C(10, 2)$. This result can be verified directly as follows:

$$C(10, 8) = \frac{P(10, 8)}{8!} = \frac{10 \cdot 9 \cdot 8 \cdot 7 \cdot 6 \cdot 5 \cdot 4 \cdot 3}{8 \cdot 7 \cdot 6 \cdot 5 \cdot 4 \cdot 3 \cdot 2 \cdot 1} = \frac{10 \cdot 9}{2 \cdot 1}$$

and

$$C(10, 2) = \frac{P(10, 2)}{2!} = \frac{10 \cdot 9}{2 \cdot 1}$$

Therefore,

$$C(10, 8) = C(10, 2)$$

In general,

$$\boxed{C(n, r) = C(n, n - r)}$$

The second formula for $C(n, r)$, which was verified in Example 4, makes this quite obvious because

$$C(n, n - r) = \frac{n!}{(n - r)![n - (n - r)]!}$$

$$= \frac{n!}{(n - r)!r!} = C(n, r)$$

EXAMPLE 8   Romano's Restaurant offers the pizza menu shown in the margin. Find how many different pizzas you can order:

a. With 1 item   **b.** With 2 items
c. With 3 items   **d.** With 4 items

Solution   **a.** Because there are exactly 6 different items available, there are 6 different pizzas with 1 item. Notice that if you use the formula for $C(6, 1)$, it gives $\frac{6}{1} = 6$.

b. The order in which the items are added is not important. (If you order pepperoni and onion, you get the same pizza as if you order onion and pepperoni.) Thus, we need to find the number of combinations of 6 things, 2 at a time:

$$C(6, 2) = \frac{P(6, 2)}{2!} = \frac{6 \cdot 5}{1 \cdot 2} = 15$$

ROMANO'S
Greek - Italian Restaurant
Menu For Lunch & Take Out
985-6666

PIZZA
OUR SPECIAL LIGHT DOUGH AND CRISPY
TO YOUR EXPECTATION - ONE SIZE ONLY
10" - 8 PIECES
PLAIN CHEESE 5.95
ANY 1 ITEM 6.95
ANY 2 COMBINATION 7.95
ANY 3 COMBINATION 8.95
SPECIAL (All Items) 9.95
ITEMS
PEPPERONI, ONION, PEPPERS, MUSHROOMS,
SAUSAGE AND MEATBALL

SOFT DRINKS
	Sm.	Lg.
COKE, 7-UP	.75	.95
TAB, ROOT BEER	.75	.95
COFFEE or ICE TEA		.95

DESSERTS
Try Our Delicious Homemade Desserts
RICE PUDDING 1.50
GALACTOBURICO 1.50
Greek Custard with Fille
BAKLAVA 2.50
Walnuts, Honey and Fille
BOUGATZA 2.50
Walnuts, honey, cinnamon and Fille
SPUMONI 2.50
Italian Flavor Ice Cream

c. Here, we need $C(6, 3)$:

$$C(6, 3) = \frac{P(6, 3)}{3!} = \frac{6 \cdot 5 \cdot 4}{3 \cdot 2 \cdot 1} = 20$$

d. Here, the answer is $C(6, 4)$, which is the same as $C(6, 2) = 15$. ◀

EXERCISE 12.3

1. Compute:
   a. $C(5, 2)$ **b.** $C(6, 4)$ **c.** $C(7, 3)$ **d.** $C(5, 0)$
   e. $C(9, 6)$

2. How many different sums of money can be formed from a penny, a nickel, a dime, a quarter, and a half-dollar, if exactly 3 coins are to be used?

3. Rework Problem 2 if 4 coins are to be used.

4. Rework Problem 2 if at least 2 coins are to be used.

5. Let $A = \{1, 2, 3, 4, 5\}$.
   a. How many subsets of 3 elements does the set A have?
   b. How many subsets of A have no more than 3 elements?

6. If 20 people all shake hands with each other, how many handshakes are there?

7. The Greek alphabet has 24 letters. In how many ways can 3 different Greek letters be selected if the order does not matter?

8. The Mathematics Department is sending 5 of its 10 members to a meeting. In how many ways can the 5 members be selected?

9. A committee is to consist of 3 members. If there are 4 men and 6 women available to serve on this committee, how many different committees can be formed?

10. The Book-of-the-Month Club offers a choice of 3 books from a list of 40. How many different selections of 3 books each can be made from this list?

11. How many different 5-card poker hands are there in a deck of 52 cards?

12. A restaurant offers 8 different kinds of sandwiches. In how many ways could you select 2 different kinds?

13. The U.S. Senate has 100 members. How many different 5-member committees can be formed from the Senate?

14. In how many ways can a committee of 7 be formed from a group of 12 eligible people?

15. Johnny has a $1 bill, a $5 bill, a $10 bill, and a $20 bill in his pocket. How many different sums of money can Johnny make with these bills if he uses at least 1 bill each time?

16. Desi has 6 coins: a penny, a nickel, a dime, a quarter, a half-dollar, and a dollar. How many different sums of money can Desi form by using 2 of these coins?

17. Refer to Problem 16. How many different sums of money can Desi form if she uses at least 1 coin each time?

18. How many different committees can be formed from 8 people if each committee must consist of at least 3 people?

19. In how many ways can 8 people be divided into 2 equal groups?

20. How many diagonals does a polygon of:
    a. 8 sides have? (A diagonal is a line segment joining two nonadjacent vertices.)
    b. n sides have?
    [*Hint:* Think of *all* the lines joining the vertices two at a time. How many of these lines are sides and not diagonals?]

DISCOVERY 12.3

Suppose a fair coin is flipped 5 times in succession. How many different outcomes are possible? Since the coin can fall in either of 2 ways (heads or tails), the number of different outcomes is $2^5 = 32$.

In how many different ways can the outcome be 2 heads and 3 tails? If you think a moment, you will realize that the answer is the number of combinations of 5 things, 2 at a time. (You simply look at the 5 tosses and determine in how many ways you can select 2 of them, the order being unimportant.) Thus, the correct answer is

$$C(5, 2) = \frac{5 \cdot 4}{2 \cdot 1} = 10$$

See if you can discover the answers to the following questions:

1. In how many different ways can the outcome be 0 heads and 5 tails? We can call this number $C(5, 0)$.

2. Rework Problem 1 for 1 head and 4 tails.

3. Rework Problem 1 for 3 heads and 2 tails.

4. Rework Problem 1 for 4 heads and 1 tail.

5. Rework Problem 1 for 5 heads and 0 tails.

6. Add the answer we found for 2 heads and 3 tails and your answers for Problems 1–5. You should come out with

$$C(5, 0) + C(5, 1) + C(5, 2) + C(5, 3) + C(5, 4) + C(5, 5) = 32$$

Explain why.

7. The result of Problem 6 is a special case of the general result

$$C(n, 0) + C(n, 1) + C(n, 2) + \cdots + C(n, n) = 2^n$$

Can you tell why this must be a correct result? [*Hint:* Think of the coin being flipped n times.]

▶ **Calculator Corner 12.3** *Some calculators can evaluate C(n, r). To do this, you must enter n and r using special keys, say* $\boxed{a}$ *and* $\boxed{b}$*, on your calculator. Thus, to evaluate C(26, 3), as in Example 6a, we press* $\boxed{26}$ $\boxed{a}$ $\boxed{3}$ $\boxed{b}$ $\boxed{2nd}$ $\boxed{_nC_r}$*. As before, the answer is 2600.*

1. Check the answer to Examples 6, 7, and 8 using your calculator.

▶ **Computer Corner 12.3** *We have two programs that deal with combinations. One will actually calculate C(n, r) and give the answer in scientific notation if the numbers are large. This program appears on page 748. The other program (page 749) will actually display (in set notation) the combinations of a set of n items when r items are selected, provided you actually list the n elements in the original set.*

1. Check the answers to Examples 6, 7, and 8 using the program on page 748.
2. Check the combinations listed in the table in Example 2 using the program on page 749.

12.4

MISCELLANEOUS COUNTING METHODS

In the preceding sections, we discussed the use of the sequential counting principle (SCP), permutations, and combinations in simple counting problems. Very often, the most difficult step in dealing with a counting problem is to decide which method or formula to use. We shall try to help you with this step in the following discussion.

As you recall, if the problem involves two or more events that are to occur in succession, you must use the SCP. For problems that involve the choosing of r items from a set of n different items, with no repetitions allowed, remember the diagram shown in Fig. 12.4a.

Figure 12.4a

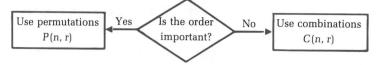

The next example will help to clarify this idea.

EXAMPLE 1 An employment agency has listed 5 highly skilled workers. Find in how many ways 2 of these workers can be selected:

a. If the first one is to be a foreman and the second one is to be a helper.
b. If they are simply to be sent to do a job.

Solution In both parts of this problem, 2 workers are to be selected from 5.

a. If the first worker is to be a foreman and the second a helper, then the *order* in which they are picked is important. Hence, we must use *permutations*. The answer is

$$P(5, 2) = 5 \cdot 4 = 20$$

b. Here the order is *not* important. (It makes no difference if Joe and Sally, or Sally and Joe are picked; both will be sent to the job.) Thus, we must use *combinations*, and the answer is

$$C(5, 2) = \frac{5 \cdot 4}{2 \cdot 1} = 10 \qquad \blacktriangleleft$$

Sometimes, we must combine more than one principle in solving a counting problem. We illustrate this in the next example.

EXAMPLE 2 A televison network has 6 different half-hour programs during "prime time" (7 P.M. to 10 P.M.). If you want to watch 3 programs in one evening:

a. How many choices do you have?
b. If one of the programs must be after 9 P.M., how many choices do you have?

Solution In this problem, you can choose the programs, but you must watch them at the times when they are presented. No permutations are allowed. Thus, to answer the questions, we must use combinations.

a. Here, we simply need to select 3 of the 6 programs. This means

$$C(6, 3) = \frac{6 \cdot 5 \cdot 4}{3 \cdot 2 \cdot 1} = 20 \text{ choices}$$

b. We divide the problem into two parts:

1. Select 1 program after 9 P.M. There are 2 choices.
2. Select 2 other programs from the 4 before 9 P.M. There are

$$C(4, 2) = \frac{4 \cdot 3}{2 \cdot 1} = 6 \text{ choices}$$

Now we use the SCP to combine the two sets of choices. This gives

$$6 \cdot 2 = 12 \text{ choices} \qquad \blacktriangleleft$$

EXAMPLE 3 Here is another view of the television problem. A local station manager has 10 different half-hour programs available and needs to schedule 6 of them in the hours from 7 P.M. to 10 P.M. The station manager feels that 4 of the programs are unsuitable for showing before 9 P.M., but is obligated to show 2 of these sometime during the evening. How many choices does this leave for the evening's schedule?

Solution   We divide this problem into two parts, as we did in Example 2b. Here, however, the station manager can control the order as well as the choice of programs. Therefore, this problem requires permutations.

1. For the hours from 7 P.M. to 9 P.M., there are 6 programs available. Since the order of showing has to be considered, the number of choices is $P(6, 4) = 6 \cdot 5 \cdot 4 \cdot 3 = 360$.
2. For the hour from 9 P.M. to 10 P.M., there are 4 programs of which 2 must be selected. The number of choices is $P(4, 2) = 4 \cdot 3 = 12$.

Now, we can use the SCP to give us the total number of choices, which is $360 \cdot 12 = 4320$. (Pity the poor manager!)   ◄

EXAMPLE 4   Roy and Rosie are eating out at an oriental restaurant. They select a special family dinner that allows an individual choice of 1 of 2 soups, 1 entree from 10 items in Group A, 1 entree from 9 items in Group B, and an individual choice of 1 of 4 desserts. How many different possibilities are there?

Solution   We consider two cases:

1. Roy picks an entree from Group A and Rosie picks one from Group B. Thus, Roy has a choice of 2 soups, 10 entrees, and 4 desserts, so by the SCP, he has $2 \cdot 10 \cdot 4 = 80$ choices. At the same time, Rosie has a choice of 2 soups, 9 entrees, and 4 desserts, so she has $2 \cdot 9 \cdot 4 = 72$ choices. Hence, by the SCP, together they have $80 \cdot 72 = 5760$ different choices available.
2. Roy picks an entree from Group B and Rosie picks one from Group A. This simply exchanges the choices we found in case 1, so the number of choices for both is again 5760.

Thus, the total number of possibilities is $2 \cdot 5760 = 11,520$.   ◄

In these examples, all the objects considered were distinct (you could tell them apart). Here is a different type of problem. If you go to Madison, Wisconsin, and look at the white pages of the phone book, you will find that the last name listed is Hero Zzyzzx (pronounced "Ziz-icks"). Can we find in how many different ways the letters in Mr. Zzyzzx's last name can be arranged? Before tackling this problem, let us look at a simpler one. It is conceivable that no one calls Mr. Zzyzzx by his proper last name; perhaps he is named zzx (zicks) for short. In how many different ways can the letters in the name zzx be arranged? To do this problem, we first rewrite the name as $z_1 z_2 x$ so that we now have three distinct things. Then we look at all the possible arrangements of z_1, z_2, and x. After this step, we erase the subscripts and look at the arrangements again. Table 12.4a shows the two sets of arrangements. Notice that with the subscripts we have 3 dis-

Table 12.4a

WITH SUBSCRIPTS		WITHOUT SUBSCRIPTS	
$z_1 z_2 x$	$z_2 z_1 x$	ZZX	ZZX
$z_1 x z_2$	$z_2 x z_1$	ZXZ	ZXZ
$x z_1 z_2$	$x z_2 z_1$	XZZ	XZZ

tinct objects, which can be ordered in $P(3, 3) = 3! = 6$ ways. The second half of the table, with the subscripts erased, shows that two permutations of z_1, z_2, and x, in which the 2 z's are simply interchanged, become identical. Hence, to find the number of distinct arrangements without subscripts, we must divide the number with subscripts by the number of ways in which the identical letters can be permuted. Because there are 2 z's, they can be permuted in 2! ways, so the number of arrangements of zzx is

$$\frac{3!}{2!} = 3$$

A similar argument leads to the general result. Suppose that a set of n objects consists of r different types, objects of the same type being indistinguishable. If there are n_1 objects of type 1, n_2 objects of type 2, . . . , n_r objects of type r, then the total number of *distinct* arrangements of the n objects is

$$\frac{n!}{n_1! n_2! \cdot \ \cdot \ \cdot \ \cdot \ n_r!}$$

With this formula, we can find the number of distinct arrangements of the letters in the name Zzyzzx. We regard the Z and the z as distinct, so there are 6 letters, 1 Z, 3 z's, 1 y, and 1 x. The formula gives

$$\frac{6!}{1! 3! 1! 1!} = 6 \cdot 5 \cdot 4 = 120$$

EXAMPLE 5   The last name in the San Francisco phone book is (are you ready?) Zachary Zzzzzzzzzzra. (Please don't ask us how to pronounce it!) In how many distinguishable ways can the letters in Zzzzzzzzzra be arranged?

Solution   Here, $n = 11$, $n_1 = 1$ (there is one Z), $n_2 = 8$ (there are 8 z's), $n_3 = 1$ (there is one r), and $n_4 = 1$ (there is one a). Thus, the number of distinct arrangements is

$$\frac{11!}{1! 8! 1! 1!} = 11 \cdot 10 \cdot 9 = 990 \qquad \blacktriangleleft$$

EXERCISE 12.4

1. Three cards are dealt in succession and without replacement from a standard deck of 52 cards.
   a. In how many different orders can the cards be dealt?
   b. How many different 3-card hands are possible?
2. An employment agency has 6 temporary workers.
   a. In how many ways could 4 of them be assigned to the research department?
   b. In how many ways could 3 of them be assigned to 3 different companies?

3. The play book for the quarterback of the Dallas Cowboys contains 50 plays.
   a. In how many ways could the quarterback select 3 plays to use in succession in the next 3 downs?
   b. In how many ways could he select a set of 3 plays to study?
4. A student must take 3 different courses on Mondays. In how many ways can the student do this:
   a. If there are 6 different courses, all available at each of the 3 hours 8 A.M., 9 A.M., and 10 A.M.? $P(6,3)$
   b. If only 1 of these courses is available each hour between 8 A.M. and 2 P.M. (6 hours)?, $C(6,3)$
5. Rework Problem 4b if the student wants to keep the hour from 12 to 1 P.M. free for lunch. $C(5,3)$
6. A student wishes to schedule mathematics, English, and science. These classes are available every hour between 9 A.M. and noon (3 hours).
   a. How many different schedules are possible?
   b. How many schedules are possible if this student wants to take mathematics at 11 A.M. with his favorite instructor, Mr. B.?
7. Peter must select 3 electives from a group of 7 courses.
   a. In how many ways can Peter do this? $C(7,3)$
   b. If all 7 of these courses are available each of the 4 hours from 8 A.M. to noon, from how many different schedules (hours and what course at each hour) can Peter choose? $P(7,4) = 840$
8. At the University of South Florida, a student must take at least 2 courses from each of 5 different areas in order to satisfy the general distribution requirement. Each of the areas has the number of courses indicated in the table.
   a. If Sandy has satisfied all the requirements except for area V, and she wishes to take 3 courses in this area, how many choices does she have?
   b. Bill has already satisfied his requirements in areas I, II, and III. Now he wishes to take the minimum number of courses in areas IV and V. How many choices does he have?
9. A class consists of 14 boys and 10 girls. They want to elect officers so that the president and secretary are girls, and the vice-president and treasurer are boys.
   a. How many possibilities are there? $P(10,2) \cdot P(14,2)$
   b. How many are there if 2 of the boys refuse to participate? $P(10,2) \cdot P(12,2)$
10. A company has 6 officers and 4 directors (10 different people). In how many ways can a committee of 4 be selected from these 10 people so that:
   a. 2 members are officers and 2 are directors?
   b. 3 members are officers and 1 is a director?
   c. All the members are officers?
   d. There are no restrictions?

AREA	COURSES
I	2
II	50
III	20
IV	40
V	100

11. There are 4 vacancies on the scholarship committee at a certain university. In order to balance the men and women on the committee, 1 woman and 3 men are to be appointed. In how many ways can this be done if there are:
    a. 7 men and 8 women available to serve?
    b. 5 men and 2 women available to serve?

12. Romano's Restaurant has the menu shown in Example 8 of Section 12.3. In how many ways can a meal consisting of a pizza with 3 items, 2 beverages, and a dessert be chosen? The menu shows that there are 6 items (toppings for the pizza), 6 beverages, and 5 desserts offered.

13. How many distinct arrangements can be made with the letters in the word TALLAHASSEE?

14. How many distinct arrangements can be made with the letters in the word MISSISSIPPI?

15. Do you know what a *palindrome* is? It is a word or phrase with the same spelling when written forwards or backwards. The longest single-word palindrome in the English language is the word REDIVIDER. How many distinct arrangements can be made with the letters in this word?

16. There is a place in Morocco with a name that has 8 vowels in a row in its spelling! Do you know what place this is? It is spelled IJOUAOUOUENE. How many distinct arrangements can be made with the letters in this name?

17. A contractor needs to buy 7 electronic components from 3 different subcontractors. The contractor wants to buy 2 of the components from the first subcontractor, 3 from the second, and 2 from the third. In how many ways can this be done?

18. An advertiser has a contract for 20 weeks that provides 3 different ads each week. If it is decided that in no 2 weeks will the same 3 ads be shown, how many different ads are necessary? [*Hint:* You need to find the least n such that $C(n, 3) \geq 20$.]

19. A cable television network wishes to show 5 movies every day for 3 weeks (21 days) without having to show the same 5 movies any 2 days in the 3 weeks. What is the least number of movies the network must have in order to do this? [See the hint in Problem 18.]

20. Repeat Problem 19 if the network wants to show the movies for 8 weeks.

*21. Polly needs to take biology, English, and history. All of these are available every hour between 9 A.M. and 3 P.M. (6 hours). If Polly must schedule 2 of these courses between 9 A.M. and 1 P.M. and 1 course between 1 P.M. and 3 P.M., how many schedules (hours and what course each hour) are available to her?

*22. Roy must elect 3 courses from among 4 courses in group I and 3 courses in group II. If he must take at least 1 of his 3 electives from each group, how many choices does he have? [*Hint:* First find how many choices he has if he elects only 1 course from group I. Then find how many choices he has if he elects 2 courses from group I. Since he must do one or the other of these, the final answer is the sum of the two answers.]

USING YOUR KNOWLEDGE 12.4

Do you know an easy way of finding how many positive integers are exact divisors of a given positive integer? For example, how many exact divisors does 4500 have? The easy way to answer this question is to write 4500 first as a product of its prime divisors. We find that

$$4500 = 2^2 3^2 5^3$$

Now you can see that every exact divisor of 4500 must be of the form $2^a 3^b 5^c$, where a is 0, 1, or 2; b is 0, 1, or 2; and c is 0, 1, 2, or 3. Because there are 3 choices for a, 3 choices for b, and 4 choices for c, the SCP tells us that the number of exact divisors of 4500 is $3 \cdot 3 \cdot 4 = 36$. Notice that the exponents in the prime factorization of 4500 are 2, 2, and 3 and the number of exact divisors is the product $(2 + 1)(2 + 1)(3 + 1)$. Try this out for a small number, say 12, where you can check the answer by writing out all the exact divisors.

1. How many exact divisors does 144 have?
2. How many exact divisors does 2520 have?
3. If the integer $N = 2^a 3^b 5^c 7^d$, how many exact divisors does N have?
4. How many exact divisors does the number $2^4 3^2 7^3$ have?

SUMMARY

Section	Item	Meaning	Example
12.1B	SCP	Sequential counting principle: If one thing can occur in m ways and a second thing can then occur in n ways, and a third thing can occur in r ways, and so on, then the sequence of things can occur in $m \times n \times r \times \cdots$ ways.	If there are 3 roads to go to the beach and 2 dates are available, then you have $3 \times 2 = 6$ different choices.
12.2	Permutation	An ordered arrangement of n distinguishable objects, taken r at a time and with no repetitions	
12.2	n! (n factorial)	$n(n - 1)(n - 2) \cdot \cdots \cdot 3 \cdot 2 \cdot 1$	$3! = 3 \cdot 2 \cdot 1 = 6$
12.2	P(n, r)	$n(n - 1)(n - 2) \cdot \cdots \cdot (n - r + 1)$ or $\dfrac{n!}{(n - r)!}$	$P(6, 2) = \dfrac{6!}{(6 - 2)!}$
12.2	0!	1	

Section	Item	Meaning	Example
12.2	Complementary counting principle	$n(A) = n(\mathcal{U}) - n(A')$	
12.2	Additive counting principle	$n(A \cup B) = n(A) + n(B) - n(A \cap B)$	
12.3	Combination	A selection of r objects without regard to order, taken from a set of n distinguishable objects	
12.3	$C(n, r)$	$\dfrac{n!}{r!(n-r)!} = \dfrac{P(n, r)}{r!}$	$C(6, 2) = \dfrac{6!}{2!4!}$
12.4	Permutations of a set of n objects, not all different	$\dfrac{n!}{n_1! n_2! \cdot \cdots \cdot n_r!}$	The number of arrangements of the letters aabbbc is $\dfrac{6!}{2!3!1!}$

PRACTICE TEST 12

1. A student wants to take two courses, A and B, both of which are available at 9, 10, and 11 A.M. Make a tree diagram to show all the possible schedules for that student. Use a notation like $(A, 9)$ to mean course A at 9 A.M., $(B, 11)$ to mean course B at 11 A.M., and so on.

2. A restaurant offers a choice of 2 soups, 3 entrees, and 5 desserts. How many different meals consisting of a soup, an entree, and a dessert are possible?

3. Suppose that one of two dice is black and the other is red.
   a. If we distinguish the result of a toss of the two dice by both numbers and colors, how many different results are possible?
   b. In how many ways could you get a sum of 5?

4. Two cards are drawn in succession and without replacement from a standard deck of 52 cards. In how many ways could these be a black jack and a red card, in that order?

5. An airline has 3 flights from city A to city B and 5 flights from city B to city C. In how many ways could you fly from city A to city C using this airline?

6. Compute:

   a. 7!   b. $\dfrac{7!}{4!}$

7. Compute:
   a. $3! \times 4!$ b. $3! + 4!$

8. Find:
   a. $P(5, 5)$ b. $P(6, 6)$

9. Find:
   a. $P(8, 2)$ b. $P(7, 3)$

10. In how many ways can 4 people be arranged in a row for a group picture?

11. Three married couples are posing for a group picture. They are to be seated in a row of 6 chairs, with each husband and wife together. In how many ways can this be done?

12. Bobby has 6 pigeons: 2 white, 2 gray, and 2 gray and white. In how many ways can Bobby select 3 of his pigeons and include *at least* 1 white bird?

13. How many counting numbers less than 50 are divisible by 2 or by 5?

14. How many different sums of money can be made from a set of coins consisting of a penny, a nickel, a dime, and a quarter if exactly 2 coins are selected?

15. Find:
    a. $C(5, 2)$ **b.** $C(6, 4)$

16. Find:

    a. $C(6, 0)$ **b.** $\dfrac{C(5, 4)}{C(5, 3)}$

17. How many subsets of 3 elements does the set $\{a, b, c, d, e, f\}$ have?

18. A student wants to schedule mathematics, English, science, and economics. These 4 classes are available every hour between 8 A.M. and noon (4 hours). How many different schedules are possible?

19. Two cards are drawn in succession and without replacement from a standard deck of 52 cards. How many different sets of 2 cards are possible?

20. Billy has 5 coins: a penny, a nickel, a dime, a quarter, and a half-dollar. How many different sums of money can Billy form by using 1, 2, or 3 of these coins?

21. On a certain night, there are 8 half-hour programs scheduled on one television station. If you want to watch 4 of these programs, how many choices do you have?

22. How many distinct arrangements can you make with the letters in the word BOOGABOO?

23. How many distinct arrangements can be made with the letters in the palindrome MADAM I'M ADAM? (Disregard the apostrophe.)

24. The A-1 Company needs 3 skilled employees, 1 to be a foreman and 2 to be helpers. If the company has 5 competent applicants, in how many ways can the employees be selected?

25. One of the biggest trading sprees on record (to that date) on the New York Stock Exchange occurred on August 26, 1982. On this day, 1977 stocks were traded, for a volume of 137,350,000 shares. Of the 1977 stocks traded, 1189 advanced (a), 460 declined (d), and 328 were unchanged (n). Suppose at the end of the day, you marked a, d, or n after each stock traded. How many distinct arrangements of all the a's, d's, and n's are possible? (Do not try to simplify your answer.)

CHAPTER 13

PROBABILITY

Blaise Pascal is undoubtedly one of the greatest might-have-beens in the history of mathematics. His most original contribution was in the theory of probability; but this was shared by the famous French mathematician, Pierre Fermat, who could easily have done it by himself. Although Pascal did first-rate mathematics, he was distracted by a morbid passion for religious subtleties, and buried his talent in a useless search for answers to unanswerable questions.

Born in the French province of Auvergne in June of 1623, Pascal moved with his father and sisters to Paris when Blaise was 7 years old. The boy was blessed with a brilliant mind and cursed with a sickly body. Protected from mathematics by a father who feared overtaxing the boy's health, Blaise, at the age of 12, finally insisted on knowing what geometry was about. Upon getting a clear explanation from his father, who was himself a mathematician, the boy plunged at once into the study of geometry, even proving a few theorems on his own without any book to help.

By the age of 16 or 17, Pascal had written an amazing essay on the conic sections, including new and deep theorems on the properties of these curves. So profound was this work that Descartes could not believe it had been done by anyone so young, and he ascribed it to the boy's father.

At the age of 18, Pascal had invented the world's first calculating machine, and had begun to work in physics and mechanics. But he continued his scientific work for only a few years, quitting at the age of 27 to devote himself to religious contemplation.

He lapsed into mathematics only a few times after this. He was 31 when the Chevalier de Méré proposed to him a problem on the division of the pot in an unfinished gambling game. Pascal wrote to Fermat about the problem, and in the ensuing correspondence these two men shared equally in establishing basic results in the theory of probability.

An interesting commentary on the influence of religion on Pascal's mind is contained in the argument known as *Pascal's wager*. The expected value of a lottery ticket is the product of the prize and the probability of winning. Even if the probability is small, the value is great if the prize is large enough. Pascal argued that even though the probability that God exists is small, the reward for true faith is eternal bliss.

BLAISE PASCAL (1623–1662)
Courtesy of IBM

It is truth very certain that, when it is not in our power to determine what is true, we ought to follow what is most probable.

RENÉ DESCARTES

SAMPLE SPACES AND PROBABILITY

The WIZARD OF ID by permission of Johnny Hart and NAS, Inc.

We often hear such statements as, "It will probably be a hot day tomorrow," "I have a good chance of getting a B in this course," and "If a penny is tossed there is a 50–50 chance that it will come up heads." Each of us has a sort of intuitive feeling as to what such a statement means, but it is not easy to give an exact mathematical formulation of this meaning. The objective of *probability theory* is to make such statements precise by giving them numerical measures.

The theory of probability is an important tool in the solution of many problems of the modern world. Although the most obvious applications are in gambling games, important applications occur in many situations involving an element of uncertainty. Probability theory is used to estimate whether a missile will hit its target, to determine premiums on insurance policies, and to make important business decisions such as where to locate a supermarket and how many checkout clerks to employ so that customers will not be kept waiting in line too long. Various sampling techniques, which are used in opinion polls and in the quality control of mass-produced items, are based on the theory of probability.

We want the **probability** of a given event to be a mathematical estimate of the likelihood that this event will occur. The following examples show how a probability may be assigned to a given event.

Photographer: Lyon, Photo Researchers

EXAMPLE 1   A fair coin is tossed; find the probability of a head coming up.

Solution   At this time we are unable to solve this problem because we have not even defined the term "probability." However, our intuition tells us the following:

1. When a fair coin is tossed, it may turn up in either of 2 ways. Assuming that the coin will not stand on edge, heads or tails are the only 2 possible **outcomes.**
2. If the coin is balanced (and this is what we mean by saying "the coin is fair"), the 2 outcomes are considered **equally likely.**

3. The probability of obtaining heads when a fair coin is tossed, denoted by $P(H)$, is 1 out of 2. That is, $P(H) = \frac{1}{2}$. ◄

A. Sample Spaces and Probability

Activities such as tossing a coin (as in Example 1), drawing a card from a deck, or rolling a pair of dice are called **experiments.** The set $\mathcal{U}$ of all possible outcomes for an experiment is called the **sample space** for the experiment. These terms are illustrated in Table 13.1a.

Table 13.1a Experiments and Sample Spaces

EXPERIMENT	POSSIBLE OUTCOMES	SAMPLE SPACE $\mathcal{U}$
A penny is tossed.	Heads or tails are equally likely outcomes.	$\{H, T\}$
There are 3 beige and 3 red balls in a box; 1 ball is drawn at random.	A beige or a red ball is equally likely to be drawn.	$\{b_1, b_2, b_3, r_1, r_2, r_3\}$
A penny and a nickel are tossed.	**Penny Nickel** H H H T T H T T	$\{(H, H), (H, T), (T, H), (T, T)\}$
One die is rolled.	The numbers from 1 to 6 are all equally likely outcomes.	$\{1, 2, 3, 4, 5, 6\}$
The pointer is spun:	The pointer is equally likely to point to 1, 2, 3, or 4.	$\{1, 2, 3, 4\}$
An integer between 1 and 50 (inclusive) is selected at random.	The integers from 1 to 50 are all equally likely to be selected.	$\{1, 2, 3, \ldots, 50\}$

Returning to Example 1, we see that the set of all possible outcomes for the experiment is $\mathcal{U} = \{H, T\}$. But there are only two subsets of $\mathcal{U}$ that can occur, namely, $\{H\}$ and $\{T\}$, and each of these is called an **event.** If heads come up, that is, if the event $E = \{H\}$ occurs, we say that we have a **favorable outcome,** or a **success.** Since there are **two** equally likely events in $\mathcal{U}$, and **one** of these is E, we assign the value $\frac{1}{2}$ to the event E.

We now enlarge upon the problem discussed in Example 1. Suppose that the coin is tossed 3 times. Can we find the probability that 3 heads appear? As before, we proceed in three steps:

1. The set of all possible outcomes for this experiment can be found by drawing a tree diagram as shown in Fig. 13.1a (page 568). As you can see, the possibilities for the first toss are labeled H and T, and likewise for the other two tosses. The number of outcomes is 8.
2. The 8 outcomes are equally likely.

3. We conclude that the probability of getting 3 heads, denoted by $P(HHH)$,* is 1 out of 8; that is, $P(HHH) = \frac{1}{8}$.

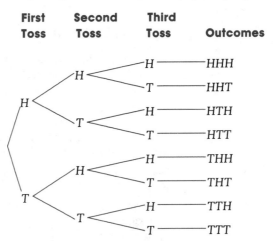

| First Toss | Second Toss | Third Toss | Outcomes |

Figure 13.1a
Tree diagram

If we want to know the probability of getting **at least** 2 heads, the 4 outcomes *HHH*, *HHT*, *HTH*, and *THH* are favorable, so the probability of getting at least 2 heads is $\frac{4}{8} = \frac{1}{2}$.

In examples such as these, in which all the possible outcomes are equally likely, the task of finding the probability of any event E can be simplified by using the following definition:

▶ **Definition 13.1a**

Suppose an experiment has $n(\mathcal{U})$ possible outcomes, **all equally likely.** Suppose further that the event E occurs in $n(E)$ of these outcomes. Then $P(E)$, the **probability** of event E, is given by

$$P(E) = \frac{\text{Number of favorable outcomes}}{\text{Number of possible outcomes}} = \frac{n(E)}{n(\mathcal{U})} \qquad (1)$$

We illustrate the use of equation (1) in the following examples.

EXAMPLE 2   A single die is rolled. Find the probability of obtaining a number greater than 4.

Solution   Let E be the event in which a number greater than 4 appears. When a die is rolled, there are 6 equally likely outcomes, so that $n(\mathcal{U}) = 6$. Two of these outcomes (5 and 6) are in E, that is, $n(E) = 2$. Hence,

$$P(E) = \frac{n(E)}{n(\mathcal{U})} = \frac{2}{6} = \frac{1}{3} \qquad \blacktriangleleft$$

* Technically, we should write $P(\{HHH\})$ instead of $P(HHH)$. However, we shall write $P(HHH)$ whenever the meaning is clear.

EXAMPLE 3   Ten balls numbered from 1 to 10 are placed in an urn. If 1 ball is selected at random, find the probability that:

a. An even-numbered ball is selected (event E)
b. Ball number 3 is chosen (event T)
c. Ball number 3 is not chosen (event T')

Solution   **a.** There are 5 outcomes (2, 4, 6, 8, 10) in E out of 10 equally likely outcomes. Hence,

$$P(E) = \tfrac{5}{10} = \tfrac{1}{2}$$

b. There is only 1 outcome (3) in the event T out of 10 equally likely outcomes. Thus, $P(T) = \tfrac{1}{10}$.
c. There are 9 outcomes (all except the 3) in T' out of the 10 possible outcomes. Hence, $P(T') = \tfrac{9}{10}$.   ◀

In Example 3 we found $P(T) = \tfrac{1}{10}$ and $P(T') = \tfrac{9}{10}$, so that $P(T') = 1 - P(T)$. This is a general result because $T \cup T' = \mathcal{U}$ and $T \cap T' = \varnothing$. Thus,

$$n(T \cup T') = n(T) + n(T') = n(\mathcal{U})$$

Therefore,

$$\frac{n(T)}{n(\mathcal{U})} + \frac{n(T')}{n(\mathcal{U})} = \frac{n(\mathcal{U})}{n(\mathcal{U})}$$

or, by Definition 13.1a,

$$P(T) + P(T') = 1$$

and

$$P(T') = 1 - P(T)$$

The next example illustrates the use of this idea.

EXAMPLE 4   A coin is thrown 3 times. Find the probability of obtaining at least 1 head.

Solution   Let E be the event that we obtain at least 1 head. Then E' is the event that we obtain 0 heads; that is, that we obtain 3 tails. From the above discussion, $P(E) = 1 - P(E')$. Here, $P(E')$ is the same as $P(TTT) = \tfrac{1}{8}$; hence, $P(E) = 1 - P(TTT) = 1 - \tfrac{1}{8} = \tfrac{7}{8}$.   ◀

EXAMPLE 5   The science of heredity uses the theory of probability to determine the likelihood of obtaining flowers of a specified color when cross-breeding. Suppose we represent with letters the genes that determine the color of an offspring flower. For example, a white offspring has genes WW, a red offspring has genes RR, and a pink offspring has genes RW or WR. When we cross-breed 2 pink flowers, each plant contributes one of its color genes to each of its offspring. The tree diagram in the margin shows the 4 possibili-

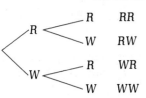

ties. Assuming that these possibilities are all equally likely, what is the probability of obtaining:

a. A white flower? **b.** A pink flower? **c.** A red flower?

Solution **a.** We see from the tree diagram that the probability of obtaining a white flower (WW) is $\frac{1}{4}$.
b. The probability of obtaining a pink flower (RW or WR) is $\frac{2}{4} = \frac{1}{2}$.
c. The probability of obtaining a red flower (RR) is $\frac{1}{4}$. ◄

EXAMPLE 6 A light and a dark die are rolled. Find:

a. The sample space for this experiment
b. The probability that the sum of the two numbers coming up is 12

Solution **a.** Figure 13.1b shows the 36 possible outcomes. The sample space for this experiment is given as follows, where the number appearing first is the number on the dark die:

$$\left\{ \begin{array}{cccccc} (1, 1) & (1, 2) & (1, 3) & (1, 4) & (1, 5) & (1, 6) \\ (2, 1) & (2, 2) & (2, 3) & (2, 4) & (2, 5) & (2, 6) \\ (3, 1) & (3, 2) & (3, 3) & (3, 4) & (3, 5) & (3, 6) \\ (4, 1) & (4, 2) & (4, 3) & (4, 4) & (4, 5) & (4, 6) \\ (5, 1) & (5, 2) & (5, 3) & (5, 4) & (5, 5) & (5, 6) \\ (6, 1) & (6, 2) & (6, 3) & (6, 4) & (6, 5) & (6, 6) \end{array} \right\}$$

Figure 13.1b

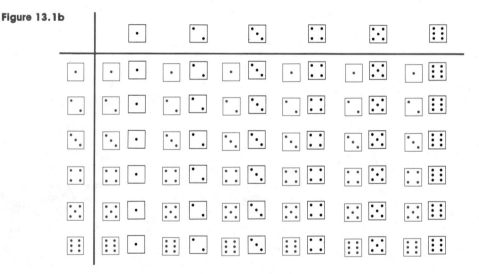

b. The probability that the sum of the two numbers coming up is 12 is $\frac{1}{36}$, because there is only 1 favorable case, (6, 6), and there are 36 possible outcomes, all equally likely. ◄

EXAMPLE 7   Find the probability of getting a king when drawing 1 card at random from a standard deck of 52 playing cards.

Solution   Since 4 of the cards are kings, the probability of drawing a king is $\frac{4}{52} = \frac{1}{13}$. ◄

B.   Events That Are Not Equally Likely

Thus far, we have considered only experiments with equally likely outcomes. Now we look at a procedure that can be used to compute probabilities of events when the outcomes are not all equally likely. We proceed in three steps:

1. We determine $\mathcal{U}$, the sample space—that is, the set of all possible outcomes for the given experiment.
2. We assign a positive number (weight) to each element of $\mathcal{U}$. (We try to assign this weight so that it measures the relative likelihood that this outcome will actually occur.)
3. Let $w(E)$ be the sum of the weights of the elements in E, and let $w(\mathcal{U})$ be the sum of the weights of the elements in $\mathcal{U}$. The probability of the event E is

$$P(E) = \frac{w(E)}{w(\mathcal{U})}$$

EXAMPLE 8   Suppose there is a race between two horses, A and B. If A is twice as likely to win as B, what is the probability that A wins the race?

Solution   1. $\mathcal{U}$, the sample space is {A wins, B wins}.
2. Because A is twice as likely to win as B, we assign a weight of 2 to {A wins} and 1 to {B wins}.
3. The weight of event E (= {A wins}) is $w(E) = 2$, and the sum of all the weights is $w(\mathcal{U}) = 3$. Thus,

$$P(E) = \frac{w(E)}{w(\mathcal{U})} = \frac{2}{3}$$ ◄

EXAMPLE 9   An accident is twice as likely to occur at noon as at 3 P.M., and 7 times as likely to occur at 6 P.M. as at 3 P.M. If an accident does occur at one of these hours, what is the probability that it occurs at noon?

Solution   1. Let N, T, and S be the events that the accident occurs at noon, at 3 P.M., and at 6 P.M., respectively. The sample space is $\mathcal{U} = \{N, T, S\}$.
2. Because event N is twice as likely as event T, and event S is 7 times as likely as event T, we assign a weight of 2 to N, 1 to T, and 7 to S.
3. The sum of all the weights is $1 + 2 + 7 = 10$, so

$$P(N) = \frac{2}{10} = \frac{1}{5}$$ ◄

A. On a single toss of a die, what is the probability of obtaining the following?

1. The number 5
2. An even number
3. A number greater than 4
4. A number less than 5

A single ball is taken at random from an urn containing 10 balls numbered 1 through 10. What is the probability of obtaining the following?

5. Ball number 8
6. An even-numbered ball
7. A ball different from 5
8. A ball whose number is less than 10
9. A ball numbered 12
10. A ball that is either less than 5 or odd

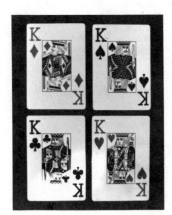

In Problems 11–16, assume that a single card is drawn from a well-shuffled deck of 52 cards. Find the probability that:

11. An ace is drawn.
12. The king of spades is drawn.
13. A spade is drawn.
14. One of the picture cards (jack, queen, or king) is drawn.
15. A picture card or a spade is drawn.
16. A red card or a picture card is drawn.

17. An executive has to visit one of his five plants for an inspection. If these plants are numbered 1, 2, 3, 4, 5 and if he is to select the plant he will visit at random, find the probability that:
    a. He will visit plant number 1.
    b. He will visit an odd-numbered plant.
    c. He will not visit plant number 4.

18. Four fair coins are tossed.
    a. Draw a tree diagram to show all the possible outcomes.
    b. Find the probability that 2 or more heads come up.
    c. Find the probability that exactly 1 head comes up.

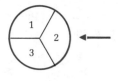

19. A disk is divided into 3 equal parts numbered 1, 2, and 3, respectively (see the figure in the margin). After the disk is spun and comes to a stop, a fixed pointer will be pointing to one of the three numbers. Suppose that the disk is spun once.
    a. Find the probability that the disk stops on the number 3.
    b. Find the probability that the disk stops on an even number.

Problems 20–23 refer to the cartoon below. Assume that all 8 travel modes listed on the spinning board are equally likely.

WORDSMITH   **by Tim Menees**

20. Find the probability that the script goes by camel train.

21. Find the probability that the script goes by sedan chair.

22. Find the probability that the script is *not* taken by a little old lady.

23. Find the probability that the script goes by life raft, hot-air balloon, or dog sled.

24. The genetic code of an organism is the self-reproducing record of the protein pattern in the organism. This code is formed by groups of small molecules that can be of 4 kinds: adenine (*A*), cytosine (*C*), guanine (*G*), and thymine (*T*).

   a. Draw a tree diagram to find all possible groups of 2 molecules. [*Note:* It is possible for both molecules to be of the same kind.]

   b. Assume that all the outcomes in part a are equally likely. Find the probability of obtaining 2 adenine molecules in a row.

   c. Find the probability of obtaining a guanine molecule and a cytosine molecule, in that order.

   d. Find the probability of obtaining two cytosine molecules in a row.

25. In a survey conducted on a Friday at Quick Shop Supermarket, it was found that 650 of 850 women who entered the supermarket bought at least 1 item. Find the probability that a woman entering the supermarket on Friday will purchase:

   a. At least 1 item   **b.** No item

The table in the margin is to be used in Problems 26–30. The table gives the numbers of males and females in a survey falling into various salary classifications. On the basis of the information in the table, find the probability that a person selected at random from those surveyed:

26. Is a female

27. Has a low income

28. Has a high income

29. Is a female with an average income

30. Is a male with a high income

SALARY	SEX M	F	TOTALS
Low	40	200	240
Average	300	160	460
High	500	300	800
TOTALS	840	660	1500

In Problems 31–35, find the probability of obtaining the following on a single toss of a pair of dice, one red and one white (see Example 6):

31. A sum of 7   **32.** A sum of 2

33. The same number on both dice

34. Different numbers on the two dice

35. An even number for the sum

B. *Use the procedure of Example 8 to solve the following problems:*

36. Stock A is twice as likely to rise in price as stock B or C, while B is as likely to rise in price as C. If it is known that exactly one of the stocks will rise in price, what is the probability that:

     a. A will rise in   **b.** B will rise in   **c.** C will rise in
        price?   price?   price?

37. A die is loaded in such a way that the probability of a particular number turning up is proportional to the number of dots showing. Find the probability that:

     a. The number 3 will turn up.

     b. An even number will turn up.

     c. A number greater than 4 will turn up.

38. Three horses, A, B, and C, are in a race. Horse A is twice as likely to win the race as horse B, and horse B is twice as likely to win as horse C. Find the probability that:

     a. A wins.   **b.** B wins.   **c.** C wins.

39. In a certain experiment, the event A can occur, or the event B can occur, or C the experiment fails. Event A is given a weight of 5, B is given a weight of 3, and C is given a weight of 2. Assuming these weights are correct, find the probability that:

     a. A occurs.   **b.** B occurs.   **c.** C occurs.

40. The weatherman predicts that it is $1\frac{1}{2}$ times as likely for it to be sunny than cloudy tomorrow, and 3 times as likely for it to be sunny than rainy. Based on these predictions, what is the probability that tomorrow will be:

     a. Sunny?   **b.** Cloudy?   **c.** Rainy?

USING YOUR KNOWLEDGE 13.1

Do you have to have surgery soon? The chances are that you will have no trouble at all! The table at the top of page 575 gives the statistics for the numbers of certain operations and the number of successes in a recent year. On the basis of these statistics, estimate the following probabilities.

1. The probability of a gallbladder operation being successful

2. The probability of an appendectomy being successful

3. The probability of a hernia operation being successful

TYPE	NUMBER OF OPERATIONS	NUMBER OF SUCCESSES
Gallbladder	472,000	465,300
Appendectomy	784,000	781,000
Hernia	508,000	506,000

13.2

COUNTING TECHNIQUES AND PROBABILITY

The counting techniques that we studied in Chapter 12 play a key role in many probability problems. In this section, we illustrate how these techniques are used in such problems.

A. Using Tree Diagrams

EXAMPLE 1 Have you heard of the "witches of Wall Street"? These are people who use astrology, tarot cards, or other "supernatural" means to predict whether a given stock will go up, go down, or stay unchanged. Not being witches, we assume that a stock is equally likely to go up, go down, or stay unchanged. A broker selects two stocks at random from the New York Stock Exchange list.

a. What is the probability that both stocks go up?
b. What is the probability that both stocks go down?
c. What is the probability that one stock goes up and one down?

Solution In order to find the total number of equally likely possibilities for the two stocks, we draw the tree diagram shown below.

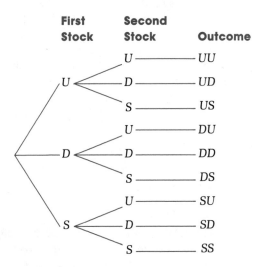

a. There is only 1 outcome (*UU*) out of 9 in which both stocks go up. Thus, the probability that both stocks go up is $\frac{1}{9}$.

b. There is only 1 outcome (*DD*) in which both stocks go down, so the probability that both go down is $\frac{1}{9}$.

c. There are 2 outcomes (*UD, DU*) in which one stock goes up and one down. Hence, the probability of this event is $\frac{2}{9}$.

(Notice that the tree diagram shows that there are 4 outcomes in which one stock stays unchanged and the other goes either up or down. The probability of this event is thus $\frac{4}{9}$.) ◀

B. Using Permutations and Combinations

Frederick Lewis, Inc.

In many games of chance, probability is used to determine "payoffs." For example, the slot machine in the picture has 3 dials with 20 symbols on each dial, as listed in Table 13.2a. In the next example, we shall find the probability of getting certain arrangements of these symbols on the 3 dials.

Table 13.2a

| | DIAL | | |
SYMBOL	1	2	3
Bar	1	3	1
Bell	1	3	3
Cherry	7	7	0
Lemon	3	0	4
Orange	3	6	7
Plum	5	1	5

EXAMPLE 2   Refer to the slot machine and to Table 13.2a to answer these questions:

a. What is the probability of getting 3 bars?

b. What is the probability of getting 3 bells?

c. What is the probability of getting 3 oranges?

d. What is the probability of getting 3 plums?

e. Based on your answers to these questions, which payoff should be the greatest and which should be the least?

Solution   **a.** We make 3 boxes representing the 3 dials:

There are 20 choices for each of the boxes (each dial has 20 symbols), so we enter a 20 in each box:

| 20 | 20 | 20 |

The total number of possibilities is $20 \times 20 \times 20 = 8000$. Now, the number of different ways to get 3 bars is $1 \times 3 \times 1 = 3$, because the first dial has 1 bar, the second has 3 bars, and the third has 1 bar. Thus,

$$P(3 \text{ bars}) = \frac{\text{Number of favorable cases}}{\text{Number of possible outcomes}} = \frac{3}{8000}$$

b. The number of ways of getting 3 bells is $1 \times 3 \times 3 = 9$. Thus,

$$P(3 \text{ bells}) = \frac{9}{8000}$$

c. The number of ways of getting 3 oranges is $3 \times 6 \times 7 = 126$. Thus,

$$P(3 \text{ oranges}) = \frac{126}{8000} = \frac{63}{4000}$$

d. The number of ways of getting 3 plums is $5 \times 1 \times 5 = 25$. Thus,

$$P(3 \text{ plums}) = \frac{25}{8000} = \frac{1}{320}$$

e. Since 3 bars is the outcome with the lowest probability and 3 oranges is the outcome with the highest probability, the greatest payoff should be for 3 bars and the least for 3 oranges. (This is how payoffs are actually determined.) ◄

The next example deals with a problem involving ordinary playing cards.

EXAMPLE 3 Two cards are drawn in succession from an ordinary deck of 52 cards. Find the probability that:

a. The cards are both aces.
b. An ace and a king, in that order, are obtained.

Solution

a. Here, the order is not important because we are simply interested in getting 2 aces. We can find this probability by using combinations. The number of ways to draw 2 aces is $C(4, 2)$, because there are 4 aces and we want a combination of any 2 of them. The number of combinations of 2 cards picked from the deck of 52 cards is $C(52, 2)$. Thus, the probability of both cards being aces is

$$\frac{C(4, 2)}{C(52, 2)} = \frac{P(4, 2)}{2!} \div \frac{P(52, 2)}{2!} = \frac{P(4, 2)}{P(52,2)}$$

$$= \frac{4 \cdot 3}{52 \cdot 51} = \frac{1}{221}$$

b. In this part of the problem, we want to consider the order in which the 2 cards are drawn, so we use permutations. The number of ways of selecting an ace is $P(4, 1)$ and the number of ways of selecting a king is $P(4, 1)$. By the SCP, the number of ways of doing these two things in succession is $P(4, 1)P(4, 1)$. The total number of ways of drawing 2 cards is $P(52, 2)$, so the probability of drawing an ace and a king, in that order, is

$$\frac{P(4, 1)P(4, 1)}{P(52, 2)} = \frac{4 \cdot 4}{52 \cdot 51} = \frac{4}{663} \qquad \blacktriangleleft$$

In part a of Example 3, we found that

$$\frac{C(4, 2)}{C(52, 2)} = \frac{P(4, 2)}{P(52, 2)}$$

This equation is a special case of a general result that can be obtained as follows:

$$\frac{C(m, r)}{C(n, r)} = C(m, r) \div C(n, r)$$

$$= \frac{P(m, r)}{r!} \div \frac{P(n, r)}{r!}$$

$$= \frac{P(m, r)}{r!} \times \frac{r!}{P(n, r)} = \frac{P(m, r)}{P(n, r)}$$

EXAMPLE 4   Suppose we take one suit, say the 13 hearts, out of a standard deck of 52 cards. Shuffle the 13 hearts and then draw 3 of them. What is the probability that none of the 3 will be an ace, king, queen, or jack?

Solution   Here the order does not matter, so we use combinations. The number of combinations of 13 things taken 3 at a time is $C(13, 3)$. There are 9 cards not including the ace, king, queen, or jack, and the number of combinations of these taken 3 at a time is $C(9, 3)$. Thus, the required probability is

$$\frac{C(9, 3)}{C(13, 3)} = \frac{P(9, 3)}{P(13, 3)} = \frac{9 \cdot 8 \cdot 7}{13 \cdot 12 \cdot 11} = \frac{42}{143} \qquad \blacktriangleleft$$

In Example 4, because we are taking 3 cards from both the 13 hearts and the 9 cards, we could use either combinations or permutations. The next example is one that requires the use of combinations.

EXAMPLE 5   A poker hand consists of 5 cards. What is the probability of getting a hand of 4 aces and a king?

Solution   Here, the order is not to be considered because any order of getting the aces and the king will result in a hand that consists of 4 aces and a king. Now, the number of ways in which 4 aces can be selected is $C(4, 4)$, and the number of ways in which 1 king can be selected is $C(4, 1)$. Hence, by the sequential counting principle (SCP), the number of ways of getting 4 aces and a king is $C(4, 4)C(4, 1)$. Furthermore, the total number of 5-card hands is $C(52, 5)$, so the required probability is

$$\frac{C(4, 4)C(4, 1)}{C(52, 5)} = \frac{1 \cdot 4}{C(52, 5)}$$

Since

$$C(52, 5) = \frac{52 \cdot 51 \cdot 50 \cdot 49 \cdot 48}{5 \cdot 4 \cdot 3 \cdot 2 \cdot 1} = 2{,}598{,}960$$

the probability of getting 4 aces and a king is

$$\frac{C(4, 4)C(4, 1)}{C(52, 5)} = \frac{4}{2{,}598{,}960} = \frac{1}{649{,}740}$$

which is very small indeed! ◀

Do we always use the SCP and/or a permutation formula and/or a combination formula in solving a probability problem? Not necessarily; sometimes it is easier to look at the actual possible outcomes or to reason the problem out directly. This is illustrated in the next example.

EXAMPLE 6   A careless commercial clerk was supposed to mail 3 bills to 3 customers. He addressed 3 envelopes, but absent-mindedly paid no attention to which bill he put in which envelope.

a. What is the probability that exactly 1 of the customers received the proper bill?
b. What is the probability that exactly 2 of the customers received the proper bills?

Solution   In the table in the margin, the headings C_1, C_2, and C_3 represent the customers; and the numbers 1, 2, and 3 below represent the bills received by the customers. Therefore, the rows represent the possible outcomes.

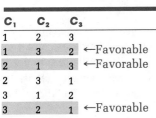

a. The table shows that there are 6 possibilities and 3 cases in which exactly 1 (that is, 1 and only 1) of the customers received the proper bill. Thus, the required probability is

$$P = \tfrac{3}{6} = \tfrac{1}{2}$$

b. Here the probability is 0, because if 2 customers received their proper bills, then the third one did also. This means that there is no case in which 2 and only 2 received the proper bills. ◀

In Example 6, if there were 4 customers and 4 bills, then there would be a total of $P(4, 4) = 24$ cases in all. (This is just the number of ways in which the bills could be permuted.) You can see that it would be quite cumbersome to list all these cases. Instead, let us draw a tree showing the possible favorable cases if 1 customer, C_1, receives the proper bill (see Fig. 13.2a). This shows the 1 under C_1; then C_2 can have only the 3 or the 4. If C_2 has the 3, then C_3 must have the 4 and C_4 the 2. (Otherwise more than 1 customer would receive the proper bill.) If C_2 has the 4, then C_3 must have the 2 and C_4 the 3. These are the only 2 favorable cases possible if C_1 gets bill 1. The same argument holds if one of the other customers gets the proper bill; there are just 2 ways in which none of the other customers gets a proper bill. Since there are 4 customers, the SCP shows that there are

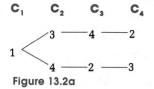

Figure 13.2a

only $4 \times 2 = 8$ favorable cases. Thus, the probability that exactly 1 of the customers gets the proper bill is $\frac{8}{24} = \frac{1}{3}$.

EXERCISE 13.2

A.

1. A man has 3 pairs of shoes, 2 suits, and 3 shirts. If he picks a pair of shoes, a suit, and a shirt at random, what is the probability that he picks his favorite shoes, suit, and shirt?

2. At the end of a meal in a restaurant, a person wants to have pie a la mode (pie and ice cream) for dessert. There are 5 flavors of ice cream —chocolate, vanilla, strawberry, peach, and coffee—and there are 2 kinds of pie—apple and cherry. If the waiter picks the pie and ice cream at random, what is the probability that the person will get apple pie with vanilla ice cream?

3. A fair die is rolled 3 times in succession. What is the probability that even numbers come up all 3 times?

B.

4. Two cards are drawn at random, in succession and without replacement, from a deck of 52 cards.
   a. Find the number of ways in which the ace of spades and a king can be selected, in that order.
   b. What is the probability of drawing the ace of spades and a king, in that order?

5. Mr. C. Nile and Mr. D. Mented agreed to meet at 8:00 P.M. in one of the Spanish restaurants in Ybor City. They were both punctual, and they both remembered the date agreed upon. Unfortunately, they forgot to specify the name of the restaurant. If there are 5 Spanish restaurants in Ybor City, and the two men each go to one of these, find the probability that:
   a. They meet each other.　　　　**b.** They miss each other.

6. P.U. University offers 100 courses, 25 of which are mathematics. All of these courses are available each hour, and a counselor selects 4 courses at random. Find the probability that the selection will not include a mathematics course. (Do not simplify your answer.)

7. A piggy bank contains 2 quarters, 3 nickels, and 2 dimes. A person takes 2 coins at random from this bank. Label the coins $Q_1, Q_2, N_1, N_2, N_3, D_1, D_2$ so that they may all be regarded as different. Then find the probability that the value of the 2 coins selected is:
   a. 35¢　　　　　　　　　　　**b.** 50¢

8. A committee of 2 is chosen at random from a population of 5 men and 6 women. What is the probability that the committee will consist of 1 man and 1 woman?

In Problems 9–13, assume that 2 cards are drawn in succession and without replacement from an ordinary deck of 52 cards. Find the probability that:

9. Two kings are obtained.   10. Two spades are obtained.
11. A spade and a king other than the king of spades (in that order) are obtained.
12. A spade and a king other than the king of spades (not necessarily in that order) are obtained.
13. Two red cards are obtained.

In Problems 14 and 15, assume that there is an urn containing five $50 bills, four $20 bills, three $10 bills, two $5 bills, and one $1 bill, and the bills all have different serial numbers so that they can be distinguished from each other. A person reaches into the urn and withdraws one bill and then another.

14. **a.** In how many ways can two $20 bills be withdrawn?
    b. How many different outcomes are possible?
    c. What is the probability of selecting two $20 bills?
15. **a.** In how many ways can a $50 bill and a $10 bill be selected, in that order?
    b. What is the probability of selecting a $50 bill and a $10 bill, in that order?
    c. What is the probability of selecting two bills, one of which is a $50 bill and the other a $10 bill?

16. An urn contains 5 white balls and 3 black balls. Two balls are drawn at random from this urn. Find the probability that:
    a. Both balls are white.   **b.** Both balls are black.
    c. One ball is white and the other is black.
17. In this problem, do not simplify your answers. What is the probability that a 5-card poker hand will contain:
    a. 2 kings, 2 aces, and 1 other card?
    b. 3 kings and 2 aces?
    c. 4 cards of a kind (same denomination)?
18. A box of light bulbs contains 95 good bulbs and 5 bad ones. If 3 bulbs are selected at random from the box, what is the probability that 2 are good and 1 is bad?
19. A plumbing company needs to hire 2 plumbers. Five people, 4 men and 1 woman, apply for the job. Since they are all equally qualified, the selection is made at random (2 names are pulled out of a hat). What is the probability that the woman is hired?
20. "Low-calorie" food is required to contain no more than 40 calories per serving. The Food and Drug Administration (FDA) suspects that a certain company is marketing illegally labeled "low-calorie" food. If an inspector selects 3 cans at random from a shelf holding 10 cans (3 legally labeled and 7 illegally labeled), what is the probability that:
    a. All 3 cans selected are legally labeled?
    b. Only 2 of the cans are legally labeled?

Have you heard of Dr. Spock? (No, not the Mr. Spock with the pointy ears in "Star Trek.") Benjamin Spock, a famous pediatrician, was accused of violating the Selective Service Act by encouraging resistance to the Vietnam War. In his trial, the defense challenged the legality of the method used to select the jury. In the Boston District Court, jurors are selected in three stages, as follows:

1. The clerk of the court selects 300 names at random from the City Directory. If a directory lists 76,000 names (40,000 women and 36,000 men), what is the probability of selecting 150 men and 150 women? (Do not simplify.)
2. The 300 names are placed in a box, and the names of 30 potential jurors are drawn. If the names in the box correspond to 160 women and 140 men, find the probability that 15 men and 15 women are selected. (Do not simplify.)
3. The subgroup of 30 is called a *venire*. From the venire, 12 jurors are selected. If the venire consists of 16 women and 14 men, what is the probability that the final jury consists of 6 men and 6 women? (Do not simplify.)

By the way, it was shown that the Spock trial judge selected only about 14.6% women, while his colleagues selected about 29% women. This showed that the trial judge systematically reduced the proportion of women jurors.

▶ **Calculator Corner 13.2**   *You can use a calculator to compute expressions such as C(52, 5) in Example 5. To do this, enter*

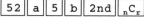

13.3

COMPUTATION OF PROBABILITIES

In this section we give four formulas, or properties, that are useful in the computation of probabilities. The letters E, A, and B stand for events in a sample space $\mathcal{U}$.

Property 1		
$P(E) = 0$	if and only if	$E = \varnothing$ (1)

Property 1 states that an **impossible event** has a probability 0. The next example illustrates this idea.

EXAMPLE 1   A die is tossed. What is the probability that a 7 turns up?

Solution   The sample space for this experiment is

$$\mathcal{U} = \{1, 2, 3, 4, 5, 6\}$$

so it is impossible for a 7 to come up. Thus, $P(7) = 0$. ◄

Property 2

$$0 \leq P(E) \leq 1 \tag{2}$$

Property 2 says that the probability of any event is a number between 0 and 1, inclusive. This follows because the number of favorable cases cannot be less than 0 or more than the total number of possible cases. Thus, $P(E) = 1$ means that the event E is **certain** to occur.

Property 3

$$P(A \cup B) = P(A) + P(B) - P(A \cap B) \tag{3}$$

Formula (3) says that the probability of A or B is the probability of A plus the probability of B, decreased by the probability of A and B. Note the key words "*or*" and "*and.*"

EXAMPLE 2   A penny and a nickel are tossed. What is the probability that one *or* the other of the coins will turn up heads?

Solution   If we use subscripts p and n for penny and nickel, respectively, we can list the possible cases as follows: (H_p, H_n), (H_p, T_n), (T_p, H_n), (T_p, T_n). Since there are 3 favorable cases out of the 4 possible,

$$P(H_p \cup H_n) = \tfrac{3}{4}$$

We can check that formula (3) gives

$$P(H_p \cup H_n) = P(H_p) + P(H_n) - P(H_p \cap H_n)$$
$$= \tfrac{1}{2} + \tfrac{1}{2} - \tfrac{1}{4} = \tfrac{3}{4}$$

as before. ◄

EXAMPLE 3   A card is drawn from a pack of 52 playing cards. Find the probability that the card is either an ace or a red card.

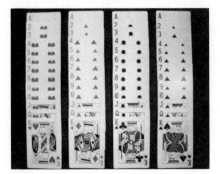

Solution Let A be the event in which the card drawn is an ace, and let R be the event in which the card drawn is red. Then by formula (3), $P(A \cup R) = P(A) + P(R) - P(A \cap R)$. Now, $P(A) = \frac{4}{52}$, $P(R) = \frac{26}{52}$, and $P(A \cap R) = \frac{2}{52}$, so $P(A \cup R) = \frac{4}{52} + \frac{26}{52} - \frac{2}{52} = \frac{7}{13}$. ◀

Formula (3), in case all outcomes in the sample space are equally likely, is derived from the fact that

$$P(E) = \frac{n(E)}{n(\mathcal{U})}$$

Hence,

$$P(A \cup B) = \frac{n(A \cup B)}{n(\mathcal{U})} = \frac{n(A) + n(B) - n(A \cap B)}{n(\mathcal{U})} \qquad \text{See Section 1.5.}$$

$$= \frac{n(A)}{n(\mathcal{U})} + \frac{n(B)}{n(\mathcal{U})} - \frac{n(A \cap B)}{n(\mathcal{U})}$$

$$= P(A) + P(B) - P(A \cap B)$$

If the outcomes are not equally likely (see Section 13.1), then the same result follows by replacing $n(E)$ and $n(\mathcal{U})$ by $w(E)$ and $w(\mathcal{U})$, where these mean the sum of the weights of the outcomes in E and $\mathcal{U}$, respectively.

EXAMPLE 4 An urn contains 5 red, 2 black, and 3 yellow balls. Find the probability that a ball selected at random from the urn will be red or yellow.

Solution By formula (3),

$$P(R \cup Y) = P(R) + P(Y) - P(R \cap Y)$$
$$= \tfrac{5}{10} + \tfrac{3}{10} - 0 = \tfrac{4}{5}$$ ◀

In Example 4, notice that $P(R \cap Y) = 0$. This means the events of selecting a red ball and selecting a yellow ball cannot occur simultaneously, so that $R \cap Y = \varnothing$. We say that A and B are **mutually exclusive** if $A \cap B = \varnothing$. For any two mutually exclusive events A and B, it follows that $P(A \cap B) = 0$, and formula (3) becomes

---

Property 4

$$P(A \cup B) = P(A) + P(B) \qquad \text{if } P(A \cap B) = 0 \qquad (4)$$

EXAMPLE 5 Show that the events R and Y of Example 4 are mutually exclusive.

Solution Because $P(R \cap Y) = 0$, $R \cap Y = \varnothing$ (property 1), so R and Y are mutually exclusive. ◀

EXAMPLE 6 In a game of blackjack (also called twenty-one), a player and the dealer each get 2 cards. Let A and B be the events defined as follows:

A: The player gets an ace and a face card for 21 points.
B: The dealer gets an ace and a 10 for 21 points.

Are A and B mutually exclusive events?

Solution   No; both player and dealer can get 21 points. (In the game of blackjack, 21 points wins, and in most casinos, the dealer would be the winner with this tie score.) ◄

EXAMPLE 7   The video games that you can attach to your television set have both sound and picture. It is estimated that the probability of the sound being defective is 0.03, the probability of at least one or the other (sound or video) being defective is 0.04, but the probability of both being defective is only 0.01. What is the probability that the video is defective?

Solution   We let S stand for the event that the sound is defective and V for the event that the video is defective. Then, using formula (3), with $P(S) = 0.03$, $P(S \cup V) = 0.04$, and $P(S \cap V) = 0.01$, we get

$$P(S \cup V) = P(S) + P(V) - P(S \cap V)$$

so that

$$0.04 = 0.03 + P(V) - 0.01$$

Thus,

$$P(V) = 0.02 \qquad ◄$$

We have used the formula $P(T') = 1 - P(T)$ to calculate the probability of the complement of an event. For example, if the probability that it will rain today is $\frac{1}{4}$, the probability that it will not rain today is $\frac{3}{4} = 1 - \frac{1}{4}$, and if the probability of a stock going up in price is $\frac{3}{8}$, the probability that the stock will not go up in price is $\frac{5}{8} = 1 - \frac{3}{8}$. We now illustrate how this property is used in the field of life insurance.

Table 13.3a is a **mortality table** for 100,000 people. According to this table, of 100,000 people alive at age 10, some 96,300 were alive at age 15; but only 3 were alive at age 95. A table like this is used to calculate a portion of the premium on life insurance policies. We use this table in the next example.

Table 13.3a   American Experience Table of Mortality

AGE IN YEARS	NUMBER ALIVE	AGE IN YEARS	NUMBER ALIVE	AGE IN YEARS	NUMBER ALIVE
10	100,000	40	78,100	70	38,600
15	96,300	45	74,200	75	26,200
20	92,600	50	69,800	80	14,500
25	89,000	55	64,600	85	5,500
30	85,400	60	57,900	90	850
35	81,800	65	49,300	95	3

EXAMPLE 8   Find the probability that a person who is alive at age 20:

a. Will still be alive at age 70
b. Will not be alive at age 70

Solution   **a.** Based on Table 13.3a, the probability that a person alive at 20 is still alive at 70 is given by

$$P(\text{Being alive at 70}) = \frac{\text{Number alive at 70}}{\text{Number alive at 20}}$$

$$= \frac{38,600}{92,600} = \frac{193}{463}$$

b. Using formula (4), we find that the probability of the person not being alive at 70 is

$$1 - \frac{193}{463} = \frac{270}{463} \qquad \blacktriangleleft$$

EXERCISE 13.3

In Problems 1–4, find the answer to the given question, and indicate which of the four properties given in this section you used.

1. A die is thrown. Find the probability that the number that turns up is a 0.

2. A die is thrown. Find the probability that an odd or an even number comes up.

3. Two dice are thrown. Find the probability that the sum of the two faces that turn up is between 0 and 13.

4. An absent-minded professor wished to mail 3 report cards to 3 of his students. He addressed 3 envelopes but, unfortunately, did not pay any attention to which card he put in which envelope. What is the probability that exactly 2 students receive their own report cards? (Assume that all 3 envelopes were delivered.)

A single ball is drawn from an urn containing ten balls numbered 1 through 10. In Problems 5–8, find the probability that the ball chosen is:

5. An even-numbered ball or a ball with a number greater than 7

6. An odd-numbered ball or a ball with a number less than 5

7. An even-numbered ball or an odd-numbered ball

8. A ball with a number that is greater than 7 or less than 5

In Problems 9–13, a single card is drawn from a deck of 52 cards. Find the probability that the card chosen is:

9. The king of hearts or a spade

10. The ace of hearts or an ace

11. The ace of diamonds or a diamond

12. The ace of clubs or a black card

13. The king of hearts or a picture card (jack, queen, or king)

14. The U.S. Weather Service reports that in a certain northern city it rains 40 days and snows 50 days in the winter. However, it rains and snows on only 10 of those days. Based on this information, what is the probability that it will rain or snow in that city on a particular winter day? (Assume there are 90 days of winter.)

15. Among the first 50 stocks listed in the New York Stock Exchange transactions on a certain day (as reported in *The Wall Street Journal*), there were 26 stocks that went down, 15 that went up, and 9 that remained unchanged. Based on this information, find the probability that a stock selected at random from this list would not have remained unchanged.

The table in the margin shows the probability that there is a given number of people waiting in line at a checkout register at Dear's Department Store. In Problems 16–20, find the probability of having:

NUMBER OF PERSONS IN LINE	PROBABILITY
0	0.10
1	0.15
2	0.20
3	0.35
4 or more	0.20

16. Exactly 2 persons in line
17. More than 3 persons in line
18. At least 1 person in line
19. More than 3 persons or fewer than 2 persons in line
20. More than 2 persons or fewer than 3 persons in line

In solving Problems 21–25, refer to Table 13.3a.

21. What is the probability that a person who is alive at age 20 will not be alive at age 65?
22. What is the probability that a person who is alive at age 25 will be alive at age 70?
23. What is the probability that a person who is alive at age 25 will not be alive at age 70?
24. What is the probability that a person who is alive at age 55 will live 80 more years? (Assume that none of the persons in the table attained 100 years of age.)
25. What is the probability that a person who is alive at 55 will live less than 80 more years? (See Problem 24.)

Problems 26–30 refer to the table below. This table shows the number of correctly and incorrectly filled out tax forms obtained from a random sample of 100 returns examined by the Internal Revenue Service (IRS) in a recent year.

	SHORT FORM (1040A)	LONG FORM (1040)		
	No Itemized Deductions	No Itemized Deductions	Itemized Deductions	TOTALS
Correct	15	40	10	65
Incorrect	5	20	10	35
TOTALS	20	60	20	100

26. Find the probability that a form was a long form (1040) or an incorrectly filled out form.
27. Find the probability that a form had no itemized deductions and was correctly filled out.
28. Find the probability that a form was not filled out incorrectly.
29. Find the probability that a form was not a short form (1040A).
30. Find the probability that a form was a long form (1040) with no itemized deductions and filled out incorrectly.

USING YOUR KNOWLEDGE 13.3

In this section we learned how to use a mortality table to calculate the probability that a person alive at a certain age will be alive at a later age. There are other tables that give the probabilities of different events. For example, many mortgage companies use a credit-scoring table to estimate the likelihood that an applicant will repay a loan. One such table appears below.

A Hypothetical Credit-Scoring Table

AGE	Under 25 — 12	25–29 — 5	30–34 — 0	35–39 — 1	40–44 — 18	45–49 — 22	50 or over — 31
TIME AT ADDRESS	Less than 1 yr — 9	1–2 yr — 0	2–3 yr — 5	3–5 yr — 0	5–9 yr — 5	10 yr or more — 21	
AGE OF AUTO	None — 0	0–1 yr — 12	2 yr — 16	3–4 yr — 13	5–7 yr — 3	8 yr or more — 0	
MONTHLY AUTO PAYMENT	None — 18	Less than $80 — 6	$80–99 — 1	$100–$139 — 4	$140 or more — 0		
HOUSING COST	Less than $125 — 0	$125–$274 — 10	$275 or more — 12	Owns clear — 12	Lives with relatives — 24		
CHECKING AND SAVINGS ACCOUNTS	Both — 15	Checking only — 2	Savings only — 2	Neither — 0			
FINANCE COMPANY REFERENCE	Yes — 0	No — 15					
MAJOR CREDIT CARDS	None — 0	1 — 5	2 or more — 15				
RATIO OF DEBT TO INCOME	No debts — 41	1–5% — 16	6–15% — 20	16% or over — 0			

In this table your score depends on the number of points you get on the nine tabulated items. To obtain your score, you add the scores (shown in color) on the individual items. For example, if your age is 21, you get 12 points. If you have lived at your present address for less than a year, you get

TOTAL SCORE	PROBABILITY OF REPAYMENT
60	0.70
65	0.74
70	0.78
75	0.81
80	0.84
85	0.87
90	0.89
95	0.91
100	0.92
105	0.93
110	0.94
115	0.95
120	0.955
125	0.96
130	0.9625

9 more points. Moreover, if your car is 2 years old, you get another 16 points. So far, your score is $12 + 9 + 16$. This should give you the idea.

A lender using the scoring table selects a cutoff point from a table, such as that shown in the margin, that gauges the probability that an applicant will repay a loan.

1. John Dough, 27 years old, living for 3 years at his present address, has a 2-year-old automobile on which he pays $200 monthly. He pays $130 per month for his apartment and has no savings account, but he does have a checking account. He has no finance company reference. He has one major credit card, and his debt-to-income ratio is 12%. Based on the credit-scoring table, what is the probability that Mr. Dough will repay a loan?
2. What is the probability in Problem 1 if John sells his car and rides the bus to work?
3. Find the probability that you will repay a loan, based on the information in the table.

DISCOVERY 13.3

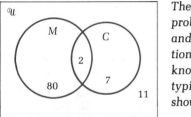

The Venn diagrams we studied in Chapter 1 can often be used to find the probability of an event by showing the number of elements in the universe and the number of elements corresponding to the event under consideration. For example, if there are 100 employees in a certain firm and it is known that 82 are males (M), 9 are clerk typists (C), and 2 of these clerk typists are male, we can draw a diagram corresponding to this situation as shown in the margin. From this diagram, we can conclude that

$$P(M) = \frac{82}{100} \qquad P(C) = \frac{9}{100} \qquad P(M \cap C) = \frac{2}{100}$$

$$P(M \cup C) = \frac{82}{100} + \frac{9}{100} - \frac{2}{100} = \frac{89}{100}$$

Using this technique, solve the following problems.

In Problems 1–3, assume that of the 100 persons in a company, 70 are married, 80 are college graduates, and 60 are both married and college graduates. Find the probability that if a person is selected at random from this group, the person will be:

1. Married and a college graduate
2. Married or a college graduate
3. Not married and not a college graduate

4. In a recent election, voters were asked to vote on two issues, A and B. A survey taken by the Gallup Poll indicated that of 1000 eligible voters, 600 persons voted in favor of A, 500 persons voted in favor of B, 200

persons voted in favor of both A and B, and 50 persons voted against both issues. If an eligible voter is selected at random, find the probability that the voter:

a. Voted for A but not B **b.** Voted for B but not A

c. Voted for both A and B **d.** Voted against both A and B

e. Did not vote at all

13.4

CONDITIONAL PROBABILITY

The WIZARD OF ID by permission of Johnny Hart and NAS, Inc.

As the *Wizard of Id* cartoon shows, it is sometimes the case that in considering the probability of an event A, we obtain additional information that may suggest a "revision" of the probability of A. For example, assume that in Getwell Hospital 70 of the patients have lung cancer (C), 60 of the patients smoke (S), and 50 have cancer and smoke. If there are 100 patients in the hospital, and 1 is selected at random, then $P(C) = \frac{70}{100}$ and $P(S) = \frac{60}{100}$. But suppose a patient selected at random tells you that he or she smokes. What is the probability that this patient has cancer? In other words, what is the probability that a patient has cancer, given that the patient smokes? The expression "given that the patient smokes" means that we must restrict our attention to those patients who smoke. We have thus added a **restrictive condition** to the problem. Essentially, the condition that the person smokes requires that we use S as our sample space.

To compute $P(C|S)$ (read, "the probability of C, given S"), we recall that there are 50 favorable outcomes (people who have lung cancer and smoke) and 60 elements in the new sample space (people who smoke). Hence, $P(C|S) = \frac{50}{60} = \frac{5}{6}$.

We note that

$$
\begin{aligned}
P(C|S) &= \frac{n(C \cap S)}{n(S)} \\
&= \frac{n(C \cap S)/n(\mathcal{U})}{n(S)/n(\mathcal{U})} = \frac{P(C \cap S)}{P(S)}
\end{aligned}
$$

This discussion suggests the next definition.

▶ **Definition 13.4a**

> If A and B are events in a sample space $\mathcal{U}$ and $P(B) \neq 0$, the **conditional probability of A, given B,** is denoted by $P(A|B)$ and is defined by
>
> $$P(A|B) = \frac{P(A \cap B)}{P(B)} \qquad (1)$$

Notice that the conditional probability of A, given B, results in a new sample space consisting of the elements in $\mathcal{U}$ for which B occurs. This gives rise to a second method of handling conditional probability, as illustrated in the following examples.

EXAMPLE 1   A die is thrown. Find the probability that a 3 came up if it is known that an odd number turned up.

Solution   **Method 1.** Let T be the event in which a 3 turns up and Q be the event in which an odd number turns up. By equation (1),

$$P(T|Q) = \frac{P(T \cap Q)}{P(Q)} = \frac{\frac{1}{6}}{\frac{3}{6}} = \frac{1}{3}$$

Method 2. We know that an odd number turned up, so our new sample space is $\mathcal{U} = \{1, 3, 5\}$. Only one outcome (3) is favorable, so $P(T|Q) = \frac{1}{3}$. ◀

EXAMPLE 2   A coin is thrown; then a die is tossed. Find the probability of obtaining a 6, given that heads came up.

Solution   **Method 1.** Let S be the event in which a 6 turns up, and let H be the event in which heads come up.

$$P(S|H) = \frac{P(S \cap H)}{P(H)} = \frac{\frac{1}{12}}{\frac{1}{2}} = \frac{1}{12} \cdot 2 = \frac{1}{6}$$

Method 2. We know that heads came up, so our new sample space is $\mathcal{U} = \{(H, 1), (H, 2), (H, 3), (H, 4), (H, 5), (H, 6)\}$. Only one outcome is favorable, $(H, 6)$, so $P(S|H) = \frac{1}{6}$. ◀

EXAMPLE 3   Two dice were thrown, and a friend tells us that the numbers that came up were different. Find the probability that the sum of the two numbers was 4.

Solution   **Method 1.** Let D be the event in which the two dice show different numbers, and let F be the event in which the sum is 4. By equation (1),

$$P(F|D) = \frac{P(F \cap D)}{P(D)}$$

Now, $P(F \cap D) = \frac{2}{36}$, because there are two outcomes, $(3, 1)$ and $(1, 3)$, in which the sum is 4 and the numbers are different, and there are 36 possible outcomes. Furthermore,

$$P(D) = \frac{36 - 6}{36} = \frac{30}{36}$$

so

$$P(F|D) = \frac{P(F \cap D)}{P(D)} = \frac{\frac{2}{36}}{\frac{30}{36}} = \frac{1}{15}$$

Method 2. We know that the numbers on the two dice were different, so we have $36 - 6 = 30$ (36 outcomes minus 6 that show the same number on both dice) elements in our sample space. Of these, only two, $(3, 1)$ and $(1, 3)$, are favorable. Hence, $P(F|D) = \frac{2}{30} = \frac{1}{15}$. ◀

EXAMPLE 4   Two dice are thrown and a friend tells you that the first die shows a 6. Find the probability that the sum of the numbers showing on the two dice is 7.

Solution   **Method 1.** Let S_1 be the event in which the first die shows a 6, and let S_2 be the event in which the sum is 7. Then

$$P(S_2|S_1) = \frac{P(S_2 \cap S_1)}{P(S_1)} = \frac{\frac{1}{36}}{\frac{6}{36}} = \frac{1}{6}$$

Method 2. We know that a 6 came up on the first die, so our new sample space is $\mathcal{U} = \{(6, 1), (6, 2), (6, 3), (6, 4), (6, 5), (6, 6)\}$. Hence, $P(S_1|S_2) = \frac{1}{6}$, because there is only one favorable outcome, $(6, 1)$. ◀

EXAMPLE 5   Suppose we represent with the letters B and b the genes that determine the color of a person's eyes. If the person has two b genes, the person has blue eyes; otherwise, the person has brown eyes. If it is known that a man has brown eyes, what is the probability that he has two B genes? (Assume that both genes are equally likely to occur.)

Solution   The tree diagram for the four possibilities appears in the margin.

B   BB   Brown eyes

b   Bb   Brown eyes

B   bB   Brown eyes

b   bb   Blue eyes

Method 1. Let T be the event in which the man has two B genes, and let B be the event in which the man has brown eyes. By Definition 13.4a,

$$P(T|B) = \frac{P(T \cap B)}{P(B)} = \frac{\frac{1}{4}}{\frac{3}{4}} = \frac{1}{3}$$

Method 2. It is known that the man has brown eyes, so we consider the three outcomes corresponding to these cases (BB, Bb, bB). Because only one of these equally likely outcomes (*BB*) is favorable, the probability that a man has two *B* genes if it is known that he has brown eyes is $\frac{1}{3}$. ◄

Other important applications also make use of conditional probability. For example, the Framingham Heart Disease Study began in 1948 and focused on strokes and heart failure. Table 13.4a, based on this study, shows the number of strokes per 1000 people for various conditions. As you can see, the incidence of stroke for people aged 45–74 increases almost fourfold as blood pressure goes from normal to high (from 8 per 1000 to 31 per 1000). Note that the numbers in the body of the table are all per 1000. This means that the table is actually giving us approximate conditional probabilities; we can interpret the last number in the table, 31, as meaning that the probability that a person has a stroke, given that this person has high blood pressure and is in the age group 45–74, is about $\frac{31}{1000}$. We look at some other aspects of this study in the next example.

Table 13.4a Strokes per 1000 People

BLOOD PRESSURE	AGES 45–74
Normal	8
Borderline	14
High	31

EXAMPLE 6   Assume the numbers in Table 13.4a are accurate, and find the probability that:

a. A person in the 45–74 age group has a stroke, given that the person has normal blood pressure.

b. A person in the 45–74 age group has a stroke, given that the person has either normal or borderline blood pressure.

c. A person in the 45–74 age group has a stroke.

d. If a person in the 45–74 age group has a stroke, what is the probability that the person has normal blood pressure?

Solution   **a.** This probability can be read directly from the table; it is $\frac{8}{1000}$.

b. Here we have two mutually exclusive sets, those with normal blood pressure and those with borderline blood pressure. Hence, the required probability is the sum of the probabilities for the two sets, that is,

$$\frac{8}{1000} + \frac{14}{1000} = \frac{22}{1000} = \frac{11}{500}$$

c. The idea is similar to that in part b; there are three mutually exclusive sets, so the required probability is the sum of the probabilities for the three sets, that is,

$$\frac{8}{1000} + \frac{14}{1000} + \frac{31}{1000} = \frac{53}{1000}$$

d. Here we know the person has a stroke, so we can use the idea of conditional probability. The population for this condition consists of the $8 + 14 + 31 = 53$ people who have a stroke. Of these, 8 have normal blood pressure. Thus, the required probability is $\frac{8}{53}$. ◄

1. A die was thrown. Find the probability that a 5 came up if it is known that an even number turned up,.

2. A coin was thrown; then a die was tossed. Find the probability of obtaining a 7, given that tails came up.

3. Two dice were thrown, and a friend tells us that the numbers that came up were identical. Find the probability that the sum of the two numbers was:
   a. 8   **b.** 9
   c. An even number   **d.** An odd number

4. Referring to Example 5 of this section, find the probability that a person has two *b* genes, given that the person has:
   a. Brown eyes   **b.** Blue eyes

5. For a family with 2 children, the sample space indicating boy (*B*) or girl (*G*) is *BB*, *BG*, *GB*, *GG*. If each of the outcomes is equally likely, find the probability that the family has 2 boys if it is known that the first child is a boy.

6. A family has 3 children. If each of the outcomes in the sample space is equally likely, find the probability that the family has 3 girls if it is known that:
   a. The first child is a girl   **b.** The first child is a boy

7. Referring to Problem 6, find the probability that the family has exactly 2 girls if it is known that the first child is a girl.

8. The table below gives the approximate number of suicides per .100,000 persons, classified according to country and age for one year:

COUNTRY	AGE (YEARS)			
	15–24	25–44	45–64	65 or over
United States	10	20	30	40
Canada	10	15	13	14
West Germany	20	30	50	50

Based on the table, find the probability that:
   a. A person between 25 and 44 years of age committed suicide, if it is known that the person lived in the United States
   b. A person between 25 and 44 years of age committed suicide, if it is known that the person lived in Canada
   c. A person committed suicide, given that the person lived in West Germany

9. The personnel director of Gadget Manufacturing Company has compiled the table on page 595, which shows the percentage of men and women employees who were absent the indicated number of days. Suppose there are as many women as men employees.

SEX	ABSENCES (DAYS)				TOTAL
	0	1-5	6-10	11 or more	
Men	20%	40%	40%	0	100%
Women	20%	20%	20%	40%	100%

    a. Find the probability that an employee missed 6–10 days, given that the employee is a woman.

    b. Find the probability that an employee is a woman, given that the employee missed 6–10 days.

In Problems 10–12, assume that 2 cards are drawn in succession and without replacement from a standard deck of 52 cards. Find the probability that:

10. The second card is the ace of hearts, if it is known that the first card was the ace of spades.

11. The second card is a king, if it is known that the first card was a king.

12. The second card is a 7, if it is known that the first card was a 6.

	LOW	HIGH
COMPUTERS	5	10
PETROLEUM	20	15

The Merrilee Brokerage House studied two groups of industries (computers and petroleum) and rated them as low risks or high risks, as shown in the margin. Use this information in Problems 13 and 14.

13. If a person selected one of these stocks at random (i.e., each stock has probability $\frac{1}{50}$ of being selected), find the probability that the person selected a computer stock, given that the person selected a low-risk stock.

14. If a person selected one of the stocks at random, find the probability that the person selected a petroleum stock, given that the person selected a high-risk stock.

15. A stock market analyst figures the probabilities that two related stocks, *A* and *B*, will go up in price. She finds the probability that *A* will go up to be 0.6 and the probability that both stocks will go up to be 0.4. What should be her estimate of the probability that stock *B* goes up, given that stock *A* does go up?

16. The Florida Tourist Commission estimates that a person visiting Florida will visit Disney World, Busch Gardens, or both with probabilities 0.5, 0.3, and 0.2, respectively. Find the probability that a person visiting Florida will visit Busch Gardens, given that the person did visit Disney World.

17. A recent survey of 400 instructors at a major university revealed the data shown in the table on page 596. Based on the data, what is the probability that:

    a. An instructor received a good evaluation, given that the instructor was tenured?

b. An instructor received a good evaluation?

	GOOD EVALUATIONS	POOR EVALUATIONS
TENURED	72	168
NONTENURED	84	76

18. Referring to the data in Problem 17, find the probability that:
    a. An instructor received a poor evaluation, given that the instructor was tenured.
    b. An instructor received a poor evaluation.

19. Billy was taking a history test, and his memory started playing tricks on him. He needed the date when Columbus reached America, and he remembered that it was 1492 or 1294 or 1249 or 1429, but was not sure which. Then he remembered that the number formed by the first three digits was not divisible by 3. What is the probability that he guessed the right date? [*Hint:* Recall that a number is divisible by 3 if the sum of its digits is divisible by 3. Use this information to find which dates this leaves Billy to choose from.]

20. Nancy was asked to guess at a preselected number between 1 and 50 (inclusive). By asking questions first, Nancy learned that the number was divisible by 2 and/or by 3. What is the probability that Nancy guessed the right number after correctly using her information? [*Hint:* Eliminate the numbers that are not divisible by 2 or by 3. This eliminates all the odd numbers that are not multiples of 3.]

USING YOUR KNOWLEDGE 13.4

The Statistical Abstracts of the United States gives the number of crime victims per 1000 persons, 12 years old and over, as follows:

SEX	ROBBERY	ASSAULT	PERSONAL LARCENY
Male	5	18	52
Female	2	9	42

Use the information in this table to do the following problems:

1. **a.** Find the probability that the victim of one of the three types of crime was a male.
   b. Find the probability that the victim of one of the three types of crime was a female.
   c. Considering your answers to parts a and b, which sex would you say is more likely to be the victim of one of these three types of crime?

2. If it is known that an assault was committed:

a. What is the probability that the victim was a male?

b. What is the probability that the victim was a female?

3. If it is known that the victim was a female, what is the probability that the crime was assault?

4. If it is known that the victim was a male, what is the probability that the crime was robbery?

13.5

INDEPENDENT EVENTS

One of the more important concepts in probability is that of *independence*. In this section we shall define what we mean when we say that two events are independent. For example, the probability of obtaining a sum of 7 when two dice are thrown *and* it is known that the first die shows a 6 is $\frac{1}{6}$, that is, $P(S|6) = \frac{1}{6}$. It is of interest that the probability of obtaining a 7 when two dice are thrown is also $\frac{1}{6}$, so $P(S|6) = P(S)$. This means that the additional information that a 6 came up on the first die does not affect the probability of the sum being 7. It can happen, in general, that the probability of an event A is not affected by the occurrence or nonoccurrence of a second event B. Hence, we state the following definition:

Definition 13.5a

> Two events A and B are said to be **independent** if and only if
>
> $$P(A|B) = P(A) \qquad (1)$$

If A and B are independent, we may substitute $P(A)$ for $P(A|B)$ in the equation

$$P(A|B) = \frac{P(A \cap B)}{P(B)} \quad \text{See equation (1), Section 13.4.}$$

to obtain

$$P(A) = \frac{P(A \cap B)}{P(B)}$$

Then, multiplying by $P(B)$, we get

$$P(A \cap B) = P(A) \cdot P(B)$$

Consequently, we see that an equivalent definition of independence is as follows:

Definition 13.5b

> Two events A and B are **independent** if and only if
>
> $$P(A \cap B) = P(A) \cdot P(B) \qquad (2)$$

A. Independent Events

The preceding ideas can be applied to experiments involving more than two events. We define **independent events** to be such that the occurrence or nonoccurrence of any one of these events does not affect the probability of any other. The most important result for applications is that if n events $E_1, E_2, \ldots, E_n$ are known to be independent, then the following multiplication rule holds:

$$P(E_1 \cap E_2 \cap E_3 \cap \cdots \cap E_n) = P(E_1) \cdot P(E_2) \cdot P(E_3) \cdot \cdots \cdot P(E_n) \qquad (3)$$

The next examples illustrate these ideas.

EXAMPLE 1   Two coins are tossed. Let E_1 be the event in which the first coin comes up tails, and let E_2 be the event in which the second coin comes up heads. Are E_1 and E_2 independent?

Solution   Because $P(E_1 \cap E_2) = \frac{1}{4}$, $P(E_1) = \frac{1}{2}$, $P(E_2) = \frac{1}{2}$, and $\frac{1}{2} \cdot \frac{1}{2} = \frac{1}{4}$, we see that $P(E_1 \cap E_2) = P(E_1) \cdot P(E_2)$. Hence, E_1 and E_2 are independent. ◀

EXAMPLE 2   We have two urns, I and II. Urn I contains 2 red and 3 black balls, while urn II contains 3 red and 2 black balls. A ball is drawn at random from each urn. What is the probability that both balls are black?

Solution   Let $P(B_1)$ be the probability of drawing a black ball from urn I, and let $P(B_2)$ be the probability of drawing a black ball from urn II. Clearly, B_1 and B_2 are independent events. Thus, $P(B_1) = \frac{3}{5}$ and $P(B_2) = \frac{2}{5}$, so $P(B_1 \cap B_2) = \frac{3}{5} \cdot \frac{2}{5} = \frac{6}{25}$. ◀

EXAMPLE 3   Bob is taking math, Spanish, and English. He estimates that his probabilities of receiving an A in these courses are $\frac{1}{10}$, $\frac{3}{10}$, and $\frac{7}{10}$, respectively. If he assumes that the grades can be regarded as independent events, find the probability that Bob makes:

a. All A's (event A)   **b.** No A's (event N)
c. Exactly two A's (event T)

Solution   **a.** $P(A) = P(M) \cdot P(S) \cdot P(E) = \frac{1}{10} \cdot \frac{3}{10} \cdot \frac{7}{10} = \frac{21}{1000}$, where M is the event in which he makes an A in math, S is the event in which he makes an A in Spanish, and E is the event in which he makes an A in English. (See the tree diagram.)

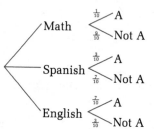

b. $P(N) = P(M') \cdot P(S') \cdot P(E') = \frac{9}{10} \cdot \frac{7}{10} \cdot \frac{3}{10} = \frac{189}{1000}$

c. There are three ways of getting exactly two A's:

   1. Getting A's in math and Spanish and not in English: The probability of this event is

$$P(M) \cdot P(S) \cdot P(E') = \frac{1}{10} \cdot \frac{3}{10} \cdot \frac{3}{10} = \frac{9}{1000}$$

   2. Getting A's in math and English and not in Spanish: The probability of this event is

$$P(M) \cdot P(S') \cdot P(E) = \tfrac{1}{10} \cdot \tfrac{7}{10} \cdot \tfrac{7}{10} = \tfrac{49}{1000}$$

3. Getting A's in Spanish and English and not in math: The probability of this event is

$$P(M') \cdot P(S) \cdot P(E) = \tfrac{9}{10} \cdot \tfrac{3}{10} \cdot \tfrac{7}{10} = \tfrac{189}{1000}$$

Since the three events we have just considered are mutually exclusive, the probability of getting exactly two A's is the sum of the probabilities we calculated. Thus,

$$P(T) = \tfrac{9}{1000} + \tfrac{49}{1000} + \tfrac{189}{1000} = \tfrac{247}{1000} \qquad \blacktriangleleft$$

EXAMPLE 4   Do you recall our mentioning the "witches of Wall Street"? (See Example 1 of Section 13.2.) The "witches" are persons who claim that they use occult powers to predict the behavior of stocks on the stock market. One of the most famous of the "witches" claims to have a 70% accuracy record. A stock broker selects 3 stocks at random from the New York Stock Exchange listing and asks this "witch" to predict their behavior. Assuming that the 70% accuracy claim is valid, find the probability that the "witch" will:

a. Correctly predict the behavior of all 3 stocks
b. Incorrectly predict the behavior of all 3 stocks
c. Correctly predict the behavior of exactly 2 of the 3 stocks

Solution   **a.** The probability of correctly predicting the behavior of all 3 stocks is the product

$$(0.70)(0.70)(0.70) = 0.343$$

b. The probability of incorrectly predicting the behavior of all 3 stocks is the product

$$(0.30)(0.30)(0.30) = 0.027$$

c. The probability of correctly predicting the behavior of 2 specific stocks and incorrectly predicting the behavior of the third stock is the product

$$(0.70)(0.70)(0.30) = 0.147$$

Because there are 3 ways of selecting the 2 specific stocks, we use the SCP and multiply the last result by 3. Thus, the probability of correctly predicting the behavior of exactly 2 of the stocks is

$$(3)(0.147) = 0.441 \qquad \blacktriangleleft$$

You can visualize the calculations in Example 4c by looking at the tree diagram in Fig. 13.5a (page 600), where C represents a correct prediction and I represents an incorrect prediction. Each branch is labeled with the probability of the event it represents. Note that if you find and add the probabilities at the ends of all the branches, the sum will be 1.

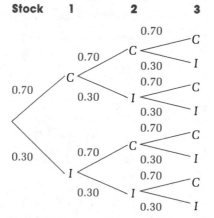

Stock   1   2   3

$P(CCI) = (0.70)(0.70)(0.30) = 0.147$

$P(CIC) = (0.70)(0.30)(0.70) = 0.147$

$P(ICC) = (0.30)(0.70)(0.70) = 0.147$

Figure 13.5a

B. Stochastic Processes

A **stochastic process** is a sequence of experiments in which the outcome of each experiment depends on chance. For example, the repeated tossing of a coin or of a die is a stochastic process. Tossing a coin and then rolling a die would also be an example of a stochastic process.

In the case of repeated tosses of a coin, we assume that on each toss there are two possible outcomes, each with probability $\frac{1}{2}$. If the coin is tossed twice, we can construct a tree diagram corresponding to this sequence of experiments (see Fig. 13.5b).

Figure 13.5b

First Toss	Second Toss	Final Outcome	Probability
H	H	(H, H)	$\frac{1}{2} \times \frac{1}{2} = \frac{1}{4}$
	T	(H, T)	$\frac{1}{2} \times \frac{1}{2} = \frac{1}{4}$
T	H	(T, H)	$\frac{1}{2} \times \frac{1}{2} = \frac{1}{4}$
	T	(T, T)	$\frac{1}{2} \times \frac{1}{2} = \frac{1}{4}$

As you can see from Fig. 13.5b, we have put on each branch the probability of the event corresponding to that branch. To obtain the probability of, say, a tail and then a head, (T, H), we multiply the probabilities on each of the branches going along the path that leads from the start to the final outcome, as indicated in the figure. This multiplication gives $(\frac{1}{2})(\frac{1}{2}) = \frac{1}{4}$, in agreement with the results we have previously obtained.

It is possible to show, by means of the sequential counting principle, that the terminal probabilities are always correctly obtained by using this multiplication technique.

EXAMPLE 5   A coin and a die are tossed. What is the probability of getting a head and a 5?

Since the outcomes depend only on chance, this is a stochastic process for which the multiplication procedure can be used. Since the probability of getting a head on the coin is $\frac{1}{2}$, and the probability of getting a 5 on the die is $\frac{1}{6}$, the probability of getting a head and a 5 is

$$P(H, 5) = \tfrac{1}{2} \times \tfrac{1}{6} = \tfrac{1}{12}$$ ◀

EXAMPLE 6 Jim has two coins, one fair (F) and the other unbalanced (U) so that the probability of its coming up heads is $\frac{2}{3}$. He picks up one of the coins, tosses it and it comes up heads. What is the probability that he picked up the unbalanced coin?

Solution We draw a tree diagram as shown. The probabilities at the ends of the branches are obtained by the multiplication technique. The starred probabilities may be taken as weights for the corresponding events. Thus, the required probability is

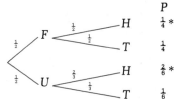

$$\frac{\frac{2}{6}}{\frac{1}{4} + \frac{2}{6}} = \frac{4}{7}$$

Note that

$$P(U|H) = \frac{P(U \cap H)}{P(U)} = \frac{\frac{2}{6}}{\frac{1}{4} + \frac{2}{6}} = \frac{4}{7}$$ ◀

EXAMPLE 7 Referring to Table 13.3a (page 585), find the probability that 2 men who are 50 and 55 years old, respectively, will both be alive at age 70.

Solution The probability that a 50-year-old man lives to 70 is $\frac{386}{698} = \frac{193}{349}$, and the probability that a 55-year-old man lives to 70 is $\frac{386}{646} = \frac{193}{323}$. Thus, the probability that both men live to 70 is

$$\frac{193}{349} \times \frac{193}{323} = \frac{37,249}{112,727} \approx 0.33$$ ◀

EXERCISE 13.5

A. 1. Two coins are tossed. Let E_1 be the event in which the first coin comes up heads, and let E_2 be the event in which the second coin comes up tails. Are E_1 and E_2 independent?

2. A bubble gum machine has 50 cherry-flavored gums, 20 grape-flavored gums, and 30 licorice-flavored gums; while a second machine has 40 cherry, 50 grape, and 10 licorice. A gum is drawn at random from each machine. Find the probability that:

   a. Both gums are cherry-flavored.
   b. Both gums are licorice-flavored.
   c. The gum from the first machine is cherry-flavored, and the one from the second machine is grape-flavored.

3. A radio repair shop has estimated the probability that a radio sent to their shop has a bad tube is $\frac{1}{4}$, the probability that the radio has a bad rectifier is $\frac{1}{8}$, and the probability that it has a bad capacitor is $\frac{1}{3}$. If we assume that tubes, rectifiers, and capacitors are independent, find the probability that:

  a. A tube, a capacitor, and a rectifier are bad in a radio sent to the shop

  b. A tube and a rectifier are bad in a radio sent to the shop

  c. None of the three parts (tubes, capacitors, and rectifiers) is bad

4. The table in the margin gives the kinds of stock available in three brokerage houses, H_1, H_2, and H_3. A brokerage house is selected at random, and one type of stock is selected. Find the probability that the stock is:

  a. A petroleum stock **b.** A computer stock

	PETROLEUM	COMPUTERS
H_1	3	2
H_2	2	3
H_3	2	2

5. A coin is tossed 3 times. Find the probability of obtaining:

  a. Heads on the first and last toss, and tails on the second toss

  b. At least 2 heads

  c. At most 2 heads

6. A die is rolled 3 times. Find the probability of obtaining:

  a. An odd number each time

  b. Two odd numbers first and an even one on the last roll

  c. At least two odd numbers

7. A card is drawn from an ordinary deck of 52 cards, and the result is recorded on paper. The card is then returned to the deck and another card is drawn and recorded. Find the probability that:

  a. The first card is a spade.

  b. The second card is a spade.

  c. Both cards are spades.

  d. Neither card is a spade.

8. Rework Problem 7, assuming that the 2 cards are drawn in succession and without replacement. [*Hint:* Make a tree diagram and assign probabilities to each of the branches.]

9. A family has 3 children. Let M be the event, "the family has at most 1 girl," and let B be the event, "the family has children of both sexes." Find:

  a. $P(M)$ **b.** $P(B)$ **c.** $P(B \cap M)$

  d. Determine whether B and M are independent.

10. Two cards are drawn in succession and without replacement from an ordinary deck of 52 cards. What is the probability that:

  a. The first card is a king and the second card is an ace?

  b. Both cards are aces? **c.** Neither card is an ace?

  d. Exactly 1 card is an ace?

11. A company has estimated that the probabilities of success for 3 products introduced in the market are $\frac{1}{4}$, $\frac{2}{3}$, and $\frac{1}{2}$, respectively. Assuming independence, find:

  a. The probability that the 3 products are successful

b. The probability that none of the products is successful

12. In Problem 11, find the probability that exactly 1 product is successful.

13. A coin is tossed. If heads come up, a die is rolled; but if tails come up, the coin is thrown again. Find the probability of obtaining:

    a. 2 tails   **b.** Heads and the number 6

    c. Heads and an even number

14. In a survey of 100 persons, the data in the table in the margin were obtained.

    a. Are S and L independent?

    b. Are S' and L' independent?

	LUNG CANCER (L)	NO LUNG CANCER (L')
SMOKER (S)	42	28
NONSMOKER (S')	18	12

15. Referring to Table 13.3a (page 585), find the probability that two persons, one 30 years old and the other 40 years old, will live to be 60.

16. In Problem 15, find the probability that both persons will live to be 70.

17. The *Apollo* module has five components: the main engine, the propulsion system, the command service module, the lunar excursion module (LEM), and the LEM engine. If each of the systems is considered independent of the others and the probability that each of the systems performs satisfactorily is 0.90, what is the probability that all the systems will perform satisfactorily?

18. A die is loaded so that 1, 2, 3, and 4 each have probability $\frac{1}{8}$ of coming up while 5 and 6 each have probability $\frac{1}{4}$ of coming up. Consider the events $A = \{1, 3, 5\}$ and $B = \{2, 4, 5\}$. Determine whether A and B are independent.

19. On one of the experimental flights of the space shuttle *Columbia*, the mission was cut short due to a malfunction of a battery aboard the ship. The batteries in the *Columbia* are guaranteed to have a failure rate of only 1 in 20. The system of 3 batteries is designed to operate as long as any one of the batteries functions properly. Find the probability that:

    a. All 3 batteries fail.

    b. Exactly 2 fail.

20. In a certain city, the probability of catching a burglar is 0.30, and the probability of convicting a caught burglar is 0.60. Find the probability that a burglar will be caught and convicted.

21. In Example 4, what is the probability of the "witch" predicting correctly the behavior of 1 of the stocks and incorrectly predicting the behavior of the other 2?

22. In Example 4, suppose the broker had selected 4 stocks. What is the probability that the "witch" would give a correct prediction for 2 of the stocks and an incorrect prediction for the other 2 stocks?

B. **23.** Three boxes, labeled A, B, and C, contain 1 red and 2 black balls, 2 red and 1 black ball, and 1 red and 1 black ball, respectively. First a box is selected at random, and then a ball is drawn at random from that box. Find the probability that the ball is red. [*Hint:* Draw a tree diagram, assign the probabilities to the separate branches, and compute the terminal probabilities by using the multiplication technique. Then add the terminal probabilities for all the outcomes in which the ball is red.]

24. There are 3 filing cabinets, each with 2 drawers. All the drawers contain letters. In one cabinet, both drawers contain airmail letters; in a second cabinet, both drawers contain ordinary letters; and in the third cabinet, one drawer contains airmail and the other contains ordinary letters. A cabinet is selected at random, and then a drawer is picked at random from this cabinet. When the drawer is opened, it is found to contain airmail letters. What is the probability that the other drawer of this cabinet also contains airmail letters? [*Hint:* Use the same procedure as in Problem 23.]

25. John has 2 coins, one fair and the other unbalanced so that the probability of its coming up heads is $\frac{3}{4}$. He picks one of the coins at random, tosses it, and it comes up heads. What is the probability that he picked the unbalanced coin?

26. A box contains 3 green balls and 2 yellow balls. Two balls are drawn at random in succession and without replacement. If the second ball is yellow, what is the probability that the first one is green?

USING YOUR KNOWLEDGE 13.5

*Suppose a fair coin is flipped 10 times in succession. What is the probability that exactly 4 of the flips will turn up heads? This is a problem in which repeated trials of the same experiment are made, and the probability of success is the same for each of the trials. This type of procedure is often called a **Bernoulli trial**, and the final probability is known as a **binomial probability**.*

Let us see if we can discover how to calculate such a probability. We represent the 10 flips and one possible success like this:

H	T	T	H	H	T	T	T	H	T
1	2	3	4	5	6	7	8	9	10

Because each flip is independent of the others, the probability of getting the particular sequence shown is $(\frac{1}{2})^{10}$. All we need do now is find in how many ways we can succeed, that is, in how many ways we can get exactly 4 heads. But this is the same as the number of ways we can select 4 of the 10 flips — that is, $C(10, 4)$. The successful ways of getting 4 heads are all mutually exclusive, so the probability of getting exactly 4 heads is

$$\frac{C(10, 4)}{2^{10}}$$

Let us suppose now that the coin is biased so that the probability of heads on any one toss is p and the probability of tails is $q = 1 - p$. The probability of getting the arrangement we have shown is now $p^4 q^6$. (Why?) Hence, the probability of getting exactly 4 heads is

$$C(10, 4)p^4 q^6$$

You should be able to convince yourself that if n is the number of trials, p is the probability of success in each trial, and $q = 1 - p$ is the probability of failure, then the probability of exactly x successes is

$$C(n, x)p^x q^{n-x}$$

1. Suppose that a fair coin is tossed 50 times in succession. What is the probability of getting exactly 25 heads? (Do not multiply out your answer.)
2. If a fair coin is tossed 6 times in succession, what is the probability of getting at least 3 heads?
3. Suppose that the coin in Problem 2 is biased 2 to 1 in favor of heads. Can you calculate the probability of getting at least 3 heads?
4. Suppose that a fair coin is tossed an even number of times, 2, 4, 6, ... What happens to the probability of getting heads in exactly half the tosses as the number of tosses increases?
5. An honest die is tossed 5 times in succession. What is the probability of getting exactly two 3's?

DISCOVERY 13.5

Figure 13.5c

Suppose you have two switches, S_1 and S_2, installed in series in an electrical circuit, and these switches have probabilities $P(S_1) = \frac{9}{10}$ and $P(S_2) = \frac{8}{10}$ of working. As you can see from Fig. 13.5c, the probability that the circuit works is the probability that S_1 and S_2 work, that is,

$$P(S_1) \cdot P(S_2) = \frac{9}{10} \times \frac{8}{10} = \frac{72}{100}$$

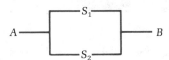

Figure 13.5d

If the same two switches are installed in parallel (see Fig. 13.5d), then we can calculate the probability that the circuit works by first calculating the probability that it does not work:

The probability that S_1 fails is $1 - \frac{9}{10} = \frac{1}{10}$.
The probability that S_2 fails is $1 - \frac{8}{10} = \frac{2}{10}$.
The probability that both S_1 and S_2 fail is $\frac{1}{10} \times \frac{2}{10} = \frac{2}{100}$.

Thus, the probability that the circuit works is $1 - \frac{2}{100} = \frac{98}{100}$. By comparing the probability that a series circuit works ($\frac{72}{100}$) with the probability that a parallel circuit works ($\frac{98}{100}$), we can see that it is better to install switches in parallel.

1. What is the probability that a series circuit with three switches, S_1, S_2, and S_3, with probabilities $\frac{1}{3}$, $\frac{1}{2}$, and $\frac{3}{4}$ of working will work?
2. What is this probability if the switches are installed in parallel?

We have just seen that, under certain circumstances, it is better to install parallel rather than series circuits to obtain maximum reliability. However, in the case of security systems, independent components in series are the most reliable. For example, the soldier in the picture is guarding a triple-threat security system that uses voice patterns, fingerprints, and handwriting to screen persons entering a maximum security area. Here is how the system operates: To enter a secure area, a person must pass through a room that has a door at each end and contains three small booths. In the first booth, the person punches in his or her four-digit identification number. This causes the machine inside the booth to intone four words, which the person must repeat. If the voice pattern matches the pattern that goes with the identification number, the machine says, "Thank you," and the person goes to the next booth. After entering his or her number there, the person signs his or her name on a Mylar sheet. If the signature is acceptable, the machine flashes a green light and the person goes to the third booth. There, he or she punches in the identification number once more, and then pokes a finger into a slot, fingerprint down. If a yellow light flashes, "IDENTITY VERIFIED," the door opens and the person can enter the high-security area.

A Pease Air Force Base guard and the triple-threat security system: Who goes there?

Courtesy of Ira Wyman

3. If each of the machines is 98% reliable, what is the probability that a person "fools" the first machine?
4. What is the probability that a person "fools" the first two machines?
5. What is the probability that a person "fools" all three machines?
6. Based on your answer to Problem 5, how would you rate the reliability of this security system?

13.6

ODDS AND MATHEMATICAL EXPECTATION

In this chapter we have several times used games of chance to illustrate the concepts of probability. In connection with these games, one often encounters such statements as "the odds are 1 to 5 for throwing a 1 with a die" or "the odds are 12 to 1 against picking an ace from a deck of cards." When a person gives you 1 to 5 odds for throwing a 1 with a die, it usually means that if a 1 does occur, you pay $5, and that the person pays $1 in case a 1 does not occur. These statements simply compare the number of favorable outcomes to the number of unfavorable outcomes. Thus, odds of 1 to 5 mean that there are one-fifth as many favorable as unfavorable outcomes.

A. Odds

▶ **Definition 13.6a**

> If an event E is such that the total number of favorable outcomes is f and the total number of unfavorable outcomes is u, then **the odds in favor of E are f to u.**

For instance, there are 4 aces in a standard deck of 52 cards. Thus, if a single card is drawn from the deck, there are 4 ways of getting an ace and 48 ways of not getting an ace. Thus, the odds in favor of drawing an ace are 4 to 48, or 1 to 12.

EXAMPLE 1 A die is rolled. What odds should a person give:

a. In favor of 1 turning up?
b. Against a 1 turning up?

Solution **a.** In this case, $f = 1$ and $u = 5$. Thus, the odds are 1 to 5.
b. There are 5 ways in which a 1 does not turn up (favorable) and 1 way in which a 1 turns up (unfavorable), so the odds are 5 to 1. ◀

EXAMPLE 2 A horse named Camarero has a record of 73 wins and 4 losses. Based on this record, what is the probability of a win for this horse?

Solution Here, $f = 73$, $u = 4$, and the probability is

$$\frac{f}{f+u} = \frac{73}{73+4} = \frac{73}{77}$$ ◀

If n is the total number of possible outcomes, and f and u are as before, then we know that

$$P(E) = \frac{f}{n} \quad \text{and} \quad P(\text{Not } E) = \frac{u}{n}$$

Therefore,

$$\frac{P(E)}{P(\text{Not } E)} = \frac{f/n}{u/n} = \frac{f}{u}$$

Thus, an equivalent definition of odds in favor of the event E is

$P(E)$ to $P(\text{Not } E)$

or, since $P(\text{Not } E) = 1 - P(E)$,

$P(E)$ to $1 - P(E)$

In the case of the die in Example 1a, the odds are $\frac{1}{6}$ to $\frac{5}{6}$, which is the same as 1 to 5. Note that if $P(E)$ and $P(\text{Not } E)$ are expressed as fractions with the same denominator, you can just compare the numerators. For example, if $P(E) = \frac{2}{7}$, then $P(\text{Not } E) = 1 - \frac{2}{7} = \frac{5}{7}$, so the odds in favor of E are 2 to 5.

EXAMPLE 3   A horse named Blue Bonnet has won 5 of her last 8 races and is thus assigned a probability of $\frac{5}{8}$ of winning her 9th race. Assuming this probability is correct, what are the odds *against* Blue Bonnet winning that race?

Solution   Since $P(\text{Winning}) = \frac{5}{8}$, we know that $P(\text{Not winning}) = 1 - \frac{5}{8} = \frac{3}{8}$. Thus, the odds in favor of Blue Bonnet are 5 to 3, and the odds *against* her are 3 to 5. ◄

EXAMPLE 4   The Florida Lottery prints 250 million tickets for each game. The prizes and the number of "instant" winning tickets are shown in the first two columns of Table 13.6a. Entry tickets are to be sent to the Lottery Department for drawings in which 14 "big prizes" are awarded (see columns 3 and 4 in the table).

Table 13.6a

PRIZE	NUMBER OF WINNERS	PRIZE	NUMBER OF WINNERS
Entry	1,000,000	$10,000	4 per game
Free ticket	25,000,000	$15,000	4 per game
$2	25,000,000	$25,000	2 per game
$5	5,000,000	$50,000	2 per game
$25	416,666	$1,000,000	2 per game
$50	208,333		
$5000	3,125		

If you buy a single Florida Lottery ticket, what is the probability that you win $50? What are the odds in favor of this?

Solution   According to Table 13.6a, of the 250 million tickets printed, 208,333 win $50. Therefore, the probability of your winning $50 with 1 ticket is

$$\frac{208,333}{250,000,000} = \frac{1}{1200}$$

Since $1 - \frac{1}{1200} = \frac{1199}{1200}$, the odds in favor of your winning $50 are 1 to 1199. ◄

B.   Expected Value

In many games of chance, we are concerned with betting. Suppose that a given event E has probability $P(E) = f/n$ of occurring and probability $P(\text{Not } E) = u/n$ of not occurring. If we agree to pay f dollars if E does not occur in exchange for receiving u dollars if E does occur, we can calculate our "expected average winnings" by multiplying $P(E)$ (the approximate proportion of the times we win) by u (the amount we win each time). Similarly, our losses will be $P(\text{Not } E) \times f$, because we lose f dollars approximately $P(\text{Not } E)$ of the times. If the bet is to be fair, the average net winnings should be 0. Let us see if this is the case. Our net winnings will be

$$P(E) \times u - P(\text{Not } E) \times f = \frac{f}{n} \times u - \frac{u}{n} \times f$$

$$= \frac{fu - uf}{n} = 0$$

as they should be. Since the odds in favor of E are f to u, we state the following definition:

▶ **Definition 13.6b**

> **Fair Bet**
>
> If the probability that event E will occur is f/n and the probability that E will not occur is u/n, where n is the total number of possible outcomes, then odds of f to u in favor of E occurring constitute a **fair bet.**

EXAMPLE 5 A woman bets that she can throw a 7 in one throw of a pair of dice. What odds should she give for the bet to be fair?

Solution $P(7) = \frac{6}{36} = \frac{1}{6}$. Here, $1 = f$ and $f + u = 6$, so $u = 5$. Hence, the odds are 1 to 5. ◀

Sometimes we wish to compute the *expected value*, or *mathematical expectation*, of a game. For example, if a woman wins $6 when she obtains a 1 in a single throw of a die and loses $12 for any other number, our intuition tells us that if she plays the game many times, she will win $6 one-sixth of the time and she will lose $12 five-sixths of the time. We might then expect her to gain $(\$6)(\frac{1}{6}) - (\$12)(\frac{5}{6}) = -\$9$; that is, to lose $9 per try on the average.

For another example, if a fair die is thrown 600 times, we would expect $(\frac{1}{6})(600) = 100$ ones to appear. This does not mean that exactly 100 ones *will appear*, but that this is the *expected* average number of ones for this experiment. In fact, if the number of ones were very far away from 100, we would have good reason to doubt the honesty of the die.

▶ **Definition 13.6c**

> **Expected Value**
>
> If the k possible outcomes of an experiment are assigned the values a_1, a_2, . . . , a_k and they occur with probabilities p_1, p_2, . . . , p_k, respectively, then the **expected value** of the experiment is given by
>
> $$E = a_1 p_1 + a_2 p_2 + a_3 p_3 + \cdot \cdot \cdot + a_k p_k$$

EXAMPLE 6 A die is thrown. If an even number comes up, a person receives $10; otherwise, the person loses $20. Find the expected value of this game.

Solution We let $a_1 = \$10$ and $a_2 = -\$20$. Now, $p_1 = \frac{3}{6} = \frac{1}{2}$ and $p_2 = \frac{1}{2}$, so $E = (\$10)(\frac{1}{2}) - (\$20)(\frac{1}{2}) = -\$5$. ◀

In Example 5, the player is expected to lose $5 per game in the long run; so we say that this game is not fair.

<table>
<tr><td>▶</td><td>**Definition 13.6d**</td><td>A game is **fair** if its expected value is 0.</td></tr>
</table>

EXAMPLE 7   A coin is thrown. If heads come up, we win $1; if tails come up, we lose $1. Is this a fair game?

Solution   Here, $a_1 = \$1$, $a_2 = -\$1$, and $p_1 = p_2 = \frac{1}{2}$, so $E = (1)(\frac{1}{2}) - 1(\frac{1}{2}) = 0$. Thus, by Definition 13.6d, the game is fair.   ◀

EXAMPLE 8   A die is thrown. A person receives double the number of dollars corresponding to the dots on the face that turns up. How much should a player pay for playing in order to make this a fair game?

Solution   The player can win $2, $4, $6, $8, $10, $12, each with probability $\frac{1}{6}$, so expected winnings (the player does not lose) are $E = 2(\frac{1}{6}) + 4(\frac{1}{6}) + 6(\frac{1}{6}) + 8(\frac{1}{6}) + 10(\frac{1}{6}) + 12(\frac{1}{6}) = \frac{42}{6} = \$7$. A person paying $7 can expect winnings of 0. Thus, $7 is a fair price to pay for playing this game.   ◀

EXAMPLE 9   Dear's Department Store wishes to open a new store in one of two locations. It is estimated that if the first location is chosen, the store will make a profit of $100,000 per year if successful and will lose $50,000 per year otherwise. For the second location, it is estimated that the annual profit will be $150,000 if successful; otherwise, the annual loss will be $80,000. If the probability of success at each location is $\frac{1}{2}$, which location should be chosen in order to maximize the expected profit?

Solution   For the first location, $a_1 = \$100,000$, $p_1 = \frac{1}{2}$, $a_2 = -\$50,000$, and $p_2 = \frac{1}{2}$. Thus, the expected profit is

$$E_1 = \$100,000(\tfrac{1}{2}) - \$50,000(\tfrac{1}{2}) = \$50,000 - \$25,000 = \$25,000$$

For the second location, $a_1 = \$150,000$, $p_1 = \frac{1}{2}$, $a_2 = -\$80,000$, and $p_2 = \frac{1}{2}$. Thus, the expected profit is

$$E_2 = \$150,000(\tfrac{1}{2}) - \$80,000(\tfrac{1}{2}) = \$75,000 - \$40,000 = \$35,000$$

The expected profit from the second location ($35,000) is greater than that for the first location ($25,000), so the second location should be chosen.   ◀

Decision problems that depend on mathematical expectation require three things for their solution: *options, values,* and *probabilities.* In Example 9 we have the information given in Table 13.6b. With this information we can find the expected value for each option and hence make the desired decision.

Table 13.6b

	OPTIONS			
	Site 1		**Site 2**	
VALUES	$100,000	−$50,000	$150,000	−$80,000
PROBABILITIES	$\frac{1}{2}$	$\frac{1}{2}$	$\frac{1}{2}$	$\frac{1}{2}$

EXERCISE 13.6

A. In Problems 1–7, find the odds in favor of obtaining:

1. A 2 in one throw of a single die

2. An even number in one throw of a single die

3. An ace when drawing 1 card from an ordinary deck of 52 cards

4. A red card when drawing 1 card from an ordinary deck of 52 cards

5. 2 tails when an ordinary coin is thrown twice

6. At least 1 tail when an ordinary coin is thrown twice

7. A vowel when 1 letter is chosen at random from among the 26 letters of the English alphabet

In Problems 8–12, find the odds against obtaining:

8. A 4 in one throw of a single die

9. An odd number in one throw of a single die

10. The king of spades when drawing 1 card from an ordinary deck of 52 cards

11. One of the picture cards (jack, queen, king) when drawing 1 card from an ordinary deck of 52 cards

12. At most 1 tail when an ordinary coin is thrown twice

13. If you buy 1 Florida Lottery ticket, what is the probability of your winning $5, according to Table 13.6a? What are the odds in favor of your winning $5?

14. If you buy 1 Florida Lottery ticket, what is the probability of your winning $5000, according to Table 13.6a? What are the odds in favor of your winning $5000?

15. If the correct odds in favor of Johnny winning a race are 3 to 2, what is the probability that Johnny wins?

B. **16.** A coin is thrown twice. If heads come up either time, we get $2; but if heads do not occur, we lose $4. What is the expected value of this game?

17. Two dice are thrown. If the sum of the dots showing is even, we get $10; otherwise, we lose $20. What is the expected value of this game?

18. A die is thrown. A person receives the number of dollars corresponding to the dots on the face that turns up. How much should a player pay in order to make this game fair?

19. In a recent charity raffle, there were 10,000 tickets in all. If the grand prize was a Lincoln Continental (priced at $21,500), what is a fair price to pay for a ticket?

20. If in Problem 19 the charity paid $10,000 for the Lincoln and they wished to make a profit of $10,000 from the raffle, for how much should each ticket sell?

21. A man offers to bet $3 against $5 that he can roll a 7 on one throw of his pair of dice. If he wins fairly consistently, are his dice fair? Explain.

22. Louie gets an "Entry" ticket in the Florida Lottery and offers to sell it to you for $10. Refer to Table 13.6a and determine whether this is a fair price. Explain.

23. If in Example 9 of this section the probabilities of success in the first and second locations are $\frac{2}{3}$ and $\frac{2}{5}$, respectively, what location should be chosen in order to have a maximum expected profit?

24. Gadget Manufacturing Company is debating whether to continue or discontinue an advertising campaign for a new product. Their research department has predicted the gain or loss to be derived from the decision to continue or discontinue the campaign, as summarized in the table below. If the president of the firm assigns odds of 4 to 1 in favor of the success of the advertising campaign, find:

   a. The expected value for the company if the advertising campaign is continued

   b. The expected value for the company if the advertising campaign is discontinued

   c. The best decision based on the answers to parts a and b

ADVERTISING CAMPAIGN	SUCCESSFUL	UNSUCCESSFUL
Continue	$20,000	−$10,000
Discontinue	$30,000	5,000

25. Repeat Problem 24 if the president of the firm assigns odds of 4 to 1 against the success of the advertising campaign.

USING YOUR KNOWLEDGE 13.6

The graph shows the probabilities that a baseball team that is ahead by 1, 2, 3, 4, 5, or 6 runs after a certain number of innings goes on to win the game. As you can see, if a team is leading by 1 run at the end of the first inning, the probability that this team wins is about 0.62. If a team is ahead by 2 runs at the end of the first inning, then the probability of this team's winning the game is about 0.72.

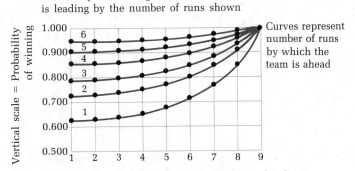

Probability of winning when team
is leading by the number of runs shown

Curves represent
number of runs
by which the
team is ahead

Vertical scale = Probability of winning

Horizontal scale = Number of innings completed

Suppose a team is ahead by 1 run at the end of the first inning. The probability that this team wins the game is

$$0.62 = \frac{62}{100} = \frac{31}{50}$$

Therefore, the odds in favor of this team winning the game should be 31 to 19.

Use the graph shown above to solve the following problems.

1. Find the probability that a team leading by 2 runs at the end of the sixth inning:
   a. Wins the game   **b.** Loses the game

2. For the same situation as in Problem 1, find:
   a. The odds in favor of this team winning the game
   b. The odds against this team winning the game

3. At the end of the sixth inning, a team is ahead by 4 runs. A man offers to bet $10 on this team. How much money should be put up against his $10 to make a fair bet?

DISCOVERY 13.6

In American roulette, the wheel has 38 compartments, 2 of which, the 0 and 00, are colored green. The rest of the compartments are numbered from 1 through 36, and half of them are red and the other half black. The wheel is spun in one direction, and a small ivory ball is spun in the other direction. If the wheel is a fair one, all the compartments are equally likely and hence the ball has probability $\frac{1}{38}$ of landing in any particular one of them. If a player bets, say, $1 on a given number, and the ball comes to rest on that number, the player receives from the croupier 36 times his or her stake, that is, $36. In this case, the player wins $35 with probability $\frac{1}{38}$ and loses $1 with probability $\frac{37}{38}$. The expected value of this game is

$$E = \$35(\tfrac{1}{38}) - \$1(\tfrac{37}{38}) = -\tfrac{\$1}{19} = -5\tfrac{5}{19} \text{ cents}$$

This may be interpreted to mean that in the long run, for every dollar that we bet in roulette, we are expected to lose $5\tfrac{5}{19}$¢.

A second way to play roulette is to bet on red or black. Suppose a player bets $1 on red. If the ball stops on a red number (there are 18 of them), the player receives twice his or her stake, thus winning $1. If a black number comes up, the player loses his or her stake. If a 0 or 00 turns up, then the wheel is spun again until it stops on a number different from 0 and 00. If this is black, the player loses the $1, but if it is red, the player receives only the original stake (gaining nothing).

1. Can you discover what is the expected value of this game?
2. Is we place 50¢ on red and 50¢ on black, what will be the expected value of the game? [*Hint:* The answer is not 0.]

SUMMARY

Section	Item	Meaning	Example
13.1A	Experiment	An activity	Tossing a coin, drawing a card from a deck
13.1A	Sample space	The set of all possible outcomes for an experiment	The sample space for tossing a coin is $\{H, T\}$
13.1A	$P(E)$	The probability of event E, $P(E) = \dfrac{n(E)}{n(\mathcal{U})}$	When tossing a coin, the probability of tails is $P(T) = \tfrac{1}{2}$.
13.1A	$P(T')$	$1 - P(T)$	The probability of a 3 when rolling a die is $\tfrac{1}{6}$. The probability of not rolling a 3 is $\tfrac{5}{6}$.
13.3	$P(E) = 0$	E is an impossible event.	Rolling a 7 on a die
13.3	$P(E) = 1$	E is a certain event.	Rolling less than 7 on a die
13.3	$P(A \cup B)$	$P(A) + P(B) - P(A \cap B)$	
13.3	Mutually exclusive events	The two events cannot occur simultaneously, $P(A \cap B) = 0$.	Throwing a 5 and a 6 on one throw of a die
13.4	$P(A|B)$, the probability of A, given B	$\dfrac{P(A \cap B)}{P(B)}$	
13.5	Independent events	$P(A \cap B) = P(A) \cdot P(B)$	Throwing a 6 with a die and tails with a coin are independent events.
13.5A	$P(E_1 \cap E_2 \cap \cdots \cap E_n)$	$P(E_1) \cdot P(E_2) \cdots \cdots P(E_n)$, when the events are independent	

Section	Item	Meaning	Example
13.5B	Stochastic process	A sequence of experiments in which the outcome of each experiment depends on chance	
13.6A	Odds in favor	The ratio of favorable to unfavorable occurrences	The odds for a 3 when tossing a die are 1 to 5.
13.6B	Expected value	$E = a_1p_1 + a_2p_2 + \cdots + a_np_n$, where the a's are the values that occur with probability p_1, p_2, and so on	

PRACTICE TEST 13

1. A single fair die is tossed. Find:
   a. The probability of obtaining a number different from 7
   b. The probability of obtaining a number greater than 2

2. A box contains 2 red balls, marked R_1 and R_2, and 3 white balls, marked W_1, W_2, and W_3. If 2 balls are drawn in succession and without replacement from this box, find the number of elements in the sample space for this experiment. (We are interested in which balls are drawn and the order in which they are drawn.)

3. A box contains 5 balls numbered from 1 to 5. If a ball is taken at random from the box, find the probability that it is:
   a. An even-numbered ball   b. Ball number 2
   c. Not ball number 2

4. Two cards are drawn at random and without replacement from a standard deck of 52 cards. Find:
   a. The probability that both cards are red
   b. The probability that neither card is an ace

5. A card is drawn at random from a standard deck of 52 cards and then replaced. Then a second card is drawn. Find:
   a. The probability that both cards are red
   b. The probability that neither card is red

6. A fair coin is tossed 5 times. What is the probability of obtaining at least 1 head?

7. An urn contains 5 white, 3 black, and 2 red balls. Find the probability of obtaining the following in a single draw:
   a. A white or a black ball   b. A ball that is not red

8. A student estimates that the probability of his passing math or English is 0.9, the probability of his passing English is 0.8, but the probability of passing both is 0.6. What should be his estimate of the probability of passing math?

9. Three cards are drawn in succession and without replacement from a standard deck of 52 cards. What is the probability that they are all face cards (jack, queen, king)?

10. Two dice are rolled. Find the probability that the sum turning up is 11, given that the first die showed a 5.

11. Two dice are rolled. Find the probability that the sum turning up is 11, given that the second die showed an even number.

12. Two dice are rolled. Find the probability that:
    a. They show a sum of 10.
    b. The first die comes up an odd number.
    c. Are these two events independent? Explain.

13. A certain drug used to reduce hypertension (high blood pressure) produces side effects in 4% of the patients. Three patients who have taken the drug are selected at random. Find the probability that:
    a. They all had side effects.
    b. None of them had side effects.

14. Roland has to take an English course and a history course, both of which are available at 8 A.M., 9 A.M., and 3 P.M. If Roland picks a schedule at random, what is the probability that he will have English at 8 A.M. and history at 3 P.M.?

15. The probability that a cassette tape is defect-free is 0.97. If 2 tapes are selected at random, what is the probability that both are defective?

16. A card is selected at random from a deck of 52 cards. What are the odds in favor of the card being:
    a. A king?   b. Not a king?

17. The probability of an event is $\frac{3}{7}$. Find:
    a. The odds in favor of this event occurring
    b. The odds against this event occurring

18. The odds in favor of an event occurring are 3 to 7.
    a. What are the odds against this event occurring?
    b. What is the probability that the event will not occur?

19. A coin is tossed twice. If exactly 1 head comes up, we receive $5, and if 2 tails come up, we receive $5; otherwise, we get nothing. How much should we be willing to pay in order to play this game?

20. The probabilities of being an "instant winner" of $2, $5, $25, or $50 in the Florida Lottery are $\frac{1}{10}$, $\frac{1}{50}$, $\frac{1}{600}$, and $\frac{1}{1200}$, respectively. What is the mathematical expectation of being an "instant winner"?

CHAPTER 14

STATISTICS

Statistical analysis was born in London, where in 1662 John Graunt published a remarkable book, *Natural and Political Observations upon the Bills of Mortality.*

At that time, London had already reached the size of approximately 100,000 inhabitants. Overcrowding, the difficulties of obtaining even the daily necessities of life, the prevalence of disease, and the many plague years all combined to make Londoners exceedingly interested in reports of births and deaths. After the great plague in 1603, these reports, which had appeared only sporadically before, became regular weekly publications. The causes of death (weird diseases such as jawfaln, King's-Evil, planet, and tissick) were reported in the *Bills of Mortality,* published regularly starting in 1629.

After this humble beginning, many mathematicians, among them such famous ones as Laplace (1749–1827) and Gauss (1777–1855), made important contributions to the basic ideas of statistics. Furthermore, the analysis of numerical data is fundamental in so many different fields that one could make a long list of scientists in such areas as biology, geology, genetics, and evolution who contributed greatly to this study. The well-known names of Charles Darwin (1809–1882), Gregor Mendel (1822–1884), and Karl Pearson (1857–1936) would surely be included in this list.

Courtesy Harper & Row, Publishers

Let the world be our laboratory and let us gather statistics on what occurs therein.

MORRIS KLINE

FREQUENCY DISTRIBUTIONS

The word "statistics" brings to the minds of most people an image of a mass of numerical data. To a statistician, **statistics** means the analysis of these data and the deduction of logical conclusions from them. It is in this sense that the science of statistics is one of the most important branches of applied mathematics.

Today we know that statistics are used throughout industry in the manufacture of goods, in the study of wages and work conditions, in insurance and investments, and in many other ways. No one can leaf through a newspaper or a news magazine without seeing evidence of the impact of statistics on our daily lives. We are constantly exhorted to buy this and not to buy that, to read this magazine, to see that movie, to eat certain foods, and not to smoke cigarettes—all on the basis of statistical evidence that seems to show the desirability of following this advice.

A. Frequency Distributions

Let us look at a statistics problem that should interest a teacher and a class who might wonder how well they are learning a certain subject. Out of 10 possible points, the class of 25 students made the following scores:

6	5	4	0	9
2	0	8	8	1
10	6	8	5	5
8	7	9	10	9
6	5	8	4	7

This listing shows at once that there were some good scores and some poor ones; but, because the scores are not arranged in any particular order, it is difficult to conclude anything else from the list.

A **frequency distribution** is often a suitable way of organizing a list of numbers to show what patterns may be present. First, the scores from 0 through 10 are listed in order in a column (see Table 14.1a). Then, by going

Table 14.1a Frequency Distribution

SCORE	TALLY MARKS	FREQUENCY
0				2
1			1
2			1
3		0
4				2
5						4
6					3
7				2
8	ℕ	5
9					3
10				2
		25 Total

through the original list in the order in which it is given, we can make tally marks on the appropriate lines of our table. Finally, in a third column we can list the number of times that each score occurs; this number is the **frequency** of the score.

It is now easier to see that a score of 8 occurred more times than any other number. This score was made by

$$\tfrac{5}{25} = \tfrac{1}{5} = 20\% \text{ of the students}$$

Ten of the students, or 40% of the class, made scores of 8 or better. Only 6, or 24%, made scores less than 5.

If there are very many items in a set of numerical data, then it is usually necessary to shorten the frequency distribution by grouping the data into intervals. For instance, in the preceding distribution we can group the scores in intervals of 2 to obtain the listing in Table 14.1b.

Of course, some of the detailed information in the first table has been lost in the second table, but for some purposes a condensed table may furnish all the information that is required.

Table 14.1b Frequency Distribution with Grouped Data

SCORE	FREQUENCY
0–1	3
2–3	1
4–5	6
6–7	5
8–9	8
10–11	2

B. Histograms

It is also possible to present the information contained in Table 14.1a by means of a special type of graph, called a **histogram,** consisting of vertical bars with no space between bars. In the histogram of Fig. 14.1a, the units on the y axis represent the frequencies, while those on the x axis indicate the scores.

Figure 14.1a Histogram

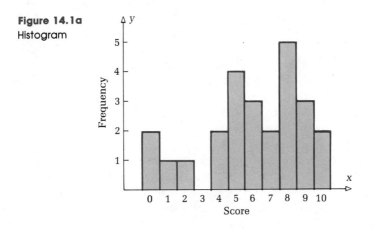

C. Frequency Polygons

From the histogram in Fig. 14.1a we can construct a **frequency polygon** (or line graph) by connecting the midpoints of the tops of the bars, as shown in

Fig. 14.1b. It is customary to extend the graph to the base line (x axis) using the midpoints of the extended intervals at both ends. This "ties the graph down," but has no predictive significance.

Figure 14.1b
Frequency polygon

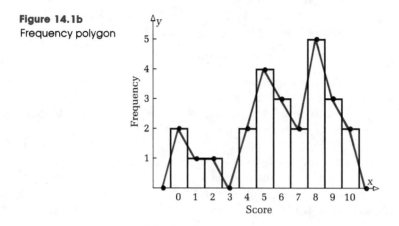

EXAMPLE 1   The following list gives the hourly wages of a group of 30 workers who are performing similar tasks, but, because of differences in seniority and skill, are paid at different rates.

$8.00	$7.90	$8.00	$8.10	$7.90	$7.90
7.90	7.80	7.90	8.00	7.80	8.00
8.10	7.70	7.90	7.80	8.10	8.00
8.00	8.10	8.20	7.80	8.20	8.10
7.70	8.00	7.80	7.70	7.80	8.00

a. Make a frequency distribution of these rates.
b. What is the most frequent rate?
c. How many workers are being paid less than $8.00/hr?
d. Make a histogram of the wage rate distribution.
e. Make a frequency polygon of the distribution.

Solution   a. The table in the margin lists the wage rates from the lowest ($7.70) to the highest ($8.20). We tally these from the given data and thus obtain the desired frequency distribution.

WAGE	TALLY MARKS	FREQUENCY
7.70	|||	3
7.80	卌|	6
7.90	卌|	6
8.00	卌|||	8
8.10	卌	5
8.20	||	2
		30 Total

b. From the frequency distribution, we read off the most frequent rate to be $8.00/hr.
c. Again, we read from the frequency distribution that 15 workers are being paid less than $8.00/hr.
d. The desired histogram appears in the figure on the next page.
e. The figure also shows the frequency polygon.

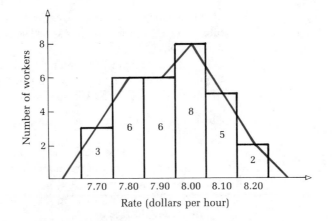

In making a frequency distribution where the data are to be grouped, we use the following procedure:

1. Decide on the classes into which the data are to be grouped. This is usually a matter of judgment and convenience, and depends on the number of items that have to be grouped. Fewer than 6 or more than 15 classes are very seldom used. If possible, we make the classes all cover equal ranges of values. If this can be done, the difference between corresponding boundaries of two adjacent classes is called the **class interval.** Classes are *never* overlapping. We always make sure that each item will go into one and only one class.
2. Sort or tally the data into the appropriate classes.
3. Count the number of items in each class.
4. Display the results in a table.
5. If desired, make a histogram and/or frequency polygon of the distribution.

The next example illustrates the procedure.

EXAMPLE 2   In a study of the voting rate in 20 cities of over 100,000 population in the United States, the following data were found:

**Turnout Rate as a Percent of the
Voting Age Population**

85.2	72.4	81.2	62.8	71.6
72.1	87.2	76.6	58.5	70.4
76.5	74.1	70.0	80.3	64.9
74.9	70.8	67.0	72.5	73.1

Inspection of the data shows that the smallest number is 58.5 and the largest is 87.2. This suggests that we go from 55 to 90, with the convenient class interval of 5 units.

a. Make a frequency distribution of the data on voting rate (r) using a class interval of 5%, so that the classes will be $55 < r \le 60, 60 < r \le 65, \ldots,$ $85 < r \le 90$.

b. Make a histogram and a frequency polygon of this distribution.

c. In what percent of the cities was the voting rate greater than 80%?

d. In what percent was the voting rate less than or equal to 70%?

Solution

a. The required frequency distribution appears in the table. (You should check this table.)

b. The histogram and frequency polygon are shown in the figure. These are constructed from the frequency distribution, just as before with ungrouped data.

c. In 4 out of 20 cities, the voting rate was greater than 80%. Thus, the required percent is $\frac{4}{20} = 20\%$.

d. In 6 out of 20 cities, the voting rate was less than or equal to 70%. Thus, in 30% of the cities, no more than 70% of the voting age population voted.

VOTING RATE r%	TALLY MARKS	FREQUENCY
$55 < r \le 60$	|	1
$60 < r \le 65$	|	1
$65 < r \le 70$	||||	4
$70 < r \le 75$	|||| |||	8
$75 < r \le 80$	||	2
$80 < r \le 85$	||	2
$85 < r \le 90$	||	2

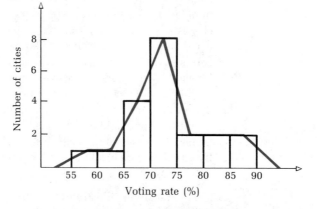

◀

EXERCISE 14.1

A.

1. Thirty students were asked to list the television programs each had watched during the preceding week. From this list, the number of hours each had spent watching televison during the week was calculated. The results are displayed in the following list:

1	5	4	7	10	8	2	3	9	6
6	12	8	14	3	4	8	7	2	1
0	3	5	8	10	12	0	15	1	4

a. Make a frequency distribution of the number of hours of television watching per student. Label the three columns "Number of hours," "Tally marks," and "Frequency."

b. What is the most frequent number of hours per student?

c. How many students watched television more than 10 hr?

d. How many students watched television 5 hr or less?

e. What percent of the students watched television more than 7 hr?

B.

WAITING TIME t min	NUMBER OF PATIENTS
0 < t ≤ 3.5	10
3.5 < t ≤ 7.0	8
7.0 < t ≤ 10.5	6
10.5 < t ≤ 14.0	16
14.0 < t ≤ 17.5	6
17.5 < t ≤ 21.0	4

2. Prepare a histogram for the data obtained in Problem 1a.

3. Have you read in the newspapers or magazines about cases in which individuals became so disgusted with the amount of time they had to wait to see a doctor or a dentist that they sued for lost wages? The waiting times for 50 patients are given in the table in the margin.

   a. Make a histogram for this set of data.

   b. What percent of the patients waited 7.0 min or less?

   c. What percent of the patients had to wait more than 10.5 min?

4. General Foods Corporation, in testing a new product, which they called Solid H, had 50 people (25 men and 25 women selected at random) taste the product and indicate their reaction on the picture ballot shown. The boxes on the ballot were then assigned scores of +3, +2, +1, 0, −1, −2, −3 in order from left to right. The table in the margin shows the frequency distribution for this taste test. (Incidentally, no significant difference was found between the men's and women's reactions.)

SCORE	FREQUENCY
−3	0
−2	1
−1	1
0	13
+1	19
+2	11
+3	5

Please check the box under the picture which expresses how you feel toward the product which you have just tasted.

   a. Make a histogram of these data.

   b. What percent of the tasters liked Solid H?

   c. What percent were undecided?

5. Would you like to be a writer? Look at the following list of authors whose books were published at the early age given:

	Age (yr)
Allen Dulles (*The Boer War: A History*)	8
Hilda Conkling (*Poems by a Little Girl*)	9
Betty Thorpe (*Fioretta*)	10
Nathalia Crane (*Janitor's Boys*)	10
David Statler (*Roaring Guns*)	9
Erlin Hogan (*The Four Funny Men*)	8
Minou Drouet (*First Poems*)	8

	Age (yr)
Dorothy Straight (*How the World Began*)	6
Kali Diana Grosvenor (*Poems by Kali*)	7
Benjamin Friedman (*The Ridiculous Book*)	9

    a. Make a frequency distribution showing the number of authors for each age.

    b. Make a histogram for the distribution in part a.

    c. What percent of the authors were less than 8 years old when they published their first books?

6. How tall are you? Here are 10 famous people and their heights:

	Height (in.)
Honoré de Balzac (French novelist)	62
Napoleon Bonaparte (French emperor)	66
Yuri Gagarin (Soviet cosmonaut)	62
Hirohito (Japanese emperor)	65
Nikita Khrushchev (Soviet leader)	63
James Madison (U.S. president)	64
Margaret Mead (U.S. anthropologist)	62
Pablo Picasso (Spanish painter)	64
Mickey Rooney (U.S. actor)	63
Tutankhamen (Egyptian king)	66

    a. Make a frequency distribution showing the number of people for each height.

    b. Make a histogram for the distribution in part a.

7. Here are 25 stocks listed on the New York Stock Exchange and their closing prices at the end of a recent year:

American Express	$22\frac{7}{8}$	Inland Steel	$30\frac{3}{8}$
American Ship	$3\frac{1}{2}$	Kellog	$52\frac{3}{8}$
Canadian Pacific	$15\frac{7}{8}$	McDermott	$14\frac{3}{4}$
Chase Manhattan	$22\frac{1}{8}$	Pan American Air	$2\frac{3}{4}$
Chrysler Motors	$22\frac{1}{8}$	Phillips Petroleum	14
Delta Airlines	$37\frac{1}{8}$	Reynolds Metal	$47\frac{5}{8}$
Exxon	$38\frac{1}{8}$	Sears	$33\frac{1}{2}$
Ford Motor Company	$75\frac{3}{8}$	Sun Company	$57\frac{3}{8}$
General Dynamics	$48\frac{3}{4}$	Texaco	$37\frac{1}{4}$
General Motors	$61\frac{3}{8}$	Transamerica	$29\frac{7}{8}$
Goodrich	$40\frac{1}{2}$	Union Pacific	54
Greyhound	$25\frac{1}{2}$	U.S. Home	3
Hewlett Packard	$58\frac{1}{4}$		

    a. Make a frequency distribution of these stocks, grouped in intervals of $10. The first two lines of your table should look like this:

PRICE	TALLY MARKS	FREQUENCY
$0 < P \leq 10$	|||	3
$10 < P \leq 20$	|||	3

   b. What is the most frequent price interval for these stocks?
   c. How many of the stocks sold for more than $40 per share?
   d. How many of the stocks sold for $30 or less per share?
   e. What percent of the stocks sold for prices between $20\frac{1}{8} and $30 per share?
   f. What percent of the stocks sold for $20 or less per share?
8. Make a histogram for the data obtained in Problem 7a.

C.

9. Do you know that certain isotopes (different forms) of the elements are used in nuclear reactors and for medical purposes such as the treatment of cancer? At the present time, about 1400 isotopes have been observed; but, of these, only 332 occur naturally. The table lists the number of elements having 1–10 naturally occurring isotopes. For instance, there are 22 elements having only 1 such isotope, but only 1 element having the maximum number, 10. Make a histogram and a frequency polygon for these data.

NUMBER OF NATURALLY OCCURRING ISOTOPES	NUMBER OF ELEMENTS	NUMBER OF NATURALLY OCCURRING ISOTOPES	NUMBER OF ELEMENTS
1	22	6	9
2	21	7	11
3	9	8	3
4	6	9	1
5	7	10	1

10. Twenty apprentices were asked to measure the diameter of a steel rod with a micrometer (an instrument that can measure to thousandths of an inch). Here are their results (in inches):

0.254	0.245	0.253	0.251
0.249	0.252	0.251	0.252
0.247	0.251	0.250	0.247
0.251	0.249	0.246	0.249
0.250	0.248	0.249	0.253

   a. Make a frequency distribution of these measurements.
   b. What single measurement has as many measurements above it as below it?
   c. What percent of the measurements are between 0.249 and 0.251 in., inclusive?
   d. What would you take as the best estimate of the diameter? Why?

11. Here is a quotation from *Robinson Crusoe*, which many of you probably have read:

> Upon the whole, here was an undoubted testimony that there was scarce any condition in the world so miserable, but was something negative or something positive, to be thankful for in it. . . .

  a. There are 151 letters in this quotation. Make a frequency distribution of the 151 letters.

  b. What letter occurs most frequently?

  c. What percent of the letters are vowels?

12. Four coins were tossed 32 times, and each time the number of heads occurring was recorded, as follows:

1	2	2	1	2	1	2	3
2	3	0	3	4	3	3	4
3	1	3	2	1	2	2	1
2	3	2	1	2	0	1	2

Label three columns "Number of heads," "Tally," and "Frequency," and prepare a frequency distribution for these data.

13. a. Make a histogram for the data in Problem 12.

  b. Now make a frequency polygon for the data of Problem 12.

14. A high school class was asked to "shoot craps" (roll dice) 3000 times. The sums of the top faces of the dice, the frequency of these sums, and the theoretical number of times the sums should have occurred appear in the table in the margin. Make a histogram for these data showing the actual frequency with a solid line and the theoretical frequency with a dotted line (perhaps of a different color).

SUM	ACTUAL FREQUENCY	THEORETICAL FREQUENCY
2	79	83
3	152	167
4	252	250
5	312	333
6	431	417
7	494	500
8	465	417
9	338	333
10	267	250
11	129	167
12	91	83

15. In a study of air pollution in a certain city, the concentration of sulfur dioxide in the air (in parts per million) was obtained for 30 days:

0.04	0.17	0.18	0.13	0.10	0.07
0.09	0.16	0.20	0.22	0.06	0.05
0.08	0.05	0.11	0.07	0.09	0.07
0.08	0.02	0.08	0.08	0.18	0.01
0.03	0.06	0.12	0.01	0.11	0.04

  a. Make a frequency distribution for these data grouped in the intervals 0.00–0.04, 0.05–0.09, 0.10–0.14, 0.15–0.19, 0.20–0.24.

  b. For what percent of the time was the concentration of sulfur dioxide more than 0.14 part per million?

16. Make a histogram and a frequency polygon for the data in Problem 15.

It was estimated in about 1940 that it would require approximately 10 years of computation to find the value of the number π (pi) to 1000 decimal places. But in the early 1960's an electronic computer calculated the value of π to more than 100,000 decimal places in less than 9 hr! Since then, several hundred thousand decimal places for π have been calculated. Here are the first 40 decimal places of π:

3.14159 26535 89793 23846 26433 83279 50288 41971

1. Make a frequency distribution of the digits after the decimal point. List the digits from 0 to 9 in your first column.
2. What are the most and the least frequently occurring digits?

Mathematicians are interested in knowing whether the digits all occur with the same frequency. This question can hardly be answered with so few decimal places. However, you should notice that only two of the digits occur with a frequency more than one unit away from what you should expect in 40 decimal places.

DISCOVERY 14.1

In this section we have shown an honest way of depicting statistical data by means of a histogram. But you can lie with statistics! Here is how. In a newspaper ad for a certain magazine, the circulation of the magazine was as shown. The bars in the diagram seem to indicate that sales in the first 9 months were tripled by the first quarter of the next year (a whopping 200% rise in sales!)

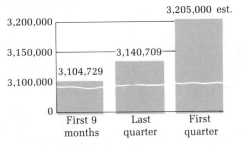

1. Can you discover what was the approximate jump in sales from the first 9 months to the first quarter of the next year?
2. Can you discover what was the approximate percent rise in sales?
3. Can you discover what is wrong with the graph?

14.2

MEASURES OF CENTRAL TENDENCY

Alberto and Barney have just gotten back their test papers. There are 9 questions and each one counted 10 points. Their scores are given in Table 14.2a on the next page.

Table 14.2a

	QUESTION									TOTAL
	1	2	3	4	5	6	7	8	9	
ALBERTO	10	7	10	7	7	10	9	10	2	72
BARNEY	10	8	10	7	7	7	10	7	7	73

Which do you think wrote a better paper? As you can see, Alberto's **average** score is $\frac{72}{9} = 8$, and Barney's **average** score is $\frac{73}{9} = 8.1$. Barney clearly has the higher average and concludes that he wrote the better paper. Do you agree?

Alberto does not agree, because he did as well as or better than Barney on 6 of the 9 questions. Alberto thinks that Barney's higher average does not tell the whole story, so he tries something else. First, he makes a frequency distribution of the two sets of scores, as shown in Table 14.2b.

Upon inspecting this list, Alberto says, "I did better than you did, Barney, because I scored 10 more often than any other number, and you scored 7 more often than any other number." Would you agree with Alberto?

The average score that we first gave is called the *mean*. It is the one that most of us think of as the average.

Table 14.2b Frequency

SCORE	ALBERTO	BARNEY
2	1	0
7	3	5
8	0	1
9	1	0
10	4	3

▶ **Definition 14.2a**

> The **mean** of a set of numbers is the sum of the numbers divided by the number of elements in the set. The mean is usually denoted by the symbol $\overline{x}$ (read, "x bar").

CROCK by Rechin, Parker, & Wilder. © 1976, Field Enterprises, Inc. Courtesy of Field Newspaper Syndicate.

Alberto used a different kind of measure that is really not an average at all. The measure he used is called the *mode*.

▶ **Definition 14.2b**

> The **mode** of a set of numbers is that number of the set that occurs most often.

If no number in the set occurs more than once, then there is no mode. However, if several numbers all occur an equal number of times and more than all the rest, then all of these several numbers are modes. Thus, it is possible for a set of numbers to have more than one mode.

The mean and the mode are useful because they give an indication of a sort of center of the set. For this reason they are called **measures of central tendency.**

EXAMPLE 1   Ten golf professionals playing a certain course scored 69, 71, 72, 68, 69, 73, 71, 70, 69, and 68. Find:

a. The mean (average) of these scores
b. The mode of these scores

Solution   **a.** $\bar{x} = \dfrac{69 + 71 + 72 + 68 + 69 + 73 + 71 + 70 + 69 + 68}{10} = \dfrac{700}{10} = 70$

b. The score that occurred most often is 69 (three times). Hence, the mode is 69.   ◀

There is a third commonly used measure of central tendency, called the *median.*

▶   **Definition 14.2c**

> The **median** of a set of numbers is the middle number when the numbers are arranged in order of magnitude. If there is no single middle number, then the median is the mean (average) of the two middle numbers.

Let us list Alberto's and Barney's scores in order of magnitude:

ALBERTO	2	7	7	7	⑨	10	10	10	10
BARNEY	7	7	7	7	⑦	8	10	10	10

Table 14.2c

MEASURE	ALBERTO	BARNEY
Mean	8	8.1
Mode	10	7
Median	9	7

The median is circled in each case.

Now, look at the three measures we have found for the scores in Table 14.2c. The mode and the median in this case would appear to some people to be evidence that perhaps Alberto did write a better paper than Barney.

EXAMPLE 2   Have you been exercising lately? You must exercise if you want to keep your weight down. Here are 10 different activities with the corresponding hourly energy expenditures (in calories) for a 150-lb person:

Fencing	300	Square dancing	350
Golf	250	Squash	600
Running	900	Swimming	300
Sitting	100	Volleyball	350
Standing	150	Wood chopping	400

a. Find the mean of these numbers.
b. Find the median number of calories spent in these activities.
c. Find the mode of these numbers.

Solution   **a.** The mean is obtained by adding all the numbers and dividing the sum by 10. The sum of the numbers in the first column is 1700, and of the numbers in the second column is 2000. Thus, the mean is

$$\frac{1700 + 2000}{10} = 370 \text{ calories per hour}$$

b. To find the median, we must first arrange the numbers in order of magnitude, as follows:

Sitting	100
Standing	150
Golf	250
Fencing	300
Swimming	300
Square dancing	350
Volleyball	350
Wood chopping	400
Squash	600
Running	900

$$\leftarrow \text{Median} = \frac{300 + 350}{2} = 325$$

Since we have an even number of items, the median is the average of the two middle items.

c. The mode is the number with the greatest frequency if there is one such number. In this case, the numbers 300 and 350 both occur twice, while all other numbers occur just once. Thus, there are two modes, 300 and 350. ◄

EXAMPLE 3   For the frequency distribution of wage rates given in Example 1 of Section 14.1, find:

a. The mean rate   **b.** The mode   **c.** The median rate

Solution   **a.** We refer to the table on page 620 and make the calculation shown in the table below, finding the mean rate to be $7.94/hr.

WAGE RATE	FREQUENCY	FREQUENCY × RATE
7.70	3	23.10
7.80	6	46.80
7.90	6	47.40
8.00	8	64.00
8.10	5	40.50
8.20	2	16.40
	30	30) 238.20
		$7.94 = \bar{x}$

b. The mode is the most frequent rate, $8.00/hr.

c. Since 15 workers get $7.90 or less and the oth
hour, the median rate is the mean of $7.90 a

    In this section, we introduced three **measur**
is how they compare:

1. The **mean** (arithmetic average) is the mos
measures. A set of data always has a uniq
account of each item of the data. On the r
takes the most calculation of the three m...
of the mean is its sensitivity to extreme values. For instanc,
of the data 2, 4, 6, 8 is $\frac{20}{4} = 5$, but the mean of 2, 4, 6, 28 is $\frac{40}{4} = 10$, a shift
of 5 units toward the extreme value 28.
2. The **mode** has the advantage of requiring no calculation. However, the
mode may not exist, as in the case of the data 2, 4, 6, 8. On the other
hand, the mode may be most useful. For example, suppose a shoe
manufacturer surveys 100 women to see which of three styles, A, B, or
C, of shoes each one prefers and finds style A selected by 30 women,
style B by 50, and style C by 20. The mode is 50 and there is not much
doubt about which style the manufacturer will feature.
3. The **median** always exists and is unique, as in the case of the mean.
However, the median requires very little computation and is not sensi-
tive to extreme values. Of course, in order to find the median, the data
must be arranged in order of magnitude, which may not be practical for
large sets of data. But the most important disadvantage of the median is
its failure to take account of each item of data. Hence, in many statistical problems, the median is not a reliable measure.

EXERCISE 14.2

1. Find the mean and the median for each set of numbers:
   a. 1, 5, 9, 13, 17   **b.** 1, 3, 9, 27, 81   **c.** 1, 4, 9, 16, 25
   d. For which of these sets are the mean and the median the same?
   Which measure is the same for all three sets? Which (if any) of the
   sets has a mode?
2. Show that the median of the set of numbers 1, 2, 4, 8, 16, 32 is 6. How
   does this compare with the mean?
3. Out of 10 possible points, a class of 20 students made the following test
   scores:

   0, 0, 1, 2, 4, 5, 5, 6, 6, 6, 7, 8, 8, 8, 8, 9, 9, 9, 10, 10

   Find the mean, the median, and the mode. Which of these three
   measures do you think is the least representative of the set of scores?
4. Find the mean and the median of the set of numbers:

   0, 3, 26, 43, 45, 60, 72, 75, 79, 82, 83

5. An instructor gave a short test to a class of 25 students and found the scores on the basis of 10 to be as follows:

SCORE	3	4	5	6	7	8	9	10
NUMBER OF STUDENTS	2	1	3	2	6	4	4	3

The instructor asked two students, Agnes and Betty, to calculate the average score. Agnes made the calculation

$$\frac{3+4+5+6+7+8+9+10}{8} = \frac{52}{8} = 6.5$$

and said the average score is 6.5. Betty calculated a *weighted average* as follows:

$$\frac{2\cdot 3+1\cdot 4+3\cdot 5+2\cdot 6+6\cdot 7+4\cdot 8+4\cdot 9+3\cdot 10}{25}$$

$$= \frac{177}{25} = 7.08$$

She then said the average was 7.08. Who is correct, Agnes or Betty? Why?

6. An investor bought 150 shares of Fly-Hi Airlines stock. He paid $60 per share for 50 shares, $50 per share for 60 shares, and $75 per share for 40 shares. What was his average cost per share? (Compare Problem 5.)

7. Make a frequency distribution of the number of letters per word in the following quotation: "For seven days seven priests with seven trumpets invested Jericho, and on the seventh day they encompassed the city seven times."
   a. Find the mode(s) of the number of letters per word.
   b. Find the median. (You can use your frequency distribution to do this.)
   c. Find the mean of the number of letters per word.
   d. Do you think your answers would give a good indication of the average length of words in ordinary English writing? Why?

8. Here are the temperatures at 1-hr intervals in Denver, Colorado, from 1 P.M. on a certain day to 9 A.M. the next day:

1 P.M.	90	8 P.M.	81	3 A.M.	66
2 P.M.	91	9 P.M.	79	4 A.M.	65
3 P.M.	92	10 P.M.	76	5 A.M.	66
4 P.M.	92	11 P.M.	74	6 A.M.	64
5 P.M.	91	12 M	71	7 A.M.	64
6 P.M.	89	1 A.M.	71	8 A.M.	71
7 P.M.	86	2 A.M.	69	9 A.M.	75

   a. What was the mean temperature? The median temperature?

b. What was the mean temperature from 1 P.M. to 9 P.M.? The median temperature?

c. What was the mean temperature from midnight to 6 A.M.? The median temperature?

9. Suppose that a dime and a nickel are tossed. They can fall in four different ways: $(H, H), (H, T), (T, H), (T, T)$, where we agree to let the first letter indicate how the dime falls and the second letter, the nickel. How many tosses do you think it would take, on the average, to get all four possibilities at least once? A good way to find out is by experimenting. Take a dime and a nickel and toss them to get your data. You can keep track of what happens with a frequency distribution like this one:

	Trial 1	Trial 2	Trial 3
(H, H)	|	|	⧄
(H, T)	|	|	|||
(T, H)	||||	|	|||
(T, T)	⧄	||||		
	11	7	12 (etc.)

You will need to make tally marks in the trials column until there is at least one mark for each possibility. Then write the total number of tosses at the bottom of the column. A new column will be needed for each trial, of course. Make 20 trials.

a. When you finish the 20 trials, make a frequency distribution of the number of tosses required to give all four possibilities.

b. Use the frequency distribution you obtained in part a to find the median number of tosses.

c. Find the mean number of tosses needed to obtain all four possibilities.

10. The mean score on a test taken by 20 students is 75; what is the sum of the 20 test scores?

11. A mathematics professor lost a test paper belonging to one of her students. She remembered that the mean score for the class of 20 was 81, and that the sum of the 19 other scores was 1560. What was the grade on the paper she lost?

12. If in Problem 11 the mean was 82, and the sum of the 19 other scores was still 1560, what was the grade on the lost paper?

13. The mean salary for the 20 workers in company A is $90 per week, while in company B the mean salary for its 30 workers is $80 per week. If the two companies merge, what is the mean salary for the 50 employees of the new company?

14. A student has a mean score of 88 on five tests taken. What score must he obtain in his next test to have a mean (average) score of 80 on all six tests?

Have you ever been in line at the counter of a supermarket or a department store for so long that you were tempted to walk out? There is a mathematical theory called queuing (pronounced, "cueing") theory that studies ways in which lines at supermarkets, department stores, and so on, can be reduced to a minimum. The following problems show how a store manager can estimate the average number of people waiting at a particular counter.

TIME	CUSTOMERS
1	A, B
2	C, D
3	
4	E, F
5	

1. Suppose that in a 5-min interval customers arrive as indicated in the table in the margin. (Arrival time is assumed to be at the beginning of each minute.) In the first minute, A and B arrive. During the second minute, B moves to the head of the line (A was gone because it took 1 min to serve him), and C and D arrive, and so on. From the figure, find:
   a. The average (mean) number of people in line
   b. The mode of the number of people in line

TIME	CUSTOMERS
1	A
2	B
3	C, D, E
4	F
5	

2. Use the ideas of Problem 1 and suppose that the list of arrivals is as shown in the table in the margin. (Assume it takes 1 min to serve the first customer in line and that customer leaves immediately.)
   a. Draw a diagram showing the line during each of the first 5 min.
   b. Find the mean of the number of people in line during the 5 min.
   c. Find the mode of the number of people in line.

We have just studied three measures of central tendency: the mean, the median, and the mode. All these measures are frequently called averages. Suppose that the chart at the top of the next page shows the salaries at Scrooge Manufacturing Company.

1. Scrooge claims that the workers should not unionize; after all, he says, the average salary is $21,000. Can you discover what average this is?
2. Manny Chevitz, the union leader, claims that Scrooge Manufacturing really needs a union. Just look at their salaries! A meager $6000 on the average. Can you discover what average he means?

$100,000 $50,000 $25,000 $10,000 $6,000 each
Boss   Boss' son   Boss' assistant   Boss' secretaries   Workers

3. B. Crooked, the politician, wants both union and management support. He says that the workers are about average as far as salary is concerned. You can just figure it out. The company's average salary is $8000. Can you discover what average B. Crooked has in mind?

▶ **Calculator Corner 14.2**   *If you have a calculator with* $\boxed{\Sigma +}$ *(read, "sigma plus") and* $\boxed{\bar{x}}$ *keys, you are in luck. The calculation for the mean is automatically done for you. First, place the calculator in the statistics mode (press* $\boxed{\text{mode}}$ $\boxed{\text{stat}}$ *or* $\boxed{\text{2nd}}$ $\boxed{\text{stat}}$*). To find the mean of the numbers in Example 2, enter*

$$\boxed{300}\ \boxed{\Sigma +}\ \boxed{250}\ \boxed{\Sigma +}\ \boxed{900}\ \boxed{\Sigma +}\ \boxed{100}\ \boxed{\Sigma +}\ \boxed{150}\ \boxed{\Sigma +}\ \boxed{350}$$
$$\boxed{\Sigma +}\ \boxed{600}\ \boxed{\Sigma +}\ \boxed{300}\ \boxed{\Sigma +}\ \boxed{350}\ \boxed{\Sigma +}\ \boxed{400}\ \boxed{\Sigma +}\ \boxed{\text{2nd}}\ \boxed{\bar{x}}$$

The display gives the mean $\bar{x} = 370$.

▶ **Computer Corner 14.2**   *The mean, median, and mode of a list of numbers can be found using the program on page 752. Simply enter the numbers and press return.*

14.3

MEASURES OF DISPERSION

Most of the time we want to know more about a set of numbers than we can learn from a measure of central tendency. For instance, the two sets of numbers {3, 5, 7} and {0, 5, 10} both have the same mean and the same median, 5, but the two sets of numbers are quite different. Clearly, some information about how the numbers vary will be useful in describing the set.

A number that describes how the numbers of a set are spread out, or dispersed, is called a **measure of dispersion.** A very simple example of such a measure is the range.

▶ **Definition 14.3a** | The **range** of a set of numbers is the difference between the greatest and the least of the numbers in the set.

The two sets {3, 5, 7} and {0, 5, 10} have ranges, $7 - 3 = 4$ and $10 - 0 = 10$, respectively. Because the range is determined by only two numbers of the set, you can see that it gives us very little information about the other numbers of the set. The range actually gives us only a general notion of the spread of the given data.

Another measure of dispersion is called the **standard deviation.** It is the most commonly used of these measures, and the only additional one that we shall consider. The easiest way to define the standard deviation is by means of a formula.

▶ **Definition 14.3b**

Let a set of n numbers be denoted by $x_1, x_2, x_3, \ldots, x_n$, and let the mean of these numbers be denoted by $\bar{x}$. Then the **standard deviation** *s* is given by

$$s = \sqrt{\frac{(x_1 - \bar{x})^2 + (x_2 - \bar{x})^2 + (x_3 - \bar{x})^2 + \cdots + (x_n - \bar{x})^2}{n}}$$

In order to find the standard deviation, we have to find:

1. The mean, $\bar{x}$, of the set of numbers
2. The difference (deviation) between each number of the set and the mean
3. The squares of these deviations
4. The mean of these squares
5. The square root of this last mean, which is the number s.

The last four steps motivate the name **root-mean-square deviation,** which is often used for the standard deviation. As we shall learn, the number s gives a good indication of how the data are spread about the mean.

EXAMPLE 1   The ages of 5 schoolchildren were found to be 7, 9, 10, 11, and 13. Find the standard deviation s for this set of ages.

Solution   We follow the five steps given above, making a table as shown below.

Calculation of the Standard Deviation

AGE x	DIFFERENCE FROM MEAN $x - \bar{x}$	SQUARE OF DIFFERENCE $(x - \bar{x})^2$
7	-3	9
9	-1	1
10	0	0
11	1	1
13	3	9

50   Sum of ages   20   Sum of squares

$\bar{x} = \frac{50}{5} = 10$ Mean of ages $\frac{20}{5} = 4$ Mean of squares

$\sqrt{4} = 2 = s$

1. The mean of the five ages is

$$\bar{x} = \frac{7 + 9 + 10 + 11 + 13}{5} = \frac{50}{5} = 10 \quad \text{Column 1}$$

2. We now find the difference (deviation) between each number and the mean (column 2).
3. We square the numbers in column 2 to get column 3.
4. We find the mean of the squares in column 3:

$$\frac{9 + 1 + 0 + 1 + 9}{5} = \frac{20}{5} = 4$$

5. The standard deviation is the square root of the number found in step 4. Thus, $s = \sqrt{4} = 2$. ◄

The number s, although it seems complicated to compute, is a most useful number to know. In many practical applications, about 68% of the data are within 1 standard deviation from the mean. That is, 68% of the numbers lie between $\bar{x} - s$ and $\bar{x} + s$. Also, about 95% of the data are within 2 standard deviations from the mean; that is, 95% of the numbers lie between $\bar{x} - 2s$ and $\bar{x} + 2s$.

For example, if the mean of a set of 1000 numbers is 200 and the standard deviation is 25, then approximately 680 of the numbers lie between 175 and 225, and all but about 50 of the numbers lie between 150 and 250. Thus, even with no further information, the number s gives a fair idea of how the data are spread about the mean. These ideas are discussed more fully in Section 14.4.

EXAMPLE 2   A consumer group checks the price of 1 dozen large eggs at 10 chain stores, with the following results:

STORE NUMBER	1	2	3	4	5	6	7	8	9	10
PRICE (Cents)	70	68	72	60	63	75	66	65	72	69

Find the mean, median, mode, and standard deviation. What percent of the data are within 1 standard deviation from the mean?

Solution   In the table on page 638, the data are arranged in order of magnitude. The mean is found to be 68¢. The median is the average of the two middle prices, 68¢ and 69¢, which is 68.5¢. The mode is 72¢. The calculation of the standard deviation is shown in the table. The result is $s = 4.34$. To find the percentage of the data within 1 standard deviation from the mean, we first find $\bar{x} - s = 63.66$ and $\bar{x} + s = 72.34$. By examining the data, we see that seven of the prices are between these two numbers. Thus, 70% of the prices are within 1 standard deviation from the mean price.

x	x − x̄	(x − x̄)²
60	−8	64
63	−5	25
65	−3	9
66	−2	4
68	0	0
69	1	1
70	2	4
72	4	16
72	4	16
75	7	49
680		188

$$\bar{x} = \frac{680}{10} = 68 \qquad \frac{188}{10} = 18.8$$

$$s = \sqrt{18.8} \approx 4.34$$

◀

Note that you are not expected to calculate square roots. Use Table I in the back of the book or a calculator.

EXERCISE 14.3

In Problems 1–10:

a. State the range. **b.** Find the standard deviation s.

1. 3, 5, 8, 13, 21
2. 1, 4, 9, 16, 25
3. 5, 10, 15, 20, 25
4. 6, 9, 12, 15, 18
5. 5, 6, 7, 8, 9
6. 4, 6, 8, 10, 12
7. 5, 9, 1, 3, 8, 7, 2
8. 2, 0, 4, 6, 8, 10, 8, 2
9. −3, −2, −1, 0, 1, 2, 3
10. −6, −4, −2, 0, 2, 4, 6

11. Out of 10 possible points, a class of 20 students made the following test scores:

0, 0, 1, 2, 4, 4, 5, 6, 6, 6, 7, 8, 8, 8, 8, 9, 9, 9, 10, 10

  a. What is the mode?
  b. What is the median?
  c. What is the mean?
  d. Calculate the standard deviation.
  e. What percent of the scores lie within 1 standard deviation from the mean?
  f. What percent of the scores lie within 2 standard deviations from the mean?

12. Suppose that the 4 students who scored lowest on the test in Problem

11 dropped the course. Answer the same questions as in Problem 11 for the remaining students.

13. Elmer Duffer plays golf on a par 75 course that is really too tough for him. His scores in his last 10 games are 103, 110, 113, 102, 105, 110, 111, 110, 106, 110.

    a. What is Elmer's mode?
    b. What is his median score?
    c. What is his mean score?
    d. Calculate the standard deviation of his scores.
    e. Which of his scores are more than 1 standard deviation from his mean score? What percent of the games is this?

14. Answer the same questions as in Problem 13 for the best 8 of Elmer's 10 games.

15. Suppose the standard deviation of a set of numbers is 0. What does this tell you about the numbers?

16. Two classes, each with 100 students, took an examination with maximum possible score 100. In the first class, the mean score was 75 and the standard deviation was 5. In the second class, the mean score was 70 and the standard deviation was 15. Which of the two classes do you think had more scores of 90 or better? Why?

USING YOUR KNOWLEDGE 14.3

A **binomial experiment** is one that consists of a number of identical trials, each trial having only two possible outcomes (like tossing a coin that must fall heads or tails). Let us consider one of the outcomes as a success and the other as a failure. If p is the probability of success, then $1 - p$ is the probability of failure.

Suppose the experiment consists of n trials; then the theoretical expected number of successes is pn. For instance, if the experiment consists of tossing a fair coin 100 times, then the expected number of heads is $(\frac{1}{2})(100) = 50$. This means that if the experiment of tossing the coin 100 times is repeated many times, then the average number of heads is theoretically 50. In general, if a binomial experiment is repeated many times, then the theoretical mean (average) number of successes is pn, where p is the probability of success in one trial and n is the number of trials in the experiment.

If we let P_k denote the probability of k successes and $n - k$ failures in a binomial experiment with n trials, then the set of numbers $P_0, P_1, P_2, \ldots, P_n$ constitutes a **binomial frequency distribution.** The following simple formula has been obtained for the standard deviation of such a distribution:

$$s = \sqrt{np(1 - p)}$$

For example, if the experiment consists of tossing a fair coin 10,000 times and tallying the number of heads, then $n = 10,000$, $p = \frac{1}{2}$, $\bar{x} = 5000$, and

$$s = \sqrt{10{,}000(\tfrac{1}{2})(1 - \tfrac{1}{2})} = \sqrt{2500} = 50$$

If this experiment (tossing the coin 10,000 times) were repeated many times, then we would expect the average number of heads to be close to 5000. Although we are not justified in expecting the number of heads in any one experiment to be exactly 5000, we may expect that about 68% of the time the number of heads will be between 4950 and 5050.

1. If a fair die is rolled, the probability that it comes up 2 is $\tfrac{1}{6}$ and the probability that it comes up *not* 2 is $\tfrac{5}{6}$. If we regard 2 as a success and any other number as a failure, what is the standard deviation for the experiment of rolling the die 180 times?
2. Suppose that in rolling a die we regard a 3 or a 4 as a success and any other number as a failure. What is the standard deviation for the experiment of rolling the die 18 times? How far away from the mean would the number of successes have to be before you became suspicious of the die's honesty?
3. Suppose a die is loaded so that the probability that a 6 comes up is $\tfrac{1}{4}$. If we regard a 6 as a success and any other number as a failure, what is the standard deviation for the experiment of rolling the die 400 times?

▶ **Calculator Corner 14.3**

If you have a calculator with a $\boxed{\sigma_\mathrm{n}}$ key, it will compute the standard deviation for a set of data at the push of a button. For example, to find the standard deviation of Example 1, set the calculator in the statistics mode and enter

$$\boxed{7}\ \boxed{\Sigma +}\ \boxed{9}\ \boxed{\Sigma +}\ \boxed{10}\ \boxed{\Sigma +}\ \boxed{11}\ \boxed{\Sigma +}\ \boxed{13}\ \boxed{\text{2nd}}\ \boxed{\sigma_\mathrm{n}}$$

The result is given as 2.

▶ **Computer Corner 14.3**

To find the standard deviation of a set of numbers requires some tedious calculations. We have a program (page 752) that will do it for you. As a matter of fact, since you need the mean to find the standard deviation, the program will also give you the mean (and the range, in case you need it). You need only press RETURN (ENTER) after each entry. When all entries are in, press RETURN (ENTER) one more time, and the mean, range, and standard deviation will be displayed.

14.4

THE NORMAL CURVE

A scientist gets 1000 students to look through a microscope that magnifies a line segment that is actually 0.1 mm long and to guess at the apparent length. He records each guess and makes a frequency polygon, which he then "smooths out" to obtain the graph shown in Fig. 14.4a. He also calcu-

Figure 14.4a
Normal curve

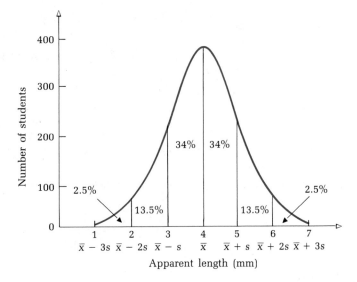

lates the mean $\bar{x}$, which he finds to be 4 mm, and the standard deviation s, which he finds to be 1 mm.

The **bell-shaped** graph in Fig. 14.4a is typical of what is called a **normal curve.** The vertical centerline marks the mean guess, $\bar{x} = 4$. The curve is symmetric with respect to this centerline; that is, if you fold the graph in half along the centerline, the two parts of the curve coincide exactly. In this experiment, the standard deviation $s = 1$, and vertical lines are drawn at intervals of 1 standard deviation, that is, at $\bar{x} + s$, $\bar{x} + 2s$, $\bar{x} + 3s$, $\bar{x} - s$, $\bar{x} - 2s$, and $\bar{x} - 3s$. The percentages shown on the graph are the approximate theoretical percentages of the guesses that are expected to fall in each interval. These percentages are obtained by advanced mathematical methods and we shall simply accept them here.

As you can see, about 68% of the guesses are within 1 standard deviation from the mean. Altogether, approximately 95% of the guesses lie within 2 standard deviations from the mean. In a normal distribution, almost all the data are within 3 standard deviations from the mean.

Refer to Fig. 14.4a, and try to answer the following questions:

1. What is the mean length that the line segment appeared to be?
2. To about how many students did the length appear to be within 1 mm of the mean?
3. To about how many students did the segment appear to be more than 5 mm long?

If your answers to these questions are 4 mm, 680 students, and 160 students, respectively, then you have a good start toward understanding the normal curve.

EXAMPLE 1   The heights of 1000 girls are measured and found to be normally distributed, with a mean of 64 in. and a standard deviation of 2 in.

a. About how many of the girls are over 68 in. tall?
b. About how many are between 60 and 64 in. tall?
c. About how many are between 62 and 66 in. tall?

Solution   We refer to Fig. 14.4a for the required percentages.

a. Because 68 in. is 2 standard deviations above the mean, about 2.5%, or
25, of the girls are over 68 in. tall.
b. We see that 64 in. is the mean, and 60 in. is exactly 2 standard deviations below the mean; hence, we add 13.5% and 34% to find that 47.5%,
or 475, girls are between 60 and 64 in. tall.
c. Because 62 in. is 1 standard deviation below the mean, and 66 in. is 1
standard deviation above the mean, we add 34% and 34% to find that
about 68%, or 680, girls will be between these two heights.   ◄

EXAMPLE 2   A standardized reading comprehension test is given to 10,000 high school
students. The scores are found to be normally distributed, with a mean of
500 and a standard deviation of 60. If a score below 440 is considered to
indicate a serious reading deficiency, about how many of the students are
rated as seriously deficient in reading comprehension?

Solution   Since 440 is exactly 1 standard deviation below the mean, scores below
440 are more than 1 standard deviation below the mean. By referring to
the percentages in Fig. 14.4a, we see that we must add 13.5% and 2.5% to
get the total percentage of students who scored more than 1 standard
deviation below the mean. Thus, 16% of the 10,000 students, or 1600, are
rated as seriously deficient in reading comprehension.   ◄

EXERCISE 14.4

1. Farmer Brown has planted a field of experimental corn. By judicious
sampling, it is estimated that there are about 20,000 plants and that a
graph of their heights looks like that shown in the figure on page 643.
a. What is the mean height of Farmer Brown's corn?
b. What is the standard deviation from the mean?
c. What percent of the cornstalks are between 90 and 110 in. tall?
d. About how many stalks are between 80 and 90 in. tall?
2. Suppose you were informed that the annual income of lawyers is
normally distributed, with a mean of $40,000 and a standard deviation
of $10,000.
a. What would you estimate for the percent of lawyers with incomes
over $50,000?

Farmer Brown's corn

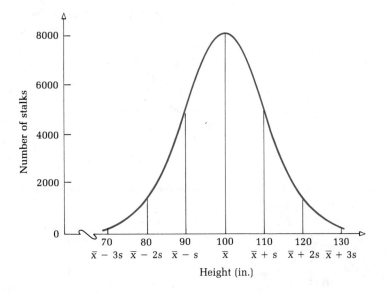

Height (in.)

b. What percent of lawyers would you estimate have an annual income of less than $20,000?

c. If a lawyer were selected at random, what would be the probability that his or her annual income is more than $60,000?

d. If the information given here were correct, would you think it very likely that 50% of all lawyers have annual incomes of over $50,000? Why?

3. Part of a test given to small children consists of putting together a simple jigsaw puzzle. Suppose that such a puzzle is given to 1000 children, each child is timed, and a graph of the times is made. Suppose the graph is a normal curve with a mean time of 120 sec and a standard deviation of 15 sec.

a. About how many of the children finished the puzzle in less than 90 sec?

b. How many took more than 150 sec?

c. If you rated as "average" all the children within 1 standard deviation from the mean, how many children would fall into this classification?

4. For a certain standardized placement test, it was found that the scores were normally distributed, with a mean of 200 and a standard deviation of 30. Suppose this test is given to 1000 students.

a. How many are expected to make scores between 170 and 230?

b. How many are expected to score above 260?

c. What is the expected range of all the scores?

5. A psychology teacher gave an objective-type test to a class of 500 students, and after seeing the results, decided that the scores were normally distributed. The mean score was 50, and the standard devia-

tion was 10. The teacher assigned a grade of A to all scores of 70 or over, B to scores of 60 to 69, C to scores 40 to 59, D to scores of 30 to 39, and F to scores below 30. About how many of each grade did the teacher assign?

6. In a study of 100 common stocks, it was found that the annual dividend rates were normally distributed, with a mean of 4.0% and a standard deviation of 0.5%.

   a. About how many of these stocks do you think paid dividends of over 5%?

   b. About how many paid between 3 and 5%?

   c. If you picked one of these stocks at random, what do you think would be the probability that it paid at least 4.5%?

7. The lifetimes of a random sample of 200 automobile tires were found to be normally distributed, with a mean of 26,000 mi and a standard deviation of 2500 mi. About how many of these tires gave out before running 21,000 mi?

8. Suppose that 100 measurements of the specific gravity of copper gave a mean of 8.8 with a standard deviation of 0.2. Between what limits did about 95% of the measurements fall?

9. Suppose that 10 measurements of the length of a wooden beam have a mean of 20 ft and a standard deviation of 0.5 in. Between what limits do almost all the measurements fall?

10. An experiment consists of tossing 100 dimes repeatedly and noting the number of heads each time. The graph of the number of heads turns out to be very nearly a normal curve, with a mean of 50 and a standard deviation of 5.

    a. Within what limits would you expect the number of heads to be 95% of the time?

    b. What percent of the time would you expect the number of heads to be between 45 and 55?

    c. Suppose that a particular one of the dimes arouses your suspicion by turning up heads too often. You toss this dime 100 times. How many times will it have to turn up heads in order for you to be almost 100% certain that it is not a fair coin? [*Hint:* Almost all the data in a normal distribution fall within 3 standard deviations of the mean.]

DISCOVERY 14.4

We have discovered that we can make fairly accurate predictions about the dispersion of the measurements in a normal distribution. For instance, 68% of the measurements fall between $\bar{x} - s$ and $\bar{x} + s$, 95% between $\bar{x} - 2s$ and $\bar{x} + 2s$, and nearly 100% between $\bar{x} - 3s$ and $\bar{x} + 3s$. But what can we say in case the distribution is not normal?

The great Russian mathematician, Pafnuti Lvovich Chebyshev (1821–1894), discovered the following remarkable result:

Chebyshev's Theorem

For any distribution with a finite number N of measurements and for any h such that $h > 1$, the number of measurements within h standard deviations of the mean is **at least** equal to

$$\left(1 - \frac{1}{h^2}\right) N$$

For example, if $h = 2$, then $1 - (1/h^2) = \frac{3}{4}$, so at least $\frac{3}{4}$N, or 75%, of the measurements fall between $\bar{x} - 2s$ and $\bar{x} + 2s$. This is not as large a percentage as for a normal distribution, but the amazing thing is that this result holds for any kind of distribution at all—as long as there is only a finite number of measurements!

Suppose we have 20 numbers with a mean of 8 and a standard deviation of 2. How many of the numbers can we **guarantee** fall between 2 and 14? Because 2 and 14 are each 3 standard deviations from the mean, we take $h = 3$ in Chebyshev's theorem and obtain

$$\left(1 - \frac{1}{3^2}\right)(20) = \frac{160}{9} \approx 17.8$$

Thus, the theorem guarantees that at least 18 of the 20 numbers fall between 2 and 14.

In the same way, by taking $h = 1.5$, we find that

$$\left(1 - \frac{1}{1.5^2}\right)(20) = \left(\frac{5}{9}\right)(20) \approx 11.1$$

Hence, we can guarantee that at least 12 of the 20 numbers fall within 1.5 standard deviations from the mean, that is, between 5 and 11.

You should notice that Chebyshev's theorem makes no claim at all for the case where $h \leq 1$.

1. If 100 measurements have a mean of 50 and a standard deviation of 5, can you discover how many of the measurements must be between:
   a. 40 and 60?   **b.** 35 and 65?   **c.** 43 and 57?
2. Can you discover the smallest value of h that is large enough to guarantee that of a set of measurements at least:
   a. 96% will be within h standard deviations from the mean?
   b. 91% will be within h standard deviations from the mean?
   c. 64% will be within h standard deviations from the mean?
3. Find the mean and the standard deviation of the following numbers: 1, 1, 1, 2, 6, 10, 11, 11, 11. How many of these numbers lie within 1 standard deviation from the mean? How many lie within 2 standard

deviations from the mean? How do these results compare with those predicted by Chebyshev's theorem?

4. Do you think it is possible for all the measurements in an experiment to be *less* than 1 standard deviation from the mean? Can you discover how to justify your answer? [*Hint:* The formula for the standard deviation shows that $ns^2 = (x_1 - \bar{x})^2 + (x_2 - \bar{x})^2 + \cdots + (x_n - \bar{x})^2.$]

5. Can Chebyshev's theorem be used to find the percent of the measurements that must fall between 2 and 3 standard deviations from the mean? What can you say about this percent?

14.5

MEASURES OF POSITION; z-SCORES

Rudy made a score of 80 in his American history test and a score of 80 in his geometry test. Which of these is the better score? Without additional information, we cannot answer this question. However, if we are told that the mean score in the American history test was 60, with a standard deviation of 25.5, and the mean score in the geometry test was 70, with a standard deviation of 14.5, then there is a way for us to compare Rudy's two scores.

A. z-Scores

In order to make a valid comparison, we have to restate the scores on a common scale. A score on this scale is known as a **standardized score,** or a **z-score.**

Definition 14.5a

> If x is a given score and $\bar{x}$ and s are the mean and standard deviation of the entire set of scores, then the **corresponding z-score** is
>
> $$z = \frac{x - \bar{x}}{s}$$

Since the numerator of z is the difference between x and the mean, *the z-score gives the number of standard deviations that x is from the mean.*

EXAMPLE 1   Compare Rudy's scores in American history and geometry, given all the preceding information.

Solution   Rudy's z-scores are

American history: $z = \dfrac{80 - 60}{25.5} = 0.78$

Geometry: $z = \dfrac{80 - 70}{14.5} = 0.69$

Thus, Rudy did better in American history than in geometry. ◀

B. Distribution of z-Scores

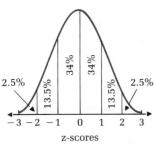

Figure 14.5a
Distribution of z-scores

For a normal distribution of scores, if we subtract $\overline{x}$ from each score, the resulting numbers will have a mean of 0. If we then divide each number by the standard deviation s, the resulting numbers will have a standard deviation of 1. Thus, the z-scores are distributed as shown in Fig. 14.5a. For instance, 34% of the z-scores lie between 0 and 1, 13.5% between 1 and 2, and 2.5% are greater than 2. For such a distribution of scores, the probabilities of randomly selecting z-scores between 0 and a given point to the right of 0 have been calculated and appear in tables such as Table II in the back of the book. To read the probability that a score falls between 0 and 0.25 standard deviation above the mean, we go down the column under z to 0.2 and then across to the column under 5; the number there is 0.099, the desired probability.

EXAMPLE 2 For a certain normally distributed set of data, the mean is $\overline{x} = 100$ and the standard deviation is 15. Find the probability that a randomly selected item of the data falls between 100 and 120.

Solution **1.** We first find the z-score for 120:

$$z = \frac{x - \overline{x}}{s} = \frac{120 - 100}{15} = \frac{4}{3} \approx 1.33$$

2. We then refer to Table II and read down the column under z to the number 1.3 and then across to the column under 3 to read the desired probability, 0.408. ◀

EXERCISE 14.5

A.

1. In a certain normal distribution of scores, the mean is 5 and the standard deviation is 1.25. Find the z-score corresponding to a score of:
  a. 6   **b.** 7   **c.** 7.5

2. In a certain normal distribution of scores, the mean is 10 and the standard deviation is 2. Find the z-score corresponding to a score of:
  a. 11   **b.** 13   **c.** 14.2

3. Gretchen scored 85 on a test in German and also on a test in English. If the mean in the German test was 75, with a standard deviation of 20, and the mean in the English test was 80, with a standard deviation of 15, which of Gretchen's 85's was the better score?

4. José scored 88 on a Spanish test and 90 on an algebra test. If the mean in the Spanish test was 78, with a standard deviation of 7.5, and the mean in the algebra test was 82, with a standard deviation of 6.5, which of José's scores was the better score?

B. In Problems 5–9, assume a normally distributed set of test scores with a mean of $\bar{x} = 100$ and a standard deviation of 15.

5. Find the probability that a person selected at random will have a score between:

   a. 100 and 110   **b.** 100 and 130

6. Find the probability that a person selected at random will have a score between 80 and 120. [Hint: In Example 2, we found the probability that the score is between 100 and 120 to be 0.408. The probability that the score is between 80 and 100 is also 0.408. (Recall the symmetry of the normal curve.)]

7. Find the probability that a person selected at random will have a score between 55 and 145.

8. Find the probability that a person selected at random will have a score between 75 and 100.

9. Find the probability that a person selected at random will have a score between 110 and 130. [Hint: In Problem 5 you found the probability that the score will be between 100 and 110 and the probability that the score will be between 100 and 130. You should be able to see how to combine these two results to get the desired probability.]

10. In Problem 10 of Exercise 14.4, it was noted that the distribution of heads if 100 dimes are tossed repeatedly is approximately a normal distribution, with a mean of 50 and a standard deviation of 5. Find the probability of getting 60 heads if 100 fair coins are tossed. [Hint: To use the normal curve, consider 60 to be between 59.5 and 60.5 and proceed as in Problem 9. This will give a very good approximation and is much easier to calculate than the exact probability, which is $100!/(60!40!2^{100})$.]

11. The heights of the male students in a large college were found to be normally distributed, with a mean of 5 ft 7 in. and a standard deviation of 3 in. Suppose these students are to be divided into five equal-sized groups according to height. Approximately what is the height of the shortest student in the tallest group?

12. If 10 of the boys who show up for the basketball team are over 6 ft 3 in. tall, about how many male students are there in the college of Problem 11?

USING YOUR KNOWLEDGE 14.5

z-Scores provide a way of making comparisons among different sets of data. To compare scores within one set of data, we use a measurement called a **percentile.** Percentiles are used extensively in educational measurements and enable us to convert raw scores into meaningful comparative scores. If you take an exam and are told that you scored in the 95th percentile, it does not mean that you scored 95% on the exam, but rather that you scored higher than 95% of the persons taking the exam. The formula used to find the percentile corresponding to a particular score is as follows:

$$\text{Percentile of score } x = \frac{\text{Number of scores less than } x}{\text{Total number of scores}} \cdot 100$$

Thus, if 80 students take a test, and 50 students score less than you do, you will be in the $\frac{50}{80} \cdot 100 = 62.5$ percentile. Use this knowledge to solve the following problems.

1. A student took a test in a class of 50 students, and 40 of the students scored less than she did. What was her percentile?
2. A student took a test in a class of 80, and only 9 students scored better than he did. What was his percentile?
3. The scores in a class were as follows:

   83, 85, 90, 90, 92, 93, 97, 97, 98, 100

   a. What percentile corresponds to a score of 90?
   b. What percentile corresponds to a score of 97?
   c. What percentile corresponds to a score of 100?
   d. What percentile corresponds to a score of 83?

▶ **Calculator Corner 14.5** You can find the probability that a randomly selected item will fall between the mean and a given z-score by using the $\boxed{\text{R(t)}}$ key on your calculator. Thus, to find the probability that a randomly selected item falls between the mean and a score of 120, as in Example 2, place the calculator in the statistics mode and press

$$\boxed{120}\ \boxed{-}\ \boxed{100}\ \boxed{=}\ \boxed{\div}\ \boxed{15}\ \boxed{\text{2nd}}\ \boxed{\text{R(t)}}$$

The result is given as 0.40878. Note that there is a slight difference (due to rounding) between this answer and the one in the text.

1-5. Use a calculator to rework Problems 5-9 in Exercise 14.5.

14.6

STATISTICAL GRAPHS

A graph of a set of data can often provide information at a glance that might be difficult and less impressive to glean from a table of numbers. No table of numbers would make the visual impact created by the graph in Fig. 14.6a, page 650, for example.

It is almost always possible to alter the appearance of a graph to make things seem better than they are. For instance, Fig. 14.6b is a portion of Fig.

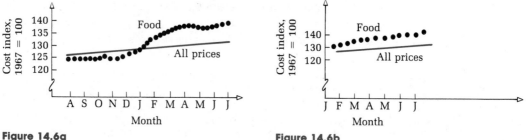

Figure 14.6a
The high cost of eating

Figure 14.6b

14.6a, but with a compressed vertical scale. Obviously, things look better on this graph! Can you see why an economist might feel it politically advantageous to publish one of these graphs rather than the other?

A. Bar Graphs

Newspapers and magazines often publish **bar graphs** of the type shown in Fig. 14.6c. These graphs again have the advantage of displaying the data in a form that is easy to understand. It would be difficult for most people to obtain the same information from a table of numbers. As in the case of line graphs, bar graphs may also be made to distort the truth. For example, consider the two graphs in Figs. 14.6d and 14.6e. The graph in Fig. 14.6d does not have the bars starting at 0, so it gives a somewhat exaggerated picture of the proportion of gasoline saved at lower speeds, even though the numerical data are the same for both graphs. The graph in Fig. 14.6e gives a correct picture of the proportion of gasoline saved. Why do you think the first bar graph rather than the second would be published?

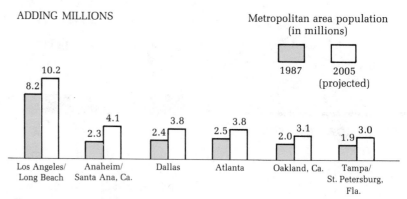

Figure 14.6c
Bar graph showing metropolitan areas that will gain more than 1 million residents by the year 2005

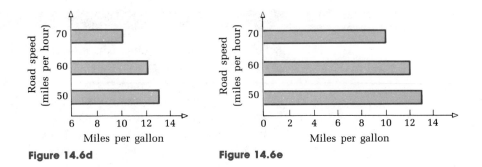

Figure 14.6d

Figure 14.6e

B. Circle Graphs

Graphs like those in Figs. 14.6f and 14.6g are called **circle graphs,** or **pie charts.** Such graphs are a very popular means of displaying data; and they are also susceptible to being drawn to make things look better than they are. For instance, compare the graph in Fig. 14.6f with the version of the same data that was published by the Internal Revenue Service. Does the visual impression of Fig. 14.6g make you feel that individual income taxes are not quite so large a chunk of the federal income as Fig. 14.6f indicates?

Circle graphs are quite easy to draw if you know how to use a simple compass and a protractor. For the graph in Fig. 14.6f, which shows where the typical dollar of federal money comes from, 23¢ is 23% of a dollar. The entire circle corresponds to 360°, so you would use 23% of 360°, or $82.8° \approx 83°$, for the slice that represents 23¢, and likewise for the other slices. This is the most accurate and honest way to present data on a circle graph.

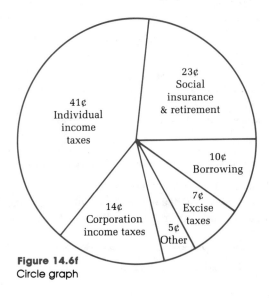

Figure 14.6f
Circle graph

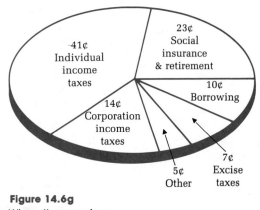

Figure 14.6g
Where it comes from
From an IRS income tax bulletin

A.

1. Following is a list of the approximate average cash income per farm in the United States from 1940 through 1985:

1940	$9,000	1965	$42,000
1945	$22,000	1970	$54,000
1950	$29,000	1975	$90,000
1955	$30,000	1980	$141,000
1960	$35,000	1985	$150,000

Make a bar graph to show these data.

2. Recently, U.S. companies had the following investments in Mideast oil and gas:

Crude oil and natural gas production facilities	$3385 million
Pipelines, refineries, and other plants	$1995 million
Marketing facilities and other related investments	$1035 million

Make a bar graph to show these data.

3. Many people grumble about the high taxes in the United States. Consider the following list, which gives the total taxes (federal, state, and local) in cents per dollar of national output, as compiled by the *Organization for Economic Cooperation and Development*:

Canada	31.2
France	36.9
Great Britain	34.4
Sweden	42.3
United States	29.9
West Germany	34.7

Arrange the data in order of decreasing amounts, and make a bar graph to illustrate these figures.

4. The trend in employee contributions to Social Security is indicated by the following figures, which give the percent deducted for selected years:

YEAR	1978	1980	1982	1984	1986	1988
PERCENT	6.05	6.13	6.7	7.0	7.15	7.51

Make a bar graph to illustrate this set of data.

5. The following figures give some idea of the growth of defense spending in the United States. Make a bar graph to illustrate these figures.

Expenditures by the Defense Department

YEAR	1983	1984	1985	1986
AMOUNT (Billions)	$208	$224	$244	$273

6. The total expenditures of the Federal government are much larger than those of the Defense Department alone. (See Problem 5.) Make a bar graph to illustrate these figures.

Expenditures of the Federal Government

YEAR	1983	1984	1985	1986
AMOUNT (Billions)	$796	$842	$937	$990

B.

7. The U.S. Department of Labor updated its theoretical budget for a retired couple. The high budget for such a couple is apportioned approximately as follows:

Food	22%	Medical care	6%
Housing	35%	Other family costs	7%
Transportation	11%	Miscellaneous	7%
Clothing	6%	Income taxes	3%
Personal care	3%		

Make a circle graph to show this budget.

8. The pie chart shown here appeared side-by-side with the one in Fig. 14.6g. Make a circle graph to present the same data. Compare the impression made by your circle graph with the pie chart.

Where it goes

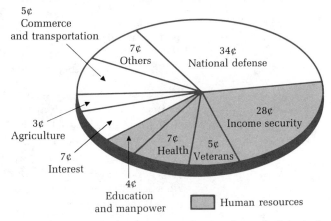

9. According to an advertisement for a color television, the top six brands of television sets were voted as best by the following percentages of about 2700 people:

BRAND	1	2	3	4	5	6
PERCENT	50.1	21.1	8.8	8.5	5.8	5.7

a. Make a circle graph to illustrate this information.
b. Make a bar graph for the same data.
c. Which of these do you think makes the stronger impression? Why?

1989   1990

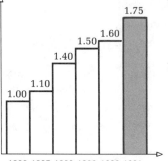

1986 1987 1988 1989 1990 1991

10. A survey made by the University of Michigan's Institute for Social Research showed that most women really enjoy keeping house. The survey found that about 67% of the women who responded had an unqualified liking for housework, while only 4% had an unqualified dislike for housework. Make a bar graph to illustrate these data.

11. The Mighty Midget Canning Company wants to impress the public with the growth of Mighty Midget business, which they claim has doubled over the previous year. They publish the pictorial graph shown in the margin.

   a. Can you see anything wrong with this? [*Hint:* Your mind compares the volumes pictured here. The volume of a cylinder is $\pi r^2 h$. What happens if you double the radius r and the height h?]

   b. Draw a bar graph that correctly represents the situation.

12. The U. B. Wary Company wants to give its stockholders a very strong impression of the rapid rate at which earnings have grown and prints the bar graph shown in the margin in its annual report. Redraw this graph to give a more honest impression.

13. Do you know that playing golf is not a particularly good way to lose weight? Here is the calorie consumption per hour for five popular activities. Make a bar graph with horizontal bars, with each bar identified at the left side of your graph.

Bicycling (15 mi/hr)	730
Running (6 mi/hr)	700
Swimming (40 yd/min)	550
Walking (4 mi/hr)	330
Golf (walking and carrying your clubs)	300

(These figures apply to a person weighing 150 lb; you have to add or subtract 10% for each 15-lb difference in weight.)

14. Women are generally lighter than men, so they require fewer calories per day to get around. Here are three occupations and their energy-per-day requirements. Make a vertical bar graph with the bars for male and female side-by-side for comparison. On the average, about what percent more calories does the male require than the female for these three occupations?

	Calories per Day
University student, male	2960
University student, female	2300
Laboratory technician, male	2850
Laboratory technician, female	2100
Office worker, male	2500
Office worker, female	1900

15. Although the best a cold remedy can do is ease the discomfort (without curing the cold), the relief seems to be worth plenty to the victims.

Here is how people in the United States spent money on cold remedies in a recent year. Make a bar graph with horizontal bars to represent the data. Be sure to identify the bars.

	Millions Spent
Cough drops and sore throat remedies	$130
Nasal sprays, drops, and vaporizers	$160
Aspirin substitutes	$275
Cold and cough syrups	$310
Aspirin	$575

16. The more you learn, the more you earn! Here are the median incomes by educational attainment for persons 25 years old or over:

	Women	**Men**
Less than 8 years	$9,800	$14,600
8 years	$10,800	$16,800
High school (1–3 years)	$11,800	$19,100
High school (4 years)	$14,600	$23,300
College (1–3 years)	$17,000	$25,800
College (4 or more years)	$21,900	$33,900

Make a side-by-side vertical bar graph for these data. Find the women's income as a percent of men's in each category.

DISCOVERY 14.6

Misleading graphs can be used in statistics to accomplish whatever deception you have in mind. The two graphs shown below, for example, give

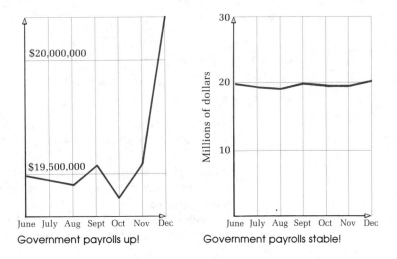

Government payrolls up!   Government payrolls stable!

exactly the same information. However, the graph on the left seems to indicate a steep increase in government payrolls, while the graph on the right shows the stability of the same payrolls!

1. Can you discover what is wrong?

MAKING PREDICTIONS

Table 14.7a

YEAR	TIME (sec)
1948	24.4
1952	23.7
1956	23.4
1960	23.2
1964	23.0
1968	22.5
1972	22.4
1976	22.37
1980	No U.S. participation
1984	21.81

Table 14.7a gives the winning times (in seconds) for the women's 200-m dash in the Olympic Games from 1948 to 1984. The points in Fig. 14.7a are the graphs of the corresponding number pairs (1948, 24.4), (1952, 23.7), and so on. As you can see, the points do not lie on a straight line. However, we can draw a straight line that goes "between" the points and seems to fit the data fairly well. Such a line has been drawn in Fig. 14.7a. Statisticians sometimes use lines like this to make predictions of unknown results. For example, the winning time for this event in the 1988 Olympics as indicated by the line in the graph is about 21.4 sec. The actual winning time (made by Florence Griffith Joyner of the United States) was 21.34 sec. (The time predicted by the graph is less than $\frac{3}{10}$ of 1% in error!)

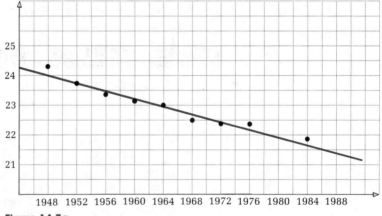

Figure 14.7a

How do we draw a line to fit data such as that in Table 14.7a? One way is just to use our best visual judgment; but a better way is to calculate a *least-squares line*. If we have n points, say $(x_1, y_1), (x_2, y_2), \ldots, (x_n, y_n)$, in the xy plane, a line $y = mx + b$ is called the **least-squares line** for these points if the sum of the squares of the differences between the actual y values of the points and the corresponding y values on the line is as small as possible. The line in Fig. 14.7a is the least-squares line for the data in Table 14.7a. If you have a calculator and wish to know the details of determining a least-squares line, then you should read Using Your Knowledge 14.7.

EXAMPLE 1 The table in the margin gives the winning times in seconds for the men's 100-m freestyle swim in the Olympic Games from 1960 to 1984. Make a graph of these data; then draw a line "between" the points and predict the winning time for this event in the 1988 Olympics.

Solution The required graph is shown in the figure below. The line in this figure is the least-squares line for the given data. From this line, we can read the predicted time for the 1988 Olympics as about 48.3 sec. The actual winning time was 48.63 sec, so the predicted time is off by less than 0.7%.

YEAR	TIME (sec)
1960	55.2
1964	53.4
1968	52.2
1972	51.22
1976	49.99
1980	50.40
1984	49.80

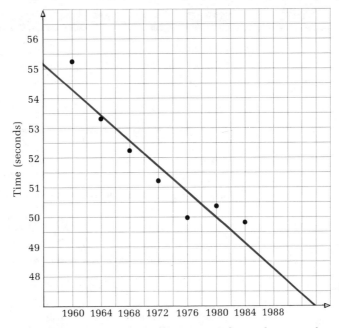

Another way in which statisticians make predictions is based on a **sampling procedure.** The idea is quite simple. The statistician takes data from a **random sample** of the population and assumes that the entire population behaves in the same way as the sample. The difficulties lie in making certain that the sample is actually random and represents the population in a satisfactory manner. We shall not discuss the different ways in which a sample is selected, but only mention that one of these ways uses a table of random numbers. (Such tables can be generated by a computer.)

EXAMPLE 2 A certain school has 3000 students. A random sample of 50 students is selected and asked to name their favorite track and field event. If 10 of these students name the 100-m dash, how many students in the school would you expect to have the same favorite event?

Solution   We would expect $\frac{10}{50}$, or 20%, of the student body to favor the 100-m dash. Thus, the required number is 20% of 3000 or (0.20)(3000) = 600.   ◄

EXAMPLE 3   In a certain county, 15% of the eleventh-grade students failed a required literacy test. If 60 of the eleventh-grade students in this county were selected at random, how many of these would we expect to have failed the test?

Solution   We would expect 15% of the 60 students to have failed the test; that is, (0.15)(60) = 9 students.   ◄

EXERCISE 14.7

In Problems 1–3, each table gives the winning results in an Olympic event from 1960 to 1984. Make a graph of the data. Draw the best line that you can "between" the points, and predict the results for the 1988 Olympics.

1. Men's 400-m hurdles, time in seconds

YEAR	1960	1964	1968	1972	1976	1980	1984
TIME	49.3	49.6	48.1	47.8	47.6	48.7	47.8

2. Women's 100-m freestyle swim, time in seconds

YEAR	1960	1964	1968	1972	1976	1980	1984
TIME	61.2	59.5	60.0	58.6	55.7	54.8	55.9

3. Women's running high jump, height in inches

YEAR	1960	1964	1968	1972	1976	1980	1984
HEIGHT	72.75	74.75	71.75	75.63	76.00	77.50	79.50

4. The table below shows the world records for the men's mile run since 1965. The times are the number of seconds over 3 min. Make a graph of the data. Draw the best line that you can "between" the points, and predict when the mile will first be run in 3 min and 45 sec.

YEAR	1965	1966	1967	1975	1980	1981	1985
TIME	53.6	51.3	51.1	49.4	48.8	47.3	46.3

5. In the school of Example 2, of the 50 students selected at random, 12 said they preferred their hamburgers plain. How many of the 3000

students in the school would you expect to prefer their hamburgers plain?

6. A state welfare department selected 150 people at random from its welfare roll of 10,000 people. Upon investigation, it was found that 9 of the 150 had gotten on the welfare roll through fraud. About how many of the 10,000 people on the roll would you expect to be guilty of fraud?

7. An automobile tire manufacturer selected a random sample of 150 tires from a batch of 10,000 tires. It was found that 3 of the 150 tires were defective. How many of the 10,000 tires should the manufacturer expect to be defective?

8. An automobile manufacturer selected a random sample of 150 cars that it manufactured and found a defective steering assembly in 5 of the 150 cars. If the manufacturer had turned out 5000 cars under the same conditions, how many of these should he expect to have defective steering assemblies?

USING YOUR KNOWLEDGE 14.7

In this section we described what is meant by a least-squares line to fit a given set of points. In order to facilitate writing the formulas for the calculation of such a line, we shall use the summation symbol Σ to indicate a sum. For example, if we have a number of x values, then Σ x stands for the sum of these values. Similarly, Σ x² stands for the sum of the squares of the x values. Thus, if the x values are 1, 2, 3, 4, then $\Sigma x = 1 + 2 + 3 + 4 = 10$ and $\Sigma x^2 = 1 + 4 + 9 + 16 = 30$.

Although the derivation of the least-squares formula is too advanced to be given here, the formulas themselves are not difficult to describe. Suppose we want the least-squares line for n given points, (x_1, y_1), $(x_2, y_2), \ldots, (x_n, y_n)$. If the equation of this line is

$$y = mx + b \tag{1}$$

then

$$m = \frac{n(\Sigma\, xy) - (\Sigma\, x)(\Sigma\, y)}{n(\Sigma\, x^2) - (\Sigma\, x)^2} \tag{2}$$

and

$$b = \frac{(\Sigma\, x^2)(\Sigma\, y) - (\Sigma\, x)(\Sigma\, xy)}{n(\Sigma\, x^2) - (\Sigma\, x)^2} \tag{3}$$

x	y	x²	xy
1	4.4	1	4.4
2	3.7	4	7.4
3	3.4	9	10.2
4	3.2	16	12.8
5	3.0	25	15.0
6	2.5	36	15.0
7	2.4	49	16.8
8	2.37	64	18.96
10	1.81	100	18.1
46	26.78	304	118.66

To use these formulas efficiently, we make a table with the headings x, y, x², and xy, as shown in the margin. The first two columns simply list the x's and y's, the third column lists the x²'s, and the last column gives the xy products. After filling out the table, we add the four columns to get Σ x, Σ y, Σ x², and Σ xy, the four sums that are required in equations (2) and (3).

We illustrate the calculation of the least-squares line for the data for the women's 200-m dash given at the beginning of this section. To simplify the arithmetic, we designate the successive Olympics starting with 1948 as $1, 2, 3, \ldots$; these are the x values. For the y values, we take the number of seconds over 20. Then our calculations are as shown in the table.

We have thus found $\Sigma x = 46$, $\Sigma y = 26.78$, $\Sigma x^2 = 304$, and $\Sigma xy = 118.66$, and we are ready to use equations (2) and (3). Since there are nine points, $n = 9$, and we get

$$m = \frac{(9)(118.66) - (46)(26.78)}{(9)(304) - (46)^2} = -0.264$$

$$b = \frac{(304)(26.78) - (46)(118.66)}{(9)(304) - (46)^2} = 4.327$$

(These answers are rounded to three decimal places.) The required line has the equation, given by equation (1),

$$y = -0.264x + 4.327$$

If we put $x = 9$ in this equation, we get $y = 1.951 \approx 1.95$. Thus, the predicted winning time for the 1980 Olympics is $20 + 1.95$, or about 22 sec. (Of course, you must not carry this type of prediction too far, because after about 36 Olympics, the winner would reach the finish line before she started the race!)

1. Find the least-squares line for the data in Example 1. Let x be the number of the Olympics with 1960 as number 1, and let the y values be the number of seconds over 50.
2. Find the least-squares line for Problem 1, Exercise 14.7.
3. Find the least-squares line for Problem 2, Exercise 14.7.
4. Find the least-squares line for Problem 3, Exercise 14.7.
5. Find the least-squares line for Problem 4, Exercise 14.7.
   [Hint: Let x be the number of years after 1965 and let y be the number of seconds over 45.]

14.8

SCATTERGRAMS AND CORRELATION

Twenty students tried out for the basketball team at West Side High. The coach listed their heights and weights as in Table 14.8a. We have graphed the ordered pairs (Height, Weight) as shown in Fig. 14.8a. This type of graph, the scattered set of points, is called a **scattergram.** The scattergram indicates how the height and weight are related. As you might expect, in any group of boys (or girls) the greater height would usually correspond to the greater weight.

Table 14.8a

HEIGHT (in.)	WEIGHT (lb)	HEIGHT (in.)	WEIGHT (lb)
61.4	106	68.9	147
62.6	108	68.9	152
63.0	101	69.3	143
63.4	114	69.7	143
63.8	112	70.1	150
65.7	123	70.9	147
66.1	121	70.9	163
67.3	136	72.8	158
67.7	143	72.8	165
68.1	143	73.2	163

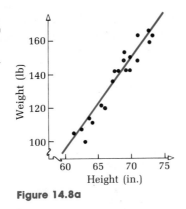

Figure 14.8a

The line drawn "between" the points in Fig. 14.8a is the least-squares line for the data in Table 14.8a. Notice that most of the points lie close to the line. Because this line slopes upward, we say that the scattergram shows a **positive correlation** between the heights and the weights of the 20 students.

Three kinds of correlation are possible. The scattergrams in Fig. 14.8b. illustrate typical cases. A good illustration of a negative correlation appears in Example 1 of Section 14.7.

Figure 14.8b

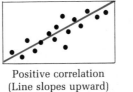
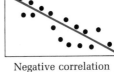
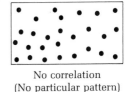

Positive correlation (Line slopes upward) Negative correlation (Line slopes downward) No correlation (No particular pattern)

EXAMPLE 1   Ten students were selected at random and a comparison was made of their high school grade-point averages (GPA) and their grade-point averages at the end of their freshman year in college (see the table). Make a scattergram and decide what kind of correlation is present.

HIGH SCHOOL GPA	COLLEGE GPA
2.2	2.0
2.4	2.0
2.5	2.7
2.7	2.3
2.9	3.0
3.0	2.5
3.2	2.8
3.5	3.4
3.9	4.0
4.0	3.9

Solution   We graph the given ordered pairs as shown in the scattergram below. This scattergram indicates a positive correlation between the high school and college GPA's. Are you surprised?

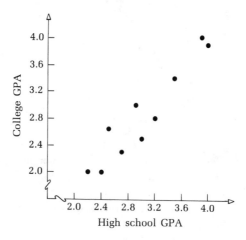

High school GPA

◄

In Problems 1–8, state what kind of correlation you would expect in a scattergram for the indicated ordered pairs.

1. (Length of person's leg, Person's height)
2. (Outdoor temperature, Cost of air-conditioning a house)
3. (Student's weight, Student's score on math test)
4. (Person's salary, Cost of person's home)
5. (Altitude, Atmospheric pressure)
6. (Weight of auto, Miles per gallon of fuel)
7. (Student's score on college aptitude test, Student's GPA)
8. (Speed of auto, Miles per gallon)

In Problems 9–16, make a scattergram for the given data, and state what kind of correlation is present.

9.

x	2	4	7	3	1	5	8	10	6	9
y	1	3	8	4	2	2	6	7	6	4

10.

x	9	10	8	8	6	4	5	5	3	2
y	4	2	3	6	4	8	7	5	8	10

11. The following table gives the weights and the highway miles per gallon for eight automobiles:

WEIGHT (lb)	2800	1900	2000	3300	3100	2900	4000	2600
MILES PER GALLON	19	34	28	19	24	23	16	24

12. The following table gives the scores of 10 students on an English exam and their corresponding scores on an economics exam:

ENGLISH	50	95	55	20	85	75	45	20	80	90
ECONOMICS	75	95	70	35	70	80	40	15	60	90

13. Here are some recent statistics on years of schooling successfully completed and average annual salaries for men over 25:

YEARS OF SCHOOL	8	12	15	16
AVERAGE SALARY	$16,800	$23,300	$25,800	$33,900

14. The following table gives the gain in reading speed for students in a speed-reading program:

WEEKS IN PROGRAM	2	3	3	4	5	6	8	9
SPEED GAIN (Words per Minute)	40	60	80	100	110	150	190	220

15. A student was curious about the effect of antifreeze on the freezing point of a water–antifreeze mixture. He went to the chemistry lab, where he made the measurements in the following table:

PERCENT ANTIFREEZE (By Volume)	10	20	30	40	50
FREEZING POINT (Degrees C)	−4	−10	−20	−24	−36

16. The following table gives the heights of students and their corresponding scores on an English test:

HEIGHT (in.)	62	67	70	64	72	68	65	61	73	67
TEST SCORE	85	60	75	70	95	35	60	80	45	100

(The following discussion requires a comprehension of the material in Using Your Knowledge 14.7.)

In order to describe how good the correlation is for a given set of ordered pairs, Karl Pearson introduced a measure called a **correlation coefficient.** In order to calculate this coefficient, we need exactly the quantities that we use to find the least-squares line in addition to one more sum, $\Sigma\, y^2$, the sum of the squares of the y's. In terms of these quantities, the correlation coefficient r is given by the formula

$$r = \frac{n(\Sigma\, xy) - (\Sigma\, x)(\Sigma\, y)}{\sqrt{n(\Sigma\, x^2) - (\Sigma\, x)^2}\ \sqrt{n(\Sigma\, y^2) - (\Sigma\, y)^2}} \tag{1}$$

If you have calculated the least-squares line, then the formula for r is quite easy to use in spite of its forbidding appearance.

Pearson devised his formula so that $-1 \le r \le 1$, with a value of r close to 1 meaning that there is a high positive correlation, and a value of r close to -1 meaning that there is a high negative correlation. It is usually agreed that a value of r between -0.2 and 0.2 indicates an insignificant correlation.

We illustrate the use of equation (1) by calculating r for the data we used to find a least-squares line in Using Your Knowledge 14.7. By adding the squares of the y values, we obtain $\Sigma\, y^2 = 84.75$ (rounded to two places), and from the table on page 659, $\Sigma\, x = 46$, $\Sigma\, y = 26.78$, $\Sigma\, x^2 = 304$, $\Sigma\, xy = 118.66$, and $n = 9$ (the number of ordered pairs). By substituting into equation (1), we get

$$r = \frac{(9)(118.66) - (46)(26.78)}{\sqrt{(9)(304) - (46)^2}\ \sqrt{(9)(84.75) - (26.78)^2}}$$

$$= \frac{-163.94}{\sqrt{620}\ \sqrt{45.58}} \approx -0.98$$

Thus, we have just found that there is a very high negative correlation between the x's and y's. This means that the data are very closely described by the least-squares line we found earlier.

You should not assume that there is any cause-and-effect relationship between two variables simply because the correlation is high. A classic example of a high positive correlation is that of the number of storks found in English villages and the number of babies born in these villages. Do you think there is a cause-and-effect relationship here?

Use your calculator to find the correlation coefficient for the following problems in Exercise 14.8:

1. Problem 11   2. Problem 12   3. Problem 13
4. Problem 15   5. Problem 16

Courtesy of PSH

Section	Item	Meaning	Example
14.1A	Frequency distribution	A way of organizing a list of numbers	
14.1A	Frequency	Number of times an entry occurs	
14.1B	Histogram	A special type of graph consisting of vertical bars with no space between the bars	
14.1C	Frequency polygon	A line graph connecting the midpoints of the tops of the bars in a histogram	
14.2	Mean, $\bar{x}$	The sum of the scores divided by the number of scores	The mean of 3, 7, and 8 is 6.
14.2	Mode	The number that occurs most often	The mode of 1, 2, 2, and 3 is 2.
14.2	Median	If the numbers are arranged in order of magnitude for an odd number of scores, the median is the middle number; for an even number of scores, the median is the average of the two middle numbers.	Median ↓ 1 3 ⑧ 15 19 1 2 5 9 11 18 $\frac{5+9}{2} = 7 \leftarrow$ Median
14.3	Range	The difference between the greatest and the least numbers in a set	The range of 2, 8, and 19 is 17.
14.3	Standard deviation, s	$\sqrt{\dfrac{(x_1 - \bar{x})^2 + \cdots + (x_n - \bar{x})^2}{n}}$, where $\bar{x}$ is the mean and n is the number of items	
14.5	z-Score	$z = \dfrac{x - \bar{x}}{s}$, where x is a score, $\bar{x}$ is the mean, and s is the standard deviation	

The following scores were made on a scholastic aptitude test by a group of 25 high school seniors:

85	65	89	83	98
67	88	87	88	90
95	77	91	73	88
99	67	91	72	86
79	83	61	70	75

Use these data for Problems 1–3.

1. Group the scores into intervals of $60 < s \le 65$, $65 < s \le 70$, $70 < s \le 75$, and so on. Then make a frequency distribution with this grouping.
2. Make a histogram for the frequency distribution in Problem 1.
3. Make a frequency polygon for the distribution in Problem 1.
4. During a certain week, the following maximum temperatures (in degrees Fahrenheit) were recorded in a large eastern city: 78, 82, 82, 71, 69, 73, 70.
   a. Find the mean of these temperatures.
   b. Find the mode of the temperature readings.
   c. Find the median high temperature for the week.
5. a. Find the range of temperatures in Problem 4.
   b. Find the standard deviation of the temperatures in Problem 4.
6. A fair coin is tossed 256 times. If this experiment is repeated many times, the numbers of heads will form an approximately normal distribution, with a mean of 128 and a standard deviation of 8.
   a. Within what limits may we be almost 100% confident that the total number of heads in 256 tosses will lie?
   b. What is the probability that heads will occur fewer than 112 times?
7. A normal distribution consists of 1000 scores, with a mean of 100 and a standard deviation of 20.
   a. About how many of the scores are above 140?
   b. About how many scores are below 80?
   c. About how many scores are between 60 and 80?
8. A testing program shows that the breaking points of fishing lines made from a certain plastic fiber are normally distributed, with a mean of 10 lb and a standard deviation of 1 lb.
   a. What is the probability that one of these lines selected at random has a breaking point of more than 10 lb?
   b. What is the probability that one of these lines selected at random has a breaking point of less than 8 lb?
9. On a multiple-choice test taken by 1000 students, the scores were normally distributed, with a mean of 50 and a standard deviation of 5.

Find the z-score corresponding to a score of:

a. 58   **b.** 62

10. Agnes scored 88 in a French test and 90 in a psychology test. The mean score in the French test was 76, with a standard deviation of 18, and the mean in the psychology test was 80, with a standard deviation of 16. If the scores were normally distributed, which of Agnes' scores was the better score?

11. With the data given in Problem 9, find the probability that a randomly selected student will have a score between 50 and 62. (Use Table II in the back of the book.)

12. Here is a list of 5 of the most active stocks on the New York Stock Exchange on February 19, 1988. Make a bar graph of the yield rates of these stocks.

Stock	Price	Dividend	Yield Rate
Fed DS	$60\frac{1}{4}$	$1.48	2.5%
Ford M	$44\frac{3}{4}$	2.00	4.5
Noes Ut	$20\frac{3}{4}$	1.76	8.5
Exxon	$42\frac{3}{4}$	2.00	4.7
Gen El	$43\frac{1}{4}$	1.40	3.2

13. In a recent poll, the features that patrons liked in a restaurant were listed as follows:

Self-service salad bar	57%
Varied portion sizes	47%
More varied menu	42%
All-you-can-eat specials	37%
Self-service soup bar	30%

Use horizontal bars and make a bar graph of this information.

14. Draw a circle graph to show the percent of measurements in a normal distribution that fall in the intervals $\bar{x}$ to $\bar{x} + s$, $\bar{x} + s$ to $\bar{x} + 2s$, and $\bar{x} + 2s$ to $\bar{x} + 3s$.

15. A marketing executive for a food manufacturer wants to show that food is an important part of a family budget. She finds that the typical budget is as follows:

Monthly Family Budget

Savings	$ 300
Housing	500
Clothing	200
Food	800
Other	200
Total	$2000

Make a circle graph for these data.

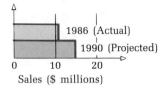

Sales ($ millions)

16. The bar graph in the figure in the margin shows the 1986 sales and the projected 1990 sales of the Wesellum Corporation. Read the graph, and estimate the percent increase that was projected for 1990 over 1986.

17. The testing department of Circle Tire Company checks a random sample of 150 of a certain type of tire that the company makes and finds a defective tread on 3 of these tires. In a batch of 10,000 of these tires, how many are expected to have defective treads?

18. In a large county, 50,000 high school students took a reading comprehension test, and 4000 of these students got a rating of excellent. In a random sample of 100 of these students, how many should be expected to have gotten an excellent rating on this test?

19. Graph the 5 points given in the table below. Draw the best line you can "between" these points, and estimate the value of y for x = 7.

x	1	2	3	4	5
y	10.2	7.6	5.8	4.4	2.0

20. What kind of correlation would you expect for the indicated ordered pairs?
    a. (Value of a family's home, Family's annual income)
    b. (Number of hours of training, Number of minutes in which runner can do the mile run)
    c. (Person's shoe size, Person's salary)
    d. (Number of children getting polio immunization, Number of children getting polio)

CHAPTER 15

CONSUMER MATHEMATICS

Have you ever wished for a money tree? The picture shows an East Indian money tree. The tin coins, used by the people of the Malay Peninsula in the nineteenth century, were broken off as needed.

The first coins were probably made about 2500 years ago in Lydia, a country that is now part of western Turkey. The coins were bean-shaped lumps of a natural mixture of gold and silver called *electrum*, and they were stamped with a design showing that the king guaranteed them to be of uniform size. These coins were accepted by traders as a convenient medium of exchange and inspired other countries to make their own coins.

According to many historians, coins were also independently invented in China and India. About 3000 years ago, the Chinese used miniature bronze tools for money, and these miniatures were later modified into coins.

The first paper money was used in China about 1400 years ago. Marco Polo, who traveled in China in the 1200's, was amazed to see the Chinese people using paper money instead of coins. In a book about his travels, he said that the Chinese accepted the paper money freely because they could easily use it to purchase whatever merchandise they might need. However, Europeans were skeptical about a piece of paper having any value, and it was not until the seventeenth century, when banks began to issue paper bills (called *bank notes*) to depositors and borrowers, that paper money was accepted.

In the early American colonies, several kinds of paper money were circulated. Some of this money was in the form of bills of credit issued by the individual colonies, and some, backed by real estate, was issued by the newly established land banks. During the Revolutionary War, the Continental Congress printed money that was called *Continental currency*. Unfortunately, this currency was overissued and soon became worthless. You may have heard the phrase, "not worth a continental," which stems from this bad experience.

Until 1863, most of the currency in the United States consisted of notes issued by various state banks. In 1863, Congress approved the establishment of national banks with the authority to issue bank notes. By taxing all new issues of state bank notes, they forced such notes out of existence. From 1863 until 1913, when the Federal Reserve System was established, the main currency was in the form of national bank notes. After 1913, Federal Reserve notes gradually replaced national bank notes and became the dominant currency of the United States.

Courtesy Chase Manhattan Archives

INTEREST

There are many consumer problems that you will probably encounter and that will directly affect your own pocketbook. Some of these problems involve calculating best buys, checking credit account statements, comparing interest rates, using credit cards, and purchasing a home or a car. In this chapter, we have tried to discuss most of these problems.

Realistic problems of these types are usually solved with the aid of a calculator. Hence, the examples show how a calculator can be used to obtain the answer. Of course, a calculator is not absolutely necessary; it is just a tool to help you do the arithmetic faster and easier. If you choose to use a calculator, we recommend one that is based on *algebraic logic*. You can tell if your calculator uses algebraic logic by doing the calculation: $2 + 3 \times 4$. Press $\boxed{2}$ $\boxed{+}$ $\boxed{3}$ $\boxed{\times}$ $\boxed{4}$ $\boxed{=}$. If the display shows 14, the logic is algebraic. If it shows 20, the calculator does not use the algebraic order of operations that gives priority to multiplication and division, and you should refer to your user's manual to determine how to proceed.

A. Simple Interest

How much would you get if you received 12% interest on $1 million for 10 years? The answer depends on how this interest is calculated! In any event, the answer will involve three things:

1. the **principal, P:** The amount of money loaned or deposited in the transaction ($1 million in this illustration)
2. The **rate, r:** The ratio of the amount of interest per period to the principal used, usually given as a percent (12% in our illustration)
3. The **time, t:** The term during which the principal is used, usually stated in years (10 years in our illustration), but may be otherwise specified

The **simple interest** for 1 year at the rate r on a principal P is just the principal times the rate—that is, Pr. The simple interest for t years is thus obtained by multiplying by t. This means that the formula for calculating the simple interest I on a principal P at the rate r for t years is

$$I = Prt$$

Notice that in the calculation of simple interest, the principal is just the original principal; the periodic interest does *not* earn further interest. Thus, if you receive 12% interest ($r = 12\% = 0.12$) on \$1 million ($P = \$1,000,000$) for 10 years ($t = 10$), the simple interest is

$$I = \quad P \quad \times \quad r \quad \times \quad t$$
$$= 1,000,000 \times 0.12 \times 10 \quad \text{dollars}$$
$$= 1,200,000 \quad \text{dollars}$$

Of course, at the end of the 10 years, you would also get your \$1 million back, so you would receive \$1,000,000 + \$1,200,000 = \$2,200,000 at simple interest. The final amount A is given as $A = P + I$.

However, if your annual interest were calculated on your original principal *plus* all previously earned interest — that is, if your interest were **compounded annually** — you would receive the much greater amount of \$3,105,848 (to the nearest dollar). We will discuss compound interest later in this section.

As a consumer, you will be interested in three important applications of simple interest: loans and deposits, taxes, and discounts. We consider these applications next.

Of course, you know that borrowing money from (or depositing money in) a bank or lending institution involves interest. Here is an example.

EXAMPLE 1   A loan company charges 32% simple interest for a 2-year, \$600 loan.

a. What is the total interest on this loan?
b. What is the interest for 3 months?
c. What is the total amount A that must be paid to the loan company at the end of 2 years?

Solution   **a.** The interest is given by $I = Prt$, where $P = 600$, $r = 32\% = 0.32$, and $t = 2$. Thus,

$$I = 600 \times 0.32 \times 2 = 384$$

On a calculator with a percent key $\boxed{\%}$, press

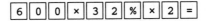

The interest for the 2 years is \$384.
b. Here, $P = 600$, $r = 32\% = 0.32$, and $t = \frac{3}{12} = \frac{1}{4}$, because 3 months is $\frac{3}{12}$ of a year. Thus,

$$I = 600 \times 0.32 \times \tfrac{1}{4}$$
$$= 600 \times 0.08 \qquad 0.32 \times \tfrac{1}{4} = 0.08$$
$$= 48.00$$

The interest for 3 months is $48.

c. At the end of 2 years, the loan company must be paid the original $600 plus the interest of $384; that is,

$$A = 600 + 384 = 984$$

The company must be paid $984. ◀

B. Taxes

You have probably heard the saying, "There is nothing certain but death and taxes." Here is a simple problem "it is certain" you can do.

EXAMPLE 2   A state has a 6% sales tax. Mary Rios buys an item priced at $84.

a. What is the sales tax on this item?
b. What is Mary's total cost for this item?

Solution   **a.** The sales tax S is simply 6% of 84; that is,

$$S = 0.06 \times 84 = 5.04$$

The tax is $5.04.

b. The total cost is the price, $84, plus the tax:

$$\$84 + \$5.04 = \$89.04$$

A calculator with a percent key $\boxed{\%}$ will give the total cost automatically if you press $\boxed{8}\,\boxed{4}\,\boxed{+}\,\boxed{6}\,\boxed{\%}\,\boxed{=}$. ◀

C. Discounts

In Examples 1 and 2, the consumer had to pay interest or taxes. But there is some hope! Sometimes, you can obtain a discount on certain purchases. Such a discount is usually stated as a percent. For instance, the coupon shown in the next example would entitle you to a 20% discount on certain purchases.

EXAMPLE 3   Ralph McWaters purchased a $42 item and used his coupon to get 20% off.

a. How much was his discount?
b. How much did he have to pay for the item?

Solution   **a.** His discount d was 20% of 42:

$$d = 0.20 \times 42 = \$8.40$$

b. Since he had a discount of $8.40, he had to pay

$42 − $8.40 = $33.60

for the item. A calculator with a percent key $\boxed{\%}$ will obtain the final price if you press $\boxed{4}\,\boxed{2}\,\boxed{-}\,\boxed{2}\,\boxed{0}\,\boxed{\%}\,\boxed{=}$. ◀

D. Compound Interest

As we mentioned earlier, when interest is **compounded,** the interest is calculated not only on the original principal but also on the earned interest. For example, if you deposit $1000 in a savings account that pays 6% interest compounded annually, then in the first year, the account will earn interest calculated as

$I = Prt$
$\quad = \$1000 \times 0.06 \times 1 = \$60.00$

If you make no withdrawal, then at the beginning of the second year the accumulated amount will be

$1000 + $60 = $1060

which is the new principal. In the second year, this new principal will earn interest

$I = Prt$
$\quad = \$1060 \times 0.06 \times 1 = \$63.60$

Thus, at the beginning of the third year, the accumulated amount will be

$1060 + $63.60 = $1123.60

and so on.

You can see that when interest is compounded, the earned interest increases each year ($60, $63.60, and so on, in the preceding illustration). This is because the interest at the end of a year is calculated on the accumulated amount (principal plus interest) at the beginning of that year. Piecewise calculation of the accumulated amount is a very time-consuming procedure, but it can be avoided by developing a general formula for the amount A_n accumulated after n interest periods and the use of special tables. To develop the formula for A_n, we let I be the compound interest, P be the original principal, i be the rate per period, and A_1 be the compound amount at the end of the first period:

$I = Pi$ Interest for the first period
$A_1 = P + I$
$\quad = P + Pi$ Substitute Pi for I.
$\quad = P(1 + i)$ Use the distributive property.

After the end of the second period, the compound amount A_2 is

$$A_2 = A_1 + A_1 i$$
$$= A_1(1 + i) \qquad \text{Use the distributive property.}$$
$$= P(1 + i)(1 + i) \qquad \text{Substitute } P(1 + i) \text{ for } A_1.$$
$$= P(1 + i)^2 \qquad \text{Substitute } (1 + i)^2 \text{ for } (1 + i)(1 + i).$$

If we continue this procedure, after n periods, A_n will be

$$A_n = P(1 + i)^n$$

Thus, if we deposit $1 at 6% compounded annually for 20 years,

$$A_{20} = \$1(1 + 0.06)^{20}$$

Fortunately, there are tables that give the value of the accumulated amount for a $1 initial deposit at compound interest i for n time periods.

Table 15.1a Amount (in dollars) to Which $1 Will Accumulate in n Periods under Compound Interest

n	1%	2%	3%	4%	5%	6%
1	1.0100	1.0200	1.0300	1.0400	1.0500	1.0600
2	1.0201	1.0404	1.0609	1.0816	1.1025	1.1236
3	1.0303	1.0612	1.0927	1.1249	1.1576	1.1910
4	1.0406	1.0824	1.1255	1.1699	1.2155	1.2625
5	1.0510	1.1041	1.1593	1.2167	1.2763	1.3382
6	1.0615	1.1262	1.1941	1.2653	1.3401	1.4185
7	1.0721	1.1487	1.2299	1.3159	1.4071	1.5036
8	1.0829	1.1717	1.2668	1.3686	1.4775	1.5938
9	1.0937	1.1951	1.3048	1.4233	1.5513	1.6895
10	1.1046	1.2190	1.3439	1.4802	1.6289	1.7908
11	1.1157	1.2434	1.3842	1.5395	1.7103	1.8983
12	1.1268	1.2682	1.4258	1.6010	1.7959	2.0122
13	1.1381	1.2936	1.4685	1.6651	1.8856	2.1329
14	1.1495	1.3195	1.5126	1.7317	1.9799	2.2609
15	1.1610	1.3459	1.5580	1.8009	2.0789	2.3966
16	1.1726	1.3728	1.6047	1.8730	2.1829	2.5404
17	1.1843	1.4002	1.6528	1.9479	2.2920	2.6928
18	1.1961	1.4282	1.7024	2.0258	2.4066	2.8543
19	1.2081	1.4568	1.7535	2.1068	2.5270	3.0256
20	1.2202	1.4859	1.8061	2.1911	2.6533	3.2071
21	1.2324	1.5157	1.8603	2.2788	2.7860	3.3996
22	1.2447	1.5460	1.9161	2.3699	2.9253	3.6035
23	1.2572	1.5769	1.9736	2.4647	3.0715	3.8198
24	1.2697	1.6084	2.0328	2.5633	3.2251	4.0489
30	1.3478	1.8114	2.4273	3.2434	4.3219	5.7434
36	1.4308	2.0399	2.8983	4.1039	5.7918	8.1473
42	1.5188	2.2972	3.4607	5.1928	7.7616	11.5570
48	1.6122	2.5870	4.1323	6.5705	10.4013	16.3939

Table 15.1a is such a table. To find the value of the accumulated amount $(1 + 0.06)^{20}$ in Table 15.1a, we go down the column headed n until we reach 20, and then go across to the column headed 6%. The accumulated amount given there is $3.2071. If we wish to know the accumulated amount for an original deposit of $1000 instead of $1, we simply multiply the $3.2071 by 1000 to obtain $3207.10.

In using Table 15.1a, here is one word of warning: The entries in this table have been rounded to four decimal places from more accurate values. Consequently, you should not expect answers to be accurate to more than the number of digits in the table entry. If more accuracy is needed, you must use a table with more decimal places or a calculator.

Many financial transactions call for interest to be compounded more often than once a year. In such cases, the interest rate is customarily stated as a nominal annual rate, it being understood that the actual rate per

n	7%	8%	9%	10%	11%	12%
1	1.0700	1.0800	1.0900	1.1000	1.1100	1.1200
2	1.1449	1.1664	1.1881	1.2100	1.2321	1.2544
3	1.2250	1.2597	1.2950	1.3310	1.3676	1.4049
4	1.3108	1.3605	1.4116	1.4641	1.5181	1.5735
5	1.4026	1.4693	1.5386	1.6105	1.6851	1.7623
6	1.5007	1.5869	1.6771	1.7716	1.8704	1.9738
7	1.6058	1.7138	1.8280	1.9487	2.0762	2.2107
8	1.7182	1.8509	1.9926	2.1436	2.3045	2.4760
9	1.8385	1.9990	2.1719	2.3579	2.5580	2.7731
10	1.9672	2.1589	2.3674	2.5937	2.8394	3.1058
11	2.1049	2.3316	2.5804	2.8531	3.1518	3.4785
12	2.2522	2.5182	2.8127	3.1384	3.4985	3.8960
13	2.4098	2.7196	3.0658	3.4523	3.8833	4.3635
14	2.5785	2.9372	3.3417	3.7975	4.3104	4.8871
15	2.7590	3.1722	3.6425	4.1772	4.7846	5.4736
16	2.9522	3.4259	3.9703	4.5950	5.3109	6.1304
17	3.1588	3.7000	4.3276	5.0545	5.8951	6.8660
18	3.3799	3.9960	4.7171	5.5599	6.5436	7.6900
19	3.6165	4.3157	5.1417	6.1159	7.2633	8.6128
20	3.8697	4.6610	5.6044	6.7275	8.0623	9.6403
21	4.1406	5.0338	6.1088	7.4002	8.9492	10.8038
22	4.4304	5.4365	6.6586	8.1403	9.9336	12.1003
23	4.7405	5.8715	7.2579	8.9543	11.0263	13.5523
24	5.0724	6.3412	7.9111	9.8497	12.2392	15.1786
30	7.6123	10.0627	13.2677	17.4494	22.8923	29.9599
36	11.4239	15.9682	22.2512	30.9127	42.8181	59.1356
42	17.1443	25.3395	37.3175	54.7637	80.0876	116.7231
48	25.7289	40.2106	62.5852	97.0172	149.7970	230.3908

interest period is the nominal rate divided by the number of periods per year. For instance, if interest is at 18%, compounded monthly, then the actual interest rate is $\frac{18}{12}\% = 1.5\%$ per month, because there are 12 months in a year.

EXAMPLE 4   Find the accumulated amount and the interest earned for:

a. $8000 at 8% compounded annually for 5 years
b. $3500 at 12% compounded semiannually for 10 years

Solution   **a.** In Table 15.1a, we go down the column under n until we come to 5 and then across to the column under 8%. The number there is 1.4693. Hence, the accumulated amount will be

$8000 × 1.4693 = $11,754   to the nearest dollar

The interest earned is the difference between the $11,754 and the original deposit, that is, $11,754 − $8000 = $3754.

b. Because *semiannually* means twice a year, interest is compounded every 6 months, and the actual rate per interest period is the nominal rate, 12% divided by 2, or 6%. In 10 years, there are 2 × 10 = 20 interest periods. Hence, we go down the column under n until we come to 20 and then across to the column headed 6% to find the accumulated amount 3.2071. Since this is the amount for $1, we multiply by $3500 to get

$3500 × 3.2071 = $11,225   to the nearest dollar

The amount of interest earned is $11,225 − $3500 = $7725.   ◀

EXERCISE 15.1

A.   In Problems 1–10, find the simple interest.

	Principal	Rate	Time
1.	$3,000	8%	1 year
2.	$4,500	7%	1 year
3.	$2,000	9%	3 years
4.	$6,200	8%	4 years
5.	$4,000	10%	6 months
6.	$6,000	12%	4 months
7.	$2,500	10%	3 months
8.	$12,000	9%	1 month
9.	$16,000	7%	5 months
10.	$30,000	8%	2 months

B. 11. The state sales tax in Florida is 6%. Desiree Cole bought $40.20 worth of merchandise.
   a. What was the tax on this purchase?
   b. What was the total price of the purchase?

12. The state sales tax in Alabama is 4%. Beto Frias bought a refrigerator priced at $666.
   a. What was the sales tax on this refrigerator?
   b. How much was the total price of the purchase?

13. Have you seen an FICA (Federal Insurance Contribution Act, better known as Social Security) deduction taken from your paycheck? For 1989, the FICA tax rate is 7.51% of your annual salary. Find the FICA tax for a person earning $24,000 a year.

14. It is projected that the FICA tax for 1990 will be 7.65% of your annual salary. Walter Snyder makes $30,000 a year. What will his FICA tax deduction be?

C. 15. An article selling for $200 was discounted 20%.
   a. What was the amount of the deduction?
   b. What was the final cost after the discount?
   c. If the sales tax was 5%, what was the final cost after the discount and including the sales tax?

16. A Sealy mattress sells regularly for $240. It is offered on sale at 50% off.
   a. What is the amount of the discount?
   b. What is the price after the discount?
   c. If the sales tax rate is 6%, what is the total price of the mattress after the discount and including the sales tax?

17. A jewelry store is selling rings at a 25% discount. If the original price of a ring was $500:
   a. What is the amount of the discount?
   b. What is the price of the ring after the discount?

18. If you have a Magic Kingdom Club Card from Disneyland, Howard Johnson's offers a 10% discount on double rooms. A family stayed at Howard Johnson's for 4 days. The rate per day was $45.
   a. What was the price of the room for the 4 days?
   b. What was the amount of the discount?
   c. If the sales tax rate was 6%, what was the total bill?

19. Some oil companies are offering a 5% discount on the gasoline you buy if you pay cash. Anita Gonzalez filled her gas tank, and the pump registered $14.20.
   a. If she paid cash, what was the amount of her discount?
   b. What did she pay after her discount?

20. U-Mart had bicycles selling regularly for $120. The bicycles were put on sale at 20% off. The manager found that some of the bicycles were dented, and offered an additional 10% discount.
   a. What was the price of a dented bicycle after the two discounts? (Careful! This is not a 30% discount, but 20%, followed by 10%.)

b. Would it be better to take the 20% discount followed by the 10% discount, or to take a single 28% discount?

D. In Problems 21–24, use Table 15.1a to find the final accumulated amount and the total interest if interest is compounded annually. (Answers to nearest five digits.)

21. $100 at 6% for 8 years

22. $1000 at 9% for 11 years

23. $2580 at 12% for 9 years

24. $6230 at 11% for 12 years

In Problems 25–30, use Table 15.1a to find the final accumulated amount and the total interest. (Give answers to the same number of digits as the table entry.)

25. $12,000 at 10% compounded semiannually for 8 years

26. $15,000 at 14% compounded semiannually for 10 years

27. $20,000 at 8% compounded quarterly for 3 years

28. $30,000 at 12% compounded quarterly for 4 years

29. $40,000 at 20% compounded semiannually for 24 years

30. $50,000 at 16% compounded semiannually for 15 years

31. When a child is born, grandparents sometimes deposit a certain amount of money that can be used to send the child to college. Mary and John Glendale deposited $1000 when their granddaughter Anna was born. The account was paying 6% compounded annually.

a. How much money will there be in the account when Anna becomes 18 years old?

b. How much money would there be in the account after 18 years if the interest had been compounded semiannually?

32. When Natasha was born, her mother deposited $100 in an account paying 6% compounded annually. After 10 years, the money was transferred into another account paying 10% compounded semiannually.

a. How much money was in the account after the first 10 years?

b. How much money was in the account at the end of 18 years? (Give answers to the nearest cent.)

33. Jack loaned Janie $3000. She promised to repay the $3000 plus interest at 10% compounded annually at the end of 3 years.

a. How much did she have to pay Jack at the end of 3 years?

b. How much interest did she pay?

34. Bank A pays 8% interest compounded quarterly, while bank B pays 10% compounded semiannually, If $1000 was deposited in each bank, how much money would there be at the end of 5 years in the:

a. Bank A account?   **b.** Bank B account?

c. In which bank would you deposit your money?

35. How much more would there be at the end of 5 years if $1000 were invested at 12% compounded quarterly rather than semiannually?

*You may have noticed banks advertising various **effective** annual interest rates. Do you know what an effective annual interest rate is? It is the simple interest rate that would result in the same total amount of interest as the compound rate yields in 1 year. For example, if interest is at 6% compounded quarterly, then a table shows that a $1 deposit would accumulate to $1.0614 in 1 year. (Use n = 4 and a periodic rate of 1½%; see Table 15.1c. Thus, the interest is $0.0614 and, because the initial deposit is $1, the effective annual rate is 0.0614, or 6.14%.*

Table 15.1b *Accumulated Amount of $1 for n Periods*

PERIODIC RATE	3%	4¼%	6%	7½%	9%
n = 2	1.0609	1.0920	1.1236	1.1556	1.1881

Table 15.1c *Accumulated Amount of $1 for n Periods*

PERIODIC RATE	1¼%	1½%	1¾%	2%	2¼%
n = 4	1.0509	1.0614	1.0719	1.0824	1.0931

Table 15.1d *Accumulated Amount of $1 for n Periods*

PERIODIC RATE	½%	¾%	1%	1¼%	1½%
n = 12	1.0617	1.0938	1.1268	1.1607	1.1956

You can see now that an easy way to find the effective annual rate is to look in a table for the accumulated amount of $1. If the nominal rate is r% and interest is compounded k times a year, then you will find the required entry by going down the column under n until you come to k and then across to the column under r/k%. By subtracting 1 from this entry, you obtain the desired effective rate. Tables 15.1b, c, and d give some selected values from compound interest tables.

Find the effective annual rate for each of the following:

1. 6%, compounded semiannually
2. 6%, compounded monthly
3. 8%, compounded quarterly
4. 12%, compounded monthly
5. 9%, compounded quarterly
6. 15%, compounded semiannually
7. 18%, compounded monthly
8. 15%, compounded monthly

▶ **Calculator Corner 15.1** *What happens if you do not have Table 15.1a? Use a calculator, of course. If your calculator has a $\boxed{y^x}$ key (a power key), then the quantity $(1.06)^{20}$ can be obtained by pressing*

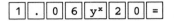

$$\boxed{1}\boxed{.}\boxed{0}\boxed{6}\boxed{y^x}\boxed{2}\boxed{0}\boxed{=}$$

This time the accumulated amount is given to nine decimal places!

15.2

CONSUMER CREDIT

One of the costs associated with credit cards is the **finance charge** that is collected if you decide to pay for your purchases later than the allowed payment period. Usually, if the entire balance is paid within a certain time (25–30 days), there is no charge. However, if you want more time, then you will have to pay the finance charge computed at the rate printed on the monthly statement you receive from the company issuing the card. Figure 15.2a shows the top portion of such a statement. As you can see, the periodic rate (monthly) is 1.5%. This rate is used to calculate the charge on $600. Where does the $600 come from? The back of the statement indicates that "The Finance Charge is computed on the Average Daily Balance, which is the sum of the Daily Balances divided by the number of days in the Billing Period." Fortunately, the computer calculated this average daily balance and came out with the correct amount. In Example 1, we shall verify only the finance charge.

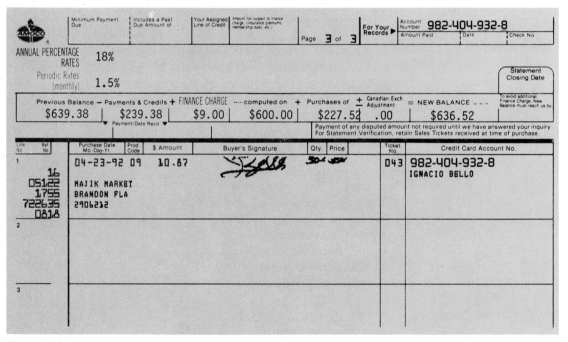

Figure 15.2a

EXAMPLE 1   Find the finance charge (interest) to be paid on the statement shown if the monthly rate is 1.5% computed on the average daily balance of $600.

Solution   The finance charge is 1.5% of $600, that is,

$$0.015 \times \$600 = \$9.00$$ ◄

Next, let us look at a different problem. Suppose you wish to obtain a national credit card. First, you have to make an application to the issuing bank for such a card. If your application is accepted, then you must pay a fee. (Some credit unions issue cards free.) After some time, the card finally arrives in your mail! Now, suppose you wish to use your card at a restaurant where these cards are accepted. Instead of collecting cash, the cashier will place your card in a stamping machine that will print on a receipt certain information that is on the card: the card number, your name, and the expiration date of the card; it will also print the name and identification numbers of the restaurant as well as the date of the transaction. The cashier will then write the amount of the purchase on the receipt and add the applicable tax. You will sign the receipt and be given a copy for your records. Figure 15.2b shows such a receipt.

At the end of the billing period, a statement is sent to you. If the balance due is $10 or less, you must pay the account *in full.* Otherwise, you must make a minimum payment of $10 or 5% of the balance due, *whichever is greater.* (Terms vary from one bank to another.)

Figure 15.2b

EXAMPLE 2   The customer who signed the receipt in Fig. 15.2b received a statement at the end of the month. The new balance was listed as $37.50. Find:

a. The minimum payment due

b. The finance charge that will be due the next month if only the minimum payment is made now

Solution   **a.** Because the new balance is $37.50 and 5% of $37.50 is $1.875, the minimum payment is $10. (Remember, you pay $10 or 5% of the new balance, whichever is greater.)

b. After paying the $10, the customer's new balance is

$37.50 − $10 = $27.50

The finance charge is 1.5% of $27.50, or

$0.015 \times 27.50 = \$0.4125$

Thus, if no additional credit card purchases are made, the finance charge will be $0.41.   ◄

Many large department stores prefer to handle their own credit card business and therefore do not accept bank credit cards. This procedure offers two main advantages to the stores:

1. The stores save the commission on sales that the national credit card companies charge for their services.
2. The interest (finance charges) collected from their customers is a welcome source of revenue to the stores.

Most charge accounts at department stores are called **revolving charge accounts.** Although the operational procedure of these accounts is similar to that employed by the national credit card companies, there may be some differences between them.

1. The interest for revolving charge accounts is $1\frac{1}{2}$% per month on the unpaid balance for balances under $500. If the balance is over $500, some accounts charge only 1% interest per month on the amount over $500.
2. The minimum monthly payment may be established by the department store, and it may or may not be similar to that of the national credit card companies.

EXAMPLE 3   Mary Lewis received her statement from Sears, where she has a revolving charge account. Her previous balance was $225.59, and she charged an additional $18.72 to her account. Find:

a. The finance charge for the month (Sears charges interest on the average daily balance as described in the form shown in Fig. 15.2c.)
b. The new balance
c. The minimum monthly payment

Sears REVOLVING CHARGE ACCOUNT

SEARS, ROEBUCK AND CO.

Please return this portion with your payment. You may pay at any Sears Store, or by mail. **Please allow 5 days to assure processing by the Billing Date.**

Please mention this number when ordering or writing.

97 11234 56789 7

MS MARY LEWIS
1234 MAIN ST
ANYWHERE IL 60606

Amount Due

INSTAL.PMT.   NONE DUE

NEW BALANCE

$_____
AMOUNT PAID

PLEASE DIRECT INQUIRIES TO YOUR NEAREST SEARS STORE AND REFER TO THIS STATEMENT WHEN MAKING AN INQUIRY.

DATE Mo. Day	Reference Number	TRANSACTION DESCRIPTION See reverse side for a more detailed description of the department numbers indicated below.	CHARGES	PAYMENTS AND CREDITS
		FINANCE CHARGE ON AVG DAILY BAL OF $222.99	3.34	
06 08	TPBT	MENS WORK & SPORTSWEAR 41	18.72	

ACCOUNT NUMBER	BILLING DATE	PREVIOUS BALANCE	NEW BALANCE	MINIMUM PAYMENT
97 11234 56789 7	JULY 06 1975	$ 225.59	$	$

If the *FINANCE CHARGE* exceeds 50¢, the *ANNUAL PERCENTAGE RATE* is 18% of the AVERAGE DAILY BALANCE excluding any purchases added during the monthly billing period and excluding any unpaid Finance Charge.

NOTICE: See reverse side for important information.

14351-610 Rev. 7/74

To avoid a Finance Charge next month, pay this amount within 30 days from Billing Date.

If you prefer to pay in installments, pay this amount or more within 30 days from Billing Date. The sooner you pay and the more you pay, the smaller your Finance Charge.

Thank You for Shopping at Sears

On a Sears Revolving Charge Account your monthly payments decrease as your account balance decreases . . . and, likewise your monthly payments increase as your account balance increases. Payments are flexible with your balance, as shown on the table below. Any premiums for Group Insurance, for which you may have contracted, other than Accidental Death and Disability Insurance, are in addition to your minimum monthly payments.

New Balance	Minimum Payment
$ .01 to $ 10.00	Balance
10.01 to 200.00	$10.00
200.01 to 250.00	15.00
250.01 to 300.00	20.00
300.01 to 350.00	25.00
350.01 to 400.00	30.00
400.01 to 450.00	35.00
450.01 to 500.00	40.00
Over $500.00	1/10 of New Bal.

Figure 15.2c

Courtesy of Sears, Roebuck and Co.

Solution   **a.** Since the average daily balance was $222.99, the finance charge is $1\frac{1}{2}\%$ (18% annual rate) of the $222.99, that is,

$$0.015 \times \$222.99 = \$3.34$$

b. The new balance is calculated as follows:

Previous balance	$225.59
Finance charge	3.34
New purchases	18.72
New balance	$247.65

c. The minimum monthly payment is found by using the information given at the bottom of Fig. 15.2c. (This is a copy of the table that appears on the back of the statement.) Because the new balance is between $200.01 and $250, the minimum payment is $15.   ◀

Some stores charge interest on the **unpaid balance.** How do we find the new balance under this method? The next example will tell you.

EXAMPLE 4   Ms. Spoto received a statement from a department store where she has a charge account. Her previous balance was $280. She made a $20 payment and charged an additional $30.12 to her account. If the store charges 1.5% of the unpaid balance, find:

a. The finance charge   **b.** The new balance

Solution   **a.** The unpaid balance is $280 − $20 = $260, so the finance charge is 1.5% of $260, which is 0.015 × 260 = $3.90.

b. The new balance is computed as follows:

Unpaid balance	$260.00
Finance charge	3.90
Purchases	30.12
New balance	$294.02

◀

There is another way of charging interest when consumers buy on credit, the **add-on interest** used by furniture stores, appliance stores, and car dealers. For example, suppose you wish to buy some furniture costing $2500 and you make a $500 down payment. The amount to be financed is $2000. If the store charges a 10% add-on rate for 5 years (60 monthly payments), the interest will be

$$I = Prt$$
$$= \$2000 \times 0.10 \times 5 = \$1000$$

Thus, the total amount to be paid is $2000 + $1000 = $3000. The monthly payment is found by dividing this total by the number of payments:

$$\text{Monthly payment} = \frac{\$3000}{60} = \$50$$

The *best* interest rate is shown in this photo.

Note that the add-on interest is charged on the *entire* $2000 for the 5 years, but the customer does *not* have the full use of the entire amount for the 5 years. It would be fairer to charge interest on the *unpaid balance* only.

EXAMPLE 5   A car costing $8500 can be bought with $2500 down and a 12% add-on interest rate to be paid in 48 monthly installments. Find:

a. The total interest charged   **b.** The monthly payment

Solution   **a.** The amount to be financed is $8500 − $2500 = $6000. The interest is 12% of $6000 for 4 years. Thus,

Interest $= 0.12 \times \$6000 \times 4 = \$2880$

b. Total amount owed $= \$6000 + \$2880 = \$8880$

$$\text{Monthly payment} = \frac{\$8880}{48} = \$185 \qquad \blacktriangleleft$$

EXERCISE 15.2

In Problems 1–5, find the new balance, assuming that the bank charges $1\frac{1}{2}$% per month on the unpaid balance.

	Previous Balance	Payment	New Purchases
1.	$100	$10	$50
2.	$300	$190	$25
3.	$134.39	$25	$73.98
4.	$145.96	$55	$44.97
5.	$378.93	$75	$248.99

In Problems 6–15, find:

a. The finance charge for the month
b. The new balance
c. The minimum monthly payment

Use the following rates and payments table:

MONTHLY RATE	UNPAID BALANCE	NEW BALANCE	MINIMUM PAYMENT
$1\frac{1}{2}$%	Up to $500	Under $200	$10
1%	Over $500	Over $200	5% of new balance

	Previous Balance	New Purchases		Previous Balance	New Purchases
6.	$50.40	$173	**7.**	$85	$150
8.	$154	$75	**9.**	$344	$60
10.	$666.80	$53.49	**11.**	$80.45	$98.73
12.	$34.97	$50	**13.**	$55.90	$35.99
14.	$98.56	$45.01	**15.**	$34.76	$87.53

16. Phyllis Phillips has a revolving charge account that charges a finance charge on the unpaid balance using the following schedule:

$1\frac{1}{2}$% per month of that portion of the balance up to $300
1% per month on that portion of the balance over $300

If the previous month's balance was $685, find the finance charge.

17. Daisy Rose has a credit card that charges a finance charge on the previous balance according to the following schedule:

2% per month on balances up to $100
$1\frac{1}{2}$% per month on balances between $100 and $200
1% per month on balances of $200 or over

If the previous month's balance was $190, find the finance charge.

18. Mr. Dan Dapper received a statement from his clothing store showing a finance charge of $1.50 on a previous balance of $100. Find the monthly finance charge rate.

19. Paul Peters received a statement from the ABC Department Store showing a previous balance of $90. If the ABC store's finance charges are 1.5% on the previous balance, find the finance charge for the month.

20. In Problem 19, if the monthly rate were 1.25%, what would be the finance charge for the month?

21. A $9000 car can be purchased with $1600 down, the balance plus a 9% add-on interest rate to be paid in 36 monthly installments. Find:
   a. The total interest charged
   b. The monthly payment, rounded to the nearest dollar

22. The Ortegas move into their first apartment and decide to buy furniture priced at $400 with $40 down, the balance plus 10% add-on interest to be paid in monthly installments in 1 year. Find:
   a. The total interest charged   **b.** The monthly payment

23. Wayne Sierpinski wishes to buy a stove and a refrigerator from an appliance dealer. The cost of the two items is $2400, and Wayne pays $400 down and finances the balance at 15% add-on interest to be paid in 18 monthly installments. Find:
   a. The total interest charged
   b. The monthly payment, to the nearest dollar

24. Bill Seeker bought a boat costing $8500 with $1500 down, the balance plus add-on interest to be paid in 36 monthly installments. If the add-on interest rate was 18%, find:
   a. The total interest charged
   b. The monthly payment, to the nearest dollar
25. Felicia Johnson bought a freezer costing $500 on the following terms: $100 down and the balance plus a 10% add-on interest rate to be paid in 18 monthly installments. Find:
   a. The total interest to be paid by Ms. Johnson
   b. The amount of her monthly payment, to the nearest dollar

USING YOUR KNOWLEDGE 15.2

A table shows that a monthly payment of $61 (to the nearest dollar) for 18 months will repay $1000 with interest at 1% per month on the unpaid balance. Can we find the equivalent add-on interest rate? Yes; here is how to do it. Eighteen payments of $61 make a total of $1098 (18 × $61), which shows that the total interest paid is $98. Since this is the interest for 18 months (1½, or ³⁄₂, years), the equivalent add-on interest rate is

$$\frac{98}{(1000)(\frac{3}{2})} = \frac{98}{1500} = 0.065$$

or 6.5%, to the nearest tenth of a percent.

A table shows that the following monthly installment payments will repay $1000 in the stated term and at the stated rate of interest on the unpaid balance. Find the equivalent add-on interest rate to the nearest tenth of a percent.

Monthly Payment	Term	Rate per Month on Unpaid Balance
1. $47	2 years	1%
2. $64	18 months	1½%
3. $50	2 years	1½%
4. $48	2 years	1¼%

15.3

ANNUAL PERCENTAGE RATE (APR) AND THE RULE OF 78

As you have seen in Sections 15.1 and 15.2, there are many ways of stating the interest rates used to compute credit costs. A few examples are 12%

simple interest, 12% compounded annually, 12% add-on interest, and 1% per month on the unpaid balance. How can you compare various credit costs? Without some help, it is difficult to do this. The Congress of the United States recognized this difficulty and enacted the Truth in Lending Act on July 1, 1969. This law helps the consumer to know exactly what credit costs. Under this law, all sellers (car dealers, banks, credit card companies, and so on) must disclose to the consumer two items:

1. The finance charge
2. The *annual percentage rate* (APR)

A. APR

Recall that the finance charge is the total dollar amount you are charged for credit. It includes interest and other charges such as service charges, loan and finder's fees, credit-related insurance, and appraisal fees. The **annual percentage rate (APR)** is the charge for credit stated as a percent.

In general, the lowest APR corresponds to the best credit buy, regardless of the amount borrowed or the period of time for repayment. For example, suppose you borrow $100 for a year and pay a finance charge of $8. If you keep the entire $100 for the whole year and then pay $108 all at one time, then you are paying an APR of 8%. On the other hand, if you repay the $100 plus the $8 finance charge in 12 equal monthly payments (8% add-on), you do not really have the use of the $100 for the whole year. What, in this case, is your APR? The formulas needed to compute the APR are rather complicated, and as a consequence, tables such as Table 15.3a have been constructed to help you find the APR. These tables are based on the cost per $100 of the amount financed. To use Table 15.3a, you must first find the finance charge per $1 of the amount financed and then multiply by 100. Thus, to find the APR on the $100 borrowed at 8% add-on interest and repaid in 12 equal payments of $9, we first find the finance charge per $100 as follows:

1. The finance charge is $108 − $100 = $8.
2. The charge per $100 financed is

$$\frac{\text{Finance charge}}{\text{Amount financed}} \times 100 = \frac{\$8}{\$100} \times 100 = \$8$$

Since there are 12 payments, we look across the row labeled 12 in Table 15.3a until we find the number closest to $8. This number is $8.03. We then read the heading of the column in which the $8.03 appears to obtain the APR. In our case, the heading is $14\frac{1}{2}$%. Thus, the 8% add-on rate is equivalent to a $14\frac{1}{2}$% APR. (Of course, Table 15.3a will give the APR only to the nearest $\frac{1}{2}$%.)

Table 15.3a True Annual Interest Rate (APR)

NUMBER OF PAYMENTS	14%	14½%	15%	15½%	16%	16½%	17%	17½%	18%
6	$ 4.12	$ 4.27	$ 4.42	$ 4.57	$ 4.72	$ 4.87	$ 5.02	$ 5.17	$ 5.32
12	7.74	8.03	8.31	8.59	8.88	9.16	9.45	9.73	10.02
18	11.45	11.87	12.29	12.72	13.14	13.57	13.99	14.42	14.85
24	15.23	15.80	16.37	16.94	17.51	18.09	18.66	19.24	19.82
30	19.10	19.81	20.54	21.26	21.99	22.72	23.45	24.18	24.92
36	23.04	23.92	24.80	25.68	26.57	27.46	28.35	29.25	30.15
42	27.06	28.10	29.15	30.19	31.25	32.31	33.37	34.44	35.51
48	31.17	32.37	33.59	34.81	36.03	37.27	38.50	39.75	41.00

Note: Numbers in body of table are finance charges per $100 of amount financed.

EXAMPLE 1   Mary Lewis bought some furniture that cost $1400. She paid $200 down and agreed to pay the balance in 30 monthly installments of $48.80 each. What was the APR for her purchase?

Solution   We first find the finance charge per $100 as follows:

Payments:	$30 \times 48.80 =$ $1464
Amount financed:	-1200
Finance charge:	$ 264

Finance charge per $100: $\dfrac{\$264}{\$1200} \times 100 = \$22$

We now turn to Table 15.3a and read across the row labeled 30 (the number of payments) until we find the number closest to $22. This number is $21.99. We then read the column heading to obtain the APR, 16%.

◄

B. The Rule of 78

In all the preceding examples, we have assumed that the consumer will faithfully make the payments until the debt is satisfied. But what if you wish to pay in full before the final due date? (Perhaps your rich aunt gave you some money.) In many cases, you are entitled to a partial refund of the finance charge! The problem is to find how much you should get back. One way of calculating the refund is to use the **rule of 78.** This rule assumes that the final payment includes a portion, say a, of the finance charge, the payment before that includes $2a$ of the finance charge, the second from the final payment includes $3a$ of the finance charge, and so on. If the total number of payments is 12, then the finance charge is paid off by the sum of $a + 2a + 3a + 4a + 5a + 6a + 7a + 8a + 9a + 10a + 11a + 12a = 78a$ dollars. If the finance charge is F, then we see that

$$78a = F$$

so $a = \frac{1}{78}F$. This is the reason for the name rule of 78. Now, suppose you borrow \$1000 for 1 year at 8% add-on interest. The interest is \$80, and the monthly payment is one-twelfth of \$1080, that is, \$90. If you wish to pay off the loan at the end of 6 months, are you entitled to a refund of half of the \$80 interest charge? Not according to the rule of 78. Your remaining finance charge payments, according to this rule, are

$$\tfrac{1}{78}F + \tfrac{2}{78}F + \tfrac{3}{78}F + \tfrac{4}{78}F + \tfrac{5}{78}F + \tfrac{6}{78}F = \tfrac{21}{78}F$$

Since $F = \$80$, you are entitled to a refund of $\frac{21}{78} \times 80$, or \$21.54. There are six payments of \$90 each for a total of \$540, so you would need to pay \$540 $-$ \$21.54 $=$ \$538.46 to cover the balance of the loan.

Notice that to obtain the numerator of the fraction $\frac{21}{78}$, we had to add $1 + 2 + 3 + 4 + 5 + 6$. If there were n payments remaining, then to find the numerator we would have to add

$$1 + 2 + 3 + \cdots + (n-2) + (n-1) + n$$

There is an easy way to do this. Let us call the sum S. Then we can write S twice, once forwards and once backwards:

$$S = 1 + \quad 2 \quad + \quad 3 \quad + \cdots + (n-2) + (n-1) + n$$
$$S = n + (n-1) + (n-2) + \cdots + \quad 3 \quad + \quad 2 \quad + 1$$

If we add these two lines, we get

$$2S = (n+1) + (n+1) + (n+1) + \cdots + (n+1) + (n+1) + (n+1)$$

and, because there are n terms on the right,

$$2S = n(n+1)$$
$$S = \frac{n(n+1)}{2}$$

Thus, for $n = 6$, we obtain

$$S = \frac{6 \times (6+1)}{2} = \frac{6 \times 7}{2} = 21$$

as before. For $n = 12$, we find $S = (12 \times 13)/2 = 78$, which again agrees with our previous result.

Now suppose the loan is for 15 months, and you wish to pay off in full after 10 payments, so that there are 5 payments remaining. By arguing in the same way as for the 12-month loan, you can see that you are entitled to a refund of a fraction a/b of the finance charge, where the numerator is

$$a = 1 + 2 + 3 + 4 + 5 = \frac{5 \times 6}{2} = 15$$

and the denominator is

$$b = 1 + 2 + 3 + \cdots + 15 = \frac{15 \times 16}{2} = 120$$

Thus, you are entitled to $\frac{15}{120} = \frac{1}{8}$ of the total finance charge.

In general, if the loan calls for a total of n payments and the loan is paid off with r payments remaining, then the unearned interest is a fraction a/b of the total finance charge, with the numerator

$$a = 1 + 2 + 3 + \cdots + r = \frac{r(r + 1)}{2}$$

and the denominator

$$b = 1 + 2 + 3 + \cdots + n = \frac{n(n + 1)}{2}$$

Thus,

$$\frac{a}{b} = \frac{r(r + 1)}{n(n + 1)} \quad \text{The 2's cancel.}$$

and the unearned interest is given by

$$\frac{r(r + 1)}{n(n + 1)} \times F$$

where F is the finance charge. For example, if an 18-month loan is paid off with 6 payments remaining, the amount of unearned interest is

$$\frac{6 \times 7}{18 \times 19} \times F = \frac{7}{57} F$$

Although the denominator is no longer 78 (except for a 12-payment loan), the rule is still called the rule of 78.

EXAMPLE 2   Cal Olleb purchased a televison set on a 15-month installment plan that included a $60 finance charge and called for payments of $25 monthly. If Cal decided to pay off the loan at the end of the eighth month, find:

a. The amount of the interest refund using the rule of 78
b. The amount needed to pay off the loan

Solution   **a.** Here, $n = 15$ and $r = 7$. We substitute into the formula to obtain

$$\frac{7 \times 8}{15 \times 16} \times \$60 = \frac{7}{30} \times \$60 = \$14$$

b. There are 7 payments of $25 each left, that is, $175. Thus, Cal needs

$$\$175 - \$14 = \$161$$

to pay off the loan.   ◀

A. In Problems 1–10, find the APR.

	Amount Financed	Finance Charge	Number of Payments
1.	$2500	$194	12
2.	$2000	$166	12
3.	$1500	$264	24
4.	$3500	$675	24
5.	$1500	$210	18
6.	$4500	$1364	36
7.	$4500	$1570	48
8.	$4000	$170.80	6
9.	$5000	$1800	48
10.	$4000	$908.80	30

B. In Problems 11–15, find:

a. The unearned finance charge

b. The amount needed to pay off the loan

	Finance Charge	Number of Payments	Frequency	Amount	Number of Payments Remaining
11.	$15.60	12	Monthly	$25	4
12.	$23.40	12	Monthly	$35	5
13.	$31.20	12	Monthly	$45	6
14.	$52.00	18	Weekly	$10	9
15.	$58.50	20	Weekly	$10	5

16. Alfreda Brown bought a car costing $6500 with $500 down and the rest to be paid in 48 equal installments of $173.
    a. What was the finance charge?
    b. What was the APR?

17. Gerardo Noriega bought a dinette set for $300, which he paid in 12 monthly payments of $27.
    a. What was the finance charge?
    b. What was the APR on this sale?

18. Yu-Feng Liang bought a car for $6500. He made a down payment of $1000 and paid off the balance in 48 monthly payments of $159 each.
    a. What was the finance charge?
    b. What was the APR?

19. A used sailboat is selling for $1500. The owner wants $500 down and 18 monthly payments of $63.
    a. What finance charge does the owner have in mind?

b. What is the APR for this transaction?

20. Natasha Gagarin paid $195 interest on a $2000 purchase. If she made 12 equal monthly payments to pay off the account, what was the APR for this purchase?

21. Virginia Osterman bought a television set on a 12-month installment plan that included a $31.20 finance charge and called for payments of $50 per month. If she decided to pay the full balance at the end of the eighth month, find:
    a. The interest refund
    b. The amount needed to pay off the loan

22. Marie Siciliano bought a washing machine on a 12-month installment plan that included a finance charge of $46.80 and called for a monthly payment of $70. If Marie wanted to pay off the loan after 7 months, find:
    a. The interest refund
    b. The amount needed to pay off the loan

23. A couple buys furniture priced at $800 with $80 down and the balance to be paid at 10% add-on interest. If the loan is to be repaid in 12 equal monthly payments, find:
    a. The finance charge   **b.** The monthly payment
    c. The interest refund if they decide to pay off the loan after 8 months
    d. The amount needed to pay off the loan

24. Dan Leizack is buying a video recorder that costs $1200. He paid $200 down and financed the balance at 15% add-on interest to be repaid in 18 monthly payments. Find:
    a. The finance charge   **b.** The monthly payment
    c. The interest refund if he pays off the loan after 9 months
    d. The amount needed to pay off the loan

25. Joe Clemente bought a stereo costing $1000 with $200 down and 10% add-on interest to be paid in 18 equal monthly installments. Find:
    a. The finance charge   **b.** The monthly payment
    c. The interest refund if he pays off the loan after 15 months
    d. The amount needed to pay off the loan

15.4

BUYING A HOUSE

For many of you, the single largest credit purchase (and investment) of your life will probably be the buying of a house. This purchase will require you to make many decisions in what may be unfamiliar areas. We will try to help you make these decisions wisely.

One of the first, if not *the* first, decision you must make when buying a house is how much to spend. Certain rules of thumb are sometimes used as guides in helping a family decide what price home to buy. Here are three such rules:

---

1. Spend no more than 2 to $2\frac{1}{2}$ times your annual income.
2. Limit housing expenses to 1 week's pay out of each month's salary.
3. Do not let the amount of the monthly payment of principal, interest, taxes, and insurance exceed one-quarter of your take-home pay.

EXAMPLE 1   John and Pat Harrell earn $30,000 annually. Their take-home pay is $26,000. Can they afford a $60,000 home with a $40,000 mortgage that requires monthly payments of $570, including principal, interest, taxes, and insurance?

Solution   Here are the maximum amounts they can spend according to the three criteria given above:

1. $2.5 \times \$30,000 = \$75,000$
2. If the Harrells earn $30,000 annually, in 1 week they earn

$$\frac{\$30,000}{52} = \$576.92$$

3. The Harrells' take-home pay is $26,000, of which one-quarter is $26,000/4 = \$6500$ annually. This comes to

$$\frac{\$6500}{12} = \$541.67 \text{ per month}$$

Thus, the Harrells qualify under the first and second criteria, but not under the third. Of course, they must have the $20,000 down payment!   ◄

Now that you know some criteria used to determine whether you can afford a certain house, let us look at some of the criteria that banks and other lending institutions use in making loans.

A **mortgage** is a loan contract in which the lender agrees to lend you money to buy a specific house or a piece of property. You, in turn, agree to repay the money according to the terms in the contract. There are many different types of mortgage loan plans, but we shall discuss only two of these: **conventional loans** and **Federal Housing Authority (FHA) loans.**

Conventional loans are arranged between you and a private lender. In these loans, the **amount of the down payment,** the **repayment period,** and the **interest rate** are agreed on by the **borrower** and the **lender.** The lender usually does not require taxes and insurance to be paid in advance through a reserve **(escrow)** account. Lenders sometimes require borrowers to pay for mortgage insurance in order to qualify for a conventional loan. In addition, they may require that the buyer:

1. Be steadily employed and a resident of the state in which the property is located
2. Have enough savings to make one or two mortgage payments
3. Have the necessary down payment in hand (not borrowed)

Maximum amounts for conventional loans are set by individual lenders. Loans up to 80% of the value of the property are quite common, and loans of 90 to 95% can often be obtained.

FHA loans are made by private lenders and are insured by the Federal Housing Administration. The FHA does *not* make loans; it simply insures the lender against loss in case you, the borrower, fail to repay the loan in full. To pay expenses and to cover this insurance, the FHA charges an insurance premium of $\frac{1}{2}$% per year on the unpaid balance of the loan. This charge is included in the monthly payment. To qualify for an FHA loan, a buyer must have:

1. A total housing expense less than 35% of the buyer's net income
2. Total monthly payments (all debts with 12 or more payments plus total housing expense) less than 50% of the net income

Interest rates on these loans are established by the Housing and Urban Development Commission (HUD), and are usually 1 or 2 percentage points lower than conventional loan rates. The FHA loan maximum for a single-family dwelling is $67,500. Down payments are as follows:

3% of the first $25,000, plus 5% of the balance over $25,000, up to the maximum loan amount

A minimum of 3% down is required.

EXAMPLE 2   A family wishes to buy a $64,000 house.

a. If a lender is willing to loan them 90% of the price of the house, what will be the amount of the loan?
b. What will be the down payment with that loan?
c. If they decide to obtain an FHA loan instead, what will be the minimum down payment?
d. What will be the maximum FHA loan they can get?

Solution   a. 90% of $64,000 = $57,600
b. $64,000 − $57,600 = $6400 (=10% of $64,000)
c. With an FHA loan, the family will have to pay down 3% of $25,000 plus 5% of the amount over $25,000 ($64,000 − $25,000 = $39,000). Thus, the minimum down payment is computed as follows:

$$3\% \text{ of } \$25,000 = \$ \ 750$$
$$5\% \text{ of } \$39,000 = \underline{\$1950}$$
$$\text{Total minimum down payment} = \$2700$$

d. $64,000 − $2700 = $61,300

[*Note:* This is less than the maximum allowed loan because of the required minimum down payment.] ◀

The last item we shall discuss in connection with mortgages is the **actual amount** of the monthly payment. This amount depends on three factors:

1. The amount borrowed
2. The interest rate
3. The number of years taken to pay off the loan

Table 15.4a shows the monthly payments for $1000 borrowed at various rates and for various times. To figure the actual monthly payment, find the appropriate interest rate and the payment period, and then multiply the amount shown in the table by the number of thousands of dollars borrowed. Thus, to figure the monthly payment on a $40,000 mortgage at 15% for 20 years, look down the column for 20 years until you come to the row labeled 15%. The amount per $1000 is $13.17. Multiply this amount by 40 (there are 40 thousands in $40,000) to obtain $526.80 for the required monthly payment.

Table 15.4a   Monthly Payments for Each $1000 Borrowed

INTEREST RATE	PAYMENT PERIOD				
	10 years	15 years	20 years	25 years	30 years
9%	12.67	10.14	9.00	8.39	8.05
$9\frac{1}{2}$%	12.94	10.44	9.32	8.74	8.41
10%	13.22	10.75	9.65	9.09	8.78
$10\frac{1}{2}$%	13.49	11.05	9.98	9.44	9.15
11%	13.78	11.37	10.32	9.80	9.52
$11\frac{1}{2}$%	14.06	11.68	10.66	10.16	9.90
12%	14.35	12.00	11.01	10.53	10.29
$12\frac{1}{2}$%	14.64	12.33	11.36	10.90	10.67
13%	14.93	12.65	11.72	11.28	11.06
$13\frac{1}{2}$%	15.23	12.98	12.07	11.66	11.45
14%	15.53	13.32	12.44	12.04	11.85
$14\frac{1}{2}$%	15.83	13.66	12.80	12.42	12.25
15%	16.13	14.00	13.17	12.81	12.64
$15\frac{1}{2}$%	16.44	14.34	13.54	13.20	13.05
16%	16.75	14.69	13.91	13.59	13.45
$16\frac{1}{2}$%	17.06	15.04	14.29	13.98	13.85
17%	17.38	15.39	14.67	14.38	14.26

EXAMPLE 3   Athanassio and Gregoria Pappas wish to obtain a 30-year loan to buy a $50,000 house.

a. If they can get a loan of 95% of the value of the house, what is the amount of the loan?

b. What will be the down payment with that loan?

c. If the interest rate is 15%, what will be the monthly payment?

d. What will be the minimum down payment with an FHA loan?

e. If the FHA loan carries 13% interest, what will be the monthly payment?

Solution

a. 95% of $50,000 = $47,500

b. $50,000 − $47,500 = $2500

c. We read from Table 15.4a that the amount per $1000 on a 30-year loan at 15% is $12.64. Thus, the monthly payment will be

$47.5 \times \$12.64 = \$600.40$ The mortgage loan is for 47.5 thousands.

d. With an FHA loan, the minimum down payment is

$$3\% \text{ of } \$25,000 = \$\ 750$$

$$\text{plus } 5\% \text{ of } \$25,000 = \underline{\$1250}$$

$$\$2000$$

e. In Table 15.4a, we find the amount per $1000 on a 30-year loan at 13% to be $11.06. The amount to be financed is $48,000 ($50,000 − $2000). Thus, the monthly payment will be

$48 \times \$11.06 = \$530.88$

Note: This is not the entire payment, because for an FHA loan, interest and taxes must be added to this amount.   ◀

EXERCISE 15.4

1. A family has take-home pay of $34,000 on a $40,000 annual salary. Can they afford an $80,000 house with a $70,000 mortgage requiring payments of $750 per month, including principal, interest, taxes, and insurance?
   a. Use the first criterion given in the text.
   b. Use the second criterion given in the text.
   c. Use the third criterion given in the text.

2. A family earns $36,000 annually and has $30,000 take-home pay. Can they afford a $95,000 house with a $60,000 mortgage requiring monthly payments of $570, including principal, interest, taxes, and insurance?
   a. Use the first criterion given in the text.
   b. Use the second criterion given in the text.
   c. Use the third criterion given in the text.

3. The Browning family wants to buy a $77,000 house.
   a. If they can get a loan of 80% of the value of the house, what is the amount of the loan?
   b. What will be the down payment on this loan?
   c. If they decide to obtain an FHA loan, what will be the down payment? (Do not forget that the maximum FHA loan is $67,500.)
4. The Scotdale family wants to buy a $60,000 house.
   a. If they can get a loan of 95% of the purchase price, what will be the amount of the loan?
   b. What will be the down payment with this loan?
   c. If they use an FHA loan, what will be the down payment?

In Problems 5–10, find the total monthly payment, including taxes and insurance, for the given mortgage loans.

	Amount	Rate	Time (Years)	Annual Taxes	Annual Insurance
5.	$30,000	13%	20	$400	$160
6.	$40,000	14%	30	$600	$180
7.	$45,000	15%	25	$540	$210
8.	$50,000	16%	20	$720	$240
9.	$73,000	17%	30	$840	$380
10.	$80,000	$15\frac{1}{2}$%	15	$1000	$390

11. The Aikido family wants to obtain a conventional loan for 30 years at 16%. Suppose they find a lender that will loan them 95% for the $60,000 house they have selected, and their taxes and insurance amount to $1500 a year.
    a. What will be their down payment on the loan?
    b. What will be their total monthly payment, including taxes and insurance?
12. The Perez family is planning to buy a $50,000 house. Suppose they get a loan of 80% of the price of the house, and this is a 25-year loan at 14%.
    a. What will be their down payment on the loan?
    b. If their taxes and insurance amount to $510 annually, what will be their monthly payment, including taxes and insurance?
13. The Green family obtained a 30-year FHA loan at 12% to buy a $30,000 house. They made the minimum required down payment, and their taxes and insurance amounted to $360 annually.
    a. What was their down payment?
    b. What was their total monthly payment?
14. A family was planning to buy a $45,000 house with an FHA loan carrying $13\frac{1}{2}$% interest over a 20-year period. If they could get the largest possible loan, and their taxes and insurance amounted to $600 annually, find:
    a. Their down payment   b. Their total monthly payment

15. The Bixley family has a $50,000 mortgage loan at 14% for 20 years.
    a. What is their monthly mortgage payment?
    b. How many payments will they have to make in all?
    c. What is the total amount they will pay for principal and interest?
    d. What is the total interest they will pay?
    e. If their loan was 80% of the price of the house, is the price more or less than the total interest?

16. The Peminides have a $35,000 mortgage loan at 12% for 30 years.
    a. What is their monthly mortgage payment?
    b. How many payments will they have to make in all?
    c. What is the total amount they will pay for principal and interest?
    d. If their loan was 80% of the price of the house, is the price more or less than the total interest?

17. If you think house prices are high, we have bad news! The costs we have mentioned in the text are not all-inclusive. When you buy a house, you also have to pay **closing costs.** These costs include various fees and are usually paid at the time of closing—that is, when the final mortgage contract is signed. They are in addition to the agreed-upon down payment. Here are some typical closing costs for a $50,000 house with the buyer making a 20% down payment:

Credit report fee	$ 20	
$\frac{1}{12}$ estimated taxes of $600	50	To escrow account
Insurance premium for 1 year	300	
$\frac{1}{12}$ insurance premium	25	To escrow account
Title insurance	220	
Mortgage recording fee	20	
1% loan fee	400	
Total closing costs		

   a. What would be the total cash payment, down payment plus closing costs, at the time of closing?
   b. If the buyer had to make escrow account deposits each month for taxes and insurance, what would be the combined monthly payment under mortgage terms of 14% for 30 years?
   c. Suppose the lender agreed to add the closing costs to the loan amount instead of asking for cash. What would be the combined monthly payment with the same terms as in part b?

18. Here are some different closing costs for a $40,000 house with a 10% down payment:

Credit report fee	$ 10
Mortgage recording fee	15
Lot survey	45
Loan fee ($1\frac{1}{2}$%)	540
Insurance premium for 1 year	180
Total closing costs	

a. What would be the total cash payment, down payment plus closing costs, at the time of closing?

b. If no escrow account was required, what would be the monthly payment for a 25-year, 15% loan?

c. If the lender added the closing costs to the loan amount instead of asking for cash, what would be the monthly payment under the same terms as in part b?

USING YOUR KNOWLEDGE 15.4

There are so many different types of loans available that we cannot discuss all the possible financial alternatives you may have when you buy a house. The accompanying article from the July 1982, Money Magazine might help you to make some sense out of the existing confusion.

A gallery of loans

This chart will help you shop for the housing loan that's best for you. Your choice should be determined by your income and your expectations about inflation and interest rates. As they rise and fall, so will rates for inflation-indexed and adjustable-rate mortgages. Fixed-rate loans don't fluctuate, but they can be expensive. All of the loans listed are widely available, except for the adjustable-balance mortgages. They have been offered so far only in Utah, but they may become more common in the future if inflation worsens. For additional details about all the loans, see the story.

Type of loan	Typical minimum down payment	Initial interest rate	Interest rate after five years	Who should consider
Fixed rate				
Conventional	10%	16% to 17%	Unchanged	High-income people who believe interest rates won't drop much
Graduated payment	5%	16% to 17%	Unchanged	People who feel certain their incomes will rise substantially
Growing equity	10%	14% to 16%	Unchanged	Borrowers who can afford high payments and want to pay off their loans early
Zero interest rate	30%	0	0	Same as above
Adjustable rate				
Typical adjustable rate	10%	15% to 16½%	Unknown	People who expect interest rates to drop
Dual rate	10%	15% to 16½%	Unknown	Borrowers who, in return for lower monthly payments, are willing to give some of their equity to lenders if interest rates rise
Balloon payment	20%	12% to 16%	Loan is usually repaid by then	Borrowers who believe they'll be able to refinance their loans at lower rates in the future
Equity sharing				
Shared equity	10%	16% to 17%	Unchanged	People who are willing to give investors part of their houses' tax benefits and future appreciation in return for help in raising down payments and making monthly loan payments
Partnership mortgage	10%	16% to 17%	Unchanged	Low-income people who are willing to give investors most of the tax deductions a home generates and some of the future appreciation in exchange for down payments and help in monthly payments
Inflation indexed				
Adjustable balance	10%	10½%	Unknown	Borrowers who are confident that their incomes will keep pace with inflation

Reprinted from the July 1982 issue of *Money Magazine* by special permission. © 1982, Time, Inc.

Section	Item	Meaning	Example
15.1A	$I = Prt$	Simple interest equals Prt, where P is principal, r is rate, and t is time.	The interest on a $500 2-year loan at 12% is $I = 500 \cdot 0.12 \cdot 2 = \$120$.
15.1A	$A = P + I$	Amount equals principal plus interest.	
15.1D	$A_n = P(1 + i)^n$	Compound amount equals $P(1 + i)^n$, where P is principal, i is rate per period, and n is the number of periods.	
15.3A	APR	Annual percentage rate	
15.3B	Rule of 78	The unearned interest on a loan of n periods with r remaining periods is $$\frac{r(r + 1)}{n(n + 1)} \times F$$ where F is the finance charge.	

PRACTICE TEST 15

1. The Ready-Money Loan Company charges 28% simple interest (annual) for a 2-year, $800 loan. Find:
   a. The total interest on this loan
   b. The interest for 3 months
   c. The total amount to be paid to the loan company at the end of 2 years
2. A state has a 6% sales tax.
   a. What is the sales tax on a microwave oven priced at $360?
   b. What is the total cost of this oven?
3. In a sale, a store offers a 20% discount on a freezer chest that is normally priced at $390.
   a. How much is the discount?
   b. What is the sale price of the freezer?
4. Here is a portion of a compound interest table to use in this problem:

Amount (in dollars) to Which $1 Will Accumulate
in n Periods under Compound Interest

n	2%	4%	6%	8%	10%
1	1.0200	1.0400	1.0600	1.0800	1.1000
2	1.0404	1.0609	1.1236	1.1664	1.2100
3	1.0612	1.1249	1.1910	1.2597	1.3310
4	1.0824	1.1699	1.2625	1.3605	1.4641
5	1.1041	1.2167	1.3382	1.4693	1.6105
6	1.1262	1.2653	1.4185	1.5869	1.7716
7	1.1487	1.3159	1.5036	1.7138	1.9487
8	1.1717	1.3686	1.5938	1.8509	2.1436

Find the accumulated amount and the interest earned for:
a. $100 at 8% compounded semiannually for 2 years
b. $100 at 8% compounded quarterly for 2 years

5. A credit card holder is obligated to pay the balance in full if it is less than $10. Otherwise, the minimum payment is $10 or 5% of the balance, whichever is more. Suppose that a customer received a statement listing the balance as $185.76.

a. Find the minimum payment due.
b. The finance charge is 1.5% per month. What will be the amount of this charge on the next statement if the customer makes only the minimum payment?

6. JoAnn Jones received a statement showing that she owed a balance of $179.64 to a department store where she had a revolving charge account. JoAnn made a payment of $50 and charged an additional $23.50. If the store charges 1.5% per month on the unpaid balance, find:

a. The finance charge for the month
b. The new balance

7. A car costing $6500 can be bought with $1500 down and a 12% add-on interest to be paid in 48 equal monthly installments.

a. What is the total interest charge?
b. What is the monthly payment?

8. Here is a table for you to use in this problem:

True Annual Interest Rate for a 12-Payment Plan

	14%	14½%	15%	15½%	16%
FINANCE CHARGE Per $100 of the Amount Financed	7.74	8.03	8.31	8.59	8.88

Sam Bearss borrows $200 and agrees to pay $18.10 per month for 12 months.

a. What is the APR for this transaction?

b. If Sam decided to pay off the balance of the loan after 5 months (with 7 payments remaining), use the rule of 78 to find the amount of the interest refund.

c. Find the amount needed to pay off the loan.

9. The Mendoza family wants to buy a $50,000 house.

    a. If a bank was willing to loan them 75% of the price of the house, what would be the amount of the loan?

    b. What would be the down payment for this house?

    c. If they decided to obtain an FHA loan instead, what would be the minimum down payment? (Recall that the FHA requires a down payment of 3% of the first $25,000 and 5% of the balance up to the maximum loan amount of $67,500.)

    d. What would be the maximum FHA loan they could get?

10. Refer to Problem 9. Suppose the Mendoza family contracted for a 15-year mortgage at 12% with the bank that loaned them 75% of the price of the house. What is their monthly payment for principal and interest? (Use the table below.)

Monthly Payment ($) for Each $1000 Borrowed

RATE	10 years	15 years	20 years
11%	13.78	11.37	10.32
12%	14.35	12.00	11.01
13%	14.93	12.65	11.72

3rd century A.D. The abacus used beads to do counting and calculations.

Courtesy of IBM

1617 Napier's bones solved multiplication problems by adding adjacent numbers.

Courtesy of IBM

1642 Pascal's adding machine did additions using wheels and cogs.

Courtesy of IBM

1673 Leibniz' calculator had movable components and a hand crank.

Courtesy of IBM

1804 Joseph Marie Jacquard built an automated loom that used cards for weaving instructions.

Courtesy of IBM

1822 Charles Babbage invented the "difference engine," which constructed mathematical tables.

Courtesy of IBM

CHAPTER 16

COMPUTERS

1834 Babbage's "analytical engine" used punched cards to give directions for computations.

Courtesy of IBM

1890 Herman Hollerith speeded up the 1890 census using punched cards to record data.

Courtesy of IBM

1930 Vannevar Bush's "differential analyzer" was able to solve complex differential equations.

Courtesy of the MIT Museum

1941 Konrad Zuse designed and built the first general-purpose computer.

Courtesy of Library of Congress

1944 Howard Aiken worked at IBM to develop the Mark I, used by the Navy to solve ballistic problems.

Courtesy of Leni Iselin for *Fortune*

1946 J. P. Eckert and John Mauchly invented ENIAC, the first electronic digital computer.

AP/Wide World Photos

GETTING STARTED

The **first generation** of modern computers started with ENIAC (Electronic Numerical Integrator and Computer) and was followed by a **second generation** (transistors replacing tubes) as well as a **third** (using integrated circuits) and a **fourth** (with miniaturized memory chips). Personal computers (PC for short) belong to the fourth generation.

A. The Components

A typical personal computer has components that include:

1. A **keyboard** (to enter information and instructions)
2. A **video monitor** (to view the instructions and results)
3. One or more **disk drives** (to store information permanently or to enter additional instructions)
4. A **system unit** (into which the other three components are plugged) housing the **central processing unit (CPU)**

When you turn on a computer, a quartz clock in the CPU emits a reset signal that clears the computer of all instructions and gets it ready to "boot," a process so named because the computer pulls itself up "by its own bootstraps." At this point, the computer will give itself a checkup by whirring, moaning, and groaning for a short time. (Some computers display on the screen the status of the equipment checked. If a specific part is functional, the display will say PASS next to the name of the part.) And then . . . nothing! (If a floppy disk is being used, some models display a screen message saying that floppy A—that is, disk drive A—is not ready and that there is a disk read error.) Why? Because we have not given the computer any specific instructions as to how to proceed. How do we do that? Let us start all over again.

Courtesy of IBM

B. Getting Restarted

This time, before we start, place the **system diskette** (or a copy of it), with its label up, in drive A (the left drive). Now, turn on the computer. After checking itself, the computer will beep and will ask you to enter a new time and date. You must do so using the format 1-2-90 or 1/2/90 for the date and then pressing the ENTER or RETURN key ⏎ . The time used in most computers is a 24-hour system, so 4:30 in the afternoon is written as 16:30. To type the colon (:), you have to press the SHIFT, up-arrow key ↑ , as on a regular typewriter. Now that you know how, here is a little secret: You can bypass entering the date and time by simply pressing ⏎ twice. The make and model of your computer, the version of the instructions (program) in the system disk (something like 1.1, 2.3, or 3.3), and the copyright holder for the **disk operating system (DOS)** being used will then appear on the screen. The computer is now ready and the next line will say

    A>_

This means that your system program is on, and drive A is waiting for action.

C. Back to Basics

We can find out what programs, or **files,** are on the system disk by asking the computer for a **directory.** Type DIR and press ↵. A list of all the files will appear on the screen. Two of these, BASIC and BASICA (which has more features), enable the computer to speak in BASIC. But how do we get started? Type:

    A:BASICA ↵   or   BASICA ↵

which means, "Load the program named BASICA from the diskette in drive A."

The monitor again indicates your computer model, the DOS version being used, the memory available (for example, 61850 bytes free, more or less depending on the memory in your machine), and the word OK, which means the computer is now really ready for BASIC. Do you see a little blinking dash after the OK? It is called a **cursor,** and it tells us where anything we type will appear on the screen. Now, one more thing. To make sure we have a clean start, we want to erase all prior programs that may be somewhere in the machine. Type NEW and press ↵. The computer then answers OK.

What do we want to do next? Whatever it is, we have to use the BASIC language to communicate with the computer. The set of instructions we give the computer is called a **program,** and the writing of these instructions is called **programming.** Let us write a very simple program as an introduction to BASIC. Type the following line:

    10 PRINT ''I AM GLAD TO MEET YOU.''

Hold it. Do *not* press the ↵ yet! First, if you made a typing error, do not worry. Use the backspace key ← as your eraser. Each time you press the ← key, the character to the left of the cursor will be erased. If your mistake was at the beginning of the line, erase to that point and retype. (If you hold the backspace key down for a few seconds, it may wipe out the entire line.) Now, study carefully what you typed.

1. Is everything after the word PRINT inside quotations?
2. Did you write any extra quotation marks?
3. Do you have any misspelled words you wish to change?

If everything is satisfactory, press ⏎. The flashing cursor will move to the left margin to let you know your instructions have been entered. (By now, you should have noticed that instructions are entered by pressing the ⏎ key. We shall omit this step from now on.) But what if you find an error after the instructions are typed? Simply retype the entire line correctly. When you type ⏎, the old line will be replaced by the new one. Finally, you have to tell **(command)** the computer to **execute,** or **run,** your program. Type RUN and press ⏎. If you have been careful, the computer will display

```
I AM GLAD TO MEET YOU.
```

EXAMPLE 1   What will be printed if you run the following programs?

a. 10   PRINT I AM A PROGRAMMER NOW.
b. 20   PRINT I AM A PROGRAMMER NOW.''
c. 30   PRINT ''I AM A PROGRAMMER NOW.'''
d. 40   PRINT ''I AM A PROGRAMMER NOW.''

Solution   **a.** Five 0's will print. There are no quotation marks after the word PRINT, so each word will be interpreted as a variable starting with a 0 value.
b. Five 0's will print. Again, there are no quotations after PRINT.
c. I AM A PROGRAMMER NOW.' Since the computer prints *everything* between the quotation marks, it will print the ' (prime) at the end of the sentence.
d. I AM A PROGRAMMER NOW.   ◄

A last word of wisdom: Always remove the diskette(s) before turning off the system—or you might end up with a "zapped" disk, with some of its files erased, unreadable, or unusable.

EXERCISE 16.1

A.   **1.** Name the main characteristic of second-generation computers.
2. Name the main characteristic of third-generation computers.
3. Name the main characteristic of fourth-generation computers.
4. To what computer generation do personal computers belong?
5. Name four components of a typical personal computer.

B.   **6.** Write two different ways in which you can enter the date June 8, 1992 in a personal computer.
7. Write two different ways in which you can enter the date August 4, 1997 in a personal computer.
8. Suppose it is 9 o'clock in the morning and you want to enter this time into your computer. What would you type?

9. Suppose it is 8 o'clock in the evening and you want to enter this time into your computer. What would you type?
10. What does the abbreviation DOS stand for?

C. 11. What happens if you type DIR and press ⏎?
12. After typing DIR and pressing ⏎, the screen showed

```
MICKEY
MINNIE
GOOFY
```

You want to use the Goofy program. What do you have to do?
13. Suppose you type

```
I DO NOT MAKE MISTE_
```

The cursor is right after the E. How would you erase the E?
14. Suppose you write the following program:

```
10 PRINT ''I DO NOT MAKE MISTEAKS''
```

Then you press the ⏎ key. How would you correct the misspelled word without using the ← key?

In Problems 15–20, what will be printed if you run the given program?

15. 10 PRINT WE NEED QUOTATION MARKS AT THE END''
16. 20 PRINT 'WE NEED QUOTATION MARKS AT THE END'
17. 30 PRINT ''WE NEED QUOTATION MARKS AT THE END''
18. 10 PRINT I AM A PROGRAMMER
     10 PRINT I AM A PROGRAMMER''
19. 20 PRINT ''I KNOW THE ANSWER''
     20 PRINT I KNOW THE ANSWER
20. 30 PRINT THIS IS THE END''
     30 PRINT ''THIS IS THE END''

16.2

PROGRAMMING A COMPUTER

Now that you know how to load BASIC, we will discuss how to communicate with your computer in this language by using a **program,** which is a list of instructions that can be stored in the computer to be executed, or run. The writing of these instructions is called **programming.** Always remember that a computer has no mind of its own; it will do exactly what you instruct it to do (which in many cases may *not* be what you really want it to do). Just remember that every detail is important. A misspelled word or a misplaced comma can upset an entire program.

Several languages have been designed to fill various programming needs. Some of these languages are:

ALGOL	**ALG**orithm-**O**riented **L**anguage
BASIC	**B**eginner's **A**ll-purpose **S**ymbolic **I**nstruction **C**ode
COBOL	**CO**mmon **B**usiness-**O**riented **L**anguage
FORTRAN	**FOR**mula **TRAN**slation
LISP	**LIS**t **P**rocessing
PASCAL	Named after the French mathematician Blaise Pascal
C	Designed by Dennis Ritchie of Bell Telephone Company to improve an earlier language developed by Ken Thompson and named—you guessed it—B!

For now, we shall concentrate on BASIC, one of the easiest languages to use because it is so close to ordinary English that it is almost self-explanatory. Where shall we start? With the mathematical operations, of course.

A. Mathematical Operations

In BASIC, the symbols for the mathematical operations are:

Symbol	Meaning	Example	
+	Plus	3+5	
−	Minus	3−5	
*	Times	3*5	3 times 5
/	Divided by	3/5	3 divided by 5
^	With exponent	3^5	3^5 or 3 with exponent 5

[Note: Some systems use ** and other systems use the upward-pointing arrow for exponentiation.]

Parentheses are used just as in ordinary mathematical notation. For example, in BASIC, the expression $4 \times 3^2 - (8 + 9)(5 \div 12)^4$ would appear as follows:

```
4*3^2-(8+9)*(5/12)^4
```

You can remember this order by memorizing:
Please
Excuse
My
Dear
Aunt
Sally

The order of operations is exactly as in algebra:

1. Operations inside **p**arentheses are done first.
2. **E**xponentiation precedes the other operations.
3. **M**ultiplications and **d**ivisions are done next.
4. **A**dditions and **s**ubtractions are done last.

Thus, $4 + 3 \cdot 5$ will give $4 + 15 = 19$, since $3 \cdot 5$ is done first. Parentheses can be used to make sure calculations are done in the order you want. For example, to add $4 + 3$ first in $4 + 3 \cdot 5$, you should write $(4 + 3) \cdot 5$. Then the result will be $7 \cdot 5 = 35$.

EXAMPLE 1   Write in BASIC:

    a. $4(2 + 3)$ **b.** $5^2 + 12^2$ **c.** $8 \div (3^4 - 4^3)$

Solution   **a.** `4*(2+3)`   **b.** `5^2+12^2`   **c.** `8/(3^4-4^3)`   ◄

EXAMPLE 2   Evaluate:

    a. `(16/4)^3`   **b.** `2^3+3^2`   **c.** `(2*5)^3`

Solution   **a.** $(16 \div 4)^3 = 4^3 = 64$ **b.** $2^3 + 3^2 = 8 + 9 = 17$
    c. $(2 \times 5)^3 = 10^3 = 1000$ ◄

If you want the computer to perform some mathematical operations or run a program that will do a certain task, you must type commands at the keyboard. In BASIC, a program consists of a set of lines that are called *statements*. A **statement** is an expression or an instruction that is meaningful to the computer. Each statement is preceded by a **line number** and followed by a **keyword (command)** that tells the computer what type of operation is to be performed. If a statement is too long for a single line, it may be split into two statements with another number assigned to the second line. If more lines are needed, they must be treated in the same manner.

Each instruction that tells a computer to do something is called a **command.** You can include several commands in one line, but they must be separated by colons (:). We shall now consider three simple commands: LET, PRINT, and END.

B. The LET, PRINT, and END Commands

A command such as LET A = 5 tells the computer to assign the name A to a storage space and to store 5 in that space. Unless a later command changes this, the computer will use the number 5 for the letter A in each calculation it is instructed to do. (You can actually shorten things by simply writing A = 5; the LET is optional.)

The PRINT command means exactly what it says. If you type the commands LET A = 5 and then PRINT A, the computer will display the number 5. If you type PRINT A^2, it will display 25 (the square of the number in storage space A), and it will keep the number 5 in storage space A. If you only type PRINT 5^2, the computer will again display 25, but it will not store a number anywhere.

Finally, to tell the computer it has come to the end of a program, enter a line number (some fussy programmers use nines, for example, 9, 99, or 999) and the word END.

EXAMPLE 3   What will be the output if the following program is run?

```
10  LET A=5
20  LET B=2
30  LET C=3
40  LET X=A*B
50  LET Y=A*B+C
60  LET Z=Y-A^B
70  LET W=B/A+C^3
80  PRINT X; Y; Z; W
90  END
```

Solution We type in the word RUN (no line number is needed), and the computer will display the values it calculated for X, Y, Z, and W. The last three lines of the completed and run program would look like this:

```
RUN
10     13     -12     27.4
```
OK Some computers omit the OK.

This means that X = 10, Y = 13, Z = −12, and W = 27.4. (You can check these answers by doing the computations indicated on lines 40, 50, 60, and 70.) The effect of the semicolon separating the variables in line 80 is to make the answers appear next to each other. If you wanted the answers displayed on separate lines, you would type a separate PRINT command for each of the variables X, Y, Z, and W. The OK on the third line tells us that the computer is ready for further instructions. ◀

Study Example 3 carefully to see what the computer was instructed to do. In the first three lines, it was commanded to store the numbers 5, 2, and 3 in spaces named A, B, and C, respectively. Then it was instructed to calculate $A \times B = 5 \times 2$, $A \times B + C = 5 \times 2 + 3$, $(A \times B + C) - A^B = (5 \times 2 + 3) - 5^2$, and finally, $B/A + C^3 = \frac{2}{5} + 3^3$. Then it was commanded to print the answers next to each other. If we had entered commas, the answers would be spaced differently. To illustrate the difference between commas and semicolons, type the following:

```
10  PRINT ''THESE NUMBERS ARE PRINTED USING COMMAS''
20  PRINT 1, 2, 3
30  PRINT ''THESE NUMBERS ARE PRINTED USING SEMICOLONS''
40  PRINT 1; 2; 3

RUN
```

You should get a result that looks like this:

```
THESE NUMBERS ARE PRINTED USING COMMAS
 1                        2                        3
THESE NUMBERS ARE PRINTED USING SEMICOLONS
 1        2        3
```

Since many computers use only capital letters, we have used capital letters throughout our programs. It is very important to distinguish the number 0 (zero) from the letter "oh" (O) and the number 1 (one) from lowercase "el" (l). The computer has separate keys for these symbols.

A computer that is equipped to use BASIC will always execute your commands in the order in which they are numbered. This is the reason for numbering the lines 10, 20, 30, and so on. If you find you have left out an instruction, you can insert it by using an intermediate line number and typing a line at the end without having to retype the entire program. For instance, suppose that in Example 3, the instruction in line 50 had been omitted. In line 60 we have

```
LET  Z=Y-A^B
```

but Y was not defined, so when we type in RUN, the computer will assign the value 0 to Y and thus give a different value for Z. We can correct this by simply typing the line

```
45  LET  Y=A*B+C
```

at the end of the program. (Recall that if you type in an incorrect instruction, you can correct it just by typing a new line with the same line number and the correct instruction. The computer will automatically replace the original line with the new one.)

EXAMPLE 4   What number will the computer display for the following program?

```
10  LET  X=2
20  LET  Y=4*X^3-3*X^2+10*X-156
30  PRINT Y
40  END

RUN
```

Solution   Since the computer does exponentiation first, multiplication and division second, and addition and subtraction last, the program asks for the value of

$$4(2^3) - 3(2^2) + (10)(2) - 156 = 32 - 12 + 20 - 156 = -116$$

Thus, the printout will be -116.   ◄

In a sequence of multiplications and divisions, the operations will be done in the order in which they occur from left to right, unless parentheses indicate otherwise.

EXAMPLE 5   Justify the two answers given for the following program:

```
10  LET  X=2*3/12*4
20  LET  Y=2*3/(12*4)
```

```
30  PRINT X; Y
40  END

RUN
2      .125
OK
```

In doing the calculation called for in line 10, the computer would first multiply 2 times 3, to get 6. It would then divide 6 by 12, to get 0.5. Then it would multiply 0.5 by 4 to get the answer 2.

In doing the calculation called for in line 20, the computer would multiply 2 times 3 and 12 times 4, getting 6 and 48, respectively. Then it would divide 6 by 48 to get 0.125. ◄

Can you determine the printout for the next program?

```
10  LET X=(6+2*3)/6*4
20  LET Y=6+2*3/(6*4)
30  PRINT X+Y; X-Y; X*Y
40  END

RUN
```

If you got the answers 14.25, 1.75, and 50, then you are beginning to understand the order of operations and the LET and PRINT commands.

Note that the printouts in the above examples were just numbers, identified only if you go back and see what you asked the computer to do. You can label the answers in the printout by using the PRINT statement and quotation marks. (Some systems use single rather than double quotation marks.) This is done by modifying the PRINT commands in the preceding program to match those shown below. The semicolon after the item in quotation marks in line 30, for example, causes the computer to print the equation $X + Y = 14.25$ without any extra space after the equals sign.

```
30  PRINT ''X+Y=''; X+Y
40  PRINT ''X-Y=''; X-Y
50  PRINT ''X*Y=''; X*Y
60  END

RUN
X+Y=14.25
X-Y=1.75
X*Y=50
```

Notice that with a single PRINT command you can ask the computer to print out more than one item, but in that case the items must be separated by semicolons or commas in the PRINT line of your program. Thus, lines 30, 40, and 50 could be made into one line reading like this:

```
25  PRINT ''X+Y=''; X+Y, ''X-Y=''; X-Y, ''X*Y=''; X*Y
```

A. In Problems 1–8, write each expression in BASIC.

1. $(3 + 4) \div (5 + 9)$ **2.** $(5 - 2)(7 + 8)$ **3.** $3^2 + 4^2$

4. $2 \times 3 \div 4 \times 6$ **5.** $2(3^3) - 5(4^2)$ **6.** $3(2^5) - 2(3^4)$

7. $\dfrac{5 \times 8}{6 \times 9}$ **8.** $\dfrac{6 - 2}{5 \times 4}$

In Problems 9–18, evaluate the given BASIC expression.

9. `4+8/2-24/6` **10.** `3*4-2` **11.** `(4+8)/2-24/6`

12. `3*4-2*3` **13.** `2^3+3*2^2` **14.** `3*5^2-4*5`

15. `3*4/6*8` **16.** `3*4/(6*8)` **17.** `4*5^3`

18. `(4*5)^3`

B. **19.** Determine what would be printed if the following program were run:

```
10  LET X=456-241+612
20  LET Y=.62/(.31+.93)
30  LET Z=2^3-6
40  PRINT X; Y; Z
50  PRINT 2*X; Y^Z
60  END
```

20. Determine what would be printed if the following program were run:

```
10  LET X=1+2*3/6*5
20  LET Y=1+2*3/(6*5)
30  LET Z=(1+2)*3/6*5
40  LET W=(1+2)*3/(6*5)
50  PRINT X; Y; Z; W
60  END
```

21. Determine what would be printed if the following program were run:

```
10  LET A=3
20  LET B=2
30  LET X=A^B+B^A
40  PRINT ''A=''; A, ''B=''; B, ''X=''; X
50  END
```

22. Determine what would be printed if the following program were run:

```
10  LET X=10/2.5
20  LET Y=X^2
30  LET Z=1*2*3*4
40  LET W=3*Z-2*Y
50  PRINT ''W=''; W
60  END
```

The spirals in the daisy shown here are seen as two distinct sets radiating clockwise and counterclockwise, with each set always made up of a predetermined number of spirals. Most daisies have 21 and 34, adjacent numbers in the Fibonacci sequence.

Photo by Rutherford Platt; courtesy PSH

23. Write a BASIC program that will calculate and print the sum of the first four counting numbers. Run your program if a computer is available.

24. Write a BASIC program that will calculate and print the sum of the squares of the first four counting numbers. Run your program if a computer is available.

25. The Fibonacci sequence is a sequence of numbers that starts with 1, 1, 2. Each following term is the sum of the two terms that precede it. Thus, the fourth term is 1 + 2, or 3, and the fifth term is 2 + 3, or 5. Write a BASIC program that will compute and print the first eight terms of the sequence. Start with

```
10  LET  A=1
20  LET  B=1
30  LET  C=A+B
```

Run your program if a computer is available.

USING YOUR KNOWLEDGE 16.2

We have discussed how to edit a line using two different methods:

1. Pressing the backspace key (which will erase any symbols to its left)
2. Retyping the line number we wish to edit and writing the information correctly

There is a third method of editing lines: using the EDIT command. You may have heard the story of a retraction that was to appear in a newspaper. The headline to be edited on the computer was:

```
10  PRINT  ''MR. X IS A SOP ON THE POLICE FORCE.''
```

We wish to change this line so that when printed it will read:

```
MR. X IS A COP ON THE POLICE FORCE
```

To do this, type:

    EDIT 10   *Remember to hit the* ⏎ *key.*

We get a flashing cursor under the 1 in 10. Since we want to change the S in the word SOP, press the right arrow key → , and make the cursor stop under the S. Press the key that says Del , and the S will be deleted. Now, to insert C, the correct letter, press Ins and the letter C. What is the rest of the story? The programmer in the newspaper did not know about the EDIT command and tried to retype the line correctly. It now read:

```
10  PRINT  ''MR. X IS A COP ON THE POLICE FARCE.''
```

1. Give the steps needed to use the EDIT command in fixing the new line 10.

THE READ, DATA, GOTO, AND INPUT COMMANDS

One of the important characteristics of the digital computer is that it can be programmed to do various types of repetitive calculations. The READ, DATA, and GOTO commands are often used in problems that require repetitive calculations.

A. The READ, DATA, and GOTO Commands

The READ command makes the computer read one or more numbers from a DATA statement. On lines between the READ and DATA statements, the computer can be commanded to perform various operations with these numbers. The GOTO command makes the computer go to the line with the line number typed after the word GOTO. For instance, GOTO 20 instructs the computer to go to line 20. The GOTO command plays an important role in a READ and DATA program; it instructs the computer to return to the READ command, pick the next number(s) from the DATA line, and continue the execution of the program by moving to line 30. All of this is illustrated by the following program.

```
10   PRINT ''A'', ''B'', ''A/B''
20   READ A, B
30   PRINT A, B, A/B
40   GOTO 20
50   DATA 1, 2, 3, 4, 5, 8, 9, 15
60   END

RUN
A   B    A/B
1   2    .5
3   4    .75
5   8    .625
9   15   .6
OUT OF DATA IN 20
```
This is one of many error messages the computer can send. There are 75 more.

Line 10 of this program simply furnishes headings for the table that is to be printed out. The READ command in line 20 tells the computer to take the first two numbers in the DATA line (line 50) and call the first number A and the second B. Line 30 instructs the computer to print the numbers A and B and their quotient A/B. The GOTO command in line 40 tells the computer to go back to line 20 and READ the next two numbers in the DATA line for A and B. The first time line 20 is executed, the computer takes A = 1, B = 2 (the first two numbers in the DATA line) and then finds their quotient and prints the values of A, B, and A/B as instructed by line 30. The second time line 20 is executed, the computer takes the next two numbers, A = 3, B = 4, from the DATA line and then goes to line 30 as before. This procedure is continued until the data are exhausted. The output is displayed as shown and the computer displays OUT OF DATA IN 20 when it has used all the numbers on the DATA line. The program may be ended in this way if we wish. If we have more data to run with the same program, we simply retype line 50 with the new data and then type RUN. The computer executes the program all over again but now with the new data.

Notice that the items in lines 10 and 30 are separated by commas. This is to give proper spacing between the items enclosed by quotation marks in line 10 and correspondingly in line 30. For most systems, the use of commas divides the page into five zones of 14 spaces each and a sixth zone 10 spaces wide. Each item is typed out in a separate zone. You should check the spacing for the system you are using.

A root of a number such as the square root, the cube root, the fourth root, and so on, is indicated by using the exponents $1/2$, $1/3$, $1/4$, and so on. Thus, $2^{(1/7)}$ stands for the number that gives 2 when raised to the seventh power; that is, $2^{(1/7)}$ is the seventh root of 2. If we write a fractional exponent such as $2/3$, we mean to take the cube root and square the result. For instance, $8^{(2/3)}$ is 4. You can check this by taking the cube root of 8, which is 2, and then squaring to get 4. This convention is used in the next example.

EXAMPLE 1   Explain the printout for the following program:

```
10   READ B
20   LET X=5^(1/B)
30   PRINT ''B=''; B, ''5^(1/B)=''; X
40   GOTO 10
50   DATA 2, 3, 4, 5, 6
60   END

RUN
B=2        5^(1/B)=2.236068
B=3        5^(1/B)=1.709976
B=4        5^(1/B)=1.495349
B=5        5^(1/B)=1.37973
B=6        5^(1/B)=1.30766
OUT OF DATA IN 10
```

Solution   When line 10 is executed for the first time, the computer takes B = 2, the first number in line 50. It then goes to line 20, and calculates $5^{(1/2)}$, that is, the square root of 5, which it calls X. Next, it prints B=2 and $5^{(1/B)}$=2.236068, according to the instruction in line 30. Line 40 tells the computer to return to line 10. The second time line 10 is executed, the computer takes B = 3, the second number in line 50. It then repeats the calculation called for in line 20, this time finding the cube root of 5 and calling that X. The printout is similar to that in the first stage. This procedure is repeated until all the numbers in line 50 are used. The program computes in succession the square root, the cube root, the fourth root, the fifth root, and the sixth root of 5. ◄

EXAMPLE 2 The five roots obtained in Example 1 may be rounded to three decimal places and checked for accuracy by raising each to the power corresponding to the root and comparing with 5. Thus, 2.236 would be squared, 1.710 would be cubed, and so on. Write a program to calculate the required powers of these decimals.

Solution

```
10   READ A, B
20   PRINT A^B
30   GOTO 10
40   DATA 2.236, 2, 1.710, 3, 1.495, 4, 1.380, 5, 1.308, 6
50   END
```

If this program is run, the resulting five powers are found to be 4.999696, 5.000211, 4.995337, 5.0049, and 5.007794. (Final digits may be different because of computer rounding.) ◄

Note that no expression involving a calculation, not even anything as simple as $1 + 2$ or $1/2$, is permitted in the DATA statement. For a number such as $1/3$, you have to use .33333333.

B. The INPUT Command

We have already mentioned that the power of the computer lies in its tremendous ability to perform repetitive tasks at amazing speeds. So far, however, the commands we have learned are run once, and a particular answer is obtained. The next command will enable us to use the same program to get as many answers as we wish. For this, we need some input.

The INPUT statement allows you to assign data to letters (variables) from the keyboard. When the INPUT command is executed, the computer displays a question mark indicating that it is waiting for data to be input and the ENTER key to be pressed. Here is a very simple program to demonstrate INPUT:

```
10   PRINT ''TYPE ANY NUMBER''
20   INPUT A

RUN
```

The printout will be:

```
TYPE ANY NUMBER
? _
```

The question mark tells you the computer is waiting! Type in any number and press ⏎ . The computer prints the number. Some give the OK sign. But there is a shortcut. BASICA lets you combine the INPUT and PRINT statements. Change line 10 to

```
10   INPUT ''TYPE ANY NUMBER'';A
```

Delete line 20 by typing

```
20   ⏎
```

```
RUN
```

The printout is now

```
TYPE ANY NUMBER?
```

If you type a number, some computers say OK. Great. We get the same result, but saved one line of typing and one line on the printout. Let us use these ideas to modify the program in Example 2. Instead of having to read the data in line 40, we write

```
10   INPUT A, B
20   PRINT A^B
30   GOTO 10
```

```
RUN
```

The computer shows us the customary question mark (our turn). Now, enter the first pair of numbers on the data line 2.236, 2 and ⏎ . The answer 4.999696 appears on the next line. If we enter the next pair of numbers on line 40, 1.710, 3 and ⏎ , we get the next answer, 5.000211. In this manner, we can get all the answers in Example 2, and as many more as we wish to run. There is only one problem. The computer is out of control. It keeps asking for more pairs of numbers. How can we stop it? Press Ctrl and Scroll Lock simultaneously. We finally get a break! We are in charge again.

EXAMPLE 3   Use the INPUT statement to obtain the first line of the printout in Example 1.

Solution

```
10   INPUT B
20   LET X=5^(1/B)
30   PRINT ''B=''; B, ''5^(1/B)=''; X
40   GOTO 10
50   END
```

If you RUN the program, the ? will ask for a number. Enter 2 and the result will be just like the first line of the printout of Example 1. Since line 40 tells the computer to go back to 10, the program is repeated automatically (you do not have to type RUN). Note that in line 30, there is a comma after the second B so that the printout is not crowded. (Try it using a semicolon.)

Since you have an endless loop (the computer will GOTO 10) after printing the answer, how would you get out of it? Do the same thing we did before. Press $\boxed{\texttt{Ctrl}}$ and $\boxed{\texttt{Scroll Lock}}$ simultaneously. Here is one last piece of advice. If you ran this program several times, you might want to "erase" the screen somewhere along the line. You can do this by typing CLS and pressing $\boxed{\hookleftarrow}$. ◄

EXERCISE 16.3

A. *In Problems 1–6, determine what would be printed if the given program were run.*

1.
```
10  READ A, B
20  PRINT A; B; (A+B)/2
30  GOTO 10
40  DATA 5, 7, 25, 35, 40, 48
50  END
```

2.
```
10  READ A, B, C
20  PRINT (A+B+C)/3
30  GOTO 10
40  DATA 75, 80, 64, 90, 100, 62
50  END
```

3.
```
10  READ P
20  PRINT ''1.05*''; P; ''=''; 1.05*P
30  GOTO 10
40  DATA 500, 750, 1500, 2000
50  END
```

4.
```
10  PRINT ''A'', ''A^2'', ''A^3''
20  READ A
30  PRINT A, A^2, A^3
40  GOTO 20
50  DATA 1, 2, 3, 4, 5
60  END
```

5.
```
10  PRINT ''X'', ''Y''
20  READ X
30  LET Y=X^2+3*X-1
40  PRINT X, Y
50  GOTO 20
60  DATA -3, -2, -1, 0, 1, 2, 3
70  END
```

6.
```
10  PRINT ''X'', ''Y''
20  READ X
30  LET Y=X^3-X^2
40  PRINT X, Y
50  GOTO 20
60  DATA -2, -1, 0, 1, 2
70  END
```

7. Write a program that computes and prints a table of the fourth powers of the integers from 1 to 8. The first lines of your table should look like this:

```
N  N^4
1  1
2  16
3  81
```

Run your program if a computer is available.

8. The "triangular" numbers are the integers obtained by starting with 1, which is the first triangular number, adding 2 to get 3, the second triangular number, then adding 3 to get 6, the third triangular number, and so on. The $(n + 1)$th triangular number is obtained by adding $n + 1$ to the nth triangular number. Write a program that will compute and print a table of the first 10 triangular numbers. Make your printout start like this:

```
N  NTH TRIANG. NO.
1  1
2  3
3  6
```

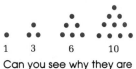

1   3   6   10

Can you see why they are called triangular numbers?

Run your program if a computer is available.

9. A formula obtained in algebra for the sum of the squares of the first N positive integers is $S = N(N + 1)(2N + 1)/6$. Write a program that will compute and print a table of the values of S for the first 10 positive integers. Your table should start like this:

```
N  S
1  1
2  5
3  14
```

Run your program if a computer is available.

10. A formula obtained in algebra for the sum of the cubes of the first N positive integers is $S = N^2(N + 1)^2/4$. Write a program that will compute and print the sum of the cubes of the first 10 positive integers. Run your program if a computer is available.

In Problems 11–16, use the INPUT statement to modify the indicated program so that the first line of the new printout is the same as before.

11. Problem 1   **12.** Problem 2   **13.** Problem 3
14. Problem 4   **15.** Problem 5   **16.** Problem 6

17. Use the INPUT statement to modify the program in Problem 8 so that the first line of the new printout is the same as before.

18. Use the INPUT statement to modify the program in Problem 9 so that the first line of the new printout is the same as before.

19. Use the INPUT statement to write a program to calculate the area of a rectangle. [*Hint:*   The area *A* of a rectangle is found by multiplying its length *L* by its width *W*. Run your program if a computer is available.]

20. Use the INPUT statement to write a program to calculate the area of a triangle. [*Hint:*   The area *A* of a triangle is found by multiplying its base *B* by its height *H* and dividing this product by 2. Run your program if a computer is available.]

16.4

THE IF. . . THEN. . . COMMAND

One of the most interesting and important features of the modern digital computer is its ability to make decisions. In BASIC, a computer is asked to make a decision by means of the IF. . . THEN. . . command. A mathematical statement using equality or inequality signs is inserted after the word IF and a command is inserted after the word THEN. *If the statement following the IF is true, the computer will go to the line named after the THEN. If the statement is not true, the computer will simply go on to the next command in the usual numerical order.*

The IF. . . THEN. . . statement is often used with the GOTO statement implied. Thus, we can type

```
10  IF A=5 THEN GOTO 50
```

or the shorter version

```
10  IF A=5 THEN 50
```

Now, let us use all this information to construct a program that finds the average of six numbers, say 0.121, 0.156, 0.162, 0.170, 0.135, and 0.174. The arithmetic involved is simple: Add the numbers and divide by 6. To tell the computer to get the sum *S* of the six numbers, we start with a sum of 0 and add the first number 0.121. The sum *S* is 0.121. Now, we add the next number, 0.156, to this sum, obtaining the new sum $S = 0.121 + 0.156 = 0.277$. We then get the next number, 0.162, and add it to 0.277.

The computer can do this by being **initialized** at S = 0 (see line 10 of the following program), reading some data (lines 20 and 60), and continuing to add new values of A to the sum (lines 40 and 50). How do we tell the computer to stop? We throw in a "flag," a number that tells the computer we are finished (the 999 in line 30). This number is written in the DATA line, and when it is reached, the computer is told to calculate the answer S/6, call it M (line 70), and then print it in the form "M = " (line 80).

```
10  LET S=0
20  READ A
30  IF A=999 THEN 70
40  LET S=S+A
50  GOTO 20
60  DATA .121, .156, .162, .170, .135, .174, 999
70  LET M=S/6
80  PRINT ''M=''; M
90  END

RUN
M=.153
```

A quicker way of doing this is to type:

```
10  INPUT ''WRITE THE NUMBERS A, B, C, D, E, F TO AVERAGE''; A,
    B, C, D, E, F
20  LET M=(A+B+C+D+E+F)/6
30  PRINT ''M=''; M
```

If you then enter the numbers in the DATA line, the answer M = 0.153 is displayed.

The flowchart in Fig. 16.4a illustrates the way in which a portion of the preceding program is recycled. The diamond-shaped decision box creates a loop in the flowchart, and the corresponding part of the program is called a **loop.** Loops are quite useful in programs that require recycling a set of operations.

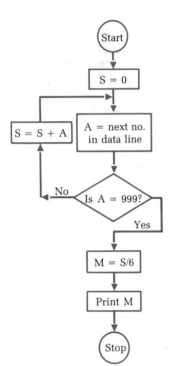

Figure 16.4a

EXAMPLE 1 Determine what will be printed if the following program is run:

```
10  LET A=0
20  READ B
30  IF B=9999 THEN 90
40  IF B>A THEN 60
50  GOTO 20
60  LET A=B
70  GOTO 20
80  DATA 761, 325, 671, 892, 346, 9999
90  PRINT A
95  END
```

Solution   In the first stage, the computer reads B = 761 ≠ 9999, so it goes to line 40. There it finds the IF statement to be true (761 > 0), and is instructed to go to line 60. Line 60 tells the computer to replace the current value of A by the current value of B. Thus, the new value of A is 761. Next, the computer goes to line 70, then to line 20 and reads the next value of B, 325 ≠ 9999. This time, the IF statement in line 40 is *not* true (325 ≯ 761), so the computer then continues its normal flow to line 50 and so back to line 20. Notice that the current value of A is replaced by the current value of B only when B is greater than A. The procedure continues through the first five numbers in the DATA line, at which time the value of A is 892, the largest number encountered so far. Then the computer reads 9999 from the DATA line, and line 30 tells it to go to line 90, where it is instructed to print the current value of A. This ends the program. The final printout is, of course, the largest of the first five numbers in the DATA line, 892. The 9999 was simply a convenient number (not one of the data numbers) used to stop the program. This little trick avoids the OUT OF DATA IN 20 error message.   ◀

The general form of the IF. . . THEN. . . instruction is

```
IF [relationship] THEN [statement]
```

The following relational symbols are permitted in the relationship:

=	Equals
<	Is less than
>	Is greater than
<=	Is less than or equal to
>=	Is greater than or equal to
<>	Is not equal to

The following instruction illustrates a correct usage of these symbols:

```
30  IF (A^2+B^2)>=C^2 THEN 45
```

If the current value of $(A^2 + B^2)$ is greater than or equal to the current value of C^2, the computer goes to line 45; but if the current value of $(A^2 + B^2)$ is less than the current value of C^2, the computer proceeds to the line with the smallest number that is greater than 30 (normal order).

In BASIC, the symbol SQR(X) causes the computer to calculate the square root of X. Any positive number (or 0) may be used for X, and the computer will calculate the nonnegative square root.

EXAMPLE 2   Determine what will be printed if the following program is run:

```
10  LET X=1
20  PRINT X; SQR(X)
30  LET X=X+1
```

```
40  IF X>10 THEN 60
50  GOTO 20
60  END
```

Solution In the first stage, the computer prints 1 and its square root. Line 30 instructs the computer to replace the current value of X (which is 1) by X + 1 (which is 2). Since the IF statement in line 40 is not true, the computer goes to line 50, which returns it to line 20 and makes it print out 2 and its square root. This procedure is repeated, at each stage the value of X being increased by 1, until the computer comes to X = 11. Then the IF statement in line 40 is true, and the computer goes to line 60, which ends the program. The printout will thus be a two-column array with the integers from 1 to 10 in the first column and decimal values for their respective square roots in the second column. ◀

Another important special function is symbolized by INT(X), which stands for the greatest integer that is less than or equal to X. If X is an integer, then INT(X) = X. For example, INT(5) = 5, INT(−3) = −3, and INT(0) = 0. If X is not an integer, however, INT(X) is the integer that just precedes X on the number line. Thus, for positive X, INT(3.571) = 3, INT(4.3172) = 4, and INT(0.678) = 0. For negative X (look at the number line in Fig. 16.4b), INT(−2.18) = −3 and INT(−1.98) = −2.

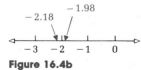

Figure 16.4b

EXAMPLE 3 What will be the printout if the following program is run?

```
10  LET A=A+5
20  PRINT A; INT(SQR(A))
30  IF A>25 THEN 50
40  GOTO 10
50  END
```

Solution Since SQR(5) is a number between 2 and 3, INT(SQR(5)) = 2. If you follow the loop in this program, you will see that the printout is a two-column array:

```
 5   2
10   3
15   3
20   4
25   5
30   5
```
◀

The next example gives an interesting application of both INT(X) and SQR(X) in finding all the exact divisors of a given integer.

EXAMPLE 4 Give the printout of the following program if it were run:

```
10   LET A=42
20   LET D=1
30   LET Q=A/D
40   IF Q=INT(Q) THEN 70
50   LET D=D+1
60   GOTO 25
25   IF D>SQR(A) THEN 90 ← Notice the convenience of not
70   PRINT D; Q              having to retype the program over
80   GOTO 50                 again because of an omitted com-
90   END                     mand.
```

Solution   1   42   Notice that line 40 instructs the computer to print out the
             2   21   divisor D and the quotient Q only if Q is an integer.
             3   14
             6   7   ◄

EXERCISE 16.4

1. Describe what would be printed if the following program were run.
You need not do the calculation called for in line 20.

```
10   LET X=1
20   PRINT X; X^(1/3)
30   LET X=X+1
40   IF X>100. THEN 60
50   GOTO 20
60   END
```

2. Determine what the following program will accomplish. You need not
compute the actual numbers that are printed out.

```
10   LET A=1
20   LET B=A^4
30   LET A=A+1
40   IF B<=100000 THEN 20
50   PRINT A-1; B
60   END
```

3. Determine what would be printed if the following program were run:

```
10   READ A, B
20   PRINT A; B; INT(A/B)
30   GOTO 10
40   DATA 9, 4, 100, 32, 500, 45
50   END
```

4. Determine what would be printed if the following program were run:

```
10  READ A, B
20  IF A=99 THEN 70
30  LET C=SQR(A^2+B^2)
40  PRINT A; B; C
50  GOTO 10
60  DATA 3, 4, 5, 12, 20, 21, 99, 1
70  END
```

5. Determine what would be printed if the following program were run using 48 for the first number and 60 for the second:

```
10   INPUT ''FIRST NUMBER''; F
20   INPUT ''SECOND NUMBER''; S
30   P=F*S
40   R=INT(F/S)
50   R=F-S*R
60   F=S
70   S=R
80   IF R<>0 THEN 40
90   PRINT ''THE GCF IS''; F
100  PRINT ''THE LCM IS''; P/F
110  END
```

6. Determine what would be printed if the following program were run using $2.58 for the cost and $3.00 for the amount tendered:

```
10   INPUT ''COST OF ITEM $'';I
20   I=INT(100*I+.5)
30   INPUT ''AMOUNT TENDERED $''; A
40   A=INT(100*A+.5)
50   IF I>A THEN PRINT ''YOU DID NOT GIVE ME ENOUGH MONEY!'': GOTO 170
60   IF I=A THEN PRINT ''THANK YOU. THERE IS NO CHANGE.'': GOTO 170
70   C=A-I
80   D=INT(C/100): C=C-100*D
90   Q=INT(C/25): C=C-25*Q
100  DI=INT(C/10): C=C-10*DI
110  N=INT(C/5): C=C-5*N
120  P=C
130  PRINT ''YOUR CHANGE IS'', ''DOLLARS'', ''QUARTERS'', ''DIMES''
140  PRINT ''      '', D, Q, DI
150  PRINT, ''NICKELS'', ''PENNIES''
160  PRINT, N, P
170  END
```

7. Write a program that will find and print the smallest positive integer whose fifth power is greater than 100,000. Run your program if a computer is available.

8. Write a program that will calculate and print a table of the first 10 positive integers and their respective fourth roots. Do not use the READ, DATA type of program. Run your program if a computer is available.

9. If $1 is invested in a savings account that earns interest at the rate R compounded annually, the total amount, A dollars, at the end of N years is given by $A = (1 + R)^N$. Write a program that finds and prints the number of years it takes for money to double at interest rates of 4, 5, 6, 7, and 8%. This means to find the smallest integer N for which $(1 + R)^N \geq 2$ and $R = 0.04, 0.05, 0.06, 0.07$, and 0.08. Ask the computer to print a two-column table with headings R and N. Run your program if a computer is available.

10. We know that the sum of the probabilities of all the possible outcomes of an experiment is 1. However, the probabilities may be given as rounded decimal approximations, so there may be round-off errors. For example, to three decimal places, the probabilities of getting 0, 1, 2, 3, 4, or 5 heads when a fair coin is tossed five times are 0.031, 0.156, 0.312, 0.312, 0.156, and 0.031, respectively (see Chapter 13). The maximum rounding error in each of these numbers is 0.0005. Hence, the sum, say S, has a maximum rounding error of 6 × 0.0005, or 0.003, and we should be willing to accept a sum that is within 0.003 of the correct value, that is, if the absolute value of $1 - S$ is less than or equal to 0.003. In BASIC, the symbol ABS(X) stands for the absolute value of X. Write a program that will add up the six given probabilities and print the message "OK" if the sum is acceptable, that is, if ABS(1−S)≤0.003, and the message "SORRY" otherwise. Run your program if a computer is available.

11. Repeat Problem 10 for N probabilities, $p_1, p_2, \ldots, p_N$, with the requirement that ABS(1−S) be less than or equal to 0.0005N. You can do this with a READ, DATA type program, where the first instruction is LET N=_____. Write $p_1, p_2, \ldots, p_N$ for the probabilities in the DATA line. It is then understood that the appropriate data will be entered when the numerical values are given and before the program is run. Run the program if a computer is available.

12. It is known that if M and N are two positive integers with $M > N$, then the numbers 2MN, $M^2 - N^2$, and $M^2 + N^2$ are integers that can be taken as the measures of the three sides of a right triangle. Thus, you can show that $(2MN)^2 + (M^2 - N^2)^2 = (M^2 + N^2)^2$. Write a program that will compute and print out the sides for all possible right triangles for M = 2, 3, 4, and 5. For convenience, let $A = 2MN$, $B = M^2 - N^2$, and $C = M^2 + N^2$. Your table should start like this:

```
A    B    C
4    3    5
6    8   10
12    5   13
```

We have used $M = 2, N = 1$ in the first line, $M = 3, N = 1$ in the second line, and $M = 3, N = 2$ in the third line. Run the program if a computer is available.

USING YOUR KNOWLEDGE 16.4

Suppose you deposit $1000 for 1 year in a savings account that pays 5% interest compounded monthly. The banker who credits your account at the end of each month would be interested in the following program, which calculates and prints the interest and the new amount (principal plus interest) for the first month:

```
10   LET  A=1000
20   LET  B=100*.05*A/12
30   LET  I=.01*INT(B+.5)
40   LET  A=A+I
50   PRINT  I; A
60   END
```

Lines 20 and 30 of this program involve the simple but clever little trick that is used to round off the interest to the nearest cent. This trick becomes clear if you carry out the computation of the interest. The actual interest is

$$0.05 \times 1000 \div 12 = 4.166666. \ . \ .$$

The banker wants to round this to 4.17, which we can do by inspection; but remember that the computer must be instructed how to do this. We first multiply the interest by 100 to get 416.6666. . . . Then we add 0.5 to get 417.1666. . . . Next, we find INT(417.1666. . .), which is 417, and finally we multiply by 0.01 to get 4.17. The remainder of the program should offer no difficulty.

Notice that by changing line 10 to let A be any desired amount, this same program can be used over and over for all the accounts that carry the same interest rate, 5% compounded monthly. Remember that a modern computer can do thousands of additions in a fraction of a second; hence, the apparent added complication needed to round off properly would not increase appreciably the time taken by the computer to run the program.

1. Modify the above program to take account of an initial deposit of $5000 and an interest rate of $5\frac{1}{4}$% compounded monthly.
2. Repeat Problem 1 if the interest is compounded daily (365 days in a year).

3. Determine what the following program will print if it is run:

```
10    PRINT ''N'', ''I'', ''A''
20    PRINT 0, 0, 1000
30    LET A=1000
40    LET N=0
50    LET B=100*.05*A/12
60    LET I=.01*INT(B+.5)
70    LET A=A+I
80    LET N=N+1
90    IF N>12 THEN 120
100   PRINT N, I, A
110   GOTO 50
120   END
```

A **perfect number** is a positive integer that is equal to the sum of its proper divisors. (Recall that a proper divisor is an exact divisor that is less than the number itself.) The smallest perfect number is 6, as you can easily verify. No odd perfect number has ever been found, and mathematicians are convinced that there is none; but no one has been able to prove this. The following program will discover the next even perfect number after 6. See if you can follow through and convince yourself of this fact. [Hint: Try taking $P = 2$ instead of 8 and carry out the instructions in the program. The printout should be the number 6.]

```
10    LET P=8
20    LET S=0
30    LET D=1
40    LET Q=P/D
50    IF Q<>INT(Q) THEN 70
60    LET S=S+D
70    LET D=D+1
80    IF D>=P THEN 100
90    GOTO 40
100   IF S=P THEN 130
110   LET P=P+2
120   GOTO 20
130   PRINT P; ''IS THE NEXT PERFECT NUMBER.''
140   END
```

1. Can you discover what modification to make in the above program if you were trying to find an odd perfect number?
2. Can you make a flowchart for the above program?

THE FOR AND NEXT COMMANDS

When a set of calculations is to be recycled several times, the FOR and NEXT commands are frequently more convenient and result in a shorter program than one constructed with the IF. . .THEN. . . command. The FOR line could have the form

```
20  FOR K=1 TO 8
```

which tells us the starting point and the ending point of the count. The NEXT line would then be of the form

```
50  NEXT K
```

Notice that the same letter, K, is used as in the FOR line. These two lines would cause the program between the lines to be performed eight times. Here, the starting value of K is 1, and with no further instruction in the FOR line, the computer automatically increases the value of K by 1 each time it goes to the NEXT line. Thus, the program calls for eight loops. The program below illustrates the FOR and NEXT commands used to calculate $8! = 1 \times 2 \times 3 \times 4 \times 5 \times 6 \times 7 \times 8$.

```
10  LET F=1
20  FOR K=1 TO 8
30  LET F=F*K
40  NEXT K
50  PRINT ''FACTORIAL 8=''; F
60  END
```

In this program, line 20 tells the computer to execute a loop eight times using $K = 1, 2, 3, 4, 5, 6, 7, 8$ in succession and doing whatever it is instructed to do between this line and the command NEXT, which closes the loop. If the program is run, the computer will automatically calculate $1 \times 1 = 1$, $1 \times 2 = 2$, $2 \times 3 = 6$, $6 \times 4 = 24$, . . . , $5040 \times 8 = 40,320$ in succession, which exhausts the K. The computer will then go to line 50, print the message FACTORIAL 8 = 40320, and end the program.

In executing the command FOR K = _____ TO _____, the computer will use a step size of 1 unless it is instructed to do otherwise. The next examples illustrate the flexibility of the FOR command.

EXAMPLE 1   Write a program that will add the odd integers from 1 to 13 and print the sum.

Solution
```
10  LET S=0
20  FOR K=1 TO 13 STEP 2
30  LET S=S+K
```

```
40  NEXT K
50  PRINT S
60  END
```

Notice that line 20 specifies a step size of 2. Thus, the computer will use K = 1, 3, 5, 7, 9, 11, and 13 in succession. ◀

The blanks in the command FOR K = _____ TO _____ STEP _____ may be filled by numbers or expressions that are evaluated in other parts of the program. For instance, the following are acceptable:

```
FOR K=5 TO 12.5 STEP .5
FOR K=15 TO -7 STEP -2.56
FOR K=N TO N+1 STEP .01*J   ← Here, the N and J would be
                              evaluated in other parts of the
                              program.
```

EXAMPLE 2   If you ran the program in Example 1, you would find the printout to be the number 49, which is 7^2. Since the program added the first seven odd positive integers, you might be curious about this coincidence and want to experiment a little. Write a program that will add the first N odd positive integers for N = 10, 15, 20, and 30, and print out N, the sum S, and SQR(S).

Solution
```
10  PRINT ''N'', ''S'', ''SQR(S)''
20  READ N
30  LET S=0
40  FOR K=1 to 2*N-1 STEP 2   ← Recall that the Nth odd
50  LET S=S+K                   positive integer is 2N − 1.
60  NEXT K
70  PRINT N, S, SQR(S)
80  GOTO 20
90  DATA 10, 15, 20, 30
95  END
```

If you run this program, the printout will be as follows:

```
N    S    SQR(S)
10   100   10
15   225   15
20   400   20
30   900   30
OUT OF DATA IN 20
```
◀

Although we have considered only the most elementary features of BASIC, the ten commands that we have discussed can be used to write many interesting and useful programs. A brief summary of these ten commands is given in Table 16.5a on the next page.

Table 16.5a

COMMAND	EXAMPLE	WHAT IT DOES
LET	LET A=5+2^3	Makes assignments and computations
PRINT	PRINT ''A=''; A	Prints messages and results
END	END	Ends a program
READ	READ X, Y, Z	Enters numbers from DATA line
DATA	DATA 2, 4, -7, 3.14	Stores data for READ line
GOTO	GOTO 40	Transfers computer to another line
INPUT	INPUT A	Assigns data to letters
IF. . . THEN. . .	IF K>8 THEN 90	Makes and acts on a decision
FOR	FOR J=1 TO 15	Starts a loop
NEXT	NEXT J	Closes a loop

EXERCISE 16.5

In Problems 1–6, figure out what would be printed if the program were run.

1.
```
10 LET S=0
20 FOR K=1 TO 5
30 LET S=S+K^2
40 PRINT S
50 NEXT K
60 END
```

2.
```
10 LET S=0
20 FOR K=1 TO 11 STEP 2
30 LET S=S+K^2
40 NEXT K
50 PRINT S
60 END
```

3.
```
10 LET P=1
20 FOR K=2 TO 10 STEP 2
30 LET P=P*K
40 NEXT K
50 PRINT P
60 END
```

4.
```
10 LET P=1
20 FOR K=2 TO 6 STEP 2
30 LET P=P*K
40 NEXT K
50 PRINT SQR(P)
60 END
```

5.
```
10 READ N
20 LET P=1
30 FOR K=1 TO N STEP 2
40 LET P=P*K
50 NEXT K
60 PRINT N; P
70 DATA 5, 9, 11
80 GOTO 10
90 END
```

6.

```
10  PRINT ''N'', ''F''          60  NEXT K
20  READ N                      70  PRINT N, F
30  LET F=1                     80  DATA 3, 5, 8
40  FOR K=1 TO N                90  GOTO 20
50  LET F=F*K                   95  END
```

7. Here is part of a payroll calculation for three employees. It is assumed that time-and-a-half is paid for work over 40 hr per week. Use the following notation:

I for employee's identification number
R for hourly rate of pay
H for hours worked
P for gross pay

Determine what the printout would be if the program were run.

```
10  PRINT ''I'', ''R'', ''H'', ''P''
20  FOR K=1 TO 3
30  READ I, R, H
40  IF H>40 THEN 70
50  LET P=R*H
60  GOTO 80
70  LET P=R*40+1.5*R*(H-40)
80  PRINT I, R, H, P
90  NEXT K
95  DATA 001, 4.00, 42, 002, 3.75, 39, 003, 3.50, 45
99  END
```

8. Suppose the total tax that the employer in Problem 7 must withhold for various purposes comes to 10% of each employee's gross pay. Modify the program given in Problem 7 so that it will calculate and print out (on a line with the employee number, I) the gross pay P, the tax T, and the net pay N.

9. Write a program that will compute and print out all the divisors of 36. You can use the instruction FOR K=1 TO 6. Do you see why? Run your program if a computer is available.

10. Write a program that will print out a table of the positive integers from 1 to 20 and their respective cube roots. Run your program if a computer is available.

11. Write a program that will print out a table of the even integers from 2 to 20 and their respective fifth powers. Run your program if a computer is available.

12. Write a program that will print out a table of the square roots of the numbers from 1 to 2 in steps of 0.1. Run your program if a computer is available.

Suppose you deposit $100 at the beginning of each month in a bank account that pays 5% interest compounded monthly. The following program will print out the number of the month, the month's interest, and the accumulated amount A at the end of each month for 1 year. The program assumes that the bank will round off to cents in the usual way. (For an explanation of line 50, see Using Your Knowledge 16.4.) Use the following notation:

$N =$ Number of month

$I =$ Interest for the month

$A =$ Accumulation at the end of the month

```
 10   PRINT ''N'', ''I'', ''A''
 20   LET A=0
 30   LET N=0
 40   FOR K=1 TO 12
 50   LET B=100*.05*(A+100)/12
 60   LET I=.01*INT(B+.5)
 70   LET A=A+I+100
 80   LET N=N+1
 90   PRINT N, I, A
100   NEXT K
110   END
```

Check this program for the first two or three values of K to assure yourself that it does what it is supposed to do. Run the program if a computer is available.

1. Suppose you wanted to continue the deposits for a second year and would like to compare the figures for the second year with those for the first year. How would you modify the program to give you these figures?
2. How would you modify the program if the interest rate were $5\frac{1}{4}$% compounded monthly?
3. Suppose you borrow $500 to pay for some wheels and you agree to repay the loan at the rate of $50 per month beginning 1 month after you get the money. Each $50 payment is to include interest on the unpaid balance at the rate of 1% per month. The final payment will be a partial payment (that is, less than $50) 1 month after the last full payment. Write a program that will calculate the total number of payments, the amount of the last payment, and the total amount of interest on the loan. (Since you do not know ahead of time how many payments are required, you will not want to use the FOR, NEXT type of program.)

*In the Discovery of Section 16.4, we gave a program to find a perfect number. If a number is not perfect, it is classified as **deficient** if the sum of its proper divisors is less than the number itself; it is classified as **abundant** if the sum of its proper divisors is greater than the number itself. To discover whether a given integer, say 18, is deficient, abundant, or perfect, we can use the following program:*

```
 10   LET N=18
 20   LET S=0
 30   FOR K=1 TO INT(N/2)
 40   LET Q=N/K
 50   IF Q<>INT(Q) THEN 70
 60   LET S=S+K
 70   NEXT K
 80   IF S<N THEN 110
 90   IF S>N THEN 130
100   IF S=N THEN 150
110   PRINT N; ''IS DEFICIENT.''
120   GOTO 160
130   PRINT N; ''IS ABUNDANT.''
140   GOTO 160
150   PRINT N; ''IS PERFECT.''
160   END
```

1. Check through this program to see if it will actually classify 18 correctly. You do not need a computer to do this.
2. Can you discover how to write a program that will classify all the integers from 2 to 30 as deficient (D), abundant (A), or perfect (P)? Here is some help for you.

   a. Start your program with the instruction

   ```
   PRINT ''D'', ''A'', ''P''
   ```

   This will give you appropriate headings for your table.

   b. Modify the program given above by replacing the three PRINT instructions by

   ```
   PRINT N, PRINT '' '', N, and PRINT '' '', '' '', N
   ```

   The effect of the quotes with an empty space between them is to make the computer move over one column before printing the value of N.

   c. Your program should have two FOR, NEXT loops in it, with the above program as the "inner" loop.

   d. GOOD LUCK!

Section	Item	Meaning	Example
16.2A	`*`	Times	4∗6 (4 times 6)
16.2A	`/`	Divided by	4∕6 (4 divided by 6)
16.2A	`^`	With exponent	4∧6 (4^6)
16.2B	`LET`	Makes assignments and computations	`LET A=3+4^2`
16.2B	`PRINT`	Prints messages and results	`PRINT ''A='';  7` prints A = 7
16.2B	`END`	Ends a program	
16.3A	`READ`	Enters numbers from DATA line	`READ X, Y, Z`
16.3A	`DATA`	Stores data for READ line	`DATA 3, 5, 6.5, -2`
16.3A	`GOTO`	Transfers computer to another line	`GOTO 20`
16.3B	`INPUT`	Assigns data to letters	`INPUT A`
16.4	`IF. . . THEN. . .`	Makes and acts on a decision	`IF A>B, THEN 90` If A is greater than B, the computer will go to line 90.
16.4	`SQR(X)`	The square root of X	`SQR(16)` The computer will calculate $\sqrt{16}$.
16.4	`INT(X)`	The greatest integer in X	`INT(5)=5, INT(4.5)=4,` `INT(-3.5)=-4`
16.5	`FOR`	Starts a loop	`FOR M=1 TO 5`
16.5	`NEXT`	Closes a loop	`NEXT M`

PRACTICE TEST 16

1. Suppose you are about to write line 80 on your program. You want your computer to write the phrase

   `THE ANSWER IS . . .`

   What instructions would you give the computer?
2. Write in BASIC:
   a. $4(2 + 5^2) \div 8$ b. $3^2 - 3 \cdot 2^3$
3. Evaluate:
   a. $(15/3)^2$ b. `2^4-3^3`   c. `3*4/2-1`

In Problems 4-12, determine what the printout would be if the given BASIC program were run.

4.
```
10  LET  A=5
20  LET  B=12
30  LET  C=SQR(A^2+B^2)
40  PRINT  2*A;  2*B;  2*C
50  END
```

5.
```
10  LET  X=9
20  LET  Y=3*X
30  LET  Z=Y^(1/3)
40  LET  U=Y-2*X/Z
50  PRINT  U
60  END
```

6.
```
10  READ  X
20  LET  Y=3*X-2*X^2
30  PRINT  X;  Y
40  GOTO  10
50  DATA  1,  2,  3
60  END
```

7. Use the INPUT statement to obtain the first line of the printout in Problem 6.

8.
```
5   LET  A=0
10  READ  B
20  LET  A=A+B
30  PRINT  A
40  GOTO  10
50  DATA  1,  2,  3,  4,  5
60  END
```

9.
```
10  LET  A=1
20  READ  B
30  IF  B=999  THEN  80
40  IF  A>B  THEN  60
50  LET  A=B
60  GOTO  20
70  DATA  25,  2001,  3260,  437,  500,  64,  999
80  PRINT  A
90  END
```

10.
```
10  FOR  K=1  TO  4
20  LET  B=2^K
30  PRINT  B
40  NEXT  K
50  END
```

11.
```
10  LET  S=0
20  FOR  K=1  TO  10
30  LET  S=S+K
40  NEXT  K
50  PRINT  S
60  END
```

12.
```
10  LET  A=18
20  FOR  K=1  TO  9
30  LET  B=18/K
40  IF  B=INT(B)  THEN  60
50  GOTO  70
60  PRINT  K
70  NEXT  K
80  END
```

13. Write a BASIC program that will compute and print out the value of $\sqrt{(38)^2 - 4(13)(17)}$.
14. Write a READ, DATA program that will compute and print out the cube roots of 20, 30, 38, and 42, and that will not print out an OUT OF DATA message.
15. Write a FOR, NEXT program that will print out a table of the fifth roots of the odd integers from 1 to 15.

In this Appendix you will find programs that will help you do exercises throughout this text. There are some new and more advanced commands in these programs. The explanation of most of them follows:

A$ Read, "A string," a type of variable that can be assigned to indicate letters, words, and/or combinations of letters, numbers, spaces, and other characters

DIM   Establishes the number of DIMensions and the number of elements per dimension in numeric and string arrays

INKEY$ Checks the keyboard looking for a pressed key

INSTR   Used to search for one STRing INside another string

LEN   Measures the LENgth of strings by counting the number of characters enclosed in quotes or assigned to string variables

LOCATE   Controls the cursor, placing it on a specific row and column on the screen

VAL   Converts numbers written as strings into their numeric value

For a nearly exhaustive list of commands used in BASIC, see David A. Lien, *The BASIC Handbook*, 3rd edition, published by Compusoft.

Statement Generator Program

This program will find a statement with a given truth table.

```
10 PRINT "PUT CAPS LOCK ON PLEASE"
20 INPUT "HOW MANY VARIABLES (<5) ";N:CLS:FOR I=1 TO N:PRINT "P";I;"    ";:NEXT
30 PRINT "DESIRED TRUTH VALUE      DESIRED CONJUNCTION "
40 FOR I=1 TO 70:PRINT "=";:NEXT:DIM B$(2^N,N+2) :PRINT
50 FOR I=1 TO N :T=0:A$="T" :FOR J=1 TO 2^N:K=2^N/2^I:T=T+1
60 IF  T>K  AND A$="T" THEN A$="F"  :T=1
70 IF T>K AND A$="F" THEN A$="T" :T=1
80 LOCATE J+3,7*I-5 :PRINT A$:B$(J,I)=A$:NEXT J:NEXT I
90 FOR J=1 TO 2^N:LOCATE J+3,30:INPUT "> ";B$(J,N+1):LOCATE J+3,30:PRINT ." "
100 LOCATE J+3,40
110 IF B$(J,N+1)="F" THEN X$="~"
120  X$=X$+"("
130 FOR I=1 TO N -1  :IF B$(J,I)="F" THEN X$=X$+"~"
140 X$=X$+"P"+STR$(I)+" ^ "   :NEXT
150 I=N :IF B$(J,I)="F" THEN X$=X$+"~"
160 X$=X$+"P"+STR$(I)+")":PRINT X$:B$(J,N+2)=X$:X$="":NEXT J
170 INPUT "PRESS RETURN FOR FUNCTION ";X$:F=0
180 FOR I=1 TO 2^N-1:IF B$(I,N+1) = "T" THEN PRINT B$(I,N+2);" v ";:F=1
190 NEXT :IF B$(2^N,N+1) ="T" THEN PRINT B$(2^N,N+2):F=1
200 IF F=0 THEN PRINT "P1^~P1" :PRINT
210 INPUT "RUN AGAIN (Y/N) ";A$:IF A$="Y" THEN RUN
```

Power Set Program

This program will list all the subsets (the power set) of a set, provided you enter the number of elements in the set and the elements themselves.

```
10 INPUT "NUMBER OF ELEMENTS ";N:DIM A$(N):DIM B(N):S$="("
20 FOR I= 1 TO N:PRINT "ENTER ELEMENT ";I;:INPUT A$(I):NEXT:CLS
30 FOR I=1 TO N:S$=S$+A$(I)+" ":NEXT:S$=S$+")":PRINT "THE POWER SET OF ";S$;"="
40 PRINT "( )  ";:FOR I=1 TO 2^N-1:J=1
50 IF B(J)=1 THEN B(J)=0:J=J+1 :GOTO 50
60 B(J)=1 :S$="":S$="("
70 FOR K=1 TO N
80 IF B(K)=1 THEN S$=S$+A$(K)+ " "
90 NEXT
100 S$=S$+")":PRINT S$;" ";
110 NEXT
120 PRINT ")"
130 INPUT "RUN AGAIN (Y/N) ";A$ :IF A$="Y" THEN RUN
```

Converting from Base 10 Program

This program will convert a number from base 10 to base X (X < 11).

```
5 PRINT "PUT CAPS LOCK ON "
10 CLS:PRINT "METHOD OF SUCCESSIVE APPROXIMATIONS -CHANGE OF BASE"
20 INPUT "ENTER NUMBER IN BASE TEN ";N :INPUT "CHANGE TO WHAT BASE (<11) ";K
30 PRINT:PRINT K;"|> ";N :T=N
40 L=INT(N/K):R=N-L*K :N=L
50 IF L<K THEN PRINT TAB(7);L;TAB(20);R:GOTO 70
60 PRINT K;"|> ";N;TAB(20);R :X$=RIGHT$(STR$(R),1)+X$:GOTO 40
70 X$=RIGHT$(STR$(L),1)+RIGHT$(STR$(R),1)+X$
80 PRINT "THUS , ";T;" = " ;X$;" BASE ";K
90  PRINT :INPUT "RUN AGAIN (Y/N) ";A$:IF A$="Y" THEN RUN
```

Converting to Base 10 Program

This program will convert a number from base X (X < 11) to base 10.

```
5 PRINT "PUT CAPS LOCK ON PLEASE"
10 CLS:PRINT "CHANGE FROM A BASE X (X<11) TO BASE 10"
15 INPUT "ENTER THE BASE OF THE NUMBER ";B
20 INPUT "ENTER NUMBER PLEASE ";X$   :K=LEN(X$):S$=X$
22 FOR I=1 TO K
24 S= VAL(RIGHT$(X$,1)):X$=LEFT$(X$,K-I)
26 T=S*B^(I-1)+T
30 NEXT
40 PRINT S$;" BASE ";B ;" = ";T;"  BASE 10"
50 INPUT "RUN AGAIN (Y/N) ";X$:IF X$="Y" THEN RUN
```

Base Converter Program (X, Y < 11)

This program will convert a number from base X to base Y, provided X and Y are less than 11.

```
5 PRINT "PUT CAPS LOCK ON PLEASE"
10 CLS:PRINT "CHANGE FROM BASE X TO BASE Y (X AND Y <11)"
20 INPUT "ENTER THE ORIGINAL BASE OF THE NUMBER ";B
30 INPUT "ENTER NUMBER PLEASE ";X$   :K=LEN(X$):S$=X$
40 FOR I=1 TO K
50 S= VAL(RIGHT$(X$,1)):X$=LEFT$(X$,K-I)
60 T=S*B^(I-1)+T
70 NEXT
80  N=T : INPUT "CHANGE TO WHAT BASE ";K   :X$=""
90 L=INT(N/K):R=N-L*K  :N=L
100 IF L<K THEN GOTO 120
110 X$=RIGHT$(STR$(R),1)+X$:GOTO 90
120 X$=RIGHT$(STR$(L),1)+RIGHT$(STR$(R),1)+X$
130 PRINT "THUS , ";S$;" IN BASE ";B;" = " ;X$;" BASE ";K
140 PRINT :INPUT "RUN AGAIN (Y/N) ";A$:IF A$="Y" THEN RUN
```

Base Converter Program (X, Y < 17)

This program will convert a number from base X to base Y, provided X and Y are less than 17.

```
10 DIM A$(16):FOR I=0 TO 9 :A$(I)=RIGHT$(STR$(I),1):NEXT
20 A$(10)="A":A$(11)="B":A$(12)="C":A$(13)="D":A$(14)="E":A$(15)="F":A$(16)="G"
30 CLS:PRINT "CHANGE FROM BASE X TO BASE Y (X,Y <17)"
40 INPUT "ENTER THE ORIGINAL BASE OF THE NUMBER ";B
50 INPUT "ENTER NUMBER IN THE ORIGINAL BASE PLEASE ";X$:A$=X$:K=LEN(X$):S$=X$
60 FOR I=1 TO K
70 S= RIGHT$(X$,1):X$=LEFT$(X$,K-I)
80 S=VAL(S$):FOR J=10 TO 16 :IF S$=A$(J) THEN S=J
90 NEXT
100 T=S*B^(I-1)+T
110 NEXT
120  N=T : INPUT "CHANGE TO WHAT BASE ";K   :X$=""
130 L=INT(N/K):R=N-L*K :N=L
140 IF L<K THEN GOTO 160
150 X$=RIGHT$(STR$(R),1)+X$:GOTO 130
160 X$=A$(L)+A$(R)+X$
170 PRINT "THUS , ";A$;" IN BASE ";B;" = " ;X$;" BASE ";K
180 PRINT :INPUT "RUN AGAIN (Y/N) ";A$:IF A$="Y" THEN RUN
```

Prime Searcher Program

This program will search and display all the prime numbers up to the number you desire.

```
10 CLS:PRINT "SEARCH FOR PRIMES : I WILL LIST ALL PRIMES UP TO THE"
20 INPUT "NUMBER YOU WOULD LIKE TO TEST UP TO ";N
30 CLS:DIM A(N)
40 FOR I=2 TO INT(N^.5)
50 IF A(I)<>0 THEN GOTO 70
60 FOR K=2 TO INT(N/I) :A(I*K)=1:NEXT
70 NEXT
80 CLS: FOR I= 2 TO N
90 IF A(I)=0 THEN PRINT TAB(L*7);I; :L=L+1
100 IF L=10 THEN PRINT :L=0   :C=C+1
110 IF C=20 THEN INPUT "PRESS RETURN FOR MORE ";A$:CLS:C=0
120 NEXT
130 PRINT:PRINT "THATS ALL"
140 INPUT "RUN AGAIN (Y/N) ";A$:IF A$="Y" THEN RUN
```

Sieve of Eratosthenes Program

This program will write the numbers up to 320 on the screen and then create a Sieve of Eratosthenes before your own eyes. The program eliminates the multiples of 2, 3, and so on, leaving only primes displayed on the screen.

```
10 CLS:PRINT "SEARCH FOR PRIMES BY SIEVE METHOD "  :DIM A$(320)
20 FOR I=2 TO 320 :A$(I) =RIGHT$(STR$(I),LEN(STR$(I))-1):NEXT:J=1
30 FOR S=2 TO INT(320^.5)
31 IF A$(S)= "   " THEN GOTO  110
40 FOR I=2 TO 320
50 IF J=16 THEN J=0:PRINT
60 PRINT TAB(J*5);A$(I); :FOR K=1 TO 200:NEXT
70 J=J+1
80 NEXT
95 LOCATE 23,5:PRINT"CROSSING OUT MULTIPLES OF ";S
96 FOR K=2*S TO 320 STEP S:A$(K)="   ":NEXT
100 LOCATE 2,1
110 LOCATE 2,1 : J=1:NEXT
120 LOCATE 23,5:PRINT "                             "
121 INPUT "RUN AGAIN (Y/N) ";A$:IF A$="Y"  THEN RUN
```

Prime Factorization Program

This program will factor any number into a product of primes.

```
10 CLS:PRINT "FACTOR A NUMBER INTO A PRODUCT OF PRIMES" :K=0
20 INPUT "ENTER NUMBER PLEASE ";N
30 IF N=1 THEN PRINT N;"=" ;1 :GOTO 90
40 PRINT N;"= ";
50 FOR I=2 TO N
60 IF K=0 AND INT(N/I)*I=N THEN PRINT I; :K=1:N=N/I:GOTO 50
70 IF INT(N/I)*I =N THEN PRINT "*";I; :N=N/I :GOTO 50
80 NEXT
90 PRINT
100 INPUT "RUN AGAIN (Y/N) ";A$:IF A$="Y" THEN RUN
```

GCF Finder Program

This program will find the GCF of two integers. Make sure you enter the larger of the two integers first.

```
10 PRINT "FIND THE GREATEST COMMON FACTOR  OF TWO INTEGERS"
20 INPUT "ENTER LARGEST INTEGER ";B:INPUT "ENTER SMALLEST INTEGER ";C
30 L=B:S=C:LR=S
40 Q=INT(B/C):R=B-Q*C
50 IF R=0 THEN GCF=LR :GOTO 70
60 LR=R :B=C:C=R :GOTO 40
70 PRINT "THE GREATEST COMMON FACTOR  OF ";L;" AND ";S;" IS
80 PRINT LR
90 INPUT "RUN AGAIN (Y/N) ";A$:IF A$="Y" THEN RUN
```

Reducing Fractions

This program will reduce a proper or improper fraction, provided you enter its numerator and denominator.

```
10 H=0:CLS:PRINT "REDUCE A PROPER OR IMPROPER FRACTION "
20 INPUT "ENTER NUMERATOR   ";B:INPUT "ENTER DENOMINATOR   ";C
30 L=B:S=C:LR=B :IF C>B THEN H=B:B=C:C=H
40 Q=INT(B/C):R=B-Q*C
50 IF R=0 THEN GCD=LR :GOTO 70
60 LR=R :B=C:C=R :GOTO 40
70  B=L/LR:C=S/LR
80 IF B>C THEN I= INT(B/C) :B=B-I*C
90 PRINT L;TAB(15);B
100 IF I<>0 THEN PRINT "------  =";TAB(10);I;
102 IF I=0 THEN PRINT "------  =";
110 PRINT TAB(15);"------"
120 PRINT    S;TAB(15);C
130 INPUT " RUN AGAIN (Y/N) ";A$ :IF A$ ="Y" THEN RUN
```

LCM Finder Program

This program will find the LCM of several numbers, provided you enter the numbers.

```
10 R=0:CLS:PRINT "FIND THE LCM":INPUT "HOW MANY NUMBERS ";N:DIM A(N)
20 FOR I=1 TO N:PRINT "NUMBER ";I;:INPUT"";A(I):NEXT:J=2:LM =1
30 PRINT "THE LCM OF ";:FOR I=1 TO N:PRINT A(I);:IF R<A(I) THEN R=A(I)
40 NEXT
50 FOR I=1 TO N
60 IF (INT(A(I)/J)*J=A(I)) AND (T=0) THEN T=T+1:A(I)=A(I)/J:L=I:GOTO 80
70 IF (INT(A(I)/J)*J=A(I)) THEN A(I)=A(I)/J :T=T+1
80 NEXT
100 LM=LM*J:IF T<=1 THEN A(L)=A(L)*J :LM=LM/J:J=J+1
110 T=0:L=0  :IF J<> R THEN GOTO 50
130 PRINT "IS : ";:FOR I=1 TO N :LM=LM*A(I):NEXT:PRINT LM
140 INPUT "RUN AGAIN (Y/N) ";A$: IF A$= "Y" THEN RUN
```

Addition of Fractions Program

This program will add or subtract fractions if you enter their numerators and denominators. Answers are not reduced.

```
10 CLS:R=0:PRINT "ADD OR SUBTRACT FRACTIONS :NOTE 3/5  - 2/5  = 3/5 + (-2)/5"
20 PRINT "ENTER (-A)/B FOR - A/B":INPUT "HOW MANY FRACTIONS ";N
25 DIM A(N),B(N),C(N)
30 FOR I=1 TO N:PRINT "NUMERATOR ";I;:INPUT"";B(I): PRINT "DENOMINATOR ";I;
35 INPUT "";A(I):C(I)=A(I):NEXT:J=2:LM=1
40 FOR I=1 TO N:IF R<A(I) THEN R=A(I)
50 NEXT
60 FOR I=1 TO N
70 IF (INT(A(I)/J)*J=A(I)) AND (T=0) THEN T=T+1:A(I)=A(I)/J:L=I:GOTO 90
80 IF (INT(A(I)/J)*J=A(I)) THEN A(I)=A(I)/J :T=T+1
90 NEXT
100 LM=LM*J:IF T<=1 THEN A(L)=A(L)*J :LM=LM/J:J=J+1
110 T=0:L=0  :IF J<> R THEN GOTO 60
120 FOR I=1 TO N :LM=LM*A(I):NEXT:FOR I=1 TO N:T=B(I)*LM/C(I)+T:NEXT
122 FOR I=1 TO N-1:PRINT B(I);"/";C(I);" + ";:NEXT:PRINT B(N);"/";C(N);" ="
130 PRINT T;"/";LM
140 INPUT "RUN AGAIN (Y/N) ";A$:IF A$="Y" THEN RUN
```

Addition of Fractions (Reduced) Program

This program will add or subtract several fractions and then reduce the answer. The program asks for the numerator and denominator of each of the fractions involved.

```
10 CLS:R=0:PRINT "ADD OR SUBTRACT FRACTIONS :NOTE 3/5  - 2/5  = 3/5 + (-2)/5"
20 PRINT "ENTER (-A)/B FOR - A/B":INPUT "HOW MANY FRACTIONS ";N
30 DIM A(N),B(N),C(N)
40 FOR I=1 TO N:PRINT "NUMERATOR ";I;:INPUT"";B(I): PRINT "DENOMINATOR ";I;
50 INPUT "";A(I):C(I)=A(I):NEXT:J=2:LM=1
60 FOR I=1 TO N:IF R<A(I) THEN R=A(I)
70 NEXT
80 FOR I=1 TO N
90 IF (INT(A(I)/J)*J=A(I)) AND (T=0) THEN T=T+1:A(I)=A(I)/J:L=I:GOTO 110
100 IF (INT(A(I)/J)*J=A(I)) THEN A(I)=A(I)/J :T=T+1
110 NEXT
120 LM=LM*J:IF T<=1 THEN A(L)=A(L)*J :LM=LM/J:J=J+1
130 T=0:L=0  :IF J<> R THEN GOTO 80
140 FOR I=1 TO N :LM=LM*A(I):NEXT:FOR I=1 TO N:T=B(I)*LM/C(I)+T:NEXT
150 FOR I=1 TO N-1:PRINT B(I);"/";C(I);" + ";:NEXT:PRINT B(N);"/";C(N);" ="
160 H=0:B=T:C=LM:I=0
170 L=B:S=C:LR=B :IF C>B THEN H=B:B=C:C=H
180 Q=INT(B/C):R=B-Q*C
190 IF R=0 THEN GCD=LR :GOTO 210
200 LR=R :B=C:C=R :GOTO 180
210  B=L/LR:C=S/LR
220 IF B>C THEN I= INT(B/C) :B=B-I*C
230 PRINT L;TAB(15);B
240 IF I<>0 THEN PRINT "------  =";TAB(10);I;
250 IF I=0 THEN PRINT "------  =";
260 PRINT TAB(15);"------"
270 PRINT     S;TAB(15);C
280 INPUT "RUN AGAIN (Y/N) ";A$:IF A$="Y" THEN RUN
```

Factorial Program for Large *n*

This program will calculate n! for large n. The actual answer will be displayed on the screen.

```
10 DIM A(300) ,B(10) :CLS:PRINT "CALCULATE FACTORIALS "
20 INPUT "ENTER NUMBER ";A$    :K=LEN(A$)
30 FOR I=1 TO K   :A(I)=VAL(MID$(A$,K-I+1,1)):NEXT
40 FOR I=     VAL(A$)-1  TO 2 STEP -1
50 FOR J=1 TO K:A(J)=A(J)*I:NEXT
60 FOR J=1 TO K:C=INT(A(J)/10):A(J+1)=A(J+1)+C:A(J)=A(J) MOD 10    :NEXT
80 IF A(K+1)<>0 THEN K=K+1
85 IF A(K)>10 THEN A(K+1)=INT(A(K)/10):A(K)=A(K) MOD 10 :K=K+1:GOTO 85
90 NEXT
92 FOR I=K TO 1 STEP -1:X$=X$+RIGHT$(STR$(A(I)),1):NEXT
100 PRINT X$
110 INPUT "RUN AGAIN (Y/N) ";A$:IF A$="Y" THEN RUN
```

Factorial Program (*n* < 30)

This program will calculate n! for a given n. If n is large, the answer is given in scientific notation.

```
10 CLS :PRINT " FIND THE FACTORIAL OF AN INTEGER " :J=1
20 INPUT "ENTER NUMBER ";N :FOR I=2 TO N:J=J*I:NEXT
30 PRINT N;" FACTORIAL IS ";J
40 INPUT "RUN AGAIN (Y/N) ";A$:IF A$="Y" THEN RUN
```

Matrix Addition or Subtraction Program

This program will add (enter a 1) or subtract (enter a 2) two matrices, provided you enter the number of rows and columns and all the entries for each matrix.

```
10 CLS:M=1:PRINT "ADD OR SUBTRACT MATRICES ":INPUT "NUMBER OF ROWS ";R
20 INPUT "NUMBER OF COLUMNS ";C:G=4:PRINT"PRESS 1 FOR ADD, PRESS 2 FOR SUBTRACT"
30 DIM A(R,C),B(R,C)
40 G=VAL(INKEY$):IF G<>1 AND G<>2 THEN GOTO 40
50 CLS:PRINT"ENTER MATRIX A ":IF G=2 THEN M=-1
60 FOR I=1 TO R:FOR J=1 TO C:LOCATE 2,3 :PRINT "ENTRY IN ROW ";I;"COLUMN ";J
70 LOCATE 2+I,J*5 :INPUT;" ",A(I,J) :NEXT J:PRINT:NEXT I :PRINT
80 PRINT "ENTER MATRIX B"
90 FOR I=1 TO R:FOR J=1 TO C:LOCATE 3+R,3 :PRINT "ENTRY IN ROW ";I;"COLUMN ";J
100 LOCATE  4+R+I,J*5 :INPUT;" ",B(I,J) :NEXT J:PRINT:NEXT I :PRINT
110 FOR I=1 TO R:FOR J=1 TO C:A(I,J)=A(I,J)+M*B(I,J):NEXT J:NEXT I
120 PRINT :PRINT"THE RESULTING MATRIX IS ":LOCATE 5+2*R,2
130 FOR I=1 TO R:FOR J=1 TO C:LOCATE 7+2*R+I,5*J:PRINT A(I,J):NEXT J:NEXT I
140 INPUT "RUN AGAIN (Y/N) ";A$:IF A$="Y" THEN RUN
```

Matrix Multiplication Program

This program will multiply two matrices, provided you enter the number of rows and columns and all the entries for each matrix.

```
10 CLS:M=1:PRINT "MULTIPLY TWO MATRICES":INPUT "NUMBER OF ROWS A";R
20 INPUT "NUMBER OF COLUMNS A";C:INPUT"NUMBER OF COLUMNS B ";B  :CLS
30 DIM A(R,C),B(C,B),C(R,B)
40 CLS:PRINT"ENTER MATRIX A "
50 FOR I=1 TO R:FOR J=1 TO C:LOCATE 2,3 :PRINT "ENTRY IN ROW ";I;"COLUMN ";J
60 LOCATE 2+I,J*5 :INPUT;" ",A(I,J) :NEXT J:PRINT:NEXT I :PRINT
70 PRINT "ENTER MATRIX B"
80 FOR I=1 TO C:FOR J=1 TO B:LOCATE 3+R,3 :PRINT "ENTRY IN ROW ";I;"COLUMN ";J
90 LOCATE  4+R+I,J*5 :INPUT;" ",B(I,J) :NEXT J:PRINT:NEXT I :PRINT
100 FOR I=1 TO R:FOR J=1 TO B:FOR K=1 TO C:C(I,J)=C(I,J)+A(I,K)*B(K,J)
110 NEXT K:NEXT J:NEXT I
120 PRINT :PRINT"THE RESULTING MATRIX IS ":LOCATE 6+2*R,2
130 FOR I=1 TO R:FOR J=1 TO B:LOCATE 8+2*R+I,5*J:PRINT C(I,J):NEXT J:NEXT I
140 INPUT "RUN AGAIN (Y/N) ";A$:IF A$="Y" THEN RUN
```

Determinants Program

This program will calculate the determinant of a 2 × 2 or 3 × 3 matrix, provided you enter the matrix.

```
10 CLS:PRINT "CALCULATE THE DETERMINANT OF A 2X2 OR 3X3 MATRIX "
20 INPUT "NUMBER OF ROWS";R:CLS:DIM A(3,3):IF R=2 THEN A(3,3)=1
40 PRINT"ENTER MATRIX A "
50 FOR I=1 TO R:FOR J=1 TO R:LOCATE 2,3 :PRINT "ENTRY IN ROW ";I;"COLUMN ";J
60 LOCATE 2+I,J*5 :INPUT;" ",A(I,J) :NEXT J:PRINT:NEXT I :PRINT
70 D=A(1,1)*A(2,2)*A(3,3)+A(1,2)*A(2,3)*A(3,1)+A(1,3)*A(2,1)*A(3,2)
80 D=D-A(3,1)*A(2,2)*A(1,3)-A(3,2)*A(2,3)*A(1,1)-A(3,3)*A(2,1)*A(1,2)
90 PRINT:PRINT "THE DETERMINANT IS ";D
140 INPUT "RUN AGAIN (Y/N) ";A$:IF A$="Y" THEN RUN
```

Permutations and Combinations Formula Program

This program will find the number of combinations and permutations of n objects taken r at a time.

```
10 CLS:PRINT "PERMUTATIONS  P(n,r) AND COMBINATIONS C(n,r)"
20 INPUT "NUMBER OF ITEMS (n) ";N
30 INPUT "NUMBER OF THINGS TO BE TAKEN AT A TIME (r) ";R
40 K= N-R+1 :P=1 :S=1
50 FOR I=N-R+1 TO N:P=P*I:NEXT
55 FOR I=2 TO R :S=S*I:NEXT
60 PRINT "P(";N;",";R;") = ";P
65 PRINT "C(";N;",";R;") =";P/S
70 INPUT "RUN AGAIN (Y/N) ";A$:IF A$="Y" THEN RUN
```

Combinations of r Items from n Items Program

This program will list as a set all possible combinations of r items picked from a set of n items. You must enter the number of items to be picked and the set from which you wish to pick them.

```
10 CLS:PRINT "FIND ALL COMBINATIONS OF R ITEMS TAKEN FROM A SET OF N ITEMS "
20 INPUT "HOW MANY ITEMS TO BE PICKED ";R:INPUT "HOW MANY ITEMS TO PICK FROM";N
30 DIM A$(N),B(N):FOR I= 1 TO N:PRINT "ENTER ELEMENT ";I;:INPUT A$(I):NEXT:CLS
40 FOR I=1 TO N:S$=S$+A$(I)+" ":NEXT:PRINT "THE COMBINATIONS ARE:"
50 FOR I=1 TO 2^N-1:J=1
60 IF B(J)=1 THEN B(J)=0:J=J+1 :GOTO 60
70 B(J)=1 :S$="":S$="(" :L=0:FOR K=1 TO N:L=L+B(K):NEXT K
80 IF L<> R THEN GOTO 130
90 FOR K=1 TO N
100 IF B(K)=1 THEN S$=S$+A$(K)+ " "
110 NEXT
120 S$=S$+")":PRINT S$;" ";
130 NEXT :PRINT
140 INPUT "RUN AGAIN (Y/N) ";A$ :IF A$="Y" THEN RUN
```

Distance Between Two Points Program

This program will find the distance between (x_1, y_1) and (x_2, y_2). You must enter each of the four numbers separately.

```
10 CLS:PRINT "CALCULATE THE DISTANCE BETWEEN TWO POINTS"
12 PRINT "ENTER FIRST POINT ":INPUT "X1= ";X1:INPUT "Y1= ";Y1
14 PRINT "ENTER SECOND POINT ":INPUT "X2= ";X2:INPUT "Y2= ";Y2
16 T=(X2-X1)^2 +(Y2-Y1)^2
18 FOR I=INT(T^.5+1 ) TO 1 STEP -1 :IF INT(T/I^2)*I^2=T THEN GOTO 20
19 NEXT
20 IF I=1 THEN PRINT "SQRT(";T;") = ";T^.5 :GOTO 25
22 PRINT I"*SQRT(";T/I^2;") = ";T^.5
25 INPUT "RUN AGAIN (Y/N) ";A$ :IF A$="Y" THEN RUN
```

Temperature Converter Program

This program will convert temperatures from Fahrenheit to Celsius (centigrade) and vice versa. Make sure you enter F or C following the temperature you wish to convert.

```
10 CLS:PRINT "FAHRENHEIT TO CELSIUS  OR  CELSIUS TO FAHRENHEIT  CONVERTER "
20 PRINT "ENTER TEMPERATURE FOLLOWED BY C FOR CELSIUS OR F FOR FAHRENHEIT"
30 INPUT " ";A$ :A=VAL(A$):IF INSTR(A$,"C")=0 THEN GOTO 50
40 F=9*A/5+32 :PRINT A;" C  = ";F;" F":GOTO 60
50 C=(A -32)*5/9: PRINT A ;"F = ";C;" C"
60 INPUT "RUN AGAIN (Y/N) ";A$:IF A$="Y" THEN RUN
10 CLS:PRINT "FAHRENHEIT TO CELSIUS  OR  CELSIUS TO FAHRENHEIT  CONVERTER "
20 PRINT "ENTER TEMPERATURE FOLLOWED BY C FOR CELSIUS OR F FOR FAHRENHEIT"
30 INPUT " ";A$ :A=VAL(A$):IF INSTR(A$,"C")=0 THEN GOTO 50
40 F=9*A/5+32 :PRINT A;" C  = ";F;" F":GOTO 60
50 C=(A -32)*5/9: PRINT A ;"F = ";C;" C"
60 INPUT "RUN AGAIN (Y/N) ";A$:IF A$="Y" THEN RUN
10 CLS:PRINT "FAHRENHEIT TO CELSIUS  OR  CELSIUS TO FAHRENHEIT  CONVERTER "
20 PRINT "ENTER TEMPERATURE FOLLOWED BY C FOR CELSIUS OR F FOR FAHRENHEIT"
30 INPUT " ";A$ :A=VAL(A$):IF INSTR(A$,"C")=0 THEN GOTO 50
40 F=9*A/5+32 :PRINT A;" C  = ";F;" F":GOTO 60
50 C=(A -32)*5/9: PRINT A ;"F = ";C;" C"
60 INPUT "RUN AGAIN (Y/N) ";A$:IF A$="Y" THEN RUN
```

Inverse Finder Program

This program will find the inverse of a 2×2 or 3×3 matrix when the matrix is entered.

```
10 DIM A(10),B(10):CLS:PRINT "FIND INVERSE OF 2X2 OR 3X3 MATRIX":PRINT
20 K=1:INPUT "NUMBER OF ROWS ";N:IF N=2 THEN B(9)=1
30 FOR I=1 TO N:FOR J=1 TO N :LOCATE 5,1:PRINT "INPUT ENTRY ROW ";I;"COLUMN ";J
40 LOCATE I+7,J*10:INPUT " ",B(K ):K=K+1:NEXT J:IF N=2 THEN K=K+1
50 NEXT I
60 A(1)=B(5)*B(9)-B(8)*B(6):A(2)=B(7)*B(6)-B(4)*B(9):A(3)=B(4)*B(8)-B(7)*B(5)
70 A(4)=B(8)*B(3)-B(2)*B(9):A(5)=B(1)*B(9)-B(7)*B(3):A(6)=B(7)*B(2)-B(1)*B(8)
80 A(7)=B(2)*B(6)-B(5)*B(3):A(8)=B(4)*B(3)-B(1)*B(6):A(9)=B(1)*B(5)-B(4)*B(2)
90 D=A(1)*B(1)+A(2)*B(2)+A(3)*B(3):IF D=0 THEN PRINT "NO INVERSE DET.0":GOTO 160
100 C=A(2):A(2)=A(4):A(4)=C:C=A(3):A(3)=A(7):A(7)=C:C=A(6):A(6)=A(8):A(8)=C
110 K=1:PRINT:PRINT "INVERSE IS":FOR I=1 TO N:FOR J=1 TO N:LOCATE I+14,J*15
120   IF INT(A(K)/D)*D=A(K) THEN PRINT A(K)/D :GOTO 140
130 PRINT A(K);"/";D
140 K=K+1:NEXT J :IF N=2 THEN K=K+1
150 NEXT I
160 INPUT "RUN AGAIN (Y/N) ";A$:IF A$="Y" THEN RUN
```

Matrix Reduction Program

This program will reduce a matrix to a desired form if you enter the number of rows and columns and the elements of the matrix.

```
10 CLS:PRINT"THIS PROGRAM WILL MANUALLY ROW REDUCE A MATRIX"
20 INPUT "NUMBER ROWS ";N:INPUT "NUMBER COLUMNS ";M:DIM A(N,M,2)
30 FOR I=1 TO N:FOR J=1 TO M:LOCATE 4,6 :PRINT "ENTER ROW ";I;"COLUMN ";J
40 LOCATE I+5,J*9:INPUT"",A$:A(I,J,1)=VAL(A$):A(I,J,2)=1:X=INSTR(A$,"/")
50 IF X<>0 THEN B$=RIGHT$(A$,LEN(A$)-X):A(I,J,2)=VAL(B$)
60 NEXT J:NEXT I
70 PRINT "1 ADD A TIMES ROW I TO ROW J":PRINT"2 MULTIPLY A TIMES ROW I"
80 PRINT "3 SWAP ROW I WITH ROW J":INPUT "ENTER 1,2 OR 3 AND RETURN ";C$
85 C=VAL(C$):IF C$="" THEN RUN "MENU"
90 IF C=3 THEN INPUT "I ";I:INPUT "J ";J :GOTO 140
100 INPUT "A ";A$:X=INSTR(A$,"/"):T=VAL(A$):B=1:IF X= 0 THEN GOTO 120
110 B=VAL(RIGHT$(A$,LEN(A$)-X))
120 INPUT "I ";I:P=I:IF C=1 THEN INPUT "J ";J :P=J:GOTO 160
130 FOR K=1 TO M:A(I,K,1)=A(I,K,1)*T:A(I,K,2)=A(I,K,2)*B:NEXT K :GOTO 180
140 FOR K=1 TO M:FOR H= 1 TO 2:C=A(I,K,H):A(I,K,H)=A(J,K,H):A(J,K,H)=C
150 NEXT H:NEXT K :GOTO 220
160 FOR K=1 TO M:A(J,K,1)=A(J,K,1)*A(I,K,2)*B+A(I,K,1)*A(J,K,2)*T
170 A(J,K,2)=A(J,K,2)*A(I,K,2)*B:NEXT K
180 FOR J=1 TO M:IF A(P,J,1) =0 THEN  A(P,J,2)=1
190 FOR K=A(P,J,1) TO 2 STEP -1:T=A(P,J,1):B=A(P,J,2)
200 IF (INT(T/K)*K=T AND INT(B/K)*K=B) THEN T=T/K:B=B/K:A(P,J,1)=T:A(P,J,2)=B
210 NEXT K :NEXT J
220 CLS:FOR I=1 TO N:FOR J=1 TO M:LOCATE I+5,J*9:T=A(I,J,1):B=A(I,J,2)
230 IF B=1 THEN PRINT T ELSE B$=STR$(B):A$=STR$(T)+"/"+B$:PRINT A$
240 NEXT J:NEXT I :PRINT :PRINT:GOTO 70
```

Solving Systems of Equations

This program will solve a system of two or three linear equations when you enter the coefficients of the variables and the numerical part of each equation. Recall that for an equation such as $x + y = 3$, you must enter a coefficient of 1 for x and y. The numerical value is 3.

```
10 A$(1)="X  +":A$(2)="Y  +":A$(3)="Z":CLS:PRINT "SOLVE SYSTEM OF EQUATIONS "
20 INPUT "NUMBER OF EQUATIONS (2 OR 3) ";N:K=1:IF N=2 THEN B(9)=1:A$(2)="Y"
30 FOR I=1 TO N:FOR J=1 TO N :LOCATE 5,1:PRINT "INPUT ENTRY ROW ";I;"COLUMN ";J
40 LOCATE I+7,J*10+4:PRINT A$(J):LOCATE I+7,J*10:INPUT " ",B(K) :K=K+1:NEXT J
50 LOCATE I+7,J*10+8:INPUT "= ",C(I):IF N=2 THEN K=K+1
60 NEXT I:FOR I=1 TO 9:D(I)=B(I):NEXT:A$(1)="X":A$(2)="Y":GOSUB 100:D1=D
70 IF D=0 THEN PRINT "I CAN'T SOLVE THIS ONE " :GOTO 120
80 FOR I=1 TO 3:B(I)=C(1):B(I+3)=C(2):B(I+6)=C(3):GOSUB 100:T(I)=D:NEXT
90 FOR I=1 TO N :PRINT A$(I);"=";T(I)/D1:NEXT:GOTO 120
100 A(1)=B(5)*B(9)-B(8)*B(6):A(2)=B(7)*B(6)-B(4)*B(9):A(3)=B(4)*B(8)-B(7)*B(5)
110 D=0:D=A(1)*B(1)+A(2)*B(2)+A(3)*B(3):FOR J=1 TO 9:B(J)=D(J):NEXT:RETURN
120 INPUT "RUN AGAIN (Y/N) ";A$:IF A$="Y" THEN RUN
```

Solving Quadratic Equations by Formula

This program will solve any quadratic equation using the quadratic formula. You enter the coefficients of x^2 and x and the numerical term.

```
10 CLS:PRINT "QUADRATIC FORMULA PROGRAM " :A$(1)="X^2 +":A$(2)="X +":P=1
20 FOR I=1 TO 3:LOCATE 4,9*I+3:PRINT A$(I):LOCATE 4,9*I:INPUT "",A(I):NEXT
30 LOCATE 4,9*I-3:PRINT " = 0":PRINT:PRINT:D=A(2)^2-4*A(1)*A(3):B=A(2)
40 FOR I=INT(ABS(D)^.5)+1 TO 2 STEP -1:IF INT(D/I^2)*I^2=D THEN P=I:I=0
50 NEXT :D=D/P^2 :S=2*A(1):FOR I=S TO 2 STEP -1:V=INT(S/I):W=INT(B/I):T=INT(P/I)
60 IF V*I=S AND W*I=B AND  T*I=P THEN S=V:B=W:P=T
70 NEXT : IF D=0 THEN PRINT "X= " ;-1*B;"/";S ;"  = ";-1*B/S:GOTO 140
80 IF D< 0 THEN D=-1*D:GOTO 120
90 PRINT "X= (";-1*B;" + ";P;" SQRT(";D;")  ) / ";S;"  = ";(-1*B+P*D^.5)/S
100 PRINT "X= (";-1*B;" - ";P;" SQRT(";D;")  ) / ";S;"  = ";(-1*B-P*D^.5)/S
110 GOTO 140
120 PRINT "X= (";-1*B;" + ";P;" i SQRT(";D;")  ) / "; S
130 PRINT "X= (";-1*B;" - ";P;" i SQRT(";D;")  ) / ";S
140 PRINT :INPUT "RUN AGAIN (Y/N) ";A$:IF A$="Y" THEN  RUN
```

Equation of a Line Program

This program will find the equation of a line, given a point and the slope of the line. You must enter the point as an ordered pair of numbers (x_1, y_1).

```
10 LINE INPUT "ENTER POINT '(X,Y)' OR SLOPE M ";A$ :R= INSTR(A$,"/")
20 LINE INPUT "ENTER POINT '(X,Y)' OR SLOPE M ";B$ :S=INSTR(B$,"/")
30 X=INSTR(A$,"(" ):Y=INSTR(B$,"(") :Z=INSTR(A$,","):W=INSTR(B$,",")
40 IF X=0 OR Y=0 THEN GOTO 130
50 X1=VAL(RIGHT$(A$,LEN(A$)-1)):Y1=VAL(RIGHT$(A$,LEN(A$)-Z))
60 X2=VAL(RIGHT$(B$,LEN(B$)-1)):Y2=VAL(RIGHT$(B$,LEN(B$)-W)):MT=Y2-Y1:MB=X2-X1
70 IF MB<0 THEN MT=-1*MT:MB=-1*MB
80 H=ABS(MT):K=MT:J=MB:IF MB=0 THEN PRINT "X= ":X2 :GOTO 170
90 B=Y1-MT*X1/MB:IF MT=0 THEN PRINT  "Y= ";Y1 :GOTO 170
100 FOR I=H TO 2 STEP -1:IF INT(K/I)*I=K AND INT(J/I)*I=J THEN MT=K/I:MB=J/I
110 NEXT:IF MB=1 THEN PRINT"Y= ";MT;"X +";B ELSE PRINT"Y= (";MT;"/";MB;") X +";B
120 GOTO 170
130 IF X<>0 THEN X1=VAL(RIGHT$(A$,LEN(A$)-1)):Y1=VAL(RIGHT$(A$,LEN(A$)-Z)):P$=B$
140 IF Y<>0 THEN X1=VAL(RIGHT$(B$,LEN(B$)-1)):Y1=VAL(RIGHT$(B$,LEN(B$)-W)):P$=A$
150 MT=VAL(P$):K=R+S:MB=1:IF K<>0 THEN MB=VAL(RIGHT$(P$,LEN(P$)-K))
160 GOTO 70
170 INPUT "RUN AGAIN (Y/N) ";A$:IF A$="Y" THEN RUN
```

Mean, Median, and Mode Program

This program will find the mean, median, and mode of a set of numbers. Press RETURN (ENTER) after each entry. When all entries are completed, press ENTER again.

```
10 DIM A(200),C(200)
20 PRINT"THIS PROGRAM FINDS THE MEAN ,MEDIAN AND MODE OF A LIST OF NUMBERS"
30 PRINT"ENTER NUMBER PRESS RETURN. WHEN DONE JUST PRESS RETURN"
40 INPUT"NUMBER AND RETURN OR JUST RETURN ";N$:P=VAL(N$)
50 T=T+P:F=F+1:IF N$="" THEN 120
60 FOR I=1 TO K
70 IF A(I)=P THEN C(I)=C(I)+1 :GOTO 100
80 NEXT I
90 K=K+1:A(K)=P:C(K)=1
100 IF C(I)>L THEN L=C(I)
110 GOTO 40
120 FOR I=1 TO K:FOR J=I+1 TO K
130 IF A(I)>=A(J) THEN B=A(I):A(I)=A(J):A(J)=B:B=C(I):C(I)=C(J):C(J)=B
140 NEXT J,I
150 PRINT"*******************************************"
160 PRINT"MEAN IS ";T/(F-1):M=INT((F-1)/2)+1
170 IF (F-1)/2 =(M-1) THEN A$="E":M1=(F-1)/2:GOTO 190
180 A$="O"
190 FOR I=1 TO K:C=C+C(I)
200 IF A$="O" AND C>=M THEN M=A(I):GOTO 240
210 IF A$="E" AND C>M1 THEN M=A(I):GOTO 240
220 IF A$="E" AND C=M1 THEN M=(A(I)+A(I+1))/2:GOTO 240
230 NEXT
240 PRINT "MEDIAN IS ";M
250 IF L=1 THEN PRINT"NO MODE ":GOTO 290
260 PRINT"MODE IS ";
270 FOR I=1 TO K :IF C(I)=L THEN PRINT"    ";A(I)
280 NEXT
290 PRINT"*******************************************"
300 INPUT"RUN AGAIN (Y/N) ";A$:IF A$="Y" THEN RUN
```

Standard Deviation, Range, and Mean Program

This program will find the standard deviation, range, and mean of a set of numbers. Press RETURN (ENTER) after each entry. When all entries are completed, press ENTER again.

```
10 DIM A(200)
20 PRINT"   <<<< STANDARD DEVIATION AND RANGE PROGRAM >>>>"
30 PRINT"ENTER NUMBER AND PRESS RETURN OR JUST PRESS RETURN WHEN DONE"
40 INPUT "NUMBER ";N$:IF N$="" THEN 100
50 IF K=0 THEN L=VAL(N$):U=VAL(N$)
60 K=K+1 :T=T+VAL(N$):A(K)=VAL(N$)
70 IF A(K)>U THEN U=A(K)
80 IF A(K)<L THEN L=A(K)
90 GOTO 40
100 PRINT"*******************************************"
110 M=T/K :T=0 :R=U-L
120 PRINT" MEAN IS            "; M
130 PRINT" RANGE IS           "; R
140 FOR I=1 TO K
150 T= T+(A(I)-M)^2
160 NEXT
170 T=T/K :S=T^.5
180 PRINT" STANDARD DEVIATION IS ";S
190 PRINT"*******************************************"
200 INPUT "RUN AGAIN (Y/N) ";A$:IF A$="Y" THEN RUN
```

TABLES

Table I Squares and Square Roots

N	N²	√N	√10N	N	N²	√N	√10N
1	1	1.00000	3.16228	51	2601	7.14143	22.5832
2	4	1.41421	4.47214	52	2704	7.21110	22.8035
3	9	1.73205	5.47723	53	2809	7.28011	23.0217
4	16	2.00000	6.32456	54	2916	7.34847	23.2379
5	25	2.23607	7.07107	55	3025	7.41620	23.4521
6	36	2.44949	7.74597	56	3136	7.48331	23.6643
7	49	2.64575	8.36660	57	3249	7.54983	23.8747
8	64	2.82843	8.94427	58	3364	7.61577	24.0832
9	81	3.00000	9.48683	59	3481	7.68115	24.2899
10	100	3.16228	10.0000	60	3600	7.74597	24.4949
11	121	3.31662	10.4881	61	3721	7.81025	24.6982
12	144	3.46410	10.9545	62	3844	7.87401	24.8998
13	169	3.60555	11.4018	63	3969	7.93725	25.0998
14	196	3.74166	11.8322	64	4096	8.00000	25.2983
15	225	3.87298	12.2474	65	4225	8.06226	25.4951
16	256	4.00000	12.6491	66	4356	8.12404	25.6905
17	289	4.12311	13.0384	67	4489	8.18535	25.8844
18	324	4.24264	13.4164	68	4624	8.24621	26.0768
19	361	4.35890	13.7840	69	4761	8.30662	26.2679
20	400	4.47214	14.1421	70	4900	8.36660	26.4575
21	441	4.58258	14.4914	71	5041	8.42615	26.6458
22	484	4.69042	14.8324	72	5184	8.48528	26.8328
23	529	4.79583	15.1658	73	5329	8.54400	27.0185
24	576	4.89898	15.4919	74	5476	8.60233	27.2029
25	625	5.00000	15.8114	75	5625	8.66025	27.3861
26	676	5.09902	16.1245	76	5776	8.71780	27.5681
27	729	5.19615	16.4317	77	5929	8.77496	27.7489
28	784	5.29150	16.7332	78	6084	8.83176	27.9285
29	841	5.38516	17.0294	79	6241	8.88819	28.1069
30	900	5.47723	17.3205	80	6400	8.94427	28.2843
31	961	5.56776	17.6068	81	6561	9.00000	28.4605
32	1024	5.65685	17.8885	82	6724	9.05539	28.6356
33	1089	5.74456	18.1659	83	6889	9.11043	28.8097
34	1156	5.83095	18.4391	84	7056	9.16515	28.9828
35	1225	5.91608	18.7083	85	7225	9.21954	29.1548
36	1296	6.00000	18.9737	86	7396	9.27362	29.3248
37	1369	6.08276	19.2354	87	7569	9.32738	29.4958
38	1444	6.16441	19.4936	88	7744	9.38083	29.6648
39	1521	6.24500	19.7484	89	7921	9.43398	29.8329
40	1600	6.32456	20.0000	90	8100	9.48683	30.0000
41	1681	6.40312	20.2485	91	8281	9.53939	30.1662
42	1764	6.48074	20.4939	92	8464	9.59166	30.3315
43	1849	6.55744	20.7364	93	8649	9.64365	30.4959
44	1936	6.63325	20.9762	94	8836	9.69536	30.6594
45	2025	6.70820	21.2132	95	9025	9.74679	30.8221
46	2116	6.78233	21.4476	96	9216	9.79796	30.9839
47	2209	6.85565	21.6795	97	9409	9.84886	31.1448
48	2304	6.92820	21.9089	98	9604	9.89949	31.3050
49	2401	7.00000	22.1359	99	9801	9.94987	31.4643
50	2500	7.07107	22.3607	100	10000	10.00000	31.6228
N	N²	√N	√10N	N	N²	√N	√10N

Table II Probabilities in a Normal Distribution

z	0	1	2	3	4	5	6	7	8	9
0.0	0.000	0.004	0.008	0.012	0.016	0.020	0.024	0.028	0.032	0.036
0.1	0.040	0.044	0.048	0.052	0.056	0.060	0.064	0.067	0.071	0.075
0.2	0.079	0.083	0.087	0.091	0.095	0.099	0.103	0.106	0.110	0.114
0.3	0.118	0.122	0.126	0.129	0.133	0.137	0.141	0.144	0.148	0.152
0.4	0.155	0.159	0.163	0.166	0.170	0.174	0.177	0.181	0.184	0.188
0.5	0.191	0.195	0.198	0.202	0.205	0.209	0.212	0.216	0.219	0.222
0.6	0.226	0.229	0.232	0.236	0.239	0.242	0.245	0.249	0.252	0.255
0.7	0.258	0.261	0.264	0.267	0.270	0.273	0.276	0.279	0.282	0.285
0.8	0.288	0.291	0.294	0.297	0.300	0.302	0.305	0.308	0.311	0.313
0.9	0.316	0.319	0.321	0.324	0.326	0.329	0.331	0.334	0.336	0.339
1.0	0.341	0.344	0.346	0.348	0.351	0.353	0.355	0.358	0.360	0.362
1.1	0.364	0.366	0.369	0.371	0.373	0.375	0.377	0.379	0.381	0.383
1.2	0.385	0.387	0.389	0.391	0.393	0.394	0.396	0.398	0.400	0.401
1.3	0.403	0.405	0.407	0.408	0.410	0.411	0.413	0.415	0.416	0.418
1.4	0.419	0.421	0.422	0.424	0.425	0.426	0.428	0.429	0.431	0.432
1.5	0.433	0.434	0.436	0.437	0.438	0.439	0.441	0.442	0.443	0.444
1.6	0.445	0.446	0.447	0.448	0.450	0.451	0.452	0.453	0.454	0.454
1.7	0.455	0.456	0.457	0.458	0.459	0.460	0.461	0.462	0.462	0.463
1.8	0.464	0.465	0.466	0.466	0.467	0.468	0.469	0.469	0.470	0.471
1.9	0.471	0.472	0.473	0.473	0.474	0.474	0.475	0.476	0.476	0.477
2.0	0.477	0.478	0.478	0.479	0.479	0.480	0.480	0.481	0.481	0.482
2.1	0.482	0.483	0.483	0.483	0.484	0.484	0.484	0.485	0.485	0.485
2.2	0.486	0.486	0.487	0.487	0.487	0.488	0.488	0.488	0.489	0.489
2.3	0.489	0.490	0.490	0.490	0.490	0.491	0.491	0.491	0.491	0.492
2.4	0.492	0.492	0.492	0.492	0.493	0.493	0.493	0.493	0.493	0.494
2.5	0.494	0.494	0.494	0.494	0.494	0.495	0.495	0.495	0.495	0.495
2.6	0.495	0.495	0.496	0.496	0.496	0.496	0.496	0.496	0.496	0.496
2.7	0.496	0.497	0.497	0.497	0.497	0.497	0.497	0.497	0.497	0.497
2.8	0.497	0.498	0.498	0.498	0.498	0.498	0.498	0.498	0.498	0.498
2.9	0.498	0.498	0.498	0.498	0.498	0.498	0.498	0.499	0.499	0.499
3.0	0.499	0.499	0.499	0.499	0.499	0.499	0.499	0.499	0.499	0.499
3.1	0.499	0.499	0.499	0.499	0.499	0.499	0.499	0.499	0.499	0.499
3.2	0.499	0.499	0.499	0.499	0.499	0.499	0.499	0.499	0.499	0.500

The number read in the body of the table is the probability that a number selected at random from a normally distributed population falls between the mean and a point that is z standard deviations greater than the mean. For instance, the entry 0.056 in the second line is the probability that the number falls between the mean and 0.14 standard deviation above the mean. For all values of z greater than 3.29, the probability is 0.500 correct to three decimal places.

ANSWERS

1. Only b and c are well-defined.   **3.** Only b and d are correct.
5. a. The set consisting of the first and last letters of the English alphabet
   b. The set of letters of the word *man*
   c. The set of names of the first Biblical man and woman
   d. The set consisting of the name of the person usually credited with the discovery of America in 1492
   e. The set of counting numbers from 1 to 7
   f. The set of products of the pairs of consecutive counting numbers, that is, $1 \cdot 2, 2 \cdot 3$, and so on to $5 \cdot 6$
7. {1, 2, 3, 4, 5, 6, 7}   **9.** {1, 2, 3, 4, 5, 6, 7}   **11.** {4, 5, 6, 7}   **13.** { } or $\varnothing$ **15.** {4, 5, 6, . . .}
17. {5, 10, 15, . . .}   **19.** { } or $\varnothing$ **21.** {1, 2}   **23.** {2, 4, 6, 8}
25. a. {WangB, Gull}   **b.** {WangB, Gull, WhrEnt, HomeSh}   **c.** {ENSCO, WDigitl, TexAir}
   d. {WhrEnt, EchBg, WDigitl}
27. a. {x|x is a stock whose last price was between 3 and 10}
   b. {x|x is a stock whose last price was more than 15}
29. a. {M, I, S, P}   **b.** {Mercury, Venus, Earth, Mars, Jupiter, Saturn, Uranus, Neptune, Pluto}
   c. $\{\frac{1}{1}, \frac{1}{2}, \frac{1}{3}, \ldots, \frac{1}{n}, \ldots\}$ **d.** { } or $\varnothing$
31. $A = B$ in c and d only.   **33. a.** =   **b.** ≠   **c.** ≠   **35.** All false except b.

USING YOUR KNOWLEDGE 1.1

1. a. If $g \in S$, then Gepetto shaves himself, which contradicts the statement that Gepetto shaves all those men and *only* those men of the village who do not shave themselves. Therefore, $g \notin S$.
   b. If $g \in D$, then Gepetto does not shave himself, and so by the same statement, he does shave himself. Thus, there is again a contradiction and $g \notin D$.

DISCOVERY 1.1

1. The word *non-self-descriptive* cannot be classified in either way without having a contradiction. If it is an element of S, then it is self-descriptive, which contradicts the definition, "non-self-descriptive is a non-self-descriptive word." On the other hand, the statement, "non-self-descriptive is a non-self-descriptive word," puts non-self-descriptive into S. So we have a contradiction either way.
3. $n + 1$

EXERCISE 1.2

1. a. $\varnothing, \{a\}, \{b\}, \{a, b\}$ **b.** $\varnothing, \{1\}, \{2\}, \{3\}, \{1, 2\}, \{1, 3\}, \{2, 3\}, \{1, 2, 3\}$
   c. $\varnothing, \{1\}, \{2\}, \{3\}, \{4\}, \{1, 2\}, \{1, 3\}, \{1, 4\}, \{2, 3\}, \{2, 4\}, \{3, 4\}, \{1, 2, 3\}, \{1, 2, 4\}, \{1, 3, 4\}, \{2, 3, 4\}, \{1, 2, 3, 4\}$
   d. $\varnothing, \{\varnothing\}$
3. a. $\varnothing, \{1\}, \{2\}, \{1, 2\}$ **b.** $\varnothing, \{x\}, \{y\}, \{z\}, \{x, y\}, \{x, z\}, \{y, z\}, \{x, y, z\}$
5. $2^4 = 16$ **7.** $2^{10} = 1024$ **9.** 5   **11.** 6
13. Yes. Every set is a subset of itself.
15. $B \subseteq A$, because every counting number that is divisible by 4 is also divisible by 2.
17. a. ⊂   **b.** ∉   **c.** ⊄
19. 8 different sums: $\varnothing, \{n\}, \{d\}, \{q\}, \{n, d\}, \{n, q\}, \{d, q\}, \{n, d, q\}$. If at least one coin must be chosen, then there are 7 possible choices.

USING YOUR KNOWLEDGE 1.2

1. {Alpine, Hindu, Mediterranean, Nordic}; note that the Ainu is *not* a subset.

3. {African Negro, Negrito, Oceanic Negro}   **5.** {Hindu, Mediterranean}   **7.** 8   **9.** 3

DISCOVERY 1.2

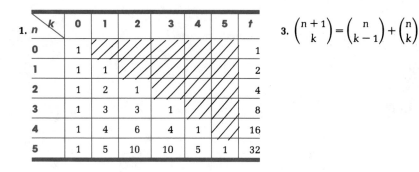

1. n / k

n \ k	0	1	2	3	4	5	t
0	1						1
1	1	1					2
2	1	2	1				4
3	1	3	3	1			8
4	1	4	6	4	1		16
5	1	5	10	10	5	1	32

3. $\binom{n+1}{k} = \binom{n}{k-1} + \binom{n}{k}$

EXERCISE 1.3

1. a. {1, 3, 4}   **b.** {1}   **c.** {1, 6}   **d.** {1, 2, 3, 4, 5, 6}   **e.** {1, 2, 3, 4, 5, 6, 7}   **f.** {1, 3, 4, 6, 7}
  g. {1, 3, 4}   **h.** {1, 2, 3, 4, 5, 6}   **i.** {1, 3, 4, 6, 7}   **j.** {1, 3, 4}
3. a. {c}   **b.** ∅   **c.** {{a, b}, a, b, c}   **d.** {{a, b}, a, b, c}
5. a. {b, d, f}   **b.** {a, c}   **c.** ∅   **d.** ∅   **e.** {a, b, c, d, f}   **f.** {a, b, c, d, f}   **g.** {c, e}   **h.** {c, e}
  i. {a, b, c, d, f}
7. a. {b, d, f}   **b.** {a, c}   **c.** {a, c}   **d.** {b, d, f}
9. 𝒰   **11.** ∅   **13.** A   **15.** ∅   **17.** A   **19.** {1, 2, 3, 4, 5}
21. {Beauty, Consideration, Kindliness, Friendliness, Helpfulness, Loyalty}
23. {Intelligence, Cheerfulness, Congeniality}   **25.** {Intelligence, Cheerfulness}
27. a. 𝒰   **b.** ∅   **c.** F   **d.** M   **e.** ∅   **29. a.** D ∩ S   **b.** F ∩ T   **c.** M ∩ D
31. a. {04, 08}, the set of full-time employees who do shop work
  b. {02, 05, 07}, the set of part-time employees who do outdoor field work or indoor office work

USING YOUR KNOWLEDGE 1.3

1. a. {Long tongue, Skin-covered horns, Native to Africa}
  b. {Long tongue, Skin-covered horns, Native to Africa}
  c. {Tall, Short, Long neck, Short neck, Long tongue, Skin-covered horns, Native to Africa}
  d. {Short, Short neck}   **e.** {Tall, Long neck}
3. {Sunbeam snakes}

DISCOVERY 1.3

5. If n = 2m − 1 or if n = 2m, the number is 2m + 1.

EXERCISE 1.4

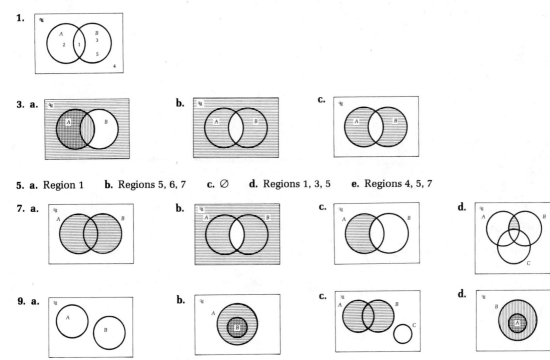

1.

3. a. **b.** **c.**

5. a. Region 1   **b.** Regions 5, 6, 7   **c.** ∅   **d.** Regions 1, 3, 5   **e.** Regions 4, 5, 7

7. a. **b.** **c.** **d.**

9. a. **b.** **c.** **d.**

11. a. A: regions 1, 3, 5, 7; $(B \cup C)$: regions 2, 3, 4, 5, 6, 7; thus, for $A \cup (B \cup C)$, we have regions 1, 2, 3, 4, 5, 6, 7.
$A \cup B$: regions 1, 2, 3, 5, 6, 7; C: regions 4, 5, 6, 7; thus, for $(A \cup B) \cup C$, we have regions 1, 2, 3, 4, 5, 6, 7.
Therefore, $A \cup (B \cup C) = (A \cup B) \cup C$.

 b. A: regions 1, 3, 5, 7; $B \cap C$: regions 6, 7; thus, for $A \cap (B \cap C)$, we have region 7.
$A \cap B$: regions 3, 7; C: regions 4, 5, 6, 7; thus, for $(A \cap B) \cap C$, we have region 7.
Therefore, $A \cap (B \cap C) = (A \cap B) \cap C$.

13. a. A: regions 1, 3, 5, 7; A': regions 2, 4, 6, 8; thus, for $A \cup A'$, we have regions 1, 2, 3, 4, 5, 6, 7, 8, the same set of
regions that represents $\mathcal{U}$. Therefore, $A \cup A' = \mathcal{U}$.

 b. From part a, we see that $A \cap A' = \emptyset$.

 c. $A - B$: regions 1, 5; $A \cap B'$: regions 1, 5; therefore, $A - B = A \cap B'$.

15. Part a, $A \cap B$

17. a. $A = \{a, b, c, e\}$, $B = \{a, b, g, h\}$, $\mathcal{U} = \{a, b, c, d, e, f, g, h\}$

 b. $A \cup B = \{a, b, c, e, g, h\}$ **c.** $(A \cap B)' = \{c, d, e, f, g, h\}$

19.

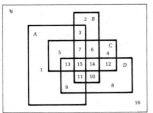

21. Arizona, California, Florida, Texas

USING YOUR KNOWLEDGE 1.4

1. AB⁺   **3.** No   **5.** No

DISCOVERY 1.4

1. $2^4 = 16$ **3. a.** Region 11   **b.** Regions 8, 16

EXERCISE 1.5

1. 30   **3.** 20   **5.** 40   **7. a.** None   **b.** 10   **c.** 10   **9. a.** 22   **b.** 36   **c.** 6
11. a. \$120,000   **b.** \$510,000   **c.** \$1,305,000   **13.** 200   **15. a.** 5   **b.** 30   **c.** 20
17. \$450   **19.** 28   **21. a.** 120   **b.** 80   **c.** 50   **23. a.** 80   **b.** 120   **c.** 50
25. a. 73   **b.** 55   **c.** 91   **d.** 38   **e.** 5

USING YOUR KNOWLEDGE 1.5

1. The Venn diagram shows that with the additional information, the statistics in the cartoon are possible.

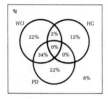

DISCOVERY 1.5

1.

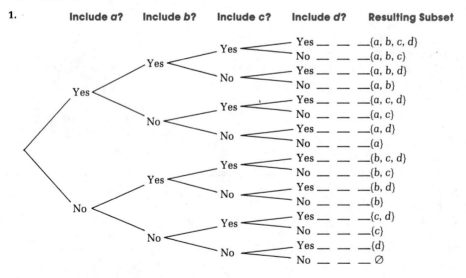

3. There are 2 choices for each element, so there are 2^n choices for an n-element set.

PRACTICE TEST 1

1. {3, 4, 5, 6, 7, 8, 9}

2. a. The set of vowels in the English alphabet; {x|x is a vowel in the English alphabet}
   b. The set of even counting numbers less than 10; {x|x is an even counting number less than 10}

3. ∅, {$}, {¢}, {%}, {$, ¢}, {$, %}, {¢, %}   **4. a.** ∈, ∈   **b.** ∈, ∉   **5. a.** ∈, ∉   **b.** ∈, ∈

6. a. {King}   **b.** ∅   **c.** {Queen}   **d.** {Queen}   **7. a.** {Ace, Queen, Jack}   **b.** {King}

8. **9.**

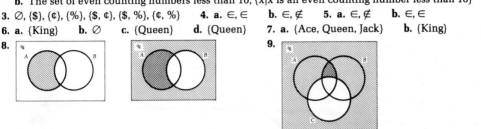

10. a. {1, 2, 5}   **b.** {2, 3, 4, 6, 8}

11. $A \cap B$: regions 5, 7; C: regions 3, 4, 6, 7; $(A \cap B) \cup C$: regions 3, 4, 5, 6, 7; $A \cup C$: regions 1, 3, 5, 6, 7; $B \cup C$: regions 2, 3, 4, 5, 6, 7; $(A \cup C) \cap (B \cup C)$: regions 3, 4, 5, 6, 7. This verifies that $(A \cap B) \cup C = (A \cup C) \cap (B \cup C)$.

12. $(A \cap B)'$: regions 1, 2, 3, 4, 6, 8; A': regions 2, 3, 6, 8; B': regions 1, 3, 4, 8; $A' \cup B'$: regions 1, 2, 3, 4, 6, 8. This verifies that $(A \cap B)' = A' \cup B'$.

13. a. ∉   **b.** ∈   **c.** ∉   **14. a.** =   **b.** ≠   **15. a.** =   **b.** ≠   **16. a.** 60   **b.** 55   **17. a.** 5   **b.** 43

18. $B - A$ or $B \cap A'$

19.

20. a. 47   **b.** 12   **c.** 26   **d.** 3   **e.** 10

a. 55
b. 30
c. 45

CHAPTER 2   EXERCISE 2.1

1. Not a statement.

3. A compound statement. Components: Jane is taking an English course. She has four themes to write.

5. Not a statement.

7. A compound statement. Components: Students at Ohio State University are required to take a course in history. Students at Ohio State University are required to take a course in economics.

9. $a \wedge f$ **11.** $d \vee f$ **13.** $b \wedge p$ **15.** $a \vee m$ **17. a.** $p \wedge q$ **b.** $p \wedge \sim q$ **c.** $\sim p \wedge \sim q$ **d.** $\sim(p \wedge q)$

19. a. It is not a long time before the end of the term.   **b.** Bill's store is not making a good profit.
   c. 10 is not a round number.

21. a. Not negations of each other.   **b–c.** Negations of each other.

23. a. He is not bald, and he does not have a 10-inch forehead.   **b.** Somebody does not like Sara Lee.
   c. No circles are round.   **d.** No men earn less than $5 an hour or more than $50 an hour.

EXERCISE 2.2

1. a. Today is Friday or Monday.   **b.** Today is Friday and Monday.   **c.** Today is not Friday.
   d. The statement in part b

3. a. $\sim(g \vee s)$ **b.** $g \wedge s$ **c.** $\sim g \wedge \sim s$

5. a. $p \wedge \sim q$ **b.** $\sim q \wedge p$ **c.** $\sim(\sim p \wedge q)$ **d.** $p \vee q$ **e.** The statements in parts c and d are both true.

7. a. $g \lor j$; true **b.** $g \lor \sim j$; false **c.** $\sim g \land \sim j$; false **d.** $(g \lor j) \land \sim(g \land j)$; true
9. $b \lor c \lor (s \land e)$ **11.** $\sim[(i \land a) \lor p]$ or $\sim(i \land a) \land \sim p$ or $(\sim i \lor \sim a) \land \sim p$
13. No. The trustee could be someone other than Sam. **15.** $(i \lor d) \lor (f \lor g \lor t)$ **17.** $i \lor r \lor p$
19. Only Boris

USING YOUR KNOWLEDGE 2.2

1. a. $(e \land g) \lor a$ **b.** h **c.** $c \lor n \lor \sim t$ **3.** $a \land h \land n$ and $a \land h \land \sim t$

DISCOVERY 2.2

1.

		a	b	c	d
g	s	$g \land s$	$g \lor s$	$\sim g \lor \sim s$	$\sim g \land \sim s$
T	T	T	T	F	F
T	F	F	T	T	F
F	T	F	T	T	F
F	F	F	F	T	T

3. The table shows that if any two of the statements are both true, then the remaining two statements are both false.
5. Statements a and c always have opposite truth values; b and d always have opposite truth values.

EXERCISE 2.3

1.

1	2	4	3
p	q	$p \lor$	$\sim q$
T	T	T	F
T	F	T	T
F	T	F	F
F	F	T	T

3.

1	2	3	4
p	q	$\sim p$	$\land q$
T	T	F	F
T	F	F	F
F	T	T	T
F	F	T	F

5.

1	2	5	4	3
p	q	$\sim(p$	$\lor$	$\sim q)$
T	T	F	T	F
T	F	F	T	T
F	T	T	F	F
F	F	F	T	T

7.

1	2	6	3	5	4
p	q	$\sim(\sim p$		$\land$	$\sim q)$
T	T	T	F	F	F
T	F	T	F	F	T
F	T	T	T	F	F
F	F	F	T	T	T

9.

1	2	3	6	4	5
p	q	$(p \land q) \lor$		$(\sim p$	$\land q)$
T	T	T	T	F	F
T	F	F	F	F	F
F	T	F	T	T	T
F	F	F	F	T	F

11.

1	2	3	5	4
p	q	r	$p \land$	$(q \lor r)$
T	T	T	T	T
T	T	F	T	T
T	F	T	T	T
T	F	F	F	F
F	T	T	F	T
F	T	F	F	T
F	F	T	F	T
F	F	F	F	F

13.

1	2	3	4	7	6	5
p	q	r	(p ∨ q)	∨	(r ∧	~q)
T	T	T	T	T	F	F
T	T	F	T	T	F	F
T	F	T	T	T	T	T
T	F	F	T	T	F	T
F	T	T	T	T	F	F
F	T	F	T	T	F	F
F	F	T	F	T	T	T
F	F	F	F	F	F	T

15. a. When p and q are both true.   **b.** When p or q or both are false.   **c.** When p or q or both are true.
    d. When p and q are both false.

17. a. Both statements have the truth values TTTTTFFF.   **b.** Both statements have the truth values FFFT.
    c. Both statements have the truth values FTTT.

19. b. The conjunction p ∧ q is true only in the first row of the table, and the conjunction ~p ∧ ~q is true only in
    the last row of the table. Hence, the disjunction of these two conjunctions will be true in the first and last
    rows of the table and will be false in the other two rows.
    c. (p ∧ ~q) ∨ (~p ∧ q) has truth values FTTF. (p ∧ ~q) ∨ (~p ∧ q) ∨ (~p ∧ ~q) has truth values FTTT;
    ~(p ∧ q) also has truth values FTTT.

21. a. Let z be "Billy goes to the zoo," let e be "Billy feeds peanuts to the elephants," and let m be "Billy feeds
    peanuts to the monkeys." Then, the given statement is z ∧ (e ∨ m), which is true if Billy goes to the zoo and
    feeds peanuts to the elephants and/or the monkeys.
    b. Let f be "I file my income tax return," let p be "I pay the tax," and let j be "I go to jail." Then, the statement
    is (f ∧ p) ∨ j, which is true if one of the following is true: (1) I file the return and pay the tax. (2) I file the
    return, do not pay the tax, and go to jail. (3) I do not file the return and do go to jail.

23. a.

1	2	3	4
p	q	(p ∧ q)	* p
T	T	T	F
T	F	F	F
F	T	F	T
F	F	F	T

b.

1	2	4	3	5
p	q	(p ∧	~q)	* q
T	T	F	F	F
T	F	T	T	F
F	T	F	F	F
F	F	F	T	T

c.

1	2	3	5	4
p	q	(p ∨ q)	*	~p
T	T	T	F	F
T	F	T	F	F
F	T	T	F	T
F	F	F	F	T

25. a. ~p ∧ ~q   **b.** p ∧ q   **c.** p ∨ q

USING YOUR KNOWLEDGE 2.3

1. e ∧ c ∧ f   **3.** e ∧ f   **5.** Fish

DISCOVERY 2.3

1. Mr. Baker is the carpenter.

EXERCISE 2.4

1.

1	2	3	5	4		6	
p	q	~q	→	~p	p	→	q
T	T	F	T	F		T	
T	F	T	F	F		F	
F	T	F	T	T		T	
F	F	T	T	T		T	

Since columns 5 and 6 are identical, $\sim q \rightarrow \sim p \Leftrightarrow p \rightarrow q$.

3. a. T **b.** F **c.** T **d.** T   **5.** x may be any number.   **7.** x may be any number except 4.

9. If you got the time and we do not have the beer.

11. a.

1	2	3		5	4
p	q	r	p	→	(q ∧ r)
T	T	T		T	T
T	T	F		F	F
T	F	T		F	F
T	F	F		F	F
F	T	T		T	T
F	T	F		T	F
F	F	T		T	F
F	F	F		T	F

b.

1	2	3	4	6	5
p	q	r	(p → q)	∧	(p → r)
T	T	T	T	T	T
T	T	F	T	F	F
T	F	T	F	F	T
T	F	F	F	F	F
F	T	T	T	T	T
F	T	F	T	T	T
F	F	T	T	T	T
F	F	F	T	T	T

c. Yes. They have identical truth tables.

13. a. $\sim s \rightarrow \sim b$ **b.** $\sim b \rightarrow \sim s$ **c.** $b \rightarrow s$

15. a. You do not work or you have to pay taxes.   **b.** You do not have the time, or we got the beer.
c. You do not find a better one, or you buy it.

17.

1	2	5	3	4		7	6
p	q	~	(~p ∨ q)		p ∧	~q
T	T	F	F	T	F	F
T	F	T	F	F	T	T
F	T	F	T	T	F	F
F	F	F	T	T	F	T

Since columns 5 and 7 are identical, $\sim(\sim p \vee q) \Leftrightarrow p \wedge \sim q$.

19. a. You earn much money, but you do not pay heavy taxes.
b. Johnny does not play quarterback and his team does not lose.
c. Alice passes the test, but does not get the job.

21. TFTT   **25. a.** $f \rightarrow a$ **b.** $p \rightarrow t$

USING YOUR KNOWLEDGE 2.4

1. $d \rightarrow r$ **3.** $d \rightarrow (r \wedge a)$

DISCOVERY 2.4

1. The desired question is: "Does the first road lead to freedom and are you telling the truth, or does the first road lead to freedom and are you lying?"

EXERCISE 2.5

1. Proof of the contrapositive, "If n is not even, then n^2 is not even": If n is not even, then $n = 2k + 1$, where k is an integer, and $n^2 = (2k + 1)^2 = 4k^2 + 4k + 1 = 2(2k^2 + 2k) + 1$, which is an odd (not an even) integer. Therefore, if n^2 is even, then n is even.

3. a., b., and c. $q \rightarrow p$ **5.**

| | | | CONVERSE | INVERSE |
| | | | | |
p	q	$p \rightarrow q$	$q \rightarrow p$	$\sim p \rightarrow \sim q$
T	T	T	T	T
T	F	F	T	T
F	T	T	F	F
F	F	T	T	T

Since the converse and the inverse have identical truth tables, they are equivalent.

7. a. *Converse:* If you are not strong, then you do not eat your spinach.
  Inverse: If you eat your spinach, then you are strong.
  Contrapositive: If you are strong, then you eat your spinach.
  b. *Converse:* If you are strong, then you eat your spinach.
  Inverse: If you do not eat your spinach, then you are not strong.
  Contrapositive: If you are not strong, then you do not eat your spinach.
  c. *Converse:* If you eat your spinach, then you are strong.
  Inverse: If you are not strong, then you do not eat your spinach.
  Contrapositive: If you do not eat your spinach, then you are not strong.
9. a. If the square of an integer is divisible by 4, then the integer is even. True.
  b. If there are clouds in the sky, then it is raining. False.
  c. If I am neat and well-dressed, then I can get a date. False.
  d. If all our problems are over, then M is elected to office. False.
  e. If you pass this course, then you get passing grades on all the tests. False.
11. a. $u \rightarrow a$ **b.** If you use this box, then your recent test scores were earned after October 1.
13. a. $f \rightarrow \sim r$ **b.** $r \rightarrow \sim f$
  c. If you need to report the income, then you rented your vacation home for 15 or more days a year.
15. a. If you are under 18, then you are not admitted without parent or guardian.
  b. If you are admitted without parent or guardian, then you are not under 18.
17. a. $k \rightarrow p$ **b.** $k \leftrightarrow p$ **c.** $p \rightarrow k$ **19. a.** $b \leftrightarrow c$ **b.** $b \rightarrow c$ **c.** $c \rightarrow b$

DISCOVERY 2.5

1. The contrapositive of $\sim q \rightarrow \sim p$ is $p \rightarrow q$.
3. The inverse of $p \rightarrow q$ is $\sim p \rightarrow \sim q$, and the contrapositive of $\sim p \rightarrow \sim q$ is $q \rightarrow p$.
5. $(\sim r \wedge \sim s) \vee (p \vee q) \Leftrightarrow (r \vee s) \rightarrow (p \vee q)$ is true because
  $(r \vee s) \rightarrow (p \vee q) \Leftrightarrow \sim (r \vee s) \vee (p \vee q) \Leftrightarrow (\sim r \wedge \sim s) \vee (p \vee q)$.
7. The direct statement   **9.** The contrapositive

EXERCISE 2.6

1.

1	2	3	4
p	**q**	**(p ∧ q)**	**→ p**
T	T	T	T
T	F	F	T
F	T	F	T
F	F	F	T

Since column 4 is all T's, the statement (p ∧ q) → p is a tautology.

5. u ⟹ v; w ⟹ v

7. The table shows the possibilities. Thus, neither one implies the other.

p	**q**
T	F
F	T

11. The table shows the possibilities. Thus, q ⟹ p.

p	**q**
T	T
T	F
F	F

3.

1	3	2
p	**p ↔**	**~p**
T	F	F
F	F	T

Since column 3 is all F's, the statement p ↔ ~p is a contradiction.

9. The table shows the possibilities. Thus, p ⟹ q.

p	**q**
T	T
F	T
F	F

13. The table shows the possibilities. Thus, q ⟹ p.

p	**q**
T	T
T	F
F	F

15. Four: Statements a, b, d, and e all can be true simultaneously; or statements b, c, d, and f all can be true simultaneously.

17. The first implies the second.   **19.** Equivalent   **21.** The second implies the first.

USING YOUR KNOWLEDGE 2.6

1.

STATEMENT LANGUAGE	SET LANGUAGE
p	P
q	Q
~p	P′
~q	Q′
p ∨ q	P ∪ Q
p ∧ q	P ∩ Q
p ⟹ q	P ⊆ Q
p ⟺ q	P = Q
t, a tautology	𝒰
c, a contradiction	∅

3. a. P ∩ (Q ∪ R)′   **b.** (P ∪ Q) ∩ (Q ∪ R)′

DISCOVERY 2.6

1. Output 1 is *T*; output 2 is *F*.

3.

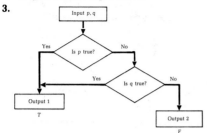

EXERCISE 2.7

1. *Premises:* "No misers are generous" and "Some old persons are not generous." *Conclusion:* "Some old persons are misers."

3. *Premises:* "All diligent students make A's" and "All lazy students are not successful." *Conclusion:* "All diligent students are lazy."

5. *Premises:* "No kitten that loves fish is unteachable" and "No kitten without a tail will play with a gorilla." *Conclusion:* "No unteachable kitten will play with a gorilla."

7. Valid

9. Invalid

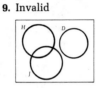

11. Valid

13. Invalid

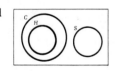

15. Invalid

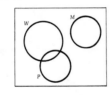

17. Invalid

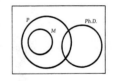

19. Valid

21. a. Invalid

b. Valid

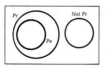

c. Invalid

1. Valid

3. Invalid

5. Invalid

7. Invalid

9. Valid

11. Only conclusion a is valid.

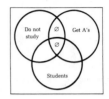

EXERCISE 2.8

1. $e \rightarrow p$
$\dfrac{\sim e}{\therefore \sim p}$
Invalid

3. $s \rightarrow e$
$\dfrac{\sim e}{\therefore \sim s}$
Valid

5. $\dfrac{g}{\therefore g \wedge r}$
Invalid

7. $w \rightarrow m$
$\dfrac{\sim w \rightarrow g}{\therefore m \vee g}$
Valid

9. $t \rightarrow b$
$\dfrac{t}{\therefore b}$
Valid

11. $s \rightarrow f$
$\dfrac{s}{\therefore f}$
Valid

13. $m \rightarrow e$
$\dfrac{\sim m}{\therefore \sim e}$
Invalid

15. $f \rightarrow s$
$\dfrac{\sim f}{\therefore \sim s}$
Invalid

17. Valid   **19.** Invalid   **21.** Valid   **23.** $p \rightarrow r$ **25.** s   **27.** q

29. All romances are well-written.   **31.** Aardvarks do not vote.

DISCOVERY 2.8

1. Kittens that play with a gorilla do not have green eyes. Or: No kitten with green eyes will play with a gorilla.

EXERCISE 2.9

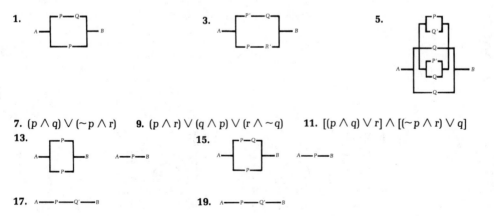

7. $(p \wedge q) \vee (\sim p \wedge r)$ **9.** $(p \wedge r) \vee (q \wedge p) \vee (r \wedge \sim q)$ **11.** $[(p \wedge q) \vee r] \wedge [(\sim p \wedge r) \vee q]$

DISCOVERY 2.9

1. In the diagram, *A* is an AND gate and *B* is a NOT gate. The inputs to *A* correspond to *p* and ~*q*, and the final output corresponds to *p* ∧ ~*q*. The table (column 4) shows that the final output voltage is low (0), except when the input voltage corresponding to *P* is high (1) and that corresponding to *Q* is low (0).

1	2	4	3
p	*q*	*p* ∧	~*q*
1	1	0	0
1	0	1	1
0	1	0	0
0	0	0	1

3. In the diagram, *A* and *D* are OR gates, *B* and *C* are NOT gates, and *E* is an AND gate. The output of *A* corresponds to *p* ∨ *q*, the output of *B* corresponds to ~*p*, and the output of *C* corresponds to ~*q*. The output of *D* corresponds to ~*p* ∨ ~*q*. Thus, the output of *E* corresponds to (*p* ∨ *q*) ∧ (~*p* ∨ ~*q*). The table (column 7) shows that the final output voltage is low (0) when the input voltages corresponding to *P* and *Q* are both high (1) or both low (0); the output voltage is high otherwise.

1	2	3	7	4	6	5
p	*q*	(*p* ∨ *q*)	∧	(~*p*	∨	~*q*)
1	1	1	0	0	0	0
1	0	1	1	0	1	1
0	1	1	1	1	1	0
0	0	0	0	1	1	1

PRACTICE TEST 2

1. b, c, d, and e are statements.

2. a. *d*: The number of the year is divisible by 4. *p*: The year is a presidential election year. The logical connective is "if . . . then . . . ." *d* → *p*

    b. *b*: I love Bill. ~*m*: Bill does not love me. The logical connective is "and." *b* ∧ ~*m*

    c. *e*: A candidate is elected president of the United States. *m*: A candidate receives a majority of the electoral college votes. The logical connective is "if and only if." *e* ↔ *m*

    d. *s*: Janet can make sense out of symbolic logic. *f*: She fails this course. The logical connective is "or." *s* ∨ *f*

    e. *s*: Janet can make sense out of symbolic logic. The logical connective is "not." ~*s*

3. a. It is not the case that he is a gentleman and a scholar.

    b. He is not a gentleman, but he is a scholar.

4. a. I will not go to the beach and I will not go to the movies.

    b. I will not stay in my room, or I will not do my homework.

    c. Pluto is a planet.

5. a. Some cats are not felines.　　**b.** No dogs are well-trained.

    c. Some dog is afraid of a mouse.

6. a. Joey does not study, and he does not fail this course.

    b. Sally studies hard, but she will not make an A in this course.

7. a. *p* ↔ *q*　　**b.** *p* ∧ *q*　　**c.** ~*p*　　**d.** *p* → *q*　　**e.** *p* ∨ *q*

8. 1	2	3	7	4	6	5
p	**q**	**(p ∨ q)**	**∧**	**(~p**	**∨**	**~q)**
T	T	T	F	F	F	F
T	F	T	T	F	T	T
F	T	T	T	T	T	F
F	F	F	F	T	T	T

9. 1	2	3	5	4
p	**q**	**(p ∨ q)**	**→**	**~p**
T	T	T	F	F
T	F	T	F	F
F	T	T	T	T
F	F	F	T	T

10. Statement b

11. When either or both of the statements, "Sally is naturally beautiful," and "Sally knows how to use makeup," is(are) true.

12. True

13. 1	2	3	6	5	4
p	**q**	**(p → q)**	**↔**	**(q ∨**	**~p)**
T	T	T	T	T	F
T	F	F	T	F	F
F	T	T	T	T	T
F	F	T	T	T	T

14. a. If you make a golf score of 62 again, then you made it once.
   b. If you did not make a golf score of 62 once, then you will not make it again.
   c. If you do not make a golf score of 62 again, then you did not make it once.

15. a. $m \rightarrow p$ **b.** $p \rightarrow m$ **c.** $p \leftrightarrow m$ **16. a.** $b \rightarrow c$ **b.** $c \rightarrow b$ **c.** $c \leftrightarrow b$

17. $b \Rightarrow a; b \Rightarrow c; c \Rightarrow a$ **18.** Statement b only

19.

Nothing in the premises tells whether John is among those who study hard. The argument is invalid.

20.

Nothing in the premises tells whether Sally is a loafer. The argument is invalid.

21. The premises may be written as: If John is a student (s), then John works hard (h). John is not a student (~s). The argument is then symbolized by:

$s \rightarrow h$
$\underline{\quad \sim s \quad}$
$\therefore \sim h$

		PREMISE	PREMISE	CONCLUSION
s	**h**	**s → h**	**~s**	**~h**
T	T	T	F	F
T	F	F	F	T
F	T	T	T	F
F	F	T	T	T

← { Premises true / Conclusion false } → Invalid

22. The premises may be written as: If Sally is a loafer (l), then Sally does not work hard ($\sim h$). Sally does not work hard ($\sim h$). The argument is then symbolized by:

$$l \to \sim h$$
$$\underline{\sim h}$$
$$\therefore l$$

			PREMISE	PREMISE	CONCLUSION
l	h	$\sim h$	$l \to \sim h$	l
T	T	F	F	T
T	F	T	T	T
F	T	F	T	F
F	F	T	T	F

$\leftarrow \left\{ \begin{array}{l} \text{Premises true} \\ \text{Conclusion false} \end{array} \right\} \to$ Invalid

23. Let r be, "You are a good runner," and let w be, "You win the race." Then the argument can be symbolized by:

$$w \to r$$
$$\underline{w}$$
$$\therefore r$$

		PREMISE	PREMISE	CONCLUSION
r	w	$w \to r$	w	r
T	T	T	T	T
T	F	T	F	T
F	T	F	T	F
F	F	T	F	F

Both premises are true in the first row only. Since the conclusion is also true there, the argument is valid.

24. If p, q, and r are all true, then the premises $p \to q$ and $\sim q \to \sim r$ are both true, but the conclusion $p \to \sim r$ is false. Hence, the argument is invalid.

25. $(p \to q) \wedge (\sim p \to \sim q) \Leftrightarrow (\sim p \vee q) \wedge (p \vee \sim q)$. Thus, a switching circuit corresponding to the given statement is

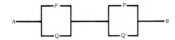

CHAPTER 3   EXERCISE 3.1

1. ∩∩||||| **3.** 9∩∩∩∩|| **5.** 9999∩∩∩||| **7.** 113   **9.** 322   **11.** 11,232
                                        9999 ||

13. ∩∩∩||||| **15.** 999⑨∩∩⓪✕✕ → 99 ✕✕∩∩∩∩∩∩∩∩⑧|||||||||
     + ∩∩|||| − 9 ✕✕∩∩✕✕| − ✕✕✕
     ‾‾‾‾‾‾‾‾‾   ‾‾‾‾‾‾‾‾‾‾‾‾‾‾
   ∩∩∩∩||||||| 99∩∩∩∩∩∩∩|||||||

17. \1   40   **19.** 1   51   **21.** 18   32   **23.** 12   51
    \2   80   \2   102   9   (64)   6   102
    \4   160   \4   204   4   128
    \8   320   8   408   2   256   3   (204)
    15   600   \16   816   1   (512)   1   (408)
                             22   1122   576   612

25. ✓✓✓✓✓✓   **27.** <<<✓✓   **29.** ✓✓ ✓✓✓   **31.** ✓✓✓✓ <✓✓✓✓✓✓✓   **33.** ✓ ✓✓ <✓✓✓   **35.** 92

37. 192   **39.** 4322   **41.** <<<✓✓   **43.** ✓✓ <✓✓✓   **45.** 126   **47.** 42,000   **49.** 90,405
                               + <<<<✓✓✓   + ✓ ✓✓✓✓✓✓✓✓
                               ‾‾‾‾‾‾‾‾‾‾‾‾   ‾‾‾‾‾‾‾‾‾‾
                               <<<<<<✓✓✓✓✓   ✓✓✓ <<✓
                               = ✓ <✓✓✓✓✓

51. LXXII   **53.** CXLV   **55.** $\overline{\text{XXXII}}$DIII

1. Assume the answer is 6: $6 + (\frac{1}{6})(6) = 7$, and $21 \div 7 = 3$. Therefore, the correct answer is $(3)(6) = 18$.

3. Assume the answer is 3: $3 + (\frac{4}{3})(3) = 5$; $5 - (\frac{1}{3})(5) = \frac{10}{3}$, and $10 \div \frac{10}{3} = 3$. Therefore, the correct answer is $(3)(3) = 9$.

DISCOVERY 3.1

1. a. $n = 9$ **b.** $n = 8$

EXERCISE 3.2

1. $(4 \times 10^2) + (3 \times 10) + (2 \times 10^0)$ **3.** $(2 \times 10^3) + (3 \times 10^2) + (7 \times 10^0)$

5. $(1 \times 10^4) + (2 \times 10^3) + (3 \times 10^2) + (4 \times 10) + (9 \times 10^0)$ **7.** 1   **9.** 45   **11.** 9071   **13.** 748,308

15. 4,000,031   **17.** $(2 \times 10) + (3 \times 10^0)$ 23   **19.** $(7 \times 10) + (1 \times 10^0)$ 71

$\dfrac{(1 \times 10) + (3 \times 10^0)}{(3 \times 10) + (6 \times 10^0)}$ $\dfrac{+13}{36}$ $\dfrac{(2 \times 10) + (3 \times 10^0)}{(9 \times 10) + (4 \times 10^0)}$ $\dfrac{+23}{94}$

21. $(7 \times 10) + (6 \times 10^0)$ 76   **23.** $(8 \times 10) + (4 \times 10^0)$ 84   **25.** 7^{11} **27.** 6^{40} **29.** 6^7 **31.** 6^{12}

$\dfrac{-(5 \times 10) + (4 \times 10^0)}{(2 \times 10) + (2 \times 10^0)}$ $\dfrac{-54}{22}$ $\dfrac{-(3 \times 10) + (1 \times 10^0)}{(5 \times 10) + (3 \times 10^0)}$ $\dfrac{-31}{53}$

33. $(2 \times 10) + (5 \times 10^0)$ 25

$\dfrac{\times (5 \times 10) + (1 \times 10^0)}{(2 \times 10) + (5 \times 10^0)}$ $\dfrac{\times 51}{25}$

$(10 \times 10^2) + (25 \times 10)$ 125

$\dfrac{10^3 \quad + (27 \times 10) + (5 \times 10^0)}{}$ 1275

$= (1 \times 10^3) + (2 \times 10^2) + (7 \times 10) + (5 \times 10^0) = 1275$

35. $(6 \times 10) + (2 \times 10^0)$ 62

$\dfrac{\times (2 \times 10) + (5 \times 10^0)}{(30 \times 10) + (10 \times 10^0)}$ $\dfrac{\times 25}{310}$

$(12 \times 10^2) + \quad (4 \times 10)$ 124

$\dfrac{12 \times 10^2 \; + (34 \times 10) + 10}{}$ 1550

$= 1 \times 10^3 + (5 \times 10^2) + (5 \times 10) = 1550$

37. $8 \times 10^0 \overline{)(6 \times 10) + (4 \times 10^0)}$ $\dfrac{(8 \times 10^0)}{}$ $8\overline{)64}$ $\dfrac{8}{}$

$\dfrac{(6 \times 10) + (4 \times 10^0)}{0}$ $\dfrac{64}{0}$

39. $6 \times 10^0 \overline{)(7 \times 10) + (2 \times 10^0)}$ $\dfrac{(1 \times 10) + (2 \times 10^0)}{}$ $6\overline{)72}$ $\dfrac{12}{}$

$\dfrac{(6 \times 10)}{(1 \times 10) + (2 \times 10^0)}$ $\dfrac{6}{12}$

$\dfrac{(1 \times 10) + (2 \times 10^0)}{0}$ $\dfrac{12}{0}$

DISCOVERY 3.2

1. There were 137,256 on the road to Rome.

EXERCISE 3.3

1. 22_{three} **3.** 31_{four} **5.** (********) ****** ; 17_{eight} **7.** (******) (******) * ; 21_{seven}

9. 22   **11.** 139   **13.** 27   **15.** 291   **17.** 30_{five} **19.** 11100_{two} **21.** $19_{sixteen}$ **23.** 41_{six} **25.** 121_{seven}

27. 46_{eight} **29.** $5BB_{sixteen}$ **31.** $73 = 1001001_{two} = 111_{eight}$

USING YOUR KNOWLEDGE 3.3

1. The trick works because the columns correspond to the binary digits in the number. For instance, $6 = 110_{two}$, and this corresponds to the number $6 = 2 + 4$, the numbers that head columns B and C. Note that 6 occurs in columns B and C, but not in column A.

3. Use the same idea, but with 5 columns.

CALCULATOR CORNER 3.3

1. 13   **3.** 113   **5.** 1914   **7.** 2620

EXERCISE 3.4

1. 67_8 **3.** 155_8 **5.** 110101_2 **7.** 11000110_2 **9.** 37_{16} **11.** $6D_{16}$ **13.** 10010101_2 **15.** 11111001101_2
17. 411_8 **19.** $15F_{16}$

USING YOUR KNOWLEDGE 3.4

1. 3   1   2   This shows that $312_4 = 110110_2$.   **3.** 1233_4 **5.** 61_8
     ↓   ↓   ↓
    11   01   10

DISCOVERY 3.4

1. 55   **3.** 000001_2 **5.** 011111_2

EXERCISE 3.5

1. 1001_2 **3.** 10011_2 **5.** 10010_2 **7.** 101_2 **9.** 1_2 **11.** 1010_2 **13.** 10010_2 **15.** 101101_2
17. 110111_2 **19.** 110_2 remainder 1_2 **21.** 100_2 remainder 10_2 **23.** 1011_2 remainder 100_2

EXERCISE 3.6

1. 600_8 **3.** 10112_8 **5.** 432_8 **7.** 7154_8 **9.** 507_8 **11.** 2306_8 **13.** 35_8 remainder 4_8
15. 250_8 remainder 5_8 **17.** 417_{16} **19.** $9B8_{16}$ **21.** $A367_{16}$ **23.** $4A451_{16}$

USING YOUR KNOWLEDGE 3.6

1. 2.625   **3.** 2.125   **5.** 58.75

PRACTICE TEST 3

1. a. ∩∩∩III
       ∩∩∩
   b. ꝯꝯꝯ ∩∩III
       ꝯꝯꝯ ∩ II

2. a. 23   **b.** 121

3. a. ٢ ٢٢٢   **b.** ⟨٢٢ ⟨٢٢٢٢٢

4. a. 82   **b.** 131

5. a.

\\1	21	**b.** 23	(21)
\\2	42	11	(42)
\\4	84	5	(84)
8	168	2	168
\\16	336	1	(336)
$\overline{23}$	$\overline{483}$		$\overline{483}$

6. a. LIII   **b.** XLII   **c.** $\overline{\text{XXII}}$ **7. a.** 67   **b.** 48,000

8. a. $(2 \times 10^3) + (5 \times 10^2) + (0 \times 10) + (7 \times 10^0)$ **b.** $(1 \times 10^2) + (8 \times 10) + (9 \times 10^0)$

9. a 3702   **b.** 59,040

10. a.

$$\begin{array}{r} 75 \\ +32 \\ \hline 107 \end{array} \qquad \begin{array}{l} (7 \times 10) + 5 \\ +(3 \times 10) + 2 \\ \hline (10 \times 10) + 7 = (1 \times 10^2) + 7 = 107 \end{array}$$

b.

$$\begin{array}{r} 56 \\ -24 \\ \hline 32 \end{array} \qquad \begin{array}{l} (5 \times 10) + 6 \\ -(2 \times 10) + 4 \\ \hline (3 \times 10) + 2 = 32 \end{array}$$

11. a. $3^4 \times 3^8 = 3^{4+8} = 3^{12}$ **b.** $2^9 \div 2^3 = 2^{9-3} = 2^6$

12. a.

$$\begin{array}{r} 83 \\ \times 21 \\ \hline 83 \\ 166 \\ \hline 1743 \end{array} \qquad \begin{array}{l} (8 \times 10) + 3 \\ \times (2 \times 10) + 1 \\ \hline (8 \times 10) + 3 \\ (16 \times 10^2) + (6 \times 10) \\ \hline (16 \times 10^2) + (14 \times 10) + 3 \end{array}$$

$$= (1 \times 10^3) + (6 \times 10^2) + (1 \times 10^2) + (4 \times 10) + 3$$
$$= (1 \times 10^3) + (7 \times 10^2) + (4 \times 10) + 3$$
$$= 1743$$

b.

$$\begin{array}{r} 7 \\ 7\overline{)54} \\ 49 \\ \hline 5 \end{array} \begin{array}{l} \leftarrow \text{Remainder} \rightarrow \end{array} \begin{array}{r} 7 \\ 7\overline{)(5 \times 10) + 4} \\ (4 \times 10) + 9 \\ \hline 5 \end{array}$$

13. a. 35   **b.** 48   **c.** 13   **14. a.** 106   **b.** 2604   **15. a.** 113_5 **b.** 53_6

16. a. 100111_2 **b.** 1000001111_2 **17. a.** 57_8 **b.** $2F_{16}$ **18. a.** 135_8 **b.** 11010111_2

19. a. $5D_{16}$ **b.** 1010111101_2 **20. a.** 53_8 **b.** 17_{16} **21. a.** 10010_2 **b.** 110_2

22. a. 100111_2 **b.** 111_2 remainder 1_2 **23. a.** 700_8 **b.** 233_8

24. a. 564_8 **b.** 77_8 **25. a.** 319_{16} **b.** $A1EC_{16}$

CHAPTER 4   EXERCISE 4.1

1. For identification   **3.** Cardinal number   **5.** For identification; ordinal number

7.

×	1	2	3	4	5	6	7	8	9	. . .
1	1	2	3	4	5	6	7	8	9	. . .
2	2	4	6	8	10	12	14	16	18	. . .
3	3	6	9	12	15	18	21	24	27	. . .
4	4	8	12	16	20	24	28	32	36	. . .
5	5	10	15	20	25	30	35	40	45	. . .
6	6	12	18	24	30	36	42	48	54	. . .
7	7	14	21	28	35	42	49	56	63	. . .
8	8	16	24	32	40	48	56	64	72	. . .
9	9	18	27	36	45	54	63	72	81	. . .
.	.	.	.	.	.	.	.	.	.	.

a. Yes. Multiplication associates a unique result with each pair of elements of N.

b. Yes. For any two natural numbers a and b, $a \times b$ is a natural number.

c. Yes. For any two natural numbers a and b, $a \times b = b \times a$.

9. a. $4(3 + 8) = 4 \times 3 + 4 \times 8 = 44$ **b.** $5(9 + 6) = 5 \times 9 + 5 \times 6 = 75$
   c. $8(3 + 8) = 8 \times 3 + 8 \times 8 = 88$ **d.** $9(4 + 9) = 9 \times 4 + 9 \times 9 = 117$

11. Yes. Yes.

13. a. Yes. The operation $*$ associates a unique result with each pair of elements of A.
   b. Yes. The result of the operation on any two elements of A is an element of A.
   c. Yes. The table is symmetric to the diagonal from upper left to lower right, so that if a and b are elements of A, then $a * b = b * a$.
   d. Yes. You can check that $(x * y) * z = x * (y * z)$ for all x, y, z that are elements of A.

15. a. Commutative property of addition **b.** Associative property of addition

17. a. Distributive property of multiplication over addition **b.** Commutative property of addition

19. a. Distributive property of multiplication over addition **b.** Commutative property of multiplication

21. a. Commutative property of addition **b.** Commutative property of multiplication

23. a. Associative property of multiplication **b.** Commutative property of multiplication

25. a. 5 **b.** 6 **c.** 8

27. No. For example, $2 + (3 \cdot 4) \neq (2 + 3)(2 + 4)$.

29. No. For example, $4 - (2 \cdot 3) \neq (4 - 2)(4 - 3)$.

31. The commutative and associative properties of multiplication

DISCOVERY 4.1

1. i. Let A, B, and C be disjoint sets such that $n(A) = a$, $n(B) = b$, and $n(C) = c$.
   ii. $n(A \cup B) = a + b$ and $n(A \cup B \cup C) = a + b + c$. Since $(A \cup B) \cup C = A \cup (B \cup C) = A \cup B \cup C$, then $n[(A \cup B) \cup C] = n[A \cup (B \cup C)]$; that is, $(a + b) + c = a + (b + c)$.

3. Since $n(A \times B) = n(B \times A)$, then $n(A) \times n(B) = n(B) \times n(A)$.

EXERCISE 4.2

1. 5̶1̶ 5̶2̶ (53) 5̶4̶ 5̶5̶ 5̶6̶ 5̶7̶ 5̶8̶ (59) 6̶0̶   The circled numbers are the desired primes: 53, 59, 61, 67, 71,
(61) 6̶2̶ 6̶3̶ 6̶4̶ 6̶5̶ 6̶6̶ (67) 6̶8̶ 6̶9̶ 7̶0̶   73, 79, 83, 89, and 97.
(71) 7̶2̶ (73) 7̶4̶ 7̶5̶ 7̶6̶ 7̶7̶ 7̶8̶ (79) 8̶0̶
8̶1̶ 8̶2̶ (83) 8̶4̶ 8̶5̶ 8̶6̶ 8̶7̶ 8̶8̶ (89) 9̶0̶
9̶1̶ 9̶2̶ 9̶3̶ 9̶4̶ 9̶5̶ 9̶6̶ (97) 9̶8̶ 9̶9̶ 1̶0̶0̶

3. 6 **5.** 4

7. a. 2, 3
   b. No. Of any two consecutive counting numbers greater than 3, one must be even (divisible by 2) and thus not a prime.

9. a. The product part of m is exactly divisible by 2, so that m divided by 2 would have a remainder of 1.
   b. The product part of m is exactly divisible by 3, so that m divided by 3 would have a remainder of 1.
   c. and d. The same reasoning as in parts a and b applies here.
   e. P was assumed to be the largest prime.
   f. m is not divisible by any of the primes from 2 to P.

11. 1, 2, 5, 10, 25, 50 **13.** 1, 2, 4, 8, 16, 32, 64, 128 **15.** 1, 7, 11, 13, 77, 91, 143, 1001 **17.** 41 is a prime.

19. $7 \cdot 13$ **21.** $2^2 \cdot 37$ **23.** 490 **25.** 1200

27. a. Divisible by 3 and by 5 **b.** Divisible by 2, by 3, and by 5 **c.** Divisible by 2, by 3, and by 5

29. a. $100 = 3 + 97 = 11 + 89 = 17 + 83 = 29 + 71 = 41 + 59$
   b. $200 = 3 + 197 = 7 + 193 = 19 + 181 = 37 + 163 = 43 + 157 = 61 + 139 = 73 + 127 = 97 + 103$

USING YOUR KNOWLEDGE 4.2

1. All the other digits are divisible by 3, so that only the sum of 2 and 7 has to be checked.
3. 999, 99, and 9 are all divisible by 9. Thus, we need to check only $2 \times 1 + 8 \times 1 + 5 \times 1 + 3$, that is, the sum of the digits. If this sum is divisible by 9, then the number is divisible by 9 and not otherwise.
5. a. Divisible by 4, not by 8 **b.** Divisible by both 4 and 8 **c.** Divisible by 4 and by 8
d. Divisible by 4, not by 8

DISCOVERY 4.2

1. 1, 2, 3, 4, and 5 are not perfect numbers. **3.** $496 = 1 + 2 + 4 + 8 + 16 + 31 + 62 + 124 + 248$
5. All primes are deficient, because they have only 1 as a proper divisor.

EXERCISE 4.3

1. GCF(14, 210) = 14 **3.** GCF(315, 350) = 35 **5.** GCF(368, 80) = 16 **7.** GCF(12, 18, 30) = 6
9. GCF(285, 315, 588) = 3 **11.** $\frac{20}{23}$ **13.** $\frac{1}{2}$ **15.** $\frac{3}{14}$ **17.** $\frac{1}{4}$ **19.** LCM(15, 55) = 165
21. LCM(32, 124) = 992 **23.** LCM(180, 240) = 720 **25.** LCM(12, 18, 30) = 180
27. LCM(285, 315, 588) = 167,580 **29.** $\frac{15}{56}$ **31.** $\frac{19}{36}$ **33.** $\frac{3}{4}$ **35.** $2\frac{5}{8}$ cups **37.** $35\frac{3}{8}$ ft **39.** $84\frac{5}{8}$ ft
41. $\frac{3}{5}$ **43.** $\frac{1}{2}$ **45.** $\frac{3}{20}$

USING YOUR KNOWLEDGE 4.3

1. Problem 1: 2|14   210   Problem 3: 5|315   350   Problem 5: 2|368   80
          7|_7   105   7|_63   70   2|184   40
             1   15   9   10   2|_92   20
      GCF(14, 210) = 2 · 7 = 14   GCF(315, 350) = 5 · 7 = 35   2|_46   10
                                                                  23   5
                                              GCF(368, 80) = 2^4 = 16

Problem 7: 2|12   18   30   Problem 9: 3|285   315   588
          3|_6   9   15   95   105   196
             2   3   5   GCF(285, 315, 588) = 3
      GCF(12, 18, 30) = 2 · 3 = 6

2. Problem 19: LCM(15, 55) = 5 · 3 · 11 = 165   Problem 21: LCM(32, 124) = 4 · 8 · 31 = 992
Problem 23: LCM(180, 240) = 60 · 3 · 4 = 720   Problem 25: LCM(12, 18, 30) = 6 · 2 · 3 · 5 = 180
Problem 27: LCM(285, 315, 388) = 3 · 5 · 7 · 19 · 3 · 28 = 167,580

CALCULATOR CORNER 4.3

1. 22 **3.** 34 **5.** 9

EXERCISE 4.4

1. 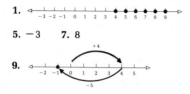 **3.**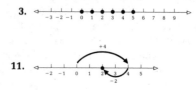

5. −3 **7.** 8

9. **11.**

13.

15.

17. $3 + (-8) = -5$ **19.** $3 + (-4) = -1$ **21.** $-5 + (-2) = -7$ **23.** $5 + (+6) = 11$ **25.** $-3 + (+4) = 1$
27. $-5 + (+3) = -2$ **29.** -15 **31.** 12   **33.** -20 **35.** 0   **37. a.** -27 **b.** 4   **39. a.** -9 **b.** -5
41. 10   **43.** -5 **45.** 11,800   **47.** $9 - (-11) = 20$ km
49. Step 1: m is assumed to be a multiplicative identity. Step 3: 1 is a multiplicative identity. Step 4: Both m and 1 equal $m \cdot 1$.
51. Step 1: Definition of subtraction. Step 3: Associative property of addition. Step 5: 0 is the additive identity.
53. b. adding a **c.** identity   **e.** q **f.** identity 0; unique
55. Step 1: 0 is the additive identity. Step 3: Distributive law. Step 5: The additive identity (0) is unique.

USING YOUR KNOWLEDGE 4.4

1. -3 **3.** 0

DISCOVERY 4.4

1. The same nomograph can be used for subtraction. To find the difference $a - b$, locate a on the top scale and $-b$ on the bottom scale, and draw a line through these two points. The point where the line crosses the middle scale gives the difference. For instance, to find $5 - 7$, locate 5 on the top scale and -7 on the bottom scale. The line through these two points crosses the middle scale at -2.

EXERCISE 4.5

1. Numerator 3, denominator 4   **3.** Numerator 3, denominator -5 **5.** $\frac{17}{41} = \frac{289}{697}$ **7.** $\frac{11}{91} = \frac{253}{2093}$ **9.** $\frac{7}{16}$
11. $\frac{4}{18} + \frac{3}{18} + \frac{7}{18} = \frac{14}{18}$ **13.** $\frac{6}{18} + \frac{3}{18} + \frac{2}{18} = \frac{11}{18}$ **15.** $\frac{16}{63}$ **17.** $\frac{19}{12}$ **19.** $\frac{176}{323}$ **21.** $\frac{2}{63}$ **23.** $-\frac{1}{12}$ **25.** $\frac{62}{323}$
27. $\frac{2}{3}$ **29.** $\frac{21}{8}$ **31.** $\frac{56}{27}$ **33.** $\frac{18}{77}$ **35.** $\frac{9}{14}$ **37.** $-\frac{1}{6}$ **39.** $\frac{1}{8}$ **41.** $\frac{1}{4}$ **43.** $\frac{49}{80}$ **45.** 16   **47.** $\frac{3}{2}$ **49.** $\frac{2}{5}$
51. a. $1\frac{9}{14}$ **b.** $3\frac{2}{3}$ **c.** $\frac{2}{7}$ **d.** -2 **53. a.** $-\frac{3}{4}$ **b.** $1\frac{1}{3}$ **c.** -18 **d.** -3
55. a. $7\frac{1}{8}$ **b.** $-1\frac{1}{8}$ **c.** $2\frac{5}{8}$ **d.** $\frac{1}{2}$

USING YOUR KNOWLEDGE 4.5

1. 126 mi   **3.** 3 in.

DISCOVERY 4.5

1. $\frac{5}{1}$ $\frac{5}{2}$ $\frac{5}{3}$ $\frac{5}{4}$ $\frac{5}{5}$ $\frac{5}{6}$ $\frac{5}{7}$ $\frac{5}{8}$ $\frac{5}{9}$ $\cdots$
$\frac{6}{1}$ $\frac{6}{2}$ $\frac{6}{3}$ $\frac{6}{4}$ $\frac{6}{5}$ $\frac{6}{6}$ $\frac{6}{7}$ $\frac{6}{8}$ $\frac{6}{9}$ $\cdots$
3. $\frac{2}{2} = 1, \frac{4}{2} = 2, \frac{3}{3} = 1, \frac{2}{4} = \frac{1}{2}$, and these have already been caught in the one-to-one correspondence.
5. Neither. The two sets have the same cardinal number.

EXERCISE 4.6

1. a. 0.9   **b.** 0.3   **c.** 1.1   **3. a.** 0.003   **b.** 0.143   **c.** 1.243   **5. a.** 0.375   **b.** 1.1666...   **c.** 0.16
7. a. 0.1875   **b.** No terminating decimal expansion   **c.** 0.015625   **9. a.** $\frac{8}{9}$ **b.** $\frac{2}{3}$

11. a. $\frac{38}{333}$ **b.** $\frac{34}{333}$ **13. a.** $\frac{137}{111}$ **b.** $\frac{17}{999}$ **15. a.** $\frac{151}{330}$ **b.** $\frac{191}{825}$
17. a. $(1 \times 10^{-3}) + (7 \times 10^{-5})$ **b.** $4 + (3 \times 10^{-1}) + (8 \times 10^{-5})$ **19.** 405.0609
21. a. 9.35×10^2 **b.** 3.72×10^{-1} **23. a.** 86,400 **b.** 90,100,000
25. 3×10^{-2} **27.** 4×10^{25} **29.** 2×10^3 hr **31.** 3×10^8 m/sec **33.** 31 yr

DISCOVERY 4.6

1. $0.\overline{4} = \frac{4}{9}$ **3. a.** $\frac{1}{3} + \frac{2}{3} = 0.999... = 0.\overline{9}$ **b.** $\frac{9}{9} = 1$ **c.** $(3)(\frac{1}{3}) = 0.999...$, or $1 = 0.\overline{9}$

PRACTICE TEST 4

1. a. Ordinal number **b.** Cardinal number **c.** Identification
2. a. Yes. Because it operates on two numbers. **b.** Yes. All items in the table are elements of A.
   c. Yes. The table is symmetric to the diagonal from upper left to lower right.
3. a. Commutative property of addition **b.** Distributive property of multiplication over addition
   c. Commutative property of multiplication
4. $1220 = 2^2 \cdot 5 \cdot 61$ **5.** Composite: $143 = 11 \cdot 13$ **6. a.** 436 and 1530 **b.** 387 and 1530 **c.** 2345 and 1530
7. GCF(18, 54, 60) = 6 **8.** $\frac{216}{254} = \frac{108}{127}$ **9.** LCM(18, 54, 60) = 540 **10. a.** $\frac{67}{504}$ **b.** $\frac{91}{264}$ **11.** $\frac{1}{8}$
12. a. -11 **b.** 27 **c.** -27 **d.** 11 **13.** 56,000 **14.** $\frac{12}{16}$ **15. a.** $\frac{3}{2}$ **b.** $-\frac{7}{4}$ **c.** $\frac{8}{21}$ **d.** $-\frac{1}{8}$
16. a. $-\frac{35}{128}$ **b.** $\frac{14}{5}$ **17. a.** 0.75 **b.** $0.0\overline{6} = 0.0666...$ **18. a.** $\frac{4}{33}$ **b.** $\frac{239}{90}$
19. a. $(2 \times 10) + (3 \times 10^0) + (5 \times 10^{-1}) + (8 \times 10^{-3})$ **b.** 803.04 **20.** 4.8×10^{-1}

CHAPTER 5 EXERCISE 5.1

1. a. Irrational **b.** Rational **3. a.** Rational **b.** Irrational **5. a.** Rational **b.** Rational
7. a. Irrational **b.** Rational **9. a.** Irrational **b.** ~~Irrational~~ Rational
11. < **13.** < **15.** < **17.** = **19.** = **21.** 0.315 (Other answers are possible.)
23. 0.3112345... (Other answers are possible.) **25.** 0.1011 (Other answers are possible.)
27. 0.101101001000... (Other answers are possible.) **29.** $\frac{7}{22}$ (Other answers are possible.)
31. 0.5101001000... (Other answers are possible.) **33.** $\frac{11}{18}$ (Other answers are possible.)
35. $0.21 < 0.2121 < 0.21211 < 0.212112111... < 0.21212$ **37.** 3.09 mi **39.** 163 mi

USING YOUR KNOWLEDGE 5.1

1. $(OB)^2 = 1^2 + (\sqrt{5})^2 = 1 + 5 = 6$, so that $OB = \sqrt{6}$. **3.** $h = 3$

EXERCISE 5.2

1. a. 4.74 **b.** -4.74 **3. a.** -4.158 **b.** -5.864 **5. a.** 0.045 **b.** 0.128 **7. a.** -0.05 **b.** 0.02
9. $3831.88 million **11.** 392.5 mi **13.** $14,422,500 **15.** 1.26×10^{19} **17.** 59.26 **19.** 78
21. a. 0.29 **b.** 0.234 **c.** 0.009 **23. a.** 0.3415 **b.** 0.9356 **c.** 0.000234
25. a. 56.7% **b.** 0.45% **c.** 900.3% **27. a.** 60% **b.** 57.1% **29.** 50.4%
31. a. 64.5 **b.** 39 **33.** $10.81 **35.** About 445,000,000 **37.** 29% **39.** 10%

USING YOUR KNOWLEDGE 5.2

1. 1371% **3.** 442% **5.** 344% **7.** 240%

EXERCISE 5.3

1. $3\sqrt{10}$ **3.** $\sqrt{122}$ is in simplest form. **5.** $6\sqrt{5}$ **7.** $10\sqrt{2}$ **9.** $8\sqrt{6}$ **11.** $3\sqrt{7}/7$ **13.** $-\sqrt{10}/5$
15. $\sqrt{2}$ **17.** $\sqrt{3}/7$ **19.** $2\sqrt{3}/3$ **21.** $2\sqrt{2}/7$ **23.** $\frac{3}{5}$ **25.** $2\sqrt{14}$ **27.** $6\sqrt{2}$ **29.** $\sqrt{14}$ **31.** $\frac{1}{5}$
33. $\sqrt{6}/2$ **35.** $5\sqrt{7}$ **37.** $-7\sqrt{7}$ **39.** $-8\sqrt{5}$ **41.** $5\sqrt{2}/4$ sec **43.** 20%

USING YOUR KNOWLEDGE 5.3

1. $6\frac{4}{13} = 6.31$ **3.** $9\frac{4}{19} = 9.21$

EXERCISE 5.4

1. $7i$ **3.** $i3\sqrt{7}$ **5.** $10i$ **7.** $i5\sqrt{2}$ **9.** $i10\sqrt{2}$ **11.** $i4\sqrt{3}$ **13.** Pure imaginary; complex
15. Real; complex **17.** Pure imaginary; complex **19.** Whole number; integer; rational; real; complex
21. Natural number; whole number; integer; rational; real; complex **23.** Complex

EXERCISE 5.5

1. a. $a_1 = 7$ **b.** $d = 6$ **c.** $a_{10} = 61$ **d.** $a_n = 6n + 1$
3. a. $a_1 = 43$ **b.** $d = -9$ **c.** $a_{10} = -38$ **d.** $a_n = 52 - 9n$
5. a. $a_1 = 2$ **b.** $d = -5$ **c.** $a_{10} = -43$ **d.** $a_n = 7 - 5n$
7. a. $a_1 = -\frac{5}{6}$ **b.** $d = \frac{1}{2}$ **c.** $a_{10} = \frac{11}{3}$ **d.** $a_n = (3n - 8)/6$
9. a. $a_1 = 0.6$ **b.** $d = -0.4$ **c.** $a_{10} = -3$ **d.** $a_n = 1 - 0.4n$
11. $S_{10} = 340$; $S_n = n(3n + 4)$ **13.** $S_{10} = 25$; $S_n = \frac{n}{2}(95 - 9n)$ **15.** $S_{10} = -205$; $S_n = \frac{n}{2}(9 - 5n)$

17. $S_{10} = 14\frac{1}{6}$; $S_n = \dfrac{n(3n - 13)}{12}$ **19.** $S_{10} = -12$; $S_n = \frac{n}{5}(4 - n)$

21. a. $a_1 = 3$ **b.** $r = 2$ **c.** $a_{10} = 1536$ **d.** $a_n = 3 \cdot 2^{n-1}$
23. a. $a_1 = \frac{1}{3}$ **b.** $r = 3$ **c.** $a_{10} = 6561$ **d.** $a_n = 3^{n-2}$

25. a. $a_1 = 16$ **b.** $r = -\frac{1}{4}$ **c.** $a_{10} = -4^{-7} = -\dfrac{1}{16,384}$ **d.** $a_n = \dfrac{(-1)^{n-1}}{4^{n-3}}$

27. $S_{10} = 3(2^{10} - 1)$; $S_n = 3(2^n - 1)$ **29.** $S_{10} = \frac{1}{6}(3^{10} - 1)$; $S_n = \frac{1}{6}(3^n - 1)$

31. $S_{10} = \dfrac{64(4^{10} - 1)}{5 \cdot 4^{10}}$; $S_n = \dfrac{64[4^n - (-1)^n]}{5 \cdot 4^n}$ **33.** $a_1 = 6$, $r = \frac{1}{2}$, $S = 12$

35. $a_1 = -8$, $r = \frac{1}{2}$, $S = -16$ **37.** $a_1 = 0.7$, $r = 0.1$, $S = \frac{7}{9}$
39. For 0.101010..., $a_1 = 0.10$, $r = 0.01$, and $S = \frac{10}{99}$, so that $2.101010... = 2 + \frac{10}{99} = \frac{208}{99}$.
41. a. $1020 **b.** $18,000 **43. a.** $95 **b.** $8625 **45.** $610.51

USING YOUR KNOWLEDGE 5.5

1. No. It is neither an arithmetic nor a geometric sequence. **3.** $a_2 + a_3 = a_4$

PRACTICE TEST 5

1. a. Rational **b.** Irrational **c.** Rational **d.** Rational **e.** Irrational **f.** Irrational
2. a. 0.24 (Other answers are possible.) **b.** 0.23456... (Other answers are possible.)
3. $C = 12.6$ in. **4. a.** 9.53 **b.** 4.63 **c.** 1.943 **d.** 5.6 **5.** 44.6 cm **6.** 4731.1 ft² **7.** 9.39 in.
8. a. 0.21 **b.** 0.0935 **c.** 0.0026 **9. a.** 52% **b.** 276.5% **c.** 60% **d.** 18.2%
10. 79.17% **11.** About 26.7 million **12. a.** $4\sqrt{6}$ **b.** Already in simplest form

13. a. $\dfrac{2\sqrt{5}}{5}$ **b.** $\dfrac{4\sqrt{3}}{7}$ **14. a.** $4\sqrt{3}$ **b.** $2\sqrt{2}$ **15. a.** $\sqrt{10}$ **b.** $2\sqrt{2}$ **16. a.** $6i$ **b.** $\sqrt{43}\,i$

17. a. Complex   **b.** Pure imaginary; complex   **c.** Real; complex   **d.** Pure imaginary; complex

18. a. Natural number; integer, rational; real; complex   **b.** Complex   **c.** Irrational; real; complex   **d.** Complex

19. a. Geometric sequence   **b.** Arithmetic sequence   **20. a.** $d = 40$ **b.** $d = -6$

21. a. $a_1 = 9$ **b.** $d = 4$ **c.** $a_{10} = 45$ **d.** $a_n = 4n + 5$ **22.** $S_{10} = 270$

23. a. $a_1 = 1$ **b.** $r = \frac{1}{2}$ **c.** $a_n = \dfrac{1}{2^{n-1}}$ **24.** $S_5 = \frac{31}{16}$ **25. a.** $\frac{4}{9}$ **b.** $\frac{7}{33}$ **c.** $2\frac{5}{9}$

CHAPTER 6   EXERCISE 6.1

1. centi   **3.** milli   **5.** kilo   **7.** 1000   **9.** 100   **11.** Liters   **13.** Centimeters   **15.** Grams
17. c. 200 cm   **19. a.** 1 cm   **21. b.** 400 m   **23. a.** 60 kg   **25. a.** 2 m   **27.** Milk   **29.** Vitamins

USING YOUR KNOWLEDGE 6.1

1. About 10 yd   **3.** 1000   **5.** About 157 cm

DISCOVERY 6.1

1. About 24   **3.** About 20   **5.** About 550,000

EXERCISE 6.2

1. About 100 cm   **3.** About 8.5 cm   **5.** About 190   **7.** About 240   **9.** 5000   **11.** 34.09   **13.** 8.413
15. 0.319   **17.** 2.1   **19.** 1500   **21.** Longer by 8.6 m   **23.** Shorter by 2.16 m
25. a. 157.5 m   **b.** 26.25 m   **c.** 15.75 m

USING YOUR KNOWLEDGE 6.2

1. d.   **3. b.**   **5. a.**

EXERCISE 6.3

1. 6300   **3.** 0.042   **5.** 1300   **7.** 4.9   **9.** 5   **11.** To heat water for instant coffee   **13.** 12 persons
15. 3.5   **17.** 10   **19.** 250

USING YOUR KNOWLEDGE 6.3

1. 60   **3.** 2   **5.** 1

DISCOVERY 6.3

1. One-third

3. *(1)* Fill the 300-ml glass twice from the pitcher, emptying the glass into the 700-ml glass. *(2)* Fill the 300-ml glass again from the pitcher and use the glass to fill the 700-ml glass. (This leaves 200 ml of milk in the 300-ml glass.) *(3)* Empty the 700-ml glass into the pitcher and then empty the 300-ml glass into the 700-ml glass. *(4)* Fill the 300-ml glass from the pitcher and then empty the glass into the 700-ml glass. This gives 500 ml of milk in the 700-ml glass.

EXERCISE 6.4

1. 1.35 **3.** 3.6 **5.** 17.6 **7.** 22 **9.** 14,000 **11.** 0.0028 **13.** 0.037 **15.** 0.041 **17.** 15
19. 30 **21.** −30 **23.** 50 **25.** 14 **27.** 260 **29.** 105.8 **31.** 37°C
33. a. 5°C **b.** 100°C
35. −108.4°F

USING YOUR KNOWLEDGE 6.4

1. g. **3.** f. **5.** a. **7.** b.

DISCOVERY 6.4

(Answers are rounded to the nearest two digits.)
1. 59 **3.** 71 **5.** 83 **7.** 61 **9.** 70

EXERCISE 6.5

(Answers are rounded using the rules on page 220.)
1. 20.32 **3.** 4.73 **5.** 46.6 **7.** 4.03 **9.** 6.44 **11.** 2.30 **13.** 2.72 **15.** 11.0 **17.** 4.73
19. 8.59 **21.** 2.46 **23.** 16.39 **25.** 183 **27.** 0.854 **29.** 64 km/hr **31.** 58 **33.** 28,900
35. 560 **37.** 48 **39.** 114 **41.** 298.3 **43.** 12.7 metric tons **45.** 17.0

USING YOUR KNOWLEDGE 6.5

1. 1.18 **3.** 96.5 **5.** 102−66−97

DISCOVERY 6.5

Across: 1. 3125 **5.** 1255 **7.** 20 **9.** 10 **13.** 2500
Down: 1. 350 **3.** 500 **7.** 218 **9.** 12 **11.** 420

PRACTICE TEST 6

1. a. 10^{-2} **b.** 10^{-3} **c.** 10^3 **2. c.** Liters **3. b.** Kilograms
4. 75 **5.** 900 **6.** 30 mm, 40 mm, and 60 mm, respectively **7.** 82 **8.** 8300 **9.** 0.92 **10.** 1.9
11. 100 **12.** 400 **13.** $\frac{1}{10}$ **14.** 0.000275 kg **15.** 0.250 **16.** 4100 **17.** 25°C **18.** 68°F
19. 24.8 (or 25 mi/hr) **20.** 114 **21. a.** 0.75 **b.** 0.795
22. 0.465 **23.** 35.56 **24.** 50.7 **25. b.** 25 cm

CHAPTER 7 EXERCISE 7.1

1. {2} **3.** {1, 2, 3, . . .} **5.** ∅ **7.** {1, 2, 3} **9.** {1, 2, 3, 4, 5} **11.** {1, 2, 3, 4, 5, 7, 8, 9, . . .} **13.** ∅
15. {23} **17.** {4} **19.** {. . . , −2, −1, 0, 1, 2, . . .} **21.** {−1} **23.** {3, 4, 5, . . .} **25.** {x|x > 1}
27. {$\frac{3}{2}$} **29.** {0} **31.** {x|x ≤ 7} **33.** Yes **35.** 150

USING YOUR KNOWLEDGE 7.1

1. $110 **3.** $202 **5.** 78 in., or 6 ft 6 in.

DISCOVERY 7.1

1. 14 dimes, 6 quarters   **3.** 84

EXERCISE 7.2

1. $\{5\}$ **3.** $\{3\}$ **5.** $\{2\}$ **7.** $\{9\}$ **9.** $\{1\}$ **11.** $\{2\}$ **13.** $\{\frac{3}{2}\}$ **15.** $\{\frac{10}{7}\}$ **17.** $\{2\}$ **19.** $\{x \mid x < 4\}$
21. $\{x \mid x > 3\}$ **23.** $\{x \mid x > 3\}$ **25.** $\{x \mid x \geq -2\}$ **27.** $\{x \mid x > -4\}$ **29.** $\{x \mid x \leq -2\}$ **31.** $\{x \mid x \leq -9\}$
33. $\{x \mid x > 3\}$ **35.** $\{x \mid x > -4\}$ **37.** $\{x \mid x \leq -\frac{2}{3}\}$ **39.** $\emptyset$ **41.** $\{x \mid x \leq 2\}$ **43.** $\{x \mid x > -2\}$ **45.** $\{x \mid x \leq \frac{2}{5}\}$

USING YOUR KNOWLEDGE 7.2

1. 32   **3.** 10%   **5.** 12.5   **7.** 200   **9.** 671 billion barrels   **11.** 67   **13.** Store *B*'s price is lower.

DISCOVERY 7.2

1. 9   **3.** 364

EXERCISE 7.3

1.   **3.**

5.   **7.**

9.   **11.**

13.   **15.**

17.   **19.**

21. $\emptyset$;   **23.**   **25.**

DISCOVERY 7.3

1. $445_{10} = 544_9$ **3.** $144_9 = 441_5$

EXERCISE 7.4

1. $\{-1, 0, 1, 2, 3, 4\}$ **3.** $\{3, 4, 5, 6\}$ **5.** $\{2\}$ **7.** $\{\ldots, -8, -7, -6, 6, 7, 8, \ldots\}$
9.   **11.** $\emptyset$ **13.**

15.   **17.** $\emptyset$ **19.**

21.

23.   **25.**

DISCOVERY 7.4

1. The snail will reach the top on the 28th day.

EXERCISE 7.5

1. 10   **3.** $\frac{1}{8}$ **5.** 3   **7.** 2   **9.** -8 **11.** {0}   **13.** $\{-5, 5\}$ **15.** $\{\ldots, -2, -1, 1, 2, \ldots\}$

17.   **19.**

21.   **23.**

25.   **27.**

29.   **31.**

33.

EXERCISE 7.6

1. $(x + 2)(x + 4)$ **3.** $(x - 4)(x + 3)$ **5.** $(x + 9)(x - 2)$ **7.** $(x - 5)^2$ **9.** $(x + 5)^2$ **11.** $\{2, 4\}$
13. $\{-2, 3\}$ **15.** $\{-1, 0, 1\}$ **17.** $\{-2, \frac{1}{2}\}$ **19.** $\{-4, 4\}$ **21.** $\{-5, 5\}$ **23.** $\{-\frac{3}{2}, \frac{8}{5}, 2\}$ **25.** $\{-6, 6\}$
27. $\{3, 9\}$ **29.** $\{-2, 10\}$ **31.** $\{-20, 1\}$ **33.** $\{-3, 4\}$ **35.** $\{-\frac{5}{2}, 1\}$ **37.** $\{-\frac{7}{2}, 1\}$
39. $\left\{\dfrac{-5 - \sqrt{13}}{2}, \dfrac{-5 + \sqrt{13}}{2}\right\}$ **41.** $\left\{\dfrac{4 - \sqrt{6}}{5}, \dfrac{4 + \sqrt{6}}{5}\right\}$ **43.** $\left\{\dfrac{3 - \sqrt{2}}{7}, \dfrac{3 + \sqrt{2}}{7}\right\}$ **45.** $\left\{\dfrac{1 - i}{3}, \dfrac{1 + i}{3}\right\}$
47. $\left\{\dfrac{-1 - i}{2}, \dfrac{-1 + i}{2}\right\}$ **49.** $\left\{\dfrac{-2 - i}{2}, \dfrac{-2 + i}{2}\right\}$

USING YOUR KNOWLEDGE 7.6

1. 2 sec   **3.** 3 sec

DISCOVERY 7.6

1. $85^2 = 7225$ was the initial number of letters.

EXERCISE 7.7

1. 6   **3.** 7   **5.** 6   **7.** -2 or 3   **9.** ± 6 **11.** 2.71 million lb   **13.** Japan: 8851 ships; Russia: 6575 ships
15. 130 mi   **17.** 14% per year   **19. a.** 62 mi   **b.** Use the mileage rate.
21. 10%   **23.** 30 mi/hr   **25.** 76.5 ft   **27.** 0.6 sec   **29.** 20 mi/hr   **31.** $x = 4, x + 1 = 5$

USING YOUR KNOWLEDGE 7.7

1. The old car should have been driven 5.22 yr.

EXERCISE 7.8

1. 7000 to 2000, 7000 : 2000, $\frac{7000}{2000}$ **3.** 70 to 4260, 70 : 4260, $\frac{70}{4260}$ **5.** $\frac{10}{3}$ **7.** 17
9. a. 6¢   **b.** 5¢   **c.** White Magic
11. $x = 12$ **13.** $x = 6$ **15.** $x = 24$ **17.** 6   **19.** 66.5 in.   **21.** 2.81   **23.** 2650
25. a. $R = kt$ **b.** $k = 45$ **c.** 2.4 min   **27. a.** $T = kh^3$ **b.** $k = 0.0005714$ **c.** 241 lb
29. a. $f = k/d$ **b.** $k = 4$ **c.** $f = 16$

PRACTICE TEST 7

1. a. $x = -5$ **b.** $x = 13$ **2. a.** $\{0, 1, 2, 3, \ldots\}$ **b.** $\{-1, 0, 1, 2, \ldots\}$ **3.** $x = 4$ **4.** $x \geq -3$

5. a.   **b.**

6. a.   **b.** The solution set is empty.

7. a.   **b.**

8. $x = \pm 3$

9.   **10.**

11.   **12.**

13.

14. a. $(x + 1)(x + 2)$ **b.** $(x + 1)(x - 4)$ **15. a.** $x = -2$ or $x = 1$ **b.** $x = 0$ or $x = 1$
16. $x = -5$ or $x = -2$ **17.** $x = -2$ or $x = 5$ **18.** $x = -\frac{5}{2}$ or $x = 1$ **19.** $x = -2$ or $x = \frac{1}{3}$
20. a. $x = \pm\frac{4}{3}$ **b.** $x = \pm\frac{2}{5}$ **21.** 200   **22.** 17   **23.** 7 and 8   **24.** 5   **25.** 4 and 5
26. a. 688 to 801, 688 : 801, $\frac{688}{801}$ **b.** $\frac{801}{1990}$
27. a. $6\frac{1}{4}$¢ per oz for the 8-oz box; $6\frac{1}{3}$¢ per oz for the 12-oz box   **b.** The 8-oz box
28. a. $x/8 = 9/5$ **b.** 14.4 ft   **29. a.** $C = ks^2$ **b.** $k = \frac{1}{225}$ **c.** $144
30. a. $t = k/I$ **b.** $k = 10$ **c.** $\frac{1}{60}$ sec

CHAPTER 8   EXERCISE 8.1

1. Domain: $\{1, 2, 3\}$; range: $\{2, 3, 4\}$ **3.** Domain: $\{1, 2, 3\}$; range: $\{1, 2, 3\}$
5. Domain: $\{x|x$ a real number$\}$; range: $\{y|y$ a real number$\}$
7. Domain: $\{x|x$ a real number$\}$; range: $\{y|y$ a real number$\}$ **9.** Domain: $\{x|x$ a real number$\}$; range: $\{y|y \geq 0\}$
11. Domain: $\{x|x \geq 0\}$; range: $\{y|y$ a real number$\}$ **13.** Domain: $\{x|x \neq 0\}$; range: $\{y|y \neq 0\}$
15. Domain: $\{-1, 0, 1, 2\}$; range: $\{-2, 0, 2, 4\}$; ordered pairs: $(-1, -2)$, $(0, 0)$, $(1, 2)$, $(2, 4)$
17. Domain: $\{0, 1, 2, 3, 4\}$; range: $\{-3, -1, 1, 3, 5\}$; ordered pairs: $(0, -3)$, $(1, -1)$, $(2, 1)$, $(3, 3)$, $(4, 5)$
19. Domain: $\{0, 1, 4, 9, 16, 25\}$; range: $\{0, 1, 2, 3, 4, 5\}$; ordered pairs: $(0, 0)$, $(1, 1)$, $(4, 2)$, $(9, 3)$, $(16, 4)$, $(25, 5)$
21. Domain: $\{1, 2, 3\}$; range: $\{2, 3, 4\}$; ordered pairs: $(1, 2)$, $(1, 3)$, $(1, 4)$, $(2, 3)$, $(2, 4)$, $(3, 4)$
23. A function. To each real x, there corresponds one real y.
25. Not a function. To each $x > 0$, there correspond two values of y.
27. A function. To each x in the domain, there corresponds one y.
29. A function. To each real x, there corresponds one real y.
31. a. 1   **b.** 7   **c.** -5 **33. a.** 0   **b.** 2   **c.** 5   **35. a.** $3x + 3h + 1$ **b.** $3h$ **c.** 3
37. a. 140 heartbeats per minute   **b.** 130 heartbeats per minute   **39. a.** 160 lb   **b.** 78 in.
41. a. 639 lb/ft²   **b.** 6390 lb/ft²

USING YOUR KNOWLEDGE 8.1

1. $\{(x, c)|c = 4(x - 40)\}$ **3.** $f(t) = 16t^2$

DISCOVERY 8.1

1. An equivalence relation   **3.** An equivalence relation   **5.** An equivalence relation

1.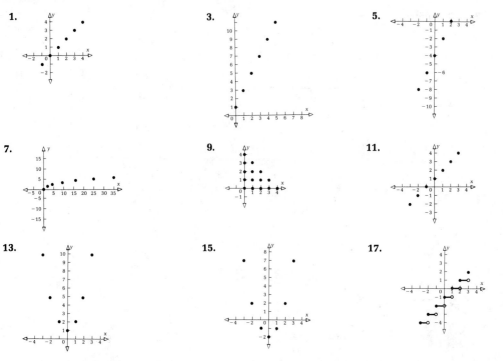

3.

5.

7.

9.

11.

13.

15.

17.

19. a. $h(x) = 2.89x + 70.64$ **b.** $h(34) = 168.9$ cm, or about 169 cm

21. a. $F(x) = 10x + 20$ **b.** $F(8) = 100$; $100
b. More than 12 but not more than 13 min.

23. a. $V = 10,000 - 2000t$

b.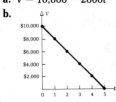

25. a.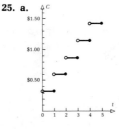

1.

WEIGHT x (oz)	POSTAGE p (cents)
$\frac{1}{2}$	25
$\frac{3}{4}$	25
1	25
$1\frac{1}{2}$	45
2	45
$2\frac{1}{4}$	65

3.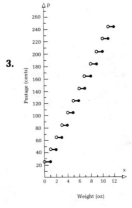

DISCOVERY 8.2

1. $f(x) = \sqrt{x}$ **3.** 10,000 units

EXERCISE 8.3

1.

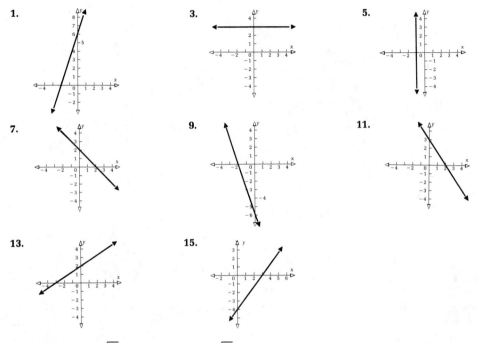

3.

5.

7.

9.

11.

13.

15.

17. 5 units   **19.** $\sqrt{73} \approx 8.54$ units   **21.** $\sqrt{58} \approx 7.62$ units   **23.** 2 units   **25.** 4 units
27. Not a right triangle; scalene   **29.** A right triangle; isosceles
31. a.   **b.** 50 mi.

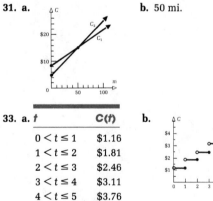

33. a.

t	$C(t)$
$0 < t \le 1$	$1.16
$1 < t \le 2$	$1.81
$2 < t \le 3$	$2.46
$3 < t \le 4$	$3.11
$4 < t \le 5$	$3.76

b.

DISCOVERY 8.3

See the answers for Exercise 8.3.

EXERCISE 8.4

1. 1 **3.** -1 **5.** $-\frac{1}{6}$ **7.** $\frac{1}{4}$ **9.** 0 **11.** $y = \frac{1}{2}x + \frac{3}{2}$ **13.** $y = -x + 6$ **15.** $y = 5$ **17. a.** 1 **b.** 2
19. a. $\frac{4}{3}$ **b.** 0 **21. a.** -1 **b.** 14 **23. a.** 0 **b.** 6
25. a. Slope not defined **b.** Does not intersect y axis **27.** $3x - y = 4$ **29.** $x + y = 5$ **31.** $10x - y = 0$
33. $w = 5h - 176$ **35.** $w = 5h - 187$ **37.** Parallel **39.** Not parallel **41.** Parallel **43.** $4x - y = 6$
45. a. $5x - 2y = 10$ **b.** $x + 2y = 3$ **c.** $2x + y = 2$ **d.** $5x - 4y = 1$
47. $m_1 = 2$, $m_2 = -\frac{415}{790}$; thus, $m_1 m_2 = -\frac{415}{395} \neq -1$ and the lines are not perpendicular.

USING YOUR KNOWLEDGE 8.4

1.

3. He will spend all his money on food.

5. a. $y = 2x + 2000$ **b.** Fixed cost $= \$1000$; marginal cost $= 50¢$ per unit

DISCOVERY 8.4

1. The intersection is at 42nd Street East and 21st Avenue North.

EXERCISE 8.5

1. $(1, 2)$ **3.** $(3, -4)$ **5.** $(2, -\frac{1}{2})$ **7.** $(1, 2)$ **9.** $(4, 2)$ **11.** $(3, -4)$ **13.** No solution **15.** $(1, 2)$
17. $(2, -\frac{1}{2})$ **19.** $(-3, 10)$ **21.** $(-2, \frac{1}{2})$ **23.** $(2, -5)$ **25.** $(\frac{2}{3}, \frac{5}{6})$ **27.** No solution **29.** $u = -\frac{5}{2}$, $v = 6$
31. a. 100 mi **b.** **c.** Company A

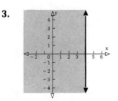

USING YOUR KNOWLEDGE 8.5

1. 33 nickels, 13 dimes **3.** 250 mi **5.** 100 adult and 200 children's tickets

EXERCISE 8.6

1.

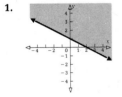

3.

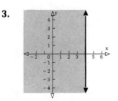

5.

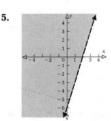

7.

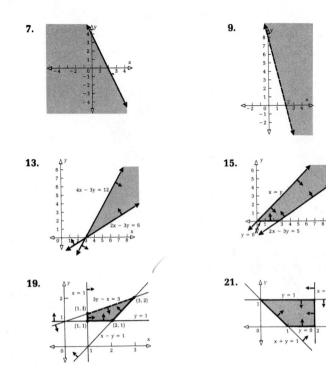

9.

11.

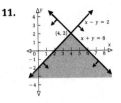

$x - y = 2$
(4, 2)
$x + y = 6$

13.

$4x - 3y = 12$
$2x - 3y = 6$

15.

$x = y$
$2x - 3y = 5$
$y = 0$

17.

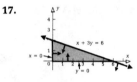

$x = 0$
$x + 3y = 6$
$y = 0$

19.

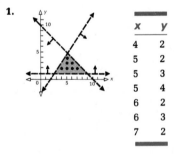

$x = 1$
$3y - x = 3$ (3, 2)
(1, $\frac{4}{3}$)
$y = 1$
(1, 1)   (2, 1)
$x - y = 1$

21.

$y = 1$ $x = 2$
$y = 0$
$x + y = 1$

23.

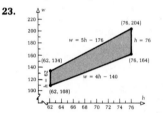

(76, 204)
$w = 5h - 176$ $h = 76$
(62, 134)   (76, 164)
$h = 62$
$w = 4h - 140$
(62, 108)

USING YOUR KNOWLEDGE 8.6

1.

x	y
4	2
5	2
5	3
5	4
6	2
6	3
7	2

DISCOVERY 8.6

1. At least two of the five points must be in or on the same small square. Hence, the distance between these two points cannot be more than $\sqrt{2}$ in., the length of the diagonal of the square.

3. Leave the first part as shown, and interchange the two shading instructions.

EXERCISE 8.7

1. Minimum value 3 at (1, 1)   **3.** Minimum value 2 at ($\frac{1}{3}$, $\frac{2}{3}$)   **5.** Minimum value 6 at (2, 2)
7. Maximum value 10 at (2, 4)   **9.** 80 cars; 20 trucks   **11.** 50 tablets   **13.** 3 oz X; 2 oz Y
15. 100 batches from I; 10 batches from II
17. 500 boxes of oranges; 100 boxes of grapefruit; 200 boxes of tangerines   **19.** 3 trees; 2 shrubs

21. a. Orange juice, 6 oz; grapefruit juice, 6 oz   **b.** 42¢
  c. 30 units vitamin A; 30 units vitamin C; 12 units vitamin D

DISCOVERY 8.7

1. $2300 is the least he can bet to win at least $100.

PRACTICE TEST 8

1. Domain: {0, 2, 3, 5}; range: {−1, 2, 3, 4}
2. Domain: the set of all real numbers; range: the set of all real numbers
3. Domain: {1, 2, 3, 4}; range: {1, 2, 3, 4}   **4.** Functions: b and c   **5. a.** 0   **b.** 0   **c.** 6   **6.** 203 mi
7.   **8.**   **9.**

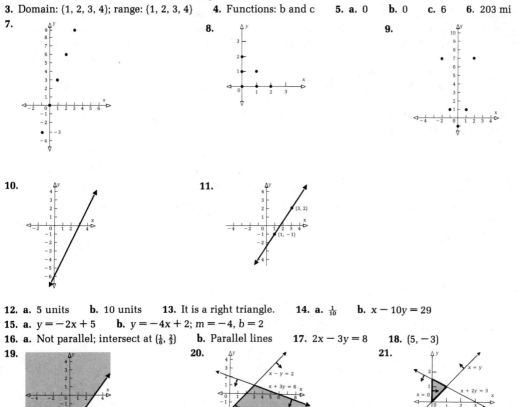

12. a. 5 units   **b.** 10 units   **13.** It is a right triangle.   **14. a.** $\frac{1}{10}$ **b.** $x - 10y = 29$
15. a. $y = -2x + 5$ **b.** $y = -4x + 2$; $m = -4$, $b = 2$
16. a. Not parallel; intersect at $(\frac{1}{6}, \frac{2}{3})$ **b.** Parallel lines   **17.** $2x - 3y = 8$ **18.** $(5, -3)$
19.   **20.**   **21.**

22. The two equations represent the same line. Solution set: $\{(a, 2a - 3) | a$ is a real number$\}$.
23. Maximum value 16 at $(0, 8)$ **24.** Minimum value -6 at $(3, 0)$
25. Run machine A 30 hr and machine B 10 hr

CHAPTER 9   EXERCISE 9.1

1. a.   **b.**   **c.**

3. The segment $\overleftrightarrow{BC}$ 5. The ray $\overrightarrow{AD}$ 7. The segment $\overrightarrow{AD}$ 9. The segment $\overrightarrow{AD}$ 11. ∅
13. The point C 15. Six; AB, AC, AD, BC, BD, CD   17. Four; ABC, ABD, ACD, BCD
19. a. AB, AC, AD, BC, BD,CD   b. AB and CD; AC and BD; AD and BC   c. No   21. True   23. True
25. True   27. False   29. False   31. Collinear   33. Coplanar   35. Yes   37. Line BD

USING YOUR KNOWLEDGE 9.1

1. 190   3. $\frac{1}{2}n(n-1)$ 5. $\frac{1}{6}n(n-1)(n-2)$

DISCOVERY 9.1

1. a. Denote the lines by a, b, c, d (property i). Denote the common points by ab, ac, ad, bc, bd, cd (property ii). None of these points can be on another line, because, suppose that point ab is also on line c. Then lines a and c would have two points in common. Therefore, there are at least six points. Now, suppose there is a seventh point, say xy. Then xy must be on two of the lines a, b, c, d (property iii). But then, by property ii, xy must coincide with one of the six points already listed. Thus, there are exactly six points.
   b. This is obvious from the listing in part a: line a, points ab, ac, ad; line b, points ab, bc, bd; line c, points ac, bc, cd; line d, points ad, bd, cd
   c. This is also obvious from the listing in part a. For instance, the only point not on the same line with ab is cd.
3. Replace "line" by "plane" and "point" by "line."

EXERCISE 9.2

1. a. ∠BAC   b. ∠β   3. ∠BAC, ∠CAD, ∠DAE, ∠EAF   5. ∠BAE, ∠CAE, ∠CAF
7. a. ∠DAE   b. ∠CAD   9. a. ∠DAF   b. ∠BAE   11. 35°   13. 15°   15. ∠B   17. 110°
19. 20°   21. 220°   23. a. 150°   b. 30°   c. 150°   25. ∠C, ∠E, ∠A, ∠G   27. a. 49°   b. 139°
29. 16   31. x = 15; 35° and 145°   33. x = 10; 30° and 60°   35. a. 30°   b. 180°   37. 90°   39. 25°

USING YOUR KNOWLEDGE 9.2

1. The sum of the angles is 0.2° too large.   3. N 50° W   5. 320°

EXERCISE 9.3

1. a. 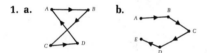   b.

3. a. C, D, G, I, J, L, M, N, O, P, R, S, U, V, W, Z   b. B, D, O   5. a. D, O   b. A, E, F, H, K, Q   7. Convex
9. Rhombus   11. Rectangle   13. Trapezoid   15. Parallelogram   17. Scalene, right
19. Scalene, acute   21. Isosceles, acute   23. Scalene, obtuse
25. a.   b.   c.

27. a.   b.   c. Impossible

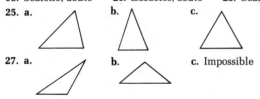

29. 65 cm   31. 12.6 yd   33. 184.6 m   35. 57.2 m   37. 30 cm   39. 10 ft   41. 90 ft
43. 716 ft by 518 ft   45. 108°   47. 135°   49. 144°

USING YOUR KNOWLEDGE 9.3

1. Either SSS or SAS
3. By the ASA statement, triangles ABC and ABT are congruent. Therefore, $\overrightarrow{AT} = \overrightarrow{AC}$. (They are corresponding sides of the two triangles.)

DISCOVERY 9.3

1. 6   **3.** 5   **5.** Impossible, because 360 divided by 25 is not a whole number.

EXERCISE 9.4

1. a and c   **3.** a and c   **5.** $x = 5\frac{1}{3}$, $y = 6\frac{2}{3}$ **7.** 14   **9.** $\frac{4}{3}$ **11.** $3\frac{1}{2}$ cm, $5\frac{1}{4}$ cm   **13.** 8 in., 12 in., 16 in.
15. $18\frac{3}{4}$ ft   **17.** 25 ft   **19.** 600 m

USING YOUR KNOWLEDGE 9.4

1. $x = 3$, $y = 4.5$ **3.** 125 ft

DISCOVERY 9.4

1. Because PQ and AB are parallel.   **3.** Corresponding sides of similar triangles are proportional.
7. Triangle ABC is similar to triangle PQC, and triangle PQC is congruent to triangle $A_1B_1C_1$. Therefore, triangle ABC is similar to triangle $A_1B_1C_1$.

EXERCISE 9.5

1. 192 cm   **3.** 50.2 cm   **5.** 37.7 in.   **7.** 15.3 in.   **9.** 80 ft

DISCOVERY 9.5

1. Construct a square inscribed in the circle. Then bisect each arc joining successive vertices. The midpoints of the arcs and the vertices of the square will be the vertices of the octagon.

EXERCISE 9.6

1. 15 in.²   **3.** 15 in.²   **5.** 22 ft²   **7.** 30 ft²   **9.** 957 in.²   **11.** 164 cm²   **13.** 3.87 cm²   **15.** 19.6 ft²
17. 8 ft   **19.** The side parallel to the base is 12 ft long. The other two sides are each 10 ft long.   **21.** 6400
23. 680 ft   **25.** About 80 ft   **27.** 12 in. by 18 in.   **29.** $3\frac{1}{2}$ ft by $4\frac{1}{2}$ ft
33. a. The area is multiplied by 4.   **b.** The area is multiplied by 9.   **c.** The area is multiplied by k^2.
35. The circle, by 6.8 cm²

USING YOUR KNOWLEDGE 9.6

1. a. $2\frac{2}{3}$ **b.** \$42   **3. a.** 15   **b.** \$60
5. a. 2.49¢ per square inch   **b.** 2.36¢ per square inch   **c.** The 10-in. pie

DISCOVERY 9.6

1. $\dfrac{1 + \sqrt{5}}{2}$

EXERCISE 9.7

1. a. A, B, C, D, E **b.** $AB, AC, AD, AE, BC, BE, CD, CE, DE$ **3.** $ABCD$

5.   **7.**

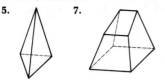

9. a. The volume is multiplied by 8.   **b.** The volume is multiplied by 27.   **11.** 50 in.³   **13.** 32 in.³
15. 1.5   **17. a.** 707 in.³   **b.** 236 in.³   **19. a.** 113 ft³   **b.** 37.7 ft³   **21.** 904 in.³
23. About 7,960,000 ft³   **25.** About 3220   **27.** About 463

DISCOVERY 9.7

3.

FIGURE	F	V	E	$F + V = E + 2$
9.7a, II	6	8	12	
9.7a, III	6	5	9	
9.7a, IV	7	10	15	
9.7b	5	6	9	
9.7c	6	6	10	

EXERCISE 9.8

1. a. 3   **b.** 0   **c.** Traversable; A, B, C **3. a.** 3   **b.** 2   **c.** Traversable; A, C
5. a. 1   **b.** 4   **c.** Not traversable   **7. a.** 5   **b.** 2   **c.** Traversable; B, F
9. a. 1   **b.** 4   **c.** Not traversable   **11.** Yes; start from B or D.
13. Yes; start from A or D. No; it is not possible to start and end outside.

DISCOVERY 9.8

1.

FIGURE	V	R	A	FIGURE	V	R	A	$V + R = A + 2$
a.	3	2	3	e.	2	3	3	
b.	4	3	5	f.	5	3	6	
c.	5	2	5	g.	6	3	7	
d.	4	3	5	h.	9	4	11	

PRACTICE TEST 9

1. a. $\overleftrightarrow{XY}$ **b.** Point Y **c.** $\overrightarrow{WZ}$ **2.** Six; $\overleftrightarrow{AB}, \overleftrightarrow{AC}, \overleftrightarrow{AD}, \overleftrightarrow{BC}, \overleftrightarrow{BD},$ and $\overleftrightarrow{CD}$. No.
3.   **a.** $\overleftrightarrow{AD}, \overleftrightarrow{BE},$ and $\overleftrightarrow{CF}, \overleftrightarrow{EF}$ and $\overleftrightarrow{BC}, \overleftrightarrow{DE}$ and $\overleftrightarrow{AB}, \overleftrightarrow{DF}$ and $\overleftrightarrow{AC}$
  b. $\overleftrightarrow{AB}$ and $\overleftrightarrow{CF}, \overleftrightarrow{AB}$ and $\overleftrightarrow{DF}, \overleftrightarrow{AB}$ and $\overleftrightarrow{EF}, \overleftrightarrow{BC}$ and $\overleftrightarrow{AD}, \overleftrightarrow{BC}$ and $\overleftrightarrow{DE}, \overleftrightarrow{BC}$ and $\overleftrightarrow{DF},$
   $\overleftrightarrow{AC}$ and $\overleftrightarrow{BE}, \overleftrightarrow{AC}$ and $\overleftrightarrow{DE}, \overleftrightarrow{AC}$ and $\overleftrightarrow{EF}, \overleftrightarrow{DE}$ and $\overleftrightarrow{CF}, \overleftrightarrow{EF}$ and $\overleftrightarrow{AD}, \overleftrightarrow{DF}$ and $\overleftrightarrow{BE}$
  c. $\overleftrightarrow{AB}, \overleftrightarrow{AC},$ and $\overleftrightarrow{AD}; \overleftrightarrow{AD}, \overleftrightarrow{DE},$ and $\overleftrightarrow{DF}; \overleftrightarrow{BE}, \overleftrightarrow{AB},$ and $\overleftrightarrow{BC}; \overleftrightarrow{BE}, \overleftrightarrow{EF},$ and $\overleftrightarrow{ED}; \overleftrightarrow{CF},$
   $\overleftrightarrow{AC},$ and $\overleftrightarrow{BC}; \overleftrightarrow{CF}, \overleftrightarrow{EF},$ and $\overleftrightarrow{DF}$

4. No; skew lines are not parallel and do not intersect, so they are not coplanar.

5. a. 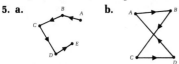 **b.**

6. a. 120° **b.** 60° **7.** $m \angle A = 135°$, $m \angle B = 45°$

8. a. $m \angle C = 130°$ **b.** $m \angle E = 50°$ **c.** $m \angle D = 130°$

9. a. $m \angle C = 99°$ **b.** $m \angle A = m \angle B = 72°$, $m \angle C = 36°$ **10.** $m \angle A = 72°$, $m \angle B = 18°$

11. a. 140° **b.** 144° **12.** 20 yd by 40 yd **13.** $XZ = 6$ in., $YZ = 4\frac{1}{2}$ in. **14.** $2\sqrt{2}\pi$ cm

15. $(2\pi - 4)$ cm² **16.** $3\frac{1}{2}$ in. **17.** 15 ft² **18.** 85 ft **19.** 15.5 ft² **20.** 0.86 in.² **21.** 4 in. **22.** $\dfrac{3\sqrt{3}}{2\pi}$

23. 30 ft³ **24.** 10 ft³

25.

Traversable   Not traversable

CHAPTER 10 EXERCISE 10.1

1. a. $\begin{bmatrix} 8 & 4 \\ 0 & -4 \end{bmatrix}$ **b.** $\begin{bmatrix} 6 & -12 \\ -9 & -3 \end{bmatrix}$ **c.** $\begin{bmatrix} 5 & 1 \\ -1 & -1 \end{bmatrix}$ **3. a.** $\begin{bmatrix} 3 & 10 \\ 5 & 0 \end{bmatrix}$ **b.** $\begin{bmatrix} -3 & 0 \\ 1 & 0 \end{bmatrix}$

5. a. $\begin{bmatrix} 0 & 13 \\ 8 & -3 \end{bmatrix}$ **b.** $\begin{bmatrix} -1 & -21 \\ -12 & -4 \end{bmatrix}$

7. a. $\begin{bmatrix} 0 & 1 & 3 \\ 7 & 3 & -3 \\ 4 & 3 & 0 \end{bmatrix}$ **b.** $\begin{bmatrix} 2 & -3 & 1 \\ -1 & -3 & -1 \\ 4 & 1 & 2 \end{bmatrix}$ **c.** $\begin{bmatrix} 1 & 0 & -1 \\ 2 & 2 & -6 \\ 1 & 5 & 1 \end{bmatrix}$

9. a. $\begin{bmatrix} -1 & 4 & 7 \\ 18 & 9 & -7 \\ 8 & 7 & -1 \end{bmatrix}$ **b.** $\begin{bmatrix} -5 & 8 & -1 \\ 6 & 9 & 1 \\ -8 & -1 & -5 \end{bmatrix}$ **11. a.** $\begin{bmatrix} 3 & -1 & 0 \\ 7 & 4 & -14 \\ 6 & 12 & 3 \end{bmatrix}$ **b.** $\begin{bmatrix} -1 & 1 & 4 \\ 5 & 1 & 3 \\ 3 & -2 & -1 \end{bmatrix}$

13. $\begin{bmatrix} 3 & 0 & 3 \\ 10 & 4 & 2 \\ 4 & -1 & 2 \end{bmatrix}$ **15.** $\begin{bmatrix} 7 & -6 & 7 \\ 4 & -3 & 0 \\ 1 & -3 & 5 \end{bmatrix}$ **17.** $\begin{bmatrix} -19 & -3 & -2 \\ 4 & 2 & 12 \\ -7 & 3 & 6 \end{bmatrix}$ **19.** $\begin{bmatrix} -15 & -9 & 2 \\ -2 & -5 & 10 \\ -10 & 1 & 9 \end{bmatrix}$

21. $I^2 = I$ **23.** $BA = I$ **25.** $BA = I$ **27.** $BA = I$

29. No; the matrices could not be conformable for both orders of multiplication.

33. a.
	E	M	L
Armchairs	20	15	10
Rockers	12	8	5

b.
	E	M	L
Armchairs	120	90	60
Rockers	72	48	30

35.
	E	M	L
Armchairs	30	15	10
Rockers	12	28	0

37.
	July	Aug.	Sept.	Oct.	Nov.
Bolts	650	1300	2100	2600	2400
Clamps	400	800	1400	1600	1800
Screws	1200	2400	4100	4800	5100

USING YOUR KNOWLEDGE 10.1

1. Reflects (a, b) across the y axis   **3.** Rotates (a, b) $180°$ around the origin

5.

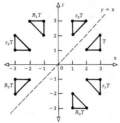

DISCOVERY 10.1

1. Yes; the sum of two 2×2 matrices is a 2×2 matrix.   **3.** Yes; the 2×2 zero matrix   **5.** Yes

7. Suppose A has an inverse, say $B = \begin{bmatrix} x & y \\ z & w \end{bmatrix}$. Then, $AB = \begin{bmatrix} 1 & 2 \\ 0 & 0 \end{bmatrix}\begin{bmatrix} x & y \\ z & w \end{bmatrix} = \begin{bmatrix} x+2z & y+2w \\ 0 & 0 \end{bmatrix}$. Since there are no

values of x, y, z, w to make this last matrix the identity matrix, A has no inverse.

EXERCISE 10.2

1. $x = 1, y = 2, z = 0$ **3.** $x = -1, y = -1, z = 3$ **5.** $x = -2, y = -1, z = 3$ **7.** No solution
9. $x = 1 - k, y = 2, z = k$, where k is any real number   **13.** 8 type I, 10 type II, 12 type III
15. 50% type I, 25% type II, 25% type III

EXERCISE 10.3

1. $\begin{bmatrix} -\frac{5}{2} & 2 \\ \frac{3}{2} & -1 \end{bmatrix}$ **3.** $\begin{bmatrix} \frac{1}{9} & \frac{1}{18} \\ -\frac{4}{9} & \frac{5}{18} \end{bmatrix}$ **5.** No inverse   **7.** $\begin{bmatrix} -1 & 0 \\ 0 & -1 \end{bmatrix}$ **9.** $x = -3, y = 5$ **11.** $x = 3, y = 2$

13. Yes, they are inverses of each other.   **15.** Yes, they are inverses of each other.

17. $\begin{bmatrix} -1 & 4 & 2 \\ -1 & 4 & 1 \\ -1 & 3 & 1 \end{bmatrix}$ **19.** $\begin{bmatrix} -\frac{3}{2} & \frac{7}{2} & -\frac{5}{2} \\ \frac{1}{2} & -\frac{1}{2} & \frac{1}{2} \\ -1 & 2 & -1 \end{bmatrix}$ **23.** $x = 2, y = 1, z = -3$ **25.** $x = -1, y = 3, z = 5$

EXERCISE 10.4

1. 1   **3.** $\frac{1}{24}$ **5.** $2ab$ **7.** -1 **9.** -4 **11.** $9x + y - 23$ **13.** $(5, -2)$ **15.** $(\frac{15}{34}, \frac{11}{17})$ **17.** $(4, 9)$
19. $(2, 3, -4)$ **21.** $(1, 0, -2)$ **23.** $(5, 2, -4)$ **25.** 8 type I, 10 type II, 12 type III
27. 50% type I, 25% type II, 25% type III

USING YOUR KNOWLEDGE 10.4

1. $2x - y + 3 = 0$ **3.** $2x + 9y - 34 = 0$ **5.** $bx + ay - ab = 0$

EXERCISE 10.5

1. Meet me tonight   **3.** Z is double agent   **5.** 30.−37.19.−70.75.−24.−14.−12.13.14.−39.22.−43.35.−7
7. Numbers are beautiful

PRACTICE TEST 10

1. $x = 3, y = 4$ **2.** $AB = BA = \begin{bmatrix} 1 & 0 \\ 0 & 1 \end{bmatrix}$ **3.** $[20 \quad 25 \quad 10] \begin{bmatrix} 1 & 2 & 1 & 1 \\ 1 & 3 & 2 & 1 \\ 4 & 4 & 2 & 2 \end{bmatrix} = \begin{matrix} \text{Frames} \ \text{Wheels} \ \text{Chains} \ \text{Paint} \\ [\quad 85 \qquad 155 \qquad 90 \qquad 65 \ ] \end{matrix}$

4. Type I, \$5.25; type II, \$7.00; type III, \$14.50   **5.** $\begin{bmatrix} 10 & -3 & 2 \\ -8 & -11 & 12 \\ 5 & -4 & 7 \end{bmatrix}$ **6.** $\begin{bmatrix} 0 & 1 & 1 \\ 6 & 2 & 1 \\ 5 & 3 & 1 \end{bmatrix}$

7. $AB = \begin{bmatrix} -3 & 4 & -1 \\ -5 & 5 & -1 \\ -2 & 11 & -4 \end{bmatrix}$; $BA = \begin{bmatrix} -2 & -1 & 1 \\ 6 & -5 & 9 \\ 2 & -3 & 5 \end{bmatrix}$ **8.** $\begin{bmatrix} 11 & 5 & 2 \\ 17 & 13 & 9 \\ 23 & 14 & 9 \end{bmatrix}$ **9.** $AA^{-1} = \begin{bmatrix} 1 & 0 & 0 \\ 0 & 1 & 0 \\ 0 & 0 & 1 \end{bmatrix}$ **10.** $\begin{bmatrix} \frac{1}{5} & \frac{1}{5} \\ \frac{2}{5} & -\frac{3}{5} \end{bmatrix}$

11. $x = -1, y = -2, z = 3$ **12.** $\frac{1}{4} \begin{bmatrix} 5 & -1 & -1 \\ -8 & 0 & 4 \\ -6 & 2 & 2 \end{bmatrix} \begin{bmatrix} 1 \\ 9 \\ 0 \end{bmatrix} = \frac{1}{4} \begin{bmatrix} -4 \\ -8 \\ 12 \end{bmatrix} = \begin{bmatrix} -1 \\ -2 \\ 3 \end{bmatrix}$ **13.** $x = -2, y = 7, z = 1$

14. $\begin{bmatrix} -1 & -1 & 2 \\ 0 & -1 & 2 \\ -2 & -4 & 7 \end{bmatrix}$ **15.** $\begin{bmatrix} -1 & -1 & 2 \\ 0 & -1 & 2 \\ -2 & -4 & 7 \end{bmatrix} \begin{bmatrix} 5 \\ 7 \\ 5 \end{bmatrix} = \begin{bmatrix} -2 \\ 3 \\ -3 \end{bmatrix}$ **16.** $\begin{bmatrix} 3 & 3 & 3 \\ 4 & 2 & 0 \\ 3 & 3 & 3 \end{bmatrix} \begin{matrix} 1 & 0 & 0 \\ 0 & 1 & 0 \\ 0 & 0 & 1 \end{matrix} \sim \begin{bmatrix} 3 & 3 & 3 \\ 4 & 2 & 0 \\ 0 & 0 & 0 \end{bmatrix} \begin{matrix} 1 & 0 & 0 \\ 0 & 1 & 0 \\ -1 & 0 & 1 \end{matrix}$

$$R_3 - R_1 \rightarrow R_3$$

The three 0's in the third row show that no row operations can reduce the given matrix to the identity matrix. Thus, the given matrix has no inverse.

17. 40 nickels, 50 dimes, 32 quarters   **18.** $x = -\frac{1}{2}, y = 0, z = 2$ **19.** The system has no solution.
20. $x = -\frac{k+1}{6}, y = \frac{4-2k}{3}, z = k$, where k is any real number   **21.** 49   **22.** -1 **23.** 0
24. $x = -2$ or $x = 4$ **25.** $x = -2, y = 3, z = -3$

CHAPTER 11 EXERCISE 11.1

1. 4   **3.** 11   **5.** 3   **7.** 7   **9.** 5   **11.** 9   **13.** 1   **15.** 2   **17.** 11   **19.** 10   **21.** 1   **23.** 12
25. 6   **27.** 4

29.

⊗	1	2	3	4	5	6	7	8	9	10	11	12
1	1	2	3	4	5	6	7	8	9	10	11	12
2	2	4	6	8	10	12	2	4	6	8	10	12
3	3	6	9	12	3	6	9	12	3	6	9	12
4	4	8	12	4	8	12	4	8	12	4	8	12
5	5	10	3	8	1	6	11	4	9	2	7	12
6	6	12	6	12	6	12	6	12	6	12	6	12
7	7	2	9	4	11	6	1	8	3	10	5	12
8	8	4	12	8	4	12	8	4	12	8	4	12
9	9	6	3	12	9	6	3	12	9	6	3	12
10	10	8	6	4	2	12	10	8	6	4	2	12
11	11	10	9	8	7	6	5	4	3	2	1	12
12	12	12	12	12	12	12	12	12	12	12	12	12

31. 3   **33.** 11   **35.** 4   **37.** 8   **39.** Impossible

DISCOVERY 11.1

1.

⊕	0	1	2	3	4
0	0	1	2	3	4
1	1	2	3	4	0
2	2	3	4	0	1
3	3	4	0	1	2
4	4	0	1	2	3

3. Yes

EXERCISE 11.2

1. False **3.** False **5.** False **7.** 2 (mod 5) **9.** 4 (mod 5) **11.** 3 (mod 5) **13.** 1 (mod 5)
15. 3 (mod 5) **17.** 3 (mod 5) **19.** 0 **21.** 0 **23.** 1 **25.** 2 **27.** 2 **29.** 2 **31.** 3 **33.** 2
35. Yes **37.** Yes; 1 **39.** Wednesday **41.** Saturday **43.** Friday

USING YOUR KNOWLEDGE 11.2

1. 5 **3.** 2 **5.** 1

DISCOVERY 11.2

1. The (5, 3) design is the same as the (5, 2) design in the figure in the text.
3. The (19, 17) and the (19, 9) designs are the same. **5.** $m = 9$

EXERCISE 11.3

1. a. a **b.** c **c.** b **3. a.** a **b.** a **c.** Yes **5. a.** c **b.** c **c.** Yes
7. Yes; if $x \in S$ and $y \in S$, then $(x @ y) \in S$.
9. a. Yes; if a and b are natural numbers, then $a \text{ F } b = a$, which is a natural number.
   b. Yes; $a \text{ F } (b \text{ F } c) = a \text{ F } b = a$ and $(a \text{ F } b) \text{ F } c = a \text{ F } c = a$. Thus, $a \text{ F } (b \text{ F } c) = (a \text{ F } b) \text{ F } c$.
   c. No; if $a \ne b$, then $a \text{ F } b = a$ and $b \text{ F } a = b$, so that $a \text{ F } b \ne b \text{ F } a$.
11. a. No; for example, $3 + 5 = 8$, which is not an odd number.
   b. Yes; the product of two odd numbers is an odd number.
   c. Yes; the sum of two even numbers is an even number.
   d. Yes; the product of two even numbers is an even number.

13.

∩	∅	{*a*}	{*b*}	{*a, b*}
∅	∅	∅	∅	∅
{*a*}	∅	{a}	∅	{a}
{*b*}	∅	∅	{b}	{b}
{*a, b*}	∅	{a}	{b}	{a, b}

15. a. ∅ **b.** ∅ **c.** Yes **17.** Yes; all elements in the table are elements of S.

19. a.

L	1	2	3	4
1	1	2	3	4
2	2	2	3	4
3	3	3	3	4
4	4	4	4	4

b. 1

21. a. No inverse **b.** No inverse **c.** No inverse **d.** 4 **23.** Yes; 1

25. a. The identity element is A, because for every B that is a subset of A, $A \cap B = B \cap A = B$.
b. No; there is no other identity element.
27. Yes; the identity element is 0. **29. a.** 3 **b.** 4
31. If a, b, c are real numbers, then a F $(b$ L $c) = a$ and $(a$ F $b)$ L $(a$ F $c) = a$ L $a = a$. Thus, this distributive property holds.
33. Yes **35.** Yes

USING YOUR KNOWLEDGE 11.3

1. $6 \times 9999 = 6(10{,}000 - 1) = 60{,}000 - 6 = 59{,}994$ **3.** $7 \times 59 = 7(60 - 1) = 420 - 7 = 413$
5. $4 \times 9995 = 4(10{,}000 - 5) = 40{,}000 - 20 = 39{,}980$

DISCOVERY 11.3

1. Here are the steps:

Think of a number:	x
Add 3 to it:	$x + 3$
Triple the result:	$3x + 9$
Subtract 9:	$3x$
Divide by the number with which you started, x:	3

EXERCISE 11.4

1. Yes (actually, a commutative group) **3.** No; no multiplicative inverses **5.** No; no identity element
7. No; no multiplicative inverses **9.** No; no multiplicative inverses **11.** Yes
13. No; no multiplicative inverse for 0 **15.** No; no identity element **17.** No; no multiplicative inverses
19. No; no multiplicative inverses

USING YOUR KNOWLEDGE 11.4

1. The system is a commutative group. It satisfies all the requirements of Definition 11.4a.
3. No; matrix multiplication is not commutative.

DISCOVERY 11.4

1.

C	1	2	3	4	5	6
1	1	2	3	4	5	6
2	2	1	4	3	6	5
3	3	6	5	2	1	4
4	4	5	6	1	2	3
5	5	4	1	6	3	2
6	6	3	2	5	4	1

3. The set is a group under the operation C.

PRACTICE TEST 11

1. a. 2 **b.** 5 **2. a.** 6 **b.** 5 **3. a.** 3 **b.** 12 **4. a.** 9 **b.** No solution
5. a. True **b.** True **c.** False **6.** 0 **7.** 2 **8.** 3 **9.** 4

10. a. $n = 2 + 7k$, k any integer   **b.** $n = 6 + 7k$, k any integer
11. a. $n = 2 + 3k$, k any integer   **b.** $n = 1 + 3k$, k any integer
12. Yes; all the entries in the table are elements of S.   **13. a.** #   **b.** #   **c.** %
14. Yes; the table is symmetrical to the diagonal from upper left to lower right.   **15.** The identity element is #.
16. a. #   **b.** ¢   **c.** %   **d.** $ **17. a.** No identity element   **b.** No   **c.** Not commutative
18. Yes; a S (b L c) = b L c and (a S b) L (a S c) = b L c. Therefore, S is distributive over L.
19. a L (b S c) = a L c and (a L b) S (a L c) = a L c. Thus, L is distributive over S.
20. Yes; the system has the five properties (closure, associative, identity, inverse, and commutative), so the system is a commutative group.
21. Yes; same explanation as for Problem 20.
22. Yes; the system has the six properties (closure, associative, commutative, identity, distributive of multiplication over addition, and inverses, except there is no inverse for 0 with respect to multiplication), so the system is a field.
23. Yes; all the requirements of Definition 11.4a are satisfied.   **24.** No; the element 0 has no inverse.
25. Yes; all the requirements of Definition 11.4b are satisfied.

CHAPTER 12   EXERCISE 12.1

1. 8 different outfits   **Suits**   **Shirts**   **3.** 8 different outcomes   **People**   **Jeans**

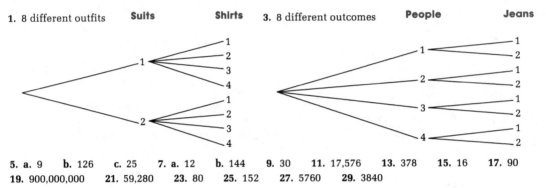

5. a. 9   **b.** 126   **c.** 25   **7. a.** 12   **b.** 144   **9.** 30   **11.** 17,576   **13.** 378   **15.** 16   **17.** 90
19. 900,000,000   **21.** 59,280   **23.** 80   **25.** 152   **27.** 5760   **29.** 3840

USING YOUR KNOWLEDGE 12.1

1. 24
3. Yes; there are not enough different sets of initials for 27,000 people. Thus, at least two people must have the same initials.

DISCOVERY 12.1

1. 12   **3.** 20,736   **5.** 12

EXERCISE 12.2

1. 24   **3.** 720   **5.** 120   **7.** 40,320   **9.** 10   **11.** 3024   **13.** 120   **15.** 5040   **17.** 1716   **19.** 650
21. 60   **23.** 9   **25.** 31   **27.** 60

USING YOUR KNOWLEDGE 12.2

1. 120   **3.** 3 (including a possible tie for fourth place)

DISCOVERY 12.2

1. $(n-1)!$ **3.** 4

EXERCISE 12.3

1. a. 10   **b.** 15   **c.** 35   **d.** 1   **e.** 84   **3.** 5   **5. a.** 10   **b.** 26   **7.** 2024   **9.** 120
11. 2,598,960   **13.** 75,287,520   **15.** 15   **17.** 63   **19.** 70

DISCOVERY 12.3

1. 1   **3.** 10   **5.** 1
7. The left side is the sum of the number of ways in which there would be 0 heads and n tails, 1 head and $n-1$ tails, 2 heads and $n-2$ tails, and so on, to n heads and 0 tails. The right side is exactly the number of ways in which n coins can fall either heads or tails. Thus, the two sides must be equal.

EXERCISE 12.4

1. a. 132,600   **b.** 22,100   **3. a.** 117,600   **b.** 19,600   **5.** 10   **7. a.** 35   **b.** 840
9. a. 16,380   **b.** 11,880   **11. a.** 280   **b.** 20   **13.** 831,600   **15.** 22,680   **17.** 210   **19.** $n=7$ **21.** 72

USING YOUR KNOWLEDGE 12.4

1. 15   **3.** $(a+1)(b+1)(c+1)(d+1)$

PRACTICE TEST 12

1.

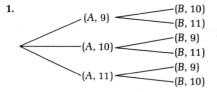

2. 30   **3. a.** 36   **b.** 4   **4.** 52   **5.** 15   **6. a.** 5040   **b.** 210   **7. a.** 144   **b.** 30
8. a. 120   **b.** 720   **9. a.** 56   **b.** 210   **10.** 24   **11.** 48   **12.** 96   **13.** 29   **14.** 6
15. a. 10   **b.** 15   **16. a.** 1   **b.** $\frac{1}{2}$ **17.** 20   **18.** 24   **19.** 1326   **20.** 25   **21.** 70   **22.** 840
23. 34,650   **24.** 30   **25.** $\dfrac{1977!}{1189!460!328!}$

CHAPTER 13   EXERCISE 13.1

1. $\frac{1}{6}$ **3.** $\frac{1}{3}$ **5.** $\frac{1}{10}$ **7.** $\frac{9}{10}$ **9.** 0   **11.** $\frac{1}{13}$ **13.** $\frac{1}{4}$ **15.** $\frac{11}{26}$ **17. a.** $\frac{1}{5}$ **b.** $\frac{3}{5}$ **c.** $\frac{4}{5}$
19. a. $\frac{1}{3}$ **b.** $\frac{1}{3}$ **21.** $\frac{1}{8}$ **23.** $\frac{3}{8}$ **25. a.** $\frac{13}{17}$ **b.** $\frac{4}{17}$ **27.** $\frac{4}{25}$ **29.** $\frac{8}{75}$ **31.** $\frac{1}{6}$ **33.** $\frac{1}{6}$ **35.** $\frac{1}{2}$
37. a. $\frac{1}{7}$ **b.** $\frac{4}{7}$ **c.** $\frac{11}{21}$ **39. a.** $\frac{1}{2}$ **b.** $\frac{3}{10}$ **c.** $\frac{1}{5}$

USING YOUR KNOWLEDGE 13.1

1. $\frac{4653}{4720}$, or about 0.986   **3.** $\frac{253}{254}$, or about 0.996

EXERCISE 13.2

1. $\frac{1}{18}$　**3.** $\frac{1}{8}$　**5. a.** $\frac{1}{5}$　**b.** $\frac{4}{5}$　**7. a.** $\frac{4}{21}$　**b.** $\frac{1}{21}$　**9.** $\frac{1}{221}$　**11.** $\frac{1}{68}$　**13.** $\frac{25}{102}$　**15. a.** 15　**b.** $\frac{1}{14}$　**c.** $\frac{1}{7}$
17. a. $\dfrac{C(4, 2)C(4, 2)C(44, 1)}{C(52, 5)}$　**b.** $\dfrac{C(4, 3)C(4, 2)}{C(52, 5)}$　**c.** $\dfrac{C(13, 1)C(48, 1)}{C(52, 5)}$　**19.** $\frac{2}{5}$

USING YOUR KNOWLEDGE 13.2

1. $\dfrac{C(40,000, 150)C(36,000, 150)}{C(76,000, 300)}$　**3.** $\dfrac{C(16, 6)C(14, 6)}{C(30, 12)}$

EXERCISE 13.3

1. 0; property (1)　**3.** 1; property (2)　**5.** $\frac{3}{5}$　**7.** 1　**9.** $\frac{7}{26}$　**11.** $\frac{1}{4}$　**13.** $\frac{3}{13}$　**15.** $\frac{41}{50}$　**17.** 0.2
19. 0.45　**21.** $\frac{433}{926}$　**23.** $\frac{252}{445}$　**25.** 1　**27.** $\frac{11}{20}$　**29.** $\frac{4}{5}$

USING YOUR KNOWLEDGE 13.3

1. About 0.83

DISCOVERY 13.3

1. $\frac{3}{5}$　**3.** $\frac{1}{10}$

EXERCISE 13.4

1. 0　**3. a.** $\frac{1}{6}$　**b.** 0　**c.** 1　**d.** 0　**5.** $\frac{1}{2}$　**7.** $\frac{1}{2}$　**9. a.** $\frac{1}{5}$　**b.** $\frac{1}{3}$　**11.** $\frac{1}{17}$　**13.** $\frac{1}{5}$　**15.** $\frac{2}{3}$
17. a. $\frac{3}{10}$　**b.** $\frac{39}{100}$　**19.** $\frac{1}{3}$

USING YOUR KNOWLEDGE 13.4

1. a. $\frac{75}{128}$　**b.** $\frac{53}{128}$　**c.** Male　**3.** $\frac{9}{53}$

EXERCISE 13.5

1. Yes　**3. a.** $\frac{1}{96}$　**b.** $\frac{1}{32}$　**c.** $\frac{7}{16}$　**5. a.** $\frac{1}{8}$　**b.** $\frac{1}{2}$　**c.** $\frac{7}{8}$　**7. a.** $\frac{1}{4}$　**b.** $\frac{1}{4}$　**c.** $\frac{1}{16}$　**d.** $\frac{9}{16}$
9. a. $\frac{1}{2}$　**b.** $\frac{3}{4}$　**c.** $\frac{3}{8}$　**d.** They are independent.　**11. a.** $\frac{1}{12}$　**b.** $\frac{1}{8}$　**13. a.** $\frac{1}{4}$　**b.** $\frac{1}{12}$　**c.** $\frac{1}{4}$
15. About 0.503　**17.** About 0.59　**19. a.** $\frac{1}{8000}$　**b.** $\frac{57}{8000}$　**21.** 0.189　**23.** $\frac{1}{2}$　**25.** $\frac{3}{5}$

USING YOUR KNOWLEDGE 13.5

1. $\dfrac{C(50, 25)}{2^{50}}$　**3.** $\frac{656}{729}$　**5.** $\frac{625}{3888}$

DISCOVERY 13.5

1. $\frac{1}{8}$　**3.** 0.02　**5.** 0.000008

1. 1 to 5 **3.** 1 to 12 **5.** 1 to 3 **7.** 5 to 21 **9.** 1 to 1 **11.** 10 to 3 **13.** $\frac{1}{50}$; 1 to 49 **15.** $\frac{3}{5}$
17. $-\$5$ **19.** \$2.15 **21.** No **23.** Build at the first location.
25. a. $-\$4000$ **b.** \$10,000 **c.** Discontinue the campaign.

USING YOUR KNOWLEDGE 13.6

1. a. 0.80 **b.** 0.20 **3.** \$1.11

DISCOVERY 13.6

1. About $-\$0.025$

PRACTICE TEST 13

1. a. 1 **b.** $\frac{2}{3}$ **2.** 20 **3. a.** $\frac{2}{5}$ **b.** $\frac{1}{5}$ **c.** $\frac{4}{5}$ **4. a.** $\frac{25}{102}$ **b.** $\frac{188}{221}$ **5. a.** $\frac{1}{4}$ **b.** $\frac{1}{4}$ **6.** $\frac{31}{32}$
7. a. $\frac{4}{5}$ **b.** $\frac{4}{5}$ **8.** 0.7 **9.** $\frac{11}{1105}$ or about 0.01 **10.** $\frac{1}{6}$ **11.** $\frac{1}{18}$
12. a. $\frac{1}{12}$ **b.** $\frac{1}{2}$ **c.** No; $P(A \cap B) = \frac{1}{36} \neq P(A)P(B)$. **13. a.** 0.000064 **b.** 0.884736 **14.** $\frac{1}{6}$ **15.** 0.0009
16. a. 1 to 12 **b.** 12 to 1 **17. a.** 3 to 4 **b.** 4 to 3 **18. a.** 7 to 3 **b.** $\frac{7}{10}$ **19.** \$3.75 **20.** $38\frac{1}{3}$ cents

CHAPTER 14 EXERCISE 14.1

1. a.

NUMBER OF HOURS	TALLY MARKS	FREQUENCY
0	||	2
1	|||	3
2	||	2
3	|||	3
4	|||	3
5	||	2
6	||	2
7	||	2
8	||||	4
9	|	1
10	||	2
11		0
12	||	2
13		0
14	|	1
15	|	1

b. 8 **c.** 4 **d.** 15 **e.** 36.7%

3. a.

b. 36% **c.** 52%

5. a.

AGE	TALLY MARKS	FREQUENCY
6	|	1
7	|	1
8	|||	3
9	|||	3
10	||	2

b.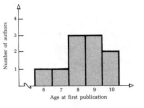

c. 20%

7. a.

PRICE	TALLY MARKS	FREQUENCY
$0 < P \le 10$	|||	3
$10 < P \le 20$	|||	3
$20 < P \le 30$	ⵌ	5
$30 < P \le 40$	ⵌ	5
$40 < P \le 50$	|||	3
$50 < P \le 60$	||||	4
$60 < P \le 70$	|	1
$70 < P \le 80$	|	1

b. $20 < P \le 30$ and $30 < P \le 40$ **c.** 9   **d.** 11
e. 20%   **f.** 24%

9.

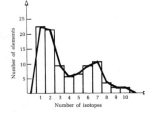

11. a.

LETTER	FREQUENCY	LETTER	FREQUENCY
a	10	n	13
b	4	o	14
c	3	p	2
d	4	q	0
e	18	r	7
f	2	s	10
g	3	t	17
h	9	u	5
i	12	v	2
j	0	w	5
k	1	x	0
l	4	y	2
m	4	z	0

b. e   **c.** 39.1%

13.

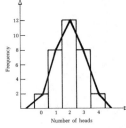

15. a.

CONCENTRATION	TALLY MARKS	FREQUENCY
0.00–0.04	ⵌ|	6
0.05–0.09	ⵌⵌ|||	13
0.10–0.14	ⵌ	5
0.15–0.19	||||	4
0.20–0.24	||	2

b. 20%

USING YOUR KNOWLEDGE 14.1

1.

DIGIT	FREQUENCY
0	1
1	4
2	5
3	6
4	4
5	4
6	3
7	3
8	5
9	5

DISCOVERY 14.1

1. 100,271   **3.** In each case, most of the bar is omitted.

EXERCISE 14.2

1. a. Mean = 9; median = 9   **b.** Mean = 24.2; median = 9   **c.** Mean = 11; median = 9
  d. Mean = median for part a only. Median = 9 for all three. None has a mode.
3. Mean = 6.05; median = 6.5; mode = 8; the mode is the least representative.
5. Betty is correct. She took account of the number of students making each score, and Agnes did not.

7.

NUMBER OF LETTERS	FREQUENCY
2	1
3	5
4	4
5	5
6	0
7	3
8	2
9	0
10	0
11	1

a. 3 and 5   **b.** 5   **c.** 5.05   **d.** No; there is too much repetition.

9. b.–c. Answers will vary, but should be approx. 8.   **11.** 60   **13.** $84 per week

USING YOUR KNOWLEDGE 14.2

1. a. 2.4   **b.** 2

DISCOVERY 14.2

1. Mean   **3.** Mean of a secretary's and a worker's salaries

EXERCISE 14.3

1. a. 18 **b.** 6.45 **3. a.** 20 **b.** 7.07 **5. a.** 4 **b.** 1.41 **7. a.** 8 **b.** 2.88 **9. a.** 6 **b.** 2
11. a. 8 **b.** 6.5 **c.** 6 **d.** 3.15 **e.** 70% **f.** 100%
13. a. 110 **b.** 110 **c.** 108 **d.** 3.52 **e.** 102, 103, 113; 30% **15.** The numbers are all the same.

USING YOUR KNOWLEDGE 14.3

1. 5 **3.** 8.66

EXERCISE 14.4

1. a. 100 in. **b.** 10 in. **c.** 68% **d.** 2700 **3. a.** 25 **b.** 25 **c.** 680
5. A, 12 or 13; B, 68 or 67; C, 340; D, 68 or 67; F, 12 or 13 **7.** 5 **9.** 19 ft 10½ in. and 20 ft 1½ in.

DISCOVERY 14.4

1. a. 75 **b.** 89 **c.** 49
3. $\bar{x} = 6$, $s = 4.5$. Three lie within 1 standard deviation from $\bar{x}$, and all nine lie within 2 standard deviations from $\bar{x}$. The theorem makes no prediction for 1 standard deviation; it predicts 75% for 2 standard deviations.
5. No; it cannot exceed 25%.

EXERCISE 14.5

1. a. 0.8 **b.** 1.6 **c.** 2 **3.** The German test score **5. a.** 0.249 **b.** 0.477 **7.** 0.998
9. 0.228 **11.** 5 ft 9½ in.

USING YOUR KNOWLEDGE 14.5

1. 80 **3. a.** 20 **b.** 60 **c.** 90 **d.** 0

EXERCISE 14.6

1.

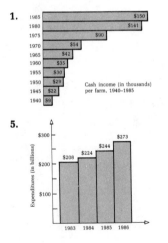

3.

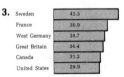

Total taxes
(in cents per dollar)

5.

7.

9. a.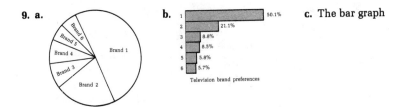

b.

Television brand preferences

c. The bar graph

11. a. Yes; to give a correct visual impression, only the height should be doubled. (If both the radius and height are doubled, the volume is multiplied by 8.)

b.

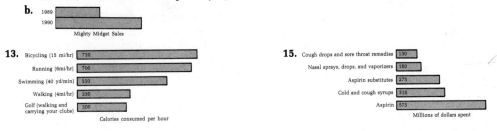

1989
1990

Mighty Midget Sales

13.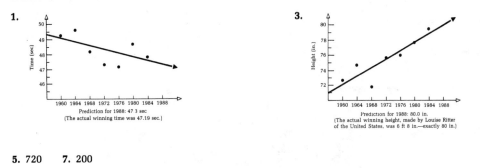

Bicycling (15 mi/hr) — 730
Running (6mi/hr) — 700
Swimming (40 yd/min) — 550
Walking (4mi/hr) — 330
Golf (walking and carrying your clubs) — 300

Calories consumed per hour

15.

Cough drops and sore throat remedies — 130
Nasal sprays, drops, and vaporizers — 160
Aspirin substitutes — 275
Cold and cough syrups — 310
Aspirin — 575

Millions of dollars spent

DISCOVERY 14.6

1. The graph on the right has a very much compressed vertical scale, which diminishes the visual effect of each increase or decrease.

EXERCISE 14.7

1.

Prediction for 1988: 47 3 sec
(The actual winning time was 47.19 sec.)

3.

Prediction for 1988: 80.0 in.
(The actual winning height, made by Louise Ritter of the United States, was 6 ft 8 in.—exactly 80 in.)

5. 720   **7.** 200

USING YOUR KNOWLEDGE 14.7

1. $y = -0.872x + 5.231$

3. $y = -1.057x + 12.186$, where $x =$ The number of the Olympics with 1960 as number 1, $y = t - 50$

5. $y = -0.293x + 8.834$, where $x =$ Year $- 1965$, $y = t - 45$

EXERCISE 14.8

1. Positive   **3.** None   **5.** Negative   **7.** Positive

9. Positive correlation

11. Negative correlation

13. Positive correlation

15. Negative correlation

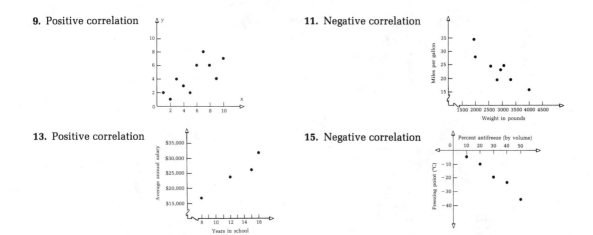

USING YOUR KNOWLEDGE 14.8

1. $r = -0.89$ **3.** $r = 0.93$ **5.** $r = -0.21$

PRACTICE TEST 14

1.

SCORE	TALLY MARKS	FREQUENCY
$60 < s \le 65$	||	2
$65 < s \le 70$	|||	3
$70 < s \le 75$	|||	3
$75 < s \le 80$	||	2
$80 < s \le 85$	|||	3
$85 < s \le 90$	ⅢⅡ	7
$90 < s \le 95$	|||	3
$95 < s \le 100$	||	2

2.–3.

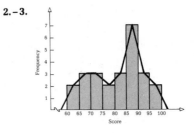

4. a. 75   **b.** 82   **c.** 73   **5. a.** 13   **b.** 5.18   **6. a.** 104 to 152   **b.** 0.025
7. a. 25   **b.** 160   **c.** 135   **8. a.** 0.5   **b.** 0.025   **9. a.** 1.6   **b.** 2.4
10. The score in French was the better score.   **11.** 0.492

12.

Fed DS	2.5%
Ford M	4.5%
Noes UT	8.5%
Exxon	4.7%
Gen El	3.2%

13.

Self-service salad bar	57%
Varied portion sizes	47%
More varied menu	42%
All-you-can-eat specials	37%
Self-service soup bar	30%

14.

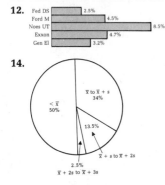

15.

16. About 25%   **17.** About 200   **18.** About 8   **19.** For $x = 7$, $y \approx -1.8$.

20. a. Positive   **b.** Positive   **c.** None   **d.** Negative

CHAPTER 15 EXERCISE 15.1

1. $240   **3.** $540   **5.** $200   **7.** $62.50   **9.** $466.67   **11. a.** $2.41   **b.** $42.61   **13.** $1802.40
15. a. $40   **b.** $160   **c.** $168   **17. a.** $125   **b.** $375   **19. a.** $0.71   **b.** $13.49
21. $159.38; $59.38   **23.** $7154.60; $4574.60   **25.** $26,195; $14,195   **27.** $25,364; $5364
29. $3,880,690; $3,840,690   **31. a.** $2854.30   **b.** $2898.30   **33. a.** $3993   **b.** $993   **35.** $15.30

USING YOUR KNOWLEDGE 15.1

1. 6.09%   **3.** 8.24%   **5.** 9.31%   **7.** 19.56%

EXERCISE 15.2

1. $141.35   **3.** $185.01   **5.** $557.48   **7. a.** $1.28   **b.** $236.28   **c.** $11.81
9. a. $5.16   **b.** $409.16   **c.** $20.46   **11. a.** $1.21   **b.** $180.39   **c.** $10.00
13. a. $0.84   **b.** $92.73   **c.** $10.00   **15. a.** $0.52   **b.** $122.81   **c.** $10.00   **17.** $2.85   **19.** $1.35
21. a. $1998   **b.** $261   **23. a.** $450   **b.** $136   **25. a.** $60   **b.** $26

USING YOUR KNOWLEDGE 15.2

1. 6.4%   **3.** 10%

EXERCISE 15.3

1. 14%   **3.** 16%   **5.** 17%   **7.** $15\frac{1}{2}$%   **9.** 16%   **11. a.** $2   **b.** $98   **13. a.** $8.40   **b.** $261.60
15. a. $4.18   **b.** $45.82   **17. a.** $24   **b.** $14\frac{1}{2}$%   **19. a.** $134   **b.** $16\frac{1}{2}$%   **21. a.** $4   **b.** $196
23. a. $72   **b.** $66   **c.** $9.23   **d.** $254.77   **25. a.** $120   **b.** $51.11   **c.** $4.21   **d.** $149.12

EXERCISE 15.4

1. a. Yes   **b.** Yes   **c.** No   **3. a.** $61,600   **b.** $15,400   **c.** $9500   **5.** $398.27   **7.** $638.95
9. $1142.65   **11. a.** $3000   **b.** $891.65   **13. a.** $1000   **b.** $328.41
15. a. $622   **b.** 240   **c.** $149,280   **d.** $99,280   **e.** The price is less than the interest.
17. a. $11,035   **b.** $549.00   **c.** $561.26

PRACTICE TEST 15

1. a. $448   **b.** $56   **c.** $1248   **2. a.** $21.60   **b.** $381.60   **3. a.** $78   **b.** $312
4. a. $116.99; $16.99   **b.** $117.17; $17.17   **5. a.** $10   **b.** $2.64   **6. a.** $1.94   **b.** $155.08
7. a. $2400   **b.** $154.17   **8. a.** $15\frac{1}{2}$%   **b.** $6.17   **c.** $120.53
9. a. $37,500   **b.** $12,500   **c.** $2000   **d.** $48,000   **10.** $450

CHAPTER 16   EXERCISE 16.1

1. Transistors replaced tubes.   **3.** Used miniaturized memory chips.
5. Keyboard, video monitor, disk drives, system unit   **7.** 8-4-97 or 8/4/97   **9.** 20:00
11. A list of all the files on the disk   **13.** Press the ⊣ key   **15.** Six 0's are printed (syntax error in 10).
17. WE NEED QUOTATION MARKS AT THE END   **19.** Four 0's are printed.

EXERCISE 16.2

1. (3+4)/(5+9)   **3.** 3^2+4^2   **5.** 2*3^3-5*4^2   **7.** 5*8/(6*9)   **9.** 4   **11.** 2   **13.** 20
15. 16   **17.** 500   **19.** 827   0.5   2   **21.** A=3   B=2   X=17
1654   0.25

23.
```
10   LET S=1+2+3+4
20   PRINT S
30   END
```
25.
```
10   LET A=1
20   LET B=1
30   LET C=A+B
40   LET D=B+C
50   LET E=C+D
60   LET F=D+E
70   LET G=E+F
80   LET H=F+G
90   PRINT A; B; C; D
100  PRINT E; F; G; H
110  END
```

USING YOUR KNOWLEDGE 16.2

1. EDIT 10 ⊣
Press the right arrow key → and stop under the A. Press Del , Ins , and the letter O. Press return ⊣ .

EXERCISE 16.3

1.
```
 5      7      6
25     35     30
40     48     44
OUT OF DATA IN 10
```
3.
```
1.05*500=525
1.05*750=787.5
1.05*1500=1575
1.05*2000=2100
OUT OF DATA IN 10
```
5.
```
 X       Y
-3      -1
-2      -3
-1      -3
 0      -1
 1       3
 2       9
 3      17
OUT OF DATA IN 20
```

7.
```
10   PRINT ''N'', ''N^4''
20   READ N
30   LET F=N^4
40   PRINT N, F
50   GOTO 20
60   DATA 1, 2, 3, 4, 5, 6, 7, 8
70   END
```

9.
```
10   PRINT ''N'', ''S''
20   READ N
30   LET S=N*(N+1)*(2*N+1)/6
40   PRINT N, S
50   GOTO 20
60   DATA 1, 2, 3, 4, 5, 6, 7, 8, 9, 10
70   END
```

11.
```
10   INPUT A,B
20   PRINT A;B;(A+B)/2
30   GOTO 10
40   END
```

13.
```
10   INPUT P
20   PRINT ''1.05*'';P;''='';1.05*P
30   GOTO 10
40   END
```

15.
```
10   INPUT X
20   PRINT ''X'', ''Y''
30   LET Y=X^2+3*X-1
40   PRINT X,Y
50   GOTO 10
60   END
```

17.
```
10   INPUT N
20   PRINT ''N'', ''NTH TRIANG. NO.''
30   LET T=T+N
40   PRINT N, T
50   GOTO 10
60   END
```

19.
```
10   INPUT ''WHAT IS THE LENGTH'';L
20   INPUT ''WHAT IS THE WIDTH'';W
30   LET A=W*L
40   PRINT ''THE AREA OF THE RECTANGLE IS'';A
50   END
```

EXERCISE 16.4

1. Would print out X and $\sqrt[3]{X}$
for X = 1, 2, 3, . . . , 100.

3.
```
9        4        2
100      32       3
500      45       11
OUT OF DATA IN 10
```

5.
```
FIRST NUMBER?  48
SECOND NUMBER?  60
THE GCF IS  12
THE LCM IS  240
```

7.
```
10   LET A=1
20   LET A=A+1
30   LET B=A^5
40   IF B<=100000 THEN 20
50   PRINT A; B
60   END
```

9.
```
10   PRINT ''R'', ''N''
20   READ R
30   IF R=1 THEN 120
40   LET N=1
50   LET A=(1 + R)^N
60   IF A>=2 THEN 90
70   LET N=N+1
80   GOTO 50
90   PRINT R, N
100  GOTO 20
110  DATA .04, .05, .06, .07, .08, 1
120  END
```

11.
```
10   LET N=    (value to be entered)
20   LET S=0
30   READ P
40   IF P=2 THEN 80
50   LET S=S+P
60   GOTO 30
70   DATA p_1, p_2, . . . p_N, 2
80   IF ABS(1-S)<=.0005*N THEN 110
90   PRINT ''SORRY''
100  GOTO 120
110  PRINT ''OK''
120  END
```

1.
```
10  LET A=5000
20  LET B=100*.0525*A/12
30  LET I=.01*INT(B+.5)
40  LET A=A+I
50  PRINT I; A
60  END
```

3. This program will print out a three-column table giving the number of the month, the interest, and the new amount at the end of the month for 12 months for an initial deposit of $1000 at 5% compounded monthly. If the program is run, the following table will be printed out:

N	I	A
0	0	1000
1	4.17	1004.17
2	4.18	1008.35
3	4.2	1012.55
4	4.22	1016.77
5	4.24	1021.01
6	4.25	1025.26
7	4.27	1029.53
8	4.29	1033.82
9	4.31	1038.13
10	4.33	1042.46
11	4.34	1046.8
12	4.36	1051.16

DISCOVERY 16.4

1. Change instruction 10 to LET P=3, and change instruction 130 to
PRINT P; ''IS THE FIRST ODD PERFECT NUMBER.''
(This program will run indefinitely; no odd perfect number has ever been found.)

EXERCISE 16.5

1.
```
1
5
14
30
55
```

3. 3840

5.
5	15
9	945
11	10395

OUT OF DATA IN 10

7.
I	R	H	P
1	4	42	172
2	3.75	39	146.25
3	3.5	45	166.25

9.
```
10  LET A=36
20  FOR K=1 TO 6
30  LET Q=A/K
40  IF Q=INT(Q) THEN 60
50  GOTO 70
60  PRINT K; Q
70  NEXT K
80  END
```

11.
```
10  PRINT ''N'', ''FIFTH POWER''
20  FOR K=2 TO 20 STEP 2
30  LET F=K^5
40  PRINT K, F
50  NEXT K
60  END
```

USING YOUR KNOWLEDGE 16.5

1. In Line 40, change the 12 to 24. No other changes are necessary.

```
3.  10  LET B=500
    20  LET N=0
    30  LET J=.01*INT(B+.5)
    40  LET P=50-J
    50  IF P>=B THEN 90
    60  LET N=N+1
    70  LET B=B-P
    80  GOTO 30
    90  LET R=B+J
   100  LET I=N*50+R-500
   110  PRINT N; R; I
   120  END
```

PRACTICE TEST 16

1. 80 PRINT ''THE ANSWER IS...''   2. a. 4*(2+5^2)/8   b. 3^2-3*2^3

3. a. 25   b. −11   c. 5   4. 10   24   26   5. 21

6.
```
1      1
2     -2
3     -9
OUT OF DATA IN 10
```

```
7. 10  INPUT X
   20  LET Y=3*X-2*X^2
   30  PRINT X;Y
   40  GOTO 10
   50  END
```

8.
```
1
3
6
10
15
OUT OF DATA IN 10
```

9. 3260

10.
```
2
4
8
16
```

11. 55

12.
```
1
2
3
6
9
```

```
13. 10  LET A=SQR(38^2-4*13*17)
    20  PRINT A
    30  END
```

```
14. 10  PRINT ''N'', ''CUBE ROOT''
    20  READ N
    30  IF N=99 THEN 80
    40  LET C=N^(1/3)
    50  PRINT N, C
    60  GOTO 20
    70  DATA 20, 30, 38, 42, 99
    80  END
```

```
15. 10  PRINT ''N'', ''FIFTH ROOT''
    20  FOR K=1 TO 15 STEP 2
    30  LET F=K^(1/5)
    40  PRINT K, F
    50  NEXT K
    60  END
```

INDEX

Boldface page numbers refer to chapter summaries.

Cylinder (continued)
  volume of a circular, 439, **451**

Dantzig, George B., 376
Darwin, Charles, 376
DATA command, 717
Decagon, 404
Decigram, 262, **273**
Deciliter, 258, **273**
Decimal(s)
  addition of, 218
  division of, 219 ff
  infinite, repeating, 194, **205**
  nonterminating, nonrepeating,
    212, **245**
  terminating, 194
Decimal form of a rational number,
  194
Decimal system, 110, 117
Decimeter, 255, **273**
Degree, 395, **448**
Dekagram, 262, **273**
Dekaliter, 258, **273**
Dekameter, 255, **273**
De Morgan, Augustus, 43
De Morgan's laws, 28, 47
Denominator, 186
  least common (LCD), 168, 188
Descartes, René, 329, 339, 565
Designs, modular, 514–515
Determinant, 487, **498**
Deviation
  root-mean-square, 636
  standard, 636, **665**
Difference
  of matrices, 459
  of rational numbers, 188
  of sets, 17
Digit(s), 117
  significant, 221
Diophantus, 281
Direct variation, 320
Discounts, 672
Disjunction(s), 46, 52, **102**
  truth table for a, 52, **102**
Disjunctive syllogism, 94, **104**
Dispersion, measure of, 655
Distance, directed, 349
Distance formula, 350, **384**
Distinct arrangements, formula for,
  559
Distribution
  binomial frequency, 639
  frequency, 618, **665**

Distribution (continued)
  normal, 641
  of z-scores, 647
Distributive laws for set operations,
  28
Distributive property, 519, **530**
  of multiplication, 153, **204**
Divisibility rules, 161 ff, 165
Division
  of approximate numbers, 222
  in bases other than 10, 126, 141
  of decimals, 219 ff
  of radicals, 230
  by zero forbidden, 186
Divisor(s), 159
  exact, number of, 562
  proper, 165
Dodecagon, 404
Dodecahedron, regular, 442
Domain of a relation, 331, **384**
Doyle, Arthur C., 495
Duplation and mediation, 111–112
Duplication, successive, 111

Echelon form, 472
Eckert, J. R., 705
Egyptian numeration system,
  109 ff, **145–146**
Elementary operations
  on equations, 282, **325**
  on inequalities, 286, **325**
Elements of a set, 2
  notation for the, 3, **39**
  number of, 29, **39**
Empty set, 9, **39**
END command, 711
ENIAC, 705, 706
Equal to, 276
Equality
  of rational numbers, 186
  of sets, 9
Equally likely outcomes, 568
Equation(s), 277, **325**
  determinant form of, of a line,
    494–495
  elementary operations on, 282
  first-degree, 285
  general form of, of a line, 358
  graph of an, 291
  linear, 285, **384**
  point–slope, of a line, 356
  procedure to solve an, 284
  quadratic, 299 ff, **326**
  slope–intercept, of a line, 357

Equivalence, 60, 77
Equivalence relation, 337
Equivalent sentences, 282
Equivalent statements, 60, **103**
Eratosthenes, 159
  sieve of, 160
Euclid, 173, 387
Euclid's algorithm, 173
Euler, Leonhard, 29, 234, 444, 503
Euler circles, 29
Euler's formula
  for networks, 447
  for polyhedrons, 442
Even vertex, 444
Event(s)
  certain, 583, **614**
  equally likely, 568
  impossible, 582, **614**
  independent, 597–598, **614**
  mutually exclusive, 584, **614**
  probability of an, 568, 571, **614**
Exact divisor, 159
Expanded form
  of decimals, 198
  of a number, 118
Expectation, mathematical, 609
Expected value, 609, **615**
Experiment(s), 567, **614**
  binomial, 639
  sample space for an, 567, **614**
Exponent(s), 118
  laws of, 119, **146**
  negative, 198, **205**
  zero, 118
Exponential form, 118
Exterior angle of a polygon, 413

Factor
  greatest common (GCF), 167, 172,
    204
  of a number, 150
Factorial, 544, 546, **562**
Factoring
  quadratics, 300
  solving quadratics by, 301
Factorization, prime, 159–160, **204**
Fahrenheit scale, 262
Fahrenheit to Celsius conversion,
  263, **273**
Fair bet, 609
Fair game, 610
False position, method of, 116
Feasible region, 377
Femur, 279

Switch(es), 97 *ff*, 605–606
Switching network, 97
Syllogism, 44, 87, 94, **104**
   disjunctive, 94, **104**
   hypothetical, 94, **104**
Symbolic logic, 43, 44
System(s)
   abstract, 469
   mathematical, 459
   positional numeral, 99
System of inequalities, graphical
   method to solve a, 372 *ff*
System of linear equations, 363 *ff*
   algebraic methods to solve a, 365 *ff*
   graphical method to solve a, 364

Target zone, 334
Tartaglia, 524
Tautology, 76, **103**
Terminating decimal, 194, 195
Tetrahedron, regular, 442
Thales, 387
Transversal(s), 397, 420, **448**
Trapezoid, 406, **450**
Tree diagram(s), 39, 534, 575
Triangle(s), 404
   acute, 404, **449**
   area of a, 427, **450**
   equiangular, 405
   equilateral, 405, 424, **449**
   isosceles, 405
   obtuse, 405

Triangle(s) *(continued)*
   Pascal's, 14
   right, 405, **449**
   scalene, 405, **449**
   similar, 414
   sum of the measures of the
      angles of a, 390
Triangular numbers, 722
Trichotomy law, 213
Truth set, 80
Truth table(s), 58 *ff*
   for a biconditional, 65
   for a conditional, 64
   of a conjunction, 51
   for the contrapositive, 71
   for the converse, 71
   of a disjunction, 52
   for the inverse, 71
   of a negation, 53
Truth values of statements, 50 *ff*
Twin primes, 163

Union of sets, 15, **39**
   number of elements in the, 30–31
Unit area, 426
Unit volume, 437
Universal set, 9, **39**

Valence, 184
Valid argument, 83, 89
Valid argument forms, 94

Variable, 276, **325**
Venn, John, 29
Venn diagrams, 23 *ff*, 87
   of a polygon, 368
   of a polyhedron, 393
   of a pyramid, 394

Weight, 261 *ff*
Well-defined set, 2
Whole numbers, set of, 176, **205**
Word problems, 308
   procedure for solving, 308, **326**

x axis, 339
x intercept, 349, **384**
x value, 330

y axis, 339
y intercept, 347, 357, **384**
y value, 330

Zero, 176
   additive property of, 180
   division by, forbidden, 186
   factorial, 546
   as the identity element for
      addition, 176, 180
   multiplication by, 176
z-score, 646, **665**
Zuse, Konrad, 705